Decision Processes by Using Bivariate Normal Quantile Pairs

N.C. Das

Decision Processes by Using Bivariate Normal Quantile Pairs

N.C. Das
Department of Statistics
Birsa Agricultural University
Ranchi, Jharkhand
India

ISBN 978-81-322-2363-4   ISBN 978-81-322-2364-1   (eBook)
DOI 10.1007/978-81-322-2364-1

Library of Congress Control Number: 2015935207

Springer New Delhi Heidelberg New York Dordrecht London
© Springer India 2015
This work is subject to copyright. All rights are reserved by the Publisher, whether the whole or part of the material is concerned, specifically the rights of translation, reprinting, reuse of illustrations, recitation, broadcasting, reproduction on microfilms or in any other physical way, and transmission or information storage and retrieval, electronic adaptation, computer software, or by similar or dissimilar methodology now known or hereafter developed.
The use of general descriptive names, registered names, trademarks, service marks, etc. in this publication does not imply, even in the absence of a specific statement, that such names are exempt from the relevant protective laws and regulations and therefore free for general use.
The publisher, the authors and the editors are safe to assume that the advice and information in this book are believed to be true and accurate at the date of publication. Neither the publisher nor the authors or the editors give a warranty, express or implied, with respect to the material contained herein or for any errors or omissions that may have been made.

Printed on acid-free paper

Springer (India) Pvt. Ltd. is part of Springer Science+Business Media (www.springer.com)

In the memory of Dr. W.F. Sheppard (Pub.1939), whose Univariate Normal Probability Integral Table (UNPIT) and its quantiles have been in common use for uncertain decision-making scenarios for more than a century

*To Sanjeev
who remains only in memory*

Preface

The most important and inevitable aspect of human life is decision-making. As civilization advances, complexities also increase, and consequently, decision-making becomes increasingly difficult. Thus, one needs newer tools and methods for decision-making. This book provides the same. The primary aim of the book was to give to the reader, particularly the one who has to make decisions, some newly developed tools and techniques for decision-making. It will help him take his decisions at his chosen level of risk, the level that brings into play three kinds of complexities, namely uncertainty, dependence and dynamism.

It is a truism to say that every person must make decisions, big or small, throughout his life. It has been assumed that for decision-makers, the risk level may vary from near-hundred to fifty-percent certainty. Decidedly, in its subtlety and power, the human mind is the most wonderful creation [Moses: 2005]. The mind must take a decision by exercising a choice or a preference and as a sequel to some action. Everyone has to act, and therefore, before acting, he has to make a decision about why, how and where he should perform that action, or even whether that action needs to be performed. He has to take a decision in such a manner that in a given situation, his action proves to be the most effective and yields the best results. This applies equally to a firm, corporation, consortium, government, organization or society. In fact, such a search for tools for making successful decisions has been the concern of mankind from the time immemorial. It is relevant to quote here a *shloka* Mahopanishad (1998)[1]:

तरवोऽपि हि जीवन्त जीवन्ति मृग पक्षिण ।
स जीवति मनोयस्य मनने नोपि जीवति ॥

"Tarvoapi hi jeewant jeevanti mriga pakshinah,
Sa jeevati manoyasya manane nopi jeevati."

---

[1]Mohopanishad Chap. 3 Shloka 13.

Roughly translated, the lines say that as compared to trees, animals and birds, man has a special faculty of the will to live and of reasoning. The mind of man is the source of his progressive development that continues from generation to generation and from epoch to epoch in a transitional chain.

Decision-making requires an exercise of these faculties. Only a properly made and result-oriented decision will be worthy of being called a "rational decision". It is a fact that in the course of a man's life, he has to face various kinds of complexities that force him to search for, and even create, alternative choices, which in turn give him the ability to make effective decisions. He has to face questions like these: What is the best decision in this situation? Or what are the criteria for preferring a particular choice. Very often, such questions are not easy to answer. This is because sometimes a decision may be good and profitable, but one also likes to know whether that decision has the potential of remaining valid if there is a change in the situation. In order to arrive at an appropriate decision, decision-makers ought to change probabilistic decision to its certainty equivalent, especially when the two choices happen to be correlated or when decisions variables are pairwise correlated, irrespective of the fact that such correlation happens to be for the observations between spaces, between times, between space and time or even between conceptual measurements or variables.

Now, making corporate decisions, involving multidisciplinary teams with varied priorities, is all the more difficult. Such situations require a much wider range of tangible alternatives apart from knowledge of risk management and also of the magnitude and direction of variable dependencies. Relative cost, market competition, customer preferences, organizational demands and the laws of the land are seen to create complexities in decision-making. This happens more on account of dependencies on frequently changing scenarios. Farming communities and agricultural scientists, in particular, have to face even greater risks because of the vagaries of the weather and also due to the fact that agricultural produce perishes faster than other kinds of products.

The present book aims to take the reader a step closer to his search for realistic solutions to the problem of decision-making. Tools, tables and techniques have been provided here that help the person making a decision, escape the drudgery of mathematics and statistics. The book focuses on application-oriented techniques that are discussed in Chaps. 6 and 7. At the same time, the reader is also informed when, where and how such decision-making concepts, ideas and methodologies are originated; what their status is as mentioned in some scriptures; and how they have changed over centuries.

The book is divided into ten chapters: Chap. 1 Introduction; Chap. 2 Decision Complexity and Methods to Meet Them; Chap. 3 Univariate Normal Distribution and Its Quantile; Chap. 4 Bivariate Normal Distribution and Heuristic-Algorithm of BIVNOR for Generating Biquantile Pairs; Chap. 5 Software Reliability Testing and Tables Explained; Chap. 6 Decision Scenario: Problems and Prospects; Chap. 7 Application Paradigms; and Chap. 8 Generated Tables by BIVNOR, which are in fact products of research done by the author, those being Table 8.1: the Table for equi-quantile values H (BIGH) and Table 8.2 containing series of four hundred tables of generated biquantile pairs for differing values of correlations and probability (risk) levels for all such grids of Table 8.1. Their need had been felt by the empirical scientists and the decision-makers since the middle of the last century. Thus, the methodology the author offers for bringing about a change in the decision-making process is dependent on the useful but judicious application of these two sets of tables.

Chapter 9 contains four sections, of which the first three are the products of software testing for the software BIVNOR, and some of its variants prepared by the author, mainly for testing validity and the reliability of the said software. Therefore, they are of interest, especially to those who are involved in the development and testing of research software, virtually a new field, as such software testing is in use mostly for commercial and utility purposes. But, the author believes that very few of them are used for testing the reliability and validity of any generated tables. Section 9.4 is on "Barycentric coordinate reading system: A tool for measuring components of mixtures and mixed-ventures". This can be adopted for guarding against risky entrepreneurship, which is an age-old technique of diversified investment. Chapter 10 is on Conclusions. Thus the tables generated in Chap. 8 are good enough to be used for such decision making problems.

The book, therefore, will prove useful to the entire group of progressive decision-makers who take decisions under conditions of uncertainty, dependence and dynamism. Presently, however, they are required to unnecessarily assume variable independence for drawing inferences. Thus, the entire group of portfolio theorists and practitioners, market analysts, insurance practitioners, actuarial scientists, agriculturists, computing and communication scientists and operation researchers would be benefited. Besides all others who are concern with either studies on particle physics or macromolecules for genome studies, and relate such studies with probability, information and correlation would gradually adopt the decision-making process recommended in the book. This they can do when they are required to change a probabilistic decision to its certainty equivalent, especially when the two choices happen to be correlated, or when decision variables are pairwise correlated irrespective of the fact that such correlation is between spaces, or times, or space and time, or even between conceptual measurements or variables.

All those who are users (which includes almost the entire group of empirical scientists) of univariate normal distribution for its quantile value for decision-making are the potential users of the tables of bivariate quantile (biquantile) pairs as an instrument of the decision-making processes, involving complexities of risk, dependence and dynamism. It will be useful for all those who require shrinkage of joint confidence interval over the existing methods, for the same confidence coefficient, including covariance analysts for smaller critical differences or L.S.D.'s, on account of variable correlation or dependence. Its usefulness for all decision-makers, who are to take decisions under risk and dependence, is irrefutable, which has been explained through numerous examples of Chap. 7. But this can have many more applications than what have been indicated in the book.

Further, the methodologies outlined in this book will facilitate decision-making under uncertainties allowing surrogation between pairs of correlated choices without change in the risk level, already fixed by the decision-maker, in terms of probability. Therefore, iso-risk technology trade off or surrogate up to a realistically feasible limit is now a possibility. Such objectives can very well be met by the use of iso-probable quantile pairs made available through series of Table 8.2 in Chap. 8, as illustrated for problems of Chap. 7. The examples given also suggest the use of equi-quantile values of Table 8.1 in Chap. 8. These tables are like any other mathematical or statistical tables, logarithmic tables, tables of mathematical functions and statistical tables. In fact, those are two-variable generalization of the inverse normal probability integral table (INPIT), wherefrom, for the given values of probability and correlation, bivariate normal quantile pairs can be obtained. These tables are of immense usefulness to entrepreneurs, decision-makers, investors and researchers in empirical sciences, and also to students of statistics, economics and econometrics, psychology, psychometrics and psychographics, social sciences, geography, geographic information system, geology, agricultural and animal husbandry-cum-veterinary sciences, medical sciences and diagnostics, remote sensing and even linguistics. Description of the tables so generated with their parameters, along with their terminologies, has been presented in Sect. 5.4 of Chap. 5.

Yet another purpose of this book is to outline in brief the process of development that took place throughout the evolutionary stages, to reach the state-of-the-art stage where it stands. There is also an attempt to give due credit to those who chiefly contributed to such phenomenal development. However, it must be added that the book includes only those aspects of decision-making process which have a very wide applicability, or which have become the landmarks in the development of statistical methodologies for decision-making.

The author realizes that the references noted may be inadequate. Further, some of those references may appear to be alien to the field covered. It is also shown how the origin of the knowledge of probabilistic thought, the concept of dependence and dynamism and the process of decision-making and their uses may be traced to the Vedic and post-Vedic era. The author has no excuse for not including many relevant and important references. He can only apologize for his ignorance and limitations.

N.C. Das

References

Mahopanishad: 108 Upanishad Gyan Khand. Pt. Shri Ram Sharma Acharya (ed.). Sanskrit Sansthan, Bareilly, U.P., India (1998)

Moses, A.M.: Last Frontiers of the Mind: Challenges of Digital Age. Prentice Hall of India, New Delhi (2005)

Acknowledgments

The author expresses his gratitude to Dr. D.P. Singh (then vice-chancellor) and Dr. K.K. Jha (then Dean of Agriculture), both of Rajendra Agricultural University, Samastipur, for granting and recommending the author's case for the International Development Agency Fellowship. The author expresses his gratitude to late Prof. Dr. A.K. Bhattacharji (the chairman of the Research Advisory Committee) and Prof. Dr. G.P. Bhattacharjee, both of IIT Kharagpur, for their learned guidance. Prof. Bhattacharjee continued his blessings and encouragement even at the manuscript stage of the book.

Author expresses his gratitude to Dr. J.K. Sengupta who taught the author the basics of chance-constrained programming. In fact, it was his initial work with Gerhard Tintner, a Nobel Laureate in Economics, that inspired the author to work on joint chance-constraints programming, for which realistic solution was need of the time. The author had an excellent opportunity to personally interact with Prof. Sengupta and receive reprints of his publications, as well as his encouragements. The author also had the opportunity of listening to a series of lectures by Leonid Hurwitz, a Polish-American mathematician, and his decision criteria in 1966 and interacting with him. The author expresses gratitude to Profs. A. Charnes, W.W. Cooper and G.H. Symonds (1958) of the Center for Cybernetic Studies, University of Texas, who shared their entire set of research publications with the author. Similar support was also received from Dr. Andras Prekopa (Hungary), Dr. Thomas Szantai (Hungary), Dr. Cristian Bereanu (Romania), Dr. William T. Ziemba (London), Dr. John P. Klein (Wiskonsin), Dr. V.S. Bawa (Bell Laboratories, U.S) and Dr. Alexey Kolbin (Russia).

Thanks are extended to Dr. S.K. Mishra (NEHU, Shillong) for his help during the FORTRAN program debugging. The author must not fail to recall late Dr. M.N. Das (former director of I.A.S.R.I., New Delhi, and the founder chair editor, *Journal of Statistics and Computer Applications*) and late Dr. D.N. Lal (former head, Department of Statistics, Patna University) not only for imparting knowledge of statistics but also for their continued affection bestowed upon the author throughout his entire career in the field of statistics and even beyond.

The author has been very thankful to one of his closest friends Prof. Dr. K.C. Prasad (former professor of mathematics, Ranchi University) for his suggestions and encouragement. That was Dr. R.N. Mishra (professor, Department of Statistics, Patna University) on whose behest the work was restarted. In return, the author has to offer his sincere thanks. The author also extends his thanks to Dr. S.N. Choudhary (Department of Physics, T.M.B. University, Bhagalpur) for his discussion on Higgs boson. The author expresses his gratitude to Dr. R.N. Sinha (head, Department of English, St. Xavier's College, Ranchi) who took all the pains of going through this book more than once and making appropriate corrections, wherever required, and to Dr. Sujit Bose of that very department for independent appraisal and making corrections. Thanks are due to Dr. Satish Sinha (I.I.C.M., Ranchi) for extending secretarial help at critical hours. The author extends his sincere thanks to Mrs. Swati Bhattacharee (librarian of I.I.M. Calcutta, Joka) for her help in renewal of his contact with Dr. J.K. Sengupta and also to Sri Chanchal Bhattacharjee (B.O.I., Varanashi) for giving typographic shape to the manuscript.

Amongst some of my esteemed friends, who directly or indirectly helped, are Dr. M.P. Jha (F.A.O. Consultant in Statistics) and Mr. P.K. Sinha (Sr. advocate, Jharkhand High Court) to whom the author's sincere thanks are duly extended.

The author fails to find suitable words to compliment his lifelong companion "LAKHI" who kept on pouring love and her immense support at the cost of her lifelong sufferings as penance during the author's studies, programme development and this work. The author extends his thanks to his respected *samadhi* Sri S.N. Sinha for reading the entire text with patience, and suggesting amendments; his eldest son-in-laws Dr. Ashok Ghose, Deepak*ji* for their good wishes; his youngest son-in-law Dr. Ashish Kumar, without whose efforts, it would not have been easy to bring it to the stage of publication; and his sons Rajiv, Rakesh and Mukesh for extending computing and internet facilities, whenever and wherever required. The author recollects help received from his grandchildren Divya, Bhavya, Rahul, Rohan, Anshu, Nishu, Shristi, Smirti, Smit, Sonal, Mohit, Shobhit, Ankit and Harshit for several cut-and-paste solutions and also for the Internet search and surfing for references. The author owes much to his loving daughters and daughter-in-laws for mental and moral support and care taken by them while working on this book. Thanks are also extended to my nephews Dr. B.K. Das (head, Department of Statistics, T.M.U., Bhagalpur) and Ashish Sinha for extending good wishes for the publication. Further, author bestowes his affection to Shourya his great-grandson.

The unknown set of referees whose valuable comments to improve and include some aspects have indeed been very helpful in revision of at least some chapters and addition of some of its sections. The author fully acknowledges their support by extending to them his gratefulness.

Acknowledgments

The editor invariably comes last in the list of getting thanks. His support is most valuable for an author, not only for mere publication but for sustained relationship. He is like one of the nearest family members for any author. Ever since, the author came in contact with Shamim Ahmad, editor at Springer, he felt in him. He gives the excellence of human touch without compromising with the publishing standards. The author extends his sincere thanks to the entire publishing group at Springer for their cooperation in the publication of this book.

N.C. Das

Contents

1	**Introduction**		1
	1.1 Decision-Making		1
	1.2 Components of Rationality and Learning Model		2
	1.3 Predicate (Stochastic) Calculus for Psycho-Kinetics of Personality Development		4
	1.4 Truth and Error		10
	1.5 The Main Users		12
	1.6 Chapter Plan		13
	References		17
2	**Decision Complexity and Methods to Meet Them**		19
	2.1 Decision Complexities: Triangular Structure		19
	2.2 Vertex of Uncertainty (V_1): Evidence of Probabilistic Thought During the Vedic and the Post-Vedic Period		21
	2.3 Probability and Its Measure as Discovered in European Continent		23
	2.4 Probability Distribution		24
	2.5 Probability Updating		25
	2.6 The Vertex of Dependence (V_2): Association and Their Measures in Vedic and Post-Vedic Era		25
	2.7 Perception of Cause and Effect as Independent and Dependent Events		26
	2.8 Development of Probabilistic Causal Algebra: Evolution of Effect as Measure of Dependence to Probabilistic Cause		29
	2.9 Emergence of Measures of Association and Dependence: Correlation and Regression Coefficients		30
	2.10 Path Coefficients, Factor Analysis and Principal Components (Karhunen–Loeve Expansion)		31
	2.11 Advancements of Concepts and Computations for Associations and Dependence		32

	2.12	Discriminant Function and Measures of Dependence	33
	2.13	The Default Correlation: Copula	35
	2.14	The Apex Vertex of Dynamism (V_3)	35
	2.15	Markov Chain and Bayesian Inference	36
	2.16	The Brownian Motion—Weiner Process	38
	2.17	Stochastic Differential Equation: ITO's Process	40
	References		41
3	**Univariate Normal Distribution and Its Quantile**		47
	3.1	Probability Distribution	47
	3.2	Normal Probability Distribution	49
	3.3	Confidence Interval	50
	3.4	Quantile and Its Role in Decision-Making	51
	3.5	Certainty Equivalent and Quantile	54
	References		57
4	**Bivariate Normal Distribution and Heuristic-Algorithm of BIVNOR for Generating Biquantile Pairs**		61
	4.1	Characterization of Joint Cumulative Distribution Function	62
	4.2	Bivariate Normal Distribution: Historical Perspectives	64
	4.3	Other Properties of Bivariate Normal Distribution	66
	4.4	Owen's Computational Scheme for Evaluating Bivariate Normal Integral	69
	4.5	Other Methods for Evaluating Bivariate Normal Integral	72
	4.6	Software to Generate Tables of Bivariate Normal Quantile (Biquantile) Pairs: Prerequisite for BIVNOR	73
	4.7	Multiplicity of Biquantile Pairs: Basics of Heuristics for Developing BIVNOR	74
	4.8	Initial Steps of Heuristic BIVNOR	74
	4.9	Problems Related to Developing BIVNOR	75
	4.10	Algorithm for BIVNOR	75
	4.11	Simultaneous (Joint) Confidence Intervals	81
		4.11.1 Roy and Bose's Multivariate Confidence Bounds	81
		4.11.2 Simultaneous Confidence Interval	81
		4.11.3 Bonferroni's Interval	83
		4.11.4 Confidence Interval Based on the Chi-Square Quantile	84
	4.12	Recent Interests in Biquantile Pairs	84
	4.13	Rizopoulos Paradox	87
	References		87

5	**Software Reliability Testing and Tables Explained**		91
	5.1	Software Reliability	91
	5.2	Evolution of Reliability Testing Criteria	92
	5.3	Procedures Adopted to Meet the Evolved Criteria	94
	5.4	Generated Tables of Chaps. 8 and 9 Explained	97
		5.4.1 Description for Table 8.1 in Chap. 8: Equi-Quantile Values (or BIGH)	97
		5.4.2 Description for Table 8.2 of Chap. 8: Iso-Probable Biquantile Pairs of 400 ($19 \times 21 + 1 = 400$) Tables	98
		5.4.3 Description of Table 9.1 of Chap. 9 (Tester Table: Owen's Table of $T(h, a)$ Function)	100
		5.4.4 Description of Table 9.2 of Chap. 9 (Tester Tables of Biquantile Pairs for Zero Correlation Against Those for Independence Hypothesis)	100
		5.4.5 Description of Table 9.3 of Chap. 9 (Tester Tables of Iso-Probable Quantile Pairs for Random Grids of Table 8.1 of Chap. 8)	101
	References		102
6	**Decision Scenario: Problems and Prospects**		103
	6.1	Decision Scenario: The Past	104
	6.2	Decision Scenario: The Present	110
	6.3	Advancements in Decision Processes: Decision Science	111
	6.4	Design of Experiment: Designing the Investigation	113
	6.5	Decision Function	114
	6.6	Discrimination and Classification	115
	6.7	Optimization: Linear Programming and Extensions	117
	6.8	Basics of Stochastic Programming and Its Chance-Constrained Version	118
	6.9	Emergence of Decision Processes Under Risk and Uncertainty	123
		6.9.1 Related References	123
		6.9.2 von Neumann and Morgenstern versus Kahneman and Tversky	124
		6.9.3 An Attempt for Compromise	125
	6.10	Developments in Uncertain Decision-Making	126
	6.11	Online Stochastic Combinatorial Optimization	130
	6.12	Fuzzy Decision	132
	References		133

7	**Application Paradigms**	137
	7.1 Introduction	138
	7.2 Comparison Between Simultaneous (Joint) Confidence Intervals	138
	7.3 Algorithmic Heuristics for Chance-Constrained Version of Discrimination/Classification	147
	7.4 Biquantile in Optimization: Bivariate Joint Chance-Constrained Linear Programming Problem	151
	7.5 Bivariate Meteorological Prediction with Equi-quantile Pairs	157
	7.6 Equi-quantiles in Bio-statistical Studies	161
	7.7 The VaR Measure (The Value at Risk)	162
	7.8 Default Correlation: Gaussian Copula Model and Biquantile Pairs	165
	7.9 Bivariate Quality Control, Six Sigma Techniques and Biquantile Pairs	167
	7.10 Bivariate Stochastic Process and Biquantile Pairs	168
	7.11 Simulation and Biquantile Pair/Equi-quantile Value	174
	7.12 From Biquantiles to Higgs Boson (God Particle)	177
	7.13 Rizopoulos's Paradox (2009)	180
	7.13.1 Basics	180
	7.13.2 The Paradox: Its Modification and Solution	181
	7.13.3 Generalization	184
	7.13.4 Reliability and Validity Tests of BIVNOR for Extreme Paradigms	184
	7.14 Further Scope for Applications of Biquantiles	184
	7.15 Information Gain and Learning	185
	7.15.1 Introduction to the Problem	185
	7.15.2 Methodology	187
	7.15.3 Results	188
	References	189
8	**Generated Tables by BIVNOR**	193
	8.1 Table of BIGH the Equi-Quantile Values	193
	8.2 Contents for Table 8.2 (Table No. 8.2-1 to 400)	196
9	**Tables Generated for Software Testing**	621
	9.1 The Table of COMP-T to Test Its Equivalence to OWEN's	621
	9.2 Tables for Testing BIVNOR COMP. PROB. with JNT. PROB Under Zero Correlation	622

	9.3	Tables Generated for Bivariate Normal Iso-Probable Quantile Pairs for TEST CASES.	630
	9.4	Barycentric Coordinate Reading System: For Analysing Mixed Activities	636
	Reference		637
10	**Conclusions**		639
	10.1	Introduction	639
	10.2	Conclusions	639
	10.3	Caveats and Cautions	642
	10.4	The Ultimate Question	642
	10.5	Feller's Dictum and Winston's Aspiration	643
	References		644
Index			645

About the Author

N.C. Das is former professor cum chief scientist at the Department of Statistics and Computer Applications, Birsa Agricultural University, Ranchi. He has over 50 years of teaching and research experience in the field of Statistics, Operations Research and Computer Applications. Earlier, Prof. Das worked as the academic secretary cum editor of the *Bihar Journal of Mathematics* for the period of 1994–1998. An active teacher and researcher, Prof. Das was a visiting professor of Statistics and Quantitative Methods at the Central University of Jharkhand (Ranchi). He has also been guiding research scholars at the Indian Institute of Coal Management (Ranchi), and Indian School of Mines (Dhanbad). Presently, Prof. Das is the president of the Jharkhand Society of Mathematical Sciences as well as of the Jharkhand Society of Statistics (Ranchi).

Chapter 1
Introduction

Human being is naturally endowed with rationality. Decision-making is an important function which makes use of rationality. A person uses his five sensors for acquiring information and generating intuition and for developing his capacity for learning, cognition and reasoning. Knowledge about these elements can be traced even in the Vedic and post-Vedic period. It is possible to develop a simple linearly additive-mock model for learning and acquiring experience that will demonstrate how psychological factors synergistically accumulate to the decision-making processes. It will also show how psycho-kinetics work to develop the kind of personality needed for decision-making and for acting on those decisions. It has to be accepted that truth is often covered with error, and it is necessary to know how truth can be reached. This problem was taken up by majority of philosophical texts. It is, therefore, important that the newly developed tools of decision-making under uncertainties, dependence and dynamism are identified and discussed.

1.1 Decision-Making

It is necessary to trace the development of decision-making processes through the ages, as could be available and considered relevant. The dictionary meaning of "decision", according to *The New International Webster's Comprehensive Dictionary of the English Language* (2004) is "The act of deciding or making up one's mind", implicitly for some action or decision. Here, the clause "making up one's mind" means the possession of the faculty of "rationality" which, in turn, means "possessing the faculty of reasoning or being judicious or sensible". Radhakrishnan (1971) wrote, "Reason is subordinate to intuition". He also says that "The philosophy of India takes its stand on the spirit, which is above mere logic, and holds that culture based on mere logic or science may be efficient but cannot be inspiring". Sir Bertrand Russell in his essay *Useless Knowledge* wrote, "An ideal life is inspired by sentiment like Buddha and guided by knowledge like Sankaracharya". Clearly, here inspiration is a derivative of intuition and reasoning that of knowledge. Though Russell's above expression appears to be factual, both these elements must not be looked upon as completely independent of each other.

Very often, they are interactive or synergic factors, because one gets generated and even enhanced by the other.

Normally, all these virtues are seen as something with which mankind is endowed naturally, through an evolutionary process. However, animals, birds, invertebrates and even plants are required to make decisions for their survival, growth and reproduction. By convention, we human beings do not consider them to be even least rational. They are rather attributed with the ability of changing their behaviour or adapting to nature. That has been exemplified by the Valmiki's (2001) *shloka* in the preface itself.

The knowledge about the awareness of such concepts to human being can be traced even in the Vedic and post-Vedic era, as ascertained from the writings of *Kathopanishad* (2004, Ch2., Vallee 1, Shloka 3):

येन रूपं रसं गन्धं शब्दान् स्पर्शाँश्च मैथुनान्।
एतेनैव विज्ञानाति किमत्र परिशिष्यते। एतद्वैतत् ॥३॥

Yen roopam rasam gandham shabdaan sparshansch maithunaan,
Etenaiv vigyaanaati kimatra parishishyate. Yetadavaitat [3].

It means "Due to whose blessing one can see the shapes and visible or taste food and drinks, smell the odour, listen to the words uttered, feel the touches and derive the pleasures of intercourse; and it is His blessings again, that it is known as to what else remains to be known and enjoyed".

The above *shloka* encompasses the cognition of features by mankind, through their info-sensors.

1.2 Components of Rationality and Learning Model

The process of decision-making is based on information acquired through undernoted five sensors, which perform corresponding orthogonal functions (there being no overlap or entanglement or confounding between each such functions), but transforming them into perceptions, for continuous learning and experience. All these being components of rationality culminating into capability of decision-making and, thereafter, taking some actions.

Info-sensors (impulses)	Function
Eyes	To observe
Ears	To listen
Nose	To smell
Tongue	To taste
Skin	To touch

1.2 Components of Rationality and Learning Model

These five sensors together generate (a) feeling, (b) thinking, (c) processing, (d) decision and (e) action/reflex action.

Similarly, there is an excerpt from a poem by William Wordsworth, quoted by Radhakrishnan (1921, 2013):

> The eye cannot choose but see,
> We cannot bid the ear be still,
> Our bodies feel where'er they will be
> Against or with our will.

Such thought in sequel naturally generates a kind of mock model for learning and of experience which has linearly additive structure.

Learning

In its simplest and accumulative form, *learning* can be explained in terms of the following model:

$$\text{Learning} = r[\text{Observe} (+) \text{Listen} (+) \text{Smell} (+) \text{Taste} (+) \text{Touch} (+) \text{Read} (+) \text{Write}] (+) \text{FORGET}$$

Here, all positive components can be enhanced through education. However, the sign (+) is not an addition, but accumulation and storage at specific points of memory cells as in pointers of C or C++ language, which is recalled at appropriate instance or necessity, and r is the number of times one repeats the activities given within parentheses, that is, observing, listening, smelling, tasting, reading and writing. This results in memorization and thus reduces the negative impact of forget-fullness.

Experience

Experience may be expressed as follows:

$$\text{Experience} = \sum_{i=0}^{\text{age}} L_i \quad \text{or} \quad \int_0^{\text{age}} dL$$

where L_i or dL stands for learning acquired in a short period of time and summation or integration for cumulative learning. Obviously, such a model is only conceptual, because till date it has not been possible to validate it for the reason that measures are not yet available for many faculties of sense-perception. Human being is naturally endowed with much less they are scale additive.

Repetition

Repetition means the number of times one repeats some activities for example r, in some order or sequence in a certain rhythm. This helps in memorization and thus the text or topics become capable of being recapitulated in the same sequence, which sometimes on forgetting comes forth on starting once again, from the beginning (ab initio). It is just like recollecting multiplication table or recitation of poem by a school child. Such psychological phenomenon may be defined as, ab initio *recollection* or *memory* ab initio. Yet another term *associative memory* is defined as recapitulating some event or phenomenon or even a name by its association with some familiar name or event, which gets pronounced with similarity. As whatever is required to be recapitulated is embedded in one's memory for a long time. Fortunately, it is possible to list or even measure the memory capacity of an individual at a particular moment of time and also of its feature of retaining in the memory. The term *associative memory* has also been explained by Colman (2009) that is somewhat similar to what has been described here.

Inspiration
Inspiration may be generated either by some impulse within or by something happening outside. Once generated, inspiration acts as a force impelling a person towards acquiring knowledge. How much knowledge a person acquires due to inspiration is dependent on differential base or background, the availability of resources and of scope. Thus, there are negative (opposing) and positive (helping) factors in the process of acquiring knowledge. The psycho-kinetics of personality development, showing an impact of such external factors or stimulus, is discussed next.

1.3 Predicate (Stochastic) Calculus for Psycho-Kinetics of Personality Development[1]

Predicate (stochastic) calculus for psycho-kinetics of personality development shows stepwise stochastic impact of external factors (stimuli) on decision orientation and result-oriented action. Now, it is necessary to formalize the structure of accumulatively advancing synergic plexus of stimuli (exposures, events or instances coming in succession) resulting (at least in probability) to a directed pathway of information acquisition, to its retaining, to cognition, to learning, and gradually contributing to action, to feedback, if needed. This finally contributes to the development of a decision-oriented personality, that of, the decision-maker. Consequently, a decision-maker achieves a state of stability, temporary if not permanent. Such decisions are empirically verifiable, at least in the sense of their probability. Such aspects can better be explained/understood through flow chart which the reader may draw for himself.

Every element of an ordered sequence of such plexus can usually be expressed as a statement, called *predicate*, which accumulates (similitude of integral calculus), say one by one, on exposure to the subject i.e., the decision-maker recursively. There is, however, the difference of a sparking synergic impact for the reason that for every such element a predicate works as catalytic to the other as it is seen to interact, that is, intersect with the other.

---

[1]Though quite interesting, some readers may skip this section until Chap. 6 is read.

1.3 Predicate (Stochastic) Calculus for Psycho-Kinetics …

Padhy (2005) and earlier Tremblay and Manohar (1997) mention a "predicate logic" as a logic that is based upon the analysis of predicates, and define, as in grammar, the object clause of a statement to be the *predicate*. They further state, "Every predicate describes something about one or more objects and able to express a statement symbolically in terms of predicates". For example, they take two statements:

(a) John is a bachelor
(b) Smith is a bachelor

They denote the predicate "is a bachelor", symbolically, by the predicate letter *B*, "John" by *j* and "Smith" by *s* and the statements (a) and (b) could be rewritten as *B*(*j*) and *B*(*s*), respectively. They, thereafter, explain the predicate formula that is a well-formed formula of the predicate calculus by using the symbols (and, or, not, and if then).

Colman (op. cit.) explains predicate calculus as "symbolic logic incorporating apart from propositions, relations between them as of (and, or, not, and if then) = axioms or rules of inference including quantifiers considered powerful form of symbolic logic".

Symbols and Operations Explained

The symbols used here are explained below, but they are not to be taken as strict mathematical operational symbols but as a relationship expressed to be the relations at least in the sense of reasonably high probability of occurrence, needing validation on empirical evidence, rather than propositions of mathematical logic. Latter, no doubt provides better understanding of underlying structure, but still keeps a little away from virtual realities.

The word *stochastic-kinetics* is meant to express the changes in personality development on exposure of *stimulus* (any event, agent, influence or instance, internal or external, which excites a sensory receptor causing response in organism stochastically), as it has been understood and defined here.

Definition Let us suppose the symbol (=) means "produces" or "is equivalent to"; (X) means "interacts" (in a mathematical sense, it is taken as "intersects"); **P** stands for "personality traits"; and **E** for external factors that affect them.

P1 (=) is defined to be, the "impact of information acquisition" on personality traits

E1 (=) is defined as "effect of observation of relevant external entity". In psychological terms, it is called the effect of *stimulus* on personality

Operations

P2(a) (=) **P1**(X) **E1**: **P2**(a) (=) "Cognized Information" (=) "Knowledge"

"The impact of information acquisition", i.e. **P1**, when interacts or intersects with the "effect of observation of relevant external entity", i.e. with **E1**, the stimulus results into acquisition of "knowledge", an entity to enhance "personality".

$$\mathbf{P2}(b) (=) r\mathbf{P2}(a)$$

The three interrelated (correlated) behavioural, emotional and informational forms of "attitude" by Luthans (2005), are the result of repeated (r-times) exposure of knowledge, reflection on personality change, if any. The same corresponds to the impact of repeated advertisement about a theme, manifesto or a product on the *subject* at the present time.[2]

Definition E2 (=) is defined to be a "state of relevant set of entities along with its change or motion if any". Such entities are motivation-causing factors, which consist of needs, drives and incentives aroused by psychological state or changes in them for the subject Luthans (op. cit.).

Operation
P3   (=) **P2**(b) (X)**E2**

Motivation is the outcome of *attitude*, which is **P2**(b) intersecting or interacting with **E2**, that is, with the state of concerned or relevant entity which modifies personality for purported decision or action.

Definition E3 (=) is defined to be trial, demonstration, exhibit or illustration during training as an element of stimulus provided by the trainer or the mentor.

Operations
P4   (=) *Practice* is the outcome of *Motivation* and of Trials: **P3**(X)**E3**(=) **P4**
P5   (=) *Skill* (=) **r.P4**: *Skill* gets developed by the repetition of *Practice*: (an effective direction to personality for purported action or decision)
P6   (=) **P5** (X) **P2**(a)(=); The resultant skill, i.e. **P5** is required to interact or intersect (X) with *Knowledge* = **P2**(a) to yield the faculty of *Ability* denoted by **P6**
P7   (=) **P6** (X) **P3** (=) **HPP**.; such created *Ability*, i.e. **P6**, is required to interact or intersect again with **P3**, the earlier obtained *Motivation* resulting into **P7**, the **Human Performance Potential** (**HPP**)

While on the other end, the external entity **E2** is required to be dichotomized into two components **E4** (=) to represent *Resources* and **E5** (=) to denote *Scope*. Here, *Resource* corresponds to *Resource Cause,* which is *Upadaan kaaran* as defined in Badrayan's (2003) Ved Vyash's *Brahmasutra* the *Vedant-darshan*.

---

[2]Rishi Sanat Kumar's discourse to Narad on the importance of *mati*, that is, "attitude" in *Chhandogyo-Upanishad*, as described further in Chap. 6. Maharishi and Kapil (2007) also in his *Sankhyacarica* discourse No. 23 lays down the basic requirements for decision-making, which includes personality components apart from possession of resources the *ashvarya*.

Operations

E6 (=) **E4**(X)**E5** (=) *opportunity*, i.e., the intersection of *resources* **E4** with *scope*, i.e. **E5**, creates *opportunity*, **E6**

E7 (=) *availed opportunity*: A part of **E6**, i.e., of *opportunity* which is actually availed gets denoted as *availed opportunity* symbolized by **E7** corresponds to *nimitta kaaran* the objective or the purposive cause of the same *Brahmasutra*

Definition P8 (=) Let **HPP** in *action*, an effective component of **P7** to be called *decision* to be represented by **P8**.

Operation

R (=) **P8** (X) **E7**; the *decision-oriented result* (=) **R**.

Thus, each of the above statements (stochastic predicates) is *stochastic modus ponens*. As the process advances, from one stage to the next, it may have to be considered as the *stochastic chain rule*. It is because the human mind does not accept any *stimulus* or its *plexus* or any *predicate* as a string that can be stored as in the memory cell of the computer. The act of learning by human mind is not the same, as putting into the memory cell of the computer. However, there exist exceptions when human mind is made to work almost like computer memory. These are situations when a school-going child is required to memorize multiplication table, sermons of scriptures or poems even without understanding their implications or intent. Still, they are not comparable because the latter are forgettable, if not made to be repeated or brought into frequent use.

Clearly, every intersection denoted by (X) is included here, and it implies union as a prerequisite. Terms such as attitude, decision, memory, motivation, performance, plexus, skill, stimulus and response as used in psychology and its cognate disciplines are available in Colman (op. cit.).

Essentially, many of these concepts are derived from those of Rishi Sanat Kumar, but they are also related to ideas that evolved over the years in modern psychology, particularly in the work of Newstrom and Keith (1997) as reported by Irudayaraj (2005). Ideas related to human psychology presented in logical statement formats above, from **P1** to **R**, should normally be seen as well-formed formulae as initiated by Hilbert (1816–62) and later developed by Ackermann (1896–1962) (see Hilbert and Ackermann (1950) and others). Recently, Tremblay and Manohar (1975), Padhy (op. cit.) and Colman (op. cit.) explained well-formed formula (WFF.) and predicate calculus as quoted above. But whatever have been said above can at best be placed in the category of stochastically well-formed formula (SWFF). So, in the sense that the relations shown above could be effective with a reasonably high probability or their contradiction be non-effective with similar probability. These bring the synthesis explained to virtual reality. Therefore, their empirical validation on availability of real-life data is of much greater relevance than deductive theorems of mathematical logic, hence the name predicate–stochastic calculus.

The concept of synergic effect existed even at the time when Valmiki wrote his *Yogavashistha*. In this text, we have a line in Volume 1, Prakarnan 1, Sarga 6, Stanza 36 that is spoken by the poet when the sage Vishvamitra arrives.

यथो चितासनगता मिथ: संवृद्ध तेजस: । ।
Yathochita ashangata mithah sambridha tejsah.

It means, "Synergic-impact was realized as soon the seers like Visvamitra and Vashistha occupied their coveted chairs in the conference hall". *Synergic* impact has also been described in greater detail by Maharishi and Kapil (2007).[3]

The external situation stated above needs to be perceived as a union of (a) resources (b) scope (c) their state, and (d) their relative law of motion, with which they might change in time and space. One may add here that the entire psychological process is actually the stochastic kinetics of interactive components of rationality, which can produce a result of a decision-oriented action. But, the largest component of rationality synergistically operative at each of the above stages of personality development happens to be the degree of curiosity that is aroused in the subject (that is, the individual human being who is to make a decision or to act, again a decision-maker) when he is exposed to some external event, happening or stimulus.

The immediate question that arises relates to measuring curiosity and that too at every stage in the development of personality. The answer lies in counting or measuring the number of questions such as, how, what, when, where, and which are raised by the subject. These questions are asked when the subject experiences the stimulus, he faces a situation in time and space at every stage of his personality development. For all these, numerous examples are available in the scriptures or texts. The entire scenario of demonstrations, examples, theorems, deductions, teaching and even preaching, done at home by parents, at schools, colleges, training centres and universities come under these activities.[4]

Considering the above-mentioned factors of personality development, it can be seen that all such union and interactions are required to pass through decision complexities.

The term *decision* implies *prima-facie* existence or generation of viable alternatives as an outcome of opportunity. This can be achieved when there is a union of resources with scope, which implies their intersection. Then, amongst them, a selection of one or some of those alternatives that are likely to attain a well-defined objective, which itself can be seen as a result of some need, requirement or preferred choice. Thus, what has been described above involves aspects of

---

[3]Maharishi Kapil was a great exponent of the Sankhya school of Indian philosophy in his *Sankhyakarica-richa* 12. He has been referred to in Sect. 6.2 (Chap. 6).

[4]An empirical study of knowledge acquisition on the exposition of exhibits (stimuli) in a farmers' fair, in terms of information gain due to Thiel (1967), as reported by Das et al. (1975), is presented in Sect. 7.15 (Chap. 7). Such an event or instance is seen to create pathway for personality change/development of the farmer as individual and of their community in decision orientation for improved farming (a risky entrepreneurship for the poor and small farmers).

decision-making capability which is dependent on knowledge, practice, attitude, skill, motivation and by HPP, as explained above, taking consideration of available resources and scope.

Developments that have been outlined above have in all probability contributed to the process of decision-making. They also take into account complexities of any kind. A person is required to handle changes in probabilities of sequence of events and their state of dependencies in terms of correlations as they change in space and proceed in time, and adjustments in decision-oriented actions. That is the main theme of evolution and application of Bayesian thought. Thus, making revision of probability on taking feedback from the system is an important component of decision-making process.

Mankind has made great strides in the development of decision-making in last century. All decisions are products of the human mind and, thus, are related to rationality and the process of thinking. The entire learning process continues from birth to death and involves experiencing, interacting and intervening at stages. These get embedded in memory and make both selectively accumulative and relevantly synergetic impact on the psychology of decision-making. Such a decision-making may need to be made in a situation of uncertainties, dependencies or dynamism, or even any two of them or all the three of them taken together. It is obvious that before any action is performed, an appropriate decision has to be taken. The entire interacting space of intuition, knowledge, learning, practice, attitude and motivation is meant for some decision and ultimately for purported action to fulfil some need or for the sake of a preferred choice. Thus, it negates the concept of "useless knowledge" as put forward by Sir Bertrand Russell and G.H. Hardy who once felt proud of the fact that his theory of numbers is not of any mundane use. Even *Brahmasutra* and similar philosophical texts are full of concepts for the mundane use of mankind. This is because the domain of applications has been expanding very rapidly to engulf any knowledge within its fold, which is complemented by equally expanding computing power and urge to derive benefit from them so that some need may be fulfilled or some choice may be made. Thus, one has to assume that man is a rational being, and with the help of that rationality, he has to make choices about what, how and when to do something. Sometimes such a choice has to be made in the absence of certain knowledge or with incomplete information.

Whatever be the situation of decision-making, it can relate to piles of fallible, peripheral and perhaps irrelevant set of information that are available at the time of decision-making. The effectiveness with which such information is scanned, assorted, processed, encrypted, stored, transmitted, decrypted, retrieved and, finally, cognized may effectively intervene or even control the appropriateness and effectiveness of the resulting decision.

In fact, cognizance of appropriate knowledge as an integral part of rationality dates back to the Vedic and Vedantic (post-Vedic) era, which gets reflected in

Chandogya Upanishad (2004), where Rishi Sanat Kumar explains to Narad about the importance of knowing or measuring mati, or the attitude.[5]

However, since the middle of the last century, considerable interest in the development of decision-making processes did arise in which mathematicians, statisticians, psychologists, economists and philosophers made a joint participation. It led to the formation of the concept of providing logical structures of empirical models of decision-making and testing their validity (Thrall et al. 1960). In this field, some well-known contributors, apart from these editors, were Bush, Debreu, Goodman, Hoffman, Marschak, Mosteller, Nash and Radner.[6]

1.4 Truth and Error

Ever since mankind became conscious of the fact that the truth behind any phenomenon remains canopied by untruths or, errors, and so efforts are made to uncover the truth. The above theme again gets reflected in *Brihadaranako Upanishad* (2004, Chap. 1, Brh.3: 28) and *Ackachhopanishada* (Part 1) (2001), wherein their seers are seen to express mankind's long cherished desire in the following words:

"असतो मा सद्गमय"
Asto-ma sada-gamaya

It means the "aspiration to come out of untruth to proceed towards the truth", implying the desired and sustained effort needed to minimize the error component to achieve the truth in its real form. In effect, it is seen to give a clarion call for breaking through the canopy of error to obtain the kernel of truth. Again in *Chandogya Upanishad* (op. cit.) (Chap. 8, Pt. 3:1), we find the following:

"त इमे सत्याः कामा अनृतापिधानास्तेषाँ सत्यानाँ सतामनृतम
पिधानम यो यो ह्यस्येत् प्रैति न तमिह दर्शनाय लभते ‖ 1 ‖"
Ta ime satyah kama anrit-apidhan-astesham satyaanaam sat-anritam, apidhanam yo yo hyasyetah praiti- na- tamih darshnaya labhate. [1]

It means that truth (*satya*) remains canopied (*api-dhana*) by the untruth; the error component (*anrita*), that is, is repeating as though it is natural for the truths to remain encapsulated by errors (untruth). For example, rice grain on the ear-head of its plants that remain covered by the husk is called *dhan*. Similarly, any natural product such as fruit, vegetable and even food grains is seen to remain covered by some kind of skin or husk, even human being at birth lies inside the placenta. They are required to be removed like error so that the truth comes out.

---

[5]As mentioned earlier as well as explained in Chap. 6.
[6]The present development, as would be seen in Chap. 6, may be regarded as continuum of the same as they thereafter continued their spate of research and publication.

1.4 Truth and Error

In fact, it is one of the greatest realizations that the truth (*satya*) remains encapsulated (*apidhana*) by untruth, the error or the transience (*anrita*), which is seen to create complexities in decision-making or in attaining the desired goal (*kaama*).

What Is Truth? and What then Is Error?

Truth: Radhakrishnan (1971, 2009) presented an analysis of what has been contended as *truth* in Indian philosophical scriptures as below:

> In the theory of *prama* or truth, the *naiyayika* sets out to inquire how the claim which we implicitly grant is justified.

Here, it must be added that such justification must be perceivable and cognizable in time and space. Clearly, such concept of truth must first come out with attribute of some sort of permanence or stability worthy of being called "constant", at least for some well-known or well-defined space within which, or alternatively, outside which, but not both simultaneously. It does not mean that any variation, or for that matter even acceleration, is not within the ambit of truth. But such features or attributes, if any, should be known or reported to mankind. Secondly, according to Radhakrisnan's interpretation of *naiyayika* "any phenomenon or concept worthy of being called true implies verifiability of its existence, acceptable to mankind, through their five sense organs". This, as explained earlier, will be acceptable to such a group that is in a position to cognize its functions or workability or differentiate its existence to its non-existence and, if possible, explain or demonstrate it. Again, he says that "It must be made clear that such function or workability is only the test of truth and not its content. Thus, the attributes of truth being verifiable, workable, being worthy of demonstration and differentiation only narrows down its domain and brings it down to the fold empiricism, implying that it is being equated with reality. It must be realized that a judgment or assertion is true not because it possesses the above stated attributes but it meets them all, because it happens to be true. Thus, it cannot and must not be contradicted. At the same time, man who is unaware of existence of such truth has to adopt tests and bring the same to the fold of empiricism, implying experimentation wherever and whenever, possible or by repeated observations as circumstances may permit".

Error: To quote Radhakrishnan again, (op. cit.), *error* can be explained as follows:

> *Pramana*, or valid knowledge, is distinguished from (*samsaya*) and erroneous knowledge (*viparyaya*), where the ideas do not lead to successful action. Illusion and hallucination fail to realize their ends, i.e. do not fulfil the expectations roused by them. We become conscious of error when the demands of our ideal past are not met by the present. We see a white object and take it to be silver, pick it up and find it to be a piece of shell. The new experience of shell contradicts the expectation of silver.

According to *Nyaya*, all error is subjective. Vatsyayana says:

> What is set aside by true knowledge is wrong apprehension, not the object... No wrong apprehension is entirely baseless. Error is the apprehension of an object as other than what it is. This view of *anyathakhyati* (error) is supported not only by the Jaina logicians, but also by Kumarila. But such apprehension is not totally unfounded. A non-existent thing

cannot produce any effect. Erroneous cognitions cannot be traced to residual impression, which are not possible without real objects.

From the above philosophical discussion, it becomes clear that the cause of error lies not in its existence or non-existence but mistaking one thing for the other, where both the *truth* and the *error* do exist simultaneously. It is also evident that one of them is nearer to the other, or similar to the other in some respect or attribute, giving a *prima-facie* reason for committing an error or apprehending one for the other. Thus, *truth* must be separated from *error* or the latter must be minimized to the extent possible.[7]

1.5 The Main Users

The use of bivariate quantile pairs, being very simple, has the great potential of offering a very wide scope of applications in an uncertain decision-making process. They can be used to tackle the complexities arising out of risk, dependence and dynamism.[8]

For example, given the probability level of, say, 95 %, one is required to use the corresponding quantile value of 1.96 for two-tailed decision for the test of significance or for building confidence interval. Similarly, for one-tailed decision or test, the corresponding quantile value is 1.645. But, what are their corresponding values when variables are more than one, and they are correlated?

Consider the following problems from the fields of pattern recognition, or classification, and its cognate disciplines such as archaeological and anthropometric classification, biometrical identification, genomics and proteomics, geographic or geological information systems, information and communication sciences, medical diagnostics, remote sensing and weather pattern studies[9]:

1. For obtaining possibly condense confidence interval with the help of tables of biquantile pairs on variables being correlated in pairs.
2. For discrimination, classification and clustering, where variable correlation is essentially required to be considered.
3. For optimization and its vast field of applications, especially joint chance constrained programming under risk and dependence.

---

[7]Further discussion on this aspect is presented in Chap. 6. The potential users of the text and the tables have been identified in Chap. 8.

[8]The potential users of the text, which contain tables generated for bivariate normal quantile pairs, are all those who presently use univariate normal quantile [Sheppard's (1939) Table] in decision-making at a given probability level. They are looking for similar quantile pairs or equi-quantile values for two variable cases when such variables are pairwise correlated (Chap. 8).

[9]Examples given in sections of Chap. 7 present several classes of problems that may draw the attention of users who may be from these fields.

1.5 The Main Users

4. For probabilistic meteorological and weather prediction, where variables are pairwise correlated.
5. For bio-statistical studies, showing pairwise variable correlation.
6. Value at risk (VaR) for two-asset risk, especially when such assets are pairwise correlated.
7. Biquantile pairs; providing alternative to Gaussian copula model.
8. Bivariate quality control charts for pairwise correlated variables, and bivariate Six Sigma limits for similar situations.
9. Bivariate stochastic process, making lag prediction; a step towards their shortest joint confidence interval considering autocorrelation or cross-auto-correlation.
10. For practitioners of simulation while seeking joint confidence cover for their results and estimates based on sequence of simulation runs as it is well established that such runs are seldom independent.
11. For seekers of condensed confidence interval for determining the size of boson (the God) particle, which are known to be highly positively correlated, the Bose–Einstein correlation (BEC).
12. For solution of Rizopoulos paradox and its generalization.
13. For any area, where stochastically condensed and valid confidence cover or such shifting cover is sought for estimate or prediction, like such cover to sliding window prediction.
14. An experiment on information gain with its impact on farmer's personality changes towards improved farming when more than one variables or personality attributes show correlation or interdependence.

1.6 Chapter Plan

Keeping the systematic process in view, chapters have been developed culminating into the development of heuristic algorithm for generation of tables for iso-probable bivariate quantile pairs that resulted in producing its tables for well-defined grids and for reliability testing of the software so prepared. Finally, it has been done keeping in view their manifold applications.

Section 2.1 introduces the concept of decision complexities and its tetrahedral structure with its triangular base. Such representation explains the phenomenon that complexities can be explained as composition of three components due to: (a) uncertainty, (b) dependence or association and (c) dynamism. Those can be represented as three vertices of the base triangle of the said tetrahedral structure of complexity. Section 2.2 deals with the vertex V_1, that is, the complexity that arises on account of uncertainty, dealing with development of the concept of uncertainty through the ages, in brief, in the Vedic era and the present time. Its use in decision-making along with its measure gives only sporadic glimpses of vast developments and their applications as aid to solving decision-making under uncertainty, which has been and is being encountered most often in real-world scenario. Section 2.3

presents probability and its measure, rediscovered in Europe. Section 2.4 introduces probability distribution. Section 2.5 outlines probability in time domain. In Sect. 2.6, the vertex of dependence, association and their measures in the Vedic and post-Vedic era have been dealt with. Section 2.7 deals with the perception of cause and effect as independent and dependent events. Reference of probabilistic causal algebra and evolution of effect as a measure of dependence to probabilistic cause have been briefly touched in Sect. 2.8. In Sect. 2.9, the emergence of correlation and regression coefficients has been briefly described. Section 2.10 refers to the analyses of correlation, as path coefficients, factor analysis, and of its equivalent Karhunen–Loeve transform. Section 2.11 introduces to the advancement made in concepts and computations for measures of association and dependence. The discriminant function and some other measures of dependence have been dealt with in Sect. 2.12. Section 2.13 introduces the concept of default correlation and *copula*. The apex vertex of dynamism, Markov chain, Brownian motion and stochastic differential equation have been briefly described in Sects. 2.14, 2.15, 2.16 and 2.17, respectively.

Chapter 3 deals with univariate probability distribution, its characterization as adopted from Ludeman (2010) and the concept of quantile. Section 3.1 presents probability distribution, its characterization and its application in decision-making under uncertainties. Sections 3.2 and 3.3 introduce to univariate normal probability distribution and confidence interval, respectively. Section 3.4, on the other hand, describes quantile and its role in decision-making. Certainty equivalence and quantile are described in Sect. 3.5.

Chapter 4 has been developed to deal the basics of joint probability distribution and its characterization and computability of biquantile pairs. The characterization of bivariate normal distribution, as adopted from Ludeman (op. cit.), has been presented in Sect. 4.1. Sections 4.2 and 4.3 present a historical perspective and problems regarding its integrability, respectively, along with other applicable properties of bivariate normal distribution. Section 4.4 traces Owen's (1956, 1962) computational scheme for evaluating bivariate normal integral, which provides the basis for heuristic algorithm for the development of software for computing the quantile tables.[10] In Sect. 4.5, some other methods for evaluating bivariate normal integral are referred to. Section 4.6 presents the development of heuristic algorithm for the software BIVNOR for generating tables of bivariate iso-probable quantile pairs. Section 4.7 identifies the problem of multiplicities of such biquantile pairs. In Sect. 4.8, lays down initial steps of heuristic for BIVNOR. Problems arising in generating tables of iso-probable biquantile pairs have been identified in Sect. 4.9. Section 4.10 presents a strategy to meet problems raised in Sect. 4.9. Sections 4.11 and 4.12 deal with simultaneous (joint) confidence intervals, whereas recent interest towards such solutions, which could be met by the use of biquantile pairs, has been quoted from Internet search. The problem posed by Rizopoulos (2009) has been stated in Sect. 4.13.

---

[10]These tables are presented in Chap. 8.

1.6 Chapter Plan

Chapter 5 is about software reliability testing. Section 5.1 introduces software reliability, while Sect. 5.2 describes evolution of criteria for standardization of research software as has been devised. Section 5.3 is about the procedures adopted to meet the requirements set in Sect. 5.2. Section 5.4 gives a description of and details about the tables.[11]

Chapter 6 is about the most important aspects of this text—decision scenarios. The chapter has twelve sections. Sections 6.1 and 6.2 deal with decision scenarios of the past and the present, respectively. The said decision scenario (past), as preached by Rishi Sanat Kumar in *Chandogya Upanishad,* is one of the principal texts of the Vedantic era. In the subsequent sections, some of the recent developments have been presented precisely. They are advancement in decision processes —decision as science, design of experiment and Wald's (1950) statistical decision function as relevant to the topics of this text in Sects. 6.3, 6.4 and 6.5, respectively. Sections 6.6, 6.7 and 6.8 contain introductory aspects of discrimination, optimization and stochastic programming and its chance constrained version with some relevant references and a short discussion about von-Neumann versus Kahnemann and Tversky. The emergence of decision-making processes, under risk and uncertainty, is the subject matter of Sect. 6.9. Section 6.10 deals with the developments in uncertain decision-making processes, and some benign comments on and excerpts from Kolbin's (2002) treatise on "Decision Making and Programming". Sections 6.11 and 6.12 have short discussion on online stochastic combinatorial optimization (OSCO) and fuzzy decision and its relevance to the topics to this text.

Chapter 7 on application paradigm is the largest section containing fifteen sections, illustrating application of biquantile pairs in different types of decision problems in its first fourteen sections leaving aside the Sect. 7.1 which is on short comment of Rao (2006) on importance of application of statistical methods. A comparison between different methods of making joint confidence interval and showing the useful role of biquantile pairs in shrinkage of such joint confidence intervals is illustrated through four examples in Sect. 7.2. Section 7.3, on the other hand, presents a new device which applies a heuristic algorithm of using biquantile pairs in deriving stochastic version of discriminant function. It has some additional advantages over the existing methods. Section 7.4 is another very important and innovative, as it illustrates how a joint chance constrained problem can be converted into its certainty equivalent by the use of biquantile pairs for specified risk (probability) level and for any given or computed correlation value as well as how optimal solution can be obtained by alternating sequence of two-stage iterative techniques. In its first stage, iterations are done for the choice of the most optimal pairs of biquantile amongst infinite, but tangible pairs. In the second stage of iteration, the regular simplex algorithm is adopted for each choice of the first-stage solutions. The said section also suggests a classification of entrepreneurs or decision-makers into four groups, starting from risk averting group to high-risk-prone group. Section 7.5 deals

---

[11] See Footnote 10.

with bivariate meteorological studies, wherein advantageous application of biquantile pairs has been demonstrated through extension of two published examples for obtaining more efficient and shorter confidence intervals, for the same confidence coefficient. Section 7.6 presents, again, the advantageous application of biquantile pairs in bio-statistical studies. Similarly, the application of such biquantile pairs, for VaR of the two-asset case, where computation of biVaR is implicit, has been demonstrated in Sect. 7.7. Section 7.8 discusses Gaussian copula model through Hull's (2007, 2009) examples and shows the advantages of using of biquantile pairs in such cases, where the correlation between two variables are directly estimable. Interestingly, the introduction to biquantile pairs for pairs of correlated characteristics in quality control problems and in much hyped Six Sigma techniques dealt with in Sect. 7.9 have been shown to be of great advantage, so much so that for such cases Six Sigma technique shows the potentiality of shrinkage of control chart to much less than even four Sigma techniques, causing a huge reduction in resources and eventually of cost. Section 7.10 is on bivariate stochastic processes in which the use of auto-correlation and cross-auto-correlation yielding, respectively, different equi-quantile pairs have been shown to be of advantage in building shorter confidence predictive intervals. Section 7.11 illustrates application of biquantile pairs/equi-quantile value to fulfil the long-awaited aspirations of simulators for valid condensed confidence interval for sequence of simulation runs in spite of the fact that as such runs are known to be correlated. In Sect. 7.12, prospect for finding highly condensed confidence interval for determining magnitude of Higgs boson (so-called God Particle) for known or estimated magnitude of BEC appears possible by using of biquantile pairs/equi-quantile value. This has been explained through an example from the Huang's (2012) text. The attention has been drawn to Rizopoulos's problem, which is named Rizopoulos paradox, because it is very likely that the problem has not met the solution till date. However, its solution has been placed in Sect. 7.13. Section 7.14 lays down the scope for some further advantageous applications of biquantile pair/equi-quantile value.

The concluding section (Sect. 7.15) presents an example of an experiment for information gain in a farmer's fair as the first step of information acquisition in quantitative terms in a series of several stages of personality development for a specific task or for some purported action to meet some specified objective.

Chapter 8 presents two sets of generated tables. The first set contains three sheets of table of equi-quantile values for probability values 0.99 through 0.5 and correlation values ranging from +0.95 to −0.95 including its value 0.0, titled Table 8.1. The second set of Table 8.2—(N = 1–400) contains four hundred tables of biquantile pairs for every grids of Table 8.1.

Chapter 9 contains three sets of tables (Tables 9.1, 9.2 and 9.3) and Sect. 9.4 contains barycentric coordinates. These tables are results of reliability tests of the software BIVNOR which has been developed to be used for generating Table 8.2—N (1–400) in Chap. 8.

Chapter 10 gives conclusions that have five sections. Section 10.1 is a small section of introduction to the chapter. Section 10.2 is on conclusions which concisely enumerates the contents of the entire text. Section 10.3 is on caveat and cautions which points out some abnormal features resulting from BIVNOR for some cases of rare use. In Sect. 10.4, some questions regarding the ultimate gain, that can be had from this text, have been posed and also answered. At last, Sect. 10.5 on Feller's (1972) dictum and Winston's (2004) aspiration has been quoted to highlight the indispensability of the consideration of dependence to achieve a long-awaited freedom from the yoke of very restrictive and unrealistic hypothesis of independence.

References

Ackachhopanishada.: Upanishad-Anka. Geeta Press, Gorakhpur (2001)
Badarayan's Brahmasutra: Vedanta-darshan. (Translated into English by Vireswaranand, S., Adidevanand, S.: Advait Ashram, Calcutta (1998). Gita Press, Gorakhpur (2003)
Brihadaranako Upanishad. Gita Press, Gorakhpur (2004)
Chandogya Upanishad Rishi Sanat Kumar's Preachings to Narad, Ch. 7, pp. 653–745., Ch. 8, Part 3, pp. 769–770. Geeta Press, Gorakhpur (2004)
Colman, A.M.: Oxford Dictionary of Psychology, 3rd edn. Oxford University Press, New Delhi (2009)
Das, N.C., Sinha, H.S.P., Ranjan, R.: Information gain in Kishan Mela. Indian J. Extension Educ. **11**(1–2), 19–24 (1975)
Feller, W.: An Introduction to Probability Theory and its Applications, vol. 1. Wiley India, New Delhi (1972)
Feller, W.: An Introduction to Probability Theory and its Applications, vol. 2. Wiley India, New Delhi (2009)
Hilbert, D., Ackermann, W.: Principles of Mathematical Logic. Chelsea Publishing Company, New York (1950)
Huang, K.: Statistical Mechanics, 2nd edn. Wiley India, New Delhi (2012)
Hull, John C.: Risk Management and Financial Institutions. Pearson Education, New Delhi (2007)
Hull, John C.: Options, Futures and other Derivatives. Pearson Education, New Delhi (2009)
Irudayaraj, I.S.F.: Understanding Organizational Behavior. X.L.R.I, Jamshedpur (2005)
Kathopanishad: Ishadi Nav Upanishad. Geeta Press, Gorakhpur (2004)
Kolbin, V.V.: Decision Making and Programming. World Scientific, Singapore (2002)
Ludeman, L.C.: Random Processes, Filtering, Estimation and Detechtion. Wiley India, New Delhi (2010)
Luthans, F.: Organizational Behavior. McGraw Hill, New Delhi (2005)
Maharishi, Kapil: Saankhyakaarikaa. Chawkhamba Krishnadas Academy, Varanasi (2007)
Newstrom, J.W., Keith, D.: Understanding Organizational Behavior: Human Behavior at Work, 10th edn, pp. 5–17 and 23. McGrawHill, India (1997)
Owen, Donald B.: Tables for computing bivariate normal probabilities. Ann. Math. Statist. **27**, 1075–1090 (1956)
Owen, D.B.: Handbook of Statistical Tables, Sec. 8.10 and 8.11. Addison-Wesley, Mass (1962)
Padhy, N.P.: Artificial Intelligence System and Intelligence System. Oxford University Press, New Delhi (2005)
Radhakrishnan, S.: Indian Philosophy, vols. 1 and 2. Oxford University Press, New Delhi (1921, 1971, 2013)
Rao, C.R.: Linear Statistical Inference and its Applications, 2nd edn. Wiley, New Jersey (2006)
Rizopoulos, D.: Quantiles for bivariate normal distribution. https://www.R-project.org/posting-guide.html/http://www.r-project.org/posting-gui (2009)

Sheppard, W.F.: The Probability Integral. British Association Mathematical Tables, vol. 7. Cambridge University Press, Cambridge (1939)

Thrall, R.M., Coombs, C.H., Davis, R.L.: Decision Processes. Wiley, New Jersey (1960)

Thiel, H.: Economics and Information Theory. North Holland Publishing Company, Amesterdam (1967)

Trembly, J.P., Manohar, R.: Discrete Mathematical Structures with Applications to Computer Science. Tata McGraw-Hill, New Delhi (1997)

Valmiki, A. (3000, B.C.).: Yogavashistha: Maha-Ramayanam, vol. 1, Ch. 2, Sarga 10. (Translated from Sanskrit to Hindi by Pandit Thakur prasad Dwedi (2001). Chaukhambha Surbharati Publication, Varanasi (2001)

Wald, A.: Statistical Decision Function. Wiley, New Jersey (1950)

Webster's: The New International Webster's Comprehensive Dictionary of the English Language. Typhoon International Corporation, Florida (2004)

Winston, W.L.: Introduction to Probability Models: Operations Research, vol. 2, 4th edn. Thompson Learning, Singapore (2004)

Chapter 2
Decision Complexity and Methods to Meet Them

In this chapter, components of decision complexity can be seen as a tetrahedral structure with complexity as vertex in third dimension. Its base is a triangle with three vertices V_1, V_2 and V_3 (see Fig. 2.1). V_1 represents the vertex of uncertainty, its conception and developments through the ages from the Vedic to the present era. These have been dealt within Sects. 2.2, 2.3, 2.4 and 2.5, which provide a genesis for such developments. The next component of complexity represented by the vertex V_2 is dependence, which has been discussed in Sects. 2.6, 2.7, 2.8, 2.9, 2.10, 2.11, 2.12 and 2.13. Mankind's awareness of its existence and the methodologies employed to explore and circumvent it have also been traced right from the Vedic and the post-Vedic era. However, only such methods have been mentioned which are popularly known and have been applied frequently and widely by empirical scientists. The third and the last component of complexity, that is dynamism, has been represented as the vertex V_3, the apex vertex of the base triangle of that tetrahedral structure, representing decision complexity. It has been summarized in Sects. 2.14, 2.15, 2.16 and 2.17, only to assert that mankind has been quite progressive on this front also, even though it has been relatively difficult to explore this aspect. This area has been difficult and is dependent on the developments in other areas. Advancements made in this field are relatively recent. Hence, they have found fewer applications, in spite of their prospective use in predictive modelling, essential in decision-making processes.

2.1 Decision Complexities: Triangular Structure

The psycho-kinetics of personality development as a pathway for decision-making processes is already explained earlier, outlining a stepwise cause–effect structure. The present section deals with the nature of complexities being encountered in decision-making and advancements made to resolve them.

Ever since the evolution of civilization, mankind's desire to explore the mysteries of nature, despite its complications and incessant intricacies, has been making a steady progress in unravelling decision complexities. The aim of such a persistent desire has been to exercise some control over and possibly to regulate the natural

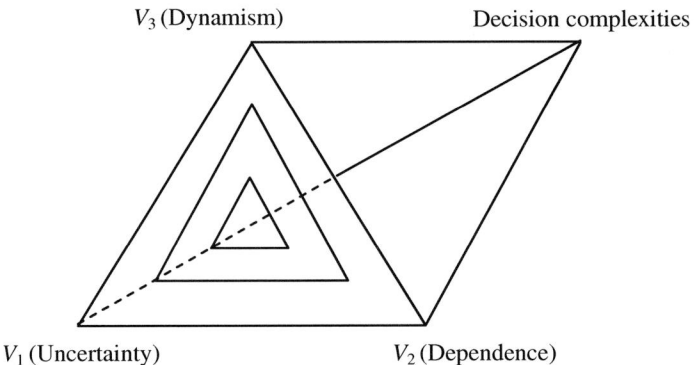

Fig. 2.1 A tetrahedral structure of decision complexities as vertex in three dimensions: vertex V_1 is uncertainty, vertex V_2 is dependence, and vertex V_3 is dynamism

system that has put them in the spiral of ever-increasing challenges at every stage of decision-making.

Decision complexities, on account of such challenges, are seen to polarize mainly around vertices of the tetrahedronal structure with its base as triangle. Each one amongst such complexities has assumingly been assigned a vertex. Factors assigned as vertices, instead of being orthogonal, are generally synergistically interdependent, which can be understood through Fig. 2.1.[1]

These three interacting components of decision complexities and challenges they pose at every stage and step of advancement have been at best as much known to mankind as unknown since the beginning of the civilization. Therefore, the mankind had to face them at every moment in their struggle for existence, subsistence and advancement. What have remained unknown are their nature, magnitude, time and sequence of their occurrences. Therefore, while on the one hand such decision complexities have remained challenges, on the other hand a step in the direction of their solution has always been a new achievement while failure on any front is bound to cause ephemeral dejection and distress, but it has also been a reason for a fresh and invigorated attempt.

It is, therefore, important to identify the three aspects of decision complexities and depict them as three vertices of the base triangle, which are uncertainty as V_1, dependence and association as V_2 and dynamism as apex vertex V_3. These three aspects of decision complexities have been considered as three interacting components. Historical developments made on each of these components can be visualized to exist even during the Vedic and the post-Vedic era. Here, the concern of the Vedic seers was to let mankind realize the degree of cognition of the appropriate concepts that were expressed by three words: uncertainty, dependence and dynamism and their connotations. This happened more than five thousand years

---

[1]This is assessed and estimated, through barycentric or areal coordinates (with coordinate reading system, as explained in Sect 9.4 (Chap 9).

2.2 Vertex of Uncertainty (V_1): Evidence of Probabilistic Thought During the Vedic and the Post-Vedic Period

Man's ability to make a decision under uncertainty grew with time. It developed on account of his training received from his immediate predecessor or from mentor and also as a result of his own experience from his successes and failures. There is enough evidence to suggest that such problems were known to man's early ancestors. There was considerable advancement made in this field in the Vedic period. To illustrate by a verse from the *Rigveda* (2003):

अक्षैर्मा दीव्यः कृषिमित्कृषस्व वित्ते रमस्व बहु मन्यमानः|
तत्र गावः किंतव तत्र जाया तन्मे वि चेष्टे सवितांय भर्यः ||
ऋ०वे० 10-34-13

Akshairma deevyah krishimit-krishasva vitte ramasva bahu manya-manah,
Tatra gavah kitav tatra jaya tanme vi chashte savitaya bharyah.

(*Rigveda* 10-34-13)

It means "Play not with dice: no, cultivate thy corn-land. Enjoy the gain and deem that wealth sufficient: there are thy cattle, there thy wife, O Gambler! So this *Savitar* himself hath told me".

This verse clearly suggests that the farmer should stop gambling and do his farming, which is obviously less risky than gambling. In this way, he can work for his own welfare and also for the sustenance of this world. It can safely be said, therefore, that even in the Vedic period, the concept of uncertainty in the event taking place was known with its consequences. It is also likely that it was being used in decision-making by grading activities on the basis of the likelihood of their happening. The implication is that the concept of probability was gradable then, if not measurable.

Further, the following verses from the *Rigveda* (op. cit.) (the celebrated *Nashadiya-Sucta*; the oracle on creation of the universe) illustrate the fact that people in that period were aware of the state of uncertainty, which is reflected in their statement about their ignorance about the birth or the creation of this universe:

को अद्धा वेद क इह प्र वोचत् कुत आऽजाता कुत इयं विसृष्टिः|
अर्वाग्देवा अस्य विसर्ज नेनाऽस्था को वेद यत आबभूव ||10-129-6||
इयं विसृष्टिर्यत आबभूव यदि वा दधे यदि वा न |
यो अस्याध्यक्षः परमे व्योमन् त्सो अंग वेद यदि वा न वेद ||10-129-7||

Ko adha ved ka pra vochat kut aajaata kut iyam vishrtishti
Arvagdeva asya visarja nenaatha ko ved yat aababhoov.
Iyam vishrishtiryata aababhuv yadi va dadhe va na
Yo asya-dhyakshah parame byoman a tso anga veda yadi va- na veda.

(Rigveda *10-129-6 & 7*)

Max Muller (2005) translates in Macdonell et al. (2005) these verses in these words: "Who knows whence this creation sprang: He from whom all this great creation came, Whether his will created or was mute, The most High Seer that is in the highest heaven, He knows it—or perchance (probably) even He knows not".

In a similar translation by Griffith and Max Muller (2007), instead of the word "perchance", the word "perhaps" has been used, connoting again the word probably.

This concept of uncertainty as expressed through the use of words such as "possibility" and/or "impossibility" was very much a matter of discussion and discourse during the Vedic (post-Vedic) period, as is evident from *Badrayan Brahmasutra* attributed to great philosopher and Badrayan (2003). Badrayan says that it was possible to attain a superhuman status such as the Gods or the seers. It was because a superhuman has a greater capacity to meditate upon the supreme power. Evidently, he uses the term "possible" in the sense of "probability", a state can be seen as gradable. He also used the term "impossibility" in a verse of his that denies the existence of any deterministic relationship between cause and effect in this creation. His use of the term "satya" (truth) implies that a particular phenomenon is deterministic.

Similarly, aadikavi Valmeek in his *Yogavashistha* (2001) in its Chap. 2 uses the word *Niyati* (fate) that clearly connotes the existence of uncertainties in the events taking place in the world. At places, he uses the word such as *Bhavitabya* and *Sambhavana* (possibility) in a similar context. Thus, there exist many references from which an awareness of the element of probability, stochasticity or uncertainty can be seen expressed in ancient Indian scriptures. Therefore, a decision had to be made and implemented after taking into account all the probabilities, as far as practicable. The concept of probability and its magnitude, apart from its empirical concept, i.e. relative frequency of occurrence, has been expressed by Harney (2003) when he says, "the interpretation of probability as of available knowledge is always possible and is thus the broader interpretation".

Device of Decision-Making Under Uncertainty:
Tulsidas's *Ramsalaka Prashnavali*

Later, Tulsidas (1574) in his *Rama charitmanas* created (15 × 15) a square calling it *Ramsalaka Prashnavali* (that is, questions and answers). He did so in order to enable an entrepreneur or a decision-maker to come to a decision under an uncertain situation. Here, a selection of apparently non-meaningful 25 components of words taken from each of the nine couplets appearing in *Ramcharitamanas* resembles today's systematic random sampling, resulting in the probability of 5/9 in favour and of 4/9 against the decision-maker in a dubious, non-informative or uncertain situation, because five out of nine times equi-probable couplets were so chosen by Tulsidas that they could convey a favourable possibility of outcomes and the remaining four of them could convey a different possibility.

That was perhaps the first attempt to conceive or grade probability as a proportion of a favourable or non-favourable number of events against its total number. Such a method is very similar to Bayesian decision taken under non-informative prior situation attaching equal probability, as expounded by Laplace (1749–1827) much later. The difference between the two (that of success to failure) being 0.055 only in terms of their probabilities. This difference in probability could very well be taken as β, the probability of committing type II error, which should be preferred, as it is sometimes more appropriate than almost the same level (0.05) of α, the probability of committing type I error for business decisions. This happens when the entrepreneur remains mentally prepared to take the risk of accepting wrong hypothesis to be true for reasonably small probability of α being considered for all uncertain statistical decisions. Such a small bias in probability of the magnitude (0.055) tilted towards a favourable decision appears justified on the ground that none of the ventures get started without any prior information.

Radhakrishnan (1971) states, "We do not start with empty minds; we possess information about the nature of the world through experience and tradition". Even empirical scientists are required to formulate hypothesis or null hypothesis on the basis of some information, which is certainly not a wild guess. Thus, the concept of non-informative prior by reformist Bayesianists is a mere formality with little reality in it. It is perhaps only to equate their dicta with those of frequencists. However, it can be interpreted to mean that, at its worst, there exists no prior information with hardly any justification as to why one should always start so pessimistically. In fact, it has been a default choice. Therefore, it is not necessary that is always to be accepted.

Incidentally, rightly or wrongly, such a practice of making a forecast persists even today when an octopus or a parrot has to forecast for the World Cup (FIFA 2010) semifinal and final results by sitting on box or drawing cards.

2.3 Probability and Its Measure as Discovered in European Continent

However, it was much later than the creation of said *Ramashalaka* with its equiprobable settings of nine couplets that Fermat (1601–1665) conceived independently the equal probability events for game situation. Others who contributed to the development of probabilistic thoughts were Pascal (1662) and Hyghens (1629–1695). It was then that the concept of probability as a measure could come into existence. After three generations of Bernoullies, J. Bernoulli (1654–1705), N. Bernoulli (1687–1759), D. Bernoulli (1700–1782), Bayes (1702–1761, 1763), deMoivre (1667–1754), Laplace (1749–1827), Gauss (1777–1866, 1821), Chebyshev (1897) and Markov (1907) the idea of probability as a measure of uncertainty was developed further and could acceptably be given a scale between 0 and 1. Even now, the controversy between frequencists and Bayesianists has not been settled fully. The central theme of such a controversy is subjectivity versus objectivity. Frequencists are hard core objectivists while bayesianists, on the other hand, are ready to accept even subjective information as prior information to revise the posterior probability on fresh objectively obtained information or evidence. However, there has never been any dispute about its limits nor about its scale in between these limits. Of late, the balance has tilted much in favour of the latter as is natural, because it provides a scope for incorporating the past with the present to project the future recursively for updating probability on the basis of fresh information or occurrence of the concerned event. There is no doubt that the evolution of calculus of probability provided a scientific (objective) basis through which one could transform a collection of relevant data into relative frequencies and take the same as an empirical approximation or estimate of corresponding probabilities scaled to lie between an open interval 0 and 1. In that calibrated form, it could be possible to use it in an uncertain decision-making process, which could be claimed to have been done on the basis of a calculated risk. Yet, there is no guarantee that such a decision can be claimed to have been made on the basis of either success or failure. A point that is often missed related to the openness of interval because of the closeness on either side is taken to be a certainty of failure or of success. Hence, it no longer remains in the domain of probability.[2]

2.4 Probability Distribution

The foundations of probability distribution theories were laid by the discovery of binomial probability distribution by Bernoulli (1713) and Poisson probability distribution by Poisson et al. (1837). But, its probability distribution was tabulated by Bortkiewicz (1898) as reported by Kruskal and Tanur (1978); Gaussian or normal probability distribution was tabulated by de Moivre (1733), Laplace (1749–1827) and Gauss (1821) who identified it as the Law of Error. This theory is being used even today. These discoveries and foundations provided the suitable basis for the works of Bienayme (1853) and Helmert (1875) who developed χ^2 (chi-square) probability distribution. They were also used later by Pearson (1948) in his discovery of the Pearsonian system of frequency curves and their fitting. After that, they were employed by Gosset (1908) (pseudonym the "Student") for his first sampling distribution of t. Then, it was Fisher (1921 and 1926) who improved the features of t and developed sampling distribution for r, the sample correlation coefficient. Pearson (1948), however, continued to contribute considerably to the development and application of chi-square statistic and used it as a criterion of the goodness of fit apart from his other contributions like statistical tables including that for bivariate normal probabilities (1930, 1931).

---

[2]It is worthwhile for interested readers to go through the article by Seneta (1981) on the history of probability.

2.5 Probability Updating

Later, Good (1950) and other Bayesianists such as Jeffreys (1967), Savage (1954), Zellner (1971), Atchison and Dunsmore (1975) and subsequently many others used subjective knowledge or that obtained from any other sources to formulate prior probability. Thereby, combining it with information obtained from empirical or experimental sources as likelihood function to revise the probability estimate and distribution function and to call the same as posterior probability and posterior probability distribution function correspondingly. Thus, the Bayesians had the advantage of absorbing prior information in the form of probability to the posterior probability. This feature had the advantage of updating probability distribution recursively. In fact, that was a major breakthrough for those who wanted to use it for predictive purposes. Thereafter, Aitchison and Dunsmore (op. cit.) obtained predictive probability distribution and exploited it for guaranteed prediction, but again such a guarantee was in terms of probabilities. Phenomenal advancements were made in other branches of probability theory and its applications such as stochastic process. A number of developments opened new vistas in the field of probabilistic decision-making in a dynamic situation for single-stage dependence. In this context, the discovery of Markov process (Markov 1907) and the branching process initiated by Galton (1908) and then developed by Feller (1972, 2009), Kendall (1947) and Harris (1963) may be mentioned. Feller (2009) and Howard et al. (1971) initiated the concept of inverse Markov process, and Das (2002) made it computable for simpler situation of (2 × 2) Markov chain and provided a method to backtrack the initial Markov transition probability matrix from its current and subsequent states. The Markov chain and its interplay with Bayesian inference have been discussed in detail in Sect. 2.15.

2.6 The Vertex of Dependence (V_2): Association and Their Measures in Vedic and Post-Vedic Era

In ancient times, man had some knowledge of, and was curious about, the complexity of phenomena and intricacies on account of dependencies. The emergence of the concept of probability as a measure of uncertainty can be cited as an example of such an attitude of curiosity. It is said that there was a time when the ancestors of man were four-footed. Gradually, he learnt to stand on his two legs leaving his forelimbs for other activities like tool making. And then, he discovered fire (अग्नि, *agni*), which he produced it by rubbing two stone chips to create sparks causing flame on pieces of dry wood. The discovery of fire was one of the greatest events in the growth of human civilization not only because man discovered a source of light and heat and learnt the art of keeping wild animals at a distance by lighting a fire, but more importantly, he discovered the principle of cause of effect, that is ignition caused by friction. It is probably because of this reason that the very first hymn of the Rigveda is invocation of the God of fire (*agni*). The hymn invokes the fire God to come to the *yagna* (a ritual to please the Gods) and make it successful. The hymn is recited before the fire is lit. It says:

"ॐ अग्निमीले पुरोहितं यज्ञस्य देवमृत्विजम | होतारं रत्नघातमम् |

Om agnimile purohitam yagyasya devamritvijam
Hotaram ratnaghatmam.

Roughly translated, it means, "Myself being enlightened by *agni*, the god of fire, I pray to you as thou art the messenger of the God, the revealer of truth, the provider of light, heat, energy, power and the mitigator of all evils, yet thou art dynamic and fulfiller of all objectives".

With the discovery of fire, man was able to light the dark night thus increasing his working time. Since producing fire was beneficial to him, he repeated the process of rubbing together stone chips or pieces of dry wood and thus came to understand the relationship between cause and effect. It was such an incident, which ushered in the concept of dependency. Such a concept of dependence is reflected in the following couplet of *Kathopanishad* (1.2.17):

एतदालम्बनँ श्रेष्ठमेतदालम्बनं परम्।
एतदालम्बनं ज्ञात्वा ब्रह्मलोके महीयते ।।17।।

Etda-alambanam shresth-metda-alambanam param,
Etda-alambanam gyatva brahma-loke maheeyate. (17)

It means, "It is He who is the most reliable source of dependence. Knowing Him as the only and ultimate source of dependence, it lets one reach the highest goal".

2.7 Perception of Cause and Effect as Independent and Dependent Events

However, when man failed to produce fire on some occasion, and similar failures in many other pursuits, he became aware of the non-deterministic relationship between cause and effect. Consequently, a desire to decipher the cause was naturally embedded in human behaviour and, in course of time, became a part of rationality. This was a process of the evolution of his rationality, because it was only through such a process that man has been able to exercise at least a partial control over the happenings in the world around him.

It was also realized during the Vedic era that very often one truth, say y, remains hidden by another truth, say x. This idea is expressed in the Rigveda:

" ऋतेन ऋतमपिहितं ध्रुवं वां
सूर्यस्य यत्र विमुचन्त्यश्रवान् " (5.62.1)

Rtena ratam apihantam dhruvam vam;
Suryasya yatra vimuchantya- ashrvaan. (5.62.1)

2.7 Perception of Cause and Effect as Independent and Dependent Events

Kashyap (2003) translates it: "There is that ever-standing truth of yours which is covered by yet another truth. It is where they unyoke the horses of the Sun". The implication is that the absolute, the eternal, the supreme and the divine is veiled by another truth. The truth of many allegorized by horses of the Lord Sun symbolizes multiple sources of energy.

This interpretation of the verse can easily be understood as a phenomenon in which one set of truth is dependent on another set of truths, identified as an independent set. This can be interpreted as

$$y = f(x)$$

where y represents the one truth, and f stands for the functional structure, providing the canopy or the covering of another truth x, or set of truths x_1, x_2, ..., which remain canopied as under:

$$y = f(x_1, x_2, \ldots, x_n)$$

where x_1, x_2, ..., etc., denote multiple energy sources represented by the horses in the chariot of the Sun God. Such a concept as Kashyap's "state upon state" and "travelling to truth" (The Rigveda: 5.19.1 and 5.12.2) suggests that even during the Vedic era, at least single-stage probabilistic dependence, now known as Markovian type, transition structure was cognizable.

These Vedic verses have not been mentioned in order to narrow down their broad philosophical import. It is only to underline the fact that from the very beginning, decision-making has been a problem and various methods were developed to handle that problem. From the very beginning, man has been trying to discover the truth as the Vedantic verse speaks of "getting out of untruth and moving towards the truth".

Similarly, Maharshi Yajnavalkya explained the concept of independence and dependence to his learned wife Maitreyi in *Brihadaranako Upanishad* (3:2:2–9) (2004). Max Muller (2007) translated it. According to Yajnavalkya, there are three *grahas* (or receptors, or info-acquirers in human beings) and their corresponding eight *atigrahas* (functions of those grahas) that are constructed on the same patterns as the syntax and semantics in a programming language (see Table 2.1).

The functions of these receptors (*grahas*) do not overlap or interfere with the working of others.[3] Therefore, in mathematical or statistical terms, it can be said that their functions are orthogonal, independent and naturally separable. But, what happens when *prana* (breathing) stops functioning? Other *grahas* and their *atigrahas* also stop working. Thus, it can be said that *prana* is an independent functionary, while the other *grahas* are dependent functionaries. Thus, those grahas that were seen to act as orthogonal functionaries are actually so only amongst themselves but they remain subservient to *prana*. In other terms, *prana* can be

---

[3] As indicated in Chap. 1.

Table 2.1 Receptors (*grahas*) and functions (*atigrahas*) of receptors

Grahas (receptors or info-acquirers)	*Atigrahas* (functions of *grahas*)
Prana (breath)	*Apana* (down breathing) with Device *nasika* (nose)
Vaak (speech)	Utterance of name, etc.
Jeevha (tongue)	Tasting
Netra (eyes)	Observing or seeing objects
Karna (ears)	Observing or seeing objects through utterances and vibrations
Buddhi (mind)	Imagination, cognition, aspiration, desire and decision
Baahu (arms)	Working
Tvacha (skin)	Touching

considered the Brahman, the supreme lord, while the others may be considered as the lower Gods.

Badrayan (1998, 2003), in his *Brahmasutra*, did classify the cause into two groups, calling one as (निमित्त कारण), that is, the "purposive cause" or efficient cause and the other (उपादान कारण) as the "resource cause". By their appropriate combinations, while taking a decision to act, that is by choice of optimal action (उपयुक्त कार्य), one can get desired result, that is produce, procure, search, transform, or transport resources to fulfil his purposive need or requirement. Further, Badrayan (2003) himself enunciated "भावे च उपलब्धे" (2.1.15), meaning that cause could be deciphered only by realization of effect through repeated action, observation sequence or experimentation. At the same time, Badrayan enunciated

नभावोऽनुपलब्धे ‖2.2.30‖

Na bhavo anuplabdhe. (2.2.30)

It means that the existence is not realizable, if it is not experienced. To be more specific, if there exists no differentiation in realization, it is not possible to decipher the specific cause of an effect. It implies that the cause remains undeciphered until effect (or differential effect) is realized. The intent was to advocate for some action (कर्म) based on gradually increasing knowledge that could be the cause for sustenance, maintenance, regeneration, development and, ultimately, satisfaction as effect, as well as of getting salvation. Thus, in philosophical terms, such a linkage between the occurrence of phenomena was conceived as a cause-and-effect relationship.

The great Indian philosopher Radhakrishnan (1971) quoted the ancient Sanskrit text from *Siddhantaa Muktavali*[4] to assert that mere antecedence is not enough to be assigned as cause. He further says that two things cannot be said to be causally related unless there is the positive or negative (*anvaya-vyatirekiki*) relation between

---

[4]pp. 19–22.

them in such a manner that the presence of the cause means the presence of effect and the absence of the cause means the absence of the effect as indicated in the text quoted above. Other philosophical texts also support of this contention, while there were others who identified three elements of causes:

1. the material causes (*upadana*)
2. the non-material or the non-inherent (*asamvahyi*) cause that is the material cause and whose efficiency is well known. The conjunction (*samyoga*) of the threads is the non-material cause,
3. efficient or the purposive cause. In support of this view, he quotes from *Vaisesika Sutra* (X. 2. 1–7), *Tarkabhabasa*, *Bhasaparichheda*, and *Tarkasamgraha* (40).

Buddhism also believed in the concept of transitive causation in which one state transmits its *paccayasatti* (matured seed) or causal energy to some newly conceived germ and seen as causal relation of the type one finds when one considers the phenomenon of a seed growing into a tree. This clearly reveals that the concept of empiricism existed even during the period of Buddhism.

Another Indian philosopher Brahma (2007) did analyse and related cause and effect as available in Indian scriptures. It may be seen in relation with the recently developed scientific approaches to the problem. Thus, the fact that raises natural question is: Why is it that mankind has always been in search of the cause–effect relationship irrespective of its level of education? The most probable answer, as already stated earlier, lies in the fact that it has become a part of man's rational behaviour to search for the probable cause so that he may be in a position to predict and ultimately to exercise control over the happenings, or on the effect, due to such causal factors by eliminating, enhancing or changing their levels or at least their probabilities of occurrences. Then, he is able to regulate the effect variable by calibrated action on the causal variable. It is this rationality that separates the human mind from those of other creatures indicating a distinct superiority in respect of reasoning that is expandable by experience and education. The synergistic impact of this rational thinking has been so great that man went on to create a vast body of knowledge and, therefore, could control and regulate many phenomena of nature. Such an intellectual activity seems to be never ending. In brief, these facts provide the essentials for research on the cause-and-effect relationship.

2.8 Development of Probabilistic Causal Algebra: Evolution of Effect as Measure of Dependence to Probabilistic Cause

The definition of "causality" as may be mentioned here is that of Suppes (1970). That may be seen as a modification of Humes's relation between cause and effect, which is cited as:

I propose to say that one event is the cause of another, if the appearance of the first is followed with a high probability by the appearance of the second, and there is no third, that we use to factor out the probability relationship between the first and the second events.

In fact, it is Suppes (op. cit.) who developed the probabilistic causal algebra and theorems related to causal events preceding and affecting the probability of occurrence of dependent event in the field of the probabilistic theory of causality. Even before that, Good (1961) developed "causal calculus" as a landmark contribution towards the mathematical formulation of philosophical thoughts to bring the same in the realm of sciences.

2.9 Emergence of Measures of Association and Dependence: Correlation and Regression Coefficients

As an initial attempt to develop a measure of the dependence of effect on cause, Gauss (1777–1866, 1821), Bravais (1846) and Edgeworth (1886, 1893, 1905, 1906, 1908, and 1909) are known to have initiated the concept of "multivariate normal distribution". Galton (1886, 1908) was the first to discover the *correlation coefficient r*, and it was he who introduced the concept of *regression* (then initially termed as *reversion*). But Weldon (1947) soon adopted and modified it by making use of *means* rather *medians* by Galton for standardizing the data and published papers on correlated variation in shrimps.

The history of developments on measures and tests for correlation and dependence are available in some of the articles published in two encyclopaedias. The first was edited by Kruskal and Tanur (1978) and the second by Kotz et al. (1986).

Pearson (1948) developed the version, called the "product moment correlation coefficient" r, now used frequently. He also developed the theory of correlation for three variables. Further, along with his associates, he pioneered the development and application of multiple regressions. Pearson even asked the contemporary philosophers to replace the idea of causation by the category of correlation as reported by Radhakrishnan (1971).

Gosset (1908) next discovered that the "Pearson's product moment correlation, r, was symmetrically distributed about zero according to Pearson type II distribution on the assumption of bivariate normal distribution with $\rho = 0$, for the data". The exact distribution of r was derived by Fisher (1915). It was this result along with his Z-transformation of r, which made it possible to make statistical test of significance for sample correlation coefficient, r, even though it is slightly biased to underestimate the value of ρ, as reported by Olkin and Pratt (1958). Thus, while the correlation coefficient could give the measure of association between two variables, the regression coefficient could be interpreted as measure of dependence of Y on X, where X be independent or *concomitant* variable and Y its dependent. The correlation coefficient could be proved to lie between −1 and +1, where positive correlation could be interpreted to indicate unidirectional association and negative correlation to indicate the association of opposite direction between two such variables. Correlation nearer to unity could be interpreted to mean strong or near perfect association between the variables. The correlation near zero was indicative of little or no association between the variables. Near perfect correlation is likely to be fused with the multi-collinearity problem.

Johnson and Wichern (2003) noted that the presence of high positive correlation causes the probability to concentrate along a line taking the direction (0, 0 to x_1, x_2) on a two-dimensional plane of x_1 and x_2, while high negative correlation causes the probability to concentrate along the line taking the direction (x_1, 0 to 0, x_2).[5]

The regression coefficient, b, of linear regression equation $Y = a + bX$ could be interpreted to mean that for unit change in X, one could expect b times change in the unit of Y. It would increase or decrease with increase in X and would depend on the sign of regression coefficient b to be positive or negative, respectively. The constant a, being the intercept, could be interpreted to mean the value of Y when the value of X was zero.

Thus, the first step of complexity on account of dependence could be solved provided there could be a valid reason for such association or dependence as may be the case and that should be linear. At the same time, such reasons were required to be provided or searched from the discipline from which the data emerged or were drawn. However, mere magnitude or sign of correlation or that of regression coefficient was not to be taken as causal phenomena. Even now, these coefficients are the most frequently used tools for modelling and prediction and also for statistical inference. Thus, devices and measures so developed were seen to solve the complexities and intricacies on account of dependence or association to stochastically considerable extent as it is in use.

2.10 Path Coefficients, Factor Analysis and Principal Components (Karhunen–Loeve Expansion)

The next stage of development in the same direction was the discovery of path coefficient by Wright (1918, 1921, 1934, 1960), Tukey et al. (1954), Vasicek (1977), Moran (1961), Duncan (1966, 1975), Blalock (1971) and Dempster (1971), which could yield a break up of correlations between causal and effect variable into several components along logically causational directed paths having similarity with the structure of weighted directed graphs. However, that could not become

---

[5]The impact of which can be seen on the length of biquantile loci to be dealt subsequently in Chap. 4 and also on effective lengths of range between BIGH and SMHL of corresponding cells in Table 8.2 of Chap. 8.

popular for many decades, but it started to play its role when structural models were used in biometrical genetics, econometrics, psychometrics and other applied and empirical sciences. For all such disciplines to explain the dynamics of cause and effect with their weights (importance) and thus reveal clues for manoeuvrability or that of controllability in decision-making processes under complexities on account of dependence. Recently, Johnson and Wichern (op. cit.) and Hair et al. (2005) have presented an excellent account of path analysis and have added further references of the works of Asher (1976) and those of Li (1955, 1956) to this aspect.

Factor analysis and principal component are other areas of analyses of dependence, providing the orthogonal factor model as an effective method to reduce the number of orthogonal sets of variables. Some early works in this area were by Burt (1938, 1941, 1948), Bordin (1941, 1943), Anderson (1946), Anderson (1984) Bartlett (1954), Darley and McNamara (1940) and Frutcher (1967). However, both the text references, i.e. those of Johnson and Wichern (op. cit.) and Hair et al. (op. cit.), give an excellent exposition of such topics.

Interestingly, Papoulis (1965), Haykin (2001) and later Ludeman (2010) and Duda et al. (2010) found a one-to-one correspondence between the well-known principal component analysis and Karhunen–Loeve expansion widely used in communication theory which exploits its eigenvalue cum eigenvector structure for dimensionality reduction of variables.

2.11 Advancements of Concepts and Computations for Associations and Dependence

Relatively, recent work on causation and dependence has been highlighted in a treatise edited by Glymour and Cooper (2004) which is a collection of papers by the editors and also by Judea Pearl, Chirstopher Meek, Peter Spirtes and Thomas Richardson, mostly in joint authorships of some other scholars. A paragraph from its preface written by Glymour and Cooper (2004) may be quoted here:

> Fully recognizing all the difficulties of causal discovery from uncontrolled and non-experimental data these developments (as uncovered in the said treatise), explore an intricate interplay between assumptions about the data generating process, patterns of association in the data, aspects of causal processes that are consistent with assumption, which can explain the emerging patterns from the data and make predictions about the outcome of interventions that can be made from incomplete causal knowledge.

These developments turn on four sets of ideas:

1. A great variety of causal structure and the patterns of association they imply can be represented by directed graph with an accompanying sets of parameters with values for its parameters specified, a graph implies a definite probability distribution over all the variables it represents.

2.11 Advancements of Concepts and Computations for Associations and Dependence

2. A graphical representation of a causal structure (usually called *model*) and values for the relevant parameters (of the models), the effects of "ideal" intervention—manipulations that fix the values one or more variables without otherwise altering the causal structure—can be calculated and similar calculations can sometimes be carried out when features of the graph and its parameters are uncertain (that is, existence of stochasticity and transformation of such parameter estimates appear to provide scope for their certainty equivalents) as indicated in 3 below:
3. New techniques permit the estimation of parameters in causal graph from appropriate sample data (supplementing the requirement envisaged in (2)).
4. The reliabilities and computational efficiencies of various algorithms for extracting features of directed graphs from sample data can be studied mathematically and tested (statistically) with simulated data and finally (compared) with real-world data, when the relevant causal structures are independently known (ascertained).

Thus, such appropriate algorithms developed for an integrated approach coupled with increasing computing power opens the door for exploring the cause–effect relationships in a much better way for the emerging patterns of interventional options for varied socio-psycho-economic strata of the society or for entrepreneurs coming from varied strata or the levels of risk zone. Nevertheless, in course of such advancements, the role of simulation that gets expressed at Glymour's idea, herein at idea 4, and many others urgently need valid and precise confidence interval which is attainable on use of biquantile pairs/equi-quantile value.[6]

2.12 Discriminant Function and Measures of Dependence

Discriminant function analysis was developed by Mahalanobis (1930 and 1936), Mahalanobis et al. (1937) and Fisher (1936, 1938 and 1940), and its theoretical developments were made by Bose and Roy (1936 and 1938). Thereafter, Rao (1950, 1962, 1966, 2006) added to the phenomena of explaining some facets, factors and features for classifications and group formations based on causally connected components, which subsequently found numerous applications in pattern recognition/classification. Rao (1950) presented the distribution of $(D^2_{p+q} - D^2_p)$ with its computational aspects. In his book (2006), he takes Fisher's (1940) example and shows the advantage of the larger dimension D^2 to have a greater discriminatory power over that of lower dimension one. It clearly shows the advantage of including

---

[6]This has been demonstrated very well through examples of Sect. 7.11 (in Chap. 7).

a greater number of variables for manifestation of more discriminatory power. Besides, the role of Rosenblatt's (1958, 1962) perceptron with its linearly separable feature is being used for a more efficient classificatory discrimination problem (Haykin, op. cit.) and also by Duda et al. (2010).

Developments, applications of correlations and regression for more than a century have become innumerable and so it is not necessary to enumerate them. However, some notable developments are of stepwise regressions, variable selection, Bayesian variable selection, ridge regression, errors in variables, adaptive regression models and their tests, multi-variate regression splines, MINMAD regression and choosing the best regression designs, nonparametric regression, fuzzy regression and their wide range applications in modelling for predictive uses for neural network and artificial intelligence for learning as well as filtering problems. These constitute the main contributions that paved the way for solving a number of problems related to dependencies.

Further, their concepts and measures have been elaborately dealt with by Kumar Jagdeo[7] (1982). Beginning with the correlation in a bivariate case, he briefly discusses Linfoot's (1957), and Schweizer and Wolff's (1976) measures of dependence, besides those analogs of measures like those of Cramer–von Mises and Kolomogorov–Smirnov's and also of several other authors.

Hollander (1982), in his encyclopaedic article, has dealt with tests for dependence and some indices of dependence in which he also deals with Cramer's index of dependence and test based on Pearson's correlation coefficient and least square estimates of regression coefficient b introduced earlier.

Special mention may be made of "Graphical models for probabilistic and causal reasoning", an article by Pearl (2004). Where the Bayesian network has been used as carriers firstly of probabilistic information and then of causal information as its sections and with their respective subsections on semantics, algorithms, system's properties, causal theories, actions, causal effect and chain of action vis-a-vis observation, action calculus with related references.

In fact, such an advancement needs a follow-up by workshops for their real-time online computability.[8] This is for the reason those who ever have examined the magnitude of advancing eigenvalues have experienced that its very first step washes away the largest amount of dependency. It being the largest root: succeeding roots decline very sharply in almost all cases. Its real-life implication is easily understood. That, by implication, reveals the phenomenon that effect of dependency is required to be studied only up to first one or two stages for most real-life cases.

---

[7]Interested readers may refer to Jagdeo (1982) in encyclopaedic article.

[8]Further advancements made through the tables generated and presented in Chap. 8 of the book and are helpful in circumventing the increase in dimensionality up to two, making the user for the first time free from assumption of independence at least by one but most significant step.

2.13 The Default Correlation: Copula

There is another measure of association named *copula*, that has emerged due to Vasicek (1977, 1987 and 2002), which, as defined by Hull (2007, 2009), is a way of defining correlation between variables with known distribution. In fact, as explained by Hull (op. cit.), *copula* is an indirect method of obtaining an estimate of correlation from the marginal distributions of two correlated variables.[9]

2.14 The Apex Vertex of Dynamism (V_3)

Like the two terms *uncertainty* and *dependence*, the concept of *dynamism* also appears to have a Vedantic origin. This is evident from this verse:

चरन्वै मधु विन्दति चरन्स्वादुमुदुम्बरम् ।
सूर्यस्य पश्य श्रेमाणं यो न तन्द्रयते चरन् ।।
चरैवेति चरैवेति ।   ऐतरेय ब्राह्मण ।। 7:3–16 ।।

Charanvai madhu vindati charan-swadu-mudumbaram,
Sooryasya pashya shremaan yo na tandrayate charan,
Charaiveti charaiveti. (Aitareya Brahman, 7:3–16).

It says, "Sweetness embraces those who are *dynamic*. It is he who is dynamic is respected like the Sun at its zenith during the noon. Hence, be on the move and keep on moving (*charaiveti, charaiveti.*)".

What then is *dynamism*? The change occurring in the value of observational or response variable or its state in a unit of time could be diagnosed as dynamic phenomena and, if measurable on a quantitative scale, is to be attributed as a measure for the impact of dynamism. Such change in course of time could be deterministic or stochastic, most often it is the latter. Undoubtedly, these three components of complication, i.e. uncertainty, dependence and dynamism, have been posing a challenge to mankind in the path of its development. However, without these aspects, this world as well as the creation, including life in it, cannot be imagined. What is fundamental to the consideration of dynamism is the concept of time lag. Bellman and Dreyfus (1962) describe the time lag as:

> In the application of the controlling force, introduce a delay or time lag. What this means is that $y(t)$ is actually dependent not upon $x(t)$ and $w(t)$, but rather upon $x(t-d)$ and $w(t-d)$ and more often upon a complex form of the past history of the process; here d stands for the period of time unit in the past on which the present is assumingly dependent.

Such a concept has existed in statistical literatures since time-series analyses and stochastic process came into existence.

---

[9]Further discussed in Chap. 4 and in examples in Sect. 7.8 (in Chap. 7).

As things stand today, a considerable progress has been made in the field of the dynamic aspect of probability through advancement in theory and application of Markov's (1907) process and, in general, Stochastic processes by Doob (1953), Bartlett (1955), Feller (1972 and 2009) and others on one hand, and dynamic programming by Bellman (1957 and 1962) and subsequently others on the other. Additionally, in the same area of further advancement and applications, a landmark contribution has been made by Howard (1960, 1971) and Dynkin and Yushkevich (1980) through respective published texts: (i) Markovian and non- Markovian decision processes and (ii) controlled Markov process. But the process of advancement is continuous and endless and, therefore, non-terminating, as explained by triangular structure in Sect. 2.1.

2.15 Markov Chain and Bayesian Inference

The essence of Markov chain lies in defining the chain of probabilities for the process, where any dynamic system passes from one state to another with assigned or estimated probabilities, provided such states are mutually exclusive and exhaustive and can be numbered ($i = 0, 1, \ldots$) to denote discrete moment of time sequence for the process to proceed. Also, such chains of probabilities have had memory of only one contiguous state, irrespective of its length. Like if x_t assumes a value i at time t, then

$$p_{ij} = P(x_{t+1} = j \,|\, x_i = i)$$

denotes the probability that process, being in state i at time t, makes the transition to state j, in the next step, which also includes the process remaining at the same state i, in its next transition. As such, this model had nothing to do with the revision of probabilities on the basis of some prior/posterior information.

Bayesian inference, based on Bayes' theorem, on the other hand, was little concerned with state-to-state transition probabilities. But, as stated earlier, it was deeply concerned with the revision of probability based on a priori information either subjective (belief etc.) or objectively observed past data or even *prior* probability distribution. Such a revision is sought through posterior information or the likelihood function.

However, the second half of the nineteenth century was a period of renaissance in which a synthesis of both the methods was seen and which yielded a new class of decision models, often called *adaptive decision* models. Though there have been numerous contributions in this area, specific mention may be made of the published texts of some contributors such as Bellman (1962), Martin (1967), Yakowitz (1969), Jazwinski (1970), Tsypkin and Nikolic (1971) and, with some similarity in approach, of Howard (1971) and Levine and Burke (1972). All these treatise, texts and monographs have had a deep impact on the development of learning and decision theories and thus became an eye opener for those who wanted to develop

2.15 Markov Chain and Bayesian Inference

algorithm for computing some practically oriented decision problems, reasonably because many of them appended computer program for their textual problems. A few recent contributions to the field are those of Kolbin (2002), Congdon (2003), Geweke (2005) and of Rossi et al. (2010). In the words of Rossi et al. (op. cit.),

> A goal of statistical inference is to use information to estimate, or to predict about unknown phenomena or feature from available data. But, there is also an undeniable role even for non-data-based information.

Whether such information is based on data or non-data or even both, as are envisaged under Bayesian inference, the inference drawn has to be probabilistic. A well-known important aspect that increases the value of inference is that it provides the scope for incorporating prior information, irrespective of those being data based, non-data based or even of no prior information, latter in the form of defused prior, that being rectangular prior. It is such an aspect of the Bayesian inference that provides a scope for revising or updating the probabilities recursively as one proceeds with decision-making and, simultaneously, updating data set at each advancing step of computation and at each stage of decision-making. Thus, the journey of synthesizing Markovian probabilistic transition with Bayesian state to state updating of probability started by Martin (1967) appears to reach its fruition stage by the works of Rossi et al. (op. cit.) by 2005 when they attempt to solve marketing problems by Markov chain Monte Carlo (MCMC) techniques for which computer programs are available. Those techniques are seen to be using simulation methods.

Similar methods of decision-making have been termed as *adaptive* decision, pioneered by Bellman (1961). Yakowitz (1969) writes,

> The adaptive control process differs from those control process model in assuming that the future state is a random variable related to the current state and controls only through probability law. Also that the said process by virtue of statistical decision theory is an appropriate model for situation demanding control in the absence of complete statistical knowledge.

Tsypkin and Nikolic (1971) also write,

> If it is not clear whether the process is deterministic or stochastic, it is logical to try the adaptive approach, that is, to solve the problem by using learning and adaptation during the process of experimentation.

He further clarifies,

> The adaptive approach is mainly related to algorithmic, or more accurately, iterative methods.

He also says that

> the adaptive approach has to be used under insufficient *a prori* information....

The restoration (estimation) of probability density function and correlation function used in solutions of optimization problem can serve as an example. One of the frequently applicable methods finding use for prediction problem is the Kalman (1969) filter, which recursively utilizes stepwise advancing auto-covariances.

However, it has to be admitted that even now these important developments could not have a significant impact on most of the decisions, related to either problems of business and economics, or of political cum social conflicts, or even of science and technology. A simple reason that can be identified is the ignorance of the main probabilistic decision tools like multi-quantile tuples of which biquantile pairs is the first but significant step.[10] Such an inference is based on the fact of the profuse application of univariate normal quantile for uncertain decision-making problems, even where it should not have been used. They should not have been used in instances of dependence or correlation between the variables. Winston (2004) clearly identified the problem of testing simulation results, as there exists a correlation in between simulation runs, whereas statistical test criteria are based on assumption of independence. Even the said MCMC techniques are not free from such shortcomings.[11]

Unfortunately, for none of the authors of texts, referenced above, the said multi-quantile multiples or even biquantile pairs have been a subject of serious concern, despite the fact that variable dependence or correlation in terms of covariance was always reflected in expressions dealt with, or derived by, almost all such celebrated authors.[12] The significant impact of single step (nearest neighbour) dependence can also be realized when Markovian (single step) decision dynamics changes its transition probabilities under Bayesian prior to posterior by likelihood loading. Such dependencies can be taken care of by auto-correlation and cross-auto-correlation which again needs the use of biquantile pairs or equi-quantile value.

2.16 The Brownian Motion—Weiner Process

Another notable area of advancement towards the study of dynamism implying the time domain is Brownian motion or Wiener's Process (1923). However, one has not been able to find a multi-variate (even bivariate) generalization of Brownian motion (Wiener process) for correlated sequence of variables or even sequence of correlated variables. For independent sequence of variables, however, Friedman (1975) has reported its time homogeneous Markov property. Ikeda and Watanbe (1981) obtained a generalized d-dimensional continuous Brownian version, but again only for mutually independent sequences of Gaussian distributions in time domain. Even its transformation has been discussed for multidimensional scenario, but only for orthogonal groups. Although assumption of independence of such preceding or

---

[10]Shown in examples given in Chap. 7.

[11]Examples in Sect. 7.11 (in Chap. 7) amply demonstrate some such features.

[12]An attempt has been made in Sect. 7.10 (in Chap. 7), to find a scope for a successful application of biquantile pairs for bivariate stochastic process of Box et al. (2004), on two auto-correlated series.

2.16 The Brownian Motion—Weiner Process

succeeding sequences is far away from reality, mankind at sometime or the other must try to free itself from such enigma, even though such an attempt may be just a single step backward or forward. This is because the advancement of civilization cannot and must not be kept hanging only for the sake of simplicity of derivation. This has been the general feelings of many researchers and applied scientists. Only two of them (Feller 2009 and Winston 2004) need to be quoted here:

1. While giving an example on branching processes and nuclear chain reactions, Feller (op. cit.) remarks,[13] "(i) physically speaking, for large numbers of particles the probabilities of fission cannot remain constant and also stochastic independence is impossible." Further, while dwelling on the survival of family names, he says, (ii) "… common inheritance and common environment are bound to produce similarities among brothers, which is contrary to our assumption of stochastic independence. Our model can be refined to take care of these objections…"
2. Discussing the "statistical analysis in simulations" Winston (op. cit.)[14] says, "Determination of the confidence interval in simulation is complicated by the fact that output data are rarely, if ever, independent. That is, the data are auto-correlated. For example, in a queuing situation, the waiting time of a customer often depends on the prior customers. Similarly, in an inventory simulation, the models are usually setup such that the beginning inventory on a given day is the ending inventory from the previous day, thereby creating a correlation. This means that classical methods of statistics, which assume independence, are not directly applicable to the analysis of simulation output data. Thus, we must modify the statistical methods to make proper inferences from simulation data".

One may, therefore, ask: "When will such methods or refinements be envisaged and what may be their basic requirements?"

Whatever be the present situation, whether deterministic, stochastic or adaptive, the mathematical principle of "recurrence relationship" is invariably applicable to move forward or to trace backward by one step or s steps, where s may not only be finite but just a few nearest steps in every considered dimension taken cautiously to avoid "large deviation" and sometimes even "chaotic state". The simplest essential link is the nearest neighbour correlation in time horizon that being a single step lag auto-correlation or cross-auto-correlation (for two parallel series), and the use of corresponding bivariate quantile pairs, as an instrument of decision-making in lieu of pairs of such univariate quantiles, is based on independent assumption.

---

[13]Refer to Feller's *An Introduction to Probability Theory and its Applications*, Vol. 1.
[14]Refer to Winston's *Introduction to Probability Model: Operations Research*.

2.17 Stochastic Differential Equation: ITO's Process

An important mathematical tool developed to handle the problem of dynamism with stochasticity is "stochastic calculus", especially the "stochastic differential equation", often known by the name of its founder as Ito's calculus. Using it, the generalized Wiener process for variable x can be defined in terms of dz as:

$$dx = a\,dt + b\,dz$$

where a and b are constants, that is $dx/dt = a$, the term $a\,dt$ represents expected drift, and the term $b\,dz$ represents additive noise, a Brownian component.

It is a well-established fact that causal phenomenon necessarily precedes effect and at least, *prima facie*, is considered responsible for the change in the state of dependent variable. Therefore, while the former qualifies to be called independent variable, the latter is known as dependent variable.

Thus, changes in attribute or characteristics of variables due to dependence can neither be free of dynamism nor of uncertainties. Nevertheless, the values of probability and of correlation between two variables considered for decision paradigm are assumed to be constant at least for limited domain of decision stage for undernoted reasons:

(a) the simplicity and
(b) the fact that dynamic part is well played by allowing change in the parameters of probability distribution, meaning means, standard deviations and correlation between variables

Therefore, the assumption of bivariate normality of the probability distribution is being considered as state, which has little to do with the principle of dynamism, unless one is required to consider auto-correlation or cross-auto-correlation with its own lagged values of variables or some corresponding auxiliary variables. The standard bivariate normal or any other standard probability distribution is being assumed, or considered, as time invariant except when, as pointed out, one is required to consider lagged values of variables. However, under bivariate set-up, one either chooses pairs of cause and effect variables or pairings with lagged ones or between spaces, but not both at a time.[15]

---

[15]Further, in Sect. 7.10 (in Chap. 7), examples of bivariate-lagged variable values have been given to illustrate the fact that useful application of biquantile pairs is possible in dynamic paradigms considered in a time-series situation.

References

Aitchison, J., Dunsmore, I.R.: Statistical Prediction Analysis. Cambridge University Press, London (1975)

Anderson, G.V.: Factor analysis of attitudes towards community problems (abstract). Am. Psychol. **1**, 462 (1946)

Anderson, T.W.: An Introduction to Multivariate Statistical Analysis. Wiley Eastern, New Delhi (1984)

Asher, H.E.: Causal modeling. Sage University paper series on quantitative applications in social sciences. 07-003. Beverly Hills and London, Sage University (1976)

Badrayan's Brahmasutra.: Vedanta Darshan: Gita Press, Gorakhpur. English Version (trans: Vireswaranand, S., Adidevanand, S). (1998), Advait Ashram. Calcutta (2003)

Bartlett, M.S.: A note on multiplying factors for various chi-squired approximations. J. Roy. Stat. Soc. Series B 16:296–98 (1954)

Bartlett, M.S.: Stochastic Processes. Cambridge University Press, Cambridge (1955)

Bayes, R.T.: An essay towards solving a problem in Doctrine of Chances. Phil. Trans. Roy. Soc. (London), 53, 370–418 (1702–1761); Reprinted in Biometrika 45, 293–315 (1958) and Facsimiles of two Papers (Commentary by W. Edwards Deming). Hafner, New York (1963)

Bellman, R.: Dynamic Programming. Princeton University Press, New Jersy (1957)

Bellman, R.: Adaptive Control Processes. Princeton University Press, Princeton (1962)

Bellman, R., Dreyfus, S.: Applied Dynamic Programming. Princeton University Press, Princeton (1962)

Bernoulli, D.: Exposition of a new theory of the measurement of risk. Econometrica 22, 23–35. First published as Specimen theoriae novae de mensura sortis (1654–1705)

Bernoulli, J.: (posthumous). Bernoulli's Theorem: Ars Conjectandi (1687–1759)

Bernoulli, N.: On the Problem of duration of play in Gambler's ruin Problem. Die Werkevon Jakov Bernoulli, B.L. van der Waerden (Ed.). Basel, Brikhauser, pp. 555–67 (1700–1782, 1975)

Bienayme.: Considerations a l' appui de la de'converte de Laplace sur loides probabilire's dans la me'thodedes moindres carre's. C.R. Acad. Sci. Paris 37. taken from Loeve, M. (1968). Prob. Theo. A E.W. Press, New Delhi (1853)

Blalock, H.W. Jr. (ed.): Casual Models in the Social Sciences. Aldine-Atherton, Chicago (1971)

Bose, R.C.: On exact distribution and moment coefficient of D^2-Statistic. Sankhya. **2**, 143–154 (1936)

Bose, R.C., Roy, S.N.: The distribution of the studentized D^2-statistic. Sankhya. **4**, 19–38 (1938)

Box, G.E.P., Jenkins, G.M., Reinsel, G.C.: Time Series Analysis: Forecasting and Control, 3rd edn. Pearson Education, New Delhi (2004)

Bordin, E.S.: Factor analysis: art or science? Psychol. Bull. **38**, 520–521 (1941)

Bordin, E.S.: Factor analysis in experimental designs in clinical and social psychology. Psychol. Rev. **50**, 415–429 (1943)

von Bortkiewicz, L.: Das Gesetz der kleinen Zahlen. Teubner, Leipzig (1898)

Brahma, N.K.: Causality and Science. PHI Learning, PHI learning (2007)

Bravais, A.: Analyse mathamatique sur les probabilites des erreus de situation d'un point. Memoires presentes par divers savants. France: l Academie Royale des Sciences de l Institute de France. 9:255–332 (1846)

Brihadaranako Upanishad. Gita Press, Gorakhpur (2004)

Burt, C.: Factor analysis by sub-matrices. J. Psychol. **6**, 339–375 (1938)

Burt, C.: The Factors of the Mind: An Introduction to Factor Analysis in Psychology. MacMillan, New York (1941)

Burt, C.: Factor analysis and canonical correlations. Br. J. Psychol. Stat. Section **1**, 95–106 (1948)

Chebyshev, P.L: J. de. Math. **12** (1897)
Congdon, P.: Applied Bayesian Modeling. Wiley, New Jersey (2003)
Darley, J.G., McNamara, W.T.: Factor analysis in the establishment of new personality test. J. Educ. Psychol. **31**, 321–334 (1940)
Das, N.C.: A method of numerical solution of inverse Markov chain. J. Bihar Math. Soc. **22**, 103–110 (2002)
Dempster, A.P.: An overview of multivariate data analysis. J. Multi. Anal. **1**, 316–346 (1971)
Doob, J.L.: Stochastic Processes. Wiley, New Jersey (1953)
Duda, R.O., Hart, P.E., Stork, D.G.: Pattern Classification, 2nd edn. Wiley India, New Delhi (2010)
Duncan, O.D.: Path analysis sociological examples. Am. J. Sociol. **72**, 1–16 (1966)
Duncan, O.D.: Introduction to Structural Equation Models. Academic Press, New York (1975)
Dynkin, Yushkevich: Controlled Markov Process. Springer, Berlin (1980)
Edgeworth, F.Y.: Philos. Mag. **5**(22):371–83 and **5**(36):98–111 (1886 and 1893)
Edgeworth, F.Y.: The law of error. Trans. Camb. Phil. Soc. 20, 36 and 113 (1905)
Edgeworth, F.Y.: The generalized law of error (or Law of large numbers). J. Roy. Stat. Soc. **69**, 497 (1906)
Edgeworth, F.Y.: On the probable errors of frequency Constants. J. Roy. Stat. Soc. **71**, 381–499, 651 and 72, 81, respectively (1908, 1909)
Feller, W.: An Introduction to Probability Theory and its Applications, vol. 1. Wiley India, New Delhi (1972, 2009)
Feller, W.: An Introduction to Probability Theory and its Applications, vol. 2. Wiley India, New Delhi (2009)
Fermat, Pierre de.: Primarily on Problem of Equitable Division of Stakes in Games of Chance (1601–1665)
FIFA.: Media and Newspaper Flash of Football Result Forecast (2010)
Fisher, R.A.: Frequency distribution of the values of correlation coefficient in sample from infinitely large population. Biometrica **10**, 507–521 (1915)
Fisher, R.A.: On probable error of the correlation coefficient deduced from small sample. Metron **1** (4), 1 (1921)
Fisher, R.A.: On application of student's distribution. Metron **5**, 90 (1926)
Fisher, R.A.: Use of multiple measurements in taxonomic problems. Ann. Eugen. **7**, 179–188 (1936)
Fisher, R.A.: The statistical utilization of multiple measurement. Ann. Eugen. **8**, 376–386 (1938)
Fisher, R.A.: The precision of discriminant functions. Ann. Eugen. **10**, 422–29 (1940)
Friedman, A.: Stochastic Differential Equations and Applications, vol. 1. Academic Press, New York (1975)
Fruchter, B.: Introduction to Factor Analysis. D. van Nostrand, Princeton (1967)
Galton, F.: Natural Inheritance. MacMillan, New York (1886)
Galton, F.: Memories of My Life. Methuen, London (1908)
Gauss, C.F.: Gauss's Work 1803–1826 and in Whittaker and Robinson 1924 on Theory of Least Squares (1777–1866:1821)
Geweke, J.: Contemporary Bayesian Econometrics and Statistics. Wiley, New Jersey (2005)
Glymour, C., Cooper, G.F. (eds.): Computation, Causation and Discovery. MIT Press, California (2004)
Good, I.J.: Probability and Weighing of Evidence. Methuen, London (1950)
Good, I.J.: Causal Calculus (1961)
Gosset, W.S.: The probable error of the mean. Biometrika **6**, 1 (1908)
Griffith, R.T.H., Max Muller, F. (eds.): English translation of the Rig Veda. English-Hindi Publisher, New Delhi (2007)

References

Hair Jr, J.F., Anderson, R.E., Tatham, R.L., Black, W.C.: Multivariate Data Analysis, 5th edn. Pearson Education, New Delhi (2005)

Harris, T.E.: The Theory of Branching Processes. Springer, Berlin (1963)

Harney, H.L.: Bayesian Inference: Parametric Estimation and Decision. Springer, Berlin (2003)

Haykin, S.: Neural Networks: A Comprehensive Foundation. Pearson Education, New Delhi (2001)

Helmert, F.R.: Uber die WahrscheinlichKiet von Potenzsummen der Beobachtun-gsfehler. Z.f. Math. u. Phys. 21 (1875)

Hollander, M.: Dependence test. Encyclopedia of Statistical Sciences Kotz et al (ed.) (op. cit.). Wiley, New Jersey (1982)

Howard, R.A.: Dynamic Programming and Markov Process. Wiley, New Jersey (1960)

Howard, R.A.: Dynamic Probabilistic System, vol. 1. Markov Models. Wiley, New Jersey (1971)

Hull, John C.: Risk Management and Financial Institutions. Dorling Kindersley, New Delhi (2007)

Hull, John C.: Options, Futures and Other Derivatives. Pearson Education, New Delhi (2009)

Hyghens, C.: De ratiocinis in ludo Aleae (1629–1695)

Ikeda, N., Watanbe, S.: Stochastic Differential Equations and Diffusions Processes. North Holland Publishing Company, Amsterdam (1981)

Jazwinski, A.H.: Stochastic Process and Filtering Theory. Academic Press, New York (1970)

Jeffrey, H.: The Theory of Probability, 3rd edn. Oxford and Clarendon Press, London (1939: 1967)

Johnson, R.A., Wichern, D.W.: Applied Multivariate Statistical Analysis, 3rd edn. PHI Learning, New Delhi (2003)

Kalman, R.E.: A new approach to linear filtering and prediction problem. ASME J Basic Eng. **91** (March), 35 (1969)

Kashyap, R.L.: Rig Veda Mantra Samhita. Sri Aurobindo Kapali Sastry Institute of Vedic Culture, Bengaluru (2003)

Kendall, M.G.: The Advanced Theory of Statistics, vol. 1. Charles Griffin, London (1947)

Kolbin, V.V.: Decision Making and Programming. World Scientific, Singapore (2002)

Kotz, S., Johnson, N.L., Read, C.B. (eds.): Encyclopedia of Statistical Sciences. Wiley, New York (1986)

Kruskal, W.H., Tanur, J.M.: International Encyclopedia of Statistics. Macmillan, New York (1978)

Kumar, J.: Concept of dependence. Encyclopedia of statistical sciences, vol. 2. deGroot et al. (eds.). Wiley, New York (1982)

Laplace, M.P.S.: A philosophical study of probabilities. First Published as Essai philosphique sure les probability's. Dover, New York (1749–1827)

Levine, G., Burke, C.J.: Mathematical Model Techniques for Learning Theories. Academic Press, New York (1972)

Li, C.C.: Population Genetics. University of Chicago Press, Chicago (1955)

Li, C.C.: The concept of path coefficient and its impact on population genetics. Biometrics **12**, 190–210 (1956)

Linfoot, E.H.: An informational measure of correlation. Inform. Control. **1**(1), 85–89 (1957)

Ludeman, L.C.: Random Processes, Filtering, Estimation and Detection. Wiley, New Jersey (2010)

Macdonell, A.A., Max Muller, F. Oldenberg, H.: The golden book of the holy vedas. Vijay Goel Publisher, Delhi, India (2005)

Mahalanobis, P.C.: On tests and measures of group divergence. J. Proc. Asiat. Soc. Beng. **26**, 541–588 (1930)

Mahalanobis, P.C.: On generalized distance in statistics. Proc. Nat. Inst. Sci. **2**, 49–55 (1936)

Mahalanobis, P.C., Bose, R.C., Roy, S.N.: Normalization of statistical variates and the use of rectangular coordinates in the theory of sampling distribution. Sankhya **3**, 1–40 (1937)

Markov, A.A.: Extension of limit theorems of probability theory to sum of variables connected to a chain: investigation of an important case of dependent trials. Izvestia Acad. Nauk. SBP VI Ser. 1: 61. (From Howard's Dynamic Probabilistic Systems, vol. 1. Markov models, Appendix B. Wiley, New Jersey) (1856–1922: 1907)

Martin, J.J.: Bayesian Decision Problems and Markov Chains. Wiley, New Jersey (1967)

Max Muller, F. (ed.): Rig Veda. Vijay Goel Publisher, New Delhi (2005)

Max Muller, F.: Upanishads. Vijay Goel Publisher, New Delhi (2007)

de Moivre, A.: The refinement of Bernoulli's theorem by obtaining normal approximation. Miscellania Analytical. Second supplement (1733)

Moran, P.A.P.: Path coefficients reconsidered. Aust. J. Stat. **3**, 87–93 (1961)

Olkin, I., Pratt, J.W.: Unbiased estimation of certain correlation coefficients. Ann. Math. Statis. **29**, 201–211 (1958)

Papoulis, A.: Probability, Random Variables and Stochastic Processes. McGraw Hill, New York (1965)

Pascal.: Logique de Port-Royal (1662)

Pearl, J.: Graphical models for probabilistic and causal reasoning. Allen N. Tucker (ed.). Computer science handbook. Chap. 70: 70–18 (2004)

Pearson, K.: Karl Pearson's Early Statistical Papers. Cambridge University, Cambridge (1948)

Poisson, S.D.: Recherches sur la probabilite des jugements en matiere criminelle et en matiere civile, precedes des regles generals du calculdespribabilites, 1837 (1781–1840)

Radhakrishnan, S.: Indian Philosophy, vol. 1. Oxford University Press, Oxford (1971)

Rao, C.R.: A note on distribution of $D_{p+q}^2-D_p^2$ and some computational aspects of D^2-statistics and distribution function. Sankhya **10**, 257–268 (1950)

Rao, C.R.: Use of distribution and allied functions in multivariate analysis. Sankhya. A **24**, 149–154 (1962)

Rao, C.R.: Discriminant function between composite hypotheses and related problems. Biometrika **53**, 315–321 (1966)

Rao, C.R.: Linear Statistical Inference and its Applications, 2nd edn. Wiley, New Jersey (2006)

Rigveda, 5-82-5.: Sarvadesik Arya Pratinidhi Sava, New Delhi (2003)

Rosenblatt, F.: The perceptron: a probabilistic model of information storage and organization in the brain. Psychol. Rev. **65**, 386–408 (1958)

Rosenblatt, F.: Principles of Neurodynamics. Spartan Books, Washington (1962)

Rossi, E., Shapzzini, F.: Model and distribution uncertainty in multivariate, GARCH estimation a Monte-Carlo analysis. Comp. Stat. Data Anal. IASC: The International Association for Scientific Computing. www.science direct com. vol. 54, pp. 2786–2800. Elsevier, Amsterdam (2010)

Savage, L.J.: Foundations of Statistics. Wiley, New Jersey (1954)

Seneta, E.: History of probability. Encyclopedia of Statistical Sciences, Vol. 7. Kotz, Johnson and Read (eds.). Wiley, New Jersey (1981)

Schweizer and Wolff.: Kumar, J. (referred by). Concept of dependence. Encyclopedia of Statistical Sciences, vol. 2 (1976)

Suppes, P.: A Probability Theory of Causality. North Holland Publishing Company, Amsterdam (1970)

Tsypkin, Y.Z., Nikolic, Z.J.: Adaptation and Learning in Automatic Systems. Academic Press, New York (1971)

Tukey, J.W.: Causation, regression and path analysis. In: Kempthrone, O et al. Biology. (ed.) Stat and Math. Hafner, New York (1954)

Tulsidas, S.G.: Shree Ram shalaka prasnavali. Ram Charit Manas, 115th edn. Geeta Press, Gorakhpur, pp. 12–14 (1574)

Valmiki, A. (3000 BC): *Yogavashistha*: *Maha-Ramayanam*; vol 1. Ch. 2, Sarga 10. tr. from Sanskrit to Hindi by Pandit Thakur, P. D (2001). Chaukhambha Surbharati Publication, Varanasi

Vasicek, O.: An equilibrium characterization of the term structure. J. Finan Econ. **5**, 177–188 (1977)

References

Vasicek, O.: Loan portfolio value (Vasicek's results). Risk. (1987)
Vasicek, O.: Probability of loss on loan portfolio. Working paper. KMV (2002)
Weldon, W.F.R.: (cited by Edgeworth, F.Y.). Encylcopedia Britannica, 11th ed. (as cited by Kendall, M.G. in his Advance Theory of Statistics, Vol. 1.). London: Charles Griffin and Company (1947)
Winston, W.L.: Introduction to Probability Models: Operations Research, vol. 2, 4th edn. Thompson Learning, Singapore (2004)
Wright, S.: Genetics **3**, 367–374 (1918)
Wright, S.: Correlation and causation. J. Agric. Res. **20**, 557–585 (1921)
Wright, S.: The method of path coefficients. Ann. Math. Statis. **5**, 161–215 (1934)
Wright, S.: Path coefficients and path regression. Biometrics **16**, 189–202 (1960)
Yakowitz, S.J.: Mathematics of Adaptive Control Processes. Elsevier, New York (1969)
Zellner, Arnold: An Introduction to Bayesian Inference in Econometrics. Wiley, New Jersey (1971)

Chapter 3
Univariate Normal Distribution and Its Quantile

The theme of this book is about quantile values of probability distribution and their applications in decision-making. This chapter highlights the univariate normal distribution, which is the most explored and exploited one. It, therefore, begins with brief characterization of such probability distribution by Ludeman (random processes, filtering, estimation and detection. Wiley India, New Delhi, 2010). These are followed by a search for definitions of quantile by early writers as well as by some later ones. Their main concern was its applications for uncertain inference in terms of confidence probability and in transforming uncertain optimization problem to its unique chance-constrained programming problem. In fact, its use as Neyman's confidence interval, as well as critical difference, happened to be of foremost importance in empirical decision-making. The chapter also touches upon the determination of nonparametric univariate quantiles. Those interested only in use of quantile may skip Sect. 3.1.

3.1 Probability Distribution

Having introduced the discovery of probability and its distribution as a measure of uncertainty in Sects. 2.3, 2.4 and 2.5, it is necessary to discuss here some basic properties of such distribution. After the discovery of normal probability distribution and also for the reason of its central limit theorem, it is logical to use it in uncertain decision-making processes. Thus, it requires to be discussed in this section.

Any continuous distribution can be defined as follows: supposing x is such a continuous variable and $f(x)$ is its density of occurrence, then probability density function $f(x)$ must satisfy that for x, $-\infty \leq x \leq +\infty$, as it has been defined in the texts of Cramer (1951), Papoulis (1965) and Feller (2009).

(i) $\qquad x = -\infty, \quad F(x) = 0$

(ii) $\qquad x = +\infty, \quad F(x) = 1$

(iii) $$F(x) = \int_{-\infty}^{x} f(x)dx. \qquad (3.1)$$

any value between 0 and 1 depends on the value of x and structure of $f(x)$. Also that

(iv) $\qquad f(x) = F'(x)$

This distribution function $F(x)$ is also called *cumulative probability density function*, denoting that $F(x)$ represents probability of x assuming any value between $-\infty$ and x. Since for the occurrence of x in the interval $x + \Delta x$, there exists a corresponding probability density function $f(x)$ between 0 and 1 as $\Delta x \to 0$. Therefore, x could qualify for being called the *continuous random variable*.

Ludeman (2010) enunciates crisply the total characterization of cumulative distribution function so as to have the undernoted important properties:

1. $F_X(x)$ is *bounded* above and below, $0 \le F_X(x) \le 1$ for all x
2. $F_X(x)$ is a *non-decreasing* function of x, $F_X(x_2) \ge F_X(x_1)$ for all $x_2 > x_1$ and all x_1
3. $F_X(x)$ is *continuous* from the right, $\lim_{a \to 0^+} F_X(x + a) = F_X(x)$
4. $F_X(x)$ can be used to *calculate probabilities* of events,

$$\left.\begin{array}{l} P[x_1 < X \le x_2] = F_x(x_2) - F_x(x_1) \\ P[x_1 \le X < x_2] = F_x(x_2^-) - F_x(x_1^-) \\ P[x_1 \le X \le x_2] = F_x(x_2) - F_x(x_1^-) \\ P[x_1 < X < x_2] = F_x(x_2^-) - F_x(x_1) \end{array}\right\} \qquad (3.2)$$

where $x^- = \lim_{\varepsilon \to 0^+} F_X(x - \varepsilon)$ (left-hand limit).

5. The relation to the probability density function $f_x(x)$ is written as

$$F_X(x) = \int_{-\infty}^{x} f_x(x)dx$$

where $f_x(x) = F'_X(x)$ as in (iv) of (3.1) called *probability density function* of x, as adopted from Ludeman (2010).

3.2 Normal Probability Distribution

For Gaussian or normal probability distribution, (i) and (ii) are the same as in (3.1), but (iii) is replaced by

$$F(t) = \int_{-\infty}^{x} f(t) dt = \frac{1}{\sqrt{2\pi}} \int_{-\infty}^{x} \exp\left(-\frac{1}{2}t^2\right) dt \qquad (3.3)$$

Clearly, t is the probability measure for variable x, where it does not exceed x, the corresponding quantile value, where t is the standard normal variate with mean $= \mu = 0$ and standard deviation $= \sigma = 1$.

More generally, t being a normally distributed variable with mean μ and standard deviation σ, its density function could be expressed as $N(t : \mu, \sigma)$ to mean

$$N(t : \mu, \sigma) = \frac{1}{\sqrt{2\pi}\sigma} \exp\left[\frac{-(t-\mu)^2}{2\sigma^2}\right] \qquad (3.4)$$

for the purpose of using probability values for corresponding values of t in the normal probability integral table. The said table, prepared by Sheppard, was remained in use for a long time but published posthumously in 1939 as reported by Kendall (1951).

Undoubtedly, the Sheppard's table has most frequently and popularly been used in uncertain decision-making problems or in finding solutions to the problems of a probabilistic nature, whenever the sample size is considered large (i.e. ≥ 30). It is well known that Sheppard's table of incomplete univariate normal integral was obtained by numerical integration, as it was not otherwise integrable. Therefore, characterization of distribution is unable to help in such an aspect. This feature would appear to be a limiting factor even while integration of incomplete bivariate normal integral, making the same also not possible. This fact is the main issue to be faced for numerical integration in obtaining the tables of biquantile pairs/equi-quantile values. Walker (1978) reported that though the normal probability function had been extensively tabulated, it was always with either the probable error (0.6745σ) or the modulus $(\sigma/\sqrt{2})$ as argument, but never as the standard error until Sheppard's (1939, posthumous) was published in its present form.

Though the discovery of normal probability distribution function is ascribed to Gauss (1803–1826) as the "law of error", earlier de Moivre (1733) and Laplace (1749–1827) had exploited this function in much greater detail in the use of such function. However, the latter had even suggested the need for the tabulation of such function as reported by Kendall (op. cit.).

It was after such advances that the use of normal probability integral became much more popular amongst the applied and empirical scientists because of its twofold applicability:

(i) For finding the probability between two values of normally distributed variable having a common mean and variance
(ii) Inversely, for finding the percentile more generally known as quantile of the normal probability distribution as the value of the variable to contain specified probability level and then interpreting such quantile value in real-life problems.

It was the second applicability, which offered the choice of value of the variable for specified risk or probability level that proved to be a great leap towards uncertain decision-making.

Abramowitz and Stegun (1972) obtained cumulative normal distribution function evaluation, i.e. $N(x: 0, 1)$ with the help of a computer program (NOR/MSDIST) based on the fifth degree polynomial evaluation. That yielded probability values correct to six places of decimal available in commonly used Microsoft® Excel® software, as reported by Hull (2009). However, the program for $G(x)$ based on Moran's formula (1980) claiming correct up to nine places of decimal was prepared. It was adopted as the result correct at least up to seven places was the necessity. That was done because it was required to satisfy the reliability test.[1]

3.3 Confidence Interval

As indicated earlier, the calculus of probability was developed to a stage where from it was possible to take a decision on the basis of bjectively calculated risk. In spite of that, it was not possible to provide any guarantee for any estimate to lie between any precise limits in probabilistic terms. It was at this juncture that the two approaches appeared:

(i) The "Fiducial inference" developed in a series of papers by Fisher (1933, 1935, 1940a, b)
(ii) Neyman's "confidence interval" (1935, 1941)

There existed a controversy between the two approaches, but the Fiducial inference approach could not continue because it required the existence of sufficient statistics and also because of the vague formulation about the concept of probability distribution for the parameter. Finally, it was due to the fact that time as a tester automatically became the best eliminator of the first approach in favour of confidence interval.

---

[1]The sample results of such tests are discussed in Chap. 5 and presented in Tables 9.2 and 9.3 in Chap. 9.

3.3 Confidence Interval

The solution to estimation problem attained desired level of celebrity and an instrument of wide applicability only after a statement of the type

(i) $$\text{Prob}[C_l \leq \theta \leq C_u] = 1 - \alpha \qquad (3.5)$$

could be possible, where θ being the parameter to be estimated and C_l and C_u the lower and upper limits, respectively, of the random interval that would believably (rather empirically) contain the parameter θ with probability $1 - \alpha$, where α could be arbitrary but reasonably small and

(ii) C_u and C_l would, respectively, be $\bar{x} \pm h s_x$
where $\bar{x}$ being the sample mean, s_x being the standard error of the sample mean, and h the quantile value of the distribution of $\bar{x}$ such that the random interval obtained between C_l and C_u would contain the parameter θ with confidence coefficient $(1 - \alpha)$, i.e. with probability $(1 - \alpha)$. This is evidenced by almost every publication on statistical methods or inference.

3.4 Quantile and Its Role in Decision-Making

Eubank (1986) in his note on quantiles says that it plays a fundamental role in statistics. They are the critical values used in hypothesis testing and interval estimation and often are the characteristics of distribution one wishes to estimate. Sample quantiles are utilized in numerous inferential settings and recently have received increased attention as a useful tool in data modelling.

Historically, the use of sample quantiles in statistics dates back to 1846, when Quetelet considered the use of the semi-interquartile range as an estimator of the probable error for a distribution. Subsequently, Galton (1889, 1908) and Edgeworth (1886, 1893) discussed the use of median as special value of quantile. Sheppard (1899) and the Pearson (1920) studied the problem of optimal quantile selection as the estimate of mean and standard deviation of the normal distribution. Smirnoff (1935) gave rigorous derivation of its limiting distribution which was generalized by Mosteller (1946), and then, Ogawa (1951) generated considerable interest in quantiles as estimation tools in location and scale parameter models. Later on, Tukey (1977) and Parzen (1979) used the same in nonparametric data modelling and also in robust statistical inference.

However, with the advent of its use in stochastic (chance-constrained) programming literature, Tintner (1955, 1960), Tintner and Sengupta (1964), Charnes and Kirby (1967) and Sengupta (1972a, b) preferred to call *quantile* as *percentile* or *fractile*. Eubank (op. cit.) defines the notion of population and sample quantiles as follows: Let F be a distribution function (DF) for a random variable x and define the associated quantile function (QF) by

$$Q(u) = F^{-1}(u) = \inf\{x : F(u) \geq u\}, \quad 0 < u < 1 \tag{3.6}$$

Clearly, u is the probability measure for variable x, where it does not exceed x, the corresponding quantile value. Thus, for a fixed p in interval $(0, 1)$, the pth population quantile for x is $Q(p)$. It follows from the definition (3.6) that the knowledge of Q is equivalent to the knowledge of F. Further, the relationships between F and Q are as follows:

(a) $FQ(u) \geq u$ with equality when F is continuous.
(b) $QF(x) \leq x$ with equality when F is continuous and strictly increasing.
(c) $F(x) \geq u$ if and only if $Q(u) \leq x$.

Thus, important property of the QF, which follows easily from (c), is that if u has a uniform distribution on $[0, 1]$, then $Q(u)$ and x have identical distributions.[2] The sample analog of Q is obtained by the use of empirical distribution function (EDF), as below.

Let $X_{1:n}, X_{2:n}, \ldots, X_{n:n}$ denote the order statistics for a random sample of size n from a distribution F. Then, the usual empirical estimator of F is

$$\hat{F}(x) = \begin{cases} 0, & x < x_{1:n} \\ \frac{j}{n}, & x_{j:n} \leq x < x_{j+1:n}, \quad j = 1, \ldots, (n-1) \\ 1, & x \geq x_{n:n} \end{cases} \tag{3.7}$$

Replacing F with $\hat{F}$ in (3.6) gives the sample or empirical quantile function (EQF).

$$\hat{Q}(u) = X_{j:n}, \quad (j-1)/n < u < j/n, \quad j = 1, \ldots, n \tag{3.8}$$

Thus, the fundamental sample statistics $\hat{Q}, \hat{F}$ and the order statistics are all closely related. Clearly, from (3.6) and (3.7), knowledge of any one implies knowledge of the other two for both continuous and discrete random variable. However, for the normal probability law is given the distribution function

$$F(x) = \int_{-\infty}^{h} \phi(x) dx$$

the quantile function is $F^{-1}(h)$ as in (3.6) by replacing h for x and the density function

$$\phi(x) = (2\pi)^{-1/2} \exp\left(\frac{-x^2}{2}\right) \tag{3.9}$$

---

[2] It is this property that has been used in defining *coupula* as may be seen latter in Sects. 4.12 and 7.8.

3.4 Quantile and Its Role in Decision-Making

Vajda (1970, 1972) uses the quantile of probability distribution function to obtain deterministic equivalent to chance constraint as

$$\text{Prob.}(ax \geq b) \geq \alpha \tag{3.10}$$

where b has a known distribution function,

$$\text{Prob.}(b \leq z) = F_b(z) \tag{3.11}$$

He finds that the smallest value B_α such that $F_b(B_\alpha) = \alpha$, which could also be written as

$$B_\alpha = F_b^{-1}(\alpha). \tag{3.12}$$

Then, the constraint above is equivalent to the non-probabilistic constraint

$$ax \geq B_\alpha \tag{3.13}$$

This is so because if $ax \geq B_\alpha$, then ax will not be smaller (but can be larger) than any of those b, which are not larger than B_α, and the probability of such b arising is α. On the other hand, if ax is smaller than B_α, then the probability $ax \geq b$ being true is less than α.

Rotar (2007) adopted a more general definition of quantile, which is instructive to be noted in the present application paradigm.

Consider a random variable X with a distribution function,

$$F(x) = P(X \leq x).$$

Let $\gamma \in [0, 1]$. We say that a number q_γ is γ quantile of X if $F(q_\gamma - \varepsilon) \leq \gamma$ and $F(q_\gamma + \varepsilon) > \gamma$ for any arbitrary small $\varepsilon > 0$. If the random variable X is continuous and its distribution function is strictly increasing, then q_γ is the unique number q for which $F(q) = \gamma$, as in Fig. 3.1a. If there are many numbers q for which $F(q) = \gamma$, then the definition above chooses the right end point of the interval where $F(x) = q$, as in Fig. 3.1b, c.

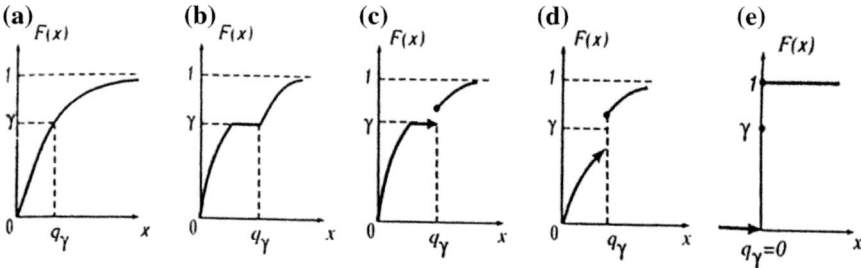

Fig. 3.1 Quantiles; adopted from Rotar (op. cit.)

In the relevant literatures, there are definitions in which γ quantile is the left end point, or the middle, or even any point from this interval. However, this difference is not essential.

If the r.v. takes on some values with positive probabilities (and the d.f. have "jumps"), it may happen that there is no number q such that $F(q) = \gamma$ (see Fig. 3.1d). Then, one may choose the point at which $F(x)$ "jumps" over level γ.

In particular, if $X = 0$, with probability one, the point "0" is the γ-quantile for all $\gamma \in [0, 1)$ (see Fig. 3.1e).

The 0.5 quantile is called the *median*. If the distribution is continuous, then $P(X \leq q_{0.5}) = P(X \geq q_{0.5}) = 0.5$. (The r.v. is likely to be larger than the median as it is to be smaller.) In general, since it may happen that $P(X = q_{0.5}) > 0$, one has

$$P(X \leq q_{0.5}) \leq 0.5 \quad \text{and} \quad P(X > q_{0.5}) \leq 0.5.$$

Another term in use for a γ-quantile is a 100γ-th percentile. The reader familiar with the notation of supremum may realize that the γ-quantile above may also be defined as

$$q_\gamma = \sup\{x : F(x) \leq \gamma\}$$

Rao and Hammed (2000) in their treatise on "Flood Frequency Analysis" gave estimates for quantiles for various probability distributions and applied the same in Flood Frequency Analysis and referred to some computer software being used for the said purpose, including some using MATLAB programming libraries.

Thus, the role of h, as it has been used in this text, that is the quantile, became apparent and implied while generating confidence interval for any estimate of the parameter for any prescribed probability level. It was indeed a great day for a decision-maker, because it was only after this discovery that a probabilistic statement could find its certainty equivalence for any prescribed probability level to which a decision-maker or an entrepreneur would prefer or like to choose with implication that he could agree/prefer or undertake the risk of facing the adversity up to the opted level of probability. This was of course on the assumption of probability distribution of the said random variable for the parameter estimate that was sought after and obtained.

3.5 Certainty Equivalent and Quantile

The terms like "deterministic equivalent" and "certainty equivalent" appear to have been introduced by Tintner (1955, 1960) in connection with the extension of the theory of linear programming to obtain the solution when its parameters were considered random variables, which he named *stochastic programming*.

Almost simultaneously, the same term "certainty equivalent" was also used by Simon (1956) and generalized by Thiel (1957), but Simon used it in the context of

3.5 Certainty Equivalent and Quantile

substitution of least square estimates of all the variables in the criterion function, which had not been observed but one then could choose the current value of the control variables to minimize the quadratic form, just as if the estimated values were the true values, as reported by Whittle (1963). Such a concept and application of the term "certainty equivalent" continues to operate, as it appears to be more general. Accordingly, Davis and Vinter (1984) consider the term as an estimate to be used if the state cannot be observed directly, as if it were the true state. They clearly make undernoted statements:

1. If the state cannot be observed directly, estimate it and use the estimate as if it were the true state.
2. The certainty equivalent principle states that, optimally, the controller acts as if the state estimate were the true state *with certainty*. Even though the controller knows that this is not the case, no other admissible strategy will give better performance.
3. To attain the self-tuning property for any stochastic control problem, if true value of the parameter is not known, one has to adopt a system controller assuming its estimate as its certainty equivalent. Take the observation and modify the estimate, iteratively for time-dependent recursive function till such process coincides with the desired result, worthy to be called *self-tuning property* for any stochastically dynamic system.

By the term "estimate", they meant either the least square estimate or minimizing well-defined loss function—generally a quadratic loss function.

Incidentally, approach suggested above by Simon (op. cit.) and Davis and Vinter (op. cit.) bear similarity with that of Aadikavi's (2001), when he considers sample mean to represent the truth.[3]

However, in programming and optimization literature, the said term "certainty equivalent" was used in a specific sense by Charnes and Cooper (1959, 1960, 1962), Tintner et al. (1963), Tintner and Sengupta (1964) and Sengupta (1969, 1972a, b, 1982, 1985a, b, 1987) and Charnes and Kirby (1967), Vajda (1970, 1972), Symonds (1967), Ziemba (1970, 1972, 1976), Ziemba and Butterworth (1974), Huang et al. (1977), Kataoka (1963), Madansky (1962), Miller and Wagner (1965), Naslund (1971), Balintfy (1970), Prekopa (1970, 1971a, b, 1973), Prekopa and Szantai (1976), Prekopa and Kelle (1978), and many others—almost all—adopted it to mean transformation of probabilistic constraint into "deterministic equivalent" by the use of quantile of the univariate probability distribution except the few.[4] Such a practice, in a way, simulates the same spirit as those of Simon as well as of Davis and Vinter (op. cit.). The essential difference was in their approach for obtaining point estimate and in building confidence interval using quantile value.

---

[3]As quoted in Sect. 6.1 in Chap. 6.
[4]Discussed in Sect. 4.6 in Chap. 4.

In spite of the advantages of the use of normal probability distribution, some developments were made to obtain quantile values for less restrictive probability distributions. To mention a few of them who initiated and later pursued the said approach, amongst which one was Theil (1957). Thereafter, Sinha (1963a, b) made use of one-sided Tebycheff's inequality developed earlier by Frechet (1937) and Uspensky (1937). Sengupta (1972a, b) preferred the use of quantile of chi-square distribution on the ground that economic and even other decision variables, though generally of stochastic nature, are seldom negative. Such applications of quantile values for obtaining certainty equivalent for nonparametric case were adopted by Disch (1978). Chandra and Chatterjee (2005) have made adequate representation and references on distribution-free covers and distribution-free confidence interval for quantile.

Recently, however, Rotar (op. cit.) introduced the same term "certainty equivalent" as any number c as a random variable (clearly univariate) taking only one value c, where it can be found as $c(X)$ such that c is equivalent to X, meaning that the decision-maker is indifferent whether to choose c or X. It may be said that decision-maker considers c, an "adequate price of X". The number $c(X)$ so defined is called a *certainty equivalent of X*.

It needs to be noted that such a unique number $c(X)$ could be easily defined but only for a univariate random variable, obviously because $c = K * \text{inv } F(X)$, where $F(X)$ is the cumulative probability value of X, which, if violated, would also violate the equality. In fact, such a value of c happens to be a multiple of quantile value for the given probability level, K being some multiplier, usually the standard deviation.

The univariate normal distribution had already become popular, first because of its limiting property on account of the central limit theorem and secondly, due to the development of a superb normal probability integral table by Sheppard (op. cit.). Here, Gottfried (1978) may be quoted in order to indicate why normal probability distribution is so frequently used: "... Nevertheless experience shows that the degree of non-normality in practice is often so small that the assumption of actual normality does not lead to erroneous conclusions" as noted from Kruskal and Tanur (1978).

Hence, it was convenient to consult the readymade table and obtain quantile value corresponding to a given or prescribed probability level and to be clear that it was the probability level to which the decision-maker could presumably show his inclination or agree to undertake as the risk level.

It appears at this stage that for every decision-maker or entrepreneur and for every decision, risk is implicit. This again reminds us of the sayings of Gaudpad (op. cit.) in the introduction and the maxim of Gauther, which says:

> To get profit without risk, experience without danger and reward without work, is as impossible as it is to live without being born.

Thus, it is clear that for any, or every, real-life activity, risk is a must and in fact it is the cost one has to pay for decision-making under incomplete information and prediction paradigms. Yet, another advantage of using quantile value obtained by the inversion of probability value while using univariate normal distribution is its uniqueness for every value of the latter and vice versa. This advantage will be lost as soon as one is required to use bivariate or multi-variate normal probability distribution. Such an apprehension created a kind of fear, like that of poet Milton (2009) felt on his paradise being lost, even without touching the forbidden fruit.[5]

References

Aadikavi, V.: Yogavasishtha Maharamayanam. Chaukhambha Sanskrit Pratisthan, Delhi (2001)
Abramowitz, M., Stegun, I.: Handbook of Mathematical Functions. Dover Publications, New York (1972)
Balintfy, J.L.: Nonlinear programming for models with joint chance constraints. In: Integer and Nonlinear Programming, pp. 337–352. Elsevier, New York (1970)
Chandra, T.K., Chatterjee, D.: A First Course in Probability, 3rd edn. Narosa Publishing House, New Delhi (2005)
Charnes, A., Cooper, W.W.: Chance constrained programming. Manage. Sci. **6**(1), 73–79 (1959)
Charnes, A., Cooper, W.W.: Chance constrained programming with normal deviates and linear decision rules. Naval Res. Logistics Q. **7**(4), 533–544 (1960)
Charnes, A., Cooper, W.W.: Chance constraints and normal deviates. J. Am. Stat. Assoc. **57**(297), 134–148 (1962)
Charnes, A., Kirby, M.: Some special P-models in chance constrained programming. Manage. Sci. **14**, 183–195 (1967)
Cramer, H.: Mathematical Methods of Statistics. Princeton University Press, Princeton (1951)
Davis, M.A.H., Vinter, R.B.: Stochastic Modelling and Control. Chapman and Hall, London (1984)
de Moivre, A.: *Miscellania Analytica*, second suppl (1733)
Disch, D.: Bayesian non-parametric inference for quantiles. A Ph.D. Dissertation. Department of Mathematics, Indiana University (1978)
Edgeworth, F.Y.: Philos. Mag., 5th series. **22**, 371–383 and **36**, 98–111 (1886, 1893)
Eubank, R.E.: Quantiles. In: Kotz, S., Johnson, N.L., Read, C.B. (eds.) Encyclopedia of Statistical Science, vol. 7. Wiley, Hoboken (1986)
Feller, W.: An Introduction to Probability Theory and its Applications, vol. 2. Wiley India, New Delhi (2009)
Fisher, R.A.: The concept of inverse probability and of fiducial probability referring to unknown parameters. Proc. Roy. Soc. A **139**, 343 (1933)
Fisher, R.A.: The fiducial argument in statistical inference. Ann. Eugen. Lond. **6**(391) (1935)
Fisher, R.A.: The precision of discriminant function. Ann. Eugen. **10**, 422–429 (1940a)
Fisher, R.A.: A note on fiducial inference. Ann. Eugen. Lond. **10**(383) (1940b)
Frechet, M.: Recherches theoriques modernes sur la theorie des probabilites. In: Forms, vol. I, Part III, Paris (1937–1938)
Galton, F.: Natural Inheritance. Macmillan, New York (1889)
Galton, F.: Memories of My Life. Methuen, London (1908)
Gauss, C.F.: Gauss's work and in Whittaker and Robinson's 1924 (1803–1826)
Gottfried, E.N.: Probability. In: Kruskal, W.H.J.M. (ed.) International Encyclopedia of Statistics. Macmillan, New York, (1978)

---

[5]Consequences of such phenomenon and loss or gain, if any, have been dealt with in Chaps. 4 and 7.

Huang, C.C., Ziemba, W.T., Ben-Tal, A.: Bounds on the expectation of a convex function of a random variable: with applications to stochastic programming. Oper. Res. **25**(2), 315–325 (1977)

Hull, J.C.: Options, Futures and Other Derivatives. Pearson Education, New Delhi (2009)

Kataoka, S.: A stochastic programming model. Econometrica **31**, 181–196 (1963)

Kendall, M.G.: The Advanced Theory of Statistics, vol. 1. Charles Griffin, London (1951)

Kruskal, W.H., Tanur, J.M.: International Encyclopedia of Statistics. Macmillan, New York (1978)

Laplace, M.P-S.: A Philosophical Study of Probabilities. Dover, New York (first published as Essai philosphique sure less probability's) (1749–1827)

Ludeman, L.C.: Random processes, filtering, estimation and detection. Wiley India, New Delhi (2010)

Madansky, A.: Methods of solutions of linear programs under uncertainty. Oper. Res. **10**, 463–470 (1962)

Millar, B.L., Wagner, H.M.: Chance-constrained programming with joint chance constraints. Oper. Res. **13**(6), 930–945 (1965)

Milton, J.: Paradise Lost. U.B.S Publishing and Distributors, New Delhi (1667, 2009)

Moran, P.A.P.: Calculation of distribution function by evaluating series. Biometrika **67**(3), 675–676 (1980)

Mosteller, M.: Ann. Math. Stat. **17**, 377–408 (1946)

Naslund, B.: A model of capital budgeting under risk. In: Byrne, R.F., et al. (eds.) Studies in Budgeting. North Holland Publishing Company, Amesterdam (1971)

Neyman, J.: On the problem of confidence intervals. Ann. Math. Stat. **6**, 111 (1935)

Neyman, J.: Fiducial argument and theory of confidence intervals. Biometrika **32**, 128 (1941)

Ogawa, J.: Osaka Math. J. **3**, 175–213 (1951)

Papoulis, A.: Probability, random variables and stochastic processes. McGraw Hill, New Delhi (1965)

Parzen, E.: J. Amer. Stat. Assoc. **74**, 105–121 (1979)

Pearson, K.: Karl Pearson's Early Statistical Papers. Cambridge University Press, Cambridge (1920–1948)

Prekopa, A.: On probabilistic constrained programming. In: Proceedings of the Princeton Symposium on Mathematical Programming, pp. 113–138. Princeton University Press, Princeton (1970)

Prekopa, A.: Logarithmic concave measures with application to stochastic programming. Acta Scientiarum Mathematicarum **32**:301–316 (1971a)

Prekopa, A.: Stochastic programming models for inventory control and water storage problems. In: Prekopa, A. (ed.). Coll. Math. Soc. Janos Bolyai, pp. 229–245 (1971b)

Prekopa, A.: Contribution to the theory of stochastic programming. Math. Prog. **4**, 202–221 (1973)

Prekopa, A., Kelle, P.: Reliability type inventory models based on stochastic programming. Math. Prog. Study **9**, 43–58 (1978)

Prekopa, A., Szantai, T.: On optimal regulation of a storage level with application to the water level regulation of a lake: a survey of mathematical programming. In: Proceedings of the 9th. International Mathematical Programming Symposium, pp. 183–210. Akademiai Kiado, Budapest (1976)

Rao, R.C., Hammed, K.: Flood Frequency Analysis. CRC Press, London (2000)

Rotar, V.I.: Actuarial Models. CRC Press, London (2007)

Sengupta, J.K.: Safety-first rules under chance-constrained linear programming. Oper. Res. **17**, 112–132 (1969)

Sengupta, J.K.: Stochastic Programming: Methods and Applications. North Holland Publishing Company, Amsterdam (1972a)

Sengupta, J.K.: Chance-constrained linear programming with Chi-square type deviation. Management Science Theory Series (1972b)

Sengupta, J.K.: Decision Models in Stochastic Programming. North Holland Publishing Company, Amsterdam (1982)

References

Sengupta, J.K.: Optimal Decision under Uncertainty-Methods, Models and Management. Springer, Berlin (1985a)

Sengupta, J.K.: Information and Efficiency in Economic Decision. Martinus Nijhoff, Dordrecht (1985b)

Sengupta, J.K.: Data envelopment analysis for efficiency measurement in stochastic case. Comput. Opns. Res. **14**, 117–129 (1987)

Sheppard, W.F.: The Probability Integral: British Association Mathematics Tables, vol. 7. Cambridge University Press, Cambridge (1939, posthumous)

Sheppard, W.F.: Philos. Trans. Roy. Soc. **192**, 101–167 (1899)

Simon, H.A.: Dynamic programming under uncertainty with a quadratic criterion function. Econometrica **24**, 74–81 (1956)

Sinha, S.M.: Stochastic Programming, vol. 24(3). A Ph.D. Dissertation. University of California, California (1963a)

Sinha, S.M.: Stochastic Programming. Operation Research Centre, University of California, Berkeley, ORC-63(22) (1963b)

Smirnoff, N.V.: Metron **12**, 59–81 (1935)

Symonds, G.: Deterministic solutions for a class of chance constrained programming problems. Oper. Res. **15**, 495–512 (1967)

Theil, H.: A note on certainty equivalence dynamic programming. Econometrica **25**, 346–349 (1957)

Tintner, G.: Stochastic linear programming with application to agricultural economics. In: Antosiewicz, H.A. (ed.) Proceedings of 2nd Symposium on Linear Programming, Washington, pp. 197–228 (1955)

Tintner, G.: A note on stochastic linear programming. Econometrica **28**, 490–495 (1960)

Tintner, G., Sengupta, J.K.: Stochastic Linear Programming and its Applications to Economic Planning: On Political Economy and Econometrics. PWN Scientific Publisher, Warsaw (1964)

Tintner, G., Millham, C., Sengupta, J.K.: A weak duality theorem for S.L.P. Unternehmensforschung **7**, 1–8 (1963)

Tukey, J.W.: Exploratory Data Analysis. Addison Wesley, Reading (1977)

Uspensky, J.V.: Introduction to Mathematical Probability. McGraw Hill, New York (1937)

Vajda, S.: Stochastic Programming. In: Abadie, J. (ed.) Integer and Nonlinear Programming. Elsevier, New York (1970)

Vajda, S.: Probabilistic Programming. Academic Press, New York (1972)

Walker, H.M.: Encyclopedic article on life and works of Karl Pearson. Kruskal and Tanur (op. cit.). Encyclopedia of Statistics 691–697 (1978)

Whittle, P.: Prediction and Regulation by Linear Least Square Methods. Applied Mathematics Series, London (1963)

Ziemba, W.T.: Computational algorithms for convex stochastic programs with simple recourse. Oper. Res. **3**, 414–431 (1970)

Ziemba, W.T.: Duality relations, and bounds for convex stochastic programs with simple recourse. Cahiers du Centre D'etudes de Recherche. Operationelle **13**(2), 85–97 (1972)

Ziemba, W.T.: Mathematical programming and portfolio selection. Instituto Nazionale di Alta Matematica Symposia Mathematica XIX (1976)

Ziemba, W.T., Butterworth, J.E.: Bounds the value of information in uncertain decision problems. J. Stochast. **4**, 1–18 (1974)

Chapter 4
Bivariate Normal Distribution and Heuristic-Algorithm of BIVNOR for Generating Biquantile Pairs

As it is the key chapter on software development, it begins with a characterization of bivariate normal distribution by Ludeman (2010). That is followed by a brief presentation of the various properties of bivariate normal distribution and its applications by Essenwagner (1976). Thereafter, Owen (1956) computational scheme for numerical integration of the bivariate normal integral is presented stepwise. The envisaged algorithm is to make iterative use of this scheme. It is very well realized that unlike univariate normal, which has a unique quantile value for a given probability level, its bivariate extension would have multiple (or infinite) quantile pairs for the same probability level. Apart from this, there arose other problems in their generation, which were sorted out and strategies to meet such problems were listed and used in perfecting the same algorithm. This was followed by finding the role of equi-quantile value BIGH for each probability level and correlation value as the initial point of such iterative scheme for the entire computational horizon of four hundred grids. In fact, the approach adopted is entirely innovative and of great economic and other consequences. Such action resulted in expansion of decision alternatives for the given or estimated correlation value without any change in probability (risk) level. Methods of forming simultaneous (joint) confidence intervals have emanated by using such biquantile pairs, having a definite edge over the existing Bonferroni's joint confidence interval. Multiplicity of biquantile pairs offers a scope even for multiple joint confidence intervals for the same confidence probability. That could yield larger number of decision alternatives and hence the scope of choice amongst them by the decision maker on some criterion function. Those who are interested only in application of quantile pairs may skip Sect. 4.1, but they also should read its concluding part.

4.1 Characterization of Joint Cumulative Distribution Function

Ludeman (2010) completely and lucidly presented "joint cumulative distribution function".[1] The said function denoted by $F_{XY}(x, y)$ for random variables $X(\zeta)$ and $Y(\zeta)$, represented by X and Y, is defined for all x and y as follows:

$$F_{XY}(x,y) = P\{\zeta : X(\zeta) \leq x \cap Y(\zeta) \leq y\} \triangleq P\{X \leq x, Y \leq y\}. \tag{4.1}$$

This provides sufficient information for computing the probabilities of all permissible events and, therefore, it qualifies for being called *total characterization*. The joint cumulative distribution function has under noted properties:

$F_{XY}(x,y)$ is *bounded* above and below,

$$0 \leq F_{XY}(x, y) \leq 1 \text{ for all } x \text{ and } y \tag{4.2}$$

1. $F_{XY}(x, y)$ is a *non-decreasing* function of x and y, meaning:

$$\begin{aligned} F_{XY}(x_2, y) &\geq F_{XY}(x_1, y) \quad \text{for all } x_2 > x_1, \quad \text{all } x_1 \text{ and all } y \\ F_{XY}(x, y_2) &\geq F_{XY}(x, y_1) \quad \text{for all } y_2 > y_1, \quad \text{all } y_1 \text{ and all } x \end{aligned} \tag{4.3}$$

2. $F_{XY}(x, y)$ is *continuous from the right* in both x and y:

$$\lim_{\substack{\varepsilon \to 0^+ \\ \delta \to 0^+}} F_{XY}(x+\varepsilon, y+\delta) = F_{XY}(x, y) \tag{4.4}$$

3. $F_{XY}(x, y)$ can be used to *calculate probability* of rectangular events as

$$P\{x_1 < X, \leq x_2, y_1 < Y \leq y_2\} = F_{XY}(x_2, y_2) - F_{XY}(x_1, y_1) \tag{4.5}$$

4. $F_{XY}(x, y)$ is related to the joint *probability density function* by

$$F_{XY}(x, y) = \int_{-\infty}^{x} \int_{-\infty}^{y} f_{XY}(x, y) \, dx \, dy \tag{4.6}$$

5. Consequently, the *joint probability density function* defined by $f_{XY}(x, y)$ is related to $F_{XY}(x, y)$ as

---

[1] Also discussed in Sect. 3.1 (in Chap. 3).

4.1 Characterization of Joint Cumulative Distribution Function

$$f_{XY}(x,y) \triangleq F''_{XY}(x_2,y_2) = \frac{\partial^2}{\partial x\,\partial y} F_{XY}(x,y) \qquad (4.7)$$

Further that the joint probability density function $f_{XY}(x, y)$ must satisfy the undernoted four relationships:

6. Positivity,

$$f_{XY}(x,y) \geq 0 \text{ for all } x \text{ and } y \qquad (4.8)$$

7. Integral over all x and y

$$F_{XY}(x,y) = \int_{-\infty}^{\infty} \int_{-\infty}^{\infty} f_{XY}(x,y)\,dx\,dy = 1 \qquad (4.9)$$

8. $f_{XY}(x, y)$ can be used to *calculate probability* of rectangular events as:

$$P\{x_1 < X \leq x_2, y_1 < Y \leq y_2\} = \int_{x_1^+}^{x_2^+} \int_{y_1^+}^{y_2^+} f_{XY}(x,y)\,dx\,dy \qquad (4.10)$$

where x_1^+, x_2^+, y_1^+ and y_2^+ are limits from the positive sides, or any event A as:

$$P(\{(x,y) \in A\}) = \iint_A f_{XY}(x,y)\,dx\,dy \qquad (4.11)$$

9. Relationship to joint cumulative distribution function,

$$\int_{-\infty}^{x^+} \int_{-\infty}^{y^+} f_{XY}(x,y)\,dx\,dy = F_{XY}(x,y) \qquad (4.12)$$

In the present context it is generally claimed to satisfy all those *characterization properties* as laid down by Ludeman (op. cit.).[2] However, it is to be noted that such total characterization as obtained for univariate normal distribution, but for non-integrability of incomplete univariate normal distribution, is also true for bivariate normal distribution. It was for this reason that Pearson (1930, 1931) had to prepare a table for bivariate normal probability for the given value of coordinate pairs, the specific values of x and y and the value of the correlation coefficient between the two variables by numerical integration. However, the problem of finding a quantile pair for a given

---

[2] For further theoretical background, Ludeman's text, or that of Papoulis (1965), should be consulted.

probability and correlation remained then untouched, which is indeed a paradox till today. Though Gupta (1963) obtained tables of equi-quantile values for multivariate normal up to twelve variables for positive correlations. Later on, and till very recently, claims have been made for developing software to generate equi-quantiles for much larger dimensions, which shall find mention subsequently. Unlike a univariate case, such quantile value is no longer unique for any given probability and correlation under bivariate setup. Instead, there are an infinite number of quantile pairs that may get generated for any tangible combination of probability level and correlation value. Such specific and important property, which needs to be exploited for creating decision alternatives, are not mentioned by anyone who ventured to work for bivariate quantiles. Such a state of affair has continued for more than six to seven decades in spite of the fact that such a set of quantile pairs was most sought after by joint optimum seekers and even by seekers of tighter confidence cover for predictive models. The same is also much sought after by simulators who are quite conscious of the fact that simulation runs are rarely independent. Therefore, tests based on independence assumption lose their validity in such cases.[3] In fact, this has been the missing link which was the dire need for the advantage of mankind. It has to be the harbinger of decision alternatives for iso-risk packages, as well as for more efficient joint confidence interval. Hence, it forms a critical point between hidden and sought after in decision processes.

The problem of finding a quantile value for a given probability is known as *inverse problem*. Milton and Hotchkiss (1969) and Majumdar and Bhattacharjee (1973, 1985, 1986) have given computer programs for obtaining quantiles for given probability values for univariate normal and incomplete beta distributions respectively, a quantity required in decision-making processes. Such a requirement was felt ever since univariate normal quantile was brought into use and was incessantly sought after Neyman (1935, 1941) who wrote on confidence interval. Such inverse for any given probability value, instead of being unique would be an infinite number of pairs for bivariate case and infinite number of such trios for trivariate case, and so on. This will be discussed further in Sects. 4.6, 4.7, 4.8, 4.9, and applications in building joint confidence intervals will be indicated.[4]

4.2 Bivariate Normal Distribution: Historical Perspectives

As reported by Bradley (1978), it was Robert Adrian who considered the bivariate normal distribution in the early nineteenth century.[5] According to Gupta (1963), it was Sheppard (1999, 1900) who perhaps was the first statistician to concern himself with the evaluation of bivariate normal probability integral given by

---

[3]The impact of these shall be seen in examples in Chap. 7.

[4]Further, their illustrations through varied examples are given of Chap. 7.

[5]The role of Galton (1886, 1908) and that of Pearson (1900) in the development and application of correlation and regression at early stages has already been dealt with in Chap. 2.

4.2 Bivariate Normal Distribution: Historical Perspectives

$$F(h,k:\rho) = \int_{-\infty}^{+h}\int_{-\infty}^{+k} f(x,y:\rho)\,dx\,dy$$

where the density function being

$$f(x,y:\rho) = \frac{1}{2\pi(1-\rho^2)^{1/2}} \exp\left(-\frac{1}{2}\frac{x^2 - 2\rho xy + y^2}{1-\rho^2}\right). \tag{4.13}$$

Historically, as reported by Gupta (op. cit.), Pearson (1901) gave a method for evaluating the integral (4.13) as a power series in ρ involving tetra-choric functions and his method provided a good approximation of the probability P_{ij} for i-th and j-th cell for small $|\rho|$.

However, from Sheppard (1900) and Pearson (1931) onward, various tables have been published for obtaining probabilities for correlated bivariate normal variables, as reported by Nicholson (1943), by Owen (1956) and subsequently by Gupta (op. cit.). Latter through his near-exhaustive bibliography, as it was at that point of time. Most of those bibliographies, including those by Nicholson (1943) and by Cadwell (1951) have been referred to.

Hymans (1968) in course of his search for simultaneous confidence intervals in economic forecasting obtained bivariate normal ellipsoid and its contour as joint interval for forecast values of both the variables to be satisfied simultaneously. Earlier, Cramer (op. cit.) obtained equi-probability curves by plotting the exponent of bivariate normal distribution, forming homothetic ellipses or ellipse of concentration as illustrated through figures by Ludeman (op. cit.).

The need for evaluating cumulative bivariate normal integral was realized by even Essenwagner (1976) in connection with several types of problems in meteorological probabilistic predictions. Such problems, similar to problems here, related to finding the probability of exceeding a threshold value of x_1 or x_2 when both were correlated. He referred to Nicholson (1943) and Cadwell (1951) amongst which Cadwell recommended the use of the tables of integral

$$F(h,k:\rho) = \frac{1}{2\pi(1-\rho^2)^{1/2}} \int_h^\infty \int_k^\infty \exp\left[-\frac{x_1^2 - 2\rho x_1 x_2 + x_2^2}{2(1-\rho^2)}\right] dx_1\,dx_2 \tag{4.14}$$

That was tabulated by Pearson (1931). Later, the same table was also published by National Bureau of Standards (1959). Incidentally, none of those authors, including Pearson, did take up the inverse problem not even for bivariate normal probability distribution.

4.3 Other Properties of Bivariate Normal Distribution

However, some interesting properties shown by Essenwagner (op. cit.) being that even if correlation coefficient $\rho \neq 0$, it can be transformed into independent variates y_1 and y_2 by coordinate transformation. That he defines as

$$y_1 = (x_1 - \bar{x}_1)\cos\varphi + (x_2 - \bar{x}_2)\sin\varphi \qquad (4.15a)$$

representing the major axis of ellipse, and

$$y_2 = (x_2 - \bar{x}_2)\cos\varphi - (x_1 - \bar{x}_1)\sin\varphi \qquad (4.15b)$$

representing the minor axis of ellipse. Here φ stands for angle between x_1 and y_1 or that between x_2 and y_2, which is obtainable as

$$\varphi = \frac{1}{2}\tan^{-1}\left(\frac{2\rho\sigma_1\sigma_2}{\sigma_1^2 - \sigma_2^2}\right). \qquad (4.16)$$

Thus y_1 and y_2 are corresponding projections of x_1 and x_2 on major and minor axes of ellipse, where $\bar{x}_1$, $\bar{x}_2$, σ_1^2, σ_2^2 and ρ denote corresponding means, variances and correlation coefficient between x_1 and x_2.

He also obtains angle between two regression lines and denotes it by

$$\theta = \tan^{-1}\left[\frac{(1-\rho^2)\sigma_1\sigma_2}{\rho(\sigma_1^2 + \sigma_2^2)}\right] \qquad (4.17)$$

and explains that when $\rho = 0$, then $\theta = \pi/2$, meaning regression lines run parallel to the axis (through point x_1, x_2). The ellipse collapses to a line when $\theta = 0$ or π. He designates σ_1 and σ_2 for the standard deviations of x_i-system and σ_a and σ_b the corresponding standard deviations for the y_i-system. Then he expresses it as

$$\begin{aligned} S_V^2 &= \sigma_1^2 + \sigma_2^2 = \sigma_a^2 + \sigma_b^2 = k_1^2 + k_2^2 \\ \sigma_a^2\sigma_b^2 &= (1-\rho^2)\sigma_1^2\sigma_2^2 = k_1 k_2 \end{aligned} \qquad (4.18)$$

which are considered roots of the matrix equation

$$\begin{pmatrix} \sigma_1^2 - k & \sigma_1\sigma_2\rho \\ \sigma_1\sigma_2\rho & \sigma_2^2 - k \end{pmatrix} = 0 \qquad (4.19)$$

k_1 and k_2 are two roots of this quadratic equation, which are obtained as

4.3 Other Properties of Bivariate Normal Distribution

$$\frac{k_1}{k_2} = \frac{1}{2}\left\{(\sigma_1^2 + \sigma_2^2) \pm [(\sigma_1^2 + \sigma_2^2)^2 - 4\sigma_1^2\sigma_2^2(1-\rho^2)]^{1/2}\right\}$$

$$= \frac{1}{2}\left\{(\sigma_1^2 + \sigma_2^2) \pm [(\sigma_1^2 - \sigma_2^2)^2 + 4\rho^2\sigma_1^2\sigma_2^2]^{1/2}\right\}$$

(4.20)

and those are equated to σ_a^2 and σ_b^2, respectively of y_i-system.

For standard bivariate normal density function, the above canonical transformation changes because both σ_1 and σ_2 are now unity, the values of k_1 and k_2, hence of σ_a^2 and σ_b^2 reduce to

$$\sigma_a^2 = k_1 = 1 + \rho \quad \text{and} \quad \sigma_b^2 = k_2 = 1 - \rho. \tag{4.21}$$

Thus the standard form of bivariate normal density reduces to undernoted canonical form

$$f(y_1, y_2) = \frac{1}{2\pi} \exp\left[-\frac{1}{2}\left(\frac{y_1^2}{1+\rho} + \frac{y_2^2}{1-\rho}\right)\right] \tag{4.22}$$

where from it is readily seen that for $\rho = 0$, the elliptical shape of the exponent becomes

$$-\frac{1}{2}(y_1^2 + y_2^2)$$

which is an equation of the circle around the origin. Also if $\rho = \pm 1$, the exponent takes the form

$$-\frac{1}{2}\left(\frac{y_1^2}{2} + \infty\right) \quad \text{or} \quad -\frac{1}{2}\left(\infty + \frac{y_2^2}{2}\right)$$

according as correlation takes its value to be positive or negative, irrespective of that the kernel of the integrand to vanish out.

The eccentricity of Eq. (4.22), a standard canonical form, is

$$\varepsilon_x = \frac{\sigma_a^2 - \sigma_b^2}{\sigma_a^2} = \frac{(1+\rho) - (1-\rho)}{1+\rho} = \frac{2\rho}{1+\rho} \tag{4.23}$$

which for $\rho = 0$, it is also zero, meaning a circular shape. For $\rho = +1$, it is also +1, meaning collapsing into straight lines on the positive side of the y_1-axis:

for $\rho = -1$, it is $-\infty$;
for $\rho = +0.5$, it is $1.0/1.5 = +0.667$;
for $\rho = -0.5$, it is $-1.0/0.5 = -2.000$.

The ellipticity, on the other hand, for this very form is

$$\varepsilon_1 = \frac{\sigma_a - \sigma_b}{\sigma_a} = \frac{(1+\rho)^{1/2} - (1-\rho)^{1/2}}{(1+\rho)^{1/2}} \qquad (4.24)$$

where from also for $\rho = 0$, it is also zero, meaning a circular shape.
For $\rho = +1$, it is also $+1$, meaning collapsing into straight line on the positive side of the y_1-axis.

For $\rho = -1$,   it is also $-\infty$.
For $\rho = 0.5$,   it is $+0.42263$.
For $\rho = -0.5$,   it is -0.7320.

Further he writes:

$$P(R) = \int_0^R f(x_1, x_2) \, dx_1 \, dx_2$$

that such bivariate normal integral has no analytical solution, and therefore suggests the use of numerical techniques as adopted by several authors referred to and also adopted here. Ludeman (op. cit.) also gave an identical statement and, therefore, suggested that Monte-Carlo sampling technique be adopted, whenever required. However, the method suggested here is not that of any simulation technique but it is an iterative technique, with the knowledge that it is almost impossible to get an estimate of error while adopting the former but obtainable by latter.

According to Essenwagner, the cumulative bivariate Gaussian integral is of use in solving more difficult problems like:

(a) What is the probability of exceeding threshold values of x_1 and x_2, pre-assigned values of x_1 and x_2?
(b) What area do we need to accommodate a certain probability density?
(c) What probability density can we expect for a circle of radius R around the origin?
(d) What is the probability density of circle segment with size of given angle, say α and size of radius R?

In fact, Essenwagner's treatise may be regarded as excellent exposition of statistical methods used in atmospheric science.[6]

---

[6]His exposition of bivariate elliptical distribution, an outcome of non-zero correlation between x_1 and x_2, where x_1 (zonal wind speed data) and x_2 (meridional wind speed data), shall be discussed in Sect. 7.5 of Chap. 7.

4.4 Owen's Computational Scheme for Evaluating Bivariate Normal Integral

Owen (1956) provided a practical solution by considering the integrand $B(h, k; \rho)$ with its ranges of integration $(-\infty, h; -\infty, k)$, which, in fact, was

$$B(h, k; \rho) = \frac{1}{2\pi(1-\rho^2)^{1/2}} \int_{-\infty}^{h} \int_{-\infty}^{k} \exp\left[-\frac{1}{2}\left(\frac{x^2 - 2\rho xy + y^2}{1-\rho^2}\right)\right] dx\, dy \quad (4.25)$$

It was he who reduced the above bivariate integral to the sum of univariate normal distribution functions and further gave expressions for it as given below: that being $B(h, k; \rho)$, in terms of $G(h)$ and $T(h, ah)$, function as

$$B(h, k; \rho) = \frac{1}{2}G(h) + \frac{1}{2}G(k) - T(h, ah) - T(k, ak) - \left[0 \text{ or } \frac{1}{2}\right] \quad (4.26)$$

where the first choice was made if $hk > 0$ or if $hk = 0$, but $(h + k) \geq 0$, and the second choice was made otherwise. Further, where

$$ah = \frac{k}{h(1-\rho^2)^{1/2}} - \frac{\rho}{(1-\rho^2)^{1/2}}, \quad ak = \frac{h}{k(1-\rho^2)^{1/2}} - \frac{\rho}{(1-\rho^2)^{1/2}} \quad (4.27)$$

Owen's $B(h, k; \rho)$ was the volume of a bivariate normal with mean zero and variance unity and correlation ρ over the lower left-hand quadrant of the xy-plane when truncated at $x = h$ and $y = k$. His $G(h)$ was the univariate normal with zero mean and unit variance integral from $-\infty$ to h. Values of function $T(h, a)$ were also tabulated by him. The T-function was tabulated only for $0 < a \leq 1$ and ∞, but it could be possible to obtain values for $1 < a < \infty$ by the use of undernoted relational formula

$$T(h, a) = \frac{1}{2}G(h) + \frac{1}{2}G(ah) - G(h)G(ah) - T\left(ah, \frac{1}{a}\right), \text{ if } a \geq 0. \quad (4.28)$$

Its values for negative a or h could be obtained by using

$$T(h, -a) = -T(h, a) \quad (4.29)$$

and

$$T(-h, a) = T(h, a) \quad (4.30)$$

noting that Eq. (4.28) would require a to be positive and hence when a was negative, first to apply Eq. (4.29) and then Eq. (4.28).

Other relational formula given by Owen (op. cit.) are:

$$T(h, 0) = 0$$
$$T(0, a) = \frac{1}{2\pi}\tan^{-1}a$$
$$T(h, 1) = G(h)[1 - G(h)]$$

and

$$T(h, \infty) = \begin{cases} \frac{1}{2}[1 - G(h)], & \text{if } h \geq 0 \\ \frac{1}{2}G(h), & \text{if } h < 0 \end{cases}. \tag{4.31}$$

Further, Owen expressed

$$G(h) = \frac{1}{(2\pi)^{1/2}}\int_a^h \exp\left(-\frac{1}{2}t^2\right)dt \tag{4.32}$$

$$T(h, a) = \frac{1}{2\pi}\int_0^{ax} \frac{\exp[-\frac{1}{2}h^2(1+x^2)]}{1+x^2}dx \tag{4.33}$$

and

$$T(h, a) = -\frac{1}{2\pi}\int_0^h \int_0^{ax} \exp\left(-\frac{x^2+y^2}{2}\right)dx\,dy + \tan^{-1}\left(\frac{a}{2\pi}\right) \tag{4.34}$$

Integrating by parts, he obtained

$$T(h, a) = \frac{1}{2}G(h) + \frac{1}{2}G(ah), -G(h)G(ah) - T\left(ah, \frac{1}{a}\right) \quad \text{for } a \geq 0 \tag{4.35}$$

and the same function -0.5 for $a < 0$.

Further, he referred Cadwell (op. cit.) and Cramer (op. cit.) for the well-known result that

$$B(0, 0; \rho) = \frac{1}{4} + \frac{1}{2\pi}\sin^{-1}\rho \tag{4.36}$$

and finally that

4.4 Owen's Computational Scheme ...

$$B(h, k; \rho) = \frac{1}{2} G(h) - T\left(h, \frac{k - \rho h}{h(1 - \rho^2)^{1/2}}\right) + \frac{1}{2} G(k) - T\left(k, \frac{h - \rho k}{k(1 - \rho^2)^{1/2}}\right) \tag{4.37}$$

for $hk > 0$ or if $hk = 0$, h or $k \geq 0$, and the same function -0.5, if $hk < 0$ or if $hk = 0$, h or $k < 0$.

For computational ease, he expressed $T(h, a)$ as undernoted series

$$T(h, a) = \frac{1}{2\pi} \left(\tan^{-1} a - \sum_{j=0}^{\infty} c_j a^{2j+1} \right)$$

where

$$c_j = (-1)^j (2j+1)^{-1} \left[1 - \exp\left(-\frac{1}{2} h^2\right) \sum_{i=0}^{j} \frac{h^{2i}}{2^i i!} \right] \tag{4.38}$$

which could converge rapidly for small values of a and h.

Thus, having used Owens T-function repeatedly, his program module had to be tested computing it for randomly selected values of h and a and to compare it with Owen's table values. Here both $G(h)$ and $T(h,a)$ are either normal distribution or expressible as its function.

However, for evaluating normal probability integral, it was preferred to use Moran's (1980) expression given by

$$\Phi(x) = \frac{1}{\sqrt{2\pi}} \int_{-\infty}^{x} \exp\left(-\frac{t^2}{2}\right) dt \tag{4.39}$$

$$\frac{1}{2} + \frac{1}{\pi} \left[\frac{x}{3\sqrt{2}} + \sum_{n=1}^{\infty} n^{-1} \exp\left(-\frac{n^2}{9}\right) \sin\left(\frac{nx\sqrt{2}}{3}\right) \right] \tag{4.40}$$

which was truncated at $n = 12$, yielding accuracy of the integral up to nine places of decimal for $|x| \leq 7$, over the series suggested by Owen (op. cit.) that claimed accuracy up to six places of decimal only. Because the objective as can be seen subsequently, was not only to find the value of $B(h, k; \rho)$ but to find bivariate quantile by the repeated use of (4.39) or (4.40). By such change, results correct up to 7–8 places of decimal could be obtained.[7] Preference to Moran's expression over those of Hill's Algorithm AS 66 (1985), Odeh and Evan's AS 70 (1974) and of

---

[7]This was needed to meet moderately high precision requirements to be explained in Sect. 4.10(8) and Chap. 5 subsequently.

Beasley and Spring's AS 111 (1977) was given due to the fact that they were based on the use of large number of larger decimal constzants.[8]

Thus the objective of having given the probability P and correlation ρ to determine the pairs of quantiles h and k that can yield the standard bivariate Gaussian incomplete integral to attain its value, equal to the pre-assigned value of P up to the desired level of approximation was met, subject to reliability and validity tests.

Since, there exist infinite number of such pairs, called for brevity biquantile pairs, the method to use them advantageously for meaningful and even optimal substitution in between members of such pairs, needs to be explained that has been done in subsequent Sects. 4.6–4.10.

4.5 Other Methods for Evaluating Bivariate Normal Integral

It is imperative to scan through some related progress made towards the computation of bivariate normal probability and also towards its quantile estimation. Since computation of bivariate normal probability also requires the computation of univariate normal probability, it would be essential to refer to the progress made in computation of the latter. Kerridge and Cook (1976) and Digvi (1975) suggested a new convergent series to evaluate univariate probability integral, whereas Young and Minder (1985) published algorithm for evaluating the bivariate normal integral. Subsequently, Borth (1973) made a modification in Owen's method for computing the bivariate normal integral. All these were followed by a series of publications by Deak (1979, 1983, 2000) for computing the univariate and multivariate normal integrals. Another related work was that of Willink (2004) which seeks to obtain bounds on bivariate normal distribution function. Later, Rose and Smith (2006) in their text described the features of Mathematica package on statistics, multi-normal distribution and gave an illustration of three dimensional pictures of bivariate normal joint probability function and of its cumulative distribution function. Illustrations of equi-quantiles for different correlation values as well as contours of iso-probable ellipsoids for probability levels 75, 90 and 99 % were shown to be obtainable by the use of the said package with a capacity to generate iso-probable ellipsoids for the combination of probabilities and correlations.

---

[8]It was not considered necessary to compare the results so obtained with those of others because results obtained met all the software reliability test criteria as presented in Chap. 5.

4.6 Software to Generate Tables of Bivariate Normal Quantile (Biquantile) Pairs: Prerequisite for BIVNOR

As we know that univariate normal quantile was needed for transforming the chance constrained programming to its certainty equivalent form for the given level of probability or the risk, apart from its several other applications.[9] Thus, obtaining the bivariate similitude of univariate quantile became the need of the hour. It became necessary, therefore, to embark on the development of software BIVNOR, which could compute bivariate quantile (biquantile) pairs that are needed in transforming a bivariate joint chance constrained programming problem into its certainty equivalents along with their other multiple uses.[10]

Such quantile pair could be obtained by the inversion of $B(h; k; \rho)$ values for the given probability, or the risk value for estimated or known correlation value. The references to the problem of inversion of probability to obtain quantile values have been scarce. Such a work in statistical computing literature is known as the *inverse problem*, an area which needs to be explored for a long time. It was Milton and Hotchkiss (1969) and Odeh and Evans (1974) that such quantile values were computed for univariate normal probability distribution, whereas Majumdar and Bhattacharjee (1973) did so for incomplete beta function, as referred to earlier, even though it is required for solving the joint chance constrained programming problem along with many other problems.

Despite the fact that there was a need for developing a joint chance constrained programming for which the basic requirement is the generation of multi-plets of multi-quantiles offering complex computational problem, the work in the area remained limited to theoretical excursions of Symonds (1968), Balintfy (1970), Armstrong and Balintfy (1975), Bereanu (1963a, b, c), Prekopa (1971, 1973), Prekopa and Szantai (1976), Prekopa and Kelle (1978), Sengupta et al. (1963), Sengupta (1969, 1971, 1982), Jagannathan and Rao (1973) and Jagannathan (1974).[11] There was hardly anyone else who showed any interest in the problem or tried to derive results of certainty equivalents for more than one variable. Even the software *Mathematica* Wolfram's (1996) do not appear near to such an inverse problem for bivariate normal probability distribution. Tables generated here are of great significance and of quicker applicability, because of their potential for having diverse uses.[12]

Here attempts have been made for generating bivariate quantile pairs. It is also realized that, in future, it is not wholly improbable that further problems may arise in this area, like those of at least trivariate normal trios.

---

[9]As discussed in Chap. 3.
[10]Some of which are discussed in Chap. 7.
[11]As mentioned in Sect. 3.5 in Chap. 3.
[12]Some of which have been illustrated through numerous examples in Chap. 7.

4.7 Multiplicity of Biquantile Pairs: Basics of Heuristics for Developing BIVNOR

Unfortunately, as mentioned earlier that unlike the unique quantile value for the given probability level, as in the case of univariate normal distribution, there cannot be any unique quantile pair for the specified probability level and given correlation value. In case of bivariate normal distribution, there exists an infinite number of such pairs of quantile values (coordinates of both the variables) which can yield and satisfy the specified probability level for given correlation value, termed as biquantile pairs. Such a situation creates the problem of selecting any one of the values and then choosing the other that satisfies the fixed criterion. However, the issue has been taken up subsequently in Sect. 4.8.

Such a set of infinite quantile pairs is named the "set of iso-probable quantile pairs", an idea parallel to equi-probability curves named by Cramer (1951), obtained by plotting the exponent of bivariate normal distribution that forms homothetic ellipses or ellipses of concentration.

It must be made clear that coordinates of iso-probable quantile pairs are meant to provide a curve for loci of equi-probable points for the bivariate normal distribution.

4.8 Initial Steps of Heuristic BIVNOR[13]

The strategy adopted here is to choose one of the quantile value h, given the value of correlation and target probability value (amounting to fixing the level of risk in terms of probability or percentage) and then to choose its pair quantile value k by iterative technique so that the target probability gets approximately attained up to its epsilon neighborhood. Here epsilon can be as small as the situation demands.

Thus, at first, the problem is to find a rationale for selecting the initial value of h. As the first step in heuristic-algorithm is to select Gupta's (op. cit.) H, which is the equi-quantile value of h, where $h = k$, its counterpart. Such a choice not only allowed the objective to be met but could cut short the table length to half because of cross-symmetry and exchangeability or swapping between the values of h and k. Here, one is reminded of the exchangeability property of dependent normal random variables dealt with in connection with stochastic majorization by Marshall and Olkin (1979) that refers to many other authors but yielding a relatively smaller gain. Another problem is such authors' pedantic style instead of indicating its practical utility. Therefore, it has been preferred to generate iso-probable quantile pairs resulting in a much larger decision alternatives for economic gain or for psychological advantages and sometimes providing a scope even for economic or political bargaining without any change in the level of risk. In fact such biquantile

---

[13]Strategy to meet problems raised in Sect. 4.7 in Chap. 4.

4.9 Problems Related to Developing BIVNOR

Problems arising in developing the software BIVNOR for generation and presentation of tables for such iso-probable quantile pairs were manifold:

1. Having an infinite number of such quantile pairs for any specified probability value as stated earlier.
2. Having another possibility of getting infinite sets of pair of quantile values for different values of ρ the correlation coefficients extending from +0.95 through 0 to −0.95 and even beyond up to near unity, if possible.
3. Forming a two-way table to provide a structure for the required tables of the biquantile pairs to be generated in order to circumvent the problem related to the above two aspects.
4. Choosing suitable seed values for generating such tables for each grid obtained in 3.
5. Selecting the initial and terminal values for each such grid obtained in 3. and providing a structural database for generating the required Tables
6. Consideration of the cross-symmetric point to be defined as cross-over point for exchanging the value of h to k and vice versa, i.e. swapping and reduction of the table to half on that count.
7. Fixing intervals for iterations and tabular input.
8. Choice of computation errors and their limits on condition in which the iteration would pass on to its next step or stop.
9. Problems related to the economics and other real-life implications and applications of multiplicity of quantile pairs for fixed probability or risk level, i.e. those of iso-probable quantile pairs so generated and of the criterion for choosing one of them at a time, if so desired.
10. Interpretation and possible real life applications of such quantile pairs generated for negative correlations for which Gupta's (op. cit.) tables of H (BIGH) were not available.

4.10 Algorithm for BIVNOR[15]

1. The problem raised in 1 in Sect. 4.9 can be solved only by choosing a finite number of equidistant probability points between some ranges that appeared to

---

[14]Economic advantages of such properties without any change in risk level are perceptible in examples given in Sect. 7.7(A)–(F), and even in other sections of Chap. 7.

[15]Strategies to solving problems raised in Sect. 4.9 in Chap. 4.

have practical significance and of applicability for which the tables are meant. The same practice has been followed in the generation of very many mathematical and statistical tables published earlier.

Probability points selected are from 0.99 to 0.90 at the interval of 0.01 and that from 0.90 to 0.50 are at the interval of 0.05 with exception of one more point, i.e. 0.9025. As this probability point is the product of 0.95 and 0.95 which is 95 % for each of the two variables of the standard normal bivariate distribution when each of them is independent, meaning zero correlation between them. It is for the reason that such probability level has found the maximum use, as the value $\alpha = 0.05$ being used as a level of significance most frequently. Following are the main reasons for the choice of differential intervals and stopping point:

(a) The number of tables to be generated must be reasonable.
(b) Poorer entrepreneurs and farmers, especially the smaller ones, would like to choose strategies with least risk, starting from probability of confidence levels 0.99 to 0.9 and would like to gradually but slowly increase their risk level, as their risk taking ability increases. Once they are used to taking a higher graded risk, they would prefer to increase the risk at the interval of 0.05. Such a decision is purely subjective but has some rationality in it aimed at reducing the number of tables.
(c) Most entrepreneurs, or decision makers, would not prefer to take a risk exceeding 0.5 excepting a few gamblers for whom the tables are not meant. Besides, it is well known that in case of univariate standard normal probability distribution for probability value lower than 0.5, the quantile value is negative. Such gambling is hardly seen to follow normal or bivariate normal distribution; rather it may be comparable to random walk.

2. The second aim was to generate such tables for differing correlation values between +0.95 and −0.95, yet to also keep them to a sizeable number. Intervals of correlation value for table generation are kept at 0.05 between the correlation 0.95 and 0.90 and also between −0.90 and −0.95. But the said interval is increased to 0.1 between the correlation values +0.9 and −0.9 to meet even the case of highly positive and negative correlations that may be encountered in practice. Correlations beyond |0.95| could be taken as near perfect positive or negative correlation and, in those cases, both the variables would face multi co-linearity problem. As a result, the bivariate distribution would gradually nearly collapse into univariate distribution along a line. However, the program BIVNOR, when required and invoked, could yield valid quantile pairs even for correlation beyond |0.95|.[16]

---

[16]Thus the problem of the size of the two-way table having been determined, its 21 columns for the range of the correlation values between +0.95 and −0.95 are required to be spread in three sheets of Table 8.1 of Chap. 8 (seven correlation values in descending order of magnitude in a sheet), whereas its rows are required to accommodate all the 19 probability levels between 0.99 and 0.50, common for all the three sheets of the said Table 8.1 with intervals as decided above. Thus, there

4.10 Algorithm for BIVNOR

3. The problem of selecting seed values for generating such tables for positive values of correlation coefficient could be solved with the help of tables generated by Gupta (op. cit.). Gupta's table is only for the positive values of correlation, viz., ρ = 0.1, 0.125, 0.2, 0.25, 0.3, 1/3, 0.375, 0.4, 0.5, 0.6, 0.625, 2/3, 0.7, 0.75, 0.8, 0.875 and 0.9, presented in major columns, whereas the number of variables or the dimensions of the multivariate distribution are shown in minor columns extending from 1 to 12. The values of H, the equi-correlated quantile values ranging between −3.5 and +3.5 at the interval of (0.1), are at its rows and entries of this three-way table are the probability values for dimensions 1–12. Since the present problem is limited to the bivariate only, it would be possible to obtain the value of H by inverse interpolation from the corresponding nearby probability values from minor Column 2. For the program BIVNOR, it has been given the name of BIGH to distinguish it from SMH, which stand for h, used as quantile value for x, one of the standard normal variates of the bivariate distribution. BIGH has been selected correct to four places of decimals. Similarly, SMK stands for k-th quantile value for y, the other member of the pair for such variables. The seed values, i.e. BIGH could be selected by inverse extrapolation for the correlation value +0.95. Intuition, inverse interpolation and iterative computation has been applied here to select the seed value, i.e. BIGH for negative values of ρ.
4. The problem of selecting initial value could be solved by rounding of BIGH to two decimal places towards its higher value. Those of terminal values is to keep SMHL, the lowest value that SMH could assume such that the highest value of SMK may not exceed 3.8, the (uppermost one sided) quantile value of the univariate normal distribution, which corresponded to probability value 0.9999 of the Sheppard's (1939) table.[17]

The graphic presentation of equi-quantile values

Such a presentation of equi-quantile values obtained from the Gupta's (op. cit.) table for positive correlations and that derived for negative correlations has been presented in Fig. 4.1. This depicts a three-dimensional coloured graph as described below:

The two extreme correlation values which are −0.95 and +0.95 have been dropped for 3-D graphic grids to make an equal number of points for x and y coordinates, both equal to 19, in order to meet the requirements of MATLAB 3-D figure. The said figure gives a very realistic view of how the value of H represented

---

(Footnote 16 continued)

could be 21 columns for differing correlation values and 19 rows for changing probability levels having 21 × 19 = 399 grids for which iso-probable quantile (biquantile) pair tables are required to be generated.

[17]Such a table of 399 grids, providing suitable data structure, is placed in Table 8.1 of Chap. 8. Its three-dimensional graph obtained by the use of MATLAB 7 with some changes in data to meet its programming requirements will be presented on a separate sheet following this page to illustrate its features.

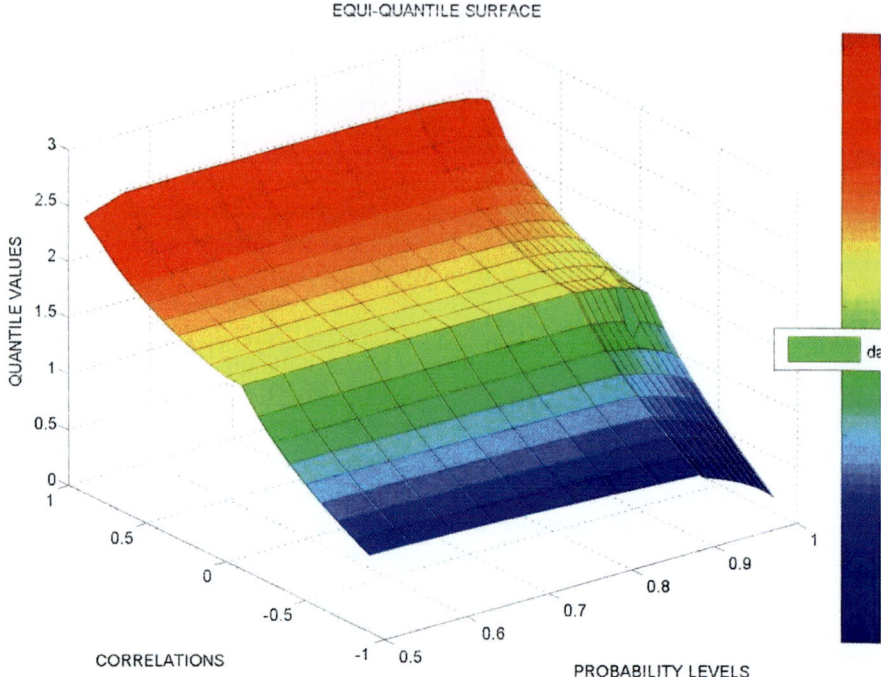

Fig. 4.1 A graphic presentation of equi-quantile values

by z-axis changes for every point and the grid of x and y axes. In the 3-D coloured graph, the probability levels ranging from 0.5 to 0.99 are shown at the x-axis, correlation values for the range −0.9 through 0 to +0.9 for the intervals +0.1 on the y-axis, and the values of H are shown on the z-axis.

5. It may be seen that as the value of h (SMH) is very close to H (BIGH), and k (SMK) nearly coincides with that, the value of SMH decreases when the corresponding value of SMK increases and vice versa showing cross-symmetry as may be expected in the case of bivariate normal situation . Hence, at such a point of coincidence, BIGH can be regarded as cross-over point; as from this point onward or backward, automatic swapping of values from SMH to SMK or vice versa takes place. This phenomenon is helpful in reducing the length or the entries (quantile pairs) of each table by nearly half.
6. The interval of iteration has initially been taken as ($H = 0.1$) and would reduce adaptively as required for iterative technique. The interval of SMH for tabular input has been kept at (DH $=$ 0.01) which is used in intervals for iterative computation as and when needed by the algorithm.
7. The next problem is to specify the magnitude of errors for termination of iteration at every intermediate stage of computation, i.e. inside the box of computational algorithm which has been kept open as below:

4.10 Algorithm for BIVNOR

EPSIL = 0.00001	For stopping when the absolute difference between the computed probability and the target one could be lower or equal to this magnitude, i.e. |BHKRO-PROB| ≤ 0.00001; where BHKRO happens to be the computed probability and PROB the pre-fixed target probability
EPSIL1 = 0.000001	For accuracy in iteration while using iterative technique
EPSIL2 = 0.00000001	For maintaining accuracy in generating univariate normal probability value $G(h)$ for the given h, the univariate normal quantile value, and those for dropping terms in the series involved in the evaluation of Owen's $T(h, a)$ function; as these have to act as the main feeders of approximation for the problem at several stages of computations and iterations

8. Thus, a two-way table of bivariate normal equi-quantile values (BIGH) at which SMH = SMK and of SMHL the lowest values of h for the combination of different probability levels and correlation values, has been prepared[18] providing a rectangular lattice type structure for data input required for generating tables of biquantile pairs, as stated earlier in Sect. 4.9: 3.

9. In a univariate case, for every probability value, there would be a unique quantile value, hence it is quite easy to get one, and only one, certainty equivalent for any probabilistic statement.[19] The situation will be different when, for every probability value, one has infinite solutions now and the question is: which among them should be chosen and why? It is the question, whose answer is not known and its implication has not been understood yet. The answer to the problem should come from the discipline from which the real-life problem and the corresponding data have emerged. The fundamental fact is of getting expansion of decision alternatives without any alteration in the risk level that is seen to have emerged. It is an unexpected feature for the decision maker who, till now, are used to getting single quantile value for a risk or probability level (for an univariate normal scenario). The implication of such unprecedented situation has to be made clear to the decision maker so that he may be in a position to take full but realistic advantages of such a change in decision scenario.

The table so generated gives the most plausible values of quantile pairs which can, in all possibility, meet the target probability values. Thus if the choice of one variable is kept at say h_1, then the choice of the other variable shall have to be kept at k_1. The other member of the pair, given the specified probability value and the value obtained for correlation coefficients because h_1 and k_1 form pairs of quantile values. It means that as one increases the value of h he will have to correspondingly decrease the value of k, or vice versa, to form

---

[18]It has been placed at the said Table 8.1 of Chap. 8 along with the corresponding SMHL values just below SMH in the same cell.

[19]It is discussed in Sect. 3.5 in Chap. 3.

a quantile pair either on the ground of feasibility or economy or for any other valid reason like conflict resolutions. This clearly provides a much wider scope to the decision maker. In this context, the following scenarios emerge:

(a) If $x = h$ is costlier compared to $y = k$ of the pair, one could decrease h and thus increase k, again forming such a pair for computed correlation value, yet the risk level would not be affected if such pairs have been chosen from the generated table of biquantile pairs, as one is on the iso-probable contour of quantile values.

(b) However, such a tradeoff between the two variables must satisfy other conditions of feasibility. For example, beyond a limit, one cannot substitute salt by sugar, tea by coffee, bread by butter or vice versa merely on the consideration of cost. Once such limits are fixed and as a result a quantile pair is made unique the solution will be found.

(c) In case, one fixes a range which also must be within the range of the table, instead of a unique value, the decision maker gets a choice of optimal solution because he converts the set of probabilistic statements to its feasible chance constrained equivalents. He optimizes it by programming (say simplex) technique, chooses the optimal solution, remembers it (keeps in memory so to say) and again does same for other chance-constrained equivalents by using other h and k pairs and compares it with the previous optimal solution, selects the better one and goes on iterating and changing h and k pairs till he obtains the best, or the global, solution within the decision space.[20]

10. For negative correlation, however, the applicability is also as clear as shown here. Nevertheless, an increase in the level of quantile value of one variable would seek the decrease in the level of the other variable on two counts:

(i) The first is the same as stated for a positive correlation. While choosing the pair of h and k, if one is higher than the other is automatically lower, so as to stick on to the iso-probable ridge of such biquantile pairs.

(ii) The second consideration is that since x and y's are negatively correlated and so are h, and k's an increase in the value of one would need a decrease in the other and vice versa. This increase and decrease is on account of negative compatibility between the two variables (on account of negative correlation) in place of substitutability or trade-off for the case of positive correlation. Varied applications in real-life paradigms may further explain the situation better.[21]

---

[20]Refer to the illustration given in Sect. 7.4 in Chap. 7.

[21]However, a negative correlation has been used in more than one example given in Sects. 7.1, 7.3 and 7.9 in Chap. 7, from which its uses and implications can easily be understood. The earliest version of the software BIVNOR, the software for generating tables of bivariate normal quantile pairs was presented by Das (1993) at the 46th Annual Conference of I.S.A.S. held at Bhubaneshwar in February 1993.

4.11 Simultaneous (Joint) Confidence Intervals

4.11.1 Roy and Bose's Multivariate Confidence Bounds

Some of the earliest researchers concerned themselves for multivariate or bivariate confidence bounds were Roy and Bose (1953), Roy (1954, 1958), Bonferroni (publication date not reported), and Millar (1966). Roy (1958) obtained such confidence bounds for the mean vector $\bar{x}$ for p-variate normal distribution as

$$\text{Prob}\left[(\bar{\mathbf{x}}' \cdot \bar{\mathbf{x}})^{1/2} - \left(\frac{T_\alpha^2}{n+1}\right)^{1/2} c_{\max}^{1/2}(\mathbf{S})\right.$$

$$\left. \leq (\boldsymbol{\mu}' \cdot \boldsymbol{\mu})^{1/2} \leq (\bar{\mathbf{x}}' \cdot \bar{\mathbf{x}})^{1/2} + \left(\frac{T_\alpha^2}{n+1}\right)^{1/2} c_{\max}^{1/2}(\mathbf{S})\right] = 1 - \alpha$$

where $\mathbf{T}^2$ was Hotelling $\mathbf{T}^{2-}$ distribution with p and $(n + 1 - p)$ d.f. and $c^2 = \mathbf{T}_\alpha^2$ was the upper α-point of the said distribution: meaning corresponding quantile value. Therefore, $c_{\max}$ being the largest amongst such c's or quantile values, Further, he also obtained expression for joint confidence interval for the difference of two bivariate normally distributed variables as

$$(\bar{x}_1 - \bar{x}_2) - \frac{st_{\alpha/2}(n-1)}{\sqrt{n}} \leq \mu_1 - \mu_2 \leq (\bar{x}_1 - \bar{x}_2) + \frac{st_{\alpha/2}(n-1)}{\sqrt{n}}$$

with confidence coefficient $1 - \alpha$.

Clearly, the limitation of such joint confidence interval was that the variables x_1 and x_2 were impliedly assumed to be scale additive. This is quite restrictive for general joint confidence interval. Besides, such confidence interval was for the difference of two means not for mean itself.

4.11.2 Simultaneous Confidence Interval

Johnson and Wichern (2003) have derived the simultaneous confidence interval for linear multivariate composition of random sample of size n from $N_p(\mu, \Sigma)$ for p-variate population of the p-vector variables $\mathbf{X}_1, \mathbf{X}_2, \ldots, \mathbf{X}_n$, where $\boldsymbol{\mu}_p$ stands for mean vector and $\Sigma_{p\times p}$ for covariance matrix, which is positive definite. Accordingly, the linear composition of the type for two variable would be

$$\mathbf{Z}_j = l_1 \mathbf{X}_{1j} + l_2 \mathbf{X}_{2j} = \mathbf{l}' \mathbf{X}_j \quad (j = 1, \ldots, n)$$

for which, naturally the sample mean and variance of Z_i ($i = 1, 2$) are

$$\bar{Z} = l'\bar{X} \quad \text{and} \quad S_z^2 = l'Sl$$

where $\bar{X}$ and S are sample mean vector and covariance matrix of X_j's, respectively. Then, they obtain joint confidence interval for $l'\mu$ having confidence coefficient $(1 - \alpha)$ as below:

$$l'\bar{X}' - c(\alpha)\sqrt{l'Sl} \leq l'\mu \leq l'\bar{X}' + c(\alpha)\sqrt{l'Sl} \quad (4.41)$$

where

$$c(\alpha) = \left[\frac{2(n-1)}{n(n-2)} F_{n,n-2}(\alpha)\right]^{1/2}$$

and $F_{2,n-2}(\alpha)$ being Snedecor's F-ratio value for 2 and $(n-2)$ degrees of freedom at $\alpha\%$ of significance level in a bivariate case. For every individual l, however, they arrived at the confidence level as below:

For $l' = [1, 0]$ and $l' = [0, 1]$ for bivariate case correspondingly to be

$$\left.\begin{array}{l}\bar{x}_1 - F_q(\alpha)\left(\frac{s_{11}}{n}\right)^{1/2} \leq \mu_1 \leq \bar{x}_1 + F_q(\alpha)\left(\frac{s_{11}}{n}\right)^{1/2} \\ \bar{x}_2 - F_q(\alpha)\left(\frac{s_{22}}{n}\right)^{1/2} \leq \mu_2 \leq \bar{x}_2 + F_q(\alpha)\left(\frac{s_{22}}{n}\right)^{1/2}\end{array}\right\} \quad (4.42)$$

where

$$F_q(\alpha) = \left[\frac{2(n-1)}{(n-2)} F_{2,n-2}(\alpha)\right]^{1/2}$$

and s_{ii} the diagonal element of dispersion matrix and n the sample size. Both of them hold simultaneously with confidence coefficient $(1 - \alpha)$ independently without modification.

There are one-at-a-time intervals, based on $t_{n-1}(\alpha/2)$, that ignore the covariance structure of bivariate distribution leading to the interval of the type

$$\bar{x}_i - t_{n-1}\left(\frac{\alpha}{2}\right)\left(\frac{s_{ii}}{n}\right)^{1/2} \leq \mu_i \leq \bar{x}_i + t_{n-1}\left(\frac{\alpha}{2}\right)\left(\frac{s_{ii}}{n}\right)^{1/2}, \quad \text{for } i = 1, 2 \quad (4.43)$$

In this approach, though each such intervals has the confidence coefficient of $(1 - \alpha)$ of covering μ_i, it is not known what to assert in general about the probability of simultaneous intervals containing μ_i's. In such a case, probability is not $(1 - \alpha)$.

Assuming each probability of such covering intervals for μ_i's to be independent, which is quite untenable, it may be $(1 - \alpha)^p$. If $(1 - \alpha)$ be 0.95 for $p = 2$, it would be

4.11 Simultaneous (Joint) Confidence Intervals

0.9025. Johnson and Wichern further assert that to guarantee a probability of $(1 - \alpha)$ to be confidence coefficient for both intervals to hold simultaneously, the individual intervals must be wider than the two separate intervals. How much wider depends on both, the number of dimension and n the sample size. For the present $p = 2$ and n is large >30, because it is required to compare each method with biquantile pairs, which corresponds only to bivariate normal distribution.

4.11.3 Bonferroni's Interval

Johnson and Wichern (op. cit.) also present Bonferroni's interval of multiple comparisons as this method has the potential of yielding better (or shorter) intervals than the simultaneous intervals dealt with in Sect. 4.11.2.

In order to keep the confidence coefficient to be $(1 - \alpha)$ even for simultaneous interval, α has been divided into two equal parts for each variable having suffix i, like the confidence coefficient $(1 - \alpha) = 1 - (\alpha_1 + \alpha_2)$ for two linear components of mean vector μ.

$$l'\mu = l_1\mu_1 + l_2\mu_2 \tag{4.44}$$

Thus it is to control the overall error rate to $(\alpha_1 + \alpha_2)$ irrespective of the correlation value, claiming equal flexibility for controlling the error for a group of important statements, but balancing it by choice for less important statements. This implies weighting of α_t with inverse of the importance of the i-th variable. However, lacking information on the relative importance of these components, they attached an equal importance to both the intervals for bivariate case.

$$\bar{x}_i - t_{n-1}\left(\frac{\alpha_i}{2}\right)\left(\frac{S_{ii}}{n}\right)^{1/2} \leq \mu_i \leq \bar{x}_i + t_{n-1}\left(\frac{\alpha_i}{2}\right)\left(\frac{S_{ii}}{n}\right)^{1/2}, \quad \text{for } i = 1, 2$$

with

$$\alpha_i = \frac{\alpha}{2} \quad \text{and} \quad 1 - \alpha = 1 - \left(\frac{\alpha}{2} + \frac{\alpha}{2}\right)$$

Therefore, with overall confidence level greater than or equal to $1 - \alpha$, like them here also two statements of a bivariate case are considered.

$$\left.\begin{array}{l}\bar{x}_1 - t_{n-1}\left(\frac{\alpha}{2\times 2}\right)\left(\frac{S_{11}}{n}\right)^{1/2} \leq \mu_1 \leq \bar{x}_1 + t_{n-1}\left(\frac{\alpha}{2\times 2}\right)\left(\frac{S_{11}}{n}\right)^{1/2} \\ \bar{x}_2 - t_{n-1}\left(\frac{\alpha}{2\times 2}\right)\left(\frac{S_{22}}{n}\right)^{1/2} \leq \mu_1 \leq \bar{x}_2 + t_{n-1}\left(\frac{\alpha}{2\times 2}\right)\left(\frac{S_{22}}{n}\right)^{1/2}\end{array}\right\} \tag{4.45}$$

which for $n \geq 30$ the distribution being normal would be

$$\left.\begin{array}{l}\bar{x}_1 - z\left(\frac{\alpha}{2\times 2}\right)\left(\frac{s_{11}}{n}\right)^{1/2} \leq \mu_1 \leq \bar{x}_1 + z\left(\frac{\alpha}{2\times 2}\right)\left(\frac{s_{11}}{n}\right)^{1/2}\\ \bar{x}_1 - z\left(\frac{\alpha}{2\times 2}\right)\left(\frac{s_{22}}{n}\right)^{1/2} \leq \mu_1 \leq \bar{x}_2 + z\left(\frac{\alpha}{2\times 2}\right)\left(\frac{s_{22}}{n}\right)^{1/2}\end{array}\right\} \quad (4.46)$$

where $z(\alpha/2\times 2)$ being naturally the quantile value of univariate normal distribution at the $(\alpha/2\times 2)$ level of significance. The statements (4.46) can be compared with (4.42). The quantile value $z(\alpha/2\times 2)$ replaces $F_q(\alpha)$, but the interval has the same structure.

4.11.4 Confidence Interval based on the Chi-Square Quantile

The third set of intervals considered by Johnson and Wichern (op. cit.) are based on the $\chi_p^2(\alpha)$ quantile, which for the bivariate case for $(n-2)$ being large based on $\chi_p^2(\alpha)$ would be

$$\left.\begin{array}{l}\bar{x}_1 - [\chi_2^2(\alpha)]^{1/2}\left(\frac{s_{11}}{n}\right)^{1/2} \leq \mu_1 \leq \bar{x}_1 + [\chi_2^2(\alpha)]^{1/2}\left(\frac{s_{11}}{n}\right)^{1/2}\\ \bar{x}_2 - [\chi_2^2(\alpha)]^{1/2}\left(\frac{s_{22}}{n}\right)^{1/2} \leq \mu_1 \leq \bar{x}_2 + [\chi_2^2(\alpha)]^{1/2}\left(\frac{s_{22}}{n}\right)^{1/2}\end{array}\right\} \quad (4.47)$$

respectively. Here $\chi_p^2(\alpha)$ means χ^2 at 1 d.f. for the level of significance 0.05 or 5 %, which takes the value 3.84 and, therefore, $[\chi_2^2(\alpha)]^{1/2} = 1.96$ and that is the quantile value of univariate independent normal distribution. Thus, the two such intervals would only be for two independent normal intervals, which cannot be considered tenable because of correlation between two such variables also needs to be taken into account. It is also for the reason that the former will reduce the confidence coefficient to[22]

$$(1-a)^2 = (0.95)^2 = 0.9025.$$

4.12 Recent Interests in Biquantile Pairs

Of late, the attention been drawn to the GHK method by Keane (1994) and of Hajivassiliou et al. (1996), as it has been referred to by Rossi et al. (2005). They write, "In many situations, the evaluation of integrals of multivariate normal distribution over a rectangular region may be desired." Further, they state that "The GHK method of Keane 1994, Hajivassiliou, et al. (op. cit.), uses an important sampling method to consider its importance to approximate this integral. The idea

---

[22]Numerical examples taken from Johnson and Wichern (op. cit.) comparing all the above methods with joint confidence intervals based on biquantile pairs have been presented in Sect. 7.2 of Chap. 7).

4.12 Recent Interests in Biquantile Pairs

of this method is to construe the importance of the technique and draw conclusions from it by using univariate truncated normal distribution."

However, the GHK algorithm may be of use for finding a higher dimension multivariate integral, but for obtaining bivariate quantile pairs of given probability and correlation that what have been achieved here are advantageous.

Further, surfing of the web for information about the problem had to be done. It is described as follows:

In the context of information dissemination through versatile use of internet, which no doubt is of great advantage to the research worker, throws the responsibility of surfing for most recent problems, queries and probable answers on the concerned and related areas of research. Thus such surfing resulted in undernoted information. Some interests for finding the bivariate normal quantile are seen to appear in the following:

1. David Shin (2003) describes his question as: "Actually, what I am looking for is something like q_{norm} to give you the quantiles for bi-variate normal distribution, if I give the function X, Y and correlation coefficient (assume $\mu = 0$, sd $= 1$). Any enlightenment will be appreciated."[23]

2. S.O.S. Mathematics CyberBoard poses the question:[24] "if I'm given the means and standard deviations, then it's possible to find the 0.95 quantile for the conditional distribution given that [gives another number]." He also answers, "I tried $F^{-1}p$ formula, but that's not giving me the right answer."

   The response is given by Royhass (20 December 2005, 12.53.37 GMT) as: "I assume you are also given the correlation. If (X, Y) have a bivariate normal distribution, then the conditional distribution of $Y/X = x$ is also normal. The quantile will be $\mu_{y/x} + \sigma^2_{y/x} z_p$, where z_p is the standard normal quantile."

To this second question, the answer given also is the same as given above. Kumar and Tripathy (2007) have worked on estimating quantiles for normal populations for two independent samples with common mean and possibly different variances. A numerical comparison of risk functions for various estimators of the quantile has been made and has been generalized to the case of bivariate normal population also.[25]

---

[23]Please refer to http://www.biostat.wustl.edu/archives/html/s-news/2003-2msg00004.html. This monographic text and especially the tables generated and placed in Chap. 8, which are Tables 8.1 and 8.2 are the answers to David's questions. It is also advised to read sections of Chap. 7 and descriptions on Tables in Chap. 5.

[24]Refer to "quantile-bivariate-normal distribution" posted on Tuesday 20 December 2005, 02:37:30 GMT.

[25]However, the abstract is released to http://atlas-conferences do not throw any light on the role of correlation values on the estimates of quantiles. Hopefully, they could have done so. Though the said work appears to be nearer to the present one, it throws no light on the actual generation of quantile pairs for a given probability level and known or estimated correlation values as has been obtained in Chap. 8 of the book.

Software named "qmvnorm" have been developed by Genz et al. (2010) and yet another by Genz et al. (2012) for obtaining equi-quantile values for multi-variate normal with the undernoted details:

Description: Computes the equi-coordinate quantile function of multivariate normal distribution for arbitrary correlation matrices based on an inversion of algorithm by Genz and Bretz (2009).

Details: Only equi-quantiles are computed i.e. the quantiles in each dimension coincide. Currently, the distribution function is inverted by using uniroot function which may result in limited accuracy of the quantiles. They deserve mention for such contribution.

Their latest announcement through Genz and Brez (2012) claims computability of multinormal quantiles as well as those of t-probs, quantiles, random deviates and densities for arbitrary correlations even for 1000 variables. Their virtual code is in FORTRAN.

However, for multivariate normal distribution for positive set of correlation values, such equi-quantile values up to 12 dimensions have long before been obtained by Gupta (1963) in the form of H. This, in the terminology of this text, is BIGH used as a starter for generating tables of biquantile pairs for positive set of correlation values. But, for a negative set of correlation such BIGH, had to be obtained. The equi-quantile values to work as starter and for other uses, by iterative technique illustrated as 3-D Fig. 4.1.[26]

Further, they report about the limited accuracy of their results. On the other hand, the book provides magnitude of error for every operation in the table itself has clearly been given. Also, the tables so derived have undergone rigorous reliability test.[27]

Another relatively recent approach is of Vasicek (1977) who introduced the concept of "copula", which has been defined by Hull (2007, 2009) as "a way of defining the correlation between variables with known distribution." It offers an innovative approach for modeling default correlation for the purpose of valuing credit derivatives and in the calculation of economic capital and offers a powerful method for handling even multivariate cases. However, the approach in this text of generating and giving tables of "inverse bivariate normal" offers no comparison to such a method, because this approach makes use of the estimates of all the five parameters of standard bivariate normal distribution, where correlation also gets estimated from user's data like other four parameters required to be used in both the approaches. Thus, it is not required to be estimated indirectly by copula approach. Though the copula approach can be easily generalized to trivariate case, it appears to be of limited use. For example, it may not be of much use in generating tables of quantile triplets that are required in transforming corresponding probabilistic statements into their certainty equivalents.

---

[26] Also presented in Table 8.1 of Chap. 8.
[27] This is reported in Chap. 5 with examples in Chap. 9.

Lo and Wilke (2010) have devised copula based estimators, which have been shown to be consistent in the presence of dependent competing risks. They suggest a computationally convenient extension of the "Copula Graphic Estimator" to a model with more than two dependent competing risks. For such method they took recourse to simulations. Further, they observed that data alone was not sufficient for proper modal identification. They aspired for possessing covariate structure which for them could reveal certain properties. In addition, they argue that if independence assumption needs to be avoided the use of the Copula function offers a recourse to model joint dependence structure. Thus, the need of covariate structure, implying knowledge about correlation estimates, was considered to be of primary importance.

It is easy, therefore, to draw the conclusion that the movement to relax an unwanted assumption of variable independence merely for introducing simplicity in the modal has become stronger with the advent of Copula.

4.13 Rizopoulos Paradox[28]

On March 29, 2009, at 16:57:27-0400, Rizopoulos floated a problem, called Rizopoulos paradox, which has so far no solution.

References

Armstrong, R.D., Balintfy, J.L.: A chance-constrained multiple choice programming algorithm. Oper. Res. **23**, 494–510 (1975)

Balintfy, J.L.: Nonlinear programming for models with joint chance-constraints. In: Integer and Nonlinear Programming, pp. 337–352. Elsevier, New York (1970)

Beasley, J.D., Springer, S.G.: Algorithm AS 111: the percentage points of the normal distribution. Appl. Statist. **26**, 118–121 (1977)

Bereanu, B.: Stochastic programming. Rev. Math. Pures et Appl. **4** (1963a)

Bereanu, B.: Stochastic transportation problem I: random consumption. Com. Acad. R.P.R **4**, 325–331 (1963b)

Bereanu, B.: On stochastic transportation problem II: random consumption. Com. Acad. R.P.R, 332–337 (1963c)

---

[28]The problem in his expression is "I know that we can find the quantile c that satisfies $P(Z1 > c, Z2 > c) = 0.05/6$ by using the R function *qmvnorm*. But, the problem here is to find the quantile c that satisfies the equation $P(Z1 > 1.975, Z2 > c) + P(Z1 > c, Z2 > c) = 0.05/6$. I think, it is different from the above-mentioned problem. I check the two R packages you mentioned. There does not seem to be a function that can directly solve c in the present problem. Please let me know if you have any other hint for me." He further clarifies that $Z1$ and $Z2$ have bivariate normal distribution with mean 0, variance 1 and correlation 0.5." The problem mentioned here has been solved in this book through tables generated by using BIVNOR and the solution is placed in Sect. 7.13 (of Chap. 7).

Bonferroni's C.E.: Publication not reported (1892–1960)
Borth, D.M.: A modification of Owen's method for computing the bivariate normal integral, Appl. Stat. **22**, 82–85 (1973)
Bradley, R.A.: Multivariate Analysis I: overview. In: Kruskal, W.H., Tanur, J.M. (eds.) International Encylopedia of Statistics. Free Press, Newyork (1978)
Cadwell, J.H.: The bivariate normal integral. Biometrika **38**, 475–479 (1951)
Cramer, H.: Mathematical Methods of Statistics. Princeton University Press, Princeton (1951)
Das, N.C.: On software for generating tables of bivariate normal quantile pairs. J. Ind. Soc. Agric. Stat. **45**(1), 130 (1993)
David Shin C.: Bivariate normal quantile function. Minbucket in part, second question p. 3 of 3, 2 Nov 2008, and p. 1 of 2, 26 Feb 2008 (2003)
Deak, I.: Computation of multiple normal probabilities. In: Prekopa, A., Kall, P. (eds.). Symposium on Stochastic Programming. Springer's Lecture Notes in Economics and Mathematical Systems, pp. 107–20 (1979)
Deak, I.: Multidimensional integration and stochastic. Wets, R. (ed.) Proceedings of the Workshop in Numerical Methods for Stochastic Optimization. IIASSA, Luxemburg (1983)
Deak, I.: Subroutine for computing normal probabilities of sets: computer Experiences. Ann. Oper. Res. **100**, 103–122 (2000)
Digvi, D.R.: Calculation of univariate and bivariate normal probability function. Ann. Stat. **7**(4), 903–910 (1975)
Essenwagner, O.: Applied Statistics in Atmospheric Science, Part A. Frequencies and Curve Fitting. Elsevier, New York (1976)
Galton, F.: Natural Inheritance. MacMillan, New York (1886)
Galton, F.: Memories of My Life. Methuen, London (1908)
Genz, A., Brez, F.: Computations of multivariate normal and t probabilities. Lecture Notes in Statistics, vol. 195. Springer-verlag, Heidelberg (2009)
Genz, A., Brez, F.: Internet google search. http://www.sci.wsu.edu/math/faculty/genz/homepage (2012) Feb 2012
Genz, A., Bretz, F., Miwa, T., Mi, X., Leisch, F., Scheipl, F., Hothorn, T.: http://www.sci.wsu.edu/math/faculty/genz/homepage (2010)
Genz, A., Bretz, F., Miwa, T., Mi, X., Leisch, F., Scheipl, F., Hothorn, T.: With Maintainer, Torsten Hothorn. Torsten.Hothorn@R-project.org (2012)
Gupta, Shanti S.: Probability integrals of multivariate normal and multi-variate—t. Ann. Math. Stat. **34**, 792–828 (1963a)
Gupta, Shanti S.: Bibliography on multivariate normal integrals and related topics. Ann. Math. Stat. **34**, 829–837 (1963b)
Hajivassiliou, V., McFadden, D., Ruud, P.: Simulation of multivariate normal rectangle probabilities and their derivatives. J. Econo. **72**, 85–114 (1996)
Hill, I.D.: The normal integral. In: Griffiths, P., Hill, I.D. (eds.) Application Statistics. Algorithms. Ellis Herwood Limited and the Royal Statistical Society, London (1985)
Hull, J.C.: Risk Management and Financial Institutions. Pearson Education, New Delhi (2007)
Hull, J.C.: Options, Futures and other Derivatives. Pearson Education, New Delhi (2009)
Hymans, S.H.: Simultaneous confidence intervals in economic forecasting. Econometrica **36**, 18–30 (1968)
Jagannathan, R., Rao, M.R.: A class of nonlinear chance-constrained programming models with joint constraint. Oper. Res. **21**(1), 360–364 (1973)
Jagannathan, R.: Chance-constrained programming models with joint chance constraints. Oper. Res. **22**(2), 358–372 (1974)
Johnson, R.A., Wichern, D.W.: Applied Multivariate Statistical Analysis. PHI Learning, New Delhi (2003)
Keane, M.: A computationally practical simulation estimator for panel data. Econometrica **62**, 96–116 (1994)
Kerridge, D.F., Cook, G.W.: Yet another series for normal integral. Biometrika **63**, 401–403 (1976)

References

Kumar, S., Tripathy, M.R.: Estimating quantile of normal Population. http://atlas-conferences.com/c/a/t/a/45.html (2007)

Lo, S.M.S., Wilke, R.A.: A copula model for dependent competing risks. J. Roy. Stat. Soc. Series C. (Appl. Stat.) **59**(2), 359–376 (2010)

Ludeman, L.C.: Random Processes, Filtering, Estimation and Detection. Wiley India, New Delhi (2010)

Majumdar, K. L., Bhattacharjee G.P.: Algorithm AS64/AS104: inverse of incomplete beta function ratio. In Griffiths, P., Hill, I.D. (eds.) Applied Statistics Algorithm. Ellis Horwood Ltd. For the Royal Statistical Society, London (1973, 1985/86)

Marshall, A.W., Olkin, I.: Inequalities: Theory of Majorization and its Applications. Academic Press, New York (1979)

Millar Jr, R.G.: Simultaneous Statistical Inference. McGraw Hill, New York (1966)

Milton, J.: Paradise Lost. New Delhi (2009, 1667)

Milton, R.C., Hotchkiss, R.: Computer evaluation of the normal and inverse normal distribution function. Technometrics **11**, 817–822 (1969)

Milton, R.C.: Computer evaluation of the multivariate normal integral. Technometrics **14**(4), 881–889 (1972)

Moran, P.A.P.: Calculation of normal distribution function. Biometrika **67**(3), 675–676 (1980)

National Bureau Standards. U.S.A (1959)

Neyman, J.: On the problem of confidence intervals. Ann. Math. Statis. **6**, 111 (1935)

Neyman, J.: Fiducial argument and theory of confidence intervals. Biometrika **32**, 128 (1941)

Nicholson, C.: The probability integral of two variables. Biometrika **33**, 59–72 (1943)

Odeh, R.E., Evans, J.O.: Algorithm AS 70: the percentage points of the normal distribution. Appl. Stat. **23**, 96–97: Ann. Math. Statis. **27**, 1075–1090 (1974)

Owen, Donald B.: Tables for computing bivariate normal probabilities. Ann. Math. Statist. **27**, 1075–1090 (1956)

Papoulis, A.: Probability, Random Variables and Stochastic Processes. McGraw-Hill Kogakusha Ltd. Tokyo, London (1965)

Pearson, K.: Karl Pearson's Early Statistical Papers. Cambridge University Press, Cambridge (1900:1948)

Pearson, K.: The Tables for Statisticians and Biometricians, vol. 2. University College, London (1914:1930–31)

Prekopa, A.: Logarithmic concave measures with application to stochastic programming. Acta Scientiarum Mathematicarum **32**, 301–316 (1971)

Prekopa, A.: Contribution to the theory of stochastic programming. Math. Prog. **4**, 202–221 (1973)

Prekopa, A., Szantai, T.: On optimal regulation of a storage level with application to the water level regulation of a lake: a survey of mathematical programming. In: Proceedings of the 9th International Mathematical Program Symposium, pp. 183–210. Akademiai Kiado, Budapest (1976)

Prekopa, A., Kelle, P.: Reliability type inventory models based on stochastic programming. Math. Prog. Study. **9,** 43–58. North Holland Publishing Company, Amsterdam (1978)

Rizopoulos, D.: Quantiles for bivariate normal distribution. https://www.R-project.org/posting-guide.html http://www.r-project.org/posting-guide.html (2009)

Rose, C., Smith, M.D.: Bivariate normal, section 6.4A. In: Mathematical Statistics with Mathematica. Springer Text in Statistics. Springer, Berlin (2006)

Rossi, P.E., Allenby, G.M., McCulloch, R.: Bayesian Statistics and Marketing. Wiley, New York (2005)

Royhaas. Answer quoted in the text. S.O.S. Mathematics Cyber Board. quantile-bivariate norm. dist (2005)

Roy, S.N., Bose, R.C.: Simultaneous confidence interval estimation. Ann. Math. Stat. **24**, 5133–5136 (1953)

Roy, S.N.: Some further results in simultaneous confidence interval estimation. Ann. Math. Stat. **25**, 752–761 (1954)

Roy, S.N.: Some Aspects of Multivariate Analysis. Asia Publishing House, New Delhi; Indian Statistical Institute Calcutta, Kolkata (1958)

Sengupta, J.K., Tintner, G., Millham, C.: On some theorems on stochastic linear programming with applications. Management Sci. **10**, 143–159 (1963)

Sengupta, J.K.: Safety first rules under chance constrained linear programming. Oper. Res. **17**, 112–132 (1969)

Sengupta, J.K.: Stochastic Programming: Methods and Applications. North Holland Publishing Company, Amsterdam (1971)

Sengupta, J.K.: Decision Models in Stochastic Programming. North Holland Publishing Company, Amsterdam (1982)

Sheppard, W.F.: On the calculation of double integral expressing normal correlation. Trans. Camb. Phil. Soc. **19**, 23–69 (1900)

Sheppard, W.F.: The probability integral. In: British Association Mathematical Tables, vol. 7. Cambridge University Press, Cambridge (1939, posthumous)

Sheppard, W.F.: Philos. Trans. Roy. Soc. Lond. (A) **192**, 101–167 (1999)

Symonds, G.: Chance-constrained equivalents of some stochastic programming problems. Oper. Res. **16**, 1152–1159 (1968)

Vasicek, O.: An equilibrium characterization of the term structure. J. Finan. Econ. **5**, 177–188 (1977)

Willink, R.: Bounds on the bivariate normal distribution function. Comm. Stat. Theory Methods **33**(10), 2281–2297 (2004)

Wolfram S.: Mathematica, Ver. 3, 3rd ed. Wolfram Media, Melbourne; Cambridge University Press, Cambridge (1996)

Young, J.C., Minder, Ch.E.: Algorithm AS-76: an integral useful in calculating non-central t and bivariate normal probabilities. In: Griffiths, P., Hills, I.D. (eds.) Applied Statistics Algorithm, pp. 145–48. [op. cit.] (1985)

Chapter 5
Software Reliability Testing and Tables Explained

There is a relatively newer discipline of software reliability testing. Therefore, it becomes necessary to first consider developing criteria for the reliability testing of each of the components of the approach suggested in this book. It is also necessary to test their reliability, as well as their validity, and decide whether or not they yield sustainable results in a fairly wide range of situations encountered by users. The way the tables developed here have been tested for their reliability and validity is new and is not reported to have been done in the past.[1] Their generation required parameters with their terminology that has been explained in Sects. 5.1–5.3. Section 5.4 on "Generated tables explained" furnishes information about the layout for such tables so as to serve a key to the user.

5.1 Software Reliability

By the time the revised version of the software BIVNOR was almost ready for generating bivariate normal iso-probable quantile pairs, a new discipline which had been in the offing for more than three decades or so came into existence. It had many of the software testing criteria, procedures and standards. There was also a host of publications about it, either as textbooks or as research papers. Some of which are *Practical Software Testing*, *Software Techniques*, *Agile Software Testing*, *The Capability Maturity Model* (*C.M.M.*) and *Software Testing: A Craftsman's Approach* and *Software Testing and Quality Assurance* written by leaders in this area. The main objective of the discipline is to lay down some criteria along with some diagnostics or test procedures to gauge whether the software newly developed meet the well laid out objectives in order to prove its worthiness to be called "standard software". A few well-known contributors in this area are Paulk et al. (2002), Perry (2005), Burnstein (2005), Loveland et al. (2005), Cockburn (2002),

---

[1]After these reliability tests, a set of tables contained in Chaps. 8 and 9 have been regenerated.

Jorgenson (2007) and Naik and Tripathy (2010) (Eds.). The books written or edited by them were consulted.

Naik and Tripathy (op. cit.) have presented these concepts in their *Theory of Program Testing Fundamental Concepts and Theorem*. They have made use of the theory of testing developed by Goodenough and Gehart and also by Goarlay. They have also included modifications in it as suggested by Weyukur and Octrand. Commenting on Goarlay's theory, Naik and Tripathy write, "An ideal goal in software development is to find out whether or not a program is correct, whereas correct program is void of faults".

An ideal process for software development consists of the following steps as applicable in present paradigm:

- A customer, or a development team, specifies the needs of the problem in hand.
- The development team considers the specifications and attempts to write a program to meet those specifications.
- A test engineer takes both the specifications and the program and selects a set of test cases. The test cases are based on those specifications and the developed program.
- The program is required to be executed with the selected test data, and the test outcome is compared with the expected outcome.
- The program is said to have faults if some tests fail. The program developed in this text passed all the tests except in some extreme cases that were of very little practical significance and applicability.[2]

5.2 Evolution of Reliability Testing Criteria

The term "validity" means getting correct and acceptable results yielded by the software under test, whereas the term reliability means getting similar results as above on repeated occasions both with varying input data from well-specified range. In this text, the above-mentioned steps were followed. The software developed was tested for validity and reliability in generating the tables.

1. The first test procedure should be to see whether the formula used and algorithms and program modules developed are based on standard publications and ranges set therein are of the standard required for the purpose in respect of computational error, which corresponds to the "unit testing", as defined by Naik and Tripathy (op. cit.).
2. The software modules that formed the essential components or units for development of BIVNOR were:

---

[2]Appropriate caveat and was obtained with satisfactory in Sect. 10.3 of Chap.10.

5.2 Evolution of Reliability Testing Criteria

(a) for generating the univariate normal probability integral values: $G(h)$, h being the quantile value[3]

(b) for generating Owen's (1956, 1962) T-function[4]

(c) for testing the equivalence of BIVNOR generated tables at $\rho = 0$, with the product of corresponding two independent univariate normal distributions which, respectively, are $G(h)$ and $G(k)$ for all probability levels considered for the generation of tables;

(d) for testing the equivalence of BIGH for each generated tables of $\rho = 0.5$ with corresponding table for H[5]

(e) to test at the point of anti-symmetry, that is, at crossover point (the point of coincidence of h and k values) as to whether $h = k$ and $H = \sqrt{h \cdot k}$? In the terminology of the module, such tests were to meet the requirements of SMH = SMK and BIGH = $\sqrt{\text{SMH} \cdot \text{SMK}}$;

(f) near such crossover point for difference in the interval of SMH = 0.01, the difference in generated SMK was also required to be very close to it, i.e. (0.01). Such nearness was set to be of the order of four places of decimal or even more; and

(g) the ensemble mean of error per cent, i.e. the mean of difference between computed probability BHKRO and the target probability PROB expressed in percentage when tested against its standard error adopting Z-test, viz. normal test of significance had to be necessarily non-significant. The ensemble being number of iso-probable quantile pairs generated for a table. That was not needed to be fixed a priori but its values obtained were mostly larger than 30. In fact, applying normal test for the case, where ensemble number was small, automatically imposed a stronger test for such case.

(h) Finally, that the module tested for release must be capable of generating such quantile pairs for any randomly chosen combination of probability levels (between 0.99 and 0.5) and correlation values (between 0.95 through 0 and −0.95) within the prescribed precision. Those were verified and were seen to follow the set norms by Hromkovich (2005), though it was not planned earlier.

3. The second task was to determine software goals, which was clearly to generate quantile pairs.[6] The criterion was prima facie fulfilled even without taking into consideration the requirements set by this new discipline. Whatever be those scientific or ethical test procedures, it was imperative to test the validity of computing another important component or unit that was of Owen's $T(h, a)$. The results of program module named OWENT were tested to match with those of

---

[3]From Eq. (4.40) in Sect. 4.3.

[4]As represented in Eqs. (4.25–4.38) in Sect. 4.3 (Chap. 4).

[5]Refer to Gupta (1963).

[6]As discussed in Chap. 4, Table 8.2 of Chap. 8 was obtained with satisfactory pairs of quantiles showing nearness of computed probabilities to targeted ones within the permitted computational error.

Owen's table for $T(h, a)$ function, for randomly selected 50 values of h and a. Such comparisons were nearly seen to satisfy the standards set by Hayes and Moulton (2009) though not planned earlier for it.

4. The third test was constructed for comparing and testing the results of BHKRO, the computed probability value for PROB, i.e. assigned probability value under correlation, that is RHO = 0. For this, sets of PROBH and PROBK, the two univariate normal probabilities, were obtained for each value of SMH, and SMK and then those of joint probability PROBJ = (PROBH * PROBK), which were tested to be true for zero correlation. Thereafter, the difference PROBD = PROBJ − BHKRO was computed to test the mean of PROBD denoted as PROBM against its standard error by taking their ratio to make Z-PROB, the standard normal test against zero for significance of computational error. The same test was performed for each PROB value, i.e. 0.99–0.50, i.e. for entire range of probabilities, for which tables were prepared and seen that each of such Z-PROB was not only non-significant but much below 1.00, even though n, the variable size of tables earlier called ensembles, were sometimes less than 30. It may be recalled that t-values are larger than corresponding Z-values adopted for normal test.

5. After this new discipline of software testing was known, the criteria and accordingly suitable tests for the same had to be set. It was required to extend the tables, and therefore, such computations were placed in Columns 5 and 6, where in Column 5, the risk per cent = (1 − Computed probability) × 100 was placed. These indicate how close the computed risks to expected ones in terms of per cent are. Column 6, on the other hand, presents the computed difference of Column 5 with target risk in per cent. The mean of Column 5 was computed to indicate ensemble mean of risk per cent and its difference with target risk to indicate the bias of ensemble mean.

5.3 Procedures Adopted to Meet the Evolved Criteria

Now, we discuss the procedures adopted to meet the evolved criteria.

To meet the very first requirement, selection of any one method out of many available for generation of univariate normal probability integral had to be done. It was preferred to select the Moran's (1980) formula based on infinite series mentioned.[7] It was not considered necessary to present the table of computed and tested values for the reason that probability values needed at step 3 of Sect. 5.2 would also require the use of the same module repeatedly and results so obtained could get reflected there itself.

---

[7]As expressed by Eq. (4.40). A small submodule G(SMH) for Eqs. (4.39) and (4.40) was also prepared.

5.3 Procedures Adopted to Meet the Evolved Criteria

Another component module (the unit test) needed was for testing Owen's $T(h, a)$ function,[8] which were required to be evaluated. The module was named as indicated earlier to be OWENT with parameters SMH for h and A for a. Sample tables of Owen's T function were computed for 50 random pairs of SMH and A, covering a wide range of SMH and A, but later arranged in increasing order of SMH values for better readability. Such computed values of $T(h, a)$ were compared with corresponding values of $T(h, a)$ function, as given in Owen's table of $T(h, a)$ function.[9] No discrepancy could be seen between the two sets up to six places of decimals. This is evident from Column 6 of the said table. However, the claim of complete lack of bias in sample selection cannot be made because only 20 of them have negative sign before zeros, coming up in set of differences, instead of 25 or its neighbouring values.

In order to test the equivalence of the results obtained by BIVNOR for correlation $\rho = 0$ with the product of corresponding $G(\text{SMH})$ and $G(\text{SMK})$, i.e. of the two independent univariate normal probability values, another software module was prepared for unit test and named BIVTEST3. Results were obtained for all probability levels considered.[10] The table heading indicates pair number in Column 1, while SMH, SMK, PROB. COMP, RISK LEVEL% and RISK% ERROR, respectively, in Columns 2–6, in the first row of each value and XH, XK, PROB.JNT, PROB-XH, PROB-XK and DIFF-PROBS, respectively, in Columns 2–6 in the second row. The main comparison was to be done between values in Column 4 of the first and second rows. Here, in the first row, there are values of PROB. COMP. Those in fact are the computed probability values, i.e. BHKRO obtained by BIVNOR for zero correlation, and in the second row are the values of the joint probabilities of two independent normal distributions, i.e. $G(\text{SMH})$ and $G(\text{SMK})$ indicated as PROB-JNT, and their differences for each pair are in Column 6 of the second row, under the heading DIFF-PROBS. It is to be seen that such differences between the two probability values, shown below DIFF-PROBS, are zero up to seven places of decimal requiring no test to be performed. Wherever differences resulted in any nonzero values, even at the seventh place of decimal, such differences had to undergo the test of significance based on Z-test (the test of significance based on normal distribution, suggested by Fisher (1950). However, for probability levels between 0.6 and 0.5, when SMH values are very low and corresponding SMK values are very high and vice versa, DIFF-PROBS indicates the appearance of nonzero values at higher decimal places. That being the seventh place of decimal: the frequency of even then none of them could be significant. Nevertheless, this phenomenon indicates that sought for equivalences are not that exact near the tail quantile pairs when the risk values exceeds 40 %. Such a feature of point-to-point coinciding or convergence of probabilities of two sets is called *convergence in distribution*, Papoulis (1965).

---

[8]Mentioned in Sect. 4.4 along with Eqs. (4.25–4.38).

[9]The said comparative values were placed in tabular format in Table 9.1 of Chap. 9.

[10]But only three such tables are placed in Table 9.2 of Chap. 9 for want of space.

The quantile values yielded by BIVNOR at correlation $\rho = 0.5$ are seen to be quite in agreement with those of Gupta's (op. cit.) table of h for the entire range of probability values considered by both.

Probabilities	0.99	0.975	0.95	0.9	0.75
*Gupta's h	2.558	2.212	1.916	1.577	1.014
Computed h	2.5571	2.2116	1.9159	1.5765	1.0138

*Courtesy Table 8.1, row 2 for $n = 2$ of the referred literature

A close examination of the main sets of tables of quantile pairs,[11] for the entire range of probability values and correlation coefficients numbering $[\{(19 \times 21) + 1\} = 400]$, clearly shows that both the conditions were very well satisfied. These conditions are that at the point of crossover (coincidence or equi-quantile, i.e. at H) between h and k, meaning between SMH and SMK, which is at BIGH:

(i) SMH = SMK
(ii) BIGH = $\sqrt{\text{SMH} \cdot \text{SMK}}$ (that BIGH is the geometric mean of SMH and SMK)

A similar examination of the said set of generated tables shows that at the initial point of iteration, which is the upper rounding off of the value at crossover point for every difference of 0.01 in SMH, the difference in computed SMK (its quantile pair), is of the same order up to four decimal places.

In order to test the difference between the computed values of the probability, i.e. Columns 5 and 6, the former for the difference between the two is expressed in per cent (to magnify it) and the latter for its standard error is also expressed in per cent. Thereafter ensemble of such differences and its standard error were computed for performing Z-test. As the ensemble size varies from table to table and not known a priori, it was decided to use a more stringent Z-test (based on normality) in place of the t-test, as mentioned earlier.

Finally, it was necessary to test the capability of the software module of generating tables of the said iso-probable quantile pairs for any random combination of probability values between 0.99 and 0.5 and for the correlation values ranging between +0.95 and −0.95. For this purpose, 37 random samples (19 for the positive and 18 for the negative correlation values) were taken. They did not coincide with the probability-correlation, combinations of Table 8.1.[12]

---

[11] As contained in Table 9.3 of Chap. 9.
[12] That is, the combination already considered earlier for such 400 tables generated and placed in Table 8.2 in Chap. 8. But for want of space, only four such tables have been given in Chap. 9.

5.3 Procedures Adopted to Meet the Evolved Criteria

Further, a successful attempt was made to generate tables for those satisfying all the conditions laid down earlier.[13] Thus, the test oracles, termed by Naik and Tripathy (2010), for testing validity and reliability of the program modules, were seen to satisfy the conditions fully.

5.4 Generated Tables of Chaps. 8 and 9 Explained

The prime objective of this text is to present the research work done with its useful applications in complex decision-making, hitherto not available. It is known by now that basically results of such research are two sets of generated tables.[14] The method was adopted as well as heuristic algorithm was developed for the generation of such tables with the help of software BIVNOR.[15] Here, while dealing with software reliability testing, such terminology has often been repeated.[16]

5.4.1 Description for Table 8.1 in Chap. 8: Equi-Quantile Values (or BIGH)

The BIGH is required as a starter for generating quantile pair, whereas SMHL is used as its terminator while running the program BIVNOR.[17] Apart from such use of it, the BIGH, being the equi-quantile value, is of use in uncertain decision-making, till the decision-maker is not in a position to decide the relative preference between two members of biquantile pair: meaning relative weight to be given to

---

[13]However, the selection of the appropriate seed values corresponding to Table 8.1 (in Chap. 8) required sometimes between 2 and 4 trials of bivariate (two-way) interpolations. Such generated tables of biquantile pairs, as rearranged for better readability, are presented in Table 9.3 of Chap. 9, as supplement to earlier ones placed in Table 8.2 of Chap. 8.

[14]Presented in Chap. 8, named "The Generated Tables", which consists of Table 8.1 of equi-quantile values and Table 8.2 containing four hundred tables, one each for biquantile pairs of bivariate normal probability integral for the given level of probability (implying the risk level) and known or estimated value of correlation coefficient between variables as per the four hundred grids of Table 8.1 of Chap. 8.

[15]These are explained in Sects. 4.6 and 4.10. The parameters, their terminologies and Owen's along with Moran's set of formulae adopted for the development of algorithms for the same and the magnitude of allowable computational errors for different stages of computations have also been presented in Sects. 4.4 and 4.10 (in Chap. 4), respectively.

[16]Thus, the present section provides a suitable niche for explaining the arrangements done to present such parameters with their corresponding terminologies through Chap. 8 along with those in Table 9.

[17]Table 8.1 contains equi-quantile values of H (BIGH) at which SMH coincides with SMK as explained in Sects. 4.8–4.10.

each member of such pair.[18] Therefore, by implication, it also stands for other values of iso-probable quantile member of the pair for any pairs of probability levels and values of correlation coefficients within the considered limits.

In Table 8.1, the said values of BIGH are printed as the upper value of each cell and SMHL as its lower value. This table has three sheets:

(1.1) The Columns of Sheet no. 1(1) show BIGH and SMHL for the range of correlation coefficients [CORR = RHO] from +0.95 to +0.4;
(1.2) The Columns of Sheet no. 1(2) show the same BIGH and SMHL for the range of correlation coefficients [CORR = RHO] from +0.3 to −0.3;
(1.3) The Columns of Sheet no. 1(3) shows those for the range of correlation coefficients [CORR = RHO] from −0.4 to −0.95.

On the other hand, the rows of each such sheet show nineteen probability levels beginning with 0.99 at the top row to 0.50 at the bottom row. Intervals of change in probability have been at 0.01 between 0.99 and 0.90 and 0.05 between 0.90 and 0.50 with the exception of including 0.9025.

5.4.2 Description for Table 8.2 of Chap. 8: Iso-Probable Biquantile Pairs of 400 (19 × 21 + 1 = 400) Tables

The series of tables contain (19 × 21 + 1 = 400) tables of iso-probable values of quantile pairs that have been generated by the software BIVNOR developed, and the same has still the potential of generating similar tables between probability range 0.99 and 0.5 and the correlation values ranging from +0.95 to −0.95, through its value zero with a few exceptions of rare use.[19]

Each such table, denoted as Table 8.2—N (with N taking its value 1–400) starts with its header giving parameter values necessary for generation of the table to its termination. Such parameter values are as follows:

1. PROB stands for the target probability value that the computed quantile pair is required to attain between 0.99 and 0.5.
2. RHO: correlation value between x and y ranging from +0.95 to −0.95 through 0 (zero) at the interval of 0.1.
3. BIGH: equi-quantile value obtained from Gupta's (1963) table for bivariate normal probability by iteration generated and presented in Table 8.1, which provides the necessary data structure. However, such values were not available in Gupta's table for negative correlations. These have been developed in this text.

---

[18]Table 8.1 also provides the necessary data structure for generating the series of tables of Table 8.2 of Chap. 8.
[19]Already explained in Sect. 10.3, "caveats and caution".

5.4 Generated Tables of Chaps. 8 and 9 Explained

4. SMHL: the lowest value of h, i.e. of SMH at which BIVNOR stops or truncates to yield further values for biquantile pairs.

However, in case the number of row entries exceeds more than 50, the table has been truncated to yield only 50 rows in order to keep its size limited to a page. For such cases, SMHL value has been raised to the appropriate level over the one given in Table 8.1. The assumption is that as many as 50 alternatives are reasonably good enough as choices for decision-making in most practical cases, even beyond as may be seen in examples of Chap. 7.

Column-wise description of the table

The column-wise description of the tables is as follows:

Column 1	Pair No. specifies the serial number of the pair of quantile SMH and corresponds SMK with their values;
Columns 2 and 3	SMH and SMK, respectively, whose SMH value is input and corresponding SMK is the computed output;
Column 4	Computed PROB.LEVEL which has been attained by the choice of pair SMH and SMK;
Column 5	RISK LEVEL% = Difference of PROB.LEVEL from unity, expressed as percentage of the latter; and
Column 6	RISK ERROR% = Difference of RISK LEVEL% to targeted risk, i.e. $(1 - \text{PROB})$, expressed in per cent.

The bottom of the table presents in sequence from left to right as given below:

1. Number of pairs of the given SMH and the corresponding computed SMK, i.e. the size or length of the table, earlier called as ensemble;
2. MEAN. COMPUT.RISK ERROR, which is only the mean of values in Column 5;
3. STD.ERR. for COMPUT.RISK ERROR computed from the values in Column 6.

The final row of Table 8.2 shows the value of Z-RATIO and its status of significance.

Notes

1. Z-RATIO being non-significant, the COMPUT.RISK ERRORS in Column 6 satisfy the condition of being called random numbers between -0.001000 and $+0.001000$. Hence, they are of use for the same Fisher and Yates (1974) and Snedecor and Cochran (1976). Thus, such random sequences are jointly stationary, i.e. for the specified probability and the given correlation, Ludeman (2010) as their mean bias stand tested for non-significance against their errors. Such non-significance of Z-RATIO is seen to amply demonstrate the

simultaneous satisfaction of Harschel's, Hagen's and Maxwell's hypotheses (Rao 2001) all the three in one, practically.[20]

2. It is important to note that in the text,[21] instruction has been given for getting the value of BIGH by interpolation. For such interpolation, it is better to adopt the geometric mean between two nearby values, because the loci of biquantile pairs or equi-quantile values are seen to generate hyperbolic or parabolic curves. The arithmetic mean was also seen to give a satisfactory result but with greater number of iterations.

5.4.3 Description of Table 9.1 of Chap. 9 (Tester Table: Owen's Table of T(h, a) Function)

Table 9.1 contains the table of Owen's $T(h, a)$ function as mentioned in Sect. 5.2, Column 1 of the table. It shows the serial number (Sl. No.) of the 50 random pairs of H and A values selected for generating COMP-T with the results of OWENTC program module. Columns 2 and 3 present such H and A values, respectively, which were selected as random pairs from Owen's T table. Column 4 presents OWEN's T taken from Owen's Table for $T(h, a)$ function.[22] Column 5 presents COMP-T values obtained as the output of the OWENTC program module. Column 6 presents the difference between corresponding values in Columns 4 and 5 and shown as DIFF-T as output of the same program module OWENTC.

5.4.4 Description of Table 9.2 of Chap. 9 (Tester Tables of Biquantile Pairs for Zero Correlation Against Those for Independence Hypothesis)

Table 9.2 presents a set of only three tables (i.e. for probability levels 0.99, 0.95 and 0.80) generated with module BIVTEST3. Here input parameters are the same as for general quantile generation tables of Table 8.2 of Chap. 8, which are PROB, RHO, BIGH and SMHL with the only exception that RHO value is taken to be zero, even though tests have been performed for all the nineteen probability levels under consideration. This has been done only to reduce the number of pages in the book.

---

[20] It is for such reason that this aspect is again referred to in Sect. 10.3(6).
[21] Especially in sections of Chap. 7.
[22] Mentioned in Sects. 4.4 and 5.2.

5.4 Generated Tables of Chaps. 8 and 9 Explained

Column 1 indicates Pair No., which is the same as ensemble no. of the general Table 8.2 of Chap. 8. Columns 2 and 3, respectively, show SMH and SMK values in Row 1 of the pair, whereas in Row 2, their nomenclature correspondingly has been changed as XH and XK to be fed as inputs for generating independent normal probabilities. Column 4 of Row 1 presents PROB.COMP obtained from BIVTEST3, whereas Row 2 of the same column presents PROB-JNT got by the product of two independent normal probabilities obtained at quantile value XH and XK, which in fact are the product of $G(XH)$ and $G(XK)$, respectively, shown in Columns 5 and 6, respectively, of Row 2. Column 5 of Row 1 presents RISK level %, which is PROB.COMPs difference with the target probability PROB expressed as percentage of the latter. Column 6 of Row 1 presents RISK % error as in general table generation, while the Row 2 of the same Column 6 shows the difference between Row 1 and Row 2 of the same Column 6, which is denoted by DIFF-PROBS. Any difference between the two would be indicative of error in computation due to BIVNOR and BIVTEST3. In case DIFF-PROBS is greater than zero, it has to undergo Z-test for its significance as is shown after Pair No. 5, whose DIFF-PROBS is of the order of 0.0000001.

For all those cases, such differences have been tested and their significance has been indicated at the end.

5.4.5 Description of Table 9.3 of Chap. 9 (Tester Tables of Iso-Probable Quantile Pairs for Random Grids of Table 8.1 of Chap. 8)

In order to test the capability of the software module, a random sample of size 37 from 399 grids formed by the combination of considered probability levels and of correlation values has been selected.[23] Iso-probable quantile pairs have been generated to see whether or not they fulfil all the conditions of reliability test laid down for the generation of the said tables. The generation of such biquantile pairs for zero correlation for three probability values considered here has already been tested and presented in Table 9.2 of Chap. 9.[24]

---

[23]From ranges already chosen for tables in Table 8.2 of Chap. 8. It is done so that they do not coincide with those of the said table.

[24]As all those conditions are seen to be met by these tables also, they are considered for inclusion as separate set of tables in Chap. 9. Thus, parameters, their nomenclatures and arrangements are the same as those of Table 8.2 of Chap. 8. For want of space, the presentation had to be limited to only four, two for the positive and other two for the negative correlation.

References

Burnstein, I.: Practical Software Testing. Springer, New Delhi (2005)
Cockburn, A.: Agile Software Development. Pearson Education, New Delhi (2002)
Fisher, R.A.: Statistical Methods for Research Workers, 11th edn. Oliver and Boyd, Edinburg (1950)
Fisher, R.A., Yates, F.: Statistical Tables for Biological, Agricultural and Medical Research, 6th edn. Oliver and Boyd, Edinburg (1974)
Gupta, S.S.: Probability integrals of multi-normal and multi t. Ann. Math. Stat. **34**, 792–828 (1963)
Hayes, R., Moulton, L.: Cluster Randomized Trials. Chapman and Hall, London (2009)
Hromkrovich, J.: Design and Algorithm of Random Algorithm: Introduction to Design Paradigm. Springer, Berlin (2005)
Jogerson, P.C.: Software Testing: A Craftsman's Approach. CRC Press, London (2007)
Loveland, S., Miller, G., Prewitt Jr, R., Shannon, M.: Software Testing Techniques: Finding the Defects that Matter. Charles River Media, New Delhi (2005)
Ludeman, L.C.: Random Processes: Filtering, Estimation and Detection. Wiley India, New Delhi (2010)
Naik, K., Tripathy, P.: Software Testing and Quality Assurance: Theory and Practice. Wiley India, New Delhi (2010)
Owen, D.B.: Tables for computing bivariate normal probabilities. Ann. Math. Statist. **27**, 1075–1090 (1956)
Owen, D.B.: Hand Book of Statistical Tables. Addison-Wesley, Reading (1962)
Papoulis, A.: Probability, Random Variables and Stochastic Processes. McGraw Hill, New York (1965)
Paulk, M.C., Weber, C.V., Curtis, B., Chrissis, M.B. (eds.): The Capability Maturity Model: Guidelines for Improving the Software Process. Pearson Education, New Delhi (2002)
Perry, W.E.: Effective Methods for Software Testing, 2nd edn. Wiley, New Jersey (2005)
Rao, C.R.: Linear Statistical Inference and its Applications. Wiley, New Jersey (2001)
Snedecor, G.W., Cochran, W.G.: Statistical Methods, 6th edn. Oxford University Press, New Delhi (1976)

Chapter 6
Decision Scenario: Problems and Prospects

A brief history of decision-making concepts and methods advocated by seers during Vedic and Vedantic (post-Vedic) eras and even later has been sketched in Sect. 6.1.[1] These, in no way, are exhaustive but quite informative. The aim is to establish at least surface connectivity with different decision tools being developed and to point out where the gaps are being felt since last six to seven decades in their applications for tangible solutions. This fulfils the aspirations of developers of such celebrated tools and of the decision-makers by the use of long-awaited biquantile pairs/equi-quantile values. This chapter has 12 sections of which Sect. 6.1, as already stated, is devoted to early periods of personality development for decision-making. The ultimate stage of the same is to reach the state of *kaivalyam*—the supremo amongst the decision-makers. The rests of the sections are on introduction to different aspects, methods and tools developed during last six to seven decades. Most of them offer only basic theme meant for mere connectivity with problems being encountered in applications of those tools. Hence, the reader is expected to consult basic-level texts and references for problem-oriented topics to have real feel of the problems and prospects of their solutions. Particularly, when the reader is required to face complexities in respect of uncertainty, dependence and dynamism, he is keen for realistic decision. Thus, the chapter aims to recollect basics of newly developed decision tools and thereby prepares for learning the artefacts of using biquantile pairs/equi-quantile value in the forthcoming application-oriented chapter.

---

[1] The chapter aims to provide an effective link between what have been achieved through previous text with what are to be presented in Chap. 7. Later half of Chap. 4 contained development of algorithm for BIVNOR, whereas contents of Chap. 5 are methods developed for their validity and reliability testers with their results in Chap. 9 of the book. Applications of tables created and presented in Chap. 8 are meant for multiple spectra of complex decision scenario as discussed in the initial chapters.

6.1 Decision Scenario: The Past

The impulse of decision-making in mankind must have been induced along with rationality.[2] However, it would be relevant to quote a couplet from the great ancient Indian philosopher Kapil, the exponent of *Sankhya* philosophy, who in his *Sankharica* (which literally means statistics at work), No. 9, deliberates:

असद्करणादुपादानग्रहणात् सर्वसम्भवाभावात् ।
शक्तस्य शक्यकरणात् कारणभावाच्च सत्कार्यम् ॥ ९ ॥

Asad-karnaat upaadaan-grahnaat sarva-sambhavaat shaktasya:
Shakya-karnaat kaaran-bhavaat cha kaaryam sat.||9||

Meaning: Decision to choose irrelevant or ineffective resource or action as probable cause cannot lead to desired objective or results. It is only the choice of effective cause when made operational could lead to desired results.

Thus, it is tacitly assumed that even in the past, the decision-maker must have had prior knowledge of the probable set of relevant causes, which may be called set of treatments, of excitations, of interventions, or of control actions, out of which the decision-maker was required to choose the most effective ones.

Normally, process of decision-making then also must would have been the choice of action known a priori. Implying that, the choice was required to be made from amongst a given set of alternative actions. The latter depended on sets of objectives, available resources and the state of mind of the decision-maker. Thus, decision-making by any individual had always been the interplay between those factors. In the good old days, every decision-maker knew his resources or capacity whatsoever reasonably well. Hence, the state of mindset, for decision-making, depended on his psychological state to face the risk of running into wrong decision, on account of uncertainty or other decision complexity elements. It was for such reasons that acquisition of appropriate knowledge had to be advised before any decision was to be made.[3]

Information acquired through the five sensors, when processed and perceived by mind, could transform into knowledge of which only the grasped ones could contribute for building conscience. It is the cumulative conscience which is conceptualized as rationality, a synergic pool, to effect human behaviour and decision, resulting in purported action.[4] An ideal example of paradigm shift (the change in attitude) that occurred in transforming Arjuna on cognition of Lord Krishna's oracle in Bhagavad Geeta is:

" नष्टो मोह: स्मृतिर्लब्धा त्वत्प्रसादान्मयाच्युत ।
स्थितोस्मि गतसंदेह: करिष्ये वचनं तव ॥७३॥ "

---

[2] Glimpses of which have been shown in Chaps. 1 and 2.

[3] In fact, this aspect has also been made clear in Chaps. 1 and 2.

[4] This is what has been stated in Chap. 1 with psycho-kinetic steps presented and explained there.

6.1 Decision Scenario: The Past

Nasto-moha-smritir-labdha, tata-prasada-na-maya-achuta
Sthito-asme-gata- sandeha-karishaye-vachanam-tava II73II:Ch-18

Meaning: Arjuna said: "My dear Krishna, O infallible one! my illusion is now gone. I have regained my memory by Your mercy (of deliberating Geeta), and now I am steady and free from doubt and am prepared to act according to Your instructions". Bhakti Vedant Swami Pravupada (1978).

It was only after that, *Gaandiva*, the famous bow of Arjuna loaded with empowered arrow got raised and the epic war began.

Here, the term *attitude* needs to be explained. McNiven (1970–1971) defines *attitude* most lucidly as "The purpose of the term *attitude* is to describe some form of human behavior, that is, generally thought of as something less than actual or final performance of an act".

Das and Singhal (2007) quote the formal definitions of *attitude* as given by Luthans (2005), Aiken (1995), respectively: "An *attitude* can be defined as a persistent tendency to feel and behave in a particular way towards some object". Whereas, Aiken defines it as "*attitudes* may be viewed as learned cognitive, effective and behavioural predispositions to respond positively or negatively to certain objects, concepts, situations, institutions or persons". Colman (2009) explains it as an enduring pattern of evaluating responses towards a person, object or issue.

Das and Singhal (op. cit.) further quote Sam Glenn while referring to the famous psychologist William James's expression "The greatest discovery of my generation is that, human beings can alter their lives by altering *attitudes* of mind".

It should be made clear that whatever information is acquired through the five sensors, gets distorted. Thus they are imperfect, incomplete, fuzzy, stochastic and variable in time and space as well as dependent on some other factors or variables or phenomena and thus have limitations.

Thus, decisions based on such information were often subject to above mentioned complexities, and those have been explained by giving them a triangular structural base.[5] The processes of decision-making under uncertainties, which get representation as Vertex V_1 in such a triangular structure of complexities, developments have been outlined, wherein it is made clear that mankind have made considerable progress in developing measure, for uncertainty. Further, a good deal of statistical methods and related techniques, for making decisions, has been developed under such situation. Thus, with calculation done, by converting such problems into its certainty equivalents and by adopting other methods, some optimal decision-making procedures have been devised. Such procedures may sometimes be preferred over risk-averting even Vapnik's (1999) risk minimization

---

[5]As discussed in Sect. 2.1 (Chap. 2).

technique or risk-prone alternatives by entrepreneur or decision-maker, irrespective of the field of application.[6]

Going back to the past in order to assess the state of contextual knowledge available, it is worthwhile to record the undernoted facts.

Apart from quantile value of the distribution, its central value—the mean is needed for obtaining the confidence interval. In this context, it is quite interesting to record its concise definition, meaningful way of its computation and its importance, believably for the first time in the history of civilization. That was written by Aadikavi Valmiki (2001) in *Yogavavashistha*, a discourse of Devarishi Vasishtha to Prince Rama, as noted below.

|| ।श्रीवसिष्ठउवाच ।।
देशाद्देशांतरं दूरं प्राप्तायाः संविदोवपुः ।
निमिषेणै वयन्मध्येतद्रुपं परमात्मनः ।।19।।

which is parsed as

देशात् तद् देशान्तरं दूरं प्राप्तायाः संविदोवपुः ।
निमिषेणै वयन्मध्येत तद् रूपं परमात्मनः ।।

Deshaat-deshantaram dooram praaptaayah samvido-vapuh:
nimisheynai vayan madhyeta tada rupam paramaatmanah.//19//

Meaning and implication: The glimpse of God (the Truth) can be had from the mean of two such observations or information (data) obtained from such spatial distance, which is recordable between a twinkling of the eyelid.

Thus, through the above expression, the controls on observations to be recorded are twofold: those being of spatial distance as well as of temporal interval (period) of only between twinkling of eyelid. Only, the mean taken from such observations (the estimate) can empirically represent the Truth (God). Thus method of calculating mean, the central value to be used for building confidence interval have emanated three to four thousand years before the Christen era. It is simple to gauge the similarity between concept of an empirically derived estimate to represent the truth expressed by Devarishi Vasistha here to those of Simon (1956), Theil (1957) and Devis and Vinter (1984).[7]

Rishi Sanat Kumar's steps for making an astute decision-maker

It is interesting to quote the discussion between Narad Muni and Rishi Sanat Kumar in *Chandogya Upanishad* (2004) (Sojourn Time Cognitive Mentoring), where the latter explains to the former that while taking decision, one has to remove three

---

[6]It has also been made clear as to how biquantile pairs/equi-quantile values generated and placed as in Tables 8.1 and 8.2 of Chap. 8 of this text are made use of while converting uncertain problems into its certainty equivalents.

[7]As quoted and explained in Sect. 3.5 in Chap. 3.

6.1 Decision Scenario: The Past

components of ignorance to attain the stage of *Kaivalyam* (the supreme state of the decision-maker), which were as follows:

(i) *Mal* → the bias
(ii) *Vic-chhepa* → the dispersion and
(iii) *Aavarna* → the canopy of error component

This is because the truth remains deviated due to bias (*mal*) and appears scattered on account of dispersion (*vic-chhepa*) as well as canopied by error component. But, their presence was considered inevitable even then.

To attain the state of *kaivalyam*, he (Rishi Sanat Kumar) enunciated all together eleven steps, as grouped herein (four stages) of cognition levels needed to be achieved by the aspirant, which according to him were in undernoted sequence:

I. Learning Stages

1. *Smarna* → awakening the memory
2. *Aasha* → rising hope or the aspiration
3. *Prana* → striving for healthy life

II. Research Stages

4. *Satya* → The truth, i.e. effort for seeking truth, recording of observations or of measurements for their perceptible features
5. *Vigyana* (*visista-gyana*) = specific knowledge → The science or of following the scientific pursuit which clearly implies analysing the data obtained at step (4), and thus getting estimates of the bias, dispersion and of error components along with cause-and-effect relationship, if any, as envisaged for the attainment of the stage of astute decision-maker (the state of reaching *kaivalyam*) in the sense of some mundane goal

III. Prerequisite for Application Stages

6. *Mati* → The attitude, meaning measurement of attitude of those concerned, on three-point scale like as tacitly meant by *asahmati* (denial or negative), *mati* (neutral) and *sahmati* (acceptance or positive); (comparable to Likert's five-point scale for attitude measurement); being adopted presently to measure the attitude, towards any predetermined objective by

the subject "the experimenter or the researcher". It is apt to compare this with "attitude" explained by Edward (1957) as personality trait or psychological state of a subject, on learning about any object with his intent to measure, the same as stated above on Likert's scale.

7. *Shradha* → The respectability, comparing such results with well-set standard results obtained at stages (4) and (5) or for products, produce or results, which command respect in public, like branded products, or for well-established facts and also for respected *rishis* (seers). This means the state of *ashradha* and mental state, between two of being undecided, also like the earlier one.

8. *Nistha* → The confidence, that of having full faith, or that of no faith for the results so obtained or towards those, for which or for whom the state of *shradha or ashradha*, could be developed at earlier stage (7). The opposite (negation) of such state, naturally be called *anistha*.

IV. Application Stages

9. *Kriti* → The stage of application of knowledge so obtained for actual working or on duties towards humanities or towards this creation

10. *Sukha* → The stage of deriving pleasure (or bearing pain), as results of stage (9). Realization of pain (*asukha*) is possibility on improper adherence of stages (1) to (9). There could be realization of pain, if stages (1) through (9) are not pursued properly but pleasure, when difficulties get removed or minimized

11. *Bhuma* → The stage of full mental, of physical and of spiritual satisfaction, as may be relevant for the occasion. The stage or the state, which once reached, does not require any scaling for measurement, as it is conceived to be a point rather than an interval. Something like trapping state for heavenly pleasure.

Naturally, all these enunciated stages were in relation to the well-set objective of the decision-maker and, at best, were expected to attain the relative efficiencies in respect of each of the eleven steps, so enunciated by Rishi Sanat Kumar, on relevant queries of Narada-muni.

Stages (4) and (5) clearly indicate how the decision-maker would then proceed after (i) removing the amount of bias and (ii) obtaining estimate of dispersion and then would get the (iii) magnitude of error component. Thus, these three components were even then identified as elements for minimizing ignorance, caused by

6.1 Decision Scenario: The Past

complexity due to uncertainty in decision-making. Having obtained estimates for these three components, resulted in to finding way for risk aversion or at least its minimization. It has some similarity with present Empirical Risk Minimization technique due to Vapnik (op cit) for a mundane god.

Rishi's concept of error (*anrita*)

It is interesting to note the divine philosophical ingenuity of Rishi Sanat Kumar, from whose enunciation one could naturally and quite elegantly get:

$$\text{Arnita} = \frac{\text{Satya} - \text{Mal}}{\text{Vic-chhepa}}$$

which corresponds to

$$z = \frac{x - \mu}{\sigma}$$

where μ was the bias required to be eliminated from the observational values x (believably true value) and σ the measure of dispersion, leaving only way to obtain measure for the error component, defined by z. Thus, it is astounding to note that how close was the approach of the said Rishi then (more than 2000 years before the Christ, probably when *Chandogya Upanishad* was enunciated/written) to that of Gauss (op. cit.), with difference that Sanat Kumar's μ was bias, whereas Gauss's μ happened to be the population mean. Nevertheless, Z retaining still its character of being the error. Which was later on identified as *Gaussian variable*.

However, it could not be ascertained as to whether ratio of type z was then obtained or not, even though it appeared easy and quite natural to obtain the same, but such remark is from the angle of the present knowledge scenario.

Philospher-Poet Rahim's concept of error (*khokha*)

In a similar context, it is interesting to recall the couplet of famous poet Rahiman for his philosophical thought during the reign of Akbar the Great, the king of the Mughal dynasty, who expressed error elimination/minimizing technique by a farm-lady at paddy threshing floor as below:

"सार सार को गहि दियो.रहिमन खोखा दियो उड़ाये।"

Saar saar ko gahi diyo, Rahiman khokha diyo uraai.

Meaning: That by winnowing the farm-lady is stacking grain kernels (symbolizing truth) and blowing away the husks (symbolizing eliminating or minimizing the untruth, meaning the error); likewise, everyone should adopt such technique in his/her life and livings. Such an expression came nearly three centuries before Gauss's (1777–1855) who gave to the world the enlightened technique of the *least squares* for parameter estimation by minimizing the error component.

Thus, it could at least be abundantly clear that even then psychological factors (often as outcome of philosophical expressions working as stimuli) like determining the levels of attitude, respectability and of confidence as acquired and learnt, were given due importance, in decision-making from day-to-day activities.

Rishi Shyavashva Atreya's concept of optimum (*Bhadram*)

The search made clearly reflects that, then also, seer Shyavashva Atreya in *Rig Vedic richa* (5-82-5) aspired to achieving *Bhadram*, hence prayed as below:

विश्वानि देव सवितर्दुरितानि परा सुव।
यद्भद्रं तन्न आ सुव ।।5।।

Vishvani deva savitar duritaani paraasuva:
Yada bhadram tanna-asuva. ||5||

Meaning: Let errors, evils, inferiors and suboptimals be kept away, as caused by the bright Sun, and let that only come, which is optimal or multi-objectively optimal, something like today's Pareto-optimal.

The above *Vedic richa,* for the reasons best known to such seers, is seen to pervade the beginnings of all *mandals* of the Rigveda and all the Adhayays (chapters) of Yajurveda.

Rishi Dadhan-gather-varna's stochastic-optimum (*shanno*)

Yet another expression of aspiration for optimality flows from the undernoted *Yajurvedic Richa by Rishi Dadhan-gathar-varna* is:

शन्नो वाता: पवताम् शन्नस्तापतु सूर्य: ।
शन्न: कनिक्रदत देव: पर्जन्योंऽ भि वर्षतु ।।10।।

Shanno vatah pavataam shannas-tapatu Suryah:
Shannah kanikra-dat devah parjanyanon avi varshatu.

Meaning: Let there be optimum flow of wind, let the Sun bestow optimum heat: optimal sound and light from thunder and lightning and optimum cum equally distributed rainwater.

It is curious to observe that happening of all the phenomena, noted in the above couplet are of stochastic nature, yet the *Vedic* seer aspires to achieve their optimum levels for the humanity, such thoughts being much broader than even Pareto-optimum.

6.2 Decision Scenario: The Present

The present scenario of decision processes is so vast that every field of knowledge claims to be their essential component, putting forward a reasonable justification. Hence, the scope of description should be kept limited to those only where, directly or indirectly, biquantile pairs find scope for application.[8]

---

[8]Refer to Chap. 7.

6.2 Decision Scenario: The Present

The complexity components of decision-making processes have been explained by a tetrahedral structure, having triangular base, wherein uncertainty, dependence and dynamism have been assigned to be the synergic components of such processes as three vertices of the base triangle.[9]

However, the history of the development of human mind, as well as of society, is suggestive of the phenomena that decision under such complexities is both individual and social (including governmental) requirements, assuming government or its implementing agency to represent the society. Hence, gains or losses on account of right or wrong decision have to be borne by the decision-maker as individual, or member of the society. Loss on account of wrong decision due to uncertainty is known as *risk*, which could be in terms of men and materials usually expressed in terms of probability but its compensation in terms of money.

It is rightly claimed that during the last half century or more, decision processes have had paradigm shift after the reintroduction of uncertainties and of risk elements required to be considered for decision-making. Broadly speaking, there have been two approaches to tackle such issues: one has been the subject matter of this monograph, where the decision-maker is required to fix up the level of risk a priori, and the other approach is that the decision-maker is required to take the risk-averting decision or risk-minimization decision.

6.3 Advancements in Decision Processes: Decision Science

The moot question is: "Is decision-making process a science?" The simple answer is "All decision-making processes cannot be placed in the category of science". Only, those decision-making processes, which pursue scientific methods, are liable to be classed as science. What then is a scientific method? A scientific method consists of the undernoted steps of decision or prescription for action:

Step 1. Formulation of hypothesis based on a prior knowledge or assumption. It requires to be categorized as *null hypothesis* (H_0) and its contradiction as *alternative hypothesis* (H_1). Thus, it is the subject matter of test on the basis of validly collected data or information quantified or at least categorized.

Step 2. Fixation of the level of significance, which is taken to mean the proportion of occasion to which the decision-maker is ready to reject the hypothesis, even if it is true. This is scientifically accepted formality that for an uncertain decision, one must make some sacrifice to lose some proportion of truth also, so that proportion of accepting wrong is kept at a minimum. A scientific inquiry accepts the hard fact of overlapping situation. Hence, a boundary line between acceptance and rejection is considered prime necessity of all empirical sciences or of entire inductive processes. It is

---

[9]Refer to Sect. 2.1 (Chap. 2).

rarely possible to get linearly separable state of space for which a decision is required to be taken in empirical sciences. Here, contention would be to consider β, the probability of accepting a wrong decision also as a criterion, at least for business decision-maker who is apt to take a reasonably small amount of such risk, and then to seek formulation to minimize the other one.

Step 3. Choice of appropriate test criterion, on which estimate is derived from sample data, is required to be tested against hypothetical value for its rejection or non-rejection. Such an estimate should accompany the estimate of its error, to be used as allowable limits of estimates deviating from hypothetically accepted value. As per requisite its probability distribution has to be known or assumed.

Step 4. Computation of the value of above-chosen test criterion and its standard error.

Step 5. Rejection or non-rejection of the null hypothesis formulated at step 1 on the basis of the test criterion value, exceeding or not exceeding the corresponding table value (representing the null hypothesis), at fixed level of significance and degree of freedom. Latter, if the sample size happens to be smaller than thirty.

Step 6. Deriving conclusion from step 5, which naturally yields decision required to be pursued.

Thus, what follows that while new technologies and methodologies do contribute to scientific decision only when they admit estimation and testing of such estimated parameters by above-enunciated scientific method, otherwise their mere quantification or even mathematical representation is to be considered merely interplay of mathematical thoughts and theorems.

It must not be taken to mean that mathematical approaches or generalization and even theorems established for environmental, social and psychological sciences are non-scientific ventures. In fact, such approaches add logical validity and exactitude in expressing as to what the phenomena are and what further could be derived (and what are impossible to be achieved) on some assumed premises. It is well known that mathematics is the language of science. However, scientific methods need to be essentially adopted when such mathematical deduction and expression are used as models, wherein empirical approaches need to be adopted, for their validation and real-time-data-based solution. Equally, what statistics offers is a scientific method. Thus, both mathematics and statistics may possibly be not regarded as physical, bio- or even social science, though they form indispensable pillars of science, as the former provides formulation and structure for the model and logically deduces the consequences in the form of theorems, deductions or their generalizations, whereas the latter adopting empirical approaches derives valid, unbiased and efficient estimates as well as draws inductive inferences resulting in validation of results, of such models, and bringing those disciplines (to which the data belong) to the level of science. Statistics, therefore, is worthy of being called "science of scientific method".

It is an admitted fact that most of the day-to-day decisions do not belong to the category of science, but only by virtue of experience and practice of the decision-maker, they are apt to be correct even most of the time, yet they cannot be considered as scientific methods. However, by training it is possible to inculcate scientific spirit for decision making.

What follows in the next section is essential component of empirical science, which needs to be followed while collecting information or data for making scientific decision.

6.4 Design of Experiment: Designing the Investigation

Prior to any recent advancement, the discipline of the "design and analysis of experiments" was discovered by Fisher (1935, 1960). The discipline helps us to study a scientific rationale for the choice of treatments and their layout on field, a device for allocating experimental units to treatments or control actions, or vice versa, and enunciates three basic principles: randomization, replication and local control offering, validity and precision, for the experimental results obtained. Tacitly, he must have thought about the four conceptual entities: (1) subject, (2) objective, (3) treatment and (4) response which are expressed here under

Subject to him must have been any entity or individual or a group on whom decision or choice of treatments or interventions could be applied or subjected to, with some objective behind the experiment or investigation.

Objective would have been to bring forth desired changes in the characteristic of the subject.

Treatment, intervention or *stimulus* would be the choice or the decision component, a set of entities of decision space, to which the subject is exposed or undergone in order to attain the well-set objective.

Response must then be to connote the cognizable change obtained in the characteristic of the subject, at least partially or stochastically, on application of treatment or imposition of intervention or of chosen elements/entities of the said decision space.

With such an onerous pursuit, Fisher (op. cit.) could provide the scientific foundation to empirical research and base for arriving at the scientific decision. Adoption of the said technique in decision-making processes definitely pushed ahead the decision process towards decision science, wherever and whenever adopted.

6.5 Decision Function

Wald (1950), on account of his classic posthumous text *Statistical Decision Function*, had the credit of architecting the edifice of a "decision function". The said famous text, used topological and measure theoretical format for providing generalization to his theoretical concepts. He, while on the one hand, established the decision function to be the generalization of both the Fisher's design of experiments and that of Neyman–Pearson theory of choice of critical region (1933, 1936, 1938) and stated that such a choice of the said theory was equivalent to the choice of decision function. On the other hand, he did also establish one-to-one correspondence between von Neumann's (1944) two-person zero sum game, to his decision function, showing the generality of the latter, in the sense that for two-person game, pure (non-randomized) strategies were adopted, whereas for decision problem, randomized strategies were adopted of which pure strategies were particular cases. However, both of them—von Neumann and Wald—used expected value criterion. And for this, the former used the utility function, while the latter preferred to define weight function. Both of them also assumed the existence of many elements or entities in decision space to get adequate choice for decision-making.

Such comparison and generalizations were natural for Wald, because those were the emerging concepts as a new discipline of scientific method, which then was termed by Neyman (1935) as *inductive behaviour*.

In the Wald's text, he initially considers a statistical problem formulated with reference to stochastic process. He further states the sample space to be a finite collection of chance variables having a joint probability distribution. But as he proceeds, he lends on to the frequent assumption made in statistical problem of the said random variables X_1, X_2, ..., etc., being independently and identically distributed. Thereafter, the distribution for each X_i's is being univariate normal. Thus, the idea of proceeding with joint distribution is dropped out, obviously because of the additional complexities on account of multivariate considerations. Both his desire and difficulties are cognizable, because such are the assumptions underlying most of the tangible results of stochastic processes.

Once again in Chap. 4 of his text, he preferred to enunciate and establish the properties of Bayes solutions, where the chance variables were assumed to be independently and identically distributed. This was required to be done for including an example computed by Milton Sobel in Chap. 4.

This is mainly because, with such an assumption, one can easily obtain tangible results. But, the reality of the fact is that random variables emerging from experiments are often such that at least some of them happen to be correlated. Hence, it could be advantageous to consider them in a real perspective. At the same time, the urge for proceeding with joint distribution amongst the researchers has been strong and long felt. But, the fear of unexpressed complexities in obtaining its quantile values has been restrictive, even in adoption of a bivariate case. It is interesting, at the same time quite challenging, to break the enigma and bring the simplest amongst the joint distribution the quantile pairs of the bivariate normal, for use in uncertain decision-making. Such an effort is likely to open a new vista for empirical research, being much more advantageous, for multidisciplinary applications.

6.6 Discrimination and Classification

Coming to yet another facet of the recent era, when decision science is seen to be blooming, for its wide applicability in every walk of life, it is apt to refer to some like pattern recognition and its application in geographical and geological information system (GIS), medical diagnostics, remote sensing, computer-oriented biometrics for identification, genomics and proteomics for synthesizing genes of choice and of oceanic temperatures and circulations around the globe for weather prediction, and also geo-morphological features of plate tectonics for earthquake prediction. These are considered cognate disciplines for collection of data from features as observed or measured mostly from photographic images and in trying to decipher some patterns in them for classification, clustering and discrimination to generate adoptable decision. It is, therefore, being considered as prerequisite to decision-making.

In such a context, it is pertinent to clarify the concepts of *features* and of *patterns*.

Features: The word "feature" connotes all characteristics that are cognizable through human sensors with or without an aid or a device. Schalkoff (1992) describes it as any extractable measurement, while Duda et al. (2010) connotes the word to signify the same by properties extracted to be used as classifiers. Thus, what emerges is the quality or property or characteristic that can be described in qualitative or quantitative or classificatory terms acquirable through seven sensors, known as *grahas*.[10] Thus, the word features try to capture all such characteristics which are cognizable through sensors man is naturally endowed with, with or without aid or devices, and not bound by the term (feature) being countable or measurable, integrable or summable, but definitely making use of them, wherever applicable. Indian philosophy is seen to connote this term to *maya*, the varied manifestations of *param Brahma*, the supreme Lord Brahma (1937, 2007).

Patterns: By the word "pattern", one usually understands a recognizable way in which something is done, arranged, organized or happened. It is a regularly repeated arrangement or design which displays some feature. In the context of pattern recognition or classification, it means trying to search any regularity or repeatability by which a feature could be represented, studied or modelled. Tsypkin and Nikolic (1971) explain "pattern recognition" as the first and most important step in

---

[10]Explained in Sect. 2.1 in Chap. 2.

processing the information, which is obtained through senses and various devices. He further says, "at first, we recognize the objects and then the relationships between objects (in fact, he means amongst the objects) and between objects and ourselves ..." Thus, his concept of pattern recognition coincides with "features", explained above. Such ambiguity in meaning arises only because he tries to put emphasis on "pattern recognition" and ignores the existence of "feature" that is primary to the existence of any pattern. But as he says in his text referred to, such ambiguity does not matter much.

With these two basic concepts of pattern recognition and its cognate disciplines, attempts are made to understand the system dynamics as it changes in time and space, offering clue for controlling, intervention, cybernation and management. These are being consequences of decision and choice as may be relevant for concerned disciplines or occasions.

A wide class of such problems requires collection of data and their statistical analysis, inference, testing and modelling, where the application of multivariable *discriminant function* dominates. It is natural that some of such variables are pairwise-correlated and inclusion of such variables often adds to the discriminatory power in classification problems. For most of the classification problems, underlying probability distribution is Gaussian or normal. This is firstly because of its having limiting property on account of the central limit theorem and secondly because of having attribute of maximum entropy H amongst all continuous density functions given mean μ and variance σ^2. The H is $\left[0.5 + \log_2\left\{(2\pi)^{1/2}\sigma\right\}\right]$ bits (Duda et al. 2010).

Thus, assumption of bivariate or multivariate normality is common features in such paradigms. Even then application of univariate quantile, while formulating the confidence interval for a group or class is frequent, is wrongly taking recourse to variable independence. Obviously, it happens due to non-availability of iso-probable biquantile pairs. But, the said malpractice kills the very spirit of multivariate concept and consideration.

As of today, for most of the bivariate normal problem, an easy recourse is to assume variable independence, which is unrealistic, often inappropriate as well as uneconomical.[11] Fukunaga (1972), in the context of pattern recognition, defines quantiles of order p as

$$\Pr\{x \leq z_i\} = P(z_i) = p,$$

where z_i is a threshold value of x, up to which the probability of attaining is p. Once he attains the value z_i, the quantile, he cleverly obtains its confidence interval, even

---

[11]With the availability of tables of iso-probable biquantile pairs, it is possible to find realistic solution, relaxing unwanted all the time-independent assumptions, needing to be shown through examples subsequently in Chap. 7.

without the knowledge of density function to which he, therefore, calls as distribution-free method of inference:

$$\Pr\{y_i < z_i < y_i\} = P\{y_i < z_i\} - P\{y_i < z_i\} = \gamma$$

But even he must have needed the use of biquantile pairs h for x and k for y to obtain bivariate normal confidence intervals to contain the required probability level, when such variables are not independent.

Schalkoff (op. cit.), while dealing with linear inequalities in the context of linear classifiers for pattern recognition problem, suggests the use of a linear programming method for arriving at the solution of suitably formulated objective function, transforming linear constrained inequalities $\mathbf{A}w > 0$ into equality $\mathbf{A}w = \mathbf{b}$, and thus, $w = \mathbf{A}^{-1}\mathbf{b}$, where $\mathbf{b}' = (b_1, b_2, ..., b_n)$ with $b_i > 0$. Such a choice of matrix $\mathbf{A}$, or for that matter, even the constrained vector $\mathbf{b}$, which are based on empirical data, in reality, have to be random, as well as, often correlated. Hence, such a problem of linear programming, requiring transformation into its certainty equivalent and forming joint chance-constrained programming problem, would need biquantile pairs.[12]

6.7 Optimization: Linear Programming and Extensions

The importance of "linear programming" by Dantzig (1947, 1951) and of "dynamic programming" by Bellman (1957) in decision scenario, as indicated earlier, is not enough, as they seem to dominate all other discoveries relating to decision processes, because they provided the following:

1. new decision-making criterion that being optimization (maximization or minimization) of objective function under operable constraints;
2. simplicity in model formulation and in interpretation of results on account of linearity assumption and of structure;
3. yet other features are its simplicity in comprehension and computation on account of simplex or revised simplex algorithms and availability of computer packages.

In fact, Dantzig's fundamental paper was in private circulation on account of the situation that it was then being used as a technique for planning the diversified activities of the US Air Force, and later, it was published in T.C. Koopman's (ed.) (1951) as *Activity Analysis of Production and Allocation*. The above information has been noted from Dorfman et al. (1958).

---

[12] An example of formulating and solving such a joint chance-constrained programming problem is given in Sect. 7.4 in Chap. 7.

It was soon realized that assumptions were too much restrictive for solving real-life problems, hence followed enormous publications, for several decades to relax such restrictions and assumptions. A series of such publications and their applications in diverse fields still continues. Early amongst some such publications were those by Dantzig himself (1955, 1963). An attempt to relax the assumption of linearity resulted in the discovery of nonlinear programming, like quadratic programming and geometric programming, whereas consideration of stochasticity or randomness in the estimates of parameters of (i) objective function, (ii) technology matrix and (iii) constraints gave rise to several versions of stochastic programming —probabilistic programming and chance-constrained programming. Consequently, there has been vast development in decision-making under uncertainty, because decision-makers of the present paradigm are fully aware of the fact that uncertainties are the realities of life, and the same intensifies often on account of incomplete irrelevant and outdated information.[13]

At this stage, it would be in fitness of the problem and also for its continuity to explain as precisely as possible the basics of stochastic programming and of its chance-constrained version.

6.8 Basics of Stochastic Programming and Its Chance-Constrained Version

Sengupta (1982) crisply defines "stochastic programming" as the method of incorporating stochasticity in parameters of the mathematical structure of such programming problem. It means, either coefficients of objective function or of technology vectors, or even the resource constraints or all of them, could be random variables. For most such models, an easy recourse was to replace such parameters by their expected values, based on information collected for them and presuming some probability distribution for them, in order to provide inferential setting often essentially required. Such practice could popularly be adopted for two reasons:

1. simplicity of getting information and of processing the expected values and their variances;
2. popularity of von Neumann's (op. cit.) expected value criterion and then newly discovered such a technique for his theory of games and economic behaviour.

However, discovery of *Prospect Theory* by Tversky and Kahneman (1981), which was based on the psychological aspects of choice-based decision-making and expected value based decisions, faced criticism. The chance-constrained version of stochastic programming, however, remained reasonably robust enough to sustain such criticism because of undernoted reasons:

---

[13]These are also discussed, in detail, in Chaps. 1 and 2.

6.8 Basics of Stochastic Programming and Its Chance-Constrained Version

1. It had the force of reasonable probability of confidence, behind its estimates of parameters and their ranges offered by the use of quantile values of univariate normal or that of t-distribution.
2. In a solution to a bivariate problem,[14] where iso-probable (iso-risk) multiple biquantile pairs, offer a wide range of viable decision alternatives. These alternatives in all possibilities could be extended to cover the varying requirements and those also that are raised through the Prospect theory, on the condition that probability distribution of involved random variables be pairwise bivariate normal. Thus, it completely relaxes the variable independence requirements. Instead of correlation between such pairs, the book provides advantageous scope for surrogation of one for the other, if found realistically viable, economically beneficial, or even psychologically preferable, meeting all the three requirements on the same platform. Clearly, these requirements are consideration for a risk in probability, dependence, in terms of correlation, and a scope of choice for the prospect theory.

The structure of chance-constrained programming problem, as explained by Vajda (1970, 1972), is as follows:

$$\text{Minimize } \mathbf{c}' \mathbf{x};$$
$$\text{Subject to } \mathbf{x} \geq \mathbf{0},$$

and to

$$\text{Prob}\left(\sum_j a_{ij} x_j \geq b_i\right) \geq \alpha_i \quad (i = 1, \ldots, m)$$

where the α_i are the assigned probability levels not equal to zero. Also, it could be even desirable to get the joint probability that all inequalities

$$\sum_j a_{ij} x_j \geq b_i$$

hold, to be at least, with probability α. The later condition is being less restrictive than the former one, if $\alpha = \Pi \alpha_i$.

While making reference to the application of quantile value,[15] for transforming probabilistic inequality, into its certainty equivalent, it is clear that

$$\sum_j a_j x_j \geq F_b^{-1}(\alpha),$$

---

[14] As can be subsequently seen through examples of Chap. 7.
[15] As explained in Sect. 3.4 in Chap. 3.

where $F_b^{-1}(\alpha)$ is the quantile value, corresponding to the chosen probability value of the distribution, which follows an assumption of randomness of such values.[16]

References to stochastic programming, its chance-constrained version

The chance-constrained programming version of stochastic programming was originated by Charnes et al. (1958), followed by several publications by those two with others like Thompson, Kirby, Raike, Stedry, Thore, Byrne, Kortanek, et al., during 1958 and 1975.

The selection of some, which has made a considerable contribution to the field of probabilistic or chance-constrained version of stochastic programming, was done, as this was the area of research where the requirement of biquantile pairs was felt most often and had to suffer most in attempts made for solving joint chance-constrained programming problem. Hence, the area was scanned for the attempt made for advancement by researchers working in this direction.

Amongst them are the group of Bereanu and his associates and the other are the group of Prekopa and associates. The criteria for scanning relate to a consideration of the joint distribution of random variables where parameters of programming models are taken to be random variables. If this is so, the question that may be asked is: how is it that the value of the correlation between them has been allowed to play a role in the adjustment of estimates of parameter values? Further in such a situation, could the necessity of biquantile be realized.

Bereanu (1963a, b, c), in his paper, considered objective function coefficients to be a random vector, whose joint probability was assumed to be known. But, while considering stochastic programming formulation of an agriculture planning problem, such random coefficients of objective function were assumed to be independently distributed. Similarly, in his other publications (1965, 1967, 1971, 1980), the phenomenon of dependence amongst random coefficient vector was not considered.

Prekopa alone and along with some associates, Szantai and Kelle during 1965 and 1978 have shown their concern to the probabilistic approach to stochastic programming models, where consideration of joint probability distribution did figure. But, they preferred logarithmic concave, measures and function, to attack such problems. The chance-constrained approach to certainty equivalent did not appear to figure at least in papers which were accessed. Thus, simultaneous consideration of probability and of correlation could not be traced in their approaches. For solving some numerical problems, Prekopa et al. (1976) preferred to assume independence of variables. However, as regards to probability distribution of such variables, they at times did consider uniform as well as exponential distribution.

It is interesting to review a publication by Katoaka (1963), wherein he formulated a stochastic programming model in certainty equivalent form. There he clearly made use of quantile as inverse of probability level α, the level of significance as probability measure, that the decision-maker would like to take as risk or

---

[16]In Sect. 7.4 (Chap. 7), advantageous applications of biquantile pairs have been shown in much sought-after joint chance-constrained version of stochastic programming.

6.8 Basics of Stochastic Programming and Its Chance-Constrained Version

alternatively $(1 - \alpha)$, the value of confidence coefficient. In his problem of maximization, Katoaka formulated the certainty equivalent of uncertain maximization problem as

$$\text{Max } f_1 = px - q(\mathbf{x}'\mathbf{V}\mathbf{x})^{1/2} \quad \text{subject to } \mathbf{A}x' = \mathbf{b}^* \text{ and } x \geq 0$$

where resource vector

$$b_i = b_i - q_i \sigma_{bi} \quad \text{and} \quad q = -I^{-1}(\alpha) \geq 0 \quad \text{and} \quad q_i = -G^{-1}(\beta_i).$$

His certainty equivalent form was transformed to

$$\text{Max } p'\bar{x} \quad \text{subject to } \mathbf{A}x \leq \mathbf{b}^*, \mathbf{b}^* \geq 0 \text{ and } x \geq 0,$$

where quantile values q and q_i were seen to have been used for changing uncertain deterministic optimization problem to its certainty equivalent.

Thus, Katoaka realistically approached the problem and, through a simple numerical example, considered the mean values of the random coefficients of objective function and of the resources, their variances as well as covariance between them. However, solution obtained by him was in terms of q and R, where q, as expected to be the value of quantile and R, being the positive definite value of the parameter $R = [\hat{x}(R)'V(x(R))]^{1/2}$, clearly implying the estimate of standard error.

However, his problem then was near to that of this text. But, there remained a gap in the solution approach on account of two features. Firstly, he did not try to obtain the value of appropriate quantile. Secondly, had he found that, he would have to face the problem of infinite such pairs and also would encounter the appropriate selection problem, as it had to be faced here, and their proper interpretation with real-life justification. In fact, its importance lies there itself.

Stancu-Minasian (1976, 1980) did consider deterministic equivalent of minimum risk problem and of Katoaka's problem, for a product of linear functions, and also Katoaka's problem, for quotient of linear function. But, the problem of correlation between such stochastic variables was not considered.

It was interesting that attention was drawn towards the publication of Ishermann (1979), which, though in German, was translated by INSDOC, New Delhi, with the hope that some clue about consideration of dependencies and thereby of correlation could perhaps be found there, while facing multi-objective optimization problem. He, while working on structuring of decision processes for multi-objective decision-making, very well expressed the importance of and consideration for interdependencies amongst objectives and thereby of partial information to be brought about in close correlation with goal forming decision-making criterion function. But, in his further discussion and numerical example, he only made use of deterministic information, knowing that partial information is mostly of stochastic nature, did not anywhere bring into picture the use and information of correlation coefficient.

Others who contributed to joint chance-constrained programming (JCCP) were Millar and Wagner (1965), Jagannathan (1974). The former considered deterministic equivalent of a JCCP model, but for independent random variable case. In such a case, the expression *joint* gets diluted and became of lesser practical utility that it deserves.[17]

Later, Jagannathan (op. cit.) considered the JCCP model with dependent right-hand elements, i.e. *resource vector*, where Jagannathan considered such a vector to have been created as a mixture of random vector ξ_r. It is too restrictive for any real-life problem. In another case, Jagannathan considered them to be dependent, but even that also to be linear combination of some sets of independent random variables. Besides, he fractionated β, the probability of exceeding the intersection of sets of inequalities, as below:

$$\text{Max} \sum_j c_j x_j$$

subject to

$$\text{Prob}\left[\bigcap_{i=1}^{k}\left\{\sum_j a_{ij}x_j \leq b_i\right\}\right] \geq \beta; \quad x \in S$$

such that

$$\sum \alpha_r \beta_r \geq \beta \quad \text{where} \quad \sum \alpha_r = 1.$$

Implying clearly that each such inequality would have to be independent, mutually exclusive, and they together could establish their exhaustiveness, so that their linear convex composition in totality be $\geq \beta$.

Such an assumption was also dampening the spirit of assumptions about dependencies of b_i's, the *resource vector* components. The numerical problem cited by him did not result in feasible solutions. Such phenomenon was natural because of non-availability of biquantile pairs, even though Gupta's (op. cit.) equi-quantile values were available for positive correlations. One probable reason may be the fact that equi-quantiles do not provide sufficient realistic appeal for practical adoption because of its unnecessary restriction on choice of making decision for giving equal weight or importance to both the variables.

The main reason for surfing through chance-constrained version of stochastic programming references has been to search for the method which could provide a solution to optimization models. That takes into consideration the joint chance-constrained problem to obtain the optimal solution, taking into consideration both the risk level in terms of probability and correlation.[18]

---

[17]Also discussed in Sect. 7.4 (in Chap. 7).
[18]This has been obtained in Sect. 7.4 (in Chap. 7).

However, most of the decision-makers have to operate with corporate decisions with multiple objectives, sometimes with conflicting ones. As a result, there may be multiple criteria for decision making. Therefore it is imperative that a precise set of criteria and methodology be chosen from a welter of criteria and methods. How this can be done has become an important question in the decision making process.

6.9 Emergence of Decision Processes Under Risk and Uncertainty

6.9.1 Related References

Apart from dealing with uncertainty, or risk, for the development and extension of optimization techniques, like stochastic and chance-constrained versions of programming and even otherwise also, there appeared several approaches for dealing with risk in decision-making.

Arrow (1951a, b, 1964, 1965, 1971) came forward with his concept of risk and risk-averse choice and their measures, which was supplemented by that of Pratt's (1964) risk-averse measure. Even though the present text is least concerned with such a risk-averse measure, it must take note of such belief which pervades probability, risk and uncertainty in decision-making. Concepts of terms like uncertainty, probability–equi-probability, possibility–impossibility and even grading choice on the basis of probability are not alien for social choice theory of Sen (1977, 2010). On the other hand, Pratt was deeply involved with these concepts.

However, such choices also have to be subset of decision set. More so, like decision under risk and uncertainty choice is also required to presume existence of the set of alternatives. Besides, choice more often is taken to mean in positive sense, whereas decision is taken to mean both in positive and in negative senses. That is the psychological relation between the two. In fact, it appears that decision is precursor to each and every choice. In other words, one may say that a decision is almost always before every choice. Having a set of alternatives, a decision tends to be stochastic, for if the decision is taken, the choice tends towards determinism. Having implication that, however uncertain, once the decision is taken, the choice is almost certain. Thus, the relation between them is a Bayesian paradigm of conditional probability.[19] Such phenomenon of decision science exhibits synergic relationship between economics and psychology with statistics providing the methods, measure and test of such synergism in decision-making, often being called *interaction*. It is such elements of synergism which accumulate to the formation of

---

[19]This aspect has cursorily been mentioned in Sects. 2.8 and 2.11 (in Chap. 2) while making references to Suppes (1970) and Pearl et al. (op. cit.).

attitude and of motivation for any specific choice, pertaining to concerned decision set.[20]

Coming to risk aversion techniques, some may be quoted such as Zeimba along with his some associates (1969, 1972, 1974, 1976 and 1977). They mainly focused on optimal portfolio choice applying mathematical and stochastic programming techniques. In their paper of 1977, they were concerned with tight lower and upper bound on the expectation of a convex function of a random variable. They realized that similar bounds were applicable in multivariate case, if the random variables were independent. For dependent case, they expressed that bounds based on Edmundson–Madansky were not available. Thus, even such bounds for bivariate normal could not be attempted then.

Byrne et al. (1971) contained papers of Hillier, Naslund, Byrne, et al., Weingartner, Ruefli, Crecine and Gerwin, and those of Davis, et al. Almost all papers of the volume considered models with optimization under risk. Models considered by Byrne et al. were on chance-constrained approach to capital budgeting and the other on capital budgeting by the application of linear programming under uncertainty. They also did not consider even two variable cases with dependency.

Impact of contributions by Markovitz, Arrow, Pratt, Klien and several others was also seen on the works of Klien, Bawa and their associates (1975, 1976, 1977, 1978, 1979) on optimal portfolio choice, its admissibility, effects of estimation risk on its choice along with diversification, safety first consideration in such choice, uncertainty in estimation on decisions, consideration of stochastic dominance, some simple rules for optimal portfolio selection and even on multivariate normal distribution, as required in chance-constrained programming for non-dominated portfolios. Bawa, in his preprint paper of 1979, has generalized chanced-constrained problem to multivariate scenario and felt the requirement of biquantile and indicated its use. However, he did not prefer to deal with the possibility of computing such quantile pairs for bivariate normal distribution. Thus, the gap of *non-availability of method to generate quantile pairs*, remained clearly *identified then itself*. For want of such quantile generation methodologies, what to ask for multivariate generalization even bivariate generalization remained alienated from the fold of application paradigm. However, many theoretical derivations and generalizations are seen to appear since then in pedantic styles by using measure theoretical notions and notations.

6.9.2 von Neumann and Morgenstern versus Kahneman and Tversky

There does not appear any demarcation between the researchers of risk averters and optimizers of risky entrepreneurships. Further, it is noteworthy to mention, as done

---

[20]This is one of the application aspects of learning model defined in Chap. 1.

earlier also, that optimization of expected utility concept appeared to dominate the scenario of the uncertain decision-making, as impact of the contribution by von Neumann and Morgenstern (1944, 1953, 1964).

On the other hand, the prospect theory propounded by Kahneman and Tversky (1979) based on psychological aspects in decision-making, which also won the Nobel Prize in Economics, could establish not only the importance of such an aspect over expected utility theory, but also of its irrelevance, thereby showing that human psychology plays an important role in decision-making irrespective of the situation being risky or not. This reminds once again of Rishi Sanat Kumar (op. cit.), where he, perhaps for the first time, in the history of mankind did disclose the role of the psychological aspects of *attitude, respectability, faith* and *sense* of pleasure to that of reaching the stage of *opulence*, as described earlier.[21]

6.9.3 An Attempt for Compromise

Soon after the publication of Kahnemann and Tversky's said paper, Sengupta (1982) came forward with a synthesis of both the approaches, wherein he explained: "The application of expected utility theory to choices between prospects is based amongst others, on two major postulates:

$$\text{Expectation: } U(\mathbf{z}, \mathbf{p}) = \sum_{i=1}^{N} p_i u(z_i)$$

Here the risk aversion, u, being a concave function.

That is, the overall utility of a prospect, denoted by U, is the expected utility of its outcomes, and the utility function of the outcome, is such that the decision-maker prefers the certain prospect z_0, to any risky prospect having its expected value to be even just $= z_0$". He then did choose the concave utility function, $u(z)$, with a form

$$u(z) = 1 - e^{\alpha z}, \quad \alpha > 0$$

and thereafter, assuming the random outcome z to be normally distributed variable, he was able to formulate the problem of maximization of

$$U(x) = -\frac{1}{2}\alpha \mathbf{x}'\mathbf{V}\mathbf{x}$$

---

[21] It is for such reasons that it was required to consider aspect like predicate (stochastic) calculus of psycho-kinetics for personality development in Chap. 1.

where $\mathbf{z} = \mathbf{c'x}$ was viewed as profits associated with n outputs, $\mathbf{x}' = (x_i: i = 1, 2, \ldots, n)$ and α being constant rate of risk aversion.

Thereafter, he listed three types of objections to the above approach and attempted to meet them, one by one, and came forward with a multi-criteria decision-making problem under a condition of risk. While dealing with *Prospect Theory*, he proposed "maximization of overall utility function OU(y) in two stages, where his OU was built up by giving different decision weight to each such decisions and then summing them up. Thereafter, reducing the problem to maximization of such weighted decision function called OU. In case such individual components were not scale additive or were qualitative in nature, the suggestion given was to follow the goal programming approach".

With all said and done, the necessity of iso-probable biquantile pairs in transforming chance-constrained programming into its certainty equivalents in case where such programming parameters were random as well as correlated was invariant. However, such an advantage of using biquantile pairs remains valid only for bivariate normal random variables till one compromises to go for equi-quantile values or may find reasonable justification for selecting any number of such correlated pairs of variables. Such possibilities are not infrequent. It is noteworthy to mention that even two-variable or one-stage dependence is capable of meeting the real-life requirement of more than eighty per cent of the problems with correlations or dependencies. It is based on the experience about obtaining magnitude for first eigenvalue in relation to that of its subsequent values, while adopting stepwise successive matrix exhaustion procedure in factor analysis for variable reduction problem.

6.10 Developments in Uncertain Decision-Making

Before the role of quantile is discussed for multivariate situation, it is appropriate to surf through some developments on the role of uncertainty in decision-making, vis-à-vis of probability modelling, which may give a sharper view of further applications and developments of the task undertaken.

Deep realization about the inevitability of uncertainty and of risk in decision-making appears to be home ground for the researchers and developers of probability modelling, so much so that the area drew the Nobel Prizes to Black and Scholes (1973) and Merton (1973) and also to Kahneman and Tversky (1979) in the field of Economics as mentioned above.

Kolbin (2002) has explained the decision-making scenario, as of recent time, from which some excerpts are quoted below:

> Decision-making (ability to argue) is the central element of administrative activities. By decision making, we mean a specific type of human activities aimed at choosing the best among available alternatives.

6.10 Developments in Uncertain Decision-Making

As quoted from Larichev (1979), this definition indicates three necessary elements in decision-making process:

- the problem to be solved, meaning identification and formulation of the problem;
- a person or a collective body, which takes a decision;
- existence or generation of several alternatives amongst which a choice will be made.

In the absence of any of these elements, the process of selection ceases to exist.

However, this definition on its surface gives a brief and crisp look, but not so simple, because each of these three elements implicates many complicated information, assertions, assumptions, and agreements, for bearing consequences in their implementations.

1. Like one "the problem to be solved" needs formulation of the problem in such a manner that *prime facie*, it be solvable, and solutions are acceptable to the body which takes the decision or empowered to take decision.
2. Laying out conditions, constraints, and specifying risk and uncertainties, availability of resources, apart from setting of objectives and its time schedule with every aspect of input required and of output achieved, from the beginning of the project to its completion.
3. Apart from the above about the resources and its funding schedule, a huge amount of relevant data need to be collected, processed through software modules, to obtain estimates of several parameters with elements of errors attached to each of them and also of magnitude of dependencies amongst them, as well as dynamics of change in them, for the projected time frame. At least some of these concepts have been highlighted in the initial chapters to the extent those were believably known to or apprehended by ancient seers, in good old days, and could reasonably be advanced through the ages and of late by Pearl and her associates (2004).
4. Besides above, software modules need to be selected, cognizing their applicability, availability, and reliability.
5. The framing of objectives requires conceptualization of utilities and their scales, risk and its measure and of optimizations of multiple criteria, and some of them may be synergic but others conflicting.

It is apt to quote Kolbin (op. cit.) again for his remark that:

> The problem of choosing alternatives or the problem of decision making in modern world becomes more important class of problems, which is a common occurrence in everyday life to businessmen and researchers, doctors, engineers, people in their life.

What has been said is no doubt an undeniable fact, but the creation of alternatives is of even greater importance, because the same expands the very decision space, providing larger scope for such choice. Further, in his essence of decision models, Kolbin (op. cit.) very realistically asserts that "basic features of social and economic decision systems are the multiplicity of persons involved in decision-making and interested in the result so obtained, the multiplicity of their interests and the multiplicity of possible developments of events".

Such features of decision processes are very much demanding sets of multiple admissible alternatives. For some scenario, there may be situation where a whole set of available viable alternatives is optimal in some sense. Again, such a situation may appear fused with incapability of defining optimality criterion itself.

Under the present paradigm, however, admissibility of decision alternatives is required to be interpreted as the iso-risk or iso-probable set of decision alternatives for a given or estimated correlation between pairs of variables.

Kolbin (op. cit.) further realizes that search problem of the entire optimal set is rather laborious and, in many cases, unsolvable, and therefore, this calls for the development of new tools for comparison amongst decision version. In such a context, the present monographic text may well be regarded as one of his aspired "new tool".

It is for such a reason that researchers from the fields of statistics, economics and psychology are seen to have been active and seriously concerned to the advancement and application of multivariate normal probability distribution together with chance-constrained programming and other methods towards tackling uncertainties in the form of risk in decision-making. Such methods included studies on risk in portfolio allocation right from the middle of the last century. This has been ushered in with the publication of Markowitz's (1952) paper which bestowed him the Nobel Prize in Economics in 1990.

This was followed by Markowitz's (1959, 1970) publications, yielding measures of efficient investment diversification. The VaR is standard measure of risk which is considered as an outcome of such contribution of Markowitz,[22] subsequently with some advancements and advantages of using biquantile pairs.

No doubt the present era has been able to formulate many aspects of decision processes, out of above enumerated problems, through mathematical formulation and modelling and able to obtain computerized and even online solutions, like that of van Hentenryck and Bent (2007), which can rightly claim wide acceptability by the decision-makers, yet infinite number of problems are seen to emerge, for the decision-makers belonging to different disciplines and almost every walk of life.

The development in decision science is seen to focus on two major aspects:

(i) Generating models, which considers uncertainties like risk management
(ii) Choice for decision alternative, which optimizes the preferred criterion, like maximizing utility or production or minimizing risk, time of completion or of wastages.

Both the aspects require generation of various alternatives, for multiple choice to the decision-maker because risk management is no less a necessity than maximization of production or profit. Literatures developed on each of these aspects are vast and beyond the scope of this book. It would be wrong to think that all

---

[22]This is illustrated through examples in Sect. 7.6 in Chap. 7.

6.10 Developments in Uncertain Decision-Making

developments in this area took place only in the second half of the twentieth century. These developments are the outcome of a long evolutionary process involving experience, thought and action which was slow but without which such developments would not have taken place.[23] The mechanism of taking decision under such scenario is the art of interventional artefact, which offers immense scope for hedging against uncertainties in decision-making, which usually is taken to be *augmented risk management technique* (ARMT).

It is interesting to note that taking two decision variables, say two investment portfolios or two cropping strategies adopted in pairs, if there happens to be correlation between them, the ARMT offers additional (synergetic) advantage of several alternatives, for choice between their levels. It means that of surrogating between their levels, called *investment-mix* (INVEST-MIX) like marketing-mix or product-mix, without affecting or change in the level of probability that being risk level, already chosen by the entrepreneur (investor). Rossi et al. (2005) have developed and applied the Bayesian approach in trying to formulate and solve some marketing-mix problems.

The scope of probability modelling in uncertain decision-making and in handling of risk and hazard is very wide.[24] Now, it would be in the fitness of situation to concern to only those which make use of quantile or percentile of some probability distribution, specially of normal distribution. For this, it would be proper to refer to Winston (2004) who recommends the use of some ready-made software such as @ Risk Excel add-in, @ Risk Functions and RISK CRIB SHEET containing modules like RISK BINOMIAL, RISK CORRMAT, RISK CUMULATIVE, RISK DISCRETE, RISK UNIFORM, RISK GENERAL, RISK MEAN and RISK NORMAL, along with few others like RISKSTDDEV and RISKWEIBULL to solve wide class of problems of uncertain decision-making paradigms, but *one and all of them for univariate probability distribution* cases.

However, the joint chance-constrained programming could be popular neither with regard to its development nor in respect of its application, for the reason to be dealt with in sequel. It was this phenomenon which caught the imagination for undertaking even a small step to advance in this direction. There was enough merit in such approach to contribute in decision-making under uncertainty when decision parameters were not only stochastic but at the same time correlated. This could be possible only when biquantile pairs are made use of, so that decision-maker may be pushed ahead at least by a step when he is required to face complexity on both counts, simultaneously.

---

[23]Section 7.4 (Chap. 7), through an example, offers the scope for increasing the depth and dimension of tackling risk management and related optimization problems like that of chance-constrained one, the present state of the art, for each stochastic but independent variable, to two such variables; that also with or without any correlation between them, irrespective of such correlation being positive or negative.

[24]It is worth recalling Sects. 3.4 and 3.5 (Chap. 3), where the terms *quantile* and its application in defining *certainty equivalent* have been introduced and discussed giving adequate references.

Basu et al. (2004) found application of model in expressing future deposits and loan repayment as jointly distributed random variables replacing the capital adequacy formula by chance constraint on meeting withdrawal claims.

Sengupta (1982) has dealt decision models, as he relates them to stochastic programming. He describes stochastic programming in a very elegant manner and states that "stochastic programming deals with methods for incorporating stochastic components in a mathematical programming framework". Thereafter, Sengupta outlines three branches of stochastic programming:

(a) active and passive approach;
(b) chance-constrained programming; and
(c) stochastic programming with recourse:

and names Kolbin as one of the reviewers of developments made in this area. Keeping the optimal decision perspective under risk and uncertainty, he outlines six steps:

1. estimation of parameter and of updating them,
2. examination of information,
3. characterization of optimal solution and of multi-criterial decision,
4. method of specifying minimax and maximax solutions in stochastic programming games,
5. evaluation of linear decision rules, and
6. defining stable solution in a differential games relating to stochastic programming.

He was fully aware of his complexities on account of uncertainty and correlation, and of difficulties to be encountered, while proceeding with both simultaneously.

Further, Sengupta (1985a, b) discussed Arrow's (op. cit.) risk-averse concepts and their certainty equivalents, where utility function of expected value was preferred over expected value of utility function, for the case of parametric measures, where he stated the Arrow–Pratt measure of absolute risk aversion. The uses of quantile for transforming risky investment into its certainty equivalents are many.[25]

6.11 Online Stochastic Combinatorial Optimization

Yet another area, emerging on decision process claiming to successfully tackle optimization problems involving uncertainty, is online stochastic combinatorial optimization (OSCO) with notable contribution by van Hentenryck and his

---

[25]In Sect. 2.1 (Chap. 2) of Sengupta, while discussing constraint equation of a linear program (8.2b), it is clearly felt that "...computing marginal probability P requires the joint probability distribution of decision variables x's, which may be difficult". Biquantile pairs made available through tables of Chap. 8, cannot solve in obtaining needed joint distribution. But can solve the problem of replacing such stochastically correlated variables into their certainty equivalents for several correlated pairs of such variables.

6.11 Online Stochastic Combinatorial Optimization

associates (2007), where their recent works get represented in book format, in which they state:

1. It is only recently that researchers have begun to study how information about input distribution may improve the performance of online algorithms.
   Accepting that:
2. OSCO pushes the direction shown by Levi et al. (2005) towards approximation algorithm with worst case performance, for a stochastic inventory problem in which the demands are correlated temporarily, further.
   They claim:
3. OSCO algorithm has at their disposal a black box to sample scenarios of their future, and they exploit past and future information to take their decisions.
   and further that:
4. OSCO may exploit complex uncertainty models including correlation between random variables.

However, they have expressed that methodologies adopted by them have black boxes, but as reader one would like to say that to a greater extent, the whole text is a set of black box for lack of solved examples.

Everyone would like online (instantaneous) decision. But, its requirement is a continuous inflow of information (data). In order to be used for decision making, the information must be such on which a prediction may be made. It means that it must relate to a period of time. In other words; it must have a history. This history must relate to the concerned phenomena about its state at initial point to its state at current point of time, giving information about its direction, magnitude and speed of advancement from one state to another, for certainty model and about the probability of such changes, if the model is stochastic. Further, if the information accumulation processes are required to yield adaptive decisions, then there must be scope for revisions of such probability under the Bayesian setup. Besides, potpourris of decision alternatives must be made richer at every stage to feed information for such online system. Not the least important is the reflexes of the decision-maker to be faster enough to keep pace with such a system in picking up and implementing optimal or suboptimal decisions; otherwise, the very purpose of online decision gets defeated. All these need samples to be reasonably large and that also to have been collected at least for some successive epochs with regular updating, with which only online decision shall yield meaningful results, at least for wider class of stochastic decision scenario.

However, these are enough, for the fact that biquantile pairs shall soon find its niche in the black box of such online ventures. For the reason most of the results of such ventures depend on reasonably large samples, as well as phenomenon of dependencies, however instantaneous it may claim.

It is apt to realize that their consensus algorithm has implicit psychological content of aggregation of attitudes of subject-concerned (promising) customers or voters, while their regret algorithm has implicit content of loss function, basically the result of economic thought.

6.12 Fuzzy Decision

Through a long series of publication, Zadeh (1962, 1992) brought about a paradigm shift in the area of decision making. He made a widespread application of fuzzy algebra, and logic. He covered the entire range of fuzzy mathematics making an extensive research which had wide technological application (Klir and Yuan 1997). It is well established that fuzziness adds to uncertainty, which is of different class than randomness, Chang and Ayyaub (2001). The essential of the fact lies in realization that fuzzicists' vision of crisp is not free from random phenomenon. Cleary, therefore, phenomenon which is classed as fuzzy happens to be random plus fuzzy.

It is an undeniable fact that such techniques are increasingly used in the development of automatic medical and other instruments and find its use in developing online technologies. Such a revolution has caused increasing number of fuzzicists apart from fuzzy technologies.

While developing the theories, initially differentiation was made between crisp parameters to their fuzzy versions. Techniques of fuzzification of crispy observations and of their defuzzification had to be developed. The concept of possibility measure was required to be introduced. Gradually, there is an increase in the trend amongst fuzzicists to introduce random elements and aspects of probability measure along with the fuzziness of parameters. Some notable contributions are being observed on fuzzy regression techniques. Chang and Ayyub (op. cit.), apart from developing fuzzy regression, have compared the same with ordinary least square regression and have developed even hybrid technique by integrating them.

However, the scope of the book is basically limited to the historical development of decision processes and complexities on account of uncertainty (due to randomness), dependence and dynamism, making use of biquantile pairs, which at least for the present do not seem to fit in the schema of fuzzy systems. It is because developments in fuzzy systems and their applications do not so far seem to face the requirements of bivariate normal distribution much less its biquantile pairs. Xu and Li (2001) are quite concerned about multidimensional least square fuzzy model. Chang (2001) attempts for bivariate regression model, but variables treated are fuzzy members belonging to the possibility class. The attempt is only mimicry of bivariate least square model, but yet to have inferential base of bivariate probability distribution. Earlier efforts by Lancu (1998), Zang et al. (1998), Yosida et al. (1998) and several others, dealing with uncertainty, frequency domain and probability like limit theorems, appear to connect fuzzicists and probabilists. The day such an emerging discipline needs the application of biquantile pairs, the same would be readily available to it, through this text, its tables in Chap. 8, as to any other class of decision processes or disciplines.

References

Aadikavi, B.: Yogavaasistha Maharamayana (Hindi commentary by Pandit Thakur Prasad Dwivedi). Chaukhamba Sanskrit Pratishthan, New Delhi (2001)
Aiken, L.R.: Attitudes and Related Constructs. Wiley, New Jersey (1995)
Atreya, S.: Rig Veda, 5-82-5. Maharishi Dayanand Sarasvati Arya Pratinidhi Sabha, New Delhi (2003)
Arrow, K.J.: Alternative approaches to the theory of choice in risk-taking situation. Econometrica **21**, 503–546 (1951a)
Arrow, K.J.: Social Choice and Individual Values. Wiley, New Jersey (1951b, 63)
Arrow, K.J.: The role of securities in the optimal allocation of risk bearing. Rev. Econ. Stud. **31**, 91–96 (1964)
Arrow, K.J.: Aspects of the Theory of Risk Bearing. Academic Book Store, Helsinki (1965)
Arrow, K.J.: The Theory of Risk Aversion (Chap. 3). In: The Essays in the Theory of Risk Bearing. Elsevier, New York (1971)
Basu, S., Bhatnagar, A., Gupta, R.: Usage of mathematical models and operations research in insurance industry. J. Inst. Risk Manage. **3**, 5 (2004)
Bawa, V.S.: On stochastically un-dominated portfolio in chance-constrained programming (pre-published memo). Bell Laboratory, New Jersey (1979)
Bellman, R. Dynamic Programming. Princeton University Press, New Jersey, (1957)
Bereanu, B.: Stochastic programming. Rev. Math. Pures et Appl. **4**, (1963a)
Bereanu, B.: Stochastic transportation problem I: random cost. Com. Acad. R.P.R. **4**, 325–331 (1963b)
Bereanu, B.: On stochastic transportation problem II: random consumption. Com. Acad. R.P.R. 332–337 (1963c)
Bereanu, B.: Distribution Problems and Minimum Risk Solution in Stochastic Programming, pp. 37–42. Publishing House of the Hungarian Academy of Sciences, Budapest (1965) (Colloq. on Application of Math. Econ. Akad. Kiado)
Bereanu, B.: On stochastic linear programming distribution problem: stochastic technology matrix. Z. Wahrscheinlichkiets theorie verw. Geb **8**, 148–152 (1967)
Bereanu, B.: The Distribution Problem in Stochastic Linear Programming: The Cartesian Integration Method. Reprint 7103. Centre of Mathematical Statistics of the Academy of Society Republic of Romania, Bucharest (1971)
Bereanu, B.: Some Numerical Methods in Stochastic Linear Programming under Risk and Uncertainty. Reprinted from M.A.H. Dempster as Stochastic Programming, pp. 169–205. Academic Press, New York (1980)
Bhakti Vedant Swami Pravupada: Bhagavat Gita: As It Is. International Society for Krishna Consciousness, Los Angeles (1978)
Black, F., Scholes, M.: The pricing of options and corporate liabilities. J. Pol. Econ. **81**, 637–659 (1973)
Brahma, N.K.: Causality and Science. PHI Learning, New Delhi (1937, 2007)
Byrne, R.F., Charnes, A., Cooper, W.W., Kortanek, K.: A chance-constrained approach to capital budgeting with portfolio type payback and liquidity constraints and horizon control posture controls. In: Theil, H. (ed.) Studies in Budgeting. Studies in Mathematical and Managerial Economics, Vol. 11, pp. 71–92, North Holland (1971) (Along with authors Davis, O.A., Guilford, D.)
Chang, Y.H.O.: Hybrid fuzzy least square regression analysis and its reliability measures. Fuzzy Sets Syst. **119**, 225–246 (2001)
Chang, Y.H.O., Ayyaub, B.M.: Fuzzy regression methods: a comparative assessment. Fuzzy Sets Syst. **119**, 187–203 (2001)

Charnes, A., Cooper, W.W., Symonds, G.H.: Cost horizons and certainty equivalents: an approach to stochastic programming of heating oil. Manage. Sci. **4**, (3): 235–263 (1958)

Chhandogyopanishad.: Rishi Sanatkumar's Preachings to Narad, Chap. 7, pp. 653–745 and Chap. 8, Part 3, pp. 769–770. Geeta Press, Gorakhpur (2004)

Colman, A.M.: Oxford Dictionary of Psychology, 3rd edn. Oxford University Press, New York (2009)

Dantzig, G.B.: Linear programming under uncertainty. Manage. Sci. **1**(3), 197–206 (1955)

Dantzig, G.B.: Linear Programming and Extensions. Princeton University Press, Princeton (1963)

Das, M.R., Singhal, M.: Attitudinal Restructuring of Channel Partners in Indian Oil Corporation for Non-Fuel Marketing Initiative. Postgraduate Certificate Dissertation. Jamshedpur: Xavier's Labour Research Institute (XLRI) (2007)

Devis, M.H.A., Vinter, R.B.: Stochastic Modelling and Control. Chapman and Hall, London (1984)

Dorfman, R., Samuelson, P., Solow, R.: Linear Programming and Economic Analysis. McGraw Hill, New York (1958)

Duda, R.O., Hart, P.E., Stork, D.G.: Pattern Classification, 2nd edn. Wiley, New Delhi (2010)

Fisher, R.A.: The fiducial argument in statistical inference. Ann. Eug. **6**, 391 (1935)

Fisher, R.A.: The statistical utilization of multiple measurement. Ann. Eugen. **8**, 376–386 (1938)

Fisher, R.A.: Design of Experiments, 7th edn, p. 44. Oliver and Byod, London (1960)

Fukunaga, K.: Introduction to Statistical Pattern Recognition. Academic Press, New York (1972)

Ishermann, H.: Strukturierung von Entscheidungsprozessen beimehrfacher Zielsetzung (tr. Structuring of decision processes for multi-objective decision making. INSDOC, New Delhi) or Spektrum. **1**, 3–26 (1979)

Jagannathan, R.: Chance-constrained programming models with joint chance-constraints. Oper. Res. **22**(2), 358–372 (1974)

Kahneman, D., Tversky, A.: Prospect theory: an analysis of decision taking under risk. Econometrica **47**(2), 263–291 (1979)

Katoaka, S.: A stochastic programming model. Econometrica **31**, 181–196 (1963)

Klir, G.J., Yuan, B.: Fuzzy Sets and Fuzzy Logic: Theory and Applications. PHI Learning, New Delhi (1997)

Kolbin, V.V.: Decision Making and Programming. World Scientific, Singapore (2002)

Lancu, I.: Propagation of uncertainty and imprecision in knowledge-based systems. **94**, 29–44 (1998)

Larichev, O.I.: Science and the Art of Decision Making. M. Nauka, Moscow (1979)

Luthans, F.: Organizational Behavior. McGraw Hill, New Delhi (2005)

Markowitz, H.M.: Portfolio selection. J. Financ. **7**, 77–91 (1952)

McNiven, M.: Attitudes and behavior: what is this all about? In: King, C.W., Tigert, D.J. (eds.) Attitude Research Reaches New Heights. The Attitude Research Committee, American Marketing Association, Marketing Research Techniques Bibliography. Series Number 14 (1970–71)

Merton, R.C.: Theory of rational option pricing. Bell J. Econ. Manage. Sci. **4**, 141–183 (1973)

Millar, B.L., Wagner, H.M.: Chance-constrained programming with joint chance constraints. Oper. Res. **13**(6), 930–945 (1965)

Neyman, J., Pearson, E.S.: The testing of statistical hypotheses in relation to probability: a priori. Proc. Camb. Phil. Soc. **29** (1933)

Neyman, J., Pearson, E.S.: Contribution to the theory of testing statistical hypotheses. Stat. Res. Memo. Parts I and II (1936, 38)

Pearl, J.: Graphical models for probabilistic and causal reasoning (Chap. 70, pp. 70–18). In: Gonzalez, T., Diaz-ferrera, J., Tucker, A.N. (eds.) Computing Handbook: Computer Science and Software Engineering, 3rd edn (2004)

References

Pratt, J.: Risk aversion in small and large sample. Econometrica **32**, 122–135 (1964)

Rossi, P.E., Allenby, G.M., McCulloch, R.: Bayesian Statistics and Marketing. Wiley, New Jersey (2005)

Schalkoff, R.J.: Pattern Recognition: Statistical, Structural and Neural Approaches. Wiley, New Jersey (1992)

Sen, A.: Choice and Preference; Choice, Welfare and Measurement. Oxford University Press, New Delhi (1982–2010)

Sengupta, J.K.: Decision Models in Stochastic Programming. North Holland Publishing Company, Amsterdam (1982)

Sengupta, J.K.: Optimal Decision Under Uncertainty: Methods, Models and Management. Springer, Berlin (1985a)

Sengupta, J.K.: Information and Efficiency in Economic Decision. Nijhoff Publisher, The Netherlands (1985b)

Simon, H.A.: Dynamic programming under uncertainty with a quadratic criterion function. Econometrica **24**, 74–81 (1956)

Stancu-Minasian, I.M.: Aspura Problemai Lui Katoaka. Studii si cercetari *Matematice* 1. Tomul. **28**, 95–109 (1976)

Stancu-Minasian, I.M.: Program stocastica cu mai multe functii obiectiv. Editura Academiei Republicii Socialiste, Bucuresti (1980)

Suppes, P.: A Probability Theory of Causality. North Holland Publishing Company, Amesterdam (1970)

Theil, H.: A note on certainty equivalence in dynamic planning. Econometrica **25**, 346–349 (1957)

Tsypkin, YaZ, Nikolic, Z.J.: Adaptation and Learning in Automatic Systems. Academic Press, New York (1971)

Tversky, A., Kahneman, D.: The framing of decision and the psychology of choice: a summary of experimental demonstration of the effects of the formulation of the decision problem on shifts of preference. Science **211**, 453–458 (1981)

Vajda, S.: Stochastic programming. In: Abadie, J. (ed.) Integer and Non-linear Programming. Elsevier, New York (1970)

Vajda, S.: Probabilistic Programming. Academic Press, New York (1972)

Van Hentenryck, P., Bent, R.: Online Stochastic Combinatorial Optimization. PHI Learning, New Delhi (2007)

Vapnik, V.N.: The nature of statistical learning theory. Springer, US (1999)

von-Neumann, J., Morgenstern, O.: Theory of Games and Economic Behavior, 2nd edn. Princeton University Press, Princeton (1944, 53 and 64)

Wald, A.: Statistical Decision Function. Wiley, New Jersey (1950, 64)

Winston, W.L.: Introduction to Probability Models, Operations Research, Vol. 2, 4th edn. Thompson Learning, Singapore, (2004)

Xu, R., Li, Chulin: Multidimensional least square fitting with fuzzy model. Fuzzy Sets Syst. **119**, 215–223 (2001)

Yoshida, Y., Yasuda, M., Nakagami, J., Kurano, M.: Limit theorem in dynamic fuzzy systems with a monotone property. Fuzzy Sets Syst. **94**, 109–119 (1998)

Zadeh, L.A.: From circuit theory to system theory. I.R.E. Proc. 856–865 (1962)

Zang, Y., Wang, C.W., Liu, S.: Frequency domain methods for solutions of n-order fuzzy differential equations. Fuzzy Sets Syst. **94**, 45–60 (1998)

Zeimba, W.T.: Solving nonlinear programming problems with stochastic objective functions. J. Financ. Quant. Anal. **7**(3), 1809–1827 (1972)

Zeimba, W.T.: Choosing investment portfolios when the returns have stable distribution. In: Proceedings of Mathematical Programming in Theory and Practice. North Holland Publishing Company, North Holland (1974)

Ziemba, W.T.: Mathematical programming and portfolio selection. Instituto Nazionale Di Alta Mathematica **19**, 269–288 (1976)

Ziemba, W.T., Parken, C., Brook-Hill, R.: Calculation of investment portfolios with risk-free borrowing and lending. Manage. Sci. **21**(2), 209–222 (1974)

Chapter 7
Application Paradigms

This chapter is the crux of the book. This chapter reveals the real power of BIVNOR which yields tables of biquantile pairs. The generated by the same are used to solve problems on joint condensed confidence interval, joint chance-constrained programming and for valid cum precise results haunting decision scientists and decision-makers for about seven decades. Thus, it is possible now to realize that a significant step of advancement has been taken in the direction of dependence and dynamism. This chapter has 15 sections of which first 14 sections exhibit the application of biquantile pairs/equi-quantile values. Though the examples more often show the use of equi-quantile values, but those are biquantile pairs which are of greater importance than the former. This is because the same offers the scope for generating larger number of alternatives for decision-making at the same level of prefixed risk or confidence coefficient and for given or computed value of correlation coefficient. However, in the absence of knowledge about relative importance, or weight of the variable, one is constrained to use the equi-quantile value. Often corporate sectors, government departments, multifaceted investors, societies with multi-objectives interested in mixed ventures seek more often larger decision alternatives for the same level of prefixed risk and known or computed value of correlation coefficient, they would prefer to go for biquantile pairs as illustrated in Sect. 7.7. Some sections in this chapter have two to four examples to amply demonstrate power for multi-spectral problems. Last but not least, important is exponentially increasing prospects for the use of such techniques and tables for similarly rising promise for results on their uses than illustrated in this chapter. A topic of application paradigm becomes clear from section headings. Further attempts have been made for computing relatively dense confidence interval for Higgs boson particle, popularly known as the God Particle, if the magnitude of Bose–Einstein correlation (BEC) is known.

7.1 Introduction

Statistical techniques developed and inference drawn must find real-world application (Rao 2006); otherwise, it carries little meaning and solves no purpose. The book does not aim to generalization of any problem by adopting pedantic style or for that matter of using measure theoretic approach for aesthetic pleasure by obtaining an explicit solution. Rather, the objective is to adopt numerical solutions to some real-life problems taken from standard literature. Thus, to fulfil the said mission, illustrations from diverse fields have been presented in subsequent subsections.

7.2 Comparison Between Simultaneous (Joint) Confidence Intervals

Example 7.1 To begin with the applications, it would be interesting to take a simple example from Kempthorne (1952). In the said example, he supposes that observation y to be made up as follows:

$$y_1 = a_1 + e_1; \quad y_2 = a_1 + a_2 + e_2; \quad y_3 = a_2 + e_3 \qquad (7.1)$$

where a_1 and, a_2 are unknown parameters, and the e's being error components are normally and independently distributed around zero with variance σ^2. He poses the question: what then are the best estimates of a_1 and a_2? For the same as natural, he minimizes

$$(y_1 - a_1)^2 + (y_2 - a_1 - a_2)^2 + (y_3 - a_2)^2.$$

Differentiating with regard to a_1 and equating the result to zero, he obtains

$$y_1 + y_2 = 2\hat{a}_1 + \hat{a}_2$$

Likewise, differentiating with respect to a_2, he gets

$$y_2 + y_3 = \hat{a}_1 + 2\hat{a}_2$$

as a result. The estimates of parameters a_1 and a_2 are obtained as

$$\hat{a}_1 = \frac{1}{3}(2y_1 + y_2 - y_3), \quad \hat{a}_2 = \frac{1}{3}(-y_1 + y_2 + 2y_3)$$

and then the minimum value of the sum of squares of deviations as

7.2 Comparison Between Simultaneous (Joint) Confidence Intervals

$$(y_1 - \hat{a}_1)^2 + (y_2 - \hat{a}_1 - \hat{a}_2)^2 + (y_3 - \hat{a}_2)^2$$

This is required to be divided by the degrees of freedom $(3 - 2 = 1)$. That is, the quantity itself is an estimate of σ^2, which he denotes as s^2. Further, he obtains minimum sum of squares simply as

$$\sum y^2 - \hat{a}_1(y_1 + y_2) + \hat{a}_2(y_2 + y_3)$$

Each such estimate, $\hat{a}_1$ and $\hat{a}_2$, is a linear function of observations, and its error therefore is a linear function of variates, each normally and independently distributed around zero with a variance σ^2. The variances of these estimates as obtained are

$$V(\hat{a}_1) = \frac{1}{9}(4 + 1 + 1)\sigma^2 = \frac{2}{3}\sigma^2, \quad V(\hat{a}_2) = \frac{1}{9}(1 + 1 + 4)\sigma^2 = \frac{2}{3}\sigma^2$$

and the covariance is

$$\text{Cov}(\hat{a}_1, \hat{a}_2) = \frac{1}{9}(-2 + 1 + 1)\sigma^2 = -\frac{1}{3}\sigma^2$$

The estimated variances are obtained by putting s^2 in place of σ^2.

Thereafter, he suggests that one may obtain confidence intervals on a_1 and a_2. It is interesting to note here that even though the estimates are correlated and, therefore, naturally have had to follow a bivariate normal distribution, he aspired to obtain two separate confidence intervals for a_1 and a_2 instead of seeking or suggesting for their joint confidence interval. It was obviously because equi-quantile value or biquantile pairs were not available then. The only way to reduce or the shrink the confidence interval, therefore, was to reduce the error for the estimate, often needing larger number of observations, designing the experiment or survey and, thereby, escalating the cost. Further, for seeking separate confidence intervals, the independence of two estimates had to be impliedly assumed even though covariance, and hence the correlation between them was amply demonstrated or obtained as nonzero. It becomes mandatory, therefore, to obtain the value of correlation between $\hat{a}_1$ and $\hat{a}_2$ and then to attempt for their joint confidence interval, as well as to examine the consequence in respect of shrinkage brought about in the width of joint confidence interval for equivalent confidence coefficient.

The correlation between the two estimates $\hat{a}_1$ and $\hat{a}_2$ for the Kempthrone's model works out to

$$\rho = -\frac{1}{3}\frac{\sigma^2}{[(2/3)\sigma^2(2/3)\sigma^2]^{1/2}} = -\frac{1}{2} = -0.5$$

Before proceeding further, it is desired to make a change in the Kempthrone's example to yield a positive correlation between the estimates a_1 and a_2 and then again to see the impact of change on the shrinkages of two joint confidence intervals for both the positive and negative correlations between the estimates a_1 and a_2 for both the scenarios, one after the other. For this, now let

$$y_1 = a_1 + e_1; \quad y_2 = a_2 + e_2; \quad y_3 = a_1 + a_2 + e_3.$$

Proceeding similarly, estimates of a_1 and a_2 are obtained as

$$\hat{a}_1 = \frac{1}{3}(-2y_1 + y_2 - y_3) \quad \text{and} \quad \hat{a}_2 = \frac{1}{3}(-y_1 + 2y_2 + y_3)$$

with their respective variances and covariance as

$$V(\hat{a}_1) = \frac{1}{9}(4+1+1)\sigma^2 = \frac{2}{3}\sigma^2, \quad V(\hat{a}_2) = \frac{1}{9}(1+4+1)\sigma^2 = \frac{2}{3}\sigma^2$$

whereas

$$\text{Cov}(\hat{a}_1, \hat{a}_2) = \frac{1}{9}(2+2-1)\sigma^2 = \frac{1}{3}\sigma^2.$$

The correlation coefficient between them is therefore

$$\rho = +\frac{1}{2} = +0.5$$

Let the confidence coefficient for the joint confidence interval be 95 %. The corresponding equi-quantile, i.e. BIGH values for the above correlation values, is needed.[1] Thus, the joint confidence interval for the estimates with the correlation value +0.5 is, therefore,

Prob $[(\hat{a}_1 - 1.9159s \le a_1 \le \hat{a}_1 + 1.9159s) \cap (\hat{a}_2 - 1.9159s \le a_2 \le \hat{a}_2 + 1.9159s)]$
$= 0.95$

and the joint confidence interval for the estimates with the correlation value -0.5 is

Prob $[(\hat{a}_1 - 1.9599s \le a_1 \le \hat{a}_1 + 1.9599s) \cap (\hat{a}_2 - 1.9599s \le a_2 \le \hat{a}_2 + 1.9599s)]$
$= 0.95$

The two independent confidence intervals for estimates a_1 and a_2 to attain the same level of confidence coefficient as that of joint one will need quantile values to be

---

[1]These are available from Table 8.1 Sheet 1, for positive correlation and Table 8.1 Sheet 3 for the negative correlation of Chap. 8. These are 1.9159 and 1.9599, respectively.

selected at $(1 - \frac{\alpha}{2 \times 2}) = 0.9875$, which is 2.24, Johnson and Wichern (2003).[2] Similar two confidence intervals with quantile value 2.24 needed to be worked out for the estimates of a_1 and a_2 each with confidence coefficient 0.95. The same confidence intervals will be taken as valid with both positive and negative correlations as the correlation being not considered then. Further, the shrinkage of confidence interval is independent of the value of estimate of parameter, and its standard error as that solely depends on the ratio of quantile values obtained for confidence coefficients, Johnson and Wichern (op. cit.).

The shrinkage brought about in confidence interval by the use of equi-quantile value in relation to that under independent assumption is

$$1.9159/2.24 = 0.8533, \quad \text{i.e.} \ (1 - 0.8599)\alpha \times 100 = 14.5\ \%,$$

when estimates are positively correlated. However, similar shrinkage when estimates are negatively correlated is

$$1.9599/2.24 = 0.8750, \quad \text{i.e.} \ (1 - 0.8750)\alpha \times 100 = 12.5\ \%.$$

On the other hand, separate confidence intervals for the estimates on assumption of independence for the confidence coefficients 95 % for each estimate shall dilute joint confidence coefficient to $0.95 \times 0.95 = 0.9025$, i.e. 90.25 %. Thus, the example clearly demonstrates relative advantage of using information about the correlation and corresponding equi-quantile value for making confidence interval.

Example 7.2

(a) For the purpose of comparison, as well as to see the relative advantages of different joint confidence intervals, an example has been taken from Johnson and Wichern (op. cit.). The said example, which considers the scores obtained by $n = 87$ college students in college-level examination program, taking only two variables, the subtest score, x_1, and college qualifying test (CQT) subtest score x_2, given below are their mean vector and covariance's in matrix form:

$$\bar{\mathbf{x}} = \begin{pmatrix} 527.74 \\ 54.69 \end{pmatrix} \quad \text{and} \quad \mathbf{S} = \begin{pmatrix} 5691.34 & 600.51 \\ 600.51 & 126.05 \end{pmatrix}$$

The computation of 95 % simultaneous confidence intervals for μ_1 and μ_2,[3] taking the table value of $F_{2,85}(0.05)$ by interpolation to be = 3.13.

$$\left[\frac{2(n-1)}{n-2} F_{2,n-2}(\alpha)\right]^{1/2} = \left[\frac{2(87-1)}{87-2} F_{2,85}(0.05)\right]^{1/2} = 2.517$$

---

[2]This is dealt in Sect. 4.11 (Chap. 4).
[3]By using formula (4.42).

and

$$\left(\frac{s_{11}}{n}\right)^{1/2} = \left(\frac{5691.34}{87}\right)^{1/2} = 8.088$$

Thus, their product to be

$$2.517 \times 8.088 = 20.358$$

and the confidence interval for μ_1, with 95 % confidence coefficient being

$$527.74 - 20.358 \leq \mu_1 \leq 527.74 + 20.358,$$

results to

$$507.382 \leq \mu_1 \leq 548.098.$$

Similarly for μ_2, it is

$$51.66 \leq \mu_2 \leq 57.72.$$

(b) For the same data set, Bonferroni's intervals are computed as below[4]:

$$\bar{x}_1 - z\frac{\alpha}{2 \times 2}\left(\frac{s_{11}}{2 \times 2}\right)^{1/2} \leq \mu_1 \leq \bar{x} + z\frac{\alpha}{2 \times 2}\left(\frac{s_{11}}{2 \times 2}\right)^{1/2}$$

and

$$\bar{x}_2 - z\frac{\alpha}{2 \times 2}\left(\frac{s_{22}}{2 \times 2}\right)^{1/2} \leq \mu_2 \leq \bar{x}_2 + z\frac{\alpha}{2 \times 2}\left(\frac{s_{22}}{2 \times 2}\right)^{1/2}$$

For $\alpha = 0.05$, $(\alpha/2 \times 2) = 0.0125$ and $(1 - \alpha/2 \times 2) = 0.9875$ at which the univariate normal quantile value is

$$z(\alpha/2 \times 2) = 2.24, \quad (s_{11}/n)^{1/2} = 8.088, \quad (s_{22}/n)^{1/2} = 1.2037.$$

The said Bonferroni's (1892–1960) intervals are

$$509.623 \leq \mu_1 \leq 545.857 \quad \text{and} \quad 51.994 \leq \mu_2 \leq 57.386,$$

respectively.

It is to be noted that even though the covariance value, implying correlation, is available, but has not been made use of in building such confidence intervals.

(c) Simultaneous (joint) confidence intervals by using biquantile equi-quantile are being considered as the best so far. The estimate of parameter being given as

---

[4]By using formula (4.45).

7.2 Comparison Between Simultaneous (Joint) Confidence Intervals

Table 7.1 Comparative interval length of joint confidence intervals

Method	l_1	l_2	Prob.	Relative % μ_1	Relative % μ_2
(a)	40.714	6.06	0.95	(d)/(a) = 74.63 %	(d)/(a) = 74.62 %
(b)	36.244	5.392	0.95	(d)/(b) = 83.83 %	(d)/(b) = 83.86 %
(c)	35.228	5.242	0.975	(d)/(c) = 86.25 %	(d)/(c) = 86.26 %
(d)	30.386	4.522	0.95		

$$\bar{x}_1 = 527.74, \quad \bar{x}_2 = 54.69$$
$$S_1 = 8.088 \quad S_2 = 1.2037$$
$$\rho = 600.51/(5691.34 \times 126.05)^{1/2} = +0.709$$

and the nearest correlation value being approximately = +0.7.
Thus, taking PROB = 0.975, RHO = +0.709, BIGH = 2.1778 and SMHL = 1.96.[5]
The difference between them being less than 0.01; one can more often depend on the use of interpolated value of the equi-quantile. The joint confidence interval for μ_1 and μ_2 being

$$P[(510.126 \leq \mu_1 \leq 545.354) \cap (52.069 \leq \mu_2 \leq 57.311)] = 0.975.$$

(d) The confidence interval obtained by equi-quantile approach for the parameter values is PROB = 0.95, RHO = +0.709, BIGH = 1.88 and SMHL = 1.65 (see Table 7.1). The equi-quantile value is $h = k = 1.8785$[6] for the nearest correlation value +0.7.
Therefore, the joint confidence interval for μ_1 and μ_2 being

$$P[(512.547 \leq \mu_1 \leq 542.933) \cap (52.429 \leq \mu_2 \leq 56.951)] = 0.95$$

Thus, the gain on account of utilizing information available for correlation between two variables[7] in terms of shrinkage in confidence interval is clearly seen. In fact, the width or shrinkage of such confidence intervals are wholly accountable to respective quantile values used for different methods as discussed below:

---

[5]The equi-quantile value $h = k = 2.187$ is obtained by interpolation between two nearest values from the said Table 8.1 of Chap. 8 (taking the mean of BIGH values for PROB levels 0.98 and 0.97 and RHO = +0.7).
[6]By Table 8.1 of Chap. 8.
[7]And thereby using equi-quantile value from Table 8.1 of Chap. 8.

The quantile values for methods (a)–(d) are 2.517, 2.24, 2.1778 and 1.8785, respectively. The percentages of (d) to (a), (d) to (b) and (d) to (c) are as expected 74.63, 83.86 and 86.25, respectively, and the corresponding shrinkage being 25.37 % over conventional confidence intervals 16.14 over the Bonferroni's confidence interval.

1. Apart from getting such shortest joint interval, the decision-maker has the scope of forming such joint intervals by choosing various alternatives of h and k from amongst the table generated for the quantile pairs of h from its value 1.88 to 1.65, and corresponding values of k or vice versa, for the same confidence coefficient, provided such intervals carry valid meaning and be of use, as it can be seen from one such example.
2. Further that, both the joint confidence intervals, (c) and (d), have been formed by the use of equi-quantile values but for different confidence coefficients, implying different risk levels, thus having different width of joint confidence interval. The interval (d) based on 95 % confidence coefficient is shorter in width to interval (c) based on 97.5 % confidence coefficient by nearly 14 % which clearly means that width of confidence interval can be bartered with confidence coefficient or the risk level.

Higher the confidence coefficient or lower the risk level, wider would be the interval, and vice versa. A wider interval would necessitate larger quantile pair value, implying greater level of investment, that being quite consistent with real-life scenario. So the long Bonferroni's interval (b), being considered as the shortest, has been in use which at best corresponds to the interval (c). But the latter enjoys superiority over it, because of reduction in risk level from 5 % to only 2.5 %, i.e. just the half.

Example 7.3 On the long-term data on wheat area and production of India,[8] it is required to find Bonferroni's joint confidence interval for 95 % confidence coefficient and also the joint confidence interval on the basis of equi-quantile.[9] Thereafter, compare them in respect of getting shrinkage percentages. In order to get the desired solution, parameter-estimates, as in Table 7.2, have been obtained.[10]

---

[8]As given in *Agricultural Research Data Book* (2007).

[9]As available from the Table 8.1 of Chap. 8.

[10]Consequently equi-quantile value from the Table 8.1 of Chap. 8 corresponding to the estimate of correlation has been obtained by interpolation: for PROB = 0.95 and CORR = +0.974; BIGH = 1.7291 which is the required equi-quantile value, and SMHL = 1.65, by iteration.

For Bonferroni's joint confidence interval, Eq. (4.46) (in Chap. 4) has been applied by using univariate normal quantile value (=2.24) to get the same for the mean value of the data in Table 7.2.

7.2 Comparison Between Simultaneous (Joint) Confidence Intervals

Table 7.2 Wheat area (in mH) and production (in mT)*

Sr. Nr.	Year	Area	Production
1	1950–51	9.75	6.46
2	1955–56	12.37	8.76
3	1960–61	12.93	11.00
4	1965–66	12.57	10.40
5	1967–68	14.99	16.54
6	1970–71	18.24	23.83
7	1975–76	20.45	28.84
8	1980–81	22.28	36.31
9	1985–86	23.00	47.05
10	1990–91	24.17	55.14
11	1995–96	25.01	62.10
12	1996–97	25.89	69.35
13	1997–98	26.70	66.35
14	1998–99	27.52	71.29
15	1999–00	27.49	76.37
16	2000–01	25.73	69.68
17	2001–02	26.34	72.77
18	2002–03	25.20	65.76
19	2003–04	26.60	72.16
20	2004–05	26.38	68.64
21	2005–06	26.48	69.35
22	2006–07	28.17	73.70

Source Agricultural Research Data Book (2007)

N	$\bar{x}$	$\bar{y}$	σx^2	σy^2	σxy	ρxy
22	22.19	49.17	34.90	667.189	148.63	+0.974

$$\text{Prob}[19.37 \leq \bar{x} \leq 25.01] = 0.975 \quad \text{and} \quad \text{Prob}[36.83 \leq \bar{y} \leq 61.65] = 0.975$$

whereas the joint confidence interval by using equi-quantile value obtained above (which is BIGH = 1.7291) is

$$\text{Prob}[(20.01 \leq \bar{x} \leq 24.37) \cap (39.65 \leq \bar{y} \leq 58.69)] = 0.95.$$

The required shrinkage percentage is

$$\left[\frac{1.729}{2.24} = 0.77; 1 - 0.77\right] = 0.23 \text{ or } 23\,\%$$

which is exclusively due the use of the equi-quantile value of BIGH obtainable on account of correlation of area under wheat crop with its production.

Table 7.3 Table of guaranteed values (in $100)

Sr. Nr.	End of policy	Cash	Paid up
1	3	4	24
2	4	14	79
3	5	24	129
4	6	35	179
5	7	45	220
6	8	56	262
7	9	68	304
8	10	80	343
9	11	93	381
10	12	110	432
11	13	129	485
12	14	148	533
13	15	168	577
14	16	187	619
15	17	207	658
16	18	227	693
17	19	248	728
18	20	269	759

It is noteworthy to mention the observation made by Johnson and Wichern (op. cit.) "that relative length (shrinkage) of such intervals does not depend on the random quantities, those being estimates of means and of standard errors. Also, that Bonferroni's intervals are always shorter than other intervals".

It is for such a reason that, in all such comparisons, what is required is to compare intervals obtained by the use of equi-quantile/biquantile pairs with Bonferroni's interval only. However, such a statement by Johnson and Wichern is not operative on the joint confidence interval based on the value of correlation coefficient as the same in randoms.

Example 7.4 Table 7.3 shows the table for guaranteed values by ABC Life Insurance Company.[11] It is required to find Bonferroni's joint confidence interval at 98 % confidence coefficient and, at the same time, also to find joint confidence interval by using equi-quantile value for the same level of confidence coefficient, and then to obtain shrinkage percentage brought by latter technique over the former.

n	$\bar{x}$	$\bar{y}$	σx^2	σy^2	σxy	ρxy
18	117.28	411.39	7118.80	54,278.84	19,385.36	+0.986

Using specified probability, i.e. risk value, and the value of the estimated correlation coefficient, the value of BIGH has been obtained[12] by interpolation. Here,

---

[11] As quoted by Rejda (2006) on p. 713.
[12] From Table 8.1 of Chap. 8.

PROB = 0.98, CORR = +0.986, BIGH = 2.1161 and SMHL = 2.06. Bonferroni's joint confidence intervals obtained by using formula (4.46) are

$$\text{Prob}\,[61.46 \leq \bar{x} \leq 173.10] = 0.995 \quad \text{and} \quad \text{Prob}\,[257 \leq \bar{y} \leq 565.53] = 0.995$$

which are based on corresponding univariate quantile value 2.808. The joint chance confidence interval by using BIGH value = 2.1161 as obtained is

$$\text{Prob}\,[(75.20 \leq \bar{x} \leq 159.36) \cap (295.19 \leq \bar{y} \leq 527.59)] = 0.98.$$

The required shrinkage, due to latter, over the former is

$$\left[\frac{2.1161}{2.808} = 0.7538\right]; \quad 1 - 0.7538 = 0.2462 \text{ or } 24.62\,\%$$

The said insurance company may be interested in knowing, overall average of cash value and of paid-up insurance claim with their most precise confidence interval, obtained on using the value of correlation coefficient, between the two.

7.3 Algorithmic Heuristics for Chance-Constrained Version of Discrimination/Classification

It was Polya (1945) who defined heuristic as the study of methods and rules of discovery and invention, Pearl (1994). The term *heuristic algorithm* has been used by Luger (2001), Russel and Norvig (2002) and Padhy (2005). Colman (2009) gives longer interpretation of the word heuristic, who writes it to mean a rough and ready procedure or rule of thumb for making decision.

In fact, the term *heuristic* has been used in the sense with which the development of this work is started, specially the development of BIVNOR, only with incomplete knowledge.

Following the above approach, the undernoted computational steps are set out through an example:

Example 7.5 Gose et al. (2003), in their example, discuss the bivariate normal density and illustrate computation of one- and two-dimensional decision boundaries for classifying two groups. But, we shall highlight the use of biquantile pair, taking correlated case only. Thus, this example, which follows here under algorithmic heuristics, has been adopted.

In the said example, they determine decision boundary for undernoted problem. They assume two bivariate clusters G and G' with parameters as below:

$$\text{Cluster } G: \bar{x}_1 = 26, \quad sx_1 = 2, \quad \bar{y}_1 = 85, \quad sy_1 = 5, \quad rx_1y_1 = +0.6$$
$$\text{Cluster } G': \bar{x}_2 = 22, \quad sx_2 = 3, \quad \bar{y}_2 = 70, \quad sy_2 = 8, \quad rx_2y_2 = +0.5 \quad (7.2)$$

and obtain decision boundary in the form of undernoted quadratic equation, which is a nonlinear separable function (Haykin 2001):

$$[-5.819x^2 + 3.167xy - y^2 + 41.89x + 97.33y - 5000.07] = 0. \qquad (7.3)$$

Equation (7.3) was converted by them to the discriminant function as

$$D = [\text{L.H.S. expression of the above equation}],$$

with decision criteria that if $D \geq 0$, the sample be classified to class G, and if $D < 0$, to class G'. They could not consider confidence bound for the discriminant function (7.3). This was so, as they had no recourse then for the joint confidence bound for the said function.

Now on the other hand, it is possible to make use of equi-quantile value assuming 0.05 or 5 % probability of misclassification leaving sample estimates of parameters unchanged.

Heuristic for chance-constrained version of discrimination

Step 1. Using whitening transformation, Gose et al. (op. cit.) obtain the following:

$$\begin{aligned} G : x_1 &= \frac{x_1 - 26}{2}, & y_1 &= \frac{y_1 - 85}{5}, & \rho_1 &= 0.6 \\ G' : x_2 &= \frac{x_2 - 22}{3}, & y_2 &= \frac{y_2 - 70}{8}, & \rho_2 &= 0.5 \end{aligned} \qquad (7.4)$$

Step 2. Consulting Tables 8.2-81 and 8.2-101, respectively, in Table 8.2 Chap. 8 for the probability (PROB) value 0.95, for both the classes but for their differing correlation values, one gets 26 quantile pairs for the class G and 28 such pairs for G'. Amongst such pairs, one was free to choose any (having larger scope for decision). However, let one prefer to choose equi-quantile value for both the classes in the absence of more information about relative preference or weight or utility for choice of the level of one variable, over that of the other. Such an equi-quantile value being ($h_1 = 1.90$, $k_1 = 1.90$) for the class G and ($h_2 = 1.92$, $k_2 = 1.92$) for the class G' get selected for further computations.

Step 3. Equi-quantile values, being 1.9 for both h and k, the joint confidence interval (JCI) for the class G is:

$$\Pr\{(x_1 \leq 26 \pm 1.9 \times 2) \cap (y_1 \leq 85 \pm 1.9 \times 5)\} = 0.95:$$

which simplifies to

$$\Pr\{(22.2 \leq x_1 \leq 29.8) \cap (75.5 \leq y_1 \leq 94.5)\} = 0.95. \qquad (7.5)$$

Similarly, taking equi-quantile value for both classes G and G' that being 1.92, the JCI simplifies to

7.3 Algorithmic Heuristics for Chance-Constrained Version ...

$$\Pr\{(16.24 \leq x_2 \leq 27.76) \cap (60.425 \leq y_2 \leq 79.57)\} = 0.95 \quad (7.6)$$

Thus, the range of overlap between two classes for the values of x lie between 22.2 and 27.76, and similar range of overlaps between two groups for the value of y lie between 75.5 and 79.57.

In case of no overlap between two classes, in respect of values of either of the two variables, the classification would be considered as perfect like linearly separable Rosenblatt's (1958) perceptron. However, the cases of such an overlap are likely to be very frequent. Therefore, the simplest method to find the solution under such a situation is to take the mean of overlapping intervals, which yields the value of $x = 24.9$ and that of $y = 77.532$ and proceed as below:

Thus if the sample mean of the variable x is higher than 24.9, the mean of x obtained above and that of y higher than 77.532, meaning those are as follows:

1. mean of $x > 24.9$ and that of $y > 77.532$
   the sample be classified to the class G, and if the sample mean of x lower than or equal to 24.9 and that of y lower than or equal to 77.532, meaning those are
2. the mean of $x < 24.9$ and that of $y < 77.532$
   the sample mean be classified to the class G'. Let these mean values of x and y be designated as x_0 and y_0, the coordinates of a point on the x-y plane.

Step 4. The sample pairs of two classes for which the mean values:

3. the mean of $x \leq 24.9$ but of $y \geq 77.532$;
4. the mean of $x \geq 24.9$ but of $y \leq 77.532$; $\quad (7.7)$

remain so far unclassified.

Step 5.

(a) Find the mean of the lower limit of x_2 in JCI in Eq. (7.6), and of x in Eq. (7.7). These are 16.24 and 24.9, respectively. The mean being 20.57 denotes the same as x_3.
(b) Find the mean of the higher limit of y_1 in JCI in Eq. (7.5), and of y in Eq. (7.7). These are 94.5 and 77.532, respectively. The mean being 86.016 denotes the same as y_3.
(c) Find the mean of higher limit of x_1 in JCI in Eq. (7.5), and of x in Eq. (7.7). These are 29.8 and 24.9, respectively. The mean being 28.35 denotes the same as x_4.
(d) Find the mean of lower limit of y_2 in JCI in Eq. (7.6), and of y in Eq. (7.7). These are 60.425 and 77.532, respectively. The mean being 68.978 denotes the same as y_4.

Step 6. Thus, one gets three points, whose coordinate pairs, respectively, are

$$\{(x_3, y_3); (x_0, y_0); (x_4, y_4)\}.$$

Joining them by fitting a quadratic least square model, one gets parabolic or the second-degree polynomial concave or convex or even a straight line, if the points are collinear. One can obtain an expression of the second-degree polynomial by fitting the same to these three points.

Step 7. A perfect-fit polynomial equation can also be obtained, which will be the line of discrimination for sample mean pairs of these two classes or even for sample mean pairs for both the classes G and G'. Because the point (x_0, y_0) falls on the line of the curve irrespective of its shape or degree of the fitted polynomial. The fitted least square quadratic polynomial is on the right-hand side of the expression D which is

$$D = y - 180.8985 + 6.1489x - 0.0785x^2$$

which, if exceeds 0, is the sample be classified into the class G, and if $D \leq 0$ into G'.

Step 8. Thus, the line of discrimination like Gose et al.'s (op. cit.) is quadratic again, but unlike them, it is explicit in x and y, quadratic in x but linear in y.

The basic difference between the two approaches is that while Gose et al. obtain a deterministic solution to a stochastic problem, the solution arrived here is a chance-constrained one, of the said problem. Hence, it deserves to be preferred. There may exist other such solutions, where the weight could be given to each of the three means, or even variances, also in place of mere sample means for classes (3) and (4) at the stage 4. However, this also is nonlinearly separable function, Haykin (2001).

Also, as biquantile pairs are infinite in numbers, one can get infinite qui-probable solutions within tangible or permissible limits or utility values, which are not obtainable otherwise even with equi-quantiles.

Remark It is notable to state the observation by Haykin (op. cit.) that for Roseblatt's perceptron to be efficient classifier, the classes G and G' must be "linearly separable". Both, those of Gose et al. and the method presented here get nonlinearly separable function, hence are more general than that of Rosenblatt's, because these are less restrictive, but these two must not claim a greater discriminating power than that of Rosenblatt's.

7.4 Biquantile in Optimization: Bivariate Joint Chance-Constrained Linear Programming Problem

Be that celebrated theorem or algorithmic computation, their ultimate goal meets in solving problems. The first stage of solving any problem lies in its formulation. A simple linear programming problem has been taken, where in resource vector **b** (with only two elements b_1 and b_2) being considered is random as well as correlated. Though it could be possible to assume coefficients of objective function, as well as coefficients of technology matrix, to be random as well as pairwise correlated with knowledge of their probability distributions. Such problems could be solvable but only with greater number of iterations.

Let the problem be

$$\text{Minimize the objective function}$$
$$z = 8x_1 + 5x_2$$
$$\text{Subject to} \tag{7.8}$$
$$7x_1 + 2x_2 \geq 16.144 \ (b_1)$$
$$x_1 + 6x_2 \geq 11.0015 \ (b_2); \quad x_1, x_2 \geq 0$$

wherein b_1 and b_2 are considered random variable and pairwise correlated, following bivariate normal distribution, with means as given in the problem which are $b_1 = 16.144$ and $b_2 = 11.0015$ and their standard deviations $sd_1 = 1.8936$ and $sd_2 = 1.7353$, having correlation coefficient between them, say RHO = +0.7. The joint confidence level is agreed to be 0.9025, incidentally taking it to be the product of two probability levels for b_1 and b_2 (0.95 and 0.95), respectively, obtainable on assumption of independence without any loss of limited generality of the problem in hand. Such values of correlation coefficient, and of the joint probability level, could have been assigned to any value within their ranges,[13] but say only those which form entries of the said tables.

The joint probability level being 0.9025 is to be interpreted that the decision-maker is in a position to take the risk to the extent of

$$1 - 0.9025 = 0.0975 \quad \text{or} \quad 9.75\,\%.$$

However, no information is supposedly available, for the choice of h (SMH) or for k (SMK). That is, about any member of the quantile pair for the given probability level 0.9025 and known correlation value +0.7, the value is initially taken as h (SMH) = 1.31, nearly lowest, to attain the given probability target with its couplet k (SMK) = 2.314 from long list of such pairs.[14] However, choosing the smallest or, for that matter, the largest value of h is not relevant with the present paradigm shift,

---

[13] As stated in Table 8.1 of Chap. 8.
[14] As shown in Table 8.2-67 of Chap. 8.

one could select any pair by cluster-randomized technique, Hayes and Moulton (2009), where the complete list of entries of the above-referred table as a list could be considered as the cluster, and any random number between such limits of h or k to be the member of the said cluster could be selected.

As the initial value of h is taken to be near the smallest, the second choice is taken as closer to the largest, from the said list. Such values of h could have been automatically generated by the software module BIVNOR to get the optimum pair of h and k. And then any compatible LP software could be embedded to get the optimum value of the objective function. Say, for example, simplex or revised simplex iterations for every pair of h and k could be adopted for the said purpose.

Thus, there could be double optimization. First on account of the choice of quantile (h and k) pairs, second on account of simplex iterations for any or every specific pairs of selected iso-probable quantile pairs. Usually the discipline, from which the problem and estimates of required parameters emerges, is required to choose individual such quantile pair or its set or range of admissible pairs, which would acquire valid and realistic physical interpretations and significance.

In fact, one LP module could be embedded in the software BIVNOR to get the solution.[15] Sometimes, when a reasonable criterion for choosing such pairs of quantile be not decidable, one could choose equi-quantile values for transforming the problem into its joint chance-constraint version and proceed with usual simplex algorithm.

The certainty equivalents for b_1 and b_2 the correlated resource constraints taking initial choice of h and k are

$$\hat{b}_1 = b_1 + hsd_1 = 16.144 + 1.31 \times 1.89 = 18.62$$
$$\hat{b}_2 = b_2 + ksd_2 = 11.0014 + 2.31 \times 1.7353 = 15.0049$$

where $h = 1.31$ and $k = 2.31$. With these changed values of b_1 and b_2, the problem entered into an LP module to get the optimum solution in four iterations. Thus, such a solution is 27.14859, as the initial minimum value of the objective function. Table 7.4 presents the detail of results obtained for advancing iterative values of h and k.

It is seen from Table 7.4 that the value of the objective function is fluctuating between initial three values of h and k pairs. Thereafter the objective value starts decreasing (as the objective function is to be minimized), for the choice of h and k pair and proceeds forward, i.e. towards iteration no. 11 (eleven), where the objective function value takes the value 26.89284*, the minimum. Further, as the iteration starts, the objective function value again starts fluctuating but with damped oscillation in search of confirmation of the globally minimal (optimal) solution. Thus, taking its value 26.89285* at iteration no. 14 (shown by "*") resulted in automatic termination of the program module on account of difference between the minimum to less than or equal to preassigned error value.

---

[15]And, therefore, the repeated consultation of Table 8.2 of Chap. 8 was not required.

7.4 Biquantile in Optimization: Bivariate Joint Chance-Constrained ...

Table 7.4 Results obtained for advancing iterative values of h and k

Iteration Nr.	Pair of values		Changed values of		Optimal value/ objective function	Nr. of simplex iteration
	SMH	SMK	$\hat{b}_1$	$\hat{b}_2$		
1	1.31	2.31	18.62	15.0049	27.14859	4
2	3.09	1.2954	22.18	13.2454	30.13596	4
3	2.20	1.318	20.40	13.2829	28.23935	4
4	1.9775	1.3461	19.955	13.5515	27.78407	4
5	1.8106	1.3864	19.6213	13.4013	27.45488	4
6	1.6355	1.4387	19.3709	13.4919	27.23241	4
7	1.5916	1.4496	19.1832	13.5973	27.08064	4
8	1.5212	1.5653	19.0424	13.712	26.98378	4
9	1.4684	1.6349	18.9368	13.8317	26.92710	4
10	1.4288	1.7049	18.8576	13.953	26.89958	4
11*	1.3991	1.7744	18.7982	14.0732	26.89284*	4
12	1.3768	1.8413	18.7537	14.19	26.90043	4
13	1.3935	1.7898	18.7871	14.0999	26.89354	4
14*	1.3977	1.778	18.7982	14.0795	26.89285*	4

Here, the minimization of the objective function is twofold: the first on account of choice of pairs of quantile values, and the second on account of the simplex iteration of LP module for each such quantile pairs of h and k, and then the optimal solution by simplex iterations of such reformulated programming problem, after reparametrization with such optimal pair of biquantiles. Again, it is to be noted that for every such reparametrization of the programming problem, with gradually optimally changing quantile pairs of h and k, four simplex iterations are required to reach the local optimum. Such numbers of simplex iterations may obviously change for every reformulation of the problem.

It is interesting to realize the function of such iterative scheme, which is its oscillatory features at the beginning with larger amplitudes, but with much smaller amplitudes at the end. It happens at the point of convergence towards the global optimum, giving a feeling of damped or dying wavelets.

It is to be recollected that the first fold operations from out of infinite such pairs, apart from giving the optimal choice of quantile pairs (h^*, k^*), offer plethora of choice for surrogating (swapping) between h and k for economic reason as well as for many other reasons. This could be administrative psychological or social. All these are attainable for prefixed level of risk and computed value of correlation, a measure of mutual dependence. Those, in the present example, correspondingly vary being 9.75 % and +0.7. At the same time, it would not be necessary for the decision-maker to stop at such an optimal level. Because the decision-maker still has the choice of stopping at any suboptimal level of quantile pair, say ($h^\wedge$, $k^\wedge$), if the same satisfies individual preference or prefixed social choice on the ground of exceeding the budget, breaking social taboos, environmental requirements or individual taste, temperament, and societal tension without a change in the risk level, until the decision-maker floats for his choice, sticking to the range of table of such biquantile pairs.

The result so obtained now needs to be perused in the light of observations made by Kall (1974) and Armstrong and Balinfy (1975), one by one to show the advantage of the method adopted here over others mentioned.

Related remarks

(I) Kall (op. cit.) remarks the following:

1. Everybody will agree that it is not true for most practical problem of LP that its coefficients shall remain fixed for the entire planning period.
2. Either the data pertaining to these coefficients are stochastic variables with known (joint) probability distribution or they are simply variables. In all such cases, the LP model does not make sense.
3. Further that stochastic linear programming (SLP) is concerned with problems arising when some or all coefficients of an LP are stochastic variables with joint probability distribution.

As Kall proceeds with the problem of (joint) probability distribution of random variables, he begins to realize the magnitude of the problem of evaluating the multi-variable probability integral of the assumed probability distribution. To simplify his problem, he looks towards assuming stochastic independence of such variables. Soon on retrospection, he expresses "such an assumption is not trivial from the practical point of view" and retorts "it seems very unlikely that these decisions do not influence each other". On the other hand, Kall finds that there are certainly many cases where assumption of stochastic independence is quite unrealistic. In his entire text, he floats into several types of difficulties even in problem formulation.

Coming to the chance-constraint version of the problem, Kall lands where convexity of formulated problem could not be guaranteed. Further, he observes "although every distribution function of one-dimensional random variable is quasi-concave, this is not true for multi-variate distribution function". In ultimate analysis, he could not be successful in formulating even two-variable joint chance-constrained programming problem, much less to find its solution.

(II) Armstrong and Balintfy (1975) expressed their view as below:

"If there is a correlation between rows (meaning rows of coefficients of LP problem), as is assumed here, the exact computation of the joint probability given by

$$\text{Min } C^T x : \; P\left\{\bigcap_{i=1}^{m}(a_i x \geq b_i)\right\} \geq \alpha \quad \text{and} \quad x \geq 0$$

will, in most cases, be impossible.

In particular, insurmountable computational difficulties arise from process of evaluating the function of multi-variate normal integral".

Equi-quantile values can be obtained by using the software **qmvnorm** for a reasonably large dimension of multi-variate normal distribution. But what needs to be realized is that the solution of only equi-quantiles cannot make their application acceptable unless it is made clear as to how and when to make use of them.[16] Besides, it cannot compensate the additional advantages of multiple options of decision-making yielded by biquantile pairs without changing the risk level. Over and above, such software does not declare about its reliability. On the other hand, the scope, for application of biquantile pairs in formulating the problem and of getting the solution, could be seen in preceding paragraphs.

It is to be believed, therefore, that all these have taken mankind a step quantitatively nearer to the realization of economists, statisticians, operation researchers and management specialists, whose one of the main concerns for last half a century or even more have been risk aversion or its minimization while playing game with uncertainties in decision-making. Such an achievement with only a reasonably cognizable assumption is that observational data must follow bivariate normal probability distribution, which in all possibility is attainable on account of bivariate central limit theorem if the sample size happens to be reasonably large. These have been traditionally taken as greater than 30, on proofs advanced by Cramer (1951). This being the centralized function of chi-square to follow a normal distribution, as its degrees of freedom exceeds 30. Thus, it is to be recollected that losing the advantage of having a unique quantile value for a probability value (level of risk) and vice versa, in case of univariate normal distribution, to that in bivariate case, was like Milton's *Paradise Lost,* that is, at its first instance. But, in fact, immense gain on account of infinite number of quantile pairs for any probability level offers very many choices for surrogating between levels of h and of k, equally for economic or for any other preference criterion or reason. This provides a scope for choice of non-dominated solutions, be any society or individual. No doubt, it is able to give a logically sound quantitative base towards a step of advancement for all those who have been concerned with risk-averting choices or choices with prefixed level of risk.

Besides this, algorithmic heuristic can be extended for the solution of joint chance-constraint to nonlinear programming problem. In fact, a successful attempt was made for a simple geometric chance-constrained programming problem to get the optimum solution.

---

[16]As discussed in Sect. 4.10 (Chap. 4).

The choice of risk level

Such choices or risks are now available not only for a bivariate paradigm but even for a multi-variate scenario, if the user could find proper interpretations for the choices and their range.

Group of decision-maker	Level of risk in terms of probability
Risk averting	0.01–0.05
Low risk prone	>0.05 and ≤0.10
Medium risk prone	>0.10 and ≤0.20
High risk prone	>0.20 and ≤0.50

Thus, the said paradise has been more than regained. There is only one difference that is in place of risk aversion. Now the decision-maker has also the choice for the level of risk. It is, however, possible to calibrate risk and classify the same as above.

There can be other similar subjectively formed groups, depending on the problems emerging from the field of application and agreeable to the consortium of decision-makers. However, the problem of non-existence of feasible region might always creep in. But, it must not be a cause of worry, as it may happen even in deterministic cases. It could possibly be due undernoted reasons:

1. Inappropriate or unrealistic values of objective function coefficients
2. Unrealistic values of coefficients of technology matrix
3. Non-matching values of resource constraints

Its answer lies in relaxing constraints which, in turns, means increasing the resource region alternatively to change the coefficients of objective function. In fact, the inclusion of information of correlation, if exists, is likely to expand such a region for any chance-constrained version of stochastic programming problem.

The problem of convexity

The problem that is seen to be haunting optimization literatures is the problem of convexity of the feasible region. In this case, such a problem is linearizable and its dimension is not very large, it must not cause much concern, in view of the fact that polygons or polyhedrons are always decomposable into non-overlapping triangles, tetrahedrons or polyhedrons, as may be the case. Those are always convex structures, like convex crystals or hyper-crystals. Of course, search for optimum shall have to be continued for every such structures, compared and then proceed to another adjacent one, or even by following without replacement random schema, till such local optimums for each structures gets on improving. With gradually increasing computing and storage power, such an algorithm appears to be feasible till it remains economical. It is hoped that such processes shall hold true and workable even for polyhedronic structures of larger but finite dimension. Such an algorithmic feature appears to be extendable even for a nonlinear cum non-convex region, by decomposing the same into a non-overlapping but finite number of such convex structures, to a satisfactory level of approximation, till it is computationally viable and economically feasible.

7.5 Bivariate Meteorological Prediction with Equi-quantile Pairs

Essenwagner (1976) has been perhaps one of the earliest to elaborately deal the application of bivariate normal distribution in meteorological parameter estimation, where various equations have been derived for estimating and making use of several parameters of the bivariate normal distribution.[17] Out of these, only one case, in which Essenwagner proceeds with wind velocity data for Thule (Greenland) in during January (1956–63) at 25-km altitude, is presented in the below-estimated statistics:

	Mean velocity (in m/s)	Estimated variance	Correlation coefficient
Zonal	$\bar{x}_1 = 14.5$	$\sigma_1^2 = 23.22$	
Meridional	$\bar{x}_2 = 8.4$	$\sigma_2^2 = 19.72$	
			$\rho = 0.428$

By making orthogonal transformation of original variables x_1 and x_2 into y_1 and y_2, estimates of the following parameters from the relational formula 4.19 and 4.20 have been obtained.

1. The variances of the transformed variables y_1 and y_2 are $\sigma_a^2 = 30.794$ and $\sigma_b^2 = 12.145$, for bivariate standard normal variables.[18]
2. The angle φ between coordinate x_1 to major axis y_1 is given by

$$\varphi = \frac{1}{2}\tan^{-1}\left(\frac{2\rho\sigma_1\sigma_2}{\sigma_1^2 - \sigma_2^2}\right) = 39.6°$$

3. The non-circular shape of lines of equal frequency density is

$$\frac{\sigma_a^2}{\sigma_b^2} = \frac{30.794}{12.145} \cong 2.5$$

---

[17] Also discussed in Sect. 4.3 (Chap. 4).
[18] The same could be changed by the expression (4.21) in Chap. 4.

4. The transformed frequency density is

$$f(y_1, y_2) = \frac{1}{2\pi \times 5.55 \times 3.48} \exp\left[-\frac{1}{2}\left(\frac{y_1^2}{30.8} + \frac{y_2^2}{12.14}\right)\right]$$

which, for the standard bivariate normal density, changes to

$$f(y_1, y_2) = \frac{1}{2\pi} \exp\left[-\frac{1}{2}\left(\frac{y_1^2}{1+\rho} + \frac{y_2^2}{1-\rho}\right)\right]$$

5. $\sigma_a^2 \sigma_b^2 = (1-\rho^2)\sigma_1^2 \sigma_2^2 = 374.0$
6. The ellipticity and eccentricity respectively are $\varepsilon_1 = 0.372$ and $\varepsilon_x = 0.778$.[19]
7. Equations of two regression lines are $x_1 = 10.5 + 0.45x_2$ and $x_2 = 2.7 + 0.39x_1$.
8. The angle between two regression lines is

$$\theta = \tan^{-1}\left[\frac{(1-\rho^2)\sigma_1\sigma_2}{\rho(\sigma_1^2 + \sigma_2^2)}\right] = 43.5°$$

9. The equations, in such transformed variables y_1 and y_2 coordinate system, with major and minor axes, are, respectively:

$$y_1 = (x_1 - 14.5)0.771 + (x_2 - 8.4)0.637$$
$$y_2 = (x_2 - 8.4)0.771 + (x_1 - 14.5)0.637.$$

10. The major and minor axes, having the following forms, in terms of original variables x_1 and x_2, respectively, are

$$x_2 = 0.827x_1 - 3.6 \quad \text{and} \quad x_2 = -1.209x_1 + 25.9.$$

Applications of biquantile pairs

Example 7.6 Adding to herein above, the computation of the joint confidence interval, for any preassigned probability value of 0.95 and for the given values of Correlation Coefficient = 0.428, is being attempted. As there being no apparent reason, for preferring the magnitude of one over the other, equi-quantile pair is chosen, which is ($h = k = 1.9257$), obtainable by interpolation from the Table 8.1 of Chap. 8. The joint confidence interval, for the variable x_1 and x_2, is thus

$$\text{Prob.}[(\bar{x}_1 - h\sigma_1 \leq x_1 \leq \bar{x}_1 + h\sigma_1) \cap (\bar{x}_2 - h\sigma_2 \leq x_2 \leq \bar{x}_2 + h\sigma_2)] = 0.95$$

or

---

[19]From expressions (4.23) and (4.24), in Chap. 4.

7.5 Bivariate Meteorological Prediction with Equi-quantile Pairs

$$\text{Prob.}[(5.22 \leq x_1 \leq 23.78) \cap (0 \leq x_2 \leq 16.95)] = 0.95$$

whereas Bonferroni's joint confidence intervals based on the corresponding quantile value = 2.24 for either variables at probability value 0.975 are

$$\text{Prob.}(3.7 \leq x_1 \leq 25.3) = 0.975 \quad \text{and} \quad \text{Prob.}(0 \leq x_2 \leq 18.35) = 0.975$$

Clearly, the shrinkage obtained for joint confidence interval due to the use of equi-quantile value BIGH = 1.9257 over Bonferroni's interval is $(1 - 1.9257/2.24) = 100$ which equals to 14.03 %. This could be achieved because of the consideration of equi-quantile value for the given correlation.

Example 7.7 There is enough scope for advantageous application of biquantile pairs in weather modelling and prediction, for the lack of data specially that of correlation or even raw data. One of the studies by Pal et al. (2003) who have reproduced satellite data of 24 classes on five characteristics related to rainfall is presented below:

1. Number of 32 × 32 pixel region,
2. Number of raining scenes,
3. Estimate of probability of rain obtained by division of (2) by (1),
4. Percentage of rain received in mm/h and
5. Mean TRMM Microwave Instrument (TMI) rain rate.

They have formed three groups of 24 classes:

(a) Group A was formed of class numbers 1, 2, 4, 10, 13 and 17, which represented six high-rain probability classes, ranging between 60 and 89 % raining scenes.
(b) Group B was formed of class numbers 3, 7, 12, 16, 20, 21 and 22, which represented seven medium-rain probability classes, ranging between 30 and 60 % raining scenes.
(c) Group C was formed of remaining 11 class numbers 5, 6, 8, 9, 11, 14, 15, 18, 19, 23 and 24, which represented low-rain probability classes ranging between 1 and 17 % raining scenes.

Table 7.5 presents the data for the purpose of studying correlation and predictive inference by three groups for only two variables, (a) probability of rain and (b) percentage of total rain received, were taken.

Thus, it is seen that Pal et al. (op. cit.) were efficient, in forming different classes on the basis of mean probability of rain % of total rain, as well as the correlations between them.

However, for the purpose of drawing predictive inference as an example consideration of only pooled estimates were taken, the number of classes in each group was small. For such a purpose, first of all biquantile pairs from BIVNOR software

Table 7.5 Classified weather data

Group and nr. of classes	Mean probability of rain	Standard deviation of probability of rain	Mean % of total rain	Standard deviation of total rain	Correlation coefficient
A (6)	74.33	9.57	13.31	11.10	+0.769
B (7)	43.94	6.07	2.49	1.01	+0.3212
C (11)	10.21	4.57	0.244	0.171	+0.9417
Pooled (24)	36.08	6.57	4.16	5.578	+0.7175

were generated for exactitude. Therefore, the probability of joint confidence region to be 0.90 and correlation value as obtained from the data +0.7175 were taken. The equi-quantile value, i.e. BIGH, was obtained in only two iterations starting from one of the probability value 0.90 and correlation value +0.7.[20]

Since both the variables, probability of rain and percentage of total rain, are considered of almost equal importance. They cannot be substituted for each other. It is reasonable to choose equi-quantile pair for joint confidence interval, which is 1.5236. Joint confidence region formed on utilizing equi-quantile value, i.e. BIGH = 1.5236, is

$$\text{Prob.}(\text{Pr}\%, \text{Rain}) = \text{Prob.}[(26.07 \le \text{Pr}\% \le 46.09) \cap (0.0 \le \text{Rain} \le 12.98)]$$
$$= 0.90.$$

Whereas, on the independence assumption, the quantile value for 95 % be 1.96 for both the variables. The Bonferroni's intervals are

$$\text{Prob.}[23.21 \le \text{Pr}\% \le 48.95] = 0.95$$

and

$$\text{Prob.}[0.0 \le \text{Rain} \le 20.09] = 0.95.$$

Thus, the reduction brought about by the former over the latter is 1.5236/1.96 = 0.7773, meaning shrinkage of 22.27 %. Comparing this result with that of Example 7.6, it is concluded that the greater the value of positive correlation, the smaller is the region of joint confidence interval.

On the other hand, greater the value of negative magnitude of correlation larger would be the region of joint confidence, but not in the same proportion, because the relationship being monotonic but not linear.

---

[20]From Table 8.1 of Chap. 8.

7.6 Equi-quantiles in Bio-statistical Studies

There is another interesting data set from the field of bio-statistics, by Daniel (2013), based on fairly large sample ($N = 155$) for pairs of data on height and CV (spine SEP measurement) both in measurement unit of centimetre. The data set is presented in Table 7.6.

Suppose that it is desired to obtain a joint confidence interval for both the variables on the basis of equi-quantile value for 95 % probability level as well as their corresponding Bonferroni's joint confidence intervals and obtain the relative shrinkage due to former over the latter. The same is obtained. Such a joint confidence interval obtained by using equi-quantile (BIGH = 1.8260 obtained by using Table 8.1 Sheet 1, and by interpolation)

$$\Pr\left[(173.296 \leq \text{Ht} \leq 176.794) \cap (16.62 \leq \text{CV} \leq 17.096)\right] = 0.95$$

and their Bonferroni's confidence intervals on independence assumption being

$$\Pr\left[(172.90 \leq \text{Ht} \leq 177.19)\right] = 0.975$$

and

$$\Pr\left[(16.566 \leq \text{CV} \leq 17.146)\right] = 0.975$$

respectively. Thus, the required shrinkage obtained is $1.8260/2.24 = 0.8152 = 81.52$, which corresponds to the shrinkage of 18.48 %.

Table 7.6 Bio-statistical data

Variable	Sample	Mean	Standard deviation	Minimum	Maximum	Correlation
Height	155	175.045	11.927	149.0	202.0	+0.848
CV	155	16.856	1.612	13.0	21.0	

7.7 The VaR Measure (The Value at Risk)

Single-asset case

Winston (2004) states that no matter where the money is invested, the value of the investment at a given date is uncertain. The concept of value at risk (VaR) is useful in determining uncertainty level of a portfolio. The VaR of a portfolio at future time, according to him, is considered to be the loss associated with the fifth percentile of the portfolio's value at that point of time. In short, it is considered to be 5 % chance that the portfolio's value would reach $80 or less, given its today's price to be $100. It then can be said that portfolio's VaR for one-year period is $20 or 20 %.

Hull (2007) reported that Morgan in the year 1994 devised VaR denoted by V as single measure of risk. It was widely accepted as important measure of risk and that has been formally defined as "A loss that will not be exceeded at some specified confidence level".

It is further explained that "about an enterprise or a venture, the decision-maker is, say $(1 - \alpha)\%$ certain, that there would not be a loss of more than V in say next N days, alternatively, it is the loss level over N days that has the probability of exceeding only $\alpha\%$. Naturally, $(1 - \alpha)\%$ has to be reasonably high for a percentile value, which corresponds to quantile used in this text".

Thus, it is a function of two parameters: the time horizon say N days, and the $(1 - \alpha)\%$ quantile value of the probability distribution for return. The computation of VaR as of today requires to make use of quantile value from univariate normal probability distribution.

The said measure has come as a great relief for the portfolio analysts and the banking industry, mainly for both, in terms of its simplicity in understanding and computation. With its popularity, it is convenient and reasonable to assume the probability distribution of return to be univariate normal, and that has been universally accepted. As a consequence, normal probability integral table (NPIT) has become a handy tool for 10-day financial as well as portfolio analysts for risk management decisions.

Hull (op. cit.) has fully explained the technique to compute and interpret the VaR. The simple formula to compute VaR is

$$\text{VaR} = \sigma N^{-1}(X)$$

where $N^{-1}(X)$ is $(1 - \alpha)\%$ quantile value obtained from normal probability integral table (NPIT). Standard time horizon set for the purpose is 10-day time and confidence level for the said period is kept at 99 %, yielding corresponding quantile value to be 2.327.

Clearly, the time horizon is a period for which values of the portfolio are required to be observed for computing the standard deviation (SD). Such computation of the SD is needed to be updated frequently (which may even be every day) for volatile scenario.

7.7 The VaR Measure (The Value at Risk)

$$N\text{-day VaR} = 1\text{-day VaR} \times \sqrt{N}$$

Two-asset case (with variants)

Very often, a company prefers to float more than one portfolio or even that by sister concerns or by competing concerns, and their market value may be found to be correlated, where computation of bi-VaR would be implicit, if the investor has invested in two such portfolios. It is natural, therefore, to seek for biquantile pairs from bivariate normal distribution to obtain bi-VaR, corresponding to the use of quantile of univariate normal distribution for obtaining VaR, as explained earlier.

It is important to note that Hull (op. cit.) explains such a concept in his both the texts referred through very same numerical example, and he worked out two-asset case assuming correlation between them to be +0.3, wherein for one-day 99 % VaR, he used univariate normal quantile value which equals 2.33.[21]

(A) The correct equi-quantile single value, out of infinite number of such pairs, happens to be 2.3157.[22] It means that for 98 % probability corresponds to 99 % and given correlation value = +0.3 for one-sided probability. On the other hand, Bonferroni's one-sided quantile value on assumption zero correlation for the same level of probability which being 0.99 is 2.813. Thus, the shrinkage obtained is 2.3157/2.813 = 0.8232, which corresponds to 17.68 %, due the use of additional knowledge of correlation value in obtaining appropriate equi-quantile value.

(B) If the investor, for some reason, attaches 10 % more preference for second choice, the corresponding quantile value would be $k = 2.5473$ and the value of corresponding h must then be reduced to 2.1699 (i.e. by 7.3 % only and not by 10 %), which is obtained by interpolation from neighbouring values of biquantile pair to maintain the same confidence level of 98 %.[23] Such substitution or of surrogation, from one to the other portfolio or vice versa, would not be possible under the assumption of both being considered independent, without affecting either the level of confidence or of loss over the additional investment made.

(C) However, in case the investor chooses, for some preferential reason, the quantile value to be 2.33, as in Hull's example, for one portfolio, he will be required to choose the quantile value of approximately 2.30 for the other portfolio in order to keep the same confidence level, as available from the same table. This means anticipated investment for the other portfolio 2.3/2.33 = 0.9871 to be kept lower by 1.29 %.

(D) Yet in another example, Hull (2009) considers options on two correlated assets. The said author considers options on two correlated assets by the use of Ito's stochastic differential equations. Nevertheless, it would be interesting to examine his problem, assuming bivariate normal distribution for a given

---

[21] Such an action was natural, because nothing could be known then about the existence of the tables of biquantile pairs before the tables (as presented in Part II) get published.
[22] As obtained through the use of Table 8.1 of Chap. 8.
[23] See Table 8.2-136 of Chap. 8.

correlation value between two such assets S_1 and S_2 to be +0.8. He is required to consider two probabilities p_1 and p_2 for exceeding some separate preassigned values for each such assets. Thus, he is required to obtain the product of such probabilities as joint probability of such occurrence, which is not realistic hence unfair. Since both the assets are considered correlated, as already assumed to have its value +0.8, one is required to consider only the joint probability distribution for which bivariate normal distribution be considered as most natural choice. Under such a situation, therefore, the equi-quantile value is 2.2437 for both portfolios at probability level 0.98.[24]

(E) Thus from such a standard form of distribution, one supposedly selects 0.75 as the joint probability of happening of both exceeding its natural equi-quantile value or any other choice of biquantile pairs adhering to above parametric requirements as of (D).[25] Then, such equi-quantile value would be (BIGH) = 0.9043. However, the corresponding biquantile pairs would range from SMH = 0.91 with SMK = 0.8985 to SMH = 0.68 with SMK = 1.8010.[26] The corresponding univariate probability value would be 0.875 by equally dividing the level of significance into two equal parts, as required for Bonferroni's intervals, having their quantile value each to be = 1.3758, assuming impliedly both to be independent and, therefore, 1.3758 × 1.3758 = 1.8928. This when compared to equi-quantile value, i.e. = 0.9043/1.8928 = 0.4778 or 47.78 % implying 52.22 % reduction in the range of quantile values (i.e. in standardized values) of decision variables due to the choice of equi-quantile value.

(F) Further, if one decides to invest 8 % more in S_1 over and above the equi-quantile value, it means an increase in SMH to 0.9043×1.08 = 0.9766. Then, the corresponding value of SMK would reduce approximately to 0.844, meaning 6.66 % reduction in investment over S_2. Such a swapping of investment between two correlated assets has been possible only by the application of biquantile pairs.

The reason obviously is the use of bivariate normal surface, where probabilities are represented by a sectional cut piece of the volume of a three-dimensional bell-shaped solid, in place of that at present, being represented by sectional area of a two-dimensional univariate normal curve, as applicable in the single-portfolio case.[27] It clearly explains the advantageous application of biquantile pairs in portfolio hedging or its alterations. This is the exchangeable property referred to in Sect. 4.8 by Marshall and Olkin (1979), which could be accruable, solely on account of the use of such biquantile pairs or equi-quantile value. Thus for all such instances, bi-Var have to be determined on the basis of biquantile pairs rather than univariate normal quantile values.

---

[24]As obtainable from Table 8.1 of Chap. 8.
[25]From Table 8.1 of Chap. 8.
[26]As obtained from Table 8.2-52 of Table 8.2 of Chap. 8.
[27]Such advantages are accruable only by the application of biquantile pairs available from the tables in Table 8.2 of Chap. 8.

7.8 Default Correlation: Gaussian Copula Model and Biquantile Pairs

Default correlation

Hull (2007) and (2009) defines the term to denote the same as measure for a tendency of two companies to default at the same moment of time. The reason, for existence of default correlation, is more than one. The companies located in the same geographic region or dealing with the same class of products are likely to be influenced by the similar events. As a result, it may experience similar constraints, taboos, resource constraints, at almost the same time.

Also default by one company may cause default by another as a result of fair or even unfair competition. The default correlation is important in the determination of probability distribution for default losses from a portfolio as a result of exposures to different counterpart. Such a default correlation is the computed value of correlation between times to default by two defaulting firms.

Gaussian copula model

The Gaussian copula model has become a practical tool and, at the same time, a standard market model for understanding time-to-default studies.

Hull (op. cit.) has presented its use quite elaborately yet lucidly in his texts referred. Others who have made considerable contributions are Vasicek (1977, 2002), Cherubini et al. (2004) and Demarta and McNeil (2005) as referred by Hull (2007, 2009). Lo and Wilke (2010), and Nelson (2006) have presented the copula model for dependent competing risks which has also been of considerable interest.[28] However, an attempt has been made to focus on Hull's texts as follows.

The model assumes that companies are bound to default eventually for competition between two companies or for other reasons. Thus, it is bound to have correlation between time to default between two companies.

Hull (2007) considers two correlated normally distributed variables t_1 and t_2 as time to default by two companies, assuming probability distribution of time to default as univariate normal. But, naturally in such a case, the joint probability distribution would be bivariate normal. But, he considers only two marginal univariate normal distributions for each of the two variables, U_1 and U_2, to which variables t_1 and t_2 are required to be transformed, respectively, assuming no knowledge about other, while making such transformation, for each one. These transformations are quantile-to-quantile transformations, say for example, value of t_1 is 0.1 with its cumulative relative frequency 0.05, corresponding to this relative frequency, treating it to be estimate of cumulative probability, the quantile value is obtained from univariate normal probability integral table which being value of

---

[28] As discussed in Chap. 4.

$U_{1,1}$, is -1.64. Similarly, for $t_2 = 0.2$, with its relative frequency, equivalent of probability being 0.20, the corresponding quantile value of $U_{1,2}$ is -0.84, and so on, for all the values of U_1, for the defaulting Firm 1 and values of U_2, for the defaulting Firm 2 get transformed on quantile-to-quantile basis.

From such transformed variate values of U_1 and U_2, he attempts to estimate the correlation coefficient to obtain a structure of correlation between the original variables t_1 and t_2, the time to default, respectively, by Firm 1 and Firm 2.

Thus, the variables U_1 and U_2 are perceived to follow bivariate normal distribution with all their five parameters, means, variances and the estimate of correlation between time to default by the said two firms.

Having estimated these, Hull advises to make use of the Excel® function for the use of cumulative bivariate normal distribution. Such a process was named the Gaussian copula model.

The Gaussian Copula model for time-to-default and biquantile pair

In fact, such a circuitous process was required to be adopted only because neither quantile values for bivariate normal distribution were known, nor existed any method to determine such a quantile pair for the given probability value. Now, since values of quantile pairs for standard bivariate normal probability distribution (biquantile) are available,[29] one should directly estimate all the five parameters from the data sets of observed variables, such as t_1 and t_2, on assumption of time-to-default data pair to follow bivariate normal distribution and can directly obtain the quantile pair for prefixed probability level and estimated correlation value. Thus, it is not required to obtain a default correlation. Clearly, by using two independent univariate quantile after establishing correlation or, for that matter, even default correlation is not only unfair but erroneous.[30]

A factor-based correlation structure, as suggested by Hull (2009), could also be computed through the use of pairs of biquantile for the correlation value between t_1 and t_2, for any preassigned value of probability or confidence level, which would be more realistic than one in practice as above.

However, the validity and importance of copulas in transforming pairs from nonnormal to normal ones, on quantile-to-quantile basis, still holds.

Copulas, whatever be the case (normal or any non-normal), are meant for deciphering the correlation structure and, once that are known, by using normal quantiles to obtain the corresponding approximate probability. On the other hand, biquantile pair (which is required to be of direct use in getting bivariate confidence volume is obtained, given the correlation and probability level, almost as easily as univariate normal quantile) is used for the given value of probability level.

---

[29]Through Table 8.2 of Chap. 8.

[30]If the need arises to compute such a quantile pair or a finer value of probability and correlation, other than those available in Tables of Part II, one can find the same value either by two-way interpolation from appropriate grids of Table 8.1 of Chap. 8.

7.9 Bivariate Quality Control, Six Sigma Techniques and Biquantile Pairs

In fact, Six Sigma and the ultimate Six Sigma techniques, as they stand today, are the natural outgrowth of statistical quality control that can be inferred by reference to the texts by Mitra (2001) and by Bhote (2007).

According to Bhote (op. cit.), Six Sigma and ultimate Six Sigma are much hyped today as "total business excellence" methodologies. Bhote, while emphasizing on the "method of achieving success", enumerates 12 Shainin–Bhote techniques of which the first one being adoption of multi-variable technique to reduce a large number of unmanageable variables to a much smaller families of related variables. This appears to be the suggested method for the variable elimination problem. But all the techniques propounded by Mitra (op. cit.), for the quality control. Bhote (op. cit.), for Six Sigma or ultimate Six Sigma variables, refers only to several sets of independent and identically distributed normal variables, making virtually no use of relation/association/dependence amongst them.

Again Bhote (op. cit.) recommends the break-up of multi-variate charts of large numbers of unmanageable list of variables into smaller and more manageable families of related causal variables, implying cause–effect analysis without clearing it further the methodologies of using measures of correlation or regression or of making use of even bivariate normal probability distribution. Similarly, Mitra (op. cit.) makes no mention of building bivariate control charts for any attributes or variables, even though at least some of them might show a reasonable degree of positive or negative correlation.

Such anomalous situations have been occurring due to the reason that consideration of even two associated or correlated variables would require iso-probable biquantile pairs not available hitherto. However, availability of such biquantile pairs[31] now opens the door for their aspired treatment correctly, at least for pairwise variable association or dependence. Additional advantages that accrue of using iso-probable biquantile pairs are as follows:

1. Reduction in the range from Six Sigma to at the most to 3.8-Sigma for both of the correlated variables and even more to either of them depending on the sign and magnitude of correlation values between them. It results in a huge cost reduction that can possibly be attained now by adopting only 3.8-Sigma.
2. Infinite number of choices to surrogate between levels of pairs of such variables within permissible limits of utility measure of the product-mix or for substitution of one by the other, on account relative cost or for the reason of taste, or for strength of material, without scaling down from the risk level (in the sense of probability) already prefixed by the entrepreneur or the decision-maker. The

---

[31]Through this book text and tables presented in Chap. 8.

phenomenon is again the exchangeability property, accruable only on account of biquantile pairs, as referred to in Sect. 7.7.[32]

3. Methodology so adopted raises possibility even for the generation and application of tri-quantile trios. In fact, such a sample of table has also been generated.

7.10 Bivariate Stochastic Process and Biquantile Pairs

Box et al.'s illustration

As an illustration, it would be interesting to quote a numerical example of bivariate stochastic process from Box et al. (2004), where they compute cross-covariance coefficient $c_{xy}(k)$ and cross-correlation coefficient $r_{xy}(k)$, utilizing undernoted formula for lag $+1$ and lag -1.

$$c_{xy}(k) = \frac{1}{n}\sum_{t=1}^{n-k}(x_t - \bar{x})(y_{t+k} - \bar{y}), \quad k = 0, 1, 2$$

and

$$c_{xy}(k) = \frac{1}{n}\sum_{t=1}^{n+k}(y_t - \bar{y})(x_{t-k} - \bar{x}), \quad k = 0, -1, -2 \tag{7.9}$$

where $\bar{x}, \bar{y}$ are the sample means of the x-series and y-series, respectively. The estimate $r_{xy}(k)$ of the cross-correlation coefficient at lag k may be obtained by substituting in (7.9)

$$\rho_{xy}(k) = \frac{\gamma_{xy}(k)}{\sigma_x \sigma_y}, \quad k = 0, \pm 1, \pm 2, \ldots \tag{7.10}$$

is called *cross-correlation coefficient* at lag k, and the function of the bivariate process, where the numerator at the right-hand side is cross-covariance coefficient between x and y at lag $+k$. Since $\rho_{xy}(k)$ is not in general equal to $\rho_{xy}(-k)$, the cross-correlation function is not symmetric about $k = 0$. The estimate $r_{xy}(k)$ of the cross-correlation coefficient $\rho_{xy}(k)$ at lag k may be obtained by substituting in (7.9) the estimate $c_{xy}(k)$ for $\gamma_{xy}(k)$,

$$s_x = [c_{xx}(0)]^{1/2} \text{ for } \sigma_x \text{ and } s_y = [c_{yy}(0)]^{1/2} \text{ for } \sigma_y$$

yielding

---

[32] Also in Sect. 4.8 (Chap. 4).

7.10 Bivariate Stochastic Process and Biquantile Pairs

$$r_{xy}(k) = \frac{c_{xy}(k)}{s_x s_y}, \quad k = 0, \pm 1, \pm 2, \ldots \tag{7.11}$$

Example 7.8 With a small sample size, the example has illustrative value only.

T	1	2	3	4	5
x_t	11	7	8	12	14
y_t	7	10	6	7	10

Since $\bar{x} = 10.4$ and $\bar{y} = 8$, taking deviations from mean, the above table reduces to

T	1	2	3	4	5
$x_t - \bar{x}$	0.6	−3.4	−2.4	1.6	3.6
$y_t - \bar{y}$	−1.0	2.0	−2.0	−1.0	2.0

Hence according to said authors, computations arrived at are

$$\sum_{t=1}^{4}(x_t - \bar{x})(y_{t+1} - \bar{y}) = (0.6)(2.0) + (-3.4)(-2.0) + (-2.4)(-1.0) + (1.6)(2.0)$$

$$= 13.6$$

and

$$c_{xy}(1) = \frac{13.60}{5} = 2.720$$

They also have obtained $s_x = 2.577$ and $s_y = 1.673$ and then

$$r_{xy} = \frac{c_{xy}(1)}{s_x s_y} = \frac{2.720}{2.577 \times 1.673} = 0.63$$

and similarly $c_{xy}(-1) = -1.640$. Therefore, $r_{xy}(-1) = -0.38$.

(a) However, interest here lie in obtaining the joint confidence interval for both x_t and y_t, for forward and backward movement by a step of input x_t and output y_t for the two correlated sequence of data, assuming their distribution to be bivariate normal. Thereby, utilizing the equi-quantile values to obtain joint confidence intervals for the input x and output y with lag +1 (i.e. of the next step for future output y) is

(b) $\text{Prob}\left[(\bar{x} - q_{+}.s_x \leq \mu_x \leq \bar{x} + q_{+}.s_x) \cap (\bar{y} - q_{+}.s_y \leq \mu_y \leq \bar{y} + q_{+}.s_y)\right] = 0.95$ (7.12)

Using for example standard deviations s_x and s_y in place of standard errors, which makes such confidence intervals for observations rather than for means, without loss of generality. For the probability value 0.95 and the value of correlation as obtained being 0.63, the equi-quantile value is 1.8936.[33] Therefore, the joint confidence interval is

$$\text{Prob}\,[(10.4 - 1.8936 \times 2.577) \le \mu_x \le 10.4 + 1.8936 \times 2.577)$$
$$\cap\,(8.0 - 1.8936 \times 1.673 \le \mu_y \le 8.0 + 1.8936 \times 1.673)] = 0.95$$

which simplifies to

$$\text{Prob}\,[(5.5202 \le \mu_x \le 15.2798) \cap (4.832 \le \mu_y \le 11.168)] = 0.95 \quad (7.13)$$

(c) The joint confidence interval for output value y and the input value x with lag -1 (i.e. next backward step), for the given probability value 0.95 and correlation value -0.38, is

$$\text{Prob}\left[(\bar{y} - q_-.s_y \le \mu_y \le \bar{y} + q_-.s_y) \cap (\bar{x} - q_-.s_x \le \mu_x \le \bar{x} + q_-.s_x)\right] = 0.95$$
$$(7.14)$$

The equi-quantile value for the given probability and correlation combination as obtained is 1.9598.[34] Substituting this value for q_- in Eq. (7.14), the said JCI is

$$\text{Prob}\,[(8.0 - 1.9598 \times 1.673) \le \mu_y \le 8.0 + 1.1.9598 \times 1.673)$$
$$\cap\,(10.4 - 1.9598 \times 2.577 \le \mu_x \le 10.4 + 1.9598 \times 2.577)] = 0.95$$

or

$$\text{Prob}\,[(4.7213 \le \mu_y \le 11.2787) \text{ and } (5.346 \le \mu_x \le 15.454)] = 0.95 \quad (7.15)$$

Box et al.'s two-variable confidence intervals and their joint confidence interval

1. Usually, for obtaining such a confidence interval, the value of correlation is either not known or not used, because biquantile pairs are not known hitherto. The values of input x and output y are taken as independent, which being unrealistic are not valid. Leaving these aspects aside, what requires to be seen is impact of such assumption on the size of confidence intervals.

---

[33] By interpolation from Table 8.1 of Chap. 8.
[34] By interpolation from two adjacent values of Table 8.1.

7.10 Bivariate Stochastic Process and Biquantile Pairs

In order to maintain same confidence coefficient of two such independent sets, confidence coefficients for each such sets have to be taken = 0.975, so that joint confidence of input x and output y taking both to be independent is 0.95. For such a value of probability, the corresponding univariate normal quantile value as obtained for Bonferroni's intervals is 2.24, consequently such intervals are

$$\text{Prob}\,[10.4 - 2.24 \times 2.577 \leq \mu_x \leq 10.4 + 2.24 \times 2.577] = 0.975;$$

and

$$\text{Prob}\,[8.0 - 2.24 \times 1.673 \leq \mu_y \leq 8.0 + 2.24 \times 1.673] = 0.975:$$

those simplify to

$$\text{Prob}\,[4.63 \leq \mu_x \leq 16.17] = 0.975$$

and

$$\text{Prob}\,[4.252 \leq \mu_y \leq 11.748] = 0.975 \tag{7.16}$$

Thus, both together would have the desired confidence coefficient = 0.95. The JCI so obtained by taking additional information of correlation coefficients and utilizing the same for equi-quantile value (1.8936) gets the advantage of shrinkage of the JCI to [1.8936/2.24 = 0.8454, i.e. 1 − 0.8454] or equals to 15.46 % of the usually adopted Bonferroni's confidence intervals based on independent assumption, for the same risk level of (1 − 0.95) = 0.05 or of 5 %.

The use of biquantile pairs has thus been able to reduce 15.46 % of the resources. This is due to the fact that expansion or reduction in value of quantile implies expansion or reduction in the value of resources (input or output), because quantiles are only standardized transform of such resources or of information.

2. Further, it is also instructive to examine the impact of backward step, i.e. on account of lag −1. Again on same count, even the negative value of the correlation coefficient, taken as (−0.38), throws no impact on JCI. Similarly, the negative value of the correlation coefficient has been obtained on independent assumption, each following univariate normal probability distribution. However, the JCI obtained on utilizing correlation value, and as a consequence based on corresponding equi-quantile value of 1.9598, would shrink to [1.9598/2.24 = 0.8748 or 87.48 %]. Thus, it has 12.52 % reduction of resources.

Auto-cross-covariance and linear predictive models

A rough scrutiny of the data of Table 7.1 raised doubts to an agricultural economist as to whether:

1. area under wheat of the present quinquennium x_t does effect even the production of the next stage of future quinquennium is y_{t+1}?
2. production of the present quinquennium y_t does effect even the area under wheat for the next of future quinquennium x_{t+1}? (Such a phenomenon is known as *area response to production*, earlier known as acreage response to production).

Supposing that to verify his hypotheses with probabilistic confidence coefficient of 95 %, the agricultural economist seeks apart from regression analyses, cross-autocorrelation estimates and likes to obtain the shortest possible joint confidence intervals and their relative shrinkages, vis-à-vis Bonferroni's joint confidence intervals, he proceeds as follows.

As per referred data set of Table 7.1, wheat growing programme was increased from quinquennium 1950–51 to 1990–91, for 10 such quinquenniums which is presented in Table 7.7.

As required, means and variances of the variables x_t for area (million hac) and y_t for production (million ton) have been computed and placed in below Table.

N	$\bar{x}$	$\bar{y}$	σx^2	σy^2	σxy	ρxy	σyx	ρyx
10	18.08	28.99	32.386	419.174	+104.49	+0.8968	+84.072	+0.7215

Thereafter, auto-cross-covariances and auto-cross-correlations between x_t and y_{t+1} and between y_t and x_{t+1} have been computed to get their respective values, as in next part of Table 7.7.

Such mean values, auto-cross-covariances and variances have been used in deriving two simple linear regression models with their R^2 values and those are placed along with their respective multiple correlation, R^2 values:

$$y_{t+1} = -29.3433 + 3.2264 x_t, \quad R^2 = 0.84$$

Table 7.7 Area and production of wheat data

t	Area in million hectare (x_t)	Production in million hectare (y_t)
1	9.75	6.46
2	12.37	8.76
3	12.39	11.00
4	12.57	10.40
5	18.24	23.83
6	20.45	22.28
7	22.28	36.31
8	23.00	47.05
9	24.17	55.14
10	25.01	62.10

Only first-half part of the data set of Table 7.2 (years 1950–51 through 1990–91)

7.10 Bivariate Stochastic Process and Biquantile Pairs

and

$$x_{t+1} = 12.224 + 0.202 y_t, \quad R^2 = 0.575$$

clearly implying that though both linear regression models are significant in making predictions envisaged, one step advance prediction of wheat production on area under the crop has shown greater reliability than a single-step advance prediction of area under wheat on production under the crop.

Joint confidence intervals for bivariate stochastic processes

Estimates of such auto-cross-correlations together with probability value, representing desired confidence level, have been used to obtain[35] respective equi-quantile values of $BIGH_1$ and $BIGH_2$, obtained by interpolation or using BIVNOR are placed below:

1. PROB $= 0.95$, $CORR_{xy} = +0.8968$, $BIGH_1 = 1.7996$ and $SMHL_1 = 1.64$.
2. PROB $= 0.95$, $CORR_{yx} = +0.7215$, $BIGH_2 = 1.8715$ and $SMHL_2 = 1.65$.

where $CORR_{xy}$ and $CORR_{yx}$ represent $\rho_{xy}(1)$ and $\rho_{yx}(-1)$, respectively, and their corresponding equi-quantile values, being those of $BIGH_1$ and $BIGH_2$. As in previous examples, these equi-quantile values $BIGH_1$ and $BIGH_2$ have been used to make joint confidence intervals, taking standard deviations in place of standard errors. Those are

$$\text{Prob}[(14.9756 \leq x_t \leq 21.1844) \cap (17.589 \leq y_{t+1} \leq 40.391)] = 0.95$$

and

$$\text{Prob}[(17.1335 \leq y_t \leq 39.9665) \cap (14.8515 \leq x_{t+1} \leq 21.3085)] = 0.95$$

respectively. The corresponding Bonferroni's intervals are

$$\text{Prob}(14.0489 \leq x_t \leq 22.1111) = 0.975$$

and

$$\text{Prob}(14.8478 \leq y_{t+1} \leq 43.4924) = 0.975;$$

for both cases, considering the variables to be independent. Hence, the values of auto-cross-correlations were not required to be used. Thus, the univariate normal quantile value, which has been used for building such Bonferroni's couple of confidence intervals, being a single value, is 2.24. Shrinkage obtained for the joint interval being

$$1.7996/2.24 = 0.8034 = 80.34 \text{ \% is } 19.66 \text{ \%}.$$

---

[35] By using Table 8.1 of Chap. 8.

whereas that for the joint interval is

$$1.8715/2.24 = 0.8356 = 83.56 \text{ is } 16.44\ \%.$$

It is interesting to observe that it is possible to make feedback recursive chain of prediction and validation by the alternate use of relation of the type

$$y_{t+1} = a + bx_t; \quad x_{t+1} = c + dy_t; \quad y_{t+2} = e + fx_{t+1}$$

until their auto-cross-correlations and their reverse auto-cross-correlations, as well as their joint confidence intervals, do not change their values. In case they change to proceed similarly with such changed values to the extent, they are realistic and meaningful.

7.11 Simulation and Biquantile Pair/Equi-quantile Value

Winston (2004) defines a simulation as a technique that imitates the operations of the real-world system that evolves over time. One of the essential components for generating simulation output is the generation of random numbers[36] with which the core result is encapsulated, which is considered mimicry of real-world scenario. For this, the quantile value of the assumed probability of error component is required. That is obtainable by a different method for univariate distribution, as can be referred to from any text on simulation including Winston (op. cit.).

The other problem, which is usually encountered in simulation, is variation in estimate with its runs executed with different set of random numbers. Thereby, there is a need for relatively compact confidence interval for prefixed confidence coefficient. But, the construction of such a confidence interval gets complicated on account of lack of independence between the results of such simulation runs. It is well established that so-generated data are correlated/auto-correlated, as can be seen from Example 7.9.

Thus, the only valid recourse to circumvent such problem lies in adoption of biquantile pair/equi-quantile value.

Example 7.9 (Gross and Harris (2010) on comparison of two system designs via queuing simulation model) The results (in Table 7.8) on mean waiting times under each design for 15 replications are obtained. The problem here is to find whether there exists a correlation between the simulation results of two designs on mean waiting times. If so, to obtain the shortest possible joint confidence interval for the confidence coefficient 0.95 and also to compare the same with Bonferroni's joint confidence interval.

For the computed correlation +0.93 and confidence coefficient 0.95, the equi-quantile value is 1.7761.[37]

---

[36]For generation of random numbers/variables refer to Degpunar (1988) and Knuth (2000).
[37]Obtained by interpolation from Table 8.1 of Chap. 8.

7.11 Simulation and Biquantile Pair/Equi-quantile Value

Table 7.8 Data on mean waiting time

Replication number	Mean waiting time	
	Design 1	Design 2
1	23.02	23.97
2	25.16	24.98
3	19.47	21.63
4	19.06	20.41
5	22.19	21.93
6	18.47	20.38
7	19.00	21.97
8	20.57	21.31
9	24.63	23.17
10	23.91	23.09
11	27.19	26.93
12	24.61	24.82
13	21.22	22.18
14	21.37	21.99
15	18.76	20.61

Design 1		Design 2		Correlation
Mean time = m_1	Sd_1	Mean time = m_2	Sd_2	
21.91	2.74	22.62	1.88	+0.93

Thus, the joint confidence interval for mean waiting time record under Design 1 and Design 2 for which confidence coefficient is 0.95 is

$$\text{Prob}\left[(21.91 - 1.7761 \times 2.74 \leq wt_1 \leq 21.91 + 1.7761 \times 2.74) \cap (22.62 - 1.7761 \times 1.88 \leq wt_2 \leq 22.62 + 1.7761 \times 1.88)\right] = 0.95.$$

which simplifies to

$$\text{Prob}[(17.04 \leq wt_1 \leq 26.78) \cap (19.28 \leq wt_2 \leq 25.96)] = 0.95 \quad (7.17)$$

whereas Bonferroni's intervals for the same Design 1 and Design 2 based on univariate normal quantile value = 2.24 are

$$\text{Prob}\left[(21.91 - 2.24 \times 2.74 \leq wt_1 \leq 21.91 + 2.24 \times 2.74) = 0.975\right]$$

and

$$\text{Prob}(22.62 - 2.24 \times 1.88 \leq wt_2 \leq 22.62 + 2.24 \times 1.88)] = 0.975$$

Those simplify to

$$\text{Prob}\,(15.77 \leq wt_1 \leq 28.5) = 0.975$$

and

$$\text{Prob}\,(18.41 \leq wt_1 \leq 26.83) = 0.975 \qquad (7.18)$$

and on account of they being considered independent, the joint confidence interval will have the confidence coefficient = $0.975 \times 0.975 = 0.95$. Here, wt_1 and wt_2 are waiting times for Design 1 and Design 2. The shrinkage brought about by Eq. (7.17) over Eq. (7.18) is $(1.7761/2.24) \times 100 = 20.21\,\%$.

It is not only the gain in shrinkage brought about by joint confidence interval (1) over (2) which matter, but it is validity of such interval because of the existence of correlation between the two.

Such shrinkage in confidence interval is quite independent of narrowing down the CI limits by variance reduction technique (VRT) by the use of common random numbers.

Example 10 (Banks et al. (2008) on simulation data obtained for job shop performing two operations) Table 7.9 presents the data on two operations milling and planing generated with random occurrences, about which the shop manager suspects to be related. Now, the problem here is to obtain the value of correlation coefficient, and if there is a correlation to obtain the joint confidence interval by using equi-quantile value, assuming that the results on two operations are following bivariate normal distribution.

Table 7.9 Banks et al.'s data on milling and planing

Order No.	Milling time (mt) (min)	Planing time (pt) (min)	Order No.	Milling time (mt) (min)	Planing time (pt) (min)
1	12.3	10.6	14	24.6	16.6
2	20.4	13.9	15	28.5	21.2
3	18.9	14.1	16	11.3	9.9
4	16.5	10.1	17	13.3	10.7
5	8.3	8.4	18	21.0	14.0
6	6.5	8.1	19	15.0	11.5
7	25.2	16.9	20	15.0	11.5
8	17.7	13.7	21	12.6	9.9
9	10.6	10.2	22	14.3	13.2
10	13.7	12.1	23	17.0	12.5
11	26.2	16.0	24	21.2	14.2
12	30.4	18.9	25	28.4	19.1
13	9.9	7.7			

7.11 Simulation and Biquantile Pair/Equi-quantile Value

No. of orders	Milling time		Planing time		Correlation
	(Mean1)	Sd1	(Mean2)	Sd2	
25	17.732	6.711	12.712	3.565	+0.96

The equi-quantile value for confidence coefficient 0.95 and correlation coefficient +0.96 is 1.7473, which yields

> Joint confidence coefficient interval for the pairs of observational records
> $= \text{Prob}[(6.006 \leq \text{mt} \leq 29.458) \cap (6.483 \leq \text{pt} \leq 18.941)]$
> $= 0.95$

(7.19)

7.12 From Biquantiles to Higgs Boson (God Particle)

In the above context, and in context to the report published in wikipedia.org/wiki/Bose-Einstein_statistics, Wikipedia.org/wiki/Higgs_boson, Wikipedia.org/wiki/Bose-Einstien correlations [7.22.2012], abbreviated as B-E statistics, H-B and BEC. This abbreviation in the particle physics is being used for both Bose–Einstein condensation and BEC. The particle condensation may be the result of quantum coherence (strong in interaction) between identical boson particles, whereas correlation provides measure of such coherence. Thus, due to phenomena of coherence, exhibiting correlation resulting in condensation, the particles are getting bunched. For example, in optics, two beams of light are said to interfere coherently forming beam.

Leaving aside other details of particle physics, as effected by correlation (BEC), the major concern here is twofold.

1. The existence of such a correlation, which is seen to play a vital role in quantum coherence, condensation and cohesive bondage of bosons:
   and of
2. The fact that search of Higgs boson lies in breaking such boson particles by LHC and determining its mass subjects to gradually thinning (condensing) its confidence interval. The state of which, as reported up to 20 July 2012, being 125.3 ± 0.6 GeV/c^2 within 4.9 sigma, where 4.9 is the value of quantile and sigma the standard deviation, and GeV refers to giga (10^9) electronic voltage per c^2 (the quantity of electricity conveyed in one second by a current of one ampere), as per Wikipedia.org/wiki/Higgs-boson.

It is to be get realized that till date such a confidence interval is being constructed on the basis of independent univariate Gaussian probability distribution and naturally on its quantile value. But now since biquantile pairs and equi-quantile value are available, the appropriate approach would be to replace the same by a biquantile pair or an equi-quantile value (which is smaller than the existing quantile value = 4.9 as reported and quoted herein. That on account of additional information, the value of BEC, the magnitude of such correlation) as may perhaps be available for every such pairs of boson particles (type-II two Higgs-doublet sector) of which such a condensation is composed of. If the same gets adopted, the resulting confidence interval shall automatically get thinner (denser).

Alternatively, the value of the confidence coefficient should increase to more than even 99 %, depending on the availability and magnitude of positive correlation coefficient (that is, the BEC). This is against 97 %, as in the report referred earlier. This result is obtained even without a change in the stochastic estimates of parameter values. Those like the mean and standard deviation as have been obtained by a fabulously costly, complex and conducted historic experiment. For example, suppose the BEC is +0.99 and the required confidence level is also kept at a probability level 0.99. The equi-quantile value comes to 2.3792 by the use of Table 8.1, as against its reported value 4.9 and confidence probability level 0.97 by Wikipedia (July 20, 2012). These are obviously inappropriate, clearly because the reported quantile value and the given probability value bear no correspondence. However, this implies that the estimate 125.3 GeV/c^2 of boson commands much greater confidence (much more than what has been reported) and even much higher than the probability of 0.99. Meaning that 99 % confidence level could possibly be attainable in less than the said quantile value, given the magnitude of BEC = +0.99. Does it not, therefore, mean that with the above magnitude of quantile value, so-called Higgs Boson (so-called the God Particle) has been achieved with the confidence coefficient of more than 99 %. Is it not a trillion-dollar question to a particle physicist, participating in such a giga experiment?

However, if the magnitude of BEC is different than what has been supposed, then the same is needed before coming to any specific conclusion. The phenomenon explained can be demonstrated through Kramers's data quoted by Huang (2012).

Here, it is worthwhile to quote the data given by Huang (2012) who has cited the undernoted experimental result due to Kramers (1955). As the number of observation is small, the example is for the sake of illustration only.

The experimental values for the specific heat of liquid He^4 are given in the accompanying Table 7.10. The values obtained are the vapour pressure curve of liquid He^4. But, one may assume that they are not very different from values of c_V at the same temperatures. (The only Bose system known to exist at low temperatures is liquid He^4, exhibiting the remarkable λ transition at which the specific heat becomes logarithmically infinite. Since He^4 atoms obey the Bose statistics, it is natural to suppose that this transition is the Bose–Einstein condensation modified by intermolecular interactions.) Here, it is required to obtain the value of correlation between the data on temperature and of specific heat, as given in Table 7.10.

7.12 From Biquantiles to Higgs Boson (God Particle)

Table 7.10 Kramer's data on temperature and specific heat of He^4

Temperature (K)	Specific heat (J/g-deg)
0.60	0.0051
0.65	0.0068
0.70	0.0098
0.75	0.0146
0.80	0.0222
0.85	0.0343
0.90	0.0510
0.95	0.0743
1.00	0.1042

Variables	Mean	Std. Dev.	Correlation
Temperature	0.80	0.433	
Specific heat	0.0358	0.043	+0.93

The next step is to construct Bonferroni's confidence intervals for both the variables on the assumption that they are independent. Then, one should obtain joint confidence interval if there is significant correlation between the two for confidence coefficients 95 %.

Bonferroni's interval based on independent assumption requires univariate normal quantile value, which is 2.24 for confidence coefficient 0.975, and 2.81 for the confidence coefficient 0.995. Bonferroni's confidence interval for the confidence coefficient 0.975 is

$$\text{Prob}\,(0.0 \leq \text{Temp.} \leq 1.77) = 0.975$$

and

$$\text{Prob}\,(0 \leq \text{Heat} \leq 0.1321) = 0.975$$

both together on independence assumption 0.975 × 0.975 = 0.95.

For joint confidence interval, what is required is to obtain first the equi-quantile value BIGH by using the Table 8.1 or BIVNOR, which is 1.7761 for the confidence coefficient = 0.95. Thus, the joint confidence interval for the data is

$$\text{Prob}[(0.03 \leq \text{Temp.} \leq 1.57) \cap (0 \leq \text{Heat} \leq 0.1121)] = 0.95$$

Level of confidence coefficient	Univariate normal quantile value on independence assumption	Equi-quantile value on correlation = 0.93	Shrinkage brought about by (3) over (2)
(1)	(2)	(3)	(4)
0.95	2.24	1.7761	$\left(1 - \frac{1.776}{2.24}\right) \times 100 = 20.71\,\%$

Thus, the advantage of high value of correlation brings shrinkage (condensation) in the confidence interval for the same level of confidence coefficient.

Looking at the exponentially rising trend in the value of specific heat for a linearly increasing value of the temperature, it was decided to transform the data of specific heat to its logarithm and then obtain the value of its correlation with temperature data. Such an act raised the value of correlation from +0.93 to +0.999, approximately. Thereafter, it was decided to obtain shrinkage brought about in confidence interval at increased value of confidence coefficient also from 0.95 to 0.99. This resulted in equi-quantile value to be 2.3437. Thus, the shrinkage obtained by equi-quantile value over Bonferroni's interval for the same confidence coefficient 0.99 happens to be $(1 - 2.3437/2.81) \times 100 = 16.60\ \%$. Thus, the said transformation increases the condensation at a much higher level of confidence coefficient, if the higher positive value of the correlation be interpreted as the said boson particle condensation.

It is evident, therefore, that any attempt for condensed confidence interval for highly correlated boson particles be made by the use of equi-quantile values, assuming variables to follow bivariate Gaussian probability distribution, equally by using biquantile pairs, if the relative importance of such particles be known and, at the same time, particle cohesiveness towards condensation at such level do not get disturbed. (The phenomenon remains to be verifiable experimentally if possible.)

7.13 Rizopoulos's Paradox (2009)

7.13.1 Basics

It is known that in case of bivariate normal distribution, the values of quantile, for any specified probability and given correlation, one is neither free to assume any of neither its value nor that value is unique like independent univariate normal quantile. Also, that of the pair, as one of the quantile for a variable increases that for the other variable decreases, forming locus of quantile curve of iso-probable values.

However, an equi-quantile value is a unique point on such quantile curve, where for the specified probability level and given correlation value, the quantile value for both the variables is equal. Hitherto, such a point has been found as only recourse for the researchers/empirical scientists by Gupta (as early as 1963) up to 12 variables. Later, Deak (2000) increased its dimensionality and speed of computation. Then only recently, Genz et al. (2012) claim to have developed the software qmvnorm as an advancement in respect of dimensionality and speed of computation only for equi-quantiles. But even for those, reliabilities remained to be tested and reported.

The knowledge of above facts alone could not carry much weight, for the decision-makers, even though they were of fundamental nature. They generally look for interval domain of quantile pairs, as they do not find reason, for both the quantile of the variables to be given equal importance or weight in decision-making processes. Such problem has been haunting more than six to seven decades, as they appeared to be offering scope for set of alternatives in decision-making.

7.13.2 The Paradox: Its Modification and Solution

The said paradox is as follows: for given two standard bivariate normal variables $Z1$ and $Z2$, with their mean values = 0, variances = 1 and correlation = 0.5, to find the value of c (the equi-quantile value), which satisfies the given probability equation:

$$P(Z1 > 1.975, Z2 < c) + P(Z1 > c, Z2 > c) = 0.05/6 = 0.00833 \quad (7.20)$$

Though the proposition is valid, it is ill-posed which is natural, because of ignorance about the existence and application of biquantile pairs. This proposition offers much wider scope for solution compared to equi-quantile value, not available as of now. That can be seen as the solution proceeds. The probability Eq. (7.20) has two parts (say, Parts 1 and 2), each of the two parts are required to be identified in probability schema, given as follows in a tabular format. Thus for any known value of c, satisfying the given probability equation appears *prima facie* to be a valid proposition. At the same time, they are mutually exclusive and also evident.

The two parts of Eq. (7.20) are expressed as below (Table 7.11):

$$\text{Part 1 (c)} + \text{Part 2 (a)} = 0.00833 \quad (7.21)$$

Now assuming at first that Part 1(c) = 0, Part 2(a) can be written as

$$P(Z1 > c, Z2 > c) = 1 - [\text{Part 2(b)} + \text{Part 2(c)} + \text{Part 2(d)}] \geq 1 \\ - P(Z1 \leq c, Z2 \leq c).$$

Equality to be attained only if each of the other two probabilities, i.e. Part 2(c) and Part 2(d) are 0, otherwise,

Table 7.11 Analysis for identification of c the quantile value

Part 1	Part 2
(a) $P(Z1 > 1.975, Z2 > c)$	(a) $P(Z1 > c, Z2 > c)$
(b) $P(Z1 \leq 1.975, Z2 > c)$	(b) $P(Z1 \leq c, Z2 \leq c)$
(c) $P(Z1 > 1.975, Z2 < c)$	(c) $P(Z1 > c, Z2 \leq c)$
(d) $P(Z1 \leq 1.975, Z2 > c)$	(d) $P(Z1 < c, Z2 > c)$

---

[38] From Table 8.1 of Chap. 8.
[39] Refer to Tables 8.2 in Chap. 8 and Table 9.9 of Chap. 9.

$$\geq 1 - 0.00833,$$

or

$$P(Z1 > c, Z2 > c) \geq 0.99167$$

It is now decided to first obtain the value of equi-quantile, c.

For such extreme case it is convenient to use BIVNOR to get BIGH = c = 2.6217 and SMH ≥ 2.40, but SMK ≤ 3.551379. The meaning of extreme pair is that the probability value outside the range from 2.40 to 3.551379 is zero, at least up to five places of decimal. Hence, the range of $1.975 < Z1 < 2.40$ is redundant, which needs removal, without effecting validity of the given equation, but not contradicting or violating the original constraint imposed by Rizopoulos.

Thus for chosen probability level and known correlation value for the stipulated combination grids starting with such arrived value of c = 2.6217 and ending with SMH = 2.40 and SMK = 3.551379, such set of solutions of Table 7.12 could be obtained. That is taken to imply that

$$P(Z1 > c, Z2 > c) = 0.00833. \tag{7.22}$$

Substituting the value of c in (7.22),

$$P(Z1 > 2.6217, Z2 > 2.6217) = 0.00833$$

which is possible only if $P(Z1 \geq 2.40, Z2 < 2.6217)$ is allowed to be 0, as assumed. Thus, the value of $Z1 = 1.975$ to less than 2.40 is redundant, because as the value of $Z1$ decreases from 2.40, the value of $Z2$ will shoot up beyond SMK = 3.551379, towards the region, where the probability density is almost zero. At the same time, the condition of equality to the set probability value = 0.00833, breaks down. Thus, the Rizopoulos paradox needs to be modified maintaining its validity (i.e. eliminating the redundant part of $Z1$ from the Part 1 which shrinks it) and Rizopoulos's said probabilistic equation is required to be changed to

$$P(Z1 > 2.40, Z2 < c) + P(Z1 > c, Z2 > c) = 0.05/6 \tag{7.23}$$

or

$$P(Z1 > 2.40, Z2 < 2.6217) + P(Z1 > 2.6217, Z2 > 2.6217) = 0.05/6 = 0.00833$$

where $Z1 > 1.975$ is required to be replaced by $Z1 > 2.40$, because the value of $Z1$ between 1.975 and 2.40 is not any way contributing probability mass at the most up to six places of decimal. Thus, Rizopoulos's Part 1 interval being lose is required to be tight-end without effecting the intended condition of the equation.

7.13 Rizopoulos's Paradox (2009)

Table 7.12 Result of computational experiment performed with BIVNOR

Partwise value of the equation			Probability	BIGH = c	SMH	SMK
Part 1	Part 2	Sum				
0.000	0.00833	0.00833	0.99167	2.6217	2.40	3.551379
0.001	0.00733		0.99267	2.6662	2.45	3.476478
0.002	0.00633		0.99367	2.7224	2.50	3.614584
0.003	0.00533		0.99467	2.7738	2.56	3.655455
0.004	0.00433		0.99567	2.8416	2.63	3.770240
0.005	0.00333		0.99667	2.9249	2.72	3.745235
0.006	0.00233		0.99767	3.0368	2.85	3.585843
0.007	0.00133		0.99867	3.2058	3.03	3.666800
0.008	0.00033		0.99967	3.5900	3.51	3.671823
0.00813	0.00020		0.99980	3.7050	3.69	3.720061
0.00819	0.00014		0.99986	3.8000	3.81	3.790026

Now, what is required to be seen as to what happens when the value of Part 1 is allowed to gradually increase from 0 to 0.00833 and how the value of c changes consequently. Graphical representation of the same is given in Fig. 7.1.

From the experimental result it is evident that the value of c is not unique, but can assume infinite values between 2.6217 to 3.80, as the value of probability component of Part 1 increases with corresponding decrease in the said value of Part 2 of the modified Rizopoulos probabilistic equation to be equal to 0.00833. It is thus clear that while the value of Part 1 may be set to zero, but that of Part 2 cannot be reduced below 0.00014, with corresponding maximum value of Part 1 not greater than 0.00819.

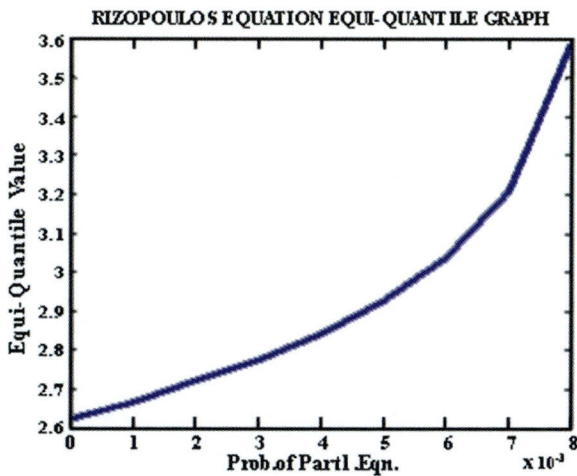

Fig. 7.1 Equi-quantile graph of computational experiment

7.13.3 *Generalization*

It is easy to visualize that such types of problem can be generalized for different values of correlations and levels of probabilities. But, the results of consequential experiments shall depend on validity test of the proposed problem, as had to be done in the present problem.

7.13.4 *Reliability and Validity Tests of BIVNOR for Extreme Paradigms*

The results of such a computational experiment to solve Rizopoulo's problem bear a witness that the software is capable of yielding a valid but approximate solutions, even for extreme probabilistic problems but only after testing their validity.

7.14 Further Scope for Applications of Biquantiles

These examples signify the applicability of biquantile pairs even for a dynamic scenario, of course, limited to a single step taken in forward or in backward direction at a time, given the values of auto-correlation and cross-auto-correlations.

Further, if pairs of such series are long and the step length is increased, even then if the same auto-correlation or cross-auto-correlations are seen to exist, the results obtained for the above bivariate stochastic processes may still hold good. For example, suppose there exist two related series of investment portfolios, one cannot only obtain JCI's as above, but even for crossover from one to the other portfolio, and then from such other to the next forming single-step dynamic; but, stochastic chain with the help of changing pair values of quantiles corresponding to the changes in values of auto-correlation/auto-cross-correlations. Therefore, the above technique offers a scope to provide condensed confidence cover, to Kalman and Bucy (1961), Kalman (1969) filterates (predictions), which are based on recursively advancing covariances or crossover covariances.

Fixed-lead prediction
It may be recalled that Ludeman's (2010) fixed-lead prediction (sliding window), where the lead period is fixed and, further if such fixation, be kept only for single-neighbouring slot at a time. It may advance recursively because auto-correlation or auto-cross-correlation between only two variables (between the two neighbouring lags), both variable values be assumed to follow bivariate Gaussian law. It has been possible to obtain biquantile pair or equi-quantile value, and hence to get shrunken/ condensed confidence cover for in-between neighbouring slots, which may also

7.14 Further Scope for Applications of Biquantiles

shift as the predicted value slides from one slot to another recursively. In fact, biquantile pairs offer the scope for providing condensed confidence covers, right from Kalman's (op. cit.) recursive filter cum prediction to Ludeman's above-mentioned fixed-lead prediction; rather, it would be capable of providing such a shifting condensed confidence covers to majority of the propositions of Ludeman (op. cit.) regarding random processes, obeying bivariate white noise, but up to a single step of movement (forward or backward), at a time, but may push to next step recursively.

Correlated Gaussian wavelet
It may also be possible to look forward for making use of such biquantile or equi-quantile value for obtaining such a condensed confidence cover to Percival and Waldens' (2000) "Correlated Gaussian Wavelet Coefficients". The importance of wavelet is many-fold as of today, it seems to represent mathematics and statistics not only of life and livings but also of economics and psychology, apart from oceanic waves to tremors of earthquakes. Such a condensed cover would, of course, be a subject to such conditions, as has been stated for others that precede. In fact, wavelets represent graphics for life and livings and its dampening towards the end of them.

Macro-molecules Condensation and Equi-quantiles
It is interesting to note the similarity between Bose Einstein condensation (correlation)[40] of particle physics with molecular bonding forces (molecular valence) which are responsible for the condensed phases i.e. for the combination of single molecules (see Holzmuller (1984)) the essentials for carrier of information from generation to generation is living system. Such phenomenon commonly disclose high positive pairwise correlation between molecules.

Further Shamir and Sharan's (2004) also explains algorithmic approaches to clustering gene expression data: where also consideration of high positive correlation between intra-cluster macro-molecules are required to play significant role in such phenomenon. Thus these appear scope for application of biquantile pairs or equi-quantile value in building condensed confidence interval if magnitude of correlation be determinable.

7.15 Information Gain and Learning[41]

7.15.1 Introduction to the Problem

It has been stated that accumulatively advancing plexus of stimuli (including exposures of events/occasions/instances, coming in successions) results with

---

[40]As discussed in Sect. 7.12.
[41]An experiment conducted in collaboration with Rajiv Ranjan.

reasonably high probability of information acquisition.[42] It cannot be deterministic phenomena for more than one reasons:

1. Efficiency and capability of mentor in respect of the manner, capacity of expression and the sequence in which the stimuli are exposed to the subject
2. Degree of simplicity or of complexity of plexus of stimuli for exposition
3. Level of prerequisite knowledge, capability and of concentration of the subject (receptor), to whom such plexus of stimuli are exposed
   Such reasons, though identifiable, are neither mutually exclusive nor independent, and together are not even exhaustive. It is for such reason that:

   (a) Mentor should assess his subject to whom proposed plexus of stimuli is to be exposed, as well as content of each such stimulus and their sequence of exposition be determined, so as to gain of information is stochastically maximum,
   while on the other hand,
   (b) The subject, who may be supposed is the decision-maker, should prepare with prerequisite knowledge, which in all probability (stochastically) make himself prepared to gain the maximum of information content of plexus of stimuli, which are likely to be explained by his mentor.
   (c) Happening of the above two aspects collectively has at a stochastically reasonably small probability, when some real-life events occur or happen in succession or simultaneously, and the subject is required to accept them as stimuli for information, learning and keep whatever learnt in practice.

Further, it is well realized that the human mind does not work as a parallel system. Rather, it is, by nature or by training, customized to acquire every element of information coming as plexus of stimulus in sequence. Even though several stimuli may be exposed to the subject at a time, but their cognizance almost always (reasonably high probability) come in succession, one by one. It is just like customer coming in a queue discipline, one by one, in a small interval of time. Such an acquired information is taken as initial milestone of the pathway of personality development, for a decision process. This is just like those of any professional or a practitioner.[43] It has been explained that there exist some more such milestones (sojourn states, similar to vertices of directed graph), to reach the said ultimate state. But, unfortunately, no investigation report appears to have been published for every psycho-states of such system kinetics.

However, it was committed to present author's work along with Das et al. (1975), on empirical findings about the information gain in a farmer's fair.

---

[42] As discussed in Sect. 1.2 (in Chap. 1).
[43] As discussed in Sect. 1.2 (Chap. 1).

7.15.2 Methodology

A systematic random sample of only 14 farmers was earmarked who got registered to visit the fair. They were interviewed through preprepared schedule containing subject matter and items of exhibits at each stalls of the said fair before their visit to well-enclosed fair. They were requested to answer as binary response (Yes or No) to questions, called *test batteries* Colman (op. cit.), about their awareness and even knowledge of exhibits to come across at the stalls of the fair. Thereafter, they were allowed to move freely into the fair with instruction that they should try to visit as much stalls as possible. They were free to make inquiries, collect leaflets, handouts and brochures about the exhibits at stalls, read them and ask questions to stall owners. After they came out of the fair, they were again exposed to the same set of schedules.

The relative frequencies, i.e. estimates of probabilities were computed by the undernoted method: X_0 and X_1 were denoted as probabilities of information about the item in question, being not known and known, by the farmer prior to his visit to the fair, respectively. Similarly, Y_0 and Y_1 stood for probabilities of information about the item in question being not known and known by the farmer after his visit to the fair, respectively. The said probabilities were computed as below for every item and all stalls of the fair.

$$\left. \begin{array}{l} X_1 = \dfrac{\text{Nr. of farmers in the sample who respond as Yes for the items before their visit to the fair}}{\text{Total nr. of farmers of the sample}} \; ; \quad X_0 = 1 - X_1 \\[1em] Y_1 = \dfrac{\text{Nr. of farmers in the sample who respond as Yes for the items after their visit to the fair}}{\text{Total nr. of farmers of the sample}} \; ; \quad Y_0 = 1 - Y_1 \end{array} \right\} \quad (7.24)$$

For measurement of information gain (IG), undernoted formula due to Thiel (1967) was used:

$$IG = Y_0 \log \frac{Y_0}{X_0} + Y_1 \log \frac{Y_1}{X_1} \quad (7.25)$$

Since such a gain of information was not comparable from item to item, because of varying number of alternatives and subitems, and also for the sake of comparison of IG, for different items, the relative IG (RIG) was computed by using the Thiel's formula:

$$RIG = \frac{IG}{Y_1 \log \frac{Y_1}{X_1}} \quad (7.26)$$

where denominator, being the maximum possible IG on the item. Thus, the maximum RIG would not exceed unity.

For measurement of stallwise and of aggregate IG, which is considered as indicator of item (stimulus) effectiveness (I.E.), again the formula suggested by Thiel (op. cit.) was used, which was

$$(I.E.)j = (p_0 \log p_0 + p_1 \log p_1)j \qquad (7.27)$$

where p_1 and p_0 denote, respectively, the probabilities of learning and non-learning individual item (stimulus) at the stall by the jth farmer. Such estimates of probability p_1 for learning about the item (stimulus) was computed as below:

$$p_1 = \frac{\text{Nr. of items learnt by the } j\text{th farmer out of total items of the stall}}{\text{Total nr. of items of the stall}}; \quad p_0 = 1 - p_1 \qquad (7.28)$$

The aggregate information so assimilated from the stall was obtained by the formula:

$$H_s = \sum_{j=1}^{n} H_j \qquad (7.29)$$

where n denotes the number of farmers of the sample and responded. The estimate of information assimilation amongst the population of farmers, who visited the fair, was obtained by the ratio method, i.e.

$$H_t = \frac{N}{n} H_s \qquad (7.30)$$

N being the size of the population of farmers who visited the said fair. Finally, the estimate of aggregate information assimilated through the fair was obtained using the formula:

$$H_a = \sum_{i=1}^{N} H_t \qquad (7.31)$$

The unit of information was in bit.

7.15.3 Results

For brevity, only relevant sample tables and their parts are presented (Table 7.13):

7.15 Information Gain and Learning

Table 7.13 (IG and RIG) on rice

Rice variety (Stimulus)	IG	RIG
I.R. 8	0.2359	0.4341
I.R.70 and Kaveri (each)	0.1983	0.4584
Jaya, Padma and Bala (each)	0.2822	0.4692

Table 7.14 Stall effectiveness (for only four stalls)

Name of the stall	Information assimilated in the sample at the stalls	Estimate of information assimilation in the population of farmers
	Hs	Ht
1. Agricultural implements	8.18179	58.79
2. Soil testing	7.00000	50.26
3. Crop varieties	6.54764	40.01
4. Fertilizers	5.26357	37.79
...	...	...
...	...	...
Ha		231.91

The total expenditure, for preparation of exhibits and stalls, was then approximately INR 550.00 only, and the estimate of total information assimilated by farmers in that fair was 231.91 bits. Thus, the estimated expenditure per bit of information assimilation then worked out to be INR 2.37 only (Table 7.14).

Thus, the study opened new vista on methodological aspect on quantitative measure of IG and of learning by the farmers, believably contributing to personality development of farmers for their purported action, for adopting improved farming technologies.

Such an effort for each stage of learning for personality development by a multi-disciplinary team of applied psychologists, economists, extension agencies, technical experts and, last but not the least, statisticians.[44]

References

Armstrong, R.D., Balintfy, J.L.: A chance-constrained multiple choice programming algorithm. Oper. Res. **23**, 494–510 (1975)

Banks, J., Carson, J.S., Nelson, B.L., Nicol, D.M.: Discrete event system simulation, 4th edn. Dorling Kindersley (India) Pvt. Ltd., licences of Pearson Education in South Asia (2008)

---

[44]As envisaged in Sect. 1.2 (Chap. 1).

Bhote, K.R.: The Ultimate Six-Sigma Beyond Quality Excellence to Total Business Excellence. PHI Learning, New Delhi (2007)

Box, G.E.P., Jenkins, G.M., Reinsel, G.C.: Time Series Analysis: Forecasting and Control, 3rd edn. Pearson Education, New Delhi (2004)

Cherubini, U., Luciano, E., Vecchiato, W.: Copula Methods in Finance. Wiley, New Jersey (2004)

Colman, A.M.: Oxford Dictionary of Psychology, 3rd edn. Oxford University Press, New Delhi (2009)

Cramer, H.: Mathematical Methods of Statistics. Princeton University Press, Princeton (1951)

Daniel, W.W.: Biostatistics: Basic Concept and Methodology for Health Sciences, 9th ed. International Student Version. Wiley Student Edition. Wiley India, New Delhi (2013)

Das, N.C., Sinha, H.S.P., Ranjan, R.: Information gain in Kishan Mela. Indian J. Extension Educ. **11**(1–2), 19–24 (1975)

Deak, I.: Subroutine for computing normal probabilities of sets: computer experiences. Ann. Oper. Res. **100**, 103–122 (2000)

Demarta, S., McNeil, A.J.: The t-Copula and Related Copulas. Department of Mathematics, ETH Zentrum, Zurich (2005)

Essenwagner, O.: Applied Statistics in Atmospheric Science, Part A. Frequencies and Curve Fitting. Elsevier, New York (1976)

Genz, A., Brez, F., Miwa, T., Mi, X., Leisch, F., Scheipl, F., Hothorn, T.: With Maintainer, Torsten Hothorn. https://Torsten-Holthorn@R-project.org (2012)

Gose, E., Johnsonbaugh, R., Jost, S.: Pattern Recognition and Image Analysis. PHI Learning, New Delhi (2003)

Gross, D., Harris, C.M.: Fundamentals of Queuing Theory, 3rd edn. Wiley India, New Delhi (2010)

Gupta, S.S.: Probability integrals of multivariate normal and multivariate t. Ann. Math. Stat. **34**, 792–828 (1963)

Hayes, R., Moulton, L.: Cluster Randomized Trials. CRC Press, New York (2009)

Haykin, S.: Neural Networks: A Comprehensive Foundation. Pearson Education, New Delhi (2001)

Holzmuller, W.: Information in biological systems: the role of macromolecules. Cambridge University Press, Cambridge, London, New York (1984)

Huang, K.: Statistical Mechanics, 2nd edn. Willy India, New Delhi (2012)

Hull, J.C.: Risk Management and Financial Institutions. Pearson Education, New Delhi (2007)

Hull, J.C.: Options, Futures and other Derivatives. Pearson Education, New Delhi (2009)

Indian Agricultural Statistics Research Institute: Agricultural Research Data Book (2007). Indian Agricultural Statistics Research Institute, New Delhi (2007)

Johnson, R.A., Wichern, D.W.: Applied Multivariate Statistical Analysis. PHI Learning, New Delhi (2003)

Kall, P.: Stochastic Linear Programming. Springer, Berlin (1974)

Kalman, R.E.: A new approach to linear filtering and prediction Problem. ASME J. Basic Eng. **91**, 35 (1969)

Kalman, R.E., Bucy, R.S.: New result in linear filtering and prediction theory. Trans. ASME J. Basic Eng. **83**, 95 (1961)

Kempthorne, O.: Design and Analysis of Experiments. John Wiley and Sons, New Jersey (1952)

Kramers, H.C.: Some properties of liquid helium below 1° K. Dissertation, Leiden (1955)

Lo, S.M.S., Wilke, R.A.: A copula model for dependent competing risks. J. R. Stat. Soc. Series C. (Appl. Stat.) **59**(2), 359–376 (2010)

Ludeman, L.C.: Random Processes, Filtering, Estimation and Detection. Wiley India, New Delhi (2010)

Luger, G.F.: Artificial Intelligence: Structure and Strategies for Complex Problem Solving. Pearson Education, New Delhi (2001)

Marshall, A.W., Olkin, I.: Inequalities: Theory of Majorization and its Applications. Academic Press, New York (1979)

References

Mitra, A.: Fundamentals of Quality Control and Improvement, 2nd edn. Pearson Education, New Delhi (2001)

Morgan, J.P.: The 1996 Amendment of Basel Committee (but Implemented in 1998): See J.C. Hall (2007), Chap. 8

Nelson, R.B.: An Introduction to Copulas, 2nd edn. Springer, Berlin (2006)

Padhy, N.P.: Artificial Intelligence System and Intelligent System. Oxford University Press, New Delhi (2005)

Pal, P.K., Kishtawal, C.M., Agrawal, N.: Multifeature classification based rainfall estimates by using visible infrared TRMM data. MAUSAM **54**, 67–74 (2003)

Pearl, J.: A probabilistic calculus in action in artificial intelligence-10. de Mantras, R.L., Poole, D. (eds.) Uncertainty in Artificial Intelligence, pp. 454–462. Morgan Kaufmann, San Matco (1994)

Pearl, J.: Graphical models for probablistic and causal reasoning. Tucker, A.N. (ed.) Computer Science Handbook, Chap. 70, pp. 70–18. CRC Press, Boca Raton (2004)

Percival, D.B., Walden, S.: Wavelet Methods for Time Series Analysis. Cambridge University Press, Cambridge (2000)

Polya, G.: How to Solve I. Princeton University Press, Princeton (1945)

Rao, C.R.: Linear Statistical Inference and its Applications, 2nd ed. Wiley, New Jersey (1973, 2006)

Rejda, G.: Principles of Risk Management and Insurance. Pearson Education, New Delhi (2006)

Rizopoulos, D.: Quantiles for bivariate normal distribution. https://www.r-project.org/posting-guide.html (2009)

Rosenblatt, F.: The perceptron: A probabilistic model of information storage and organization in the brain. Psychol. Rev. **65**, 386–408 (1958)

Russel, S.J., Norvig, P.: Artificial Intelligence: A Modern Approach. Pearson Education, New Delhi (2002)

Shamir, R., Sharan, R.: Algorithmic approaches to clustering gene expression data. In: Jiang, T., Xu, Y., Zhang, M.Q. (eds.) Current Topics in Computational Molecular Biology. Ane Books, New Delhi, India (2004)

Theil, H.: Economics and Information Theory. North Holland Publishing Company, Amsterdam (1967)

Vasicek, O.A.: An equilibrium characterization of the term structure. J. Finan. Econ. **5**, 177–188 (1977)

Vasicek, O.A.: Load portfolio value. Risk (2002)

Wikipedia: Bose–Einstein statistics, Higgs–Boson, and Bose-Einstein correlation (2012; 20th July)

Winston, W.L.: Introduction to Probability Models, Operations Research, vol. 2, 4th edn. Thompson Learning, Singapore (2004)

Chapter 8
Generated Tables by BIVNOR

8.1 Table of BIGH the Equi-Quantile Values

See Table 8.1.

Table 8.1 Bivariate normal equi-quantile values (BIGH) at which (SMH = SMK) and of (SMHL) the lowest tangible values of (h or k) for different probability levels (PROB.) and correlations (RHO.)

Sheet No. 1: For correlations between 0.95 and +0.4

PROB.	RHO.							
	0.95	0.9	0.8	0.7	0.6	0.5	0.4
0.99	2.4346	2.4694	2.5088	2.5324	2.548	2.5529	2.5648
	2.32	2.33	2.33	2.33	2.33	2.33	2.33
0.98	2.1641	2.2007	2.244	2.2702	2.288	2.3003	2.3093
	2.06	2.06	2.06	2.06	2.06	2.06	2.06
0.97	1.9925	2.0302	2.0753	2.1038	2.1235	2.1376	2.1481
	1.89	1.89	1.89	1.89	1.89	1.89	1.89
0.96	1.8632	1.9018	1.9488	1.9789	2	2.0155	2.0271
	1.76	1.76	1.76	1.76	1.76	1.76	1.76
0.95	1.7582	1.7976	1.846	1.8785	1.8996	1.9159	1.929
	1.65	1.65	1.65	1.65	1.65	1.65	1.65
0.94	1.6687	1.7089	1.7585	1.7908	1.8143	1.832	1.8455
	1.56	1.56	1.56	1.56	1.56	1.56	1.56
0.93	1.5903	1.631	1.6817	1.7152	1.7396	1.7582	1.7726
	1.48	1.48	1.48	1.48	1.48	1.48	1.48
0.92	1.5201	1.5613	1.613	1.6474	1.6728	1.6922	1.7074
	1.41	1.41	1.41	1.41	1.41	1.41	1.41

(continued)

Table 8.1 (continued)

Sheet No. 1: For correlations between 0.95 and +0.4

PROB.	RHO.						
	0.95	0.9	0.8	0.7	0.6	0.5	0.4
0.91	1.4563	1.4979	1.5506	1.5859	1.612	1.6322	1.6481
	1.35	1.35	1.35	1.35	1.35	1.35	1.35
0.9025	1.4128	1.4538	1.5071	1.5429	1.5696	1.5903	1.6068
	1.3	1.3	1.3	1.3	1.3	1.3	1.3
0.9	1.3976	1.4396	1.4931	1.5291	1.5561	1.577	1.5936
	1.29	1.29	1.29	1.29	1.29	1.29	1.29
0.85	1.1544	1.1982	1.2552	1.2947	1.3246	1.349	1.3688
	1.04	1.04	1.04	1.04	1.04	1.04	1.04
0.8	0.961	1.0065	1.0662	1.1085	1.1415	1.1684	1.1908
	0.85	0.85	0.85	0.85	0.85	0.85	0.85
0.75	0.795	0.8418	0.9043	0.949	0.9845	1.0138	1.0387
	0.69	0.68	0.68	0.68	0.68	0.68	0.68
0.7	0.6463	0.6942	0.7589	0.8059	0.8437	0.8754	0.9025
	0.55	0.54	0.53	0.53	0.53	0.53	0.53
0.65	0.5083	0.5572	0.6242	0.6734	0.7133	0.7473	0.7767
	0.43	0.42	0.39	0.39	0.39	0.39	0.39
0.6	0.3774	0.4274	0.4962	0.5478	0.5898	0.6259	0.6575
	0.31	0.31	0.27	0.26	0.26	0.26	0.26
0.55	0.2505	0.3016	0.3727	0.4262	0.4704	0.5087	0.5425
	0.21	0.21	0.18	0.13	0.13	0.13	0.13
0.5	0.1259	0.178	0.2511	0.3067	0.3531	0.3935	0.4296
	0.12	0.12	0.11	0.01	0.01	0.01	0.01

Sheet No. 2: For correlations between +0.3 and −0.3

PROB.	RHO.						
	0.3	0.2	0.1	0	−0.1	−0.2	−0.3
0.99	2.569	2.5722	2.574	2.5751	2.5757	2.5758	2.5758
	2.33	2.33	2.33	2.34	2.34	2.34	2.34
0.98	2.3157	2.3198	2.3228	2.3244	2.3255	2.3361	2.3261
	2.06	2.06	2.06	2.06	2.06	2.06	2.06
0.97	2.1555	2.161	2.1646	2.1672	2.1687	2.1695	2.1698
	1.89	1.89	1.89	1.89	1.89	1.89	1.89
0.96	2.0357	2.0419	2.0464	2.0496	2.0516	2.0528	2.0533
	1.76	1.76	1.76	1.76	1.76	1.76	1.76
0.95	1.9384	1.9456	1.9507	1.9546	1.957	1.9586	1.9594
	1.65	1.65	1.65	1.65	1.65	1.65	1.65
0.94	1.856	1.8639	1.8698	1.8741	1.8769	1.8789	1.88
	1.56	1.56	1.56	0.156	1.56	1.56	1.56
0.93	1.7838	1.7924	1.799	1.8037	1.8072	1.8094	1.8108
	1.48	1.48	1.48	1.48	1.48	1.48	1.48

(continued)

8.1 Table of BIGH the Equi-Quantile Values

Table 8.1 (continued)

Sheet No. 2: For correlations between +0.3 and −0.3

PROB.	RHO.						
	0.3	0.2	0.1	0	−0.1	−0.2	−0.3
0.92	1.7193	1.7285	1.7358	1.7411	1.7449	1.7476	1.7492
	1.41	1.41	1.41	1.41	1.41	1.41	1.41
0.91	1.6607	1.6707	1.6784	1.6844	1.6886	1.6916	1.6935
	1.35	1.35	1.35	1.35	1.35	1.35	1.35
0.9025	1.62	1.6304	1.6386	1.645	1.6495	1.6528	1.655
	1.3	1.3	1.3	1.3	1.3	1.3	1.3
0.9	1.6069	1.6175	1.6258	1.6322	1.637	1.6404	1.6426
	1.29	1.29	1.29	1.29	1.29	1.29	1.29
0.85	1.3851	1.3985	1.4094	1.4184	1.4254	1.4307	1.4346
	1.04	1.04	1.04	1.04	1.04	1.04	1.04
0.8	1.2097	1.2257	1.2392	1.2504	1.2597	1.267	1.2727
	0.85	0.85	0.85	0.85	0.85	0.85	0.85
0.75	1.06	1.0785	1.0943	1.1078	1.1193	1.1288	1.1364
	0.68	0.68	0.68	0.68	0.68	0.68	0.68
0.7	0.9262	0.947	0.965	0.9808	0.9946	1.0062	1.0159
	0.53	0.53	0.53	0.53	0.53	0.53	0.53
0.65	0.8026	0.5255	0.846	0.8641	0.8801	0.894	0.9062
	0.39	0.39	0.39	0.39	0.39	0.39	0.39
0.6	0.6857	0.7109	0.7336	0.7541	0.7724	0.7887	0.8032
	0.26	0.26	0.26	0.26	0.26	0.26	0.26
0.55	0.5729	0.6005	0.6255	0.6484	0.6692	0.688	0.7048
	0.13	0.13	0.13	0.13	0.13	0.13	0.13
0.5	0.4622	0.4921	0.5196	0.545	0.5683	0.5898	0.6093
	0.01	0.01	0.01	0.01	0.01	0.01	0.01

Sheet No. 3: For correlations between −0.4 and −0.95

PROB.	RHO.						
	−0.4	−0.5	−0.6	−0.7	−0.8	−0.9	−0.95
0.99	2.5759	2.5761	2.5761	2.5761	2.5758	2.4534	2.5757
	2.34	2.34	2.34	2.34	2.33	1.79	2.34
0.98	2.3262	2.3264	2.3265	2.3265	2.3265	2.3155	2.3265
	2.06	2.06	2.06	2.06	2.06	1.75	2.07
0.97	2.1701	2.1702	2.1702	2.1701	2.1701	2.1694	2.1702
	1.88	1.89	1.89	1.89	1.89	1.69	1.89
0.96	2.0537	2.0538	2.0538	2.0538	2.0538	2.0537	2.0538
	1.76	1.76	1.76	1.76	1.76	1.63	1.78
0.95	1.9597	1.9599	1.96	1.96	1.9599	1.96	1.9599
	1.65	1.65	1.65	1.65	1.65	1.57	1.68
0.94	1.8805	1.8807	1.8808	1.8808	1.8808	1.8808	1.8808
	1.56	1.56	1.56	1.56	1.56	1.51	1.59

(continued)

Table 8.1 (continued)

Sheet No. 3: For correlations between −0.4 and −0.95

PROB.	RHO.						
	−0.4	−0.5	−0.6	−0.7	−0.8	−0.9	−0.95
0.93	1.8115	1.8118	1.8119	1.8119	1.8119	1.8119	1.8119
	1.48	1.48	1.48	1.48	1.48	1.45	1.53
0.92	1.7501	1.7505	1.7507	1.7507	1.7507	1.7507	1.7507
	1.41	1.41	1.41	1.41	1.41	1.39	1.46
0.91	1.6947	1.6952	1.6954	1.6954	1.6954	1.6954	1.6954
	1.35	1.34	1.35	1.35	1.35	1.33	1.41
0.9025	1.6562	1.6568	1.657	1.6571	1.6571	1.6571	1.6571
	1.3	1.3	1.3	1.3	1.3	1.29	1.37
0.9	1.6439	1.6447	1.6448	1.6448	1.645	1.645	1.645
	1.29	1.29	1.29	1.29	1.29	1.28	1.36
0.85	1.4371	1.4386	1.4393	1.4395	1.4395	1.4395	1.4395
	1.04	1.04	1.04	1.04	1.04	1.04	0.51
0.8	1.2768	1.2794	1.2809	1.2815	1.2815	1.2815	1.2815
	0.85	0.85	0.85	0.85	0.85	0.85	0.47
0.75	1.1421	1.1463	1.1488	1.15	1.1504	1.1504	1.1504
	0.68	0.68	0.68	0.68	0.68	0.68	0.39
0.7	1.0237	1.0296	1.0335	1.0357	1.0363	1.0364	1.0364
	0.53	0.53	0.53	0.53	0.53	0.53	0.32
0.65	0.9159	0.9238	0.9294	0.9329	0.9343	0.9345	0.9346
	0.39	0.39	0.39	0.39	0.39	0.39	0.26
0.6	0.8153	0.8255	0.8332	0.8385	0.841	0.8416	0.8416
	0.26	0.26	0.26	0.26	0.26	0.26	0.17
0.55	0.7196	0.7323	0.7425	0.7499	0.754	0.7553	0.7554
	0.13	0.13	0.13	0.13	0.13	0.13	0.08
0.5	0.6269	0.6424	0.6554	0.6655	0.672	0.6744	0.6745
	0.01	0.01	0.01	0.01	0.01	0.01	0.01

BIGH is printed as upper value and SMHL as lower value
For description of Generated Table 8.1 consult Sect. 5.4.1 of Chap. 5

8.2 Contents for Table 8.2 (Table No. 8.2-1 to 400)

Contents of Table 8.2: Series of generated tables for biquantile pairs for differing values of correlations and probability (risk) levels.

The following tables are generated for bivariate normal iso-probable quantile pairs.

8.2 Contents for Table 8.2 (Table No. 8.2-1 to 400)

Table No.	PROB.	RHO	Page no.
8.2-1	0.99	0.95	207
8.2-2	0.98		207
8.2-3	0.97		207
8.2-4	0.96		208
8.2-5	0.95		208
8.2-6	0.94		209
8.2-7	0.93		209
8.2-8	0.92		210
8.2-9	0.91		210
8.2-10	0.9025		211
8.2-11	0.9		211
8.2-12	0.85		212
8.2-13	0.8		212
8.2-14	0.75		213
8.2-15	0.7		213
8.2-16	0.65		214
8.2-17	0.6		214
8.2-18	0.55		214
8.2-19	0.5		215
8.2-20	0.99	0.9	215
8.2-21	0.98		216
8.2-22	0.97		216
8.2-23	0.96		217
8.2-24	0.95		217
8.2-25	0.94		218
8.2-26	0.93		218
8.2-27	0.92		219
8.2-28	0.91		220
8.2-29	0.9025		220
8.2-30	0.9		221
8.2-31	0.85		221
8.2-32	0.8		222
8.2-33	0.75		223
8.2-34	0.7		223
8.2-35	0.65		224
8.2-36	0.6		224
8.2-37	0.55		225
8.2-38	0.5		225
8.2-39	0.99	0.8	226
8.2-40	0.98		226
8.2-41	0.97		227

(continued)

Table No.	PROB.	RHO	Page no.
8.2-42	0.96		228
8.2-43	0.95		228
8.2-44	0.94		229
8.2-45	0.93		230
8.2-46	0.92		231
8.2-47	0.91		231
8.2-48	0.9025		232
8.2-49	0.9		233
8.2-50	0.85		233
8.2-51	0.8		234
8.2-52	0.75		235
8.2-53	0.7		236
8.2-54	0.65		237
8.2-55	0.6		237
8.2-56	0.55		238
8.2-57	0.5		239
8.2-58	0.99	0.7	239
8.2-59	0.98		240
8.2-60	0.97		241
8.2-61	0.96		242
8.2-62	0.95		242
8.2-63	0.94		243
8.2-64	0.93		244
8.2-65	0.92		245
8.2-66	0.91		246
8.2-67	0.9025		246
8.2-68	0.9		247
8.2-69	0.85		248
8.2-70	0.8		249
8.2-71	0.75		250
8.2-72	0.7		251
8.2-73	0.65		252
8.2-74	0.6		253
8.2-75	0.55		254
8.2-76	0.5		255
8.2-77	0.99	0.6	256
8.2-78	0.98		257
8.2-79	0.97		257
8.2-80	0.96		258
8.2-81	0.95		259
8.2-82	0.94		260

(continued)

8.2 Contents for Table 8.2 (Table No. 8.2-1 to 400)

Table No.	PROB.	RHO	Page no.
8.2-83	0.93		261
8.2-84	0.92		262
8.2-85	0.91		263
8.2-86	0.9025		264
8.2-87	0.9		265
8.2-88	0.85		266
8.2-89	0.8		267
8.2-90	0.75		268
8.2-91	0.7		269
8.2-92	0.65		270
8.2-93	0.6		271
8.2-94	0.55		272
8.2-95	0.5		273
8.2-96	0.99	0.5	274
8.2-97	0.98		275
8.2-98	0.975		276
8.2-99	0.97		277
8.2-100	0.96		277
8.2-101	0.95		278
8.2-102	0.94		279
8.2-103	0.93		280
8.2-104	0.92		281
8.2-105	0.91		282
8.2-106	0.9025		283
8.2-107	0.9		284
8.2-108	0.85		285
8.2-109	0.8		286
8.2-110	0.75		287
8.2-111	0.7		288
8.2-112	0.65		289
8.2-113	0.6		290
8.2-114	0.55		291
8.2-115	0.5		292
8.2-116	0.99	0.4	293
8.2-117	0.98		294
8.2-118	0.97		295
8.2-119	0.96		296
8.2-120	0.95		297
8.2-121	0.94		298
8.2-122	0.93		299
8.2-123	0.92		300

(continued)

Table No.	PROB.	RHO	Page no.
8.2-124	0.91		301
8.2-125	0.9025		302
8.2-126	0.9		303
8.2-127	0.85		304
8.2-128	0.8		305
8.2-129	0.75		306
8.2-130	0.7		307
8.2-131	0.65		308
8.2-132	0.6		309
8.2-133	0.55		310
8.2-134	0.5		311
8.2-135	0.99	0.3	313
8.2-136	0.98		314
8.2-137	0.97		315
8.2-138	0.96		316
8.2-139	0.95		317
8.2-140	0.94		318
8.2-141	0.93		319
8.2-142	0.92		320
8.2-143	0.91		321
8.2-144	0.9025		322
8.2-145	0.9		323
8.2-146	0.85		324
8.2-147	0.8		325
8.2-148	0.75		326
8.2-149	0.7		327
8.2-150	0.65		328
8.2-151	0.6		329
8.2-152	0.55		331
8.2-153	0.5		332
8.2-154	0.99	0.2	333
8.2-155	0.98		334
8.2-156	0.97		335
8.2-157	0.96		336
8.2-158	0.95		337
8.2-159	0.94		338
8.2-160	0.93		339
8.2-161	0.92		340
8.2-162	0.91		341
8.2-163	0.9025		342
8.2-164	0.9		343

(continued)

8.2 Contents for Table 8.2 (Table No. 8.2-1 to 400)

Table No.	PROB.	RHO	Page no.
8.2-165	0.85		344
8.2-166	0.8		345
8.2-167	0.75		346
8.2-168	0.7		347
8.2-169	0.65		348
8.2-170	0.6		350
8.2-171	0.55		351
8.2-172	0.5		352
8.2-173	0.99	0.1	354
8.2-174	0.98		355
8.2-175	0.97		356
8.2-176	0.96		357
8.2-177	0.95		358
8.2-178	0.94		359
8.2-179	0.93		360
8.2-180	0.92		361
8.2-181	0.91		362
8.2-182	0.9025		363
8.2-183	0.9		364
8.2-184	0.85		365
8.2-185	0.8		366
8.2-186	0.75		367
8.2-187	0.7		368
8.2-188	0.65		369
8.2-189	0.6		371
8.2-190	0.55		372
8.2-191	0.5		373
8.2-192	0.99	0	375
8.2-193	0.98		376
8.2-194	0.97		377
8.2-195	0.96		378
8.2-196	0.95		379
8.2-197	0.94		380
8.2-198	0.93		381
8.2-199	0.92		382
8.2-200	0.91		383
8.2-201	0.9025		384
8.2-202	0.9		385
8.2-203	0.85		386
8.2-204	0.8		387
8.2-205	0.75		388

(continued)

Table No.	PROB.	RHO	Page no.
8.2-206	0.7		389
8.2-207	0.65		391
8.2-208	0.6		392
8.2-209	0.55		393
8.2-210	0.5		395
8.2-211	0.99	−0.1	396
8.2-212	0.98		397
8.2-213	0.97		398
8.2-214	0.96		399
8.2-215	0.95		400
8.2-216	0.94		401
8.2-217	0.93		402
8.2-218	0.92		403
8.2-219	0.91		404
8.2-220	0.9025		405
8.2-221	0.9		406
8.2-222	0.85		407
8.2-223	0.8		408
8.2-224	0.75		409
8.2-225	0.7		410
8.2-226	0.65		412
8.2-227	0.6		413
8.2-228	0.55		415
8.2-229	0.5		416
8.2-230	0.99	−0.2	417
8.2-231	0.98		418
8.2-232	0.97		419
8.2-233	0.96		420
8.2-234	0.95		421
8.2-235	0.94		422
8.2-236	0.93		423
8.2-237	0.92		424
8.2-238	0.91		425
8.2-239	0.9025		426
8.2-240	0.9		427
8.2-241	0.85		428
8.2-242	0.8		429
8.2-243	0.75		430
8.2-244	0.7		432
8.2-245	0.65		433
8.2-246	0.6		434

(continued)

8.2 Contents for Table 8.2 (Table No. 8.2-1 to 400)

Table No.	PROB.	RHO	Page no.
8.2-247	0.55		436
8.2-248	0.5		437
8.2-249	0.99	−0.3	439
8.2-250	0.98		440
8.2-251	0.97		441
8.2-252	0.96		442
8.2-253	0.95		443
8.2-254	0.94		444
8.2-255	0.93		445
8.2-256	0.92		446
8.2-257	0.91		447
8.2-258	0.9025		448
8.2-259	0.9		449
8.2-260	0.85		450
8.2-261	0.8		451
8.2-262	0.75		452
8.2-263	0.7		454
8.2-264	0.65		455
8.2-265	0.6		456
8.2-266	0.55		458
8.2-267	0.5		459
8.2-268	0.99	−0.4	461
8.2-269	0.98		462
8.2-270	0.97		463
8.2-271	0.96		464
8.2-272	0.95		465
8.2-273	0.94		466
8.2-274	0.93		467
8.2-275	0.92		468
8.2-276	0.91		469
8.2-277	0.9025		470
8.2-278	0.9		471
8.2-279	0.85		472
8.2-280	0.8		473
8.2-281	0.75		474
8.2-282	0.7		476
8.2-283	0.65		477
8.2-284	0.6		479
8.2-285	0.55		480
8.2-286	0.5		482
8.2-287	0.99	−0.5	483

(continued)

Table No.	PROB.	RHO	Page no.
8.2-288	0.98		484
8.2-289	0.97		485
8.2-290	0.96		486
8.2-291	0.95		487
8.2-292	0.94		488
8.2-293	0.93		489
8.2-294	0.92		490
8.2-295	0.91		491
8.2-296	0.9025		492
8.2-297	0.9		493
8.2-298	0.85		494
8.2-299	0.8		495
8.2-300	0.75		496
8.2-301	0.7		498
8.2-302	0.65		499
8.2-303	0.6		501
8.2-304	0.55		502
8.2-305	0.5		504
8.2-306	0.99	-0.6	505
8.2-307	0.98		506
8.2-308	0.97		507
8.2-309	0.96		508
8.2-310	0.95		509
8.2-311	0.94		510
8.2-312	0.93		511
8.2-313	0.92		512
8.2-314	0.91		513
8.2-315	0.9025		514
8.2-316	0.9		515
8.2-317	0.85		516
8.2-318	0.8		517
8.2-319	0.75		518
8.2-320	0.7		520
8.2-321	0.65		521
8.2-322	0.6		523
8.2-323	0.55		524
8.2-324	0.5		526
8.2-325	0.99	-0.7	527
8.2-326	0.98		528
8.2-327	0.97		529
8.2-328	0.96		530

(continued)

8.2 Contents for Table 8.2 (Table No. 8.2-1 to 400)

Table No.	PROB.	RHO	Page no.
8.2-329	0.95		531
8.2-330	0.94		532
8.2-331	0.93		533
8.2-332	0.92		534
8.2-333	0.91		535
8.2-334	0.9025		536
8.2-335	0.9		537
8.2-336	0.85		538
8.2-337	0.8		540
8.2-338	0.75		541
8.2-339	0.7		542
8.2-340	0.65		544
8.2-341	0.6		545
8.2-342	0.55		547
8.2-343	0.5		548
8.2-344	0.99	−0.8	550
8.2-345	0.98		551
8.2-346	0.97		552
8.2-347	0.96		553
8.2-348	0.95		554
8.2-349	0.94		555
8.2-350	0.93		556
8.2-351	0.92		557
8.2-352	0.91		558
8.2-353	0.9025		559
8.2-354	0.9		560
8.2-355	0.85		561
8.2-356	0.8		562
8.2-357	0.75		564
8.2-358	0.7		565
8.2-359	0.65		566
8.2-360	0.6		568
8.2-361	0.55		569
8.2-362	0.5		571
8.2-363	0.99	−0.9	572
8.2-364	0.98		574
8.2-365	0.97		575
8.2-366	0.96		577
8.2-367	0.95		578
8.2-368	0.94		579
8.2-369	0.93		580

(continued)

Table No.	PROB.	RHO	Page no.
8.2-370	0.92		581
8.2-371	0.91		583
8.2-372	0.9025		584
8.2-373	0.9		585
8.2-374	0.85		586
8.2-375	0.8		587
8.2-376	0.75		588
8.2-377	0.7		590
8.2-378	0.65		591
8.2-379	0.6		593
8.2-380	0.55		594
8.2-381	0.5		596
8.2-382	0.99	−0.95	597
8.2-383	0.98		598
8.2-384	0.97		599
8.2-385	0.96		600
8.2-386	0.95		601
8.2-387	0.94		602
8.2-388	0.93		603
8.2-389	0.92		604
8.2-390	0.91		605
8.2-391	0.9025		606
8.2-392	0.9		607
8.2-393	0.85		608
8.2-394	0.8		609
8.2-395	0.75		611
8.2-396	0.7		612
8.2-397	0.65		614
8.2-398	0.6		615
8.2-399	0.55		617
8.2-400	0.5		618

8.2 Contents for Table 8.2 (Table No. 8.2-1 to 400)

Table 8.2 N (N = 1–400) of Biquantile pairs

Parameter values					
PROB. = 0.99	RHO = 0.95		BIGH = 2.4346		SMHL = 2.33
Pair No.	SMH	SMK	PROB. level	Risk level (%)	Risk error (%)
Table No. 8.2-1					
1	2.44	2.429212	0.989998	1.000166	−0.000167
2	2.43	2.439209	0.989999	1.000100	−0.000101
3	2.42	2.450851	0.990004	0.999635	0.000364
4	2.41	2.462576	0.989991	1.000941	−0.000942
5	2.40	2.479074	0.990009	0.999141	0.000858
6	2.39	2.495658	0.990001	0.999945	0.000054
7	2.38	2.515453	0.989993	1.000696	−0.000697
8	2.37	2.541586	0.989999	1.000059	−0.000060
9	2.36	2.574058	0.990001	0.999945	0.000054
10	2.35	2.615996	0.989992	1.000798	−0.000799
11	2.34	2.683025	0.989994	1.000637	−0.000638
12	2.33	2.831396	0.989994	1.000601	−0.000602
No. of pairs		Mean COMPUT. risk errors		STD. ERR. errors	
12		−0.0002058653		0.0001420024	
Z-ratio				Status of significance	
1.44973				N.S.—not significant	
Parameter values					
PROB. = 0.98	RHO = 0.95		BIGH = 2.1641		SMHL = 2.06
Pair No.	SMH	SMK	PROB. level	Risk level (%)	Risk error (%)
Table No. 8.2-2					
1	2.17	2.158216	0.980002	1.999849	0.000149
2	2.16	2.168208	0.980004	1.999605	0.000393
3	2.15	2.179074	0.979997	2.000272	−0.000274
4	2.14	2.191596	0.979994	2.000582	−0.000584
5	2.13	2.206558	0.980002	1.999784	0.000215
6	2.12	2.223179	0.979996	2.000398	−0.000399
7	2.11	2.243024	0.979993	2.000749	−0.000751
8	2.10	2.267656	0.979997	2.000332	−0.000334
9	2.09	2.298639	0.980001	1.999927	0.000072
10	2.08	2.339100	0.979998	2.000177	−0.000179
11	2.07	2.399977	0.980004	1.999617	0.000381
12	2.06	2.510960	0.979992	2.000773	−0.000775
No. of pairs		Mean COMPUT. risk errors		STD. ERR. errors	
12		−0.0001604740		0.0001110136	
Z-ratio				Status of significance	
1.44554				N.S.—not significant	
Parameter values					
PROB. = 0.97	RHO = 0.95		BIGH = 1.9925		SMHL = 1.89
Pair No.	SMH	SMK	PROB. level	Risk level (%)	Risk error (%)
Table No. 8.2-3					
1	2.00	1.985028	0.970004	2.999628	0.000370
2	1.99	1.994612	0.970000	3.000045	−0.000048

(continued)

Table 8.2 (continued)

Parameter values					
PROB. = 0.97	RHO = 0.95		BIGH = 1.9925		SMHL = 1.89
Pair No.	SMH	SMK	PROB. level	Risk level (%)	Risk error (%)
3	1.98	2.005469	0.969998	3.000212	−0.000215
4	1.97	2.017601	0.969991	3.000885	−0.000888
5	1.96	2.031788	0.969993	3.000718	−0.000721
6	1.95	2.048426	0.970000	3.000045	−0.000048
7	1.94	2.067514	0.969995	3.000516	−0.000519
8	1.93	2.091399	0.970008	2.999175	0.000823
9	1.92	2.119300	0.969992	3.000784	−0.000787
10	1.91	2.156688	0.969992	3.000784	−0.000787
11	1.90	2.211378	0.970009	2.999115	0.000882
12	1.89	2.300559	0.969998	3.000194	−0.000197
No. of pairs		Mean COMPUT. risk errors		STD. ERR. errors	
12		−0.0001641420		0.0001630283	
Z-ratio				Status of significance	
1.00683				N.S.—Not significant	
Parameter values					
PROB. = 0.96	RHO = 0.95		BIGH = 1.8632		SMHL = 1.76
Pair No.	SMH	SMK	PROB. level	Risk level (%)	Risk error (%)
Table No. 8.2-4					
1	1.87	1.856425	0.959993	4.000670	−0.000668
2	1.86	1.866405	0.960001	3.999865	0.000137
3	1.85	1.877275	0.959996	4.000449	−0.000447
4	1.84	1.889818	0.959998	4.000163	−0.000161
5	1.83	1.904033	0.959997	4.000336	−0.000334
6	1.82	1.920706	0.960002	3.999823	0.000179
7	1.81	1.939838	0.959993	4.000676	−0.000674
8	1.80	1.963775	0.960008	3.999198	0.000805
9	1.79	1.992518	0.960002	3.999776	0.000226
10	1.78	2.029976	0.959999	4.000145	−0.000143
11	1.77	2.083182	0.959999	4.000068	−0.000066
12	1.76	2.172451	0.959993	4.000688	−0.000685
No. of pairs		Mean COMPUT. risk errors		STD. ERR. errors	
12		−0.0001407587		0.0001193745	
Z-ratio				Status of significance	
1.17914				N.S.—not significant	
Parameter values					
PROB. = 0.95	RHO = 0.95		BIGH = 1.7582		SMHL = 1.65
Pair No.	SMH	SMK	PROB. level	Risk level (%)	Risk error (%)
Table No. 8.2-5					
1	1.76	1.756402	0.950008	4.999209	0.000793
2	1.75	1.766439	0.949991	5.000920	−0.000918
3	1.74	1.778153	0.949993	5.000740	−0.000739

(continued)

8.2 Contents for Table 8.2 (Table No. 8.2-1 to 400)

Table 8.2 (continued)

Parameter values					
PROB. = 0.95	RHO = 0.95		BIGH = 1.7582		SMHL = 1.65
Pair No.	SMH	SMK	PROB. level	Risk level (%)	Risk error (%)
4	1.73	1.791547	0.949998	5.000216	−0.000215
5	1.72	1.806624	0.949991	5.000901	−0.000900
6	1.71	1.824556	0.949998	5.000228	−0.000226
7	1.70	1.845736	0.950002	4.999823	0.000179
8	1.69	1.871340	0.950002	4.999841	0.000161
9	1.68	1.903321	0.949995	5.000496	−0.000495
10	1.67	1.946371	0.949995	5.000532	−0.000530
11	1.66	2.010647	0.949995	5.000496	−0.000495
12	1.65	2.142245	0.950004	4.999644	0.000358
No. of pairs		Mean COMPUT. risk errors		STD. ERR. errors	
12		−0.0002329166		0.0001426868	
Z-ratio				Status of significance	
1.63236				N.S.—not significant	
Parameter values					
PROB. = 0.94	RHO = 0.95		BIGH = 1.6687		SMHL = 1.56
Pair No.	SMH	SMK	PROB. level	Risk level (%)	Risk error (%)
Table No. 8.2-6					
1	1.67	1.667401	0.939998	6.000197	−0.000197
2	1.66	1.677836	0.939998	6.000221	−0.000221
3	1.65	1.689565	0.939996	6.000388	−0.000387
4	1.64	1.702981	0.939999	6.000078	−0.000077
5	1.63	1.718475	0.940006	5.999375	0.000626
6	1.62	1.736052	0.939994	6.000561	−0.000560
7	1.61	1.757275	0.939998	6.000251	−0.000250
8	1.60	1.783319	0.940009	5.999118	0.000882
9	1.59	1.815358	0.939999	6.000102	−0.000101
10	1.58	1.858473	0.939997	6.000269	−0.000268
11	1.57	1.923605	0.940008	5.999244	0.000757
12	1.56	2.053724	0.940006	5.999428	0.000572
No. of pairs		Mean COMPUT. risk errors		STD. ERR. errors	
12		0.0000596046		0.0001321992	
Z-ratio				Status of significance	
0.45087				N.S.—not significant	
Parameter values					
PROB. = 0.93	RHO = 0.95		BIGH = 1.5903		SMHL = 1.48
Pair No.	SMH	SMK	PROB. level	Risk level (%)	Risk error (%)
Table No. 8.2-7					
1	1.60	1.581049	0.929993	7.000703	−0.000703
2	1.59	1.590600	0.929997	7.000345	−0.000346
3	1.58	1.601253	0.929998	7.000202	−0.000203
4	1.57	1.613206	0.929996	7.000375	−0.000376
5	1.56	1.627048	0.930009	6.999147	0.000852

(continued)

Table 8.2 (continued)

Parameter values					
PROB. = 0.93	RHO = 0.95		BIGH = 1.5903		SMHL = 1.48
Pair No.	SMH	SMK	PROB. level	Risk level (%)	Risk error (%)
6	1.55	1.642585	0.930002	6.999815	0.000185
7	1.54	1.660993	0.930009	6.999076	0.000924
8	1.53	1.682664	0.930009	6.999130	0.000870
9	1.52	1.708773	0.929998	7.000166	−0.000167
10	1.51	1.742058	0.929996	7.000363	−0.000364
11	1.50	1.786817	0.929994	7.000566	−0.000566
12	1.49	1.855164	0.930001	6.999904	0.000095
13	1.48	2.002570	0.930006	6.999379	0.000620
No. of pairs		Mean COMPUT. risk errors		STD. ERR. errors	
13		0.0000587532		0.0001481285	
Z-ratio				Status of significance	
0.39664				N.S.—not significant	
Parameter values					
PROB. = 0.92	RHO = 0.95		BIGH = 1.5201		SMHL = 1.41
Pair No.	SMH	SMK	PROB. level	Risk level (%)	Risk error (%)
Table No. 8.2-8					
1	1.53	1.510850	0.920004	7.999563	0.000435
2	1.52	1.520200	0.919994	8.000595	−0.000596
3	1.51	1.530854	0.919997	8.000255	−0.000256
4	1.50	1.542813	0.919999	8.000136	0.000137
5	1.49	1.556277	0.919993	8.000750	−0.000751
6	1.48	1.571833	0.919992	8.000773	−0.000775
7	1.47	1.590267	0.920009	7.999110	0.000888
8	1.46	1.611580	0.920001	7.999909	0.000089
9	1.45	1.637730	0.920003	7.999677	0.000322
10	1.44	1.670281	0.919992	8.000761	−0.000763
11	1.43	1.714314	0.919996	8.000428	−0.000429
12	1.42	1.780381	0.920002	7.999820	0.000179
13	1.41	1.913797	0.919994	8.000649	−0.000650
No. of pairs		Mean COMPUT. risk errors		STD. ERR. errors	
13		−0.0001745565		0.0001384289	
Z-ratio				Status of significance	
1.26098				N.S.—not significant	
Parameter values					
PROB. = 0.91	RHO = 0.95		BIGH = 1.4563		SMHL = 1.35
Pair No.	SMH	SMK	PROB. level	Risk level (%)	Risk error (%)
Table No. 8.2-9					
1	1.46	1.452609	0.909993	9.000725	−0.000727
2	1.45	1.462823	0.909998	9.000230	−0.000232
3	1.44	1.474347	0.910007	8.999264	0.000733
4	1.43	1.487185	0.910003	8.999670	0.000328
5	1.42	1.501731	0.909992	9.000789	−0.000793
6	1.41	1.518964	0.910005	8.999461	0.000536

(continued)

8.2 Contents for Table 8.2 (Table No. 8.2-1 to 400)

Table 8.2 (continued)

Parameter values					
PROB. = 0.91	RHO = 0.95		BIGH = 1.4563		SMHL = 1.35
Pair No.	SMH	SMK	PROB. level	Risk level (%)	Risk error (%)
7	1.40	1.538692	0.909992	9.000844	−0.000846
8	1.39	1.562872	0.910001	8.999872	0.000125
9	1.38	1.592287	0.909991	9.000939	−0.000942
10	1.37	1.630849	0.909993	9.000694	−0.000697
11	1.36	1.685200	0.909995	9.000481	−0.000483
12	1.35	1.777220	0.909995	9.000481	−0.000483
No. of pairs		Mean COMPUT. risk errors		STD. ERR. errors	
12		−0.0002677624		0.0001557320	
Z-ratio				Status of significance	
1.71938				N.S.—not significant	
Parameter values					
PROB. = 0.9025	RHO = 0.95		BIGH = 1.4128		SMHL = 1.30
Pair No.	SMH	SMK	PROB. level	Risk level (%)	Risk error (%)
Table No. 8.2-10					
1	1.42	1.404074	0.902506	9.749376	0.000626
2	1.41	1.413652	0.902499	9.750152	−0.000149
3	1.40	1.424545	0.902505	9.749472	0.000530
4	1.39	1.436755	0.902508	9.749240	0.000763
5	1.38	1.450481	0.902500	9.750044	−0.000042
6	1.37	1.466312	0.902496	9.750414	−0.000411
7	1.36	1.484837	0.902498	9.750247	−0.000244
8	1.35	1.506646	0.902490	9.750974	−0.000972
9	1.34	1.533696	0.902505	9.749454	0.000548
10	1.33	1.567161	0.902492	9.750784	−0.000781
11	1.32	1.612905	0.902501	9.749926	0.000077
12	1.31	1.681480	0.902493	9.750711	−0.000709
13	1.30	1.832263	0.902504	9.749579	0.000423
No. of pairs		Mean COMPUT. risk errors		STD. ERR. errors	
13		−0.0000242676		0.0001490305	
Z-ratio				Status of significance	
0.16284				N.S.—not significant	
Parameter values					
PROB. = 0.9	RHO = 0.95		BIGH = 1.3976		SMHL = 1.29
Pair No.	SMH	SMK	PROB. level	Risk level (%)	Risk error (%)
Table No. 8.2-11					
1	1.40	1.395204	0.900005	9.999496	0.000507
2	1.39	1.405437	0.899999	10.000060	−0.000060
3	1.38	1.416987	0.899999	10.000130	−0.000131
4	1.37	1.430053	0.899998	10.000230	−0.000232
5	1.36	1.445029	0.900000	10.000010	−0.000012
6	1.35	1.462308	0.899999	10.000100	−0.000101
7	1.34	1.482676	0.900001	9.999919	0.000083
8	1.33	1.506917	0.899991	10.000940	−0.000942

(continued)

Table 8.2 (continued)

Parameter values					
PROB. = 0.9	RHO = 0.95		BIGH = 1.3976		SMHL = 1.29
Pair No.	SMH	SMK	PROB. level	Risk level (%)	Risk error (%)
9	1.32	1.537575	0.900004	9.999585	0.000417
10	1.31	1.576996	0.899998	10.000180	−0.000173
11	1.30	1.633778	0.900009	9.999102	0.000900
12	1.29	1.731363	0.900000	9.999991	0.000012
No. of pairs		Mean COMPUT. risk errors		STD. ERR. errors	
12		0.0000206324		0.0001199456	
Z-ratio				Status of significance	
0.17201				N.S.—not significant	
Parameter values					
PROB. = 0.85	RHO = 0.95		BIGH = 1.1544		SMHL = 1.04
Pair No.	SMH	SMK	PROB. level	Risk level (%)	Risk error (%)
Table No. 8.2-12					
1	1.16	1.148827	0.850004	14.999650	0.000352
2	1.15	1.158817	0.850010	14.999010	0.000983
3	1.14	1.169763	0.849992	15.000770	−0.000769
4	1.13	1.182257	0.849992	15.000780	−0.000787
5	1.12	1.196497	0.849997	15.000280	−0.000286
6	1.11	1.212881	0.850005	14.999520	0.000477
7	1.10	1.231803	0.850000	14.999990	0.000006
8	1.09	1.254246	0.849995	15.000510	−0.000507
9	1.08	1.281582	0.849990	15.000960	−0.000960
10	1.07	1.316551	0.850003	14.999660	0.000340
11	1.06	1.363457	0.850004	14.999600	0.000399
12	1.05	1.434805	0.849996	15.000370	−0.000376
13	1.04	1.597009	0.850004	14.999560	0.000441
No. of pairs		Mean COMPUT. risk errors		STD. ERR. errors	
13		−0.0000489610		0.0001548640	
Z-ratio				Status of significance	
0.31615				N.S.—not significant	
Parameter values					
PROB. = 0.8	RHO = 0.95		BIGH = 0.961		SMHL = 0.85
Pair No.	SMH	SMK	PROB. level	Risk level (%)	Risk error (%)
Table No. 8.2-13					
1	0.97	0.952474	0.800008	19.999250	0.000751
2	0.96	0.962001	0.799998	20.000210	−0.000209
3	0.95	0.972713	0.800002	19.999810	0.000185
4	0.94	0.984618	0.799993	20.000700	−0.000703
5	0.93	0.998209	0.800001	19.999930	0.000072
6	0.92	1.013593	0.799995	20.000490	−0.000489
7	0.91	1.031460	0.799999	20.000090	−0.000095
8	0.90	1.052306	0.799992	20.000850	−0.000852
9	0.89	1.077508	0.799998	20.000230	−0.000226
10	0.88	1.108831	0.800010	19.999050	0.000948

(continued)

8.2 Contents for Table 8.2 (Table No. 8.2-1 to 400) 213

Table 8.2 (continued)

Parameter values					
PROB. = 0.8	RHO = 0.95		BIGH = 0.961		SMHL = 0.85
Pair No.	SMH	SMK	PROB. level	Risk level (%)	Risk error (%)
11	0.87	1.149018	0.799990	20.000970	−0.000972
12	0.86	1.206674	0.799996	20.000430	−0.000429
13	0.85	1.306808	0.799998	20.000160	−0.000161
No. of pairs		Mean COMPUT. risk errors		STD. ERR. errors	
13		−0.0001558236		0.0001459994	
Z-ratio				Status of significance	
1.06729				N.S.—not significant	

Table No. 8.2-14

Parameter values					
PROB. = 0.75	RHO = 0.95		BIGH = 0.795		SMHL = 0.69
Pair No.	SMH	SMK	PROB. level	Risk level (%)	Risk error (%)
1	0.80	0.790422	0.749993	25.000720	−0.000715
2	0.79	0.800520	0.750005	24.999550	0.000453
3	0.78	0.811656	0.749998	25.000200	−0.000197
4	0.77	0.824230	0.750002	24.999830	0.000167
5	0.76	0.838448	0.750003	24.999680	0.000316
6	0.75	0.854614	0.749993	25.000680	−0.000679
7	0.74	0.873424	0.749991	25.000930	−0.000930
8	0.73	0.895671	0.749993	25.000660	−0.000656
9	0.72	0.922734	0.750009	24.999070	0.000930
10	0.71	0.956192	0.749993	25.000670	−0.000674
11	0.70	1.000940	0.749999	25.000100	−0.000101
12	0.69	1.066759	0.749996	25.000390	−0.000387
No. of pairs		Mean COMPUT. risk errors		STD. ERR. errors	
12		−0.0001902764		0.0001529045	
Z-ratio				Status of significance	
1.24441				N.S.—not significant	

Parameter values					
PROB. = 0.7	RHO = 0.95		BIGH = 0.6463		SMHL = 0.55
Pair No.	SMH	SMK	PROB. level	Risk level (%)	Risk error (%)

Table No. 8.2-15

1	0.65	0.642621	0.700006	29.999380	0.000626
2	0.64	0.652760	0.700006	29.999430	0.000578
3	0.63	0.663998	0.699996	30.000430	−0.000429
4	0.62	0.676645	0.699992	30.000850	−0.000846
5	0.61	0.691010	0.699995	30.000470	−0.000471
6	0.60	0.707501	0.700008	29.999230	0.000775
7	0.59	0.726429	0.699999	30.000060	−0.000054
8	0.58	0.748890	0.700006	29.999390	0.000608
9	0.57	0.775782	0.699991	30.000910	−0.000906
10	0.56	0.809767	0.700002	29.999850	0.000155
11	0.55	0.854774	0.700010	29.999010	0.000995

(continued)

Table 8.2 (continued)

Parameter values					
PROB. = 0.7	RHO = 0.95		BIGH = 0.6463		SMHL = 0.55
Pair No.	SMH	SMK	PROB. level	Risk level (%)	Risk error (%)
No. of pairs		Mean COMPUT. risk errors		STD. ERR. errors	
11		0.0000859300		0.0001860308	
Z-ratio				Status of significance	
0.46191				N.S.—not significant	
Parameter values					
PROB. = 0.65	RHO = 0.95		BIGH = 0.5083		SMHL = 0.43
Pair No.	SMH	SMK	PROB. level	Risk level (%)	Risk error (%)
Table No. 8.2-16					
1	0.51	0.506606	0.650009	34.999090	0.000918
2	0.50	0.516933	0.650004	34.999620	0.000387
3	0.49	0.528455	0.650001	34.999930	0.000072
4	0.48	0.541394	0.649999	35.000120	−0.000113
5	0.47	0.556069	0.650002	34.999830	0.000179
6	0.46	0.572804	0.649998	35.000220	−0.000215
7	0.45	0.592317	0.650007	34.999300	0.000703
8	0.44	0.615132	0.649992	35.000830	−0.000829
9	0.43	0.643045	0.650005	34.999530	0.000477
No. of pairs		Mean COMPUT. risk errors		STD. ERR. errors	
9		0.0001579523		0.0001580299	
Z-ratio				Status of significance	
0.99951				N.S.—not significant	
Parameter values					
PROB. = 0.6	RHO = 0.95		BIGH = 0.3774		SMHL = 0.31
Pair No.	SMH	SMK	PROB. level	Risk level (%)	Risk error (%)
Table No. 8.2-17					
1	0.38	0.374720	0.600001	39.999940	0.000060
2	0.37	0.384948	0.599996	40.000380	−0.000381
3	0.36	0.396422	0.600009	39.999150	0.000852
4	0.35	0.409093	0.599991	40.000910	−0.000912
5	0.34	0.423602	0.600004	39.999570	0.000429
6	0.33	0.440007	0.599997	40.000300	−0.000298
7	0.32	0.458963	0.599990	40.000990	−0.000989
8	0.31	0.481329	0.599993	40.000670	−0.000674
No. of pairs		Mean COMPUT. risk errors		STD. ERR. errors	
8		−0.0002125899		0.0002043314	
Z-ratio				Status of significance	
1.04042				N.S.—not significant	
Parameter values					
PROB. = 0.55	RHO = 0.95		BIGH = 0.2505		SMHL = 0.21
Pair No.	SMH	SMK	PROB. level	Risk level (%)	Risk error (%)
Table No. 8.2-18					
1	0.26	0.241738	0.549995	45.000470	−0.000471
2	0.25	0.251294	0.549991	45.000930	−0.000930

(continued)

8.2 Contents for Table 8.2 (Table No. 8.2-1 to 400)

Table 8.2 (continued)

Parameter values					
PROB. = 0.55	RHO = 0.95		BIGH = 0.2505		SMHL = 0.21
Pair No.	SMH	SMK	PROB. level	Risk level (%)	Risk error (%)
3	0.24	0.261948	0.550000	45.000040	−0.000036
4	0.23	0.273804	0.550007	44.999320	0.000685
5	0.22	0.286986	0.549995	45.000480	−0.000483
6	0.21	0.302033	0.550005	44.999540	0.000465
No. of pairs		Mean COMPUT. risk errors		STD. ERR. errors	
6		−0.0001098429		0.0002139327	
Z-ratio				Status of significance	
0.51345				N.S.—not significant	
Parameter values					
PROB. = 0.5	RHO = 0.95		BIGH = 0.1259		SMHL = 0.12
Pair No.	SMH	SMK	PROB. level	Risk level (%)	Risk error (%)
Table No. 8.2-19					
1	0.13	0.122125	0.500005	49.999500	0.000507
2	0.12	0.132188	0.499996	50.000420	−0.000414
No. of pairs		Mean COMPUT. risk errors		STD. ERR. errors	
2		0.0000307957		0.0002662843	
Z-ratio				Status of significance	
0.11565				N.S.—not significant	
Parameter values					
PROB. = 0.99	RHO = 0.9		BIGH = 2.4694		SMHL = 2.33
Pair No.	SMH	SMK	PROB. level	Risk level (%)	Risk error (%)
Table No. 8.2-20					
1	2.47	2.468800	0.990001	0.999880	0.000119
2	2.46	2.478836	0.989996	1.000381	−0.000381
3	2.45	2.490516	0.989999	1.000083	−0.000083
4	2.44	2.503842	0.990007	0.999308	0.000691
5	2.43	2.517251	0.989999	1.000083	−0.000083
6	2.42	2.532308	0.989992	1.000810	−0.000811
7	2.41	2.550577	0.989996	1.000357	−0.000358
8	2.40	2.572056	0.990006	0.999427	0.000572
9	2.39	2.595188	0.990001	0.999916	0.000083
10	2.38	2.623096	0.990000	1.000017	−0.000018
11	2.37	2.657344	0.990001	0.999928	0.000072
12	2.36	2.699496	0.989996	1.000381	−0.000381
13	2.35	2.757366	0.989996	1.000392	−0.000393
14	2.34	2.843455	0.989991	1.000887	−0.000888
15	2.33	3.042140	0.989995	1.000464	−0.000465
No. of pairs		Mean COMPUT. risk errors		STD. ERR. errors	
15		−0.0001452863		0.0001069261	
Z-ratio				Status of significance	
1.35875				N.S.—not significant	

(continued)

Table 8.2 (continued)

Parameter values					
PROB. = 0.98	RHO = 0.9		BIGH = 2.2007		SMHL = 2.06
Pair No.	SMH	SMK	PROB. level	Risk level (%)	Risk error (%)
Table No. 8.2-21					
1	2.21	2.191439	0.979997	2.000302	−0.000304
2	2.20	2.201400	0.980005	1.999480	0.000519
3	2.19	2.211452	0.979994	2.000588	−0.000590
4	2.18	2.223159	0.979998	2.000189	−0.000191
5	2.17	2.235740	0.979995	2.000481	−0.000483
6	2.16	2.249979	0.979998	2.000201	−0.000203
7	2.15	2.265877	0.980001	1.999939	0.000060
8	2.14	2.283435	0.979997	2.000296	−0.000298
9	2.13	2.303434	0.979994	2.000618	−0.000620
10	2.12	2.326659	0.979993	2.000737	−0.000739
11	2.11	2.354674	0.980001	1.999933	0.000066
12	2.10	2.387478	0.979998	2.000183	−0.000185
13	2.09	2.429763	0.980008	1.999229	0.000769
14	2.08	2.484654	0.980009	1.999092	0.000906
15	2.07	2.564652	0.980008	1.999229	0.000769
16	2.06	2.713510	0.979998	2.000219	−0.000221
No. of pairs		Mean COMPUT. risk errors		STD. ERR. errors	
16		−0.0000438269		0.0001227835	
Z-ratio				Status of significance	
0.35694				N.S.—not significant	
Parameter values					
PROB. = 0.97	RHO = 0.9		BIGH = 2.0302		SMHL = 1.89
Pair No.	SMH	SMK	PROB. level	Risk level (%)	Risk error (%)
Table No. 8.2-22					
1	2.04	2.020447	0.969994	3.000623	−0.000626
2	2.03	2.030400	0.970006	2.999365	0.000632
3	2.02	2.040452	0.969993	3.000742	−0.000745
4	2.01	2.052166	0.970001	2.999902	0.000095
5	2.00	2.064762	0.970001	2.999926	0.000072
6	1.99	2.079024	0.970010	2.999038	0.000960
7	1.98	2.094173	0.969999	3.000093	−0.000095
8	1.97	2.111771	0.970003	2.999687	0.000310
9	1.96	2.131040	0.969992	3.000826	−0.000829
10	1.95	2.154323	0.970004	2.999580	0.000417
11	1.94	2.180844	0.970004	2.999604	0.000393
12	1.93	2.212165	0.969999	3.000122	−0.000125
13	1.92	2.251412	0.970003	2.999723	0.000274
14	1.91	2.301715	0.970002	2.999824	0.000173

(continued)

8.2 Contents for Table 8.2 (Table No. 8.2-1 to 400)

Table 8.2 (continued)

Parameter values					
PROB. = 0.97	RHO = 0.9		BIGH = 2.0302		SMHL = 1.89
Pair No.	SMH	SMK	PROB. level	Risk level (%)	Risk error (%)
15	1.90	2.372447	0.970001	2.999908	0.000089
16	1.89	2.493300	0.970003	2.999693	0.000304
No. of pairs		Mean COMPUT. risk errors		STD. ERR. errors	
16		0.0000764342		0.0001138128	
Z-ratio				Status of significance	
0.67158				N.S.—not significant	
Parameter values					
PROB. = 0.96	RHO = 0.9		BIGH = 1.9018		SMHL = 1.76
Pair No.	SMH	SMK	PROB. level	Risk level (%)	Risk error (%)
Table No. 8.2-23					
1	1.91	1.894026	0.960000	4.000008	−0.000006
2	1.90	1.903602	0.959993	4.000735	−0.000733
3	1.89	1.914455	0.960003	3.999734	0.000268
4	1.88	1.926197	0.960007	3.999269	0.000733
5	1.87	1.938828	0.960001	3.999853	0.000149
6	1.86	1.953133	0.960008	3.999239	0.000763
7	1.85	1.968722	0.960003	3.999728	0.000274
8	1.84	1.985988	0.959991	4.000884	−0.000882
9	1.83	2.006105	0.959995	4.000461	−0.000459
10	1.82	2.029464	0.960006	3.999370	0.000632
11	1.81	2.056068	0.960002	3.999782	0.000221
12	1.80	2.087482	0.959992	4.000754	−0.000751
13	1.79	2.126833	0.959996	4.000444	−0.000441
14	1.78	2.177247	0.959994	4.000646	−0.000644
15	1.77	2.249664	0.960008	3.999239	0.000763
16	1.76	2.367525	0.959994	4.000640	−0.000638
No. of pairs		Mean COMPUT. risk errors		STD. ERR. errors	
16		−0.0000441776		0.0001407918	
Z-ratio				Status of significance	
0.31378				N.S.—not significant	
Parameter values					
PROB. = 0.95	RHO = 0.9		BIGH = 1.7976		SMHL = 1.65
Pair No.	SMH	SMK	PROB. level	Risk level (%)	Risk error (%)
Table No. 8.2-24					
1	1.80	1.795203	0.950000	4.999972	0.000030
2	1.79	1.805232	0.949990	5.000967	−0.000966
3	1.78	1.816546	0.949998	5.000174	−0.000173
4	1.77	1.828755	0.949998	5.000252	−0.000250
5	1.76	1.842254	0.949999	5.000127	−0.000125
6	1.75	1.857432	0.950009	4.999089	0.000912
7	1.74	1.873904	0.950001	4.999858	0.000143
8	1.73	1.892842	0.950008	4.999214	0.000787

(continued)

Table 8.2 (continued)

Parameter values					
PROB. = 0.95	RHO = 0.9		BIGH = 1.7976		SMHL = 1.65
Pair No.	SMH	SMK	PROB. level	Risk level (%)	Risk error (%)
9	1.72	1.913858	0.949997	5.000335	−0.000334
10	1.71	1.938906	0.950009	4.999119	0.000882
11	1.70	1.967210	0.949991	5.000866	−0.000864
12	1.69	2.002676	0.950005	4.999542	0.000459
13	1.68	2.046088	0.949999	5.000055	−0.000054
14	1.67	2.103700	0.949995	5.000484	−0.000483
15	1.66	2.190356	0.950003	4.999733	0.000268
16	1.65	2.364653	0.950004	4.999561	0.000441
No. of pairs		Mean COMPUT. risk errors		STD. ERR. errors	
16		0.0000396196		0.0001333340	
Z-ratio				Status of significance	
0.29715				N.S.—not significant	
Parameter values					
PROB. = 0.94	RHO = 0.9		BIGH = 1.7089		SMHL = 1.56
Pair No.	SMH	SMK	PROB. level	Risk level (%)	Risk error (%)
Table No. 8.2-25					
1	1.71	1.707801	0.940009	5.999124	0.000876
2	1.70	1.717847	0.939992	6.000817	−0.000817
3	1.69	1.729183	0.939996	6.000418	−0.000417
4	1.68	1.741422	0.939990	6.000954	−0.000954
5	1.67	1.755347	0.940008	5.999250	0.000751
6	1.66	1.770178	0.939997	6.000352	−0.000352
7	1.65	1.787090	0.940002	5.999816	0.000185
8	1.64	1.806085	0.940006	5.999446	0.000554
9	1.63	1.827557	0.940004	5.999601	0.000399
10	1.62	1.852288	0.940001	5.999863	0.000137
11	1.61	1.881063	0.939991	6.000919	−0.000918
12	1.60	1.916618	0.940004	5.999649	0.000352
13	1.59	1.960129	0.939996	6.000424	−0.000423
14	1.58	2.018629	0.940002	5.999768	0.000232
15	1.57	2.103839	0.939993	6.000728	−0.000727
16	1.56	2.278262	0.940005	5.999536	0.000465
No. of pairs		Mean COMPUT. risk errors		STD. ERR. errors	
16		−0.0000385677		0.0001440550	
Z-ratio				Status of significance	
0.26773				N.S.—not significant	
Parameter values					
PROB. = 0.93	RHO = 0.9		BIGH = 1.631		SMHL = 1.48
Pair No.	SMH	SMK	PROB. level	Risk level (%)	Risk error (%)
Table No. 8.2-26					
1	1.64	1.622440	0.930008	6.999243	0.000757
2	1.63	1.632001	0.929999	7.000089	−0.000089

(continued)

8.2 Contents for Table 8.2 (Table No. 8.2-1 to 400)

Table 8.2 (continued)

Parameter values					
PROB. = 0.93	RHO = 0.9		BIGH = 1.631		SMHL = 1.48
Pair No.	SMH	SMK	PROB. level	Risk level (%)	Risk error (%)
3	1.62	1.642465	0.929995	7.000458	−0.000459
4	1.61	1.654032	0.930001	6.999886	0.000113
5	1.60	1.666507	0.929995	7.000489	−0.000489
6	1.59	1.680479	0.930002	6.999779	0.000221
7	1.58	1.695756	0.929999	7.000149	−0.000149
8	1.57	1.712729	0.929991	7.000870	−0.000870
9	1.56	1.732184	0.930000	7.000041	−0.000042
10	1.55	1.754123	0.929999	7.000119	−0.000119
11	1.54	1.779721	0.930008	6.999236	0.000763
12	1.53	1.809370	0.930000	6.999958	0.000042
13	1.52	1.845418	0.930000	6.999970	0.000030
14	1.51	1.890602	0.929999	7.000101	−0.000101
15	1.50	1.951566	0.930008	6.999189	0.000811
16	1.49	2.041593	0.929995	7.000476	−0.000477
17	1.48	2.234906	0.929993	7.000691	−0.000691
No. of pairs		Mean COMPUT. risk errors		STD. ERR. errors	
17		−0.0000417233		0.0001116370	
Z-ratio				Status of significance	
0.37374				N.S.—not significant	
Parameter values					
PROB. = 0.92	RHO = 0.9		BIGH = 1.5613		SMHL = 1.41
Pair No.	SMH	SMK	PROB. level	Risk level (%)	Risk error (%)
Table No. 8.2-27					
1	1.57	1.553039	0.920006	7.999402	0.000596
2	1.56	1.562601	0.919995	8.000528	−0.000530
3	1.55	1.573268	0.920004	7.999641	0.000358
4	1.54	1.584848	0.920009	7.999099	0.000900
5	1.53	1.597342	0.920002	7.999772	0.000226
6	1.52	1.611144	0.919999	8.000118	−0.000119
7	1.51	1.626452	0.919997	8.000303	−0.000304
8	1.50	1.643465	0.919992	8.000791	−0.000793
9	1.49	1.662965	0.920005	7.999456	0.000542
10	1.48	1.684566	0.919992	8.000780	−0.000781
11	1.47	1.709833	0.919994	8.000588	−0.000590
12	1.46	1.739550	0.919997	8.000255	−0.000256
13	1.45	1.775284	0.919998	8.000172	−0.000173
14	1.44	1.820161	0.920004	7.999599	0.000399

(continued)

Table 8.2 (continued)

Parameter values					
PROB. = 0.92	RHO = 0.9		BIGH = 1.5613		SMHL = 1.41
Pair No.	SMH	SMK	PROB. level	Risk level (%)	Risk error (%)
15	1.43	1.879656	0.920007	7.999283	0.000715
16	1.42	1.966660	0.919991	8.000869	−0.000870
17	1.41	2.147585	0.920003	7.999659	0.000340
No. of pairs		Mean COMPUT. risk errors		STD. ERR. errors	
17		−0.0000188748		0.0001309347	
Z-ratio				Status of significance	
0.14415				N.S.—not significant	
Parameter values					
PROB. = 0.91	RHO = 0.9		BIGH = 1.4979		SMHL = 1.35
Pair No.	SMH	SMK	PROB. level	Risk level (%)	Risk error (%)
Table No. 8.2-28					
1	1.50	1.495998	0.910003	8.999681	0.000316
2	1.49	1.506233	0.910002	8.999824	0.000173
3	1.48	1.517384	0.910001	8.999866	0.000131
4	1.47	1.529455	0.909993	9.000725	−0.000727
5	1.46	1.543034	0.910007	8.999311	0.000685
6	1.45	1.557734	0.910002	8.999806	0.000191
7	1.44	1.574144	0.910002	8.999830	0.000167
8	1.43	1.592462	0.909998	9.000194	−0.000197
9	1.42	1.613277	0.909999	9.000081	−0.000083
10	1.41	1.636983	0.909994	9.000599	−0.000602
11	1.40	1.664755	0.909996	9.000361	−0.000364
12	1.39	1.697769	0.909997	9.000265	−0.000268
13	1.38	1.738373	0.910000	8.999962	0.000036
14	1.37	1.790865	0.910008	8.999241	0.000757
15	1.36	1.863845	0.910004	8.999592	0.000405
16	1.35	1.987003	0.910002	8.999842	0.000155
No. of pairs		Mean COMPUT. risk errors		STD. ERR. errors	
16		0.0000455800		0.0000968096	
Z-ratio				Status of significance	
0.47082				N.S.—not significant	
Parameter values					
PROB. = 0.9025	RHO = 0.9		BIGH = 1.4538		SMHL = 1.30
Pair No.	SMH	SMK	PROB. level	Risk level (%)	Risk error (%)
Table No. 8.2-29					
1	1.46	1.447822	0.902504	9.749609	0.000393
2	1.45	1.457610	0.902493	9.750658	−0.000656
3	1.44	1.468514	0.902505	9.749466	0.000536
4	1.43	1.480145	0.902497	9.750282	−0.000280
5	1.42	1.493092	0.902505	9.749514	0.000489
6	1.41	1.507164	0.902500	9.750009	−0.000006
7	1.40	1.522949	0.902509	9.749150	0.000852

(continued)

8.2 Contents for Table 8.2 (Table No. 8.2-1 to 400)

Table 8.2 (continued)

Parameter values					
PROB. = 0.9025	RHO = 0.9		BIGH = 1.4538		SMHL = 1.30
Pair No.	SMH	SMK	PROB. level	Risk level (%)	Risk error (%)
8	1.39	1.540255	0.902499	9.750152	−0.000149
9	1.38	1.560062	0.902508	9.749246	0.000757
10	1.37	1.582179	0.902494	9.750562	−0.000560
11	1.36	1.607976	0.902494	9.750562	−0.000560
12	1.35	1.638237	0.902491	9.750867	−0.000864
13	1.34	1.674921	0.902494	9.750616	−0.000614
14	1.33	1.721155	0.902502	9.749776	0.000226
15	1.32	1.782412	0.902499	9.750139	−0.000137
16	1.31	1.874323	0.902491	9.750902	−0.000900
17	1.30	2.075795	0.902505	9.749544	0.000459
No. of pairs		Mean COMPUT. risk errors		STD. ERR. errors	
17		−0.0000562933		0.0001315690	
Z-ratio				Status of significance	
0.42786				N.S.—not significant	
Parameter values					
PROB. = 0.9	RHO = 0.9		BIGH = 1.4396		SMHL = 1.29
Pair No.	SMH	SMK	PROB. level	Risk level (%)	Risk error (%)
Table No. 8.2-30					
1	1.44	1.439200	0.899991	10.000920	−0.000918
2	1.43	1.449655	0.899997	10.000340	−0.000340
3	1.42	1.461033	0.900002	9.999776	0.000226
4	1.41	1.473337	0.899998	10.000210	−0.000209
5	1.40	1.486961	0.900002	9.999794	0.000209
6	1.39	1.501907	0.900000	10.000050	−0.000048
7	1.38	1.518571	0.900000	9.999985	0.000018
8	1.37	1.537345	0.900007	9.999317	0.000685
9	1.36	1.558234	0.899993	10.000670	−0.000668
10	1.35	1.582412	0.899992	10.000800	−0.000799
11	1.34	1.610666	0.899994	10.000610	−0.000608
12	1.33	1.644560	0.900004	9.999621	0.000381
13	1.32	1.685661	0.899993	10.000720	−0.000715
14	1.31	1.739834	0.900008	9.999174	0.000829
15	1.30	1.815284	0.900001	9.999889	0.000113
16	1.29	1.947174	0.900008	9.999180	0.000823
No. of pairs		Mean COMPUT. risk errors		STD. ERR. errors	
16		−0.0000599553		0.0001354974	
Z-ratio				Status of significance	
0.44248				N.S.—not significant	
Parameter values					
PROB. = 0.85	RHO = 0.9		BIGH = 1.1982		SMHL = 1.04
Pair No.	SMH	SMK	PROB. level	Risk level (%)	Risk error (%)
Table No. 8.2-31					
1	1.20	1.196403	0.849993	15.000680	−0.000679
2	1.19	1.206652	0.849992	15.000820	−0.000823

(continued)

Table 8.2 (continued)

Parameter values					
PROB. = 0.85	RHO = 0.9		BIGH = 1.1982		SMHL = 1.04
Pair No.	SMH	SMK	PROB. level	Risk level (%)	Risk error (%)
3	1.18	1.217853	0.849999	15.000110	−0.000113
4	1.17	1.230009	0.850001	14.999870	0.000131
5	1.16	1.243322	0.850005	14.999510	0.000489
6	1.15	1.257795	0.849993	15.000720	−0.000721
7	1.14	1.274020	0.849999	15.000080	−0.000077
8	1.13	1.292000	0.849997	15.000280	−0.000286
9	1.12	1.312329	0.850003	14.999700	0.000298
10	1.11	1.335205	0.849992	15.000820	−0.000817
11	1.10	1.361807	0.849994	15.000580	−0.000584
12	1.09	1.393312	0.850008	14.999180	0.000817
13	1.08	1.430899	0.850000	15.000020	−0.000024
14	1.07	1.478478	0.850006	14.999350	0.000644
15	1.06	1.541918	0.850004	14.999600	0.000393
16	1.05	1.637630	0.849992	15.000800	−0.000799
17	1.04	1.855465	0.850007	14.999320	0.000679
No. of pairs		Mean COMPUT. risk errors		STD. ERR. errors	
17		−0.0000817908		0.0001319510	
Z-ratio				Status of significance	
0.61986				N.S.—not significant	
Parameter values					
PROB. = 0.8	RHO = 0.9		BIGH = 1.0065		SMHL = 0.85
Pair No.	SMH	SMK	PROB. level	Risk level (%)	Risk error (%)
Table No. 8.2-32					
1	1.01	1.002817	0.800000	20.000030	−0.000030
2	1.00	1.012945	0.800002	19.999810	0.000191
3	0.99	1.023861	0.799998	20.000230	−0.000232
4	0.98	1.035767	0.799999	20.000090	−0.000089
5	0.97	1.048670	0.799990	20.000970	−0.000972
6	0.96	1.062967	0.799999	20.000090	−0.000089
7	0.95	1.078665	0.800001	19.999890	0.000113
8	0.94	1.096064	0.800001	19.999890	0.000113
9	0.93	1.115465	0.799995	20.000480	−0.000477
10	0.92	1.137461	0.799996	20.000410	−0.000411
11	0.91	1.162647	0.799998	20.000240	−0.000238
12	0.90	1.192009	0.800003	19.999660	0.000334
13	0.89	1.226921	0.800007	19.999270	0.000727
14	0.88	1.269544	0.799997	20.000260	−0.000262
15	0.87	1.324963	0.800004	19.999580	0.000423

(continued)

8.2 Contents for Table 8.2 (Table No. 8.2-1 to 400)

Table 8.2 (continued)

Parameter values					
PROB. = 0.8	RHO = 0.9		BIGH = 1.0065		SMHL = 0.85
Pair No.	SMH	SMK	PROB. level	Risk level (%)	Risk error (%)
16	0.86	1.402956	0.800006	19.999420	0.000584
17	0.85	1.537127	0.799998	20.000170	−0.000167
No. of pairs		Mean COMPUT. risk errors		STD. ERR. errors	
17		−0.0000268221		0.0000952190	
Z-ratio				Status of significance	
0.28169				N.S.—not significant	
Parameter values					
PROB. = 0.75	RHO = 0.9		BIGH = 0.8418		SMHL = 0.68
Pair No.	SMH	SMK	PROB. level	Risk level (%)	Risk error (%)
Table No. 8.2-33					
1	0.85	0.833874	0.749995	25.000550	−0.000548
2	0.84	0.843702	0.750004	24.999560	0.000441
3	0.83	0.854158	0.749991	25.000920	−0.000924
4	0.82	0.865644	0.750001	24.999920	0.000083
5	0.81	0.878071	0.750002	24.999760	0.000238
6	0.80	0.891643	0.750005	24.999510	0.000495
7	0.79	0.906469	0.749999	25.000080	−0.000077
8	0.78	0.922949	0.750008	24.999220	0.000781
9	0.77	0.941193	0.750008	24.999180	0.000823
10	0.76	0.961506	0.749995	25.000540	−0.000542
11	0.75	0.984680	0.749997	25.000340	−0.000340
12	0.74	1.011315	0.750000	24.999970	0.000036
13	0.73	1.042402	0.750000	25.000030	−0.000030
14	0.72	1.079517	0.749991	25.000940	−0.000942
15	0.71	1.125801	0.749992	25.000760	−0.000757
16	0.70	1.186934	0.750004	24.999590	0.000411
17	0.69	1.276214	0.750006	24.999420	0.000584
18	0.68	1.449911	0.749998	25.000230	−0.000232
No. of pairs		Mean COMPUT. risk errors		STD. ERR. errors	
18		−0.0000263515		0.0001253360	
Z-ratio				Status of significance	
0.21025				N.S.—not significant	
Parameter values					
PROB. = 0.7	RHO = 0.9		BIGH = 0.6942		SMHL = 0.54
Pair No.	SMH	SMK	PROB. level	Risk level (%)	Risk error (%)
Table No. 8.2-34					
1	0.70	0.688253	0.699995	30.000550	−0.000548
2	0.69	0.698230	0.700002	29.999790	0.000215
3	0.68	0.708892	0.699992	30.000770	−0.000769
4	0.67	0.720544	0.700000	29.999980	0.000024
5	0.66	0.733102	0.699993	30.000720	−0.000721
6	0.65	0.746874	0.699996	30.000420	−0.000423

(continued)

Table 8.2 (continued)

Parameter values					
PROB. = 0.7	RHO = 0.9		BIGH = 0.6942		SMHL = 0.54
Pair No.	SMH	SMK	PROB. level	Risk level (%)	Risk error (%)
7	0.64	0.761974	0.699999	30.000140	−0.000137
8	0.63	0.778614	0.700000	30.000040	−0.000036
9	0.62	0.797007	0.699990	30.000980	−0.000978
10	0.61	0.817757	0.699997	30.000260	−0.000262
11	0.60	0.841275	0.700007	29.999310	0.000697
12	0.59	0.868170	0.700007	29.999270	0.000733
13	0.58	0.899440	0.699996	30.000400	−0.000399
14	0.57	0.937064	0.699999	30.000140	−0.000137
15	0.56	0.983607	0.699994	30.000650	−0.000650
16	0.55	1.044956	0.699999	30.000100	−0.000101
17	0.54	1.134620	0.700004	29.999570	0.000429
No. of pairs		Mean COMPUT. risk errors		STD. ERR. errors	
17		−0.0001702044		0.0001140368	
Z-ratio				Status of significance	
1.49254				N.S.—not significant	
Parameter values					
PROB. = 0.65	RHO = 0.9		BIGH = 0.5572		SMHL = 0.42
Pair No.	SMH	SMK	PROB. level	Risk level (%)	Risk error (%)
Table No. 8.2-35					
1	0.56	0.554414	0.650005	34.999520	0.000483
2	0.55	0.564592	0.650007	34.999320	0.000685
3	0.54	0.575534	0.650002	34.999800	0.000203
4	0.53	0.587359	0.649994	35.000570	−0.000566
5	0.52	0.600284	0.650000	35.000050	−0.000042
6	0.51	0.614335	0.649997	35.000280	−0.000280
7	0.50	0.629733	0.649994	35.000630	−0.000626
8	0.49	0.646800	0.650002	34.999840	0.000167
9	0.48	0.665762	0.650010	34.999020	0.000978
10	0.47	0.686848	0.649999	35.000120	−0.000113
11	0.46	0.710876	0.650004	34.999640	0.000364
12	0.45	0.738375	0.649999	35.000130	−0.000125
13	0.44	0.770657	0.650004	34.999620	0.000387
14	0.43	0.809137	0.649991	35.000890	−0.000888
15	0.42	0.857383	0.650000	34.999970	0.000030
No. of pairs		Mean COMPUT. risk errors		STD. ERR. errors	
15		0.0000409782		0.0001223637	
Z-ratio				Status of significance	
0.33489				N.S.—not significant	
Parameter values					
PROB. = 0.6	RHO = 0.9		BIGH = 0.4274		SMHL = 0.31
Pair No.	SMH	SMK	PROB. level	Risk level (%)	Risk error (%)
Table No. 8.2-36					
1	0.43	0.424620	0.600002	39.999830	0.000167
2	0.42	0.434735	0.599991	40.000940	−0.000942

(continued)

8.2 Contents for Table 8.2 (Table No. 8.2-1 to 400)

Table 8.2 (continued)

Parameter values					
PROB. = 0.6	RHO = 0.9		BIGH = 0.4274		SMHL = 0.31
Pair No.	SMH	SMK	PROB. level	Risk level (%)	Risk error (%)
3	0.41	0.445734	0.599998	40.000240	−0.000244
4	0.40	0.457653	0.600010	39.999040	0.000954
5	0.39	0.470437	0.599997	40.000330	−0.000334
6	0.38	0.484423	0.599993	40.000670	−0.000668
7	0.37	0.499759	0.599994	40.000620	−0.000626
8	0.36	0.516696	0.600001	39.999860	0.000137
9	0.35	0.535393	0.599998	40.000160	−0.000161
10	0.34	0.556407	0.600009	39.999060	0.000942
11	0.33	0.579915	0.599994	40.000570	−0.000566
12	0.32	0.606979	0.599996	40.000360	−0.000358
13	0.31	0.638479	0.599998	40.000250	−0.000250
No. of pairs		Mean COMPUT. risk errors		STD. ERR. errors	
13		−0.0001392194		0.0001483270	
Z-ratio				Status of significance	
0.93860				N.S.—not significant	
Parameter values					
PROB. = 0.55	RHO = 0.9		BIGH = 0.3015		SMHL = 0.21
Pair No.	SMH	SMK	PROB. level	Risk level (%)	Risk error (%)
Table No. 8.2-37					
1	0.31	0.293526	0.550001	44.999950	0.000048
2	0.30	0.303300	0.550002	44.999780	0.000221
3	0.29	0.313749	0.549993	45.000750	−0.000745
4	0.28	0.325139	0.550007	44.999350	0.000650
5	0.27	0.337359	0.550003	44.999720	0.000286
6	0.26	0.350601	0.549996	45.000430	−0.000429
7	0.25	0.365074	0.549995	45.000530	−0.000530
8	0.24	0.380908	0.549990	45.000990	−0.000983
9	0.23	0.398450	0.549999	45.000140	−0.000137
10	0.22	0.417879	0.550001	44.999880	0.000125
11	0.21	0.439606	0.550000	45.000000	0.000000
No. of pairs		Mean COMPUT. risk errors		STD. ERR. errors	
11		−0.0001246731		0.0001345790	
Z-ratio				Status of significance	
0.92639				N.S.—not significant	
Parameter values					
PROB. = 0.5	RHO = 0.9		BIGH = 0.178		SMHL = 0.13
Pair No.	SMH	SMK	PROB. level	Risk level (%)	Risk error (%)
Table No. 8.2-38					
1	0.18	0.176022	0.500005	49.999540	0.000459
2	0.17	0.186181	0.499993	50.000730	−0.000733
3	0.16	0.197244	0.500006	49.999400	0.000608
4	0.15	0.209078	0.500001	49.999900	0.000101

(continued)

Table 8.2 (continued)

Parameter values					
PROB. = 0.5	RHO = 0.9		BIGH = 0.178		SMHL = 0.13
Pair No.	SMH	SMK	PROB. level	Risk level (%)	Risk error (%)
5	0.14	0.221920	0.500001	49.999900	0.000101
6	0.13	0.235911	0.500006	49.999360	0.000644
No. of pairs		Mean COMPUT. risk errors			
6		0.0001685960		0.0001795300	
Z-ratio				Status of significance	
0.93910				N.S.—not significant	
Parameter values					
PROB. = 0.99	RHO = 0.8		BIGH = 2.5088		SMHL = 2.33
Pair No.	SMH	SMK	PROB. level	Risk level (%)	Risk error (%)
Table No. 8.2-39					
1	2.51	2.507601	0.989992	1.000798	−0.000799
2	2.50	2.519194	0.990009	0.999063	0.000936
3	2.49	2.529305	0.989997	1.000345	−0.000346
4	2.48	2.541059	0.989994	1.000595	−0.000596
5	2.47	2.554459	0.989999	1.000077	−0.000077
6	2.46	2.567943	0.989992	1.000774	−0.000775
7	2.45	2.584636	0.990006	0.999415	0.000584
8	2.44	2.601415	0.990004	0.999618	0.000381
9	2.43	2.619843	0.990001	0.999939	0.000060
10	2.42	2.641483	0.990007	0.999308	0.000691
11	2.41	2.664775	0.990005	0.999516	0.000483
12	2.40	2.691282	0.990003	0.999713	0.000286
13	2.39	2.721005	0.989995	1.000458	−0.000459
14	2.38	2.757070	0.989994	1.000559	−0.000560
15	2.37	2.802604	0.990002	0.999761	0.000238
16	2.36	2.857607	0.990001	0.999951	0.000048
17	2.35	2.934580	0.990007	0.999302	0.000697
18	2.34	3.046026	0.990000	0.999987	0.000012
19	2.33	3.301320	0.989998	1.000172	−0.000173
No. of pairs		Mean COMPUT. risk errors		STD. ERR. errors	
19		0.0000315905		0.0001142271	
Z-ratio				Status of significance	
0.27656				N.S.—not significant	
Parameter values					
PROB. = 0.98	RHO = 0.8		BIGH = 2.244		SMHL = 2.06
Pair No.	SMH	SMK	PROB. level	Risk level (%)	Risk error (%)
Table No. 8.2-40					
1	2.25	2.237234	0.979993	2.000725	−0.000727
2	2.24	2.247226	0.979995	2.000505	−0.000507
3	2.23	2.258088	0.980001	1.999927	0.000072
4	2.22	2.269822	0.980008	1.999199	0.000799
5	2.21	2.281648	0.979997	2.000266	−0.000268

(continued)

8.2 Contents for Table 8.2 (Table No. 8.2-1 to 400)

Table 8.2 (continued)

Parameter values					
PROB. = 0.98	RHO = 0.8		BIGH = 2.244		SMHL = 2.06
Pair No.	SMH	SMK	PROB. level	Risk level (%)	Risk error (%)
6	2.20	2.295130	0.980001	1.999879	0.000119
7	2.19	2.309488	0.980000	2.000034	−0.000036
8	2.18	2.325504	0.980005	1.999497	0.000501
9	2.17	2.342399	0.979999	2.000141	−0.000143
10	2.16	2.360954	0.979991	2.000874	−0.000876
11	2.15	2.382735	0.980002	1.999760	0.000238
12	2.14	2.406179	0.980000	2.000046	−0.000048
13	2.13	2.432851	0.979999	2.000147	−0.000149
14	2.12	2.464315	0.980007	1.999283	0.000715
15	2.11	2.499009	0.979994	2.000594	−0.000596
16	2.10	2.541624	0.979991	2.000862	−0.000864
17	2.09	2.596847	0.980005	1.999462	0.000536
18	2.08	2.664680	0.979991	2.000880	−0.000882
19	2.07	2.770126	0.980005	1.999545	0.000453
20	2.06	2.969434	0.980008	1.999247	0.000751
No. of pairs		Mean COMPUT. risk errors		STD. ERR. errors	
20		−0.0000434262		0.0001199754	
Z-ratio				Status of significance	
0.36196				N.S.—not significant	
Parameter values					
PROB. = 0.97	RHO = 0.8		BIGH = 2.0753		SMHL = 1.89
Pair No.	SMH	SMK	PROB. level	Risk level (%)	Risk error (%)
Table No. 8.2-41					
1	2.08	2.071001	0.970003	2.999687	0.000310
2	2.07	2.081004	0.970002	2.999789	0.000209
3	2.06	2.091495	0.969994	3.000605	−0.000608
4	2.05	2.103256	0.970002	2.999795	0.000203
5	2.04	2.115898	0.970010	2.999020	0.000978
6	2.03	2.128642	0.969991	3.000885	−0.000888
7	2.02	2.143442	0.970001	2.999950	0.000048
8	2.01	2.159128	0.969999	3.000122	−0.000125
9	2.00	2.176091	0.969993	3.000736	−0.000739
10	1.99	2.195506	0.970005	2.999544	0.000453
11	1.98	2.215812	0.969990	3.000963	−0.000966
12	1.97	2.239353	0.969995	3.000540	−0.000542
13	1.96	2.266133	0.970003	2.999711	0.000286
14	1.95	2.296151	0.970002	2.999842	0.000155
15	1.94	2.330974	0.970000	3.000009	−0.000012
16	1.93	2.372164	0.969997	3.000343	−0.000346
17	1.92	2.422849	0.969996	3.000420	−0.000423

(continued)

Table 8.2 (continued)

Parameter values					
PROB. = 0.97	RHO = 0.8		BIGH = 2.0753		SMHL = 1.89
Pair No.	SMH	SMK	PROB. level	Risk level (%)	Risk error (%)
18	1.91	2.489281	0.970005	2.999497	0.000501
19	1.90	2.579273	0.969995	3.000552	−0.000554
20	1.89	2.735017	0.969999	3.000152	−0.000155
No. of pairs		Mean COMPUT. risk errors		STD. ERR. errors	
20		−0.0001055854		0.0001095773	
Z-ratio				Status of significance	
0.96357				N.S.—not significant	
Parameter values					
PROB. = 0.96	RHO = 0.8		BIGH = 1.9488		SMHL = 1.76
Pair No.	SMH	SMK	PROB. level	Risk level (%)	Risk error (%)
Table No. 8.2-42					
1	1.95	1.947601	0.959991	4.000897	−0.000894
2	1.94	1.958031	0.959999	4.000145	−0.000143
3	1.93	1.968955	0.959997	4.000348	−0.000346
4	1.92	1.980766	0.959999	4.000139	−0.000137
5	1.91	1.993466	0.960001	3.999937	0.000066
6	1.90	2.007056	0.959999	4.000127	−0.000125
7	1.89	2.021929	0.960002	3.999758	0.000244
8	1.88	2.038086	0.960006	3.999418	0.000584
9	1.87	2.055139	0.959991	4.000938	−0.000936
10	1.86	2.074652	0.959999	4.000139	−0.000137
11	1.85	2.095845	0.959996	4.000425	−0.000423
12	1.84	2.119502	0.959994	4.000616	−0.000614
13	1.83	2.146406	0.959998	4.000223	−0.000221
14	1.82	2.177340	0.960007	3.999353	0.000650
15	1.81	2.212306	0.959998	4.000211	−0.000209
16	1.80	2.254432	0.960000	3.999955	0.000048
17	1.79	2.306063	0.960004	3.999639	0.000364
18	1.78	2.371107	0.959993	4.000693	−0.000691
19	1.77	2.464411	0.960003	3.999740	0.000262
20	1.76	2.620352	0.960003	3.999728	0.000274
No. of pairs		Mean COMPUT. risk errors		STD. ERR. errors	
20		−0.0001135327		0.0000954884	
Z-ratio				Status of significance	
1.18897				N.S.—not significant	
Parameter values					
PROB. = 0.95	RHO = 0.8		BIGH = 1.846		SMHL = 1.65
Pair No.	SMH	SMK	PROB. level	Risk level (%)	Risk error (%)
Table No. 8.2-43					
1	1.85	1.842009	0.949996	5.000430	−0.000429
2	1.84	1.852019	0.949993	5.000716	−0.000715
3	1.83	1.862921	0.950001	4.999912	0.000089
4	1.82	1.874325	0.949996	5.000395	−0.000393

(continued)

8.2 Contents for Table 8.2 (Table No. 8.2-1 to 400)

Table 8.2 (continued)

Parameter values					
PROB. = 0.95	RHO = 0.8		BIGH = 1.846		SMHL = 1.65
Pair No.	SMH	SMK	PROB. level	Risk level (%)	Risk error (%)
5	1.81	1.886622	0.949994	5.000562	−0.000560
6	1.80	1.900207	0.950008	4.999155	0.000846
7	1.79	1.914299	0.949998	5.000222	−0.000221
8	1.78	1.930072	0.950008	4.999214	0.000787
9	1.77	1.946747	0.949999	5.000139	−0.000137
10	1.76	1.965108	0.949994	5.000621	−0.000620
11	1.75	1.985547	0.949997	5.000329	−0.000328
12	1.74	2.008457	0.950007	4.999286	0.000715
13	1.73	2.033840	0.950010	4.999012	0.000989
14	1.72	2.062090	0.950000	4.999984	0.000018
15	1.71	2.095160	0.950006	4.999357	0.000644
16	1.70	2.133445	0.950005	4.999548	0.000453
17	1.69	2.178900	0.949995	5.000508	−0.000507
18	1.68	2.236214	0.949997	5.000270	−0.000268
19	1.67	2.312423	0.950004	4.999554	0.000447
20	1.66	2.424715	0.950007	4.999352	0.000650
21	1.65	2.646532	0.949994	5.000610	−0.000608
No. of pairs		Mean COMPUT. risk errors		STD. ERR. errors	
21		0.0000387430		0.0001182909	
Z-ratio				Status of significance	
0.32752				N.S.—not significant	
Parameter values					
PROB. = 0.94	RHO = 0.8		BIGH = 1.7585		SMHL = 1.56
Pair No.	SMH	SMK	PROB. level	Risk level (%)	Risk error (%)
Table No. 8.2-44					
1	1.76	1.757001	0.940002	5.999768	0.000232
2	1.75	1.767041	0.939991	6.000889	−0.000888
3	1.74	1.777978	0.939993	6.000716	−0.000715
4	1.73	1.789813	0.940002	5.999780	0.000221
5	1.72	1.802159	0.939993	6.000686	−0.000685
6	1.71	1.815797	0.940003	5.999744	0.000256
7	1.70	1.830341	0.940003	5.999661	0.000340
8	1.69	1.846183	0.940008	5.999184	0.000817
9	1.68	1.863324	0.940009	5.999124	0.000876
10	1.67	1.881768	0.939997	6.000323	−0.000322
11	1.66	1.902688	0.940009	5.999083	0.000918
12	1.65	1.925306	0.940001	5.999929	0.000072
13	1.64	1.950797	0.939999	6.000078	−0.000077
14	1.63	1.979943	0.940007	5.999261	0.000739
15	1.62	2.012747	0.939999	6.000150	−0.000149

(continued)

Table 8.2 (continued)

Parameter values					
PROB. = 0.94	RHO = 0.8		BIGH = 1.7585		SMHL = 1.56
Pair No.	SMH	SMK	PROB. level	Risk level (%)	Risk error (%)
16	1.61	2.051165	0.939993	6.000728	−0.000727
17	1.60	2.097545	0.939995	6.000525	−0.000525
18	1.59	2.155794	0.940006	5.999387	0.000614
19	1.58	2.232165	0.940009	5.999094	0.000906
20	1.57	2.344631	0.940008	5.999178	0.000823
21	1.56	2.569757	0.940005	5.999488	0.000513
No. of pairs		Mean COMPUT. risk errors		STD. ERR. errors	
21		0.0001471151		0.0001272338	
Z-ratio				Status of significance	
1.15626				N.S.—not significant	
Parameter values					
PROB. = 0.93	RHO = 0.8		BIGH = 1.6817		SMHL = 1.49
Pair No.	SMH	SMK	PROB. level	Risk level (%)	Risk error (%)
Table No. 8.2-45					
1	1.69	1.673831	0.930010	6.999004	0.000995
2	1.68	1.683402	0.929994	7.000625	−0.000626
3	1.67	1.693873	0.929996	7.000375	−0.000376
4	1.66	1.705051	0.929999	7.000095	−0.000095
5	1.65	1.717134	0.930010	6.999010	0.000989
6	1.64	1.729734	0.929999	7.000065	−0.000066
7	1.63	1.743633	0.930009	6.999117	0.000882
8	1.62	1.758250	0.929997	7.000268	−0.000268
9	1.61	1.774171	0.929991	7.000947	−0.000948
10	1.60	1.791790	0.929998	7.000160	−0.000161
11	1.59	1.810720	0.929990	7.000989	−0.000989
12	1.58	1.832133	0.930007	6.999302	0.000697
13	1.57	1.855253	0.929998	7.000172	−0.000173
14	1.56	1.881644	0.930009	6.999100	0.000900
15	1.55	1.910918	0.929999	7.000065	−0.000066
16	1.54	1.945032	0.930005	6.999463	0.000536
17	1.53	1.984378	0.929996	7.000375	−0.000376
18	1.52	2.032477	0.930006	6.999374	0.000626
19	1.51	2.091673	0.930000	7.000041	−0.000042
20	1.50	2.171347	0.930007	6.999302	0.000697
21	1.49	2.288688	0.929994	7.000649	−0.000650
No. of pairs		Mean COMPUT. risk errors		STD. ERR. errors	
21		0.0000677326		0.0001334631	
Z-ratio				Status of significance	
0.50750				N.S.—not significant	

(continued)

8.2 Contents for Table 8.2 (Table No. 8.2-1 to 400)

Table 8.2 (continued)

Parameter values					
PROB. = 0.92	RHO = 0.8		BIGH = 1.613		SMHL = 1.41
Pair No.	SMH	SMK	PROB. level	Risk level (%)	Risk error (%)
Table No. 8.2-46					
1	1.62	1.606226	0.919999	8.000136	−0.000137
2	1.61	1.616006	0.919991	8.000916	−0.000918
3	1.60	1.626692	0.920004	7.999653	0.000346
4	1.59	1.637895	0.920002	7.999760	0.000238
5	1.58	1.649814	0.919997	8.000255	−0.000256
6	1.57	1.662647	0.919996	8.000374	−0.000376
7	1.56	1.676394	0.919992	8.000761	−0.000763
8	1.55	1.691451	0.920002	7.999778	0.000221
9	1.54	1.707429	0.919993	8.000655	−0.000656
10	1.53	1.725112	0.920002	7.999838	0.000161
11	1.52	1.744112	0.919993	8.000738	−0.000739
12	1.51	1.765213	0.919994	8.000613	−0.000614
13	1.50	1.788809	0.920005	7.999540	0.000459
14	1.49	1.814904	0.920004	7.999653	0.000346
15	1.48	1.844280	0.919998	8.000165	−0.000167
16	1.47	1.878114	0.919999	8.000136	−0.000137
17	1.46	1.917971	0.920009	7.999110	0.000888
18	1.45	1.964636	0.919991	8.000862	−0.000864
19	1.44	2.023971	0.920002	7.999766	0.000232
20	1.43	2.102231	0.920009	7.999087	0.000912
21	1.42	2.216606	0.919999	8.000100	−0.000101
22	1.41	2.451475	0.920002	7.999778	0.000221
No. of pairs		Mean COMPUT. risk errors		STD. ERR. errors	
22		−0.0000741171		0.0001087552	
Z-ratio				Status of significance	
0.68150				N.S.—not significant	
Parameter values					
PROB. = 0.91	RHO = 0.8		BIGH = 1.5506		SMHL = 1.35
Pair No.	SMH	SMK	PROB. level	Risk level (%)	Risk error (%)
Table No. 8.2-47					
1	1.56	1.541452	0.910001	8.999872	0.000125
2	1.55	1.551200	0.910002	8.999795	0.000203
3	1.54	1.561468	0.909995	9.000486	−0.000489
4	1.53	1.572649	0.910007	8.999264	0.000733
5	1.52	1.584355	0.910002	8.999795	0.000203
6	1.51	1.596979	0.910004	8.999628	0.000370
7	1.50	1.610329	0.909992	9.000832	−0.000834
8	1.49	1.624993	0.909999	9.000081	−0.000083
9	1.48	1.640583	0.909991	9.000951	−0.000954
10	1.47	1.657885	0.910005	8.999472	0.000525
11	1.46	1.676510	0.910006	8.999389	0.000608
12	1.45	1.696852	0.910003	8.999729	0.000268
13	1.44	1.719304	0.909999	9.000069	−0.000072

(continued)

Table 8.2 (continued)

Parameter values					
PROB. = 0.91	RHO = 0.8		BIGH = 1.5506		SMHL = 1.35
Pair No.	SMH	SMK	PROB. level	Risk level (%)	Risk error (%)
14	1.43	1.744262	0.909995	9.000534	−0.000536
15	1.42	1.772509	0.909998	9.000158	−0.000161
16	1.41	1.804439	0.909994	9.000647	−0.000650
17	1.40	1.841619	0.909997	9.000343	−0.000346
18	1.39	1.885615	0.910000	9.000039	−0.000042
19	1.38	1.939165	0.909998	9.000158	−0.000161
20	1.37	2.008133	0.910004	8.999610	0.000387
21	1.36	2.103850	0.910003	8.999741	0.000256
22	1.35	2.265383	0.910005	8.999502	0.000495
No. of pairs		Mean COMPUT. risk errors		STD. ERR. errors	
22		−0.0000067379		0.0000962514	
Z-ratio				Status of significance	
0.07000				N.S.—not significant	
Parameter values					
PROB. = 0.9025	RHO = 0.8		BIGH = 1.5071		SMHL = 1.30
Pair No.	SMH	SMK	PROB. level	Risk level (%)	Risk error (%)
Table No. 8.2-48					
1	1.51	1.504206	0.902505	9.749496	0.000507
2	1.50	1.514234	0.902496	9.750378	−0.000376
3	1.49	1.524982	0.902494	9.750593	−0.000590
4	1.48	1.536649	0.902509	9.749132	0.000870
5	1.47	1.548847	0.902502	9.749788	0.000215
6	1.46	1.561969	0.902499	9.750056	−0.000054
7	1.45	1.576214	0.902507	9.749306	0.000697
8	1.44	1.591389	0.902502	9.749824	0.000179
9	1.43	1.607888	0.902501	9.749878	0.000125
10	1.42	1.625714	0.902494	9.750622	−0.000620
11	1.41	1.645262	0.902490	9.750992	−0.000989
12	1.40	1.666924	0.902495	9.750462	−0.000459
13	1.39	1.691096	0.902509	9.749097	0.000906
14	1.38	1.717781	0.902506	9.749389	0.000614
15	1.37	1.747764	0.902495	9.750480	−0.000477
16	1.36	1.782610	0.902502	9.749794	0.000209
17	1.35	1.823107	0.902500	9.749973	0.000030
18	1.34	1.871600	0.902495	9.750456	−0.000453
19	1.33	1.932782	0.902506	9.749365	0.000638
20	1.32	2.012907	0.902495	9.750504	−0.000501

(continued)

8.2 Contents for Table 8.2 (Table No. 8.2-1 to 400)

Table 8.2 (continued)

Parameter values					
PROB. = 0.9025	RHO = 0.8		BIGH = 1.5071		SMHL = 1.30
Pair No.	SMH	SMK	PROB. level	Risk level (%)	Risk error (%)
21	1.31	2.133855	0.902495	9.750504	−0.000501
22	1.30	2.397192	0.902506	9.749436	0.000566
No. of pairs		Mean COMPUT. risk errors		STD. ERR. errors	
22		0.0000233236		0.0001137826	
Z-ratio				Status of significance	
0.20498				N.S.—not significant	
Parameter values					
PROB. = 0.9	RHO = 0.8		BIGH = 1.4931		SMHL = 1.29
Pair No.	SMH	SMK	PROB. level	Risk level (%)	Risk error (%)
Table No. 8.2-49					
1	1.50	1.486427	0.900008	9.999246	0.000757
2	1.49	1.496207	0.899998	10.000250	−0.000250
3	1.48	1.506707	0.899997	10.000330	−0.000328
4	1.47	1.517930	0.899999	10.000080	−0.000077
5	1.46	1.529880	0.899999	10.000090	−0.000083
6	1.45	1.542559	0.899991	10.000940	−0.000942
7	1.44	1.556361	0.899997	10.000350	−0.000346
8	1.43	1.571094	0.899994	10.000650	−0.000644
9	1.42	1.587151	0.900000	10.000010	−0.000006
10	1.41	1.604535	0.900004	9.999627	0.000376
11	1.40	1.623251	0.899993	10.000730	−0.000727
12	1.39	1.644082	0.900000	10.000040	−0.000042
13	1.38	1.667032	0.900003	9.999657	0.000346
14	1.37	1.692495	0.900003	9.999716	0.000286
15	1.36	1.721258	0.900008	9.999246	0.000757
16	1.35	1.753712	0.899999	10.000070	−0.000066
17	1.34	1.791817	0.900010	9.999048	0.000954
18	1.33	1.836358	0.900000	10.000030	−0.000030
19	1.32	1.891243	0.900000	9.999991	0.000012
20	1.31	1.961948	0.900002	9.999842	0.000161
21	1.30	2.061758	0.900008	9.999239	0.000763
22	1.29	2.231301	0.899992	10.000820	−0.000817
No. of pairs		Mean COMPUT. risk errors		STD. ERR. errors	
22		0.0000023324		0.0001071770	
Z-ratio				Status of significance	
0.02176				N.S.—not significant	
Parameter values					
PROB. = 0.85	RHO = 0.8		BIGH = 1.2552		SMHL = 1.04
Pair No.	SMH	SMK	PROB. level	Risk level (%)	Risk error (%)
Table No. 8.2-50					
1	1.26	1.250418	0.849993	15.000660	−0.000662
2	1.25	1.260422	0.849992	15.000770	−0.000769

(continued)

Table 8.2 (continued)

Parameter values					
PROB. = 0.85	RHO = 0.8		BIGH = 1.2552		SMHL = 1.04
Pair No.	SMH	SMK	PROB. level	Risk level (%)	Risk error (%)
3	1.24	1.271172	0.850007	14.999290	0.000703
4	1.23	1.282479	0.850008	14.999160	0.000834
5	1.22	1.294345	0.849991	15.000930	−0.000936
6	1.21	1.307362	0.850009	14.999110	0.000888
7	1.20	1.320947	0.849992	15.000760	−0.000763
8	1.19	1.335691	0.849991	15.000890	−0.000894
9	1.18	1.351599	0.849992	15.000810	−0.000811
10	1.17	1.368870	0.849999	15.000150	−0.000155
11	1.16	1.387510	0.849995	15.000510	−0.000513
12	1.15	1.407914	0.849995	15.000510	−0.000507
13	1.14	1.430479	0.850005	14.999490	0.000513
14	1.13	1.455209	0.849998	15.000170	−0.000173
15	1.12	1.482892	0.849995	15.000500	−0.000501
16	1.11	1.514315	0.850000	15.000040	−0.000042
17	1.10	1.550266	0.850000	14.999990	0.000012
18	1.09	1.592313	0.850002	14.999820	0.000179
19	1.08	1.642806	0.850004	14.999590	0.000411
20	1.07	1.705658	0.850000	14.999990	0.000012
21	1.06	1.789471	0.849999	15.000090	−0.000095
22	1.05	1.916127	0.849999	15.000140	−0.000143
23	1.04	2.199304	0.849997	15.000260	−0.000262
No. of pairs		Mean COMPUT. risk errors		STD. ERR. errors	
23		−0.0001529853		0.0001105774	
Z-ratio				Status of significance	
1.38351				N.S.—not significant	
Parameter values					
PROB. = 0.8	RHO = 0.8		BIGH = 1.0662		SMHL = 0.85
Pair No.	SMH	SMK	PROB. level	Risk level (%)	Risk error (%)
Table No. 8.2-51					
1	1.07	1.062511	0.799997	20.000350	−0.000352
2	1.06	1.072632	0.800003	19.999660	0.000340
3	1.05	1.083236	0.799994	20.000580	−0.000578
4	1.04	1.094622	0.800003	19.999700	0.000304
5	1.03	1.106602	0.799995	20.000460	−0.000459
6	1.02	1.119473	0.800001	19.999900	0.000095
7	1.01	1.133144	0.799997	20.000290	−0.000292
8	1.00	1.147915	0.800008	19.999230	0.000769
9	0.99	1.163695	0.800008	19.999230	0.000769
10	0.98	1.180685	0.800005	19.999520	0.000477
11	0.97	1.199089	0.800002	19.999800	0.000203

(continued)

8.2 Contents for Table 8.2 (Table No. 8.2-1 to 400)

Table 8.2 (continued)

Parameter values					
PROB. = 0.8	RHO = 0.8		BIGH = 1.0662		SMHL = 0.85
Pair No.	SMH	SMK	PROB. level	Risk level (%)	Risk error (%)
12	0.96	1.219109	0.799999	20.000140	−0.000137
13	0.95	1.241144	0.800006	19.999420	0.000584
14	0.94	1.265202	0.799995	20.000540	−0.000542
15	0.93	1.292268	0.800009	19.999110	0.000888
16	0.92	1.322352	0.800002	19.999800	0.000203
17	0.91	1.356633	0.800001	19.999920	0.000077
18	0.90	1.396294	0.800001	19.999880	0.000119
19	0.89	1.443299	0.800007	19.999320	0.000679
20	0.88	1.500392	0.800000	20.000010	−0.000012
21	0.87	1.573834	0.799999	20.000140	−0.000137
22	0.86	1.676527	0.799992	20.000810	−0.000811
23	0.85	1.854578	0.800004	19.999570	0.000429
No. of pairs		Mean COMPUT. risk errors		STD. ERR. errors	
23		0.0001090268		0.0000948162	
Z-ratio				Status of significance	
1.14988				N.S.—not significant	
Parameter values					
PROB. = 0.75	RHO = 0.8		BIGH = 0.9043		SMHL = 0.68
Pair No.	SMH	SMK	PROB. level	Risk level (%)	Risk error (%)
Table No. 8.2-52					
1	0.91	0.898538	0.750000	24.999980	0.000018
2	0.90	0.908523	0.750004	24.999570	0.000429
3	0.89	0.919025	0.750000	25.000020	−0.000018
4	0.88	0.930150	0.749996	25.000440	−0.000441
5	0.87	0.942003	0.749999	25.000130	−0.000131
6	0.86	0.954593	0.750000	25.000000	0.000000
7	0.85	0.967928	0.749990	25.000980	−0.000978
8	0.84	0.982311	0.749998	25.000160	−0.000161
9	0.83	0.997653	0.749998	25.000210	−0.000215
10	0.82	1.014161	0.749999	25.000100	−0.000095
11	0.81	1.032039	0.750008	24.999220	0.000781
12	0.80	1.051299	0.750004	24.999620	0.000381
13	0.79	1.072246	0.749997	25.000340	−0.000340
14	0.78	1.095283	0.749998	25.000230	−0.000226
15	0.77	1.120618	0.749991	25.000900	−0.000900
16	0.76	1.149045	0.750003	24.999720	0.000286
17	0.75	1.180969	0.750009	24.999070	0.000930
18	0.74	1.217188	0.750002	24.999830	0.000173
19	0.73	1.259279	0.749996	25.000420	−0.000417
20	0.72	1.309604	0.750001	24.999890	0.000113
21	0.71	1.371694	0.750004	24.999560	0.000441
22	0.70	1.452601	0.749998	25.000250	−0.000250

(continued)

Table 8.2 (continued)

Parameter values					
PROB. = 0.75	RHO = 0.8		BIGH = 0.9043		SMHL = 0.68
Pair No.	SMH	SMK	PROB. level	Risk level (%)	Risk error (%)
23	0.69	1.571094	0.750006	24.999450	0.000548
24	0.68	1.801023	0.750002	24.999760	0.000244
No. of pairs		Mean COMPUT. risk errors		STD. ERR. errors	
24		0.0000069141		0.0000907817	
Z-ratio				Status of significance	
0.07616				N.S.—not significant	
Parameter values					
PROB. = 0.7	RHO = 0.8		BIGH = 0.7589		SMHL = 0.53
Pair No.	SMH	SMK	PROB. level	Risk level (%)	Risk error (%)
Table No. 8.2-53					
1	0.76	0.757606	0.700006	29.999370	0.000632
2	0.75	0.767808	0.700005	29.999510	0.000489
3	0.74	0.778576	0.700001	29.999870	0.000137
4	0.73	0.790018	0.700005	29.999470	0.000530
5	0.72	0.802050	0.699993	30.000710	−0.000709
6	0.71	0.814976	0.700002	29.999760	0.000238
7	0.70	0.828615	0.699993	30.000720	−0.000721
8	0.69	0.843274	0.699997	30.000310	−0.000310
9	0.68	0.858966	0.700001	29.999940	0.000060
10	0.67	0.875807	0.700002	29.999830	0.000173
11	0.66	0.893909	0.699996	30.000430	−0.000429
12	0.65	0.913584	0.699998	30.000200	−0.000197
13	0.64	0.935045	0.700005	29.999470	0.000530
14	0.63	0.958509	0.700008	29.999160	0.000840
15	0.62	0.984191	0.699993	30.000710	−0.000709
16	0.61	1.012896	0.699990	30.000990	−0.000983
17	0.60	1.045429	0.700009	29.999150	0.000852
18	0.59	1.082206	0.700003	29.999660	0.000340
19	0.58	1.125012	0.700007	29.999310	0.000691
20	0.57	1.175636	0.699994	30.000560	−0.000554
21	0.56	1.238601	0.700008	29.999180	0.000817
22	0.55	1.320191	0.699998	30.000160	−0.000155
23	0.54	1.439192	0.700002	29.999840	0.000161
24	0.53	1.667909	0.699991	30.000880	−0.000876
No. of pairs		Mean COMPUT. risk errors		STD. ERR. errors	
24		0.0000338554		0.0001148318	
Z-ratio				Status of significance	
0.29483				N.S.—not significant	

(continued)

8.2 Contents for Table 8.2 (Table No. 8.2-1 to 400)

Table 8.2 (continued)

Parameter values					
PROB. = 0.65	RHO = 0.8		BIGH = 0.6242		SMHL = 0.39
Pair No.	SMH	SMK	PROB. level	Risk level (%)	Risk error (%)
Table No. 8.2-54					
1	0.63	0.618258	0.650007	34.999290	0.000715
2	0.62	0.628135	0.649993	35.000720	−0.000715
3	0.61	0.638633	0.649993	35.000710	−0.000703
4	0.60	0.649767	0.650001	34.999880	0.000119
5	0.59	0.661457	0.649994	35.000630	−0.000626
6	0.58	0.673917	0.649998	35.000240	−0.000232
7	0.57	0.687167	0.650004	34.999620	0.000381
8	0.56	0.701229	0.650003	34.999710	0.000292
9	0.55	0.716223	0.650000	35.000050	−0.000048
10	0.54	0.732271	0.649996	35.000430	−0.000429
11	0.53	0.749596	0.650004	34.999610	0.000393
12	0.52	0.768225	0.650005	34.999460	0.000548
13	0.51	0.788386	0.650005	34.999460	0.000542
14	0.50	0.810306	0.650003	34.999740	0.000262
15	0.49	0.834314	0.650001	34.999950	0.000054
16	0.48	0.860743	0.649993	35.000660	−0.000656
17	0.47	0.890319	0.650002	34.999760	0.000238
18	0.46	0.923574	0.650010	34.999050	0.000954
19	0.45	0.961342	0.650004	34.999640	0.000370
20	0.44	1.005239	0.649999	35.000070	−0.000072
21	0.43	1.057668	0.650002	34.999790	0.000215
22	0.42	1.122211	0.649991	35.000940	−0.000936
23	0.41	1.207337	0.649994	35.000610	−0.000608
24	0.40	1.332657	0.649996	35.000400	−0.000393
25	0.39	1.630290	0.649994	35.000650	−0.000644
No. of pairs		Mean COMPUT. risk errors		STD. ERR. errors	
25		−0.0000375968		0.0001005985	
Z-ratio				Status of significance	
0.37373				N.S.—not significant	
Parameter values					
PROB. = 0.6	RHO = 0.8		BIGH = 0.4962		SMHL = 0.27
Pair No.	SMH	SMK	PROB. level	Risk level (%)	Risk error (%)
Table No. 8.2-55					
1	0.50	0.492624	0.600007	39.999260	0.000739
2	0.49	0.502674	0.600004	39.999640	0.000358
3	0.48	0.513240	0.599995	40.000520	−0.000519
4	0.47	0.524446	0.599996	40.000380	−0.000381
5	0.46	0.536323	0.600003	39.999740	0.000262
6	0.45	0.548901	0.600009	39.999120	0.000882
7	0.44	0.562117	0.599993	40.000730	−0.000727
8	0.43	0.576303	0.599997	40.000260	−0.000262
9	0.42	0.591401	0.599998	40.000240	−0.000238

(continued)

Table 8.2 (continued)

Parameter values					
PROB. = 0.6	RHO = 0.8		BIGH = 0.4962		SMHL = 0.27
Pair No.	SMH	SMK	PROB. level	Risk level (%)	Risk error (%)
10	0.41	0.607554	0.599999	40.000100	−0.000101
11	0.40	0.624911	0.600004	39.999570	0.000423
12	0.39	0.643526	0.599999	40.000080	−0.000083
13	0.38	0.663753	0.600007	39.999320	0.000679
14	0.37	0.685561	0.599994	40.000640	−0.000644
15	0.36	0.709515	0.599994	40.000610	−0.000608
16	0.35	0.735891	0.599996	40.000440	−0.000441
17	0.34	0.765175	0.599998	40.000200	−0.000203
18	0.33	0.798057	0.600004	39.999650	0.000352
19	0.32	0.835435	0.600009	39.999060	0.000936
20	0.31	0.878419	0.599998	40.000190	−0.000191
21	0.30	0.929503	0.600000	40.000050	−0.000054
22	0.29	0.991984	0.599995	40.000490	−0.000489
23	0.28	1.130508	0.599996	40.000440	−0.000441
24	0.27	1.235342	0.599993	40.000670	−0.000668
No. of pairs		Mean COMPUT. risk errors		STD. ERR. errors	
24		−0.0000567436		0.0001008700	
Z-ratio				Status of significance	
0.56254				N.S.—not significant	
Parameter values					
PROB. = 0.55	RHO = 0.8		BIGH = 0.3727		SMHL = 0.18
Pair No.	SMH	SMK	PROB. level	Risk level (%)	Risk error (%)
Table No. 8.2-56					
1	0.38	0.365540	0.550009	44.999060	0.000936
2	0.37	0.375420	0.550008	44.999200	0.000805
3	0.36	0.385750	0.549994	45.000640	−0.000644
4	0.35	0.396775	0.550006	44.999450	0.000554
5	0.34	0.408350	0.550006	44.999380	0.000626
6	0.33	0.420534	0.549998	45.000250	−0.000250
7	0.32	0.433493	0.549998	45.000200	−0.000197
8	0.31	0.447300	0.550008	44.999220	0.000775
9	0.30	0.461846	0.549994	45.000590	−0.000590
10	0.29	0.477421	0.549990	45.000960	−0.000960
11	0.28	0.494137	0.549995	45.000470	−0.000471
12	0.27	0.512120	0.550006	44.999380	0.000626
13	0.26	0.531321	0.549994	45.000620	−0.000620
14	0.25	0.552301	0.550008	44.999210	0.000793
15	0.24	0.574866	0.549991	45.000860	−0.000864
16	0.23	0.599834	0.550007	44.999290	0.000709

(continued)

8.2 Contents for Table 8.2 (Table No. 8.2-1 to 400)

Table 8.2 (continued)

Parameter values					
PROB. = 0.55	RHO = 0.8		BIGH = 0.3727		SMHL = 0.18
Pair No.	SMH	SMK	PROB. level	Risk level (%)	Risk error (%)
17	0.22	0.627090	0.549995	45.000510	−0.000507
18	0.21	0.657547	0.549998	45.000180	−0.000179
19	0.20	0.691792	0.550002	44.999780	0.000221
20	0.19	0.798463	0.550007	44.999290	0.000709
21	0.18	0.834977	0.549997	45.000260	−0.000262
No. of pairs		Mean COMPUT. risk errors		STD. ERR. errors	
21		0.0000549988		0.0001342465	
Z-ratio				Status of significance	
0.40969				N.S.—not significant	
Parameter values					
PROB. = 0.5	RHO = 0.8		BIGH = 0.2511		SMHL = 0.11
Pair No.	SMH	SMK	PROB. level	Risk level (%)	Risk error (%)
Table No. 8.2-57					
1	0.26	0.242407	0.500009	49.999090	0.000906
2	0.25	0.252205	0.500007	49.999350	0.000650
3	0.24	0.262518	0.500007	49.999330	0.000668
4	0.23	0.273354	0.500003	49.999680	0.000322
5	0.22	0.284741	0.499993	50.000700	−0.000697
6	0.21	0.296826	0.499995	50.000530	−0.000530
7	0.20	0.309592	0.499994	50.000640	−0.000629
8	0.19	0.323157	0.499999	50.000060	−0.000063
9	0.18	0.337589	0.500010	49.999050	0.000954
10	0.17	0.352823	0.499999	50.000090	−0.000089
11	0.16	0.369168	0.500002	49.999830	0.000173
12	0.15	0.386650	0.500002	49.999810	0.000191
13	0.14	0.405444	0.500004	49.999630	0.000376
14	0.13	0.425635	0.499995	50.000550	−0.000548
15	0.12	0.525915	0.499996	50.000400	−0.000390
16	0.11	0.539795	0.500004	49.999600	0.000405
No. of pairs		Mean COMPUT. risk errors		STD. ERR. errors	
16		0.0000997501		0.0001290839	
Z-ratio				Status of significance	
0.77275				N.S.—not significant	
Parameter values					
PROB. = 0.99	RHO = 0.7		BIGH = 2.5324		SMHL = 2.33
Pair No.	SMH	SMK	PROB. level	Risk level (%)	Risk error (%)
Table No. 8.2-58					
1	2.54	2.524823	0.989997	1.000345	−0.000346
2	2.53	2.534802	0.989999	1.000148	−0.000149
3	2.52	2.544861	0.989993	1.000714	−0.000715
4	2.51	2.556562	0.989999	1.000059	−0.000060
5	2.50	2.568344	0.989997	1.000327	−0.000328

(continued)

Table 8.2 (continued)

Parameter values					
PROB. = 0.99	RHO = 0.7		BIGH = 2.5324		SMHL = 2.33
Pair No.	SMH	SMK	PROB. level	Risk level (%)	Risk error (%)
6	2.49	2.581771	0.990003	0.999671	0.000328
7	2.48	2.595282	0.989999	1.000094	−0.000095
8	2.47	2.610438	0.990001	0.999916	0.000083
9	2.46	2.625681	0.989990	1.000977	−0.000978
10	2.45	2.644134	0.989998	1.000226	−0.000226
11	2.44	2.664236	0.990004	0.999618	0.000381
12	2.43	2.685990	0.990006	0.999391	0.000608
13	2.42	2.709396	0.990002	0.999773	0.000226
14	2.41	2.736016	0.990001	0.999892	0.000107
15	2.40	2.765854	0.989998	1.000243	−0.000244
16	2.39	2.802034	0.990005	0.999546	0.000453
17	2.38	2.841433	0.989995	1.000476	−0.000477
18	2.37	2.893428	0.990006	0.999445	0.000554
19	2.36	2.954893	0.990001	0.999910	0.000089
20	2.35	3.041457	0.990007	0.999308	0.000691
21	2.34	3.165619	0.989998	1.000220	−0.000221
22	2.33	3.452381	0.989997	1.000327	−0.000328
No. of pairs		Mean COMPUT. risk errors		STD. ERR. errors	
22		−0.0000279883		0.0000877515	
Z-ratio				Status of significance	
0.31895				N.S.—not significant	
Parameter values					
PROB. = 0.98	RHO = 0.7		BIGH = 2.2702		SMHL = 2.06
Pair No.	SMH	SMK	PROB. level	Risk level (%)	Risk error (%)
Table No. 8.2-59					
1	2.28	2.260442	0.980001	1.999885	0.000113
2	2.27	2.270400	0.980007	1.999271	0.000727
3	2.26	2.280446	0.980001	1.999939	0.000060
4	2.25	2.291363	0.979999	2.000076	−0.000077
5	2.24	2.303151	0.980001	1.999855	0.000143
6	2.23	2.315812	0.980005	1.999503	0.000495
7	2.22	2.329348	0.980008	1.999211	0.000787
8	2.21	2.342978	0.979992	2.000767	−0.000769
9	2.20	2.359046	0.980003	1.999664	0.000334
10	2.19	2.375212	0.979992	2.000821	−0.000823
11	2.18	2.393819	0.979998	2.000159	−0.000161
12	2.17	2.414089	0.970004	1.999623	0.000376
13	2.16	2.436022	0.980004	1.999652	0.000346
14	2.15	2.459620	0.979994	2.000642	−0.000644
15	2.14	2.486446	0.979990	2.000988	−0.000989

(continued)

8.2 Contents for Table 8.2 (Table No. 8.2-1 to 400)

Table 8.2 (continued)

Parameter values					
PROB. = 0.98	RHO = 0.7		BIGH = 2.2702		SMHL = 2.06
Pair No.	SMH	SMK	PROB. level	Risk level (%)	Risk error (%)
16	2.13	2.518066	0.980002	1.999772	0.000226
17	2.12	2.552916	0.979999	2.000070	−0.000072
18	2.11	2.594125	0.980002	1.999837	0.000161
19	2.10	2.641694	0.979991	2.000952	−0.000954
20	2.09	2.703436	0.979999	2.000141	−0.000143
21	2.08	2.784042	0.980003	1.999688	0.000310
22	2.07	2.902262	0.980008	1.999182	0.000817
23	2.06	3.126848	0.980007	1.999301	0.000697
No. of pairs		Mean COMPUT. risk errors		STD. ERR. errors	
23		0.0000399848		0.0001103150	
Z-ratio				Status of significance	
0.36246				N.S.—not significant	
Parameter values					
PROB. = 0.97	RHO = 0.7		BIGH = 2.1038		SMHL = 1.89
Pair No.	SMH	SMK	PROB. level	Risk level (%)	Risk error (%)
Table No. 8.2-60					
1	2.11	2.097619	0.969996	3.000379	−0.000381
2	2.10	2.107607	0.969998	3.000164	−0.000167
3	2.09	2.118473	0.970010	2.999032	0.000966
4	2.08	2.129435	0.970002	2.999812	0.000185
5	2.07	2.141277	0.970000	3.000027	−0.000030
6	2.06	2.154000	0.970000	2.999973	0.000024
7	2.05	2.167606	0.970001	2.999938	0.000060
8	2.04	2.182095	0.969998	3.000188	−0.000191
9	2.03	2.197861	0.970000	2.999962	0.000036
10	2.02	2.214514	0.969993	3.000701	−0.000703
11	2.01	2.233227	0.970001	2.999854	0.000143
12	2.00	2.253612	0.970009	2.999056	0.000942
13	1.99	2.274889	0.969994	3.000569	−0.000572
14	1.98	2.299403	0.970000	2.999997	0.000000
15	1.97	2.326375	0.970000	3.000015	−0.000018
16	1.96	2.356588	0.969998	3.000236	−0.000238
17	1.95	2.391606	0.970005	2.999520	0.000477
18	1.94	2.431430	0.970001	2.999854	0.000143
19	1.93	2.479189	0.970005	2.999497	0.000501
20	1.92	2.536445	0.969997	3.000277	−0.000280
21	1.91	2.611014	0.969998	3.000182	−0.000185
22	1.90	2.716960	0.970007	2.999288	0.000709

(continued)

Table 8.2 (continued)

Parameter values					
PROB. = 0.97	RHO = 0.7		BIGH = 2.1038		SMHL = 1.89
Pair No.	SMH	SMK	PROB. level	Risk level (%)	Risk error (%)
23	1.89	2.891785	0.969999	3.000111	−0.000113
No. of pairs		Mean COMPUT. risk errors		STD. ERR. errors	
23		0.0000543892		0.0000853938	
Z-ratio				Status of significance	
0.63692				N.S.—not significant	
Parameter values					
PROB. = 0.96	RHO = 0.7		BIGH = 1.9789		SMHL = 1.76
Pair No.	SMH	SMK	PROB. level	Risk level (%)	Risk error (%)
Table No. 8.2-61					
1	1.98	1.977801	0.960003	3.999663	0.000340
2	1.97	1.987840	0.959995	4.000473	−0.000471
3	1.96	1.998763	0.959999	4.000080	−0.000077
4	1.95	2.010181	0.959996	4.000414	−0.000411
5	1.94	2.022486	0.959999	4.000127	−0.000125
6	1.93	2.035679	0.960005	3.999543	0.000459
7	1.92	2.049763	0.960010	3.999031	0.000972
8	1.91	2.064348	0.959997	4.000342	−0.000340
9	1.90	2.080608	0.960003	3.999686	0.000316
10	1.89	2.097763	0.959997	4.000301	−0.000298
11	1.88	2.116596	0.959998	4.000181	−0.000179
12	1.87	2.137110	0.959999	4.000062	−0.000060
13	1.86	2.159307	0.959993	4.000682	−0.000679
14	1.85	2.183969	0.959992	4.000836	−0.000834
15	1.84	2.211879	0.960001	3.999871	0.000131
16	1.83	2.243040	0.960006	3.999365	0.000638
17	1.82	2.278236	0.960006	3.999400	0.000602
18	1.81	2.318248	0.959994	4.000604	−0.000602
19	1.80	2.366205	0.959992	4.000801	−0.000799
20	1.79	2.425234	0.959996	4.000425	−0.000423
21	1.78	2.500025	0.959991	4.000902	−0.000900
22	1.77	2.606204	0.959999	4.000062	−0.000060
23	1.76	2.781275	0.959992	4.000807	−0.000805
No. of pairs		Mean COMPUT. risk errors		STD. ERR. errors	
23		−0.0001502534		0.0001048221	
Z-ratio				Status of significance	
1.43341				N.S.—not significant	
Parameter values					
PROB. = 0.95	RHO = 0.7		BIGH = 1.8785		SMHL = 1.65
Pair No.	SMH	SMK	PROB. level	Risk level (%)	Risk error (%)
Table No. 8.2-62					
1	1.88	1.874657	0.950002	4.999781	0.000221
2	1.87	1.884695	0.949997	5.000282	−0.000280

(continued)

8.2 Contents for Table 8.2 (Table No. 8.2-1 to 400)

Table 8.2 (continued)

Parameter values					
PROB. = 0.95	RHO = 0.7		BIGH = 1.8785		SMHL = 1.65
Pair No.	SMH	SMK	PROB. level	Risk level (%)	Risk error (%)
3	1.86	1.895621	0.950008	4.999209	0.000793
4	1.85	1.906658	0.949992	5.000842	−0.000840
5	1.84	1.918977	0.950004	4.999584	0.000417
6	1.83	1.931801	0.950003	4.999674	0.000328
7	1.82	1.945521	0.950004	4.999578	0.000423
8	1.81	1.960139	0.950003	4.999715	0.000286
9	1.80	1.975658	0.949995	5.000520	−0.000519
10	1.79	1.992469	0.949991	5.000890	−0.000888
11	1.78	2.010966	0.950000	5.000014	−0.000012
12	1.77	2.030760	0.949999	5.000144	−0.000143
13	1.76	2.052635	0.950006	4.999381	0.000620
14	1.75	2.076201	0.950000	5.000007	−0.000006
15	1.74	2.102243	0.949993	5.000663	−0.000662
16	1.73	2.131934	0.950004	4.999638	0.000364
17	1.72	2.164887	0.950000	4.999984	0.000018
18	1.71	2.202666	0.949998	5.000162	−0.000161
19	1.70	2.246836	0.949999	5.000103	−0.000101
20	1.69	2.299743	0.950001	4.999948	0.000054
21	1.68	2.366078	0.950010	4.999018	0.000983
22	1.67	2.452093	0.950003	4.999727	0.000274
23	1.66	2.578885	0.949996	5.000395	−0.000393
24	1.65	2.835518	0.949997	5.000317	−0.000316
No. of pairs		Mean COMPUT. risk errors		STD. ERR. errors	
24		0.0000183582		0.0000945581	
Z-ratio				Status of significance	
0.19415				N.S.—not significant	
Parameter values					
PROB. = 0.94	RHO = 0.7		BIGH = 1.7908		SMHL = 1.65
Pair No.	SMH	SMK	PROB. level	Risk level (%)	Risk error (%)
Table No. 8.2-63					
1	1.80	1.782038	0.940001	5.999953	0.000048
2	1.79	1.791796	0.939999	6.000102	−0.000101
3	1.78	1.802056	0.939995	6.000519	−0.000519
4	1.77	1.813016	0.939997	6.000304	−0.000304
5	1.76	1.824483	0.939991	6.000912	−0.000912
6	1.75	1.836848	0.939996	6.000442	−0.000441
7	1.74	1.850114	0.940006	5.999398	0.000602
8	1.73	1.863893	0.939998	6.000209	−0.000209
9	1.72	1.878967	0.940007	5.999333	0.000668
10	1.71	1.894949	0.940007	5.999321	0.000679
11	1.70	1.911840	0.939994	6.000650	−0.000650
12	1.69	1.930425	0.939995	6.000513	−0.000513

(continued)

Table 8.2 (continued)

Parameter values					
PROB. = 0.94	RHO = 0.7		BIGH = 1.7908		SMHL = 1.65
Pair No.	SMH	SMK	PROB. level	Risk level (%)	Risk error (%)
13	1.68	1.950704	0.940001	5.999905	0.000095
14	1.67	1.972682	0.940002	5.999833	0.000167
15	1.66	1.996750	0.940001	5.999929	0.000072
16	1.65	2.023302	0.939998	6.000173	−0.000173
17	1.64	2.053122	0.940002	5.999804	0.000197
18	1.63	2.086213	0.939991	6.000889	−0.000888
19	1.62	2.124920	0.940002	5.999816	0.000185
20	1.61	2.169247	0.939995	6.000453	−0.000453
21	1.60	2.223103	0.940005	5.999458	0.000542
22	1.59	2.288834	0.939998	6.000191	−0.000191
23	1.58	2.376599	0.940005	5.999458	0.000542
24	1.57	2.505152	0.940005	5.999542	0.000459
25	1.56	2.761995	0.940001	5.999875	0.000125
No. of pairs		Mean COMPUT. risk errors		STD. ERR. errors	
25		0.0000373675		0.0000907381	
Z-ratio				Status of significance	
0.41182				N.S.—not significant	
Parameter values					
PROB. = 0.93	RHO = 0.7		BIGH = 1.7152		SMHL = 1.48
Pair No.	SMH	SMK	PROB. level	Risk level (%)	Risk error (%)
Table No. 8.2-64					
1	1.72	1.710413	0.929998	7.000220	−0.000221
2	1.71	1.720416	0.929997	7.000292	−0.000292
3	1.70	1.730926	0.929993	7.000685	−0.000685
4	1.69	1.742338	0.930009	6.999135	0.000864
5	1.68	1.754067	0.930001	6.999856	0.000143
6	1.67	1.766701	0.930006	6.999433	0.000566
7	1.66	1.780048	0.930004	6.999559	0.000441
8	1.65	1.794304	0.930005	6.999517	0.000483
9	1.64	1.809473	0.930001	6.999869	0.000131
10	1.63	1.825947	0.930009	6.999117	0.000882
11	1.62	1.843338	0.930000	7.000017	−0.000018
12	1.61	1.862430	0.930005	6.999457	0.000542
13	1.60	1.882835	0.929997	7.000280	−0.000280
14	1.59	1.905336	0.930000	7.000041	−0.000042
15	1.58	1.929937	0.929998	7.000190	−0.000191
16	1.57	1.957422	0.930007	6.999332	0.000668
17	1.56	1.987403	0.929992	7.000780	−0.000781
18	1.55	2.022225	0.930008	6.999183	0.000817
19	1.54	2.061113	0.929997	7.000327	−0.000328
20	1.53	2.107192	0.930002	6.999755	0.000244
21	1.52	2.162030	0.930000	7.000005	−0.000006

(continued)

8.2 Contents for Table 8.2 (Table No. 8.2-1 to 400)

Table 8.2 (continued)

Parameter values					
PROB. = 0.93	RHO = 0.7		BIGH = 1.7152		SMHL = 1.48
Pair No.	SMH	SMK	PROB. level	Risk level (%)	Risk error (%)
22	1.51	2.231097	0.930010	6.999028	0.000972
23	1.50	2.320649	0.929993	7.000751	−0.000751
24	1.49	2.457249	0.929999	7.000059	−0.000060
25	1.48	2.750277	0.930004	6.999559	0.000441
No. of pairs		Mean COMPUT. risk errors		STD. ERR. errors	
25		0.0001361737		0.0000995318	
Z-ratio				Status of significance	
1.36814				N.S.—not significant	
Parameter values					
PROB. = 0.92	RHO = 0.7		BIGH = 1.6474		SMHL = 1.41
Pair No.	SMH	SMK	PROB. level	Risk level (%)	Risk error (%)
Table No. 8.2-65					
1	1.65	1.644804	0.919994	8.000583	−0.000584
2	1.64	1.655029	0.920001	7.999861	0.000137
3	1.63	1.665767	0.920005	7.999516	0.000483
4	1.62	1.677021	0.920001	7.999861	0.000137
5	1.61	1.688989	0.920002	7.999838	0.000161
6	1.60	1.701673	0.920001	7.999897	0.000101
7	1.59	1.715075	0.919995	8.000488	−0.000489
8	1.58	1.729394	0.919992	8.000821	−0.000823
9	1.57	1.745022	0.920008	7.999176	0.000823
10	1.56	1.761181	0.919991	8.000916	−0.000918
11	1.55	1.779045	0.920001	7.999951	0.000048
12	1.54	1.798227	0.920005	7.999522	0.000477
13	1.53	1.818730	0.919995	8.000546	−0.000548
14	1.52	1.841338	0.919997	8.000338	−0.000340
15	1.51	1.866052	0.919995	8.000458	−0.000459
16	1.50	1.893659	0.920006	7.999355	0.000644
17	1.49	1.923771	0.919993	8.000713	−0.000715
18	1.48	1.958343	0.920001	7.999897	0.000101
19	1.47	1.997771	0.920006	7.999415	0.000584
20	1.46	2.043229	0.919998	8.000183	−0.000185
21	1.45	2.098236	0.920006	7.999373	0.000626
22	1.44	2.165921	0.920003	7.999707	0.000292
23	1.43	2.255663	0.920007	7.999325	0.000674
24	1.42	2.386215	0.919991	8.000875	−0.000876
25	1.41	2.656020	0.920000	8.000004	−0.000006
No. of pairs		Mean COMPUT. risk errors		STD. ERR. errors	
25		−0.0000252174		0.0001032734	
Z-ratio				Status of significance	
0.24418				N.S.—not significant	

(continued)

Table 8.2 (continued)

Parameter values						
PROB. = 0.91	RHO = 0.7		BIGH = 1.5859		SMHL = 1.35	
Pair No.	SMH	SMK	PROB. level	Risk level (%)	Risk error (%)	
Table No. 8.2-66						
1	1.59	1.581810	0.910009	8.999080	0.000918	
2	1.58	1.591822	0.910006	8.999384	0.000614	
3	1.57	1.602351	0.910001	8.999902	0.000095	
4	1.56	1.613402	0.909991	9.000945	−0.000948	
5	1.55	1.625366	0.910001	8.999878	0.000119	
6	1.54	1.637855	0.909998	9.000194	−0.000197	
7	1.53	1.651264	0.910006	8.999443	0.000554	
8	1.52	1.665204	0.909990	9.000969	−0.000972	
9	1.51	1.680459	0.910001	8.999944	0.000054	
10	1.50	1.696641	0.910002	8.999801	0.000197	
11	1.49	1.713949	0.910000	9.000015	−0.000018	
12	1.48	1.732580	0.909997	9.000283	−0.000286	
13	1.47	1.752735	0.909995	9.000551	−0.000554	
14	1.46	1.774610	0.909990	9.000963	−0.000966	
15	1.45	1.798599	0.909990	9.000987	−0.000989	
16	1.44	1.825098	0.909994	9.000582	−0.000584	
17	1.43	1.854499	0.909998	9.000212	−0.000215	
18	1.42	1.887588	0.910006	8.999371	0.000626	
19	1.41	1.924759	0.910001	8.999909	0.000089	
20	1.40	1.967578	0.909994	9.000635	−0.000638	
21	1.39	2.018784	0.910004	8.999562	0.000435	
22	1.38	2.080333	0.909997	9.000338	−0.000340	
23	1.37	2.159261	0.909994	9.000588	−0.000590	
24	1.36	2.269635	0.910000	9.000051	−0.000054	
25	1.35	2.453645	0.909994	9.000564	−0.000566	
No. of pairs		Mean COMPUT. risk errors		STD. ERR. errors		
25		−0.0001620788		0.0001056869		
Z-ratio				Status of significance		
1.53357				N.S.—not significant		
Parameter values						
PROB. = 0.9025	RHO = 0.7		BIGH = 1.5429		SMHL = 1.30	
Pair No.	SMH	SMK	PROB. level	Risk level (%)	Risk error (%)	
Table No. 8.2-67						
1	1.55	1.536028	0.902508	9.749240	0.000763	
2	1.54	1.545806	0.902497	9.750307	−0.000304	
3	1.53	1.556300	0.902503	9.749681	0.000322	
4	1.52	1.567317	0.902505	9.749496	0.000507	
5	1.51	1.578861	0.902499	9.750128	−0.000125	
6	1.50	1.591129	0.902497	9.750342	−0.000340	
7	1.49	1.604319	0.902508	9.749174	0.000829	

(continued)

Table 8.2 (continued)

Parameter values					
PROB. = 0.9025	RHO = 0.7		BIGH = 1.5429		SMHL = 1.30
Pair No.	SMH	SMK	PROB. level	Risk level (%)	Risk error (%)
8	1.48	1.618043	0.902498	9.750176	−0.000173
9	1.47	1.632696	0.902491	9.750932	−0.000930
10	1.46	1.648476	0.902492	9.750807	−0.000805
11	1.45	1.665580	0.902506	9.749365	0.000638
12	1.44	1.683622	0.902499	9.750109	−0.000107
13	1.43	1.703385	0.902509	9.749090	0.000912
14	1.42	1.724484	0.902501	9.749926	0.000077
15	1.41	1.747701	0.902504	9.749567	0.000435
16	1.40	1.773042	0.902502	9.749794	0.000209
17	1.39	1.800900	0.902494	9.750647	−0.000644
18	1.38	1.832451	0.902506	9.749407	0.000596
19	1.37	1.867308	0.902491	9.750944	−0.000942
20	1.36	1.907819	0.902498	9.750247	−0.000244
21	1.35	1.954769	0.902496	9.750438	−0.000435
22	1.34	2.010898	0.902492	9.750801	−0.000799
23	1.33	2.081286	0.902500	9.750051	−0.000048
24	1.32	2.173752	0.902493	9.750748	−0.000745
25	1.31	2.314081	0.902505	9.749514	0.000489
26	1.30	2.612434	0.902496	9.750432	−0.000429
No. of pairs		Mean COMPUT. risk errors		STD. ERR. errors	
26		−0.0000479045		0.0001090298	
Z-ratio				Status of significance	
0.43937				N.S.—not significant	
Parameter values					
PROB. = 0.9	RHO = 0.7		BIGH = 1.5291		SMHL = 1.29
Pair No.	SMH	SMK	PROB. level	Risk level (%)	Risk error (%)
Table No. 8.2-68					
1	1.53	1.528201	0.899992	10.000840	−0.000840
2	1.52	1.538450	0.899995	10.000540	−0.000542
3	1.51	1.549223	0.899995	10.000530	−0.000525
4	1.50	1.560718	0.900005	9.999478	0.000525
5	1.49	1.572742	0.900005	9.999519	0.000483
6	1.48	1.585493	0.900005	9.999460	0.000542
7	1.47	1.598975	0.900002	9.999824	0.000179
8	1.46	1.613384	0.900003	9.999704	0.000298
9	1.45	1.628726	0.900002	9.999800	0.000203
10	1.44	1.645197	0.900005	9.999502	0.000501
11	1.43	1.662802	0.900003	9.999662	0.000340
12	1.42	1.681738	0.900000	9.999991	0.000012
13	1.41	1.702401	0.900007	9.999275	0.000727
14	1.40	1.724597	0.900000	10.000040	−0.000042
15	1.39	1.748917	0.899995	10.000520	−0.000519
16	1.38	1.775950	0.900003	9.999752	0.000250

(continued)

Table 8.2 (continued)

Parameter values					
PROB. = 0.9	RHO = 0.7		BIGH = 1.5291		SMHL = 1.29
Pair No.	SMH	SMK	PROB. level	Risk level (%)	Risk error (%)
17	1.37	1.805895	0.900007	9.999293	0.000709
18	1.36	1.839147	0.899998	10.000190	−0.000185
19	1.35	1.877273	0.900002	9.999806	0.000197
20	1.34	1.921448	0.900009	9.999061	0.000942
21	1.33	1.972848	0.899994	10.000600	−0.000602
22	1.32	2.036167	0.899993	10.000710	−0.000709
23	1.31	2.117657	0.899995	10.000460	−0.000453
24	1.30	2.232949	0.900009	9.999144	0.000858
25	1.29	2.428141	0.899994	10.000610	−0.000608
No. of pairs		Mean COMPUT. risk errors		STD. ERR. errors	
25		0.0000669406		0.0001037128	
Z-ratio				Status of significance	
0.64544				N.S.—not significant	
Parameter values					
PROB. = 0.85	RHO = 0.7		BIGH = 1.2947		SMHL = 1.04
Pair No.	SMH	SMK	PROB. level	Risk level (%)	Risk error (%)
Table No. 8.2-69					
1	1.30	1.289422	0.850003	14.999690	0.000310
2	1.29	1.299417	0.850004	14.999580	0.000423
3	1.28	1.309960	0.850010	14.999010	0.000989
4	1.27	1.320857	0.849993	15.000680	−0.000685
5	1.26	1.332504	0.849995	15.000550	−0.000548
6	1.25	1.344905	0.850008	14.999250	0.000751
7	1.24	1.357868	0.850005	14.999510	0.000489
8	1.23	1.371592	0.850002	14.999840	0.000155
9	1.22	1.386083	0.849991	15.000950	−0.000948
10	1.21	1.401735	0.850001	14.999860	0.000137
11	1.20	1.418358	0.850005	14.999490	0.000513
12	1.19	1.435956	0.849992	15.000750	−0.000757
13	1.18	1.455120	0.850002	14.999830	0.000173
14	1.17	1.475659	0.850001	14.999860	0.000143
15	1.16	1.497776	0.849992	15.000810	−0.000811
16	1.15	1.522060	0.849997	15.000310	−0.000310
17	1.14	1.548518	0.849992	15.000770	−0.000769
18	1.13	1.577937	0.850001	14.999930	0.000072
19	1.12	1.610712	0.850009	14.999100	0.000900
20	1.11	1.647243	0.849997	15.000270	−0.000268
21	1.10	1.689487	0.850008	14.999230	0.000769
22	1.09	1.738233	0.849999	15.000150	−0.000155
23	1.08	1.796613	0.849991	15.000880	−0.000882
24	1.07	1.869712	0.849996	15.000430	−0.000435
25	1.06	1.967304	0.850008	14.999190	0.000805

(continued)

8.2 Contents for Table 8.2 (Table No. 8.2-1 to 400)

Table 8.2 (continued)

Parameter values					
PROB. = 0.85	RHO = 0.7		BIGH = 1.2947		SMHL = 1.04
Pair No.	SMH	SMK	PROB. level	Risk level (%)	Risk error (%)
26	1.05	2.113614	0.850006	14.999440	0.000554
27	1.04	2.443027	0.850009	14.999120	0.000876
No. of pairs		Mean COMPUT. risk errors		STD. ERR. errors	
27		0.0000532184		0.0001156538	
Z-ratio				Status of significance	
0.46015				N.S.—not significant	
Parameter values					
PROB. = 0.8	RHO = 0.7		BIGH = 1.1085		SMHL = 0.85
Pair No.	SMH	SMK	PROB. level	Risk level (%)	Risk error (%)
Table No. 8.2-70					
1	1.11	1.107100	0.800003	19.999730	0.000274
2	1.10	1.117261	0.800001	19.999870	0.000131
3	1.09	1.127900	0.799996	20.000360	−0.000364
4	1.08	1.139119	0.799997	20.000300	−0.000298
5	1.07	1.150924	0.799998	20.000240	−0.000238
6	1.06	1.163321	0.799993	20.000730	−0.000733
7	1.05	1.176509	0.800001	19.999930	0.000066
8	1.04	1.190399	0.800002	19.999830	0.000173
9	1.03	1.205092	0.800000	19.999990	0.000006
10	1.02	1.220694	0.799998	20.000180	−0.000185
11	1.01	1.237309	0.799997	20.000320	−0.000322
12	1.00	1.255139	0.800005	19.999530	0.000471
13	0.99	1.274192	0.800008	19.999190	0.000811
14	0.98	1.294474	0.799993	20.000720	−0.000721
15	0.97	1.316580	0.799997	20.000350	−0.000346
16	0.96	1.340518	0.799997	20.000350	−0.000346
17	0.95	1.366687	0.800001	19.999930	0.000072
18	0.94	1.395486	0.800008	19.999170	0.000829
19	0.93	1.427120	0.799998	20.000200	−0.000203
20	0.92	1.462575	0.799992	20.000820	−0.000823
21	0.91	1.503033	0.800001	19.999890	0.000113
22	0.90	1.549287	0.799992	20.000830	−0.000829
23	0.89	1.604080	0.799994	20.000570	−0.000566
24	0.88	1.670941	0.800002	19.999810	0.000191
25	0.87	1.756132	0.799997	20.000280	−0.000286

(continued)

Table 8.2 (continued)

Parameter values					
PROB. = 0.8	RHO = 0.7		BIGH = 1.1085		SMHL = 0.85
Pair No.	SMH	SMK	PROB. level	Risk level (%)	Risk error (%)
26	0.86	1.875680	0.800001	19.999900	0.000101
27	0.85	2.079989	0.799994	20.000590	−0.000590
No. of pairs		Mean COMPUT. risk errors		STD. ERR. errors	
27		−0.0001290015		0.0000825959	
Z-ratio				Status of significance	
1.56184				N.S.—not significant	
Parameter values					
PROB. = 0.75	RHO = 0.7		BIGH = 0.949		SMHL = 0.68
Pair No.	SMH	SMK	PROB. level	Risk level (%)	Risk error (%)
Table No. 8.2-71					
1	0.95	0.948001	0.749994	25.000650	−0.000644
2	0.94	0.958281	0.750009	24.999120	0.000882
3	0.93	0.968877	0.749996	25.000420	−0.000423
4	0.92	0.980086	0.749997	25.000260	−0.000262
5	0.91	0.991820	0.749993	25.000720	−0.000721
6	0.90	1.004183	0.749991	25.000880	−0.000876
7	0.89	1.017282	0.750000	25.000000	0.000006
8	0.88	1.031027	0.749998	25.000190	−0.000191
9	0.87	1.045525	0.749991	25.000870	−0.000870
10	0.86	1.061078	0.750009	24.999060	0.000942
11	0.85	1.077304	0.749991	25.000870	−0.000864
12	0.84	1.094800	0.750000	25.000020	−0.000024
13	0.83	1.113382	0.749996	25.000360	−0.000358
14	0.82	1.133352	0.750000	25.000000	0.000006
15	0.81	1.154822	0.750003	24.999730	0.000274
16	0.80	1.177900	0.749994	25.000560	−0.000560
17	0.79	1.203087	0.749999	25.000130	−0.000131
18	0.78	1.230593	0.750002	24.999760	0.000244
19	0.77	1.260823	0.750004	24.999620	0.000381
20	0.76	1.294376	0.750006	24.999420	0.000578
21	0.75	1.331856	0.749997	25.000290	−0.000292
22	0.74	1.374646	0.749999	25.000110	−0.000107
23	0.73	1.424325	0.750007	24.999270	0.000733
24	0.72	1.482866	0.749996	25.000420	−0.000423
25	0.71	1.555171	0.749995	25.000510	−0.000507
26	0.70	1.649854	0.750005	24.999520	0.000483

(continued)

8.2 Contents for Table 8.2 (Table No. 8.2-1 to 400)

Table 8.2 (continued)

Parameter values					
PROB. = 0.75	RHO = 0.7		BIGH = 0.949		SMHL = 0.68
Pair No.	SMH	SMK	PROB. level	Risk level (%)	Risk error (%)
27	0.69	1.786469	0.749994	25.000560	−0.000560
28	0.68	2.052538	0.749995	25.000470	−0.000465
No. of pairs		Mean COMPUT. risk errors		STD. ERR. errors	
28		−0.0001292804		0.0000966095	
Z-ratio				Status of significance	
1.33817				N.S.—not significant	
Parameter values					
PROB. = 0.7	RHO = 0.7		BIGH = 0.8059		SMHL = 0.53
Pair No.	SMH	SMK	PROB. level	Risk level (%)	Risk error (%)
Table No. 8.2-72					
1	0.81	0.801821	0.700001	29.999880	0.000119
2	0.80	0.811843	0.699998	30.000220	−0.000215
3	0.79	0.822315	0.699992	30.000770	−0.000769
4	0.78	0.833344	0.699997	30.000260	−0.000262
5	0.77	0.844841	0.699991	30.000860	−0.000858
6	0.76	0.857013	0.700002	29.999790	0.000209
7	0.75	0.869677	0.699992	30.000830	−0.000829
8	0.74	0.883137	0.700001	29.999920	0.000083
9	0.73	0.897309	0.700006	29.999390	0.000614
10	0.72	0.912204	0.700000	29.999990	0.000012
11	0.71	0.928034	0.700002	29.999780	0.000221
12	0.70	0.944813	0.700001	29.999860	0.000143
13	0.69	0.962654	0.699999	30.000090	−0.000089
14	0.68	0.981672	0.699994	30.000570	−0.000566
15	0.67	1.002178	0.700006	29.999370	0.000632
16	0.66	1.024091	0.700005	29.999550	0.000453
17	0.65	1.047629	0.699991	30.000940	−0.000942
18	0.64	1.073398	0.700000	29.999980	0.000024
19	0.63	1.101420	0.700000	29.999980	0.000018
20	0.62	1.132305	0.700007	29.999260	0.000745
21	0.61	1.166275	0.699993	30.000720	−0.000721
22	0.60	1.204723	0.700006	29.999400	0.000602
23	0.59	1.248070	0.699997	30.000310	−0.000310
24	0.58	1.298299	0.699994	30.000650	−0.000644
25	0.57	1.358179	0.700006	29.999390	0.000608
26	0.56	1.431260	0.699997	30.000320	−0.000316
27	0.55	1.526957	0.700006	29.999410	0.000596

(continued)

Table 8.2 (continued)

Parameter values					
PROB. = 0.7	RHO = 0.7		BIGH = 0.8059		SMHL = 0.53
Pair No.	SMH	SMK	PROB. level	Risk level (%)	Risk error (%)
28	0.54	1.665231	0.700005	29.999540	0.000459
29	0.53	1.931674	0.700000	30.000040	−0.000042
No. of pairs		Mean COMPUT. risk errors		STD. ERR. errors	
29		−0.0000341733		0.0000922949	
Z-ratio				Status of significance	
0.37026				N.S.—not significant	
Parameter values					
PROB. = 0.65	RHO = 0.7		BIGH = 0.6734		SMHL = 0.39
Pair No.	SMH	SMK	PROB. level	Risk level (%)	Risk error (%)
Table No. 8.2-73					
1	0.68	0.666766	0.649991	35.000900	−0.000894
2	0.67	0.676720	0.649996	35.000440	−0.000435
3	0.66	0.687072	0.649992	35.000830	−0.000829
4	0.65	0.697935	0.649995	35.000520	−0.000513
5	0.64	0.709324	0.650001	34.999880	0.000119
6	0.63	0.721255	0.650006	34.999360	0.000644
7	0.62	0.733743	0.650006	34.999360	0.000638
8	0.61	0.746807	0.649996	35.000370	−0.000364
9	0.60	0.760662	0.650003	34.999670	0.000334
10	0.59	0.775230	0.650004	34.999590	0.000411
11	0.58	0.790630	0.650007	34.999330	0.000674
12	0.57	0.806885	0.650003	34.999720	0.000280
13	0.56	0.824119	0.649997	35.000270	−0.000268
14	0.55	0.842455	0.649993	35.000740	−0.000739
15	0.54	0.862118	0.650001	34.999910	0.000089
16	0.53	0.883138	0.650006	34.999410	0.000590
17	0.52	0.905647	0.650002	34.999840	0.000167
18	0.51	0.929972	0.650001	34.999860	0.000149
19	0.50	0.956349	0.650000	34.999990	0.000018
20	0.49	0.985209	0.650006	34.999380	0.000620
21	0.48	1.016794	0.650000	35.000050	−0.000042
22	0.47	1.051933	0.650003	34.999750	0.000250
23	0.46	1.091267	0.650001	34.999940	0.000060
24	0.45	1.136025	0.650001	34.999860	0.000143
25	0.44	1.187834	0.650001	34.999930	0.000072
26	0.43	1.249497	0.650006	34.999410	0.000596
27	0.42	1.325387	0.650005	34.999530	0.000477

(continued)

8.2 Contents for Table 8.2 (Table No. 8.2-1 to 400)

Table 8.2 (continued)

Parameter values					
PROB. = 0.65	RHO = 0.7		BIGH = 0.6734		SMHL = 0.39
Pair No.	SMH	SMK	PROB. level	Risk level (%)	Risk error (%)
28	0.41	1.424768	0.650004	34.999650	0.000358
29	0.40	1.570387	0.649996	35.000410	−0.000405
30	0.39	1.903362	0.649995	35.000480	−0.000477
No. of pairs		Mean COMPUT. risk errors		STD. ERR. errors	
30		0.0000555669		0.0000818283	
Z-ratio				Status of significance	
0.67907				N.S.—not significant	
Parameter values					
PROB. = 0.6	RHO = 0.7		BIGH = 0.5478		SMHL = 0.26
Pair No.	SMH	SMK	PROB. level	Risk level (%)	Risk error (%)
Table No. 8.2-74					
1	0.55	0.545413	0.600004	39.999560	0.000441
2	0.54	0.555517	0.600002	39.999810	0.000191
3	0.53	0.566100	0.600006	39.999360	0.000638
4	0.52	0.577086	0.599997	40.000300	−0.000298
5	0.51	0.588597	0.599991	40.000890	−0.000888
6	0.50	0.600756	0.600004	39.999580	0.000417
7	0.49	0.613395	0.599998	40.000240	−0.000244
8	0.48	0.626739	0.600004	39.999640	0.000358
9	0.47	0.640724	0.600001	39.999860	0.000143
10	0.46	0.655483	0.600004	39.999640	0.000358
11	0.45	0.671054	0.600005	39.999490	0.000507
12	0.44	0.687480	0.600001	39.999950	0.000048
13	0.43	0.704903	0.599998	40.000200	−0.000197
14	0.42	0.723472	0.600003	39.999670	0.000334
15	0.41	0.743242	0.600006	39.999370	0.000626
16	0.40	0.764274	0.599996	40.000390	−0.000387
17	0.39	0.787026	0.600008	39.999170	0.000834
18	0.38	0.811376	0.600001	39.999880	0.000119
19	0.37	0.837797	0.600000	40.000030	−0.000030
20	0.36	0.866576	0.599998	40.000180	−0.000185
21	0.35	0.898205	0.600002	39.999750	0.000244
22	0.34	0.933188	0.600006	39.999380	0.000620
23	0.33	0.972238	0.600006	39.999430	0.000566
24	0.32	1.016280	0.599992	40.000810	−0.000817
25	0.31	1.067234	0.599992	40.000840	−0.000846
26	0.30	1.127626	0.600005	39.999510	0.000495
27	0.29	1.201181	0.600008	39.999220	0.000775
28	0.28	1.344387	0.600003	39.999740	0.000262

(continued)

Table 8.2 (continued)

Parameter values					
PROB. = 0.6	RHO = 0.7		BIGH = 0.5478		SMHL = 0.26
Pair No.	SMH	SMK	PROB. level	Risk level (%)	Risk error (%)
29	0.27	1.470018	0.599992	40.000760	−0.000763
30	0.26	1.705734	0.600001	39.999930	0.000066
No. of pairs		Mean COMPUT. risk errors		STD. ERR. errors	
30		0.0001092111		0.0000867581	
Z-ratio				Status of significance	
1.25880				N.S.—not significant	
Parameter values					
PROB. = 0.55	RHO = 0.7		BIGH = 0.4262		SMHL = 0.13
Pair No.	SMH	SMK	PROB. level	Risk level (%)	Risk error (%)
Table No. 8.2-75					
1	0.43	0.422434	0.550007	44.999290	0.000709
2	0.42	0.432492	0.550009	44.999150	0.000852
3	0.41	0.442942	0.550003	44.999740	0.000262
4	0.40	0.453823	0.549992	45.000840	−0.000840
5	0.39	0.465272	0.549996	45.000380	−0.000381
6	0.38	0.477236	0.549999	45.000090	−0.000089
7	0.37	0.489764	0.550002	44.999790	0.000215
8	0.36	0.502816	0.549990	45.000980	−0.000978
9	0.35	0.516646	0.550000	44.999960	0.000042
10	0.34	0.531129	0.550001	44.999950	0.000054
11	0.33	0.546342	0.549994	45.000620	−0.000620
12	0.32	0.562469	0.549999	45.000150	−0.000149
13	0.31	0.579511	0.550001	44.999870	0.000131
14	0.30	0.597480	0.549990	45.000960	−0.000960
15	0.29	0.616699	0.549997	45.000320	−0.000322
16	0.28	0.637214	0.550007	44.999300	0.000703
17	0.27	0.659092	0.550010	44.999040	0.000960
18	0.26	0.682429	0.549994	45.000590	−0.000590
19	0.25	0.707835	0.550007	44.999300	0.000703
20	0.24	0.735180	0.549999	45.000080	−0.000077
21	0.23	0.765157	0.550008	44.999190	0.000817
22	0.22	0.797931	0.550003	44.999720	0.000286
23	0.21	0.834318	0.550004	44.999570	0.000435
24	0.20	0.938310	0.550008	44.999240	0.000757
25	0.19	0.977127	0.549994	45.000640	−0.000638
26	0.18	1.023599	0.549994	45.000620	−0.000614
27	0.17	1.080422	0.550000	44.999960	0.000042

(continued)

8.2 Contents for Table 8.2 (Table No. 8.2-1 to 400)

Table 8.2 (continued)

Parameter values					
PROB. = 0.55	RHO = 0.7		BIGH = 0.4262		SMHL = 0.13
Pair No.	SMH	SMK	PROB. level	Risk level (%)	Risk error (%)
28	0.16	1.152086	0.549998	45.000160	−0.000155
29	0.15	1.248475	0.550006	44.999360	0.000638
30	0.14	1.392786	0.549996	45.000360	−0.000364
31	0.13	1.700404	0.550007	44.999270	0.000733
No. of pairs		Mean COMPUT. risk errors		STD. ERR. errors	
31		0.0000488013		0.0001013899	
Z-ratio				Status of significance	
0.48132				N.S.—not significant	
Parameter values					
PROB. = 0.5	RHO = 0.7		BIGH = 0.3067		SMHL = 0.01
Pair No.	SMH	SMK	PROB. level	Risk level (%)	Risk error (%)
Table No. 8.2-76					
1	0.31	0.303386	0.500000	50.000000	−0.000003
2	0.30	0.313452	0.499999	50.000080	−0.000077
3	0.29	0.323971	0.500005	49.999470	0.000530
4	0.28	0.334872	0.499998	50.000190	−0.000185
5	0.27	0.346338	0.500008	49.999170	0.000829
6	0.26	0.358175	0.499992	50.000840	−0.000837
7	0.25	0.370693	0.500000	49.999960	0.000042
8	0.24	0.383734	0.499997	50.000300	−0.000298
9	0.23	0.397454	0.500002	49.999850	0.000155
10	0.22	0.411845	0.500003	49.999690	0.000316
11	0.21	0.426932	0.499997	50.000340	−0.000334
12	0.20	0.442883	0.499999	50.000120	−0.000113
13	0.19	0.459727	0.500002	49.999830	0.000167
14	0.18	0.477465	0.499992	50.000790	−0.000787
15	0.17	0.496389	0.499998	50.000160	−0.000158
16	0.16	0.516421	0.499992	50.000800	−0.000802
17	0.15	0.537841	0.499992	50.000790	−0.000790
18	0.14	0.560759	0.499991	50.000930	−0.000930
19	0.13	0.656291	0.500000	50.000050	−0.000048
20	0.12	0.674499	0.500008	49.999160	0.000846
21	0.11	0.695174	0.499994	50.000610	−0.000611
22	0.10	0.719164	0.500004	49.999600	0.000405
23	0.09	0.746728	0.500004	49.999630	0.000370
24	0.08	0.778545	0.499991	50.000910	−0.000912
25	0.07	0.816049	0.500003	49.999700	0.000298

(continued)

Table 8.2 (continued)

Parameter values					
PROB. = 0.5	RHO = 0.7		BIGH = 0.3067		SMHL = 0.01
Pair No.	SMH	SMK	PROB. level	Risk level (%)	Risk error (%)
26	0.06	0.860326	0.500010	49.999040	0.000960
27	0.05	0.913328	0.500001	49.999890	0.000113
28	0.04	0.978965	0.500004	49.999620	0.000381
29	0.03	1.063230	0.499998	50.000180	−0.000176
30	0.02	1.180587	0.500006	49.999370	0.000632
31	0.01	1.372099	0.499996	50.000410	−0.000405
No. of pairs			Mean COMPUT. risk errors	STD. ERR. errors	
31			−0.0000444241	0.0000940488	
Z-ratio				Status of significance	
0.47235				N.S.—not significant	
Parameter values					
PROB. = 0.99	RHO = 0.6		BIGH = 2.548		SMHL = 2.33
Pair No.	SMH	SMK	PROB. level	Risk level (%)	Risk error (%)
Table No. 8.2-77					
1	2.55	2.545220	0.989999	1.000065	−0.000066
2	2.54	2.556025	0.990008	0.999177	0.000823
3	2.53	2.566128	0.989999	1.000112	−0.000113
4	2.52	2.577874	0.990002	0.999773	0.000226
5	2.51	2.589701	0.989997	1.000291	−0.000292
6	2.50	2.603171	0.990002	0.999838	0.000161
7	2.49	2.616726	0.989996	1.000392	−0.000393
8	2.48	2.631927	0.989997	1.000261	−0.000262
9	2.47	2.648776	0.990003	0.999665	0.000334
10	2.46	2.665711	0.989996	1.000381	−0.000381
11	2.45	2.684295	0.989991	1.000941	−0.000942
12	2.44	2.706093	0.989998	1.000220	−0.000221
13	2.43	2.729543	0.990000	0.999975	0.000024
14	2.42	2.754645	0.989996	1.000440	−0.000441
15	2.41	2.784527	0.990003	0.999683	0.000316
16	2.40	2.817627	0.990005	0.999457	0.000542
17	2.39	2.853945	0.989998	1.000226	−0.000226
18	2.38	2.896609	0.989991	1.000911	−0.000912
19	2.37	2.951868	0.990000	0.999993	0.000006
20	2.36	3.019726	0.990003	0.999725	0.000274
21	2.35	3.112682	0.990009	0.999123	0.000876
22	2.34	3.249488	0.990004	0.999558	0.000441
23	2.33	3.561396	0.990001	0.999945	0.000054

(continued)

8.2 Contents for Table 8.2 (Table No. 8.2-1 to 400)

Table 8.2 (continued)

Parameter values					
PROB. = 0.99	RHO = 0.6		BIGH = 2.548		SMHL = 2.33
Pair No.	SMH	SMK	PROB. level	Risk level (%)	Risk error (%)
No. of pairs		Mean COMPUT. risk errors		STD. ERR. errors	
23		−0.0000072022		0.0000930889	
Z-ratio				Status of significance	
0.07737				N.S.—not significant	
Parameter values					
PROB. = 0.98	RHO = 0.6		BIGH = 2.288		SMHL = 2.06
Pair No.	SMH	SMK	PROB. level	Risk level (%)	Risk error (%)
Table No. 8.2-78					
1	2.29	2.286002	0.980005	1.999545	0.000453
2	2.28	2.296028	0.980001	1.999903	0.000095
3	2.27	2.306924	0.980004	1.999569	0.000429
4	2.26	2.317910	0.979995	2.000552	−0.000554
5	2.25	2.330548	0.980007	1.999343	0.000656
6	2.24	2.343279	0.980003	1.999700	0.000298
7	2.23	2.356884	0.980000	2.000010	−0.000012
8	2.22	2.371364	0.979996	2.000451	−0.000453
9	2.21	2.387503	0.980003	1.999730	0.000268
10	2.20	2.404520	0.980003	1.999676	0.000322
11	2.19	2.423198	0.980009	1.999122	0.000876
12	2.18	2.442757	0.980002	1.999772	0.000226
13	2.17	2.463979	0.979994	2.000570	−0.000572
14	2.16	2.487648	0.979992	2.000803	−0.000805
15	2.15	2.514545	0.979999	2.000058	−0.000060
16	2.14	2.544673	0.980008	1.999223	0.000775
17	2.13	2.576471	0.979992	2.000845	−0.000846
18	2.12	2.616188	0.980009	1.999068	0.000930
19	2.11	2.659141	0.979997	2.000296	−0.000298
20	2.10	2.711580	0.979994	2.000558	−0.000560
21	2.09	2.776633	0.979993	2.000701	−0.000703
22	2.08	2.863675	0.980000	2.000010	−0.000012
23	2.07	2.991458	0.980008	1.999247	0.000751
24	2.06	3.228734	0.980001	1.999897	0.000101
No. of pairs		Mean COMPUT. risk errors		STD. ERR. errors	
24		0.0000522137		0.0001079185	
Z-ratio				Status of significance	
0.48382				N.S.—not significant	
Parameter values					
PROB. = 0.97	RHO = 0.6		BIGH = 2.1235		SMHL = 1.89
Pair No.	SMH	SMK	PROB. level	Risk level (%)	Risk error (%)
Table No. 8.2-79					
1	2.13	2.117020	0.969997	3.000301	−0.000304
2	2.12	1.127006	0.969999	3.000063	−0.000066

(continued)

Table 8.2 (continued)

Parameter values					
PROB. = 0.97	RHO = 0.6		BIGH = 2.1235		SMHL = 1.89
Pair No.	SMH	SMK	PROB. level	Risk level (%)	Risk error (%)
3	2.11	2.137477	0.970000	3.000051	−0.000054
4	2.10	2.148826	0.970009	2.999091	0.000906
5	2.09	2.160271	0.970000	2.999967	0.000030
6	2.08	2.172597	0.969998	3.000248	−0.000250
7	2.07	2.186195	0.970010	2.999014	0.000983
8	2.06	2.199895	0.969999	3.000075	−0.000077
9	2.05	2.215260	0.970009	2.999056	0.000942
10	2.04	2.230731	0.969993	3.000689	−0.000691
11	2.03	2.247869	0.969990	3.000969	−0.000972
12	2.02	2.266678	0.969996	3.000426	−0.000429
13	2.01	2.287159	0.970004	2.999652	0.000346
14	2.00	2.309313	0.970008	2.999187	0.000811
15	1.99	2.333143	0.970004	2.999586	0.000411
16	1.98	2.359431	0.970001	2.999938	0.000060
17	1.97	2.388570	0.969996	3.000391	−0.000393
18	1.96	2.421733	0.969999	3.000116	−0.000119
19	1.95	2.459312	0.969999	3.000081	−0.000083
20	1.94	2.502481	0.969996	3.000391	−0.000393
21	1.93	2.553588	0.969994	3.000623	−0.000626
22	1.92	2.617319	0.970003	2.999669	0.000328
23	1.91	2.698365	0.970006	2.999413	0.000584
24	1.90	2.810790	0.970003	2.999687	0.000310
25	1.89	2.998347	0.969994	3.000629	−0.000632
No. of pairs		Mean COMPUT. risk errors		STD. ERR. errors	
25		0.0000238419		0.0001041527	
Z-ratio				Status of significance	
0.22891				N.S.—not significant	
Parameter values					
PROB. = 0.96	RHO = 0.6		BIGH = 2.00		SMHL = 1.76
Pair No.	SMH	SMK	PROB. level	Risk level (%)	Risk error (%)
Table No. 8.2-80					
1	2.01	1.990050	0.959990	4.000962	−0.000960
2	2.00	2.000000	0.960000	4.000032	−0.000030
3	1.99	2.010050	0.959990	4.000974	−0.000972
4	1.98	2.020983	0.959995	4.000479	−0.000477
5	1.97	2.032410	0.959995	4.000461	−0.000459
6	1.96	2.044723	0.960005	3.999543	0.000459
7	1.95	2.057532	0.960005	3.999549	0.000453
8	1.94	2.071230	0.960008	3.999198	0.000805
9	1.93	2.085430	0.959998	4.000223	−0.000221
10	1.92	2.100911	0.959999	4.000074	−0.000072
11	1.91	2.117678	0.960007	3.999269	0.000733

(continued)

8.2 Contents for Table 8.2 (Table No. 8.2-1 to 400)

Table 8.2 (continued)

Parameter values					
PROB. = 0.96	RHO = 0.6		BIGH = 2.00		SMHL = 1.76
Pair No.	SMH	SMK	PROB. level	Risk level (%)	Risk error (%)
12	1.90	2.134950	0.959992	4.000783	−0.000781
13	1.89	2.154293	0.960000	4.000050	−0.000048
14	1.88	2.175316	0.960009	3.999066	0.000936
15	1.87	2.197631	0.960004	3.999591	0.000411
16	1.86	2.222412	0.960009	3.999138	0.000864
17	1.85	2.248881	0.959994	4.000640	−0.000638
18	1.84	2.279381	0.960003	3.999663	0.000340
19	1.83	2.313136	0.960004	3.999645	0.000358
20	1.82	2.350927	0.959994	4.000628	−0.000626
21	1.81	2.395882	0.960005	3.999502	0.000501
22	1.80	2.447222	0.959991	4.000860	−0.000858
23	1.79	2.511199	0.959995	4.000545	−0.000542
24	1.78	2.594066	0.960005	3.999454	0.000548
25	1.77	2.706761	0.959996	4.000378	−0.000376
26	1.76	2.897727	0.959995	4.000545	−0.000542
No. of pairs		Mean COMPUT. risk errors		STD. ERR. errors	
26		−0.0000441516		0.0001155236	
Z-ratio				Status of significance	
0.38219				N.S.—not significant	
Parameter values					
PROB. = 0.95	RHO = 0.6		BIGH = 1.8996		SMHL = 1.65
Pair No.	SMH	SMK	PROB. level	Risk level (%)	Risk error (%)
Table No. 8.2-81					
1	1.90	1.899200	0.949992	5.000764	−0.000763
2	1.89	1.909639	0.950003	4.999697	0.000304
3	1.88	1.920186	0.949991	5.000920	−0.000918
4	1.87	1.931622	0.949994	5.000562	−0.000560
5	1.86	1.943559	0.949991	5.000883	−0.000882
6	1.85	1.956389	0.949998	5.000252	−0.000250
7	1.84	1.969724	0.949992	5.000830	−0.000829
8	1.83	1.984347	0.950006	4.999388	0.000614
9	1.82	1.999478	0.950002	4.999817	0.000185
10	1.81	2.015510	0.949993	5.000735	−0.000733
11	1.80	2.033227	0.950005	4.999507	0.000495
12	1.79	2.051848	0.950002	4.999834	0.000167
13	1.78	2.071767	0.949993	5.000735	−0.000733
14	1.77	2.093767	0.949998	5.000228	−0.000226
15	1.76	2.117460	0.949994	5.000568	−0.000566
16	1.75	2.144020	0.950008	4.999191	0.000811
17	1.74	2.172277	0.949993	5.000716	−0.000715
18	1.73	2.204577	0.949999	5.000150	−0.000149
19	1.72	2.240922	0.950003	4.999715	0.000286

(continued)

Table 8.2 (continued)

Parameter values					
PROB. = 0.95	RHO = 0.6		BIGH = 1.8996		SMHL = 1.65
Pair No.	SMH	SMK	PROB. level	Risk level (%)	Risk error (%)
20	1.71	2.282097	0.950000	5.000031	−0.000030
21	1.70	2.330447	0.950004	4.999590	0.000411
22	1.69	2.388320	0.950009	4.999054	0.000948
23	1.68	2.458842	0.949999	5.000097	−0.000095
24	1.67	2.552953	0.950001	4.999871	0.000131
25	1.66	2.692533	0.950007	4.999328	0.000674
26	1.65	2.974457	0.950008	4.999185	0.000817
No. of pairs		Mean COMPUT. risk errors		STD. ERR. errors	
26		−0.0000596046		0.0001124740	
Z-ratio				Status of significance	
0.52994				N.S.—not significant	
Parameter values					
PROB. = 0.94	RHO = 0.6		BIGH = 1.8143		SMHL = 1.56
Pair No.	SMH	SMK	PROB. level	Risk level (%)	Risk error (%)
Table No. 8.2-82					
1	1.82	1.808813	0.940002	5.999840	0.000161
2	1.81	1.818610	0.939991	6.000919	−0.000918
3	1.80	1.829104	0.939992	6.000799	−0.000799
4	1.79	1.840102	0.939990	6.000954	−0.000954
5	1.78	1.851995	0.940006	5.999375	0.000626
6	1.77	1.864005	0.939992	6.000823	−0.000823
7	1.76	1.877306	0.940010	5.999035	0.000966
8	1.75	1.890728	0.939993	6.000746	−0.000745
9	1.74	1.905444	0.939999	6.000090	−0.000089
10	1.73	1.921067	0.940004	5.999643	0.000358
11	1.72	1.937598	0.940002	5.999840	0.000161
12	1.71	1.955430	0.940006	5.999422	0.000578
13	1.70	1.974566	0.940009	5.999070	0.000930
14	1.69	1.995008	0.940005	5.999488	0.000513
15	1.68	2.017148	0.940001	5.999863	0.000137
16	1.67	2.041381	0.940002	5.999780	0.000221
17	1.66	2.068098	0.940009	5.999124	0.000876
18	1.65	2.097304	0.940007	5.999333	0.000668
19	1.64	2.129780	0.940004	5.999649	0.000352
20	1.63	2.166313	0.940000	6.000018	−0.000018
21	1.62	2.208466	0.940006	5.999375	0.000626
22	1.61	2.257024	0.940003	5.999732	0.000268
23	1.60	2.315115	0.940002	5.999822	0.000179
24	1.59	2.387428	0.940006	5.999446	0.000554

(continued)

8.2 Contents for Table 8.2 (Table No. 8.2-1 to 400)

Table 8.2 (continued)

Parameter values					
PROB. = 0.94	RHO = 0.6		BIGH = 1.8143		SMHL = 1.56
Pair No.	SMH	SMK	PROB. level	Risk level (%)	Risk error (%)
25	1.58	2.481781	0.940001	5.999941	0.000060
26	1.57	2.621613	0.940003	5.999714	0.000286
27	1.56	2.897553	0.939993	6.000680	−0.000679
No. of pairs		Mean COMPUT. risk errors		STD. ERR. errors	
27		0.0001247440		0.0001085871	
Z-ratio				Status of significance	
1.14879				N.S.—not significant	
Parameter values					
PROB. = 0.93	RHO = 0.6		BIGH = 1.7396		SMHL = 1.48
Pair No.	SMH	SMK	PROB. level	Risk level (%)	Risk error (%)
Table No. 8.2-83					
1	1.74	1.739200	0.929998	7.000166	−0.000167
2	1.73	1.749449	0.930000	7.000023	−0.000024
3	1.72	1.760204	0.930000	6.999958	0.000042
4	1.71	1.771470	0.929998	7.000244	−0.000244
5	1.70	1.783443	0.930001	6.999869	0.000131
6	1.69	1.796124	0.930008	6.999207	0.000793
7	1.68	1.809127	0.929990	7.000971	−0.000972
8	1.67	1.823429	0.930003	6.999654	0.000346
9	1.66	1.838251	0.929996	7.000369	−0.000370
10	1.65	1.854378	0.930008	6.999171	0.000829
11	1.64	1.871030	0.929992	7.000840	−0.000840
12	1.63	1.889382	0.930001	6.999856	0.000143
13	1.62	1.908655	0.929991	7.000923	−0.000924
14	1.61	1.930023	0.930006	6.999362	0.000638
15	1.60	1.952708	0.930003	6.999743	0.000256
16	1.59	1.977494	0.930002	6.999773	0.000226
17	1.58	2.004384	0.929992	7.000757	−0.000757
18	1.57	2.034552	0.929999	7.000071	−0.000072
19	1.56	2.068002	0.930001	6.999880	0.000119
20	1.55	2.105517	0.929998	7.000196	−0.000197
21	1.54	2.148663	0.930002	6.999815	0.000185
22	1.53	2.199007	0.930007	6.999326	0.000674
23	1.52	2.258114	0.929990	7.000977	−0.000978
24	1.51	2.333798	0.930008	6.999171	0.000829

(continued)

Table 8.2 (continued)

Parameter values					
PROB. = 0.93	RHO = 0.6		BIGH = 1.7396		SMHL = 1.48
Pair No.	SMH	SMK	PROB. level	Risk level (%)	Risk error (%)
25	1.50	2.431533	0.929994	7.000572	−0.000572
26	1.49	2.581011	0.930007	6.999284	−0.000715
27	1.48	2.894734	0.929996	7.000423	−0.000423
No. of pairs		Mean COMPUT. risk errors		STD. ERR. errors	
27		−0.0000219260		0.0001059813	
Z-ratio				Status of significance	
0.20689				N.S.—not significant	
Parameter values					
PROB. = 0.92	RHO = 0.6		BIGH = 1.6728		SMHL = 1.41
Pair No.	SMH	SMK	PROB. level	Risk level (%)	Risk error (%)
Table No. 8.2-84					
1	1.68	1.665631	0.920003	7.999730	0.000268
2	1.67	1.675605	0.920009	7.999110	0.000888
3	1.66	1.685894	0.920002	7.999831	0.000167
4	1.65	1.696696	0.919994	8.000570	−0.000572
5	1.64	1.708209	0.919998	8.000220	−0.000221
6	1.63	1.720240	0.919995	8.000523	−0.000525
7	1.62	1.733180	0.920009	7.999129	0.000870
8	1.61	1.746643	0.920009	7.999129	0.000870
9	1.60	1.760631	0.919992	8.000833	−0.000834
10	1.59	1.775927	0.920003	7.999718	0.000280
11	1.58	1.791949	0.919999	8.000064	−0.000066
12	1.57	1.809089	0.919999	8.000100	−0.000101
13	1.56	1.827350	0.919995	8.000499	−0.000501
14	1.55	1.847126	0.920000	7.999987	0.000012
15	1.54	1.868223	0.919995	8.000488	−0.000489
16	1.53	1.891428	0.920007	7.999254	0.000745
17	1.52	1.916351	0.920005	7.999468	0.000530
18	1.51	1.943777	0.920009	7.999075	0.000924
19	1.50	1.973709	0.920001	7.999873	0.000125
20	1.49	2.006933	0.919991	8.000887	−0.000888
21	1.48	2.044622	0.919991	8.000875	−0.000876
22	1.47	2.087953	0.920000	7.999975	0.000024
23	1.46	2.137710	0.919993	8.000660	−0.000662
24	1.45	2.197803	0.920002	7.999790	0.000209
25	1.44	2.271360	0.919995	8.000517	−0.000519
26	1.43	2.369325	0.920003	7.999701	0.000298

(continued)

Table 8.2 (continued)

Parameter values					
PROB. = 0.92	RHO = 0.6		BIGH = 1.6728		SMHL = 1.41
Pair No.	SMH	SMK	PROB. level	Risk level (%)	Risk error (%)
27	1.42	2.512792	0.919999	8.000070	−0.000072
28	1.41	2.803331	0.919994	8.000601	−0.000602
No. of pairs		Mean COMPUT. risk errors		STD. ERR. errors	
28		−0.0000246640		0.0001040142	
Z-ratio				Status of significance	
0.23712				N.S.—not significant	
Parameter values					
PROB. = 0.91	RHO = 0.6		BIGH = 1.612		SMHL = 1.35
Pair No.	SMH	SMK	PROB. level	Risk level (%)	Risk error (%)
Table No. 8.2-85					
1	1.62	1.604039	0.909998	9.000212	−0.000215
2	1.61	1.614002	0.910007	8.999329	0.000668
3	1.60	1.624285	0.910002	8.999777	0.000221
4	1.59	1.635086	0.909998	9.000170	−0.000173
5	1.58	1.646601	0.910008	8.999211	0.000787
6	1.57	1.658444	0.909996	9.000385	−0.000387
7	1.56	1.671202	0.910005	8.999461	0.000536
8	1.55	1.684488	0.910001	8.999866	0.000131
9	1.54	1.698694	0.910009	8.999151	0.000846
10	1.53	1.713629	0.910007	8.999264	0.000733
11	1.52	1.729490	0.910006	8.999378	0.000620
12	1.51	1.746280	0.909999	9.000087	−0.000089
13	1.50	1.764394	0.910003	8.999681	0.000316
14	1.49	1.783833	0.910009	8.999073	0.000924
15	1.48	1.804601	0.910008	8.999193	0.000805
16	1.47	1.826701	0.909990	9.000963	−0.000966
17	1.46	1.851309	0.910004	8.999628	0.000370
18	1.45	1.877646	0.909994	9.000653	−0.000656
19	1.44	1.906888	0.909997	9.000349	−0.000352
20	1.43	1.939038	0.909990	9.000981	−0.000983
21	1.42	1.975273	0.909994	9.000647	−0.000650
22	1.41	2.016376	0.909998	9.000206	−0.000209
23	1.40	2.063915	0.910007	8.999294	0.000703
24	1.39	2.119456	0.910004	8.999598	0.000399
25	1.38	2.186909	0.910000	9.000021	−0.000024

(continued)

Table 8.2 (continued)

Parameter values					
PROB. = 0.91	RHO = 0.6		BIGH = 1.612		SMHL = 1.35
Pair No.	SMH	SMK	PROB. level	Risk level (%)	Risk error (%)
26	1.37	2.273309	0.910001	8.999931	0.000066
27	1.36	2.393506	0.910003	8.999706	0.000292
28	1.35	2.593597	0.909994	9.000576	−0.000578
No. of pairs		Mean COMPUT. risk errors		STD. ERR. errors	
28		0.0001081105		0.0001034278	
Z-ratio				Status of significance	
1.04527				N.S.—not significant	
Parameter values					
PROB. = 0.9025	RHO = 0.6		BIGH = 1.5696		SMHL = 1.30
Pair No.	SMH	SMK	PROB. level	Risk level (%)	Risk error (%)
Table No. 8.2-86					
1	1.57	1.569200	0.902497	9.750348	−0.000346
2	1.56	1.579454	0.902500	9.750014	−0.000012
3	1.55	1.590229	0.902505	9.749508	0.000495
4	1.54	1.601331	0.902492	9.750766	−0.000763
5	1.53	1.613350	0.902508	9.749156	0.000846
6	1.52	1.625701	0.902500	9.749979	0.000024
7	1.51	1.638779	0.902496	9.750378	−0.000376
8	1.50	1.652586	0.902492	9.750825	−0.000823
9	1.49	1.667319	0.902496	9.750378	−0.000376
10	1.48	1.682984	0.902503	9.749675	0.000328
11	1.47	1.699386	0.902493	9.750676	−0.000674
12	1.46	1.717115	0.902499	9.750080	−0.000077
13	1.45	1.735979	0.902499	9.750069	−0.000066
14	1.44	1.756176	0.902497	9.750301	−0.000298
15	1.43	1.777906	0.902493	9.750664	−0.000662
16	1.42	1.801757	0.902508	9.749233	0.000769
17	1.41	1.827343	0.902504	9.749615	0.000387
18	1.40	1.855449	0.902505	9.749526	0.000477
19	1.39	1.886468	0.902507	9.749341	0.000662
20	1.38	1.920796	0.902501	9.749901	0.000101
21	1.37	1.959608	0.902502	9.749824	0.000179
22	1.36	2.003690	0.902495	9.750498	−0.000495
23	1.35	2.055390	0.902499	9.750074	−0.000072
24	1.34	2.116665	0.902491	9.750872	−0.000870
25	1.33	2.193770	0.902502	9.749770	0.000232
26	1.32	2.294521	0.902493	9.750748	−0.000745

(continued)

8.2 Contents for Table 8.2 (Table No. 8.2-1 to 400)

Table 8.2 (continued)

Parameter values					
PROB. = 0.9025	RHO = 0.6		BIGH = 1.5696		SMHL = 1.30
Pair No.	SMH	SMK	PROB. level	Risk level (%)	Risk error (%)
27	1.31	2.446269	0.902495	9.750551	−0.000548
28	1.30	2.776360	0.902506	9.749449	0.000554
No. of pairs		Mean COMPUT. risk errors		STD. ERR. errors	
28		−0.0000739920		0.0000943877	
Z-ratio				Status of significance	
0.78392				N.S.—not significant	
Parameter values					
PROB. = 0.9	RHO = 0.6		BIGH = 1.5561		SMHL = 1.29
Pair No.	SMH	SMK	PROB. level	Risk level (%)	Risk error (%)
Table No. 8.2-87					
1	1.56	1.552210	0.900008	9.999168	0.000834
2	1.55	1.562224	0.900005	9.999514	0.000489
3	1.54	1.572759	0.900005	9.999502	0.000501
4	1.53	1.583817	0.900006	9.999448	0.000554
5	1.52	1.595401	0.900003	9.999681	0.000322
6	1.51	1.607514	0.899995	10.000520	−0.000513
7	1.50	1.620548	0.900008	9.999168	0.000834
8	1.49	1.634117	0.900007	9.999287	0.000715
9	1.48	1.648418	0.900003	9.999752	0.000250
10	1.47	1.663649	0.900004	9.999633	0.000370
11	1.46	1.679619	0.899991	10.000940	−0.000936
12	1.45	1.696917	0.899997	10.000260	−0.000256
13	1.44	1.715349	0.900002	9.999811	0.000191
14	1.43	1.735117	0.900008	9.999233	0.000769
15	1.42	1.756026	0.899994	10.000650	−0.000644
16	1.41	1.779057	0.900005	9.999532	0.000471
17	1.40	1.803824	0.900004	9.999609	0.000393
18	1.39	1.830720	0.899997	10.000310	−0.000304
19	1.38	1.860531	0.900002	9.999848	0.000155
20	1.37	1.893261	0.899994	10.000650	−0.000644
21	1.36	1.930476	0.900007	9.999264	0.000739
22	1.35	1.971790	0.899991	10.000950	−0.000948
23	1.34	2.020331	0.900001	9.999884	0.000119
24	1.33	2.076887	0.899993	10.000740	−0.000739
25	1.32	2.146930	0.900008	9.999216	0.000787

(continued)

Table 8.2 (continued)

Parameter values					
PROB. = 0.9	RHO = 0.6		BIGH = 1.5561		SMHL = 1.29
Pair No.	SMH	SMK	PROB. level	Risk level (%)	Risk error (%)
26	1.31	2.235933	0.900007	9.999299	0.000703
27	1.30	2.359526	0.899993	10.000710	−0.000709
28	1.29	2.573965	0.899995	10.000490	−0.000483
No. of pairs		Mean COMPUT. risk errors		STD. ERR. errors	
28		0.0001042054		0.0001092814	
Z-ratio				Status of significance	
0.95355				N.S.—not significant	
Parameter values					
PROB. = 0.85	RHO = 0.6		BIGH = 1.3246		SMHL = 1.04
Pair No.	SMH	SMK	PROB. level	Risk level (%)	Risk error (%)
Table No. 8.2-88					
1	1.33	1.319808	0.849999	15.000080	−0.000083
2	1.32	1.329802	0.849999	15.000090	−0.000095
3	1.31	1.340339	0.850010	14.999010	0.000983
4	1.30	1.351228	0.850005	14.999530	0.000471
5	1.29	1.362667	0.850004	14.999630	0.000364
6	1.28	1.374660	0.850002	14.999780	0.000221
7	1.27	1.387211	0.849996	15.000380	−0.000381
8	1.26	1.400520	0.850002	14.999830	0.000173
9	1.25	1.414394	0.849993	15.000730	−0.000733
10	1.24	1.429230	0.850003	14.999690	0.000310
11	1.23	1.444835	0.850006	14.999390	0.000608
12	1.22	1.461215	0.849995	15.000500	−0.000501
13	1.21	1.478765	0.849998	15.000220	−0.000226
14	1.20	1.497489	0.850004	14.999630	0.000364
15	1.19	1.517393	0.850002	14.999810	0.000191
16	1.18	1.538677	0.849996	15.000360	−0.000358
17	1.17	1.561737	0.850003	14.999710	0.000292
18	1.16	1.586384	0.849990	15.000970	−0.000972
19	1.15	1.613599	0.850006	14.999390	0.000608
20	1.14	1.642998	0.850000	15.000020	−0.000018
21	1.13	1.675369	0.849993	15.000690	−0.000691
22	1.12	1.711497	0.849994	15.000620	−0.000620
23	1.11	1.752173	0.849997	15.000300	−0.000304
24	1.10	1.798575	0.849999	15.000060	−0.000060
25	1.09	1.852661	0.850006	14.999440	0.000560
26	1.08	1.916785	0.849996	15.000410	−0.000411

(continued)

8.2 Contents for Table 8.2 (Table No. 8.2-1 to 400)

Table 8.2 (continued)

Parameter values					
PROB. = 0.85	RHO = 0.6		BIGH = 1.3246		SMHL = 1.04
Pair No.	SMH	SMK	PROB. level	Risk level (%)	Risk error (%)
27	1.07	1.996811	0.849995	15.000480	−0.000477
28	1.06	2.103687	0.850008	14.999250	0.000745
29	1.05	2.263201	0.849999	15.000080	−0.000083
30	1.04	2.618331	0.849990	15.000990	−0.000989
No. of pairs		Mean COMPUT. risk errors		STD. ERR. errors	
30		−0.0000359551		0.0000911795	
Z-ratio				Status of significance	
0.39433				N.S.—not significant	
Parameter values					
PROB. = 0.8	RHO = 0.6		BIGH = 1.1415		SMHL = 0.85
Pair No.	SMH	SMK	PROB. level	Risk level (%)	Risk error (%)
Table No. 8.2-89					
1	1.15	1.133258	0.800006	19.999390	0.000608
2	1.14	1.143100	0.800004	19.999650	0.000352
3	1.13	1.153313	0.799994	20.000630	−0.000626
4	1.12	1.164096	0.800002	19.999840	0.000155
5	1.11	1.175261	0.799995	20.000480	−0.000483
6	1.10	1.187007	0.799998	20.000180	−0.000185
7	1.09	1.199340	0.800005	19.999460	0.000536
8	1.08	1.212166	0.799999	20.000070	−0.000072
9	1.07	1.225688	0.800000	19.999990	0.000006
10	1.06	1.239911	0.800001	19.999860	0.000143
11	1.05	1.254841	0.799997	20.000320	−0.000316
12	1.04	1.270680	0.800002	19.999800	0.000197
13	1.03	1.287336	0.799997	20.000280	−0.000286
14	1.02	1.305012	0.799995	20.000510	−0.000513
15	1.01	1.323910	0.800005	19.999520	0.000477
16	1.00	1.343842	0.799995	20.000500	−0.000501
17	0.99	1.365403	0.800009	19.999110	0.000888
18	0.98	1.388208	0.799994	20.000580	−0.000584
19	0.97	1.413048	0.800002	19.999770	0.000232
20	0.96	1.439737	0.799994	20.000550	−0.000554
21	0.95	1.468868	0.799992	20.000780	−0.000781
22	0.94	1.501037	0.800006	19.999420	0.000584
23	0.93	1.536255	0.799998	20.000240	−0.000238
24	0.92	1.575703	0.799997	20.000320	−0.000322
25	0.91	1.620564	0.800008	19.999190	0.000811
26	0.90	1.671630	0.799995	20.000530	−0.000530
27	0.89	1.732039	0.799995	20.000470	−0.000471
28	0.88	1.805707	0.800004	19.999600	0.000399

(continued)

Table 8.2 (continued)

Parameter values					
PROB. = 0.8	RHO = 0.6		BIGH = 1.1415		SMHL = 0.85
Pair No.	SMH	SMK	PROB. level	Risk level (%)	Risk error (%)
29	0.87	1.899289	0.799997	20.000260	−0.000256
30	0.86	2.030767	0.800006	19.999420	0.000584
31	0.85	2.254842	0.799998	20.000240	−0.000238
No. of pairs		Mean COMPUT. risk errors		STD. ERR. errors	
31		−0.0000307336		0.0000828340	
Z-ratio				Status of significance	
0.37103				N.S.—not significant	
Parameter values					
PROB. = 0.75	RHO = 0.6		BIGH = 0.9845		SMHL = 0.68
Pair No.	SMH	SMK	PROB. level	Risk level (%)	Risk error (%)
Table No. 8.2-90					
1	0.99	0.979031	0.749994	25.000630	−0.000626
2	0.98	0.989021	0.749995	25.000470	−0.000465
3	0.97	0.999412	0.749995	25.000510	−0.000507
4	0.96	1.010309	0.750005	24.999500	0.000501
5	0.95	1.021620	0.750006	24.999390	0.000608
6	0.94	1.033450	0.750010	24.999030	0.000972
7	0.93	1.045710	0.749997	25.000300	−0.000304
8	0.92	1.058600	0.749993	25.000720	−0.000715
9	0.91	1.072131	0.749992	25.000850	−0.000852
10	0.90	1.086406	0.750000	24.999970	0.000030
11	0.89	1.101339	0.749999	25.000090	−0.000089
12	0.88	1.117034	0.749994	25.000570	−0.000572
13	0.87	1.133600	0.749990	25.000970	−0.000966
14	0.86	1.151242	0.750002	24.999820	0.000185
15	0.85	1.169775	0.749995	25.000540	−0.000542
16	0.84	1.189599	0.750002	24.999770	0.000232
17	0.83	1.210630	0.750000	24.999990	0.000012
18	0.82	1.233172	0.750004	24.999620	0.000381
19	0.81	1.257335	0.750005	24.999530	0.000471
20	0.80	1.283425	0.750010	24.999020	0.000978
21	0.79	1.311457	0.749995	25.000510	−0.000507
22	0.78	1.342225	0.749998	25.000210	−0.000209
23	0.77	1.375941	0.749996	25.000360	−0.000358
24	0.76	1.413207	0.749991	25.000940	−0.000942
25	0.75	1.455211	0.750003	24.999730	0.000274
26	0.74	1.502753	0.750007	24.999300	0.000697
27	0.73	1.557414	0.749998	25.000250	−0.000250
28	0.72	1.622338	0.749996	25.000400	−0.000399
29	0.71	1.702236	0.750000	25.000020	−0.000018

(continued)

Table 8.2 (continued)

Parameter values					
PROB. = 0.75	RHO = 0.6		BIGH = 0.9845		SMHL = 0.68
Pair No.	SMH	SMK	PROB. level	Risk level (%)	Risk error (%)
30	0.70	1.806504	0.750009	24.999100	0.000906
31	0.69	1.956258	0.749991	25.000940	−0.000936
32	0.68	2.250353	0.750007	24.999340	0.000662
No. of pairs		Mean COMPUT. risk errors		STD. ERR. errors	
32		−0.0000711643		0.0001011268	
Z-ratio				Status of significance	
0.70371				N.S.—not significant	
Parameter values					
PROB. = 0.7	RHO = 0.6		BIGH = 0.8437		SMHL = 0.53
Pair No.	SMH	SMK	PROB. level	Risk level (%)	Risk error (%)
Table No. 8.2-91					
1	0.85	0.837447	0.700000	29.999960	0.000042
2	0.84	0.847416	0.700004	29.999600	0.000399
3	0.83	0.857724	0.699997	30.000330	−0.000328
4	0.82	0.868476	0.699994	30.000630	−0.000626
5	0.81	0.879779	0.700008	29.999160	0.000846
6	0.80	0.891447	0.700004	29.999650	0.000352
7	0.79	0.903589	0.699992	30.000760	−0.000757
8	0.78	0.916411	0.700003	29.999690	0.000316
9	0.77	0.929727	0.699999	30.000100	−0.000101
10	0.76	0.943649	0.699991	30.000950	−0.000948
11	0.75	0.958383	0.700002	29.999850	0.000155
12	0.74	0.973748	0.699996	30.000370	−0.000370
13	0.73	0.989953	0.699996	30.000420	−0.000417
14	0.72	1.007011	0.699992	30.000800	−0.000799
15	0.71	1.025135	0.700001	29.999860	0.000143
16	0.70	1.044243	0.700000	30.000000	0.000000
17	0.69	1.064547	0.700001	29.999910	0.000089
18	0.68	1.086066	0.699990	30.000990	−0.000983
19	0.67	1.109307	0.700007	29.999280	0.000721
20	0.66	1.133998	0.700000	30.000010	−0.000006
21	0.65	1.160747	0.700008	29.999200	0.000799
22	0.64	1.189577	0.700004	29.999590	0.000417
23	0.63	1.220904	0.699995	30.000540	−0.000536
24	0.62	1.255534	0.700007	29.999290	0.000715
25	0.61	1.293496	0.699994	30.000650	−0.000650
26	0.60	1.336382	0.700009	29.999080	0.000924
27	0.59	1.384615	0.700001	29.999910	0.000095
28	0.58	1.440573	0.700008	29.999220	0.000781
29	0.57	1.506636	0.700008	29.999240	0.000763
30	0.56	1.587530	0.700002	29.999760	0.000238
31	0.55	1.692673	0.700000	30.000040	−0.000042

(continued)

Table 8.2 (continued)

Parameter values					
PROB. = 0.7	RHO = 0.6		BIGH = 0.8437		SMHL = 0.53
Pair No.	SMH	SMK	PROB. level	Risk level (%)	Risk error (%)
32	0.54	1.844765	0.699998	30.000230	−0.000226
33	0.53	2.136824	0.699991	30.000930	−0.000930
No. of pairs		Mean COMPUT. risk errors		STD. ERR. errors	
33		0.0000022790		0.0000973457	
Z-ratio				Status of significance	
0.02341				N.S.—not significant	
Parameter values					
PROB. = 0.65	RHO = 0.6		BIGH = 0.7133		SMHL = 0.39
Pair No.	SMH	SMK	PROB. level	Risk level (%)	Risk error (%)
Table No. 8.2-92					
1	0.72	0.706760	0.649992	35.000780	−0.000781
2	0.71	0.716713	0.649995	35.000500	−0.000501
3	0.70	0.727048	0.649997	35.000270	−0.000268
4	0.69	0.737777	0.649997	35.000290	−0.000286
5	0.68	0.748914	0.649992	35.000770	−0.000769
6	0.67	0.760570	0.649998	35.000170	−0.000167
7	0.66	0.772662	0.649995	35.000540	−0.000536
8	0.65	0.785303	0.649995	35.000460	−0.000453
9	0.64	0.798511	0.649997	35.000280	−0.000280
10	0.63	0.812301	0.649996	35.000430	−0.000429
11	0.62	0.826792	0.650003	34.999750	0.000256
12	0.61	0.841906	0.649997	35.000310	−0.000304
13	0.60	0.857858	0.650004	34.999640	0.000370
14	0.59	0.874574	0.650001	34.999890	0.000107
15	0.58	0.892275	0.650010	34.999010	0.000995
16	0.57	0.910790	0.649995	35.000520	−0.000519
17	0.56	0.930538	0.649999	35.000070	−0.000072
18	0.55	0.951452	0.649998	35.000210	−0.000209
19	0.54	0.973661	0.649990	35.000960	−0.000954
20	0.53	0.997494	0.649996	35.000400	−0.000393
21	0.52	1.022986	0.649995	35.000480	−0.000477
22	0.51	1.050570	0.650008	34.999200	0.000799
23	0.50	1.080289	0.650007	34.999280	0.000727
24	0.49	1.112579	0.650000	34.999980	0.000024
25	0.48	1.148079	0.650000	34.999960	0.000042
26	0.47	1.187234	0.649991	35.000870	−0.000870
27	0.46	1.231275	0.649999	35.000100	−0.000101
28	0.45	1.281050	0.649995	35.000520	−0.000519
29	0.44	1.338778	0.650002	34.999780	0.000226
30	0.43	1.407076	0.650003	34.999690	0.000316
31	0.42	1.491108	0.650004	34.999620	0.000387
32	0.41	1.601124	0.650010	34.999020	0.000978

(continued)

8.2 Contents for Table 8.2 (Table No. 8.2-1 to 400)

Table 8.2 (continued)

Parameter values					
PROB. = 0.65	RHO = 0.6		BIGH = 0.7133		SMHL = 0.39
Pair No.	SMH	SMK	PROB. level	Risk level (%)	Risk error (%)
33	0.40	1.761835	0.650003	34.999690	0.000316
34	0.39	2.120231	0.650005	34.999490	0.000513
No. of pairs		Mean COMPUT. risk errors		STD. ERR. errors	
34		−0.0000808920		0.0000864363	
Z-ratio				Status of significance	
0.93586				N.S.—not significant	
Parameter values					
PROB. = 0.6	RHO = 0.6		BIGH = 0.5898		SMHL = 0.26
Pair No.	SMH	SMK	PROB. level	Risk level (%)	Risk error (%)
Table No. 8.2-93					
1	0.59	0.589698	0.600005	39.999470	0.000530
2	0.58	0.599863	0.600002	39.999770	0.000226
3	0.57	0.610385	0.599993	40.000670	−0.000674
4	0.56	0.621381	0.599997	40.000290	−0.000292
5	0.55	0.632773	0.599994	40.000650	−0.000650
6	0.54	0.644681	0.600000	39.999990	0.000012
7	0.53	0.657031	0.599997	40.000270	−0.000274
8	0.52	0.669946	0.600002	39.999830	0.000167
9	0.51	0.683356	0.599994	40.000580	−0.000584
10	0.50	0.697486	0.600007	39.999320	0.000679
11	0.49	0.712173	0.600004	39.999630	0.000370
12	0.48	0.727549	0.599999	40.000120	−0.000119
13	0.47	0.743749	0.600005	39.999550	0.000453
14	0.46	0.760718	0.600002	39.999840	0.000155
15	0.45	0.778598	0.600001	39.999950	0.000048
16	0.44	0.797436	0.599996	40.000430	−0.000429
17	0.43	0.817384	0.599995	40.000520	−0.000525
18	0.42	0.838599	0.600002	39.999750	0.000244
19	0.41	0.861144	0.600009	39.999150	0.000852
20	0.40	0.885089	0.600003	39.999740	0.000256
21	0.39	0.910709	0.599995	40.000510	−0.000513
22	0.38	0.938283	0.599990	40.000960	−0.000960
23	0.37	0.968298	0.600008	39.999190	0.000811
24	0.36	1.000664	0.600001	39.999870	0.000131
25	0.35	1.036084	0.599995	40.000520	−0.000525
26	0.34	1.075277	0.599997	40.000340	−0.000340
27	0.33	1.118976	0.599998	40.000240	−0.000238
28	0.32	1.168324	0.599996	40.000410	−0.000405
29	0.31	1.225266	0.600006	39.999440	0.000554
30	0.30	1.291968	0.599997	40.000260	−0.000262
31	0.29	1.421405	0.600001	39.999870	0.000131
32	0.28	1.519324	0.599999	40.000090	−0.000095

(continued)

Table 8.2 (continued)

Parameter values					
PROB. = 0.6	RHO = 0.6		BIGH = 0.5898		SMHL = 0.26
Pair No.	SMH	SMK	PROB. level	Risk level (%)	Risk error (%)
33	0.27	1.661040	0.600004	39.999590	0.000405
34	0.26	1.922312	0.599994	40.000630	−0.000638
No. of pairs		Mean COMPUT. risk errors		STD. ERR. errors	
34		−0.0000427450		0.0000781744	
Z-ratio				Status of significance	
0.54679				N.S.—not significant	
Parameter values					
PROB. = 0.55	RHO = 0.6		BIGH = 0.4704		SMHL = 0.13
Pair No.	SMH	SMK	PROB. level	Risk level (%)	Risk error (%)
Table No. 8.2-94					
1	0.48	0.461090	0.550004	44.999610	0.000387
2	0.47	0.470898	0.549997	45.000310	−0.000310
3	0.46	0.481133	0.550005	44.999550	0.000453
4	0.45	0.491725	0.550008	44.999200	0.000805
5	0.44	0.502705	0.550009	44.999060	0.000936
6	0.43	0.514010	0.549992	45.000850	−0.000852
7	0.42	0.525871	0.549995	45.000470	−0.000471
8	0.41	0.538233	0.550003	44.999660	0.000340
9	0.40	0.551042	0.550000	44.999990	0.000012
10	0.39	0.564445	0.550006	44.999360	0.000638
11	0.38	0.578399	0.550007	44.999260	0.000739
12	0.37	0.592965	0.550006	44.999360	0.000644
13	0.36	0.608211	0.550007	44.999290	0.000709
14	0.35	0.624112	0.549997	45.000320	−0.000316
15	0.34	0.640851	0.549996	45.000440	−0.000441
16	0.33	0.658424	0.549992	45.000840	−0.000840
17	0.32	0.677035	0.550004	44.999650	0.000352
18	0.31	0.696606	0.550006	44.999380	0.000626
19	0.30	0.717274	0.550005	44.999500	0.000507
20	0.29	0.739193	0.550004	44.999560	0.000441
21	0.28	0.762537	0.550008	44.999170	0.000829
22	0.27	0.787314	0.549997	45.000330	−0.000328
23	0.26	0.813952	0.549997	45.000320	−0.000316
24	0.25	0.842721	0.550009	44.999110	0.000894
25	0.24	0.873546	0.549993	45.000730	−0.000727
26	0.23	0.907382	0.550005	44.999460	0.000536
27	0.22	0.944277	0.550002	44.999800	0.000197
28	0.21	1.044125	0.549994	45.000590	−0.000590
29	0.20	1.083529	0.549996	45.000440	−0.000435
30	0.19	1.129259	0.550001	44.999890	0.000113
31	0.18	1.183217	0.550008	44.999190	0.000817
32	0.17	1.248108	0.550000	44.999980	0.000018
33	0.16	1.329459	0.549997	45.000350	−0.000352

(continued)

8.2 Contents for Table 8.2 (Table No. 8.2-1 to 400)

Table 8.2 (continued)

Parameter values					
PROB. = 0.55	RHO = 0.6		BIGH = 0.4704		SMHL = 0.13
Pair No.	SMH	SMK	PROB. level	Risk level (%)	Risk error (%)
34	0.15	1.437673	0.549996	45.000450	−0.000447
35	0.14	1.599293	0.549996	45.000400	−0.000399
36	0.13	1.939623	0.550001	44.999920	0.000077
No. of pairs		Mean COMPUT. risk errors		STD. ERR. errors	
36		0.0001146987		0.0000893928	
Z-ratio				Status of significance	
1.28309				N.S.—not significant	
Parameter values					
PROB. = 0.5	RHO = 0.6		BIGH = 0.3531		SMHL = 0.01
Pair No.	SMH	SMK	PROB. level	Risk level (%)	Risk error (%)
Table No. 8.2-95					
1	0.36	0.346235	0.499994	50.000650	−0.000644
2	0.35	0.356227	0.500006	49.999400	0.000602
3	0.34	0.366509	0.500005	49.999510	0.000489
4	0.33	0.377133	0.499998	50.000230	−0.000224
5	0.32	0.388159	0.499993	50.000720	−0.000715
6	0.31	0.399653	0.499999	50.000120	−0.000119
7	0.30	0.411595	0.500006	49.999410	0.000596
8	0.29	0.423973	0.500007	49.999290	0.000715
9	0.28	0.436788	0.499997	50.000280	−0.000271
10	0.27	0.450253	0.500009	49.999060	0.000942
11	0.26	0.464107	0.499990	50.001000	−0.000995
12	0.25	0.478699	0.499994	50.000600	−0.000596
13	0.24	0.493912	0.499993	50.000720	−0.000715
14	0.23	0.509858	0.499997	50.000320	−0.000322
15	0.22	0.526588	0.500004	49.999580	0.000423
16	0.21	0.544103	0.500005	49.999450	0.000548
17	0.20	0.562460	0.499999	50.000130	−0.000134
18	0.19	0.581794	0.499992	50.000840	−0.000831
19	0.18	0.602332	0.500003	49.999740	0.000262
20	0.17	0.624034	0.500009	49.999060	0.000942
21	0.16	0.646923	0.499999	50.000160	−0.000152
22	0.15	0.744088	0.499990	50.001000	−0.000995
23	0.14	0.762736	0.500000	50.000050	−0.000048
24	0.13	0.783488	0.500001	49.999930	0.000072
25	0.12	0.806770	0.500008	49.999160	0.000846
26	0.11	0.832669	0.499992	50.000840	−0.000840
27	0.10	0.862030	0.499991	50.000870	−0.000867
28	0.09	0.895484	0.500007	49.999320	0.000679

(continued)

Table 8.2 (continued)

Parameter values					
PROB. = 0.5	RHO = 0.6		BIGH = 0.3531		SMHL = 0.01
Pair No.	SMH	SMK	PROB. level	Risk level (%)	Risk error (%)
29	0.08	0.933494	0.500008	49.999190	0.000817
30	0.07	0.977229	0.500005	49.999480	0.000525
31	0.06	1.028381	0.500008	49.999180	0.000823
32	0.05	1.088901	0.499994	50.000560	−0.000563
33	0.04	1.163079	0.499992	50.000850	−0.000846
34	0.03	1.258322	0.500009	49.999100	0.000900
35	0.02	1.388649	0.500001	49.999900	0.000107
36	0.01	1.602232	0.500009	49.999060	0.000948
No. of pairs		Mean COMPUT. risk errors		STD. ERR. errors	
36		0.0000366488		0.0001078902	
Z-ratio				Status of significance	
0.33969				N.S.—not significant	
Parameter values					
PROB. = 0.99	RHO = 0.5		BIGH = 2.5529		SMHL = 2.33
Pair No.	SMH	SMK	PROB. level	Risk level (%)	Risk error (%)
Table No. 8.2-96					
1	2.56	2.555195	0.989994	1.000637	−0.000638
2	2.55	2.565179	0.989991	1.000875	−0.000876
3	2.54	2.576803	0.990003	0.999689	0.000310
4	2.53	2.588508	0.990007	0.999332	0.000668
5	2.52	2.600292	0.990002	0.999802	0.000197
6	2.51	2.613721	0.990008	0.999212	0.000787
7	2.50	2.627232	0.990004	0.999606	0.000393
8	2.49	2.642389	0.990008	0.999218	0.000781
9	2.48	2.657631	0.990001	0.999951	0.000048
10	2.47	2.674520	0.989998	1.000196	−0.000197
11	2.46	2.693059	0.989998	1.000214	−0.000215
12	2.45	2.713247	0.989998	1.000190	−0.000191
13	2.44	2.736649	0.990009	0.999057	0.000942
14	2.43	2.760141	0.990003	0.999743	0.000256
15	2.42	2.786849	0.990001	0.999939	0.000060
16	2.41	2.816773	0.989999	1.000100	−0.000101
17	2.40	2.849916	0.989993	1.000708	−0.000709
18	2.39	2.889403	0.989995	1.000530	−0.000530
19	2.38	2.935236	0.989994	1.000589	−0.000590
20	2.37	2.993665	0.990006	0.999421	0.000578
21	2.36	3.061567	0.989999	1.000130	−0.000131
22	2.35	3.154568	0.989997	1.000273	−0.000274

(continued)

Table 8.2 (continued)

Parameter values					
PROB. = 0.99	RHO = 0.5		BIGH = 2.5529		SMHL = 2.33
Pair No.	SMH	SMK	PROB. level	Risk level (%)	Risk error (%)
23	2.34	3.297670	0.989998	1.000226	−0.000226
24	2.33	3.647123	0.990008	0.999212	0.000787
No. of pairs		Mean COMPUT. risk errors		STD. ERR. errors	
24		0.0000450611		0.0001033363	
Z-ratio				Status of significance	
0.43606				N.S.—not significant	
Parameter values					
PROB. = 0.98	RHO = 0.5		BIGH = 2.3003		SMHL = 2.06
Pair No.	SMH	SMK	PROB. level	Risk level (%)	Risk error (%)
Table No. 8.2-97					
1	2.31	2.291422	0.980008	1.999199	0.000799
2	2.30	2.300600	0.979993	2.000725	−0.000727
3	2.29	2.311427	0.980006	1.999372	0.000626
4	2.28	2.322343	0.980007	1.999259	0.000739
5	2.27	2.333348	0.979996	2.000409	−0.000411
6	2.26	2.346006	0.980008	1.999247	0.000751
7	2.25	2.358755	0.980004	1.999605	0.000393
8	2.24	2.372379	0.980002	1.999795	0.000203
9	2.23	2.386878	0.980000	2.000028	−0.000030
10	2.22	2.402254	0.979995	2.000475	−0.000477
11	2.21	2.419289	0.980001	1.999879	0.000119
12	2.20	2.437204	0.980000	2.000010	−0.000012
13	2.19	2.456780	0.980003	1.999748	0.000250
14	2.18	2.477238	0.979993	2.000713	−0.000715
15	2.17	2.500924	0.980004	1.999563	0.000435
16	2.16	2.526275	0.980007	1.999337	0.000662
17	2.15	2.553294	0.979996	2.000392	−0.000393
18	2.14	2.585107	0.980006	1.999390	0.000608
19	2.13	2.618590	0.979990	2.000982	−0.000983
20	2.12	2.658434	0.979991	2.000952	−0.000954
21	2.11	2.706200	0.980003	1.999736	0.000262
22	2.10	2.760329	0.979993	2.000690	−0.000691
23	2.09	2.828635	0.979993	2.000731	−0.000733
24	2.08	2.918932	0.979996	2.000409	−0.000411

(continued)

Table 8.2 (continued)

Parameter values					
PROB. = 0.98	RHO = 0.5		BIGH = 2.3003		SMHL = 2.06
Pair No.	SMH	SMK	PROB. level	Risk level (%)	Risk error (%)
25	2.07	3.049971	0.979997	2.000254	−0.000256
26	2.06	3.293630	0.979991	2.000898	−0.000900
No. of pairs		Mean COMPUT. risk errors		STD. ERR. errors	
26		−0.0000684350		0.0001124796	
Z-ratio				Status of significance	
0.60842				N.S.—not significant	
Parameter values					
PROB. = 0.975	RHO = 0.5		BIGH = 2.212		SMHL = 1.97
Pair No.	SMH	SMK	PROB. level	Risk level (%)	Risk error (%)
Table No. 8.2-98					
1	2.22	2.204419	0.975000	2.500016	−0.000018
2	2.21	2.214002	0.974991	2.500862	−0.000864
3	2.20	2.224846	0.975006	2.499437	0.000560
4	2.19	2.235783	0.975005	2.499455	0.000542
5	2.18	2.246813	0.974991	2.500939	−0.000942
6	2.17	2.259500	0.975004	2.499604	0.000393
7	2.16	2.272282	0.975000	2.500034	−0.000036
8	2.15	2.285944	0.974998	2.500218	−0.000221
9	2.14	2.300484	0.974996	2.500397	−0.000399
10	2.13	2.315907	0.974992	2.500773	−0.000775
11	2.12	2.332992	0.975002	2.499831	0.000167
12	2.11	2.350962	0.975003	2.499688	0.000310
13	2.10	2.369817	0.974994	2.500582	−0.000584
14	2.09	2.391121	0.975003	2.499688	0.000310
15	2.08	2.413314	0.974994	2.500594	−0.000596
16	2.07	2.438741	0.975005	2.499473	0.000525
17	2.06	2.465840	0.975002	2.499813	0.000185
18	2.05	2.496176	0.975002	2.499771	0.000226
19	2.04	2.529751	0.974997	2.500302	−0.000304
20	2.03	2.568129	0.974995	2.500492	−0.000495
21	2.02	2.612874	0.974998	2.500194	−0.000197
22	2.01	2.665550	0.975000	2.500045	−0.000048
23	2.00	2.730846	0.975010	2.499044	0.000954
24	1.99	2.811890	0.975004	2.499604	0.000393
25	1.98	2.924308	0.975000	2.500034	−0.000036
26	1.97	3.108727	0.974994	2.500618	−0.000620
No. of pairs		Mean COMPUT. risk errors		STD. ERR. errors	
26		−0.0000580593		0.0000941356	
Z-ratio				Status of significance	
0.61676				N.S.—not significant	

(continued)

8.2 Contents for Table 8.2 (Table No. 8.2-1 to 400)

Table 8.2 (continued)

Parameter values					
PROB. = 0.97	RHO = 0.5		BIGH = 2.1376		SMHL = 1.89
Pair No.	SMH	SMK	PROB. level	Risk level (%)	Risk error (%)
Table No. 8.2-99					
1	2.14	2.135203	0.969993	3.000665	−0.000668
2	2.13	2.145618	0.970004	2.999652	0.000346
3	2.12	2.156127	0.969998	3.000224	−0.000226
4	2.11	2.167514	0.970003	2.999723	0.000274
5	2.10	2.178998	0.969990	3.000969	−0.000972
6	2.09	2.191752	0.969997	3.000277	−0.000280
7	2.08	2.205388	0.970008	2.999187	0.000811
8	2.07	2.219126	0.969997	3.000289	−0.000292
9	2.06	2.234529	0.970009	2.999127	0.000870
10	2.05	2.250037	0.969995	3.000498	−0.000501
11	2.04	2.267213	0.969997	3.000307	−0.000310
12	2.03	2.286059	0.970009	2.999073	0.000924
13	2.02	2.305796	0.970008	2.999210	0.000787
14	2.01	2.326425	0.969991	3.000951	−0.000954
15	2.00	2.350292	0.970004	2.999598	0.000399
16	1.99	2.375835	0.970006	2.999383	0.000614
17	1.98	2.403057	0.969992	3.000820	−0.000823
18	1.97	2.435083	0.970008	2.999157	0.000840
19	1.96	2.468792	0.969991	3.000951	−0.000954
20	1.95	2.508873	0.969998	3.000194	−0.000197
21	1.94	2.554545	0.969997	3.000265	−0.000268
22	1.93	2.608155	0.969993	3.000695	−0.000697
23	1.92	2.675173	0.970003	2.999699	0.000298
24	1.91	2.757946	0.969990	3.000963	−0.000966
25	1.90	2.879912	0.970010	2.999008	0.000989
26	1.89	3.080136	0.970005	2.999514	0.000483
No. of pairs		Mean COMPUT. risk errors		STD. ERR. errors	
26		−0.0000174399		0.0001285097	
Z-ratio				Status of significance	
0.13571				N.S.—not significant	
Parameter values					
PROB. = 0.96	RHO = 0.5		BIGH = 2.0155		SMHL = 1.76
Pair No.	SMH	SMK	PROB. level	Risk level (%)	Risk error (%)
Table No. 8.2-100					
1	2.02	2.011010	0.959996	4.000407	−0.000405
2	2.01	2.021015	0.959995	4.000503	−0.000501
3	2.00	2.031511	0.959994	4.000646	−0.000644
4	1.99	2.042890	0.960007	3.999317	0.000685
5	1.98	2.054371	0.959999	4.000086	−0.000083
6	1.97	2.066738	0.960002	3.999800	0.000203
7	1.96	2.079603	0.959997	4.000294	−0.000292
8	1.95	2.093356	0.959997	4.000264	−0.000262

(continued)

Table 8.2 (continued)

Parameter values					
PROB. = 0.96	RHO = 0.5		BIGH = 2.0155		SMHL = 1.76
Pair No.	SMH	SMK	PROB. level	Risk level (%)	Risk error (%)
9	1.94	2.108001	0.960000	4.000026	−0.000024
10	1.93	2.123538	0.960001	3.999889	0.000113
11	1.92	2.139969	0.959998	4.000175	−0.000173
12	1.91	2.157687	0.960000	3.999961	0.000042
13	1.90	2.176303	0.959991	4.000908	−0.000906
14	1.89	2.196990	0.960001	3.999937	0.000066
15	1.88	2.218969	0.960000	4.000026	−0.000024
16	1.87	2.242634	0.959993	4.000658	−0.000656
17	1.86	2.269157	0.960004	3.999627	0.000376
18	1.85	2.297368	0.959992	4.000807	−0.000805
19	1.84	2.329615	0.960001	3.999937	0.000066
20	1.83	2.365116	0.959997	4.000294	−0.000292
21	1.82	2.405438	0.959994	4.000592	−0.000590
22	1.81	2.452144	0.959994	4.000580	−0.000578
23	1.80	2.508362	0.960008	3.999192	0.000811
24	1.79	2.575657	0.960007	3.999281	0.000721
25	1.78	2.660282	0.959994	4.000592	−0.000590
26	1.77	2.782550	0.960009	3.999090	0.000912
27	1.76	2.983091	0.960000	3.999955	0.000048
No. of pairs		Mean COMPUT. risk errors		STD. ERR. errors	
27		−0.0000994120		0.0000923325	
Z-ratio				Status of significance	
1.07667				N.S.—not significant	
Parameter values					
PROB. = 0.95	RHO = 0.5		BIGH = 1.9159		SMHL = 1.65
Pair No.	SMH	SMK	PROB. level	Risk level (%)	Risk error (%)
Table No. 8.2-101					
1	1.92	1.912590	0.949994	5.000586	−0.000584
2	1.91	1.922599	0.949991	5.000872	−0.000870
3	1.90	1.933496	0.950010	4.999048	0.000954
4	1.89	1.944499	0.950005	4.999501	0.000501
5	1.88	1.956001	0.949997	5.000282	−0.000280
6	1.87	1.968395	0.950003	4.999697	0.000304
7	1.86	1.981292	0.950001	4.999948	0.000054
8	1.85	1.995085	0.950006	4.999429	0.000572
9	1.84	2.009384	0.949998	5.000252	−0.000250
10	1.83	2.024582	0.949991	5.000943	−0.000942
11	1.82	2.041072	0.949997	5.000282	−0.000280
12	1.81	2.058465	0.949996	5.000371	−0.000370
13	1.80	2.077153	0.949999	5.000139	−0.000137
14	1.79	2.097139	0.949999	5.000150	−0.000149
15	1.78	2.118426	0.949990	5.000979	−0.000978

(continued)

8.2 Contents for Table 8.2 (Table No. 8.2-1 to 400)

Table 8.2 (continued)

Parameter values					
PROB. = 0.95	RHO = 0.5		BIGH = 1.9159		SMHL = 1.65
Pair No.	SMH	SMK	PROB. level	Risk level (%)	Risk error (%)
16	1.77	2.141795	0.949993	5.000723	−0.000721
17	1.76	2.167641	0.950008	4.999209	0.000793
18	1.75	2.195184	0.950002	4.999841	0.000161
19	1.74	2.225988	0.950007	4.999340	0.000662
20	1.73	2.260057	0.950006	4.999364	0.000638
21	1.72	2.298175	0.950002	4.999841	0.000161
22	1.71	2.341905	0.950001	4.999871	0.000131
23	1.70	2.392813	0.950002	4.999799	0.000203
24	1.69	2.453245	0.949998	5.000186	−0.000185
25	1.68	2.528674	0.950001	4.999882	0.000119
26	1.67	2.627694	0.950001	4.999871	0.000131
27	1.66	2.773748	0.950002	4.999841	0.000161
28	1.65	3.068399	0.950002	4.999834	0.000167
No. of pairs		Mean COMPUT. risk errors		STD. ERR. errors	
28		−0.0000012332		0.0000937939	
Z-ratio				Status of significance	
0.01315				N.S.—not significant	
Parameter values					
PROB. = 0.94	RHO = 0.5		BIGH = 1.832		SMHL = 1.56
Pair No.	SMH	SMK	PROB. level	Risk level (%)	Risk error (%)
Table No. 8.2-102					
1	1.84	1.824230	0.940003	5.999750	0.000250
2	1.83	1.834002	0.939997	6.000352	−0.000352
3	1.82	1.844470	0.940005	5.999512	0.000489
4	1.81	1.855244	0.940001	5.999863	0.000137
5	1.80	1.866522	0.939996	6.000400	−0.000399
6	1.79	1.878501	0.939998	6.000221	−0.000221
7	1.78	1.890988	0.939993	6.000710	−0.000709
8	1.77	1.904375	0.940000	5.999995	0.000006
9	1.76	1.918274	0.939995	6.000525	−0.000525
10	1.75	1.933076	0.939994	6.000572	−0.000572
11	1.74	1.948786	0.939995	6.000537	−0.000536
12	1.73	1.965795	0.940010	5.999053	0.000948
13	1.72	1.983324	0.939997	6.000257	−0.000256
14	1.71	2.002548	0.940006	5.999369	0.000632
15	1.70	2.022687	0.939996	6.000436	−0.000435
16	1.69	2.044525	0.939991	6.000895	−0.000894
17	1.68	2.068455	0.939998	6.000191	−0.000191
18	1.67	2.094090	0.939992	6.000817	−0.000817
19	1.66	2.122603	0.940001	5.999923	0.000077
20	1.65	2.153606	0.939997	6.000304	−0.000304
21	1.64	2.188665	0.940009	5.999089	0.000912

(continued)

Table 8.2 (continued)

Parameter values					
PROB. = 0.94	RHO = 0.5		BIGH = 1.832		SMHL = 1.56
Pair No.	SMH	SMK	PROB. level	Risk level (%)	Risk error (%)
22	1.63	2.227002	0.939994	6.000596	−0.000596
23	1.62	2.271352	0.939994	6.000602	−0.000602
24	1.61	2.323673	0.940009	5.999136	0.000864
25	1.60	2.385140	0.940007	5.999285	0.000715
26	1.59	2.460832	0.940001	5.999935	0.000066
27	1.58	2.561692	0.940009	5.999094	0.000906
28	1.57	2.706472	0.939992	6.000841	−0.000840
29	1.56	3.001425	0.939998	6.000203	−0.000203
No. of pairs		Mean COMPUT. risk errors		STD. ERR. errors	
29		−0.0000816584		0.0001038303	
Z-ratio				Status of significance	
0.78646				N.S.—not significant	
Parameter values					
PROB. = 0.93	RHO = 0.5		BIGH = 1.7582		SMHL = 1.48
Pair No.	SMH	SMK	PROB. level	Risk level (%)	Risk error (%)
Table No. 8.2-103					
1	1.76	1.756402	0.930000	6.999970	0.000030
2	1.75	1.766439	0.929993	7.000745	−0.000745
3	1.74	1.777176	0.930001	6.999934	0.000066
4	1.73	1.788422	0.930008	6.999165	0.000834
5	1.72	1.799983	0.930000	6.999975	0.000024
6	1.71	1.812251	0.930000	7.000017	−0.000018
7	1.70	1.825033	0.929992	7.000834	−0.000834
8	1.69	1.838918	0.930008	6.999177	0.000823
9	1.68	1.853321	0.930010	6.999016	0.000983
10	1.67	1.868246	0.929994	7.000602	−0.000602
11	1.66	1.884475	0.930000	6.999970	0.000030
12	1.65	1.901620	0.930001	6.999869	0.000131
13	1.64	1.919685	0.929992	7.000786	−0.000787
14	1.63	1.939452	0.930005	6.999505	0.000495
15	1.62	1.960143	0.929995	7.000536	−0.000536
16	1.61	1.982933	0.930007	6.999350	0.000650
17	1.60	2.007042	0.929996	7.000375	−0.000376
18	1.59	2.033646	0.930001	6.999898	0.000101
19	1.58	2.062748	0.930005	6.999463	0.000536
20	1.57	2.094741	0.930007	6.999350	0.000650
21	1.56	2.130019	0.929998	7.000160	−0.000161
22	1.55	2.170147	0.930002	6.999779	0.000221
23	1.54	2.215910	0.930005	6.999529	0.000471
24	1.53	2.268873	0.930001	6.999928	0.000072
25	1.52	2.332165	0.929997	7.000280	−0.000280
26	1.51	2.411259	0.930001	6.999952	0.000048

(continued)

8.2 Contents for Table 8.2 (Table No. 8.2-1 to 400)

Table 8.2 (continued)

Parameter values					
PROB. = 0.93	RHO = 0.5		BIGH = 1.7582		SMHL = 1.48
Pair No.	SMH	SMK	PROB. level	Risk level (%)	Risk error (%)
27	1.50	2.515532	0.929999	7.000149	−0.000149
28	1.49	2.671550	0.929998	7.000172	−0.000173
29	1.48	3.007443	0.930003	6.999743	0.000256
No. of pairs		Mean COMPUT. risk errors		STD. ERR. errors	
29		0.0000586112		0.0000884722	
Z-ratio				Status of significance	
0.66248				N.S.—not significant	
Parameter values					
PROB. = 0.92	RHO = 0.5		BIGH = 1.6922		SMHL = 1.41
Pair No.	SMH	SMK	PROB. level	Risk level (%)	Risk error (%)
Table No. 8.2-104					
1	1.70	1.684436	0.919998	8.000172	−0.000173
2	1.69	1.694403	0.920005	7.999456	0.000542
3	1.68	1.704684	0.920002	7.999808	0.000191
4	1.67	1.715476	0.920001	7.999897	0.000101
5	1.66	1.726782	0.920000	7.999993	0.000006
6	1.65	1.738604	0.919997	8.000338	−0.000340
7	1.64	1.751139	0.920001	7.999855	0.000143
8	1.63	1.764195	0.919998	8.000236	−0.000238
9	1.62	1.778164	0.920008	7.999206	0.000793
10	1.61	1.792659	0.920003	7.999736	0.000262
11	1.60	1.808072	0.920003	7.999730	0.000268
12	1.59	1.824406	0.920003	7.999701	0.000298
13	1.58	1.841664	0.919998	8.000189	−0.000191
14	1.57	1.860239	0.920004	7.999593	0.000405
15	1.56	1.879743	0.919992	8.000755	−0.000757
16	1.55	1.900961	0.919997	8.000321	−0.000322
17	1.54	1.923895	0.920006	7.999408	0.000590
18	1.53	1.948157	0.919992	8.000845	−0.000846
19	1.52	1.974924	0.919995	8.000528	−0.000530
20	1.51	2.004197	0.919998	8.000172	−0.000173
21	1.50	2.036370	0.919999	8.000070	−0.000072
22	1.49	2.071839	0.919991	8.000887	−0.000888
23	1.48	2.112168	0.919998	8.000236	−0.000238
24	1.47	2.158143	0.920004	7.999599	0.000399
25	1.46	2.211329	0.920005	7.999516	0.000483
26	1.45	2.274074	0.919993	8.000684	−0.000685
27	1.44	2.352632	0.919998	8.000225	−0.000226
28	1.43	2.455600	0.919998	8.000231	−0.000232

(continued)

Table 8.2 (continued)

Parameter values					
PROB. = 0.92	RHO = 0.5		BIGH = 1.6922		SMHL = 1.41
Pair No.	SMH	SMK	PROB. level	Risk level (%)	Risk error (%)
29	1.42	2.607202	0.919995	8.000458	−0.000459
30	1.41	2.918379	0.920004	7.999623	0.000376
No. of pairs		Mean COMPUT. risk errors		STD. ERR. errors	
30		−0.0000488374		0.0000787727	
Z-ratio				Status of significance	
0.61998				N.S.—not significant	
Parameter values					
PROB. = 0.91	RHO = 0.5		BIGH = 1.6322		SMHL = 1.35
Pair No.	SMH	SMK	PROB. level	Risk level (%)	Risk error (%)
Table No. 8.2-105					
1	1.64	1.624437	0.909998	9.000236	−0.000238
2	1.63	1.634403	0.910005	8.999479	0.000519
3	1.62	1.644687	0.910002	8.999782	0.000215
4	1.61	1.655487	0.910003	8.999706	0.000292
5	1.60	1.666806	0.910005	8.999514	0.000483
6	1.59	1.678645	0.910005	8.999491	0.000507
7	1.58	1.691007	0.910001	8.999938	0.000060
8	1.57	1.704091	0.910003	8.999681	0.000316
9	1.56	1.717898	0.910009	8.999133	0.000864
10	1.55	1.732236	0.909999	9.000111	−0.000113
11	1.54	1.747498	0.909998	9.000248	−0.000250
12	1.53	1.763688	0.909999	9.000140	−0.000143
13	1.52	1.780807	0.909997	9.000343	−0.000346
14	1.51	1.799055	0.909997	9.000272	−0.000274
15	1.50	1.818629	0.910005	8.999527	0.000471
16	1.49	1.839338	0.909999	9.000058	−0.000060
17	1.48	1.861770	0.910005	8.999538	0.000459
18	1.47	1.885735	0.909999	9.000140	−0.000143
19	1.46	1.911819	0.909999	9.000147	−0.000149
20	1.45	1.940419	0.910006	8.999401	0.000596
21	1.44	1.971538	0.910003	8.999735	0.000262
22	1.43	2.005959	0.909999	9.000063	−0.000066
23	1.42	2.044470	0.909998	9.000194	−0.000197
24	1.41	2.087853	0.909991	9.000910	−0.000912
25	1.40	2.138068	0.909992	9.000778	−0.000781
26	1.39	2.197851	0.910009	8.999062	0.000936
27	1.38	2.269552	0.910007	8.999294	0.000703
28	1.37	2.360206	0.909994	9.000617	−0.000620

(continued)

8.2 Contents for Table 8.2 (Table No. 8.2-1 to 400)

Table 8.2 (continued)

Parameter values					
PROB. = 0.91	RHO = 0.5		BIGH = 1.6322		SMHL = 1.35
Pair No.	SMH	SMK	PROB. level	Risk level (%)	Risk error (%)
29	1.36	2.487004	0.909994	9.000599	−0.000602
30	1.35	2.701514	0.910005	8.999538	0.000459
No. of pairs		Mean COMPUT. risk errors		STD. ERR. errors	
30		0.0000724869		0.0000844947	
Z-ratio				Status of significance	
0.85789				N.S.—not significant	
Parameter values					
PROB. = 0.9025	RHO = 0.5		BIGH = 1.5903		SMHL = 1.30
Pair No.	SMH	SMK	PROB. level	Risk level (%)	Risk error (%)
Table No. 8.2-106					
1	1.60	1.581049	0.902508	9.749186	0.000817
2	1.59	1.590795	0.902502	9.749764	0.000238
3	1.58	1.601058	0.902505	9.749532	0.000471
4	1.57	1.611644	0.902495	9.750504	−0.000501
5	1.56	1.622946	0.902505	9.749484	0.000519
6	1.55	1.634577	0.902498	9.750176	−0.000173
7	1.54	1.646930	0.902504	9.749585	0.000417
8	1.53	1.659812	0.902504	9.749651	0.000352
9	1.52	1.673226	0.902493	9.750724	−0.000721
10	1.51	1.687565	0.902498	9.750181	−0.000179
11	1.50	1.702637	0.902500	9.750027	−0.000024
12	1.49	1.718445	0.902493	9.750729	−0.000727
13	1.48	1.735383	0.902499	9.750091	−0.000089
14	1.47	1.753257	0.902498	9.750176	−0.000173
15	1.46	1.772463	0.902509	9.749138	0.000864
16	1.45	1.792612	0.902498	9.750247	−0.000244
17	1.44	1.814491	0.902503	9.749693	0.000310
18	1.43	1.837710	0.902492	9.750832	−0.000829
19	1.42	1.863055	0.902494	9.750634	−0.000632
20	1.41	1.890530	0.902493	9.750711	−0.000709
21	1.40	1.920529	0.902491	9.750927	−0.000924
22	1.39	1.953838	0.902500	9.750021	−0.000018
23	1.38	1.990460	0.902494	9.750629	−0.000626
24	1.37	2.031962	0.902499	9.750062	−0.000060
25	1.36	2.079130	0.902500	9.750038	−0.000036
26	1.35	2.134310	0.902509	9.749121	0.000882
27	1.34	2.199853	0.902509	9.749061	0.000942
28	1.33	2.281231	0.902510	9.749008	0.000995
29	1.32	2.387824	0.902496	9.750378	−0.000376

(continued)

Table 8.2 (continued)

Parameter values					
PROB. = 0.9025	RHO = 0.5		BIGH = 1.5903		SMHL = 1.30
Pair No.	SMH	SMK	PROB. level	Risk level (%)	Risk error (%)
30	1.31	2.549325	0.902504	9.749615	0.000387
31	1.30	2.895425	0.902500	9.749985	0.000018
No. of pairs		Mean COMPUT. risk errors		STD. ERR. errors	
31		0.0000054017		0.0000983747	
Z-ratio				Status of significance	
0.05491				N.S.—not significant	
Parameter values					
PROB. = 0.9	RHO = 0.5		BIGH = 1.577		SMHL = 1.29
Pair No.	SMH	SMK	PROB. level	Risk level (%)	Risk error (%)
Table No. 8.2-107					
1	1.58	1.574006	0.900001	9.999924	0.000077
2	1.57	1.584031	0.899995	10.000500	−0.000501
3	1.56	1.594576	0.899996	10.000380	−0.000381
4	1.55	1.605642	0.900001	9.999889	0.000113
5	1.54	1.617233	0.900007	9.999311	0.000691
6	1.53	1.629155	0.899994	10.000600	−0.000596
7	1.52	1.641997	0.900009	9.999138	0.000864
8	1.51	1.655176	0.899998	10.000170	−0.000167
9	1.50	1.669281	0.900007	9.999352	0.000650
10	1.49	1.683923	0.899998	10.000210	−0.000203
11	1.48	1.699498	0.899998	10.000210	−0.000203
12	1.47	1.716007	0.900000	10.000020	−0.000018
13	1.46	1.733454	0.899998	10.000230	−0.000226
14	1.45	1.752037	0.899998	10.000210	−0.000203
15	1.44	1.771956	0.900005	9.999538	0.000465
16	1.43	1.793017	0.899997	10.000320	−0.000322
17	1.42	1.815811	0.899999	10.000060	−0.000060
18	1.41	1.840342	0.899999	10.000060	−0.000060
19	1.40	1.867003	0.900003	9.999657	0.000346
20	1.39	1.895798	0.899995	10.000470	−0.000465
21	1.38	1.927903	0.900009	9.999090	0.000912
22	1.37	1.962933	0.900003	9.999734	0.000268
23	1.36	2.002061	0.899995	10.000470	−0.000465
24	1.35	2.046856	0.900004	9.999609	0.000393
25	1.34	2.098104	0.900002	9.999829	0.000173
26	1.33	2.158152	0.899992	10.000840	−0.000834
27	1.32	2.232474	0.900007	9.999311	0.000691
28	1.31	2.326543	0.900002	9.999818	0.000185

(continued)

8.2 Contents for Table 8.2 (Table No. 8.2-1 to 400)

Table 8.2 (continued)

Parameter values					
PROB. = 0.9	RHO = 0.5		BIGH = 1.577		SMHL = 1.29
Pair No.	SMH	SMK	PROB. level	Risk level (%)	Risk error (%)
29	1.30	2.458334	0.899999	10.000130	−0.000131
30	1.29	2.684101	0.899992	10.000830	−0.000823
No. of pairs		Mean COMPUT. risk errors		STD. ERR. errors	
30		0.0000055759		0.0000835266	
Z-ratio				Status of significance	
0.06676				N.S.—not significant	
Parameter values					
PROB. = 0.85	RHO = 0.5		BIGH = 1.349		SMHL = 1.04
Pair No.	SMH	SMK	PROB. level	Risk level (%)	Risk error (%)
Table No. 8.2-108					
1	1.35	1.348098	0.850002	14.999800	0.000203
2	1.34	1.358256	0.850001	14.999890	0.000107
3	1.33	1.368857	0.850003	14.999720	0.000274
4	1.32	1.379809	0.849992	15.000750	−0.000757
5	1.31	1.391310	0.849991	15.000950	−0.000948
6	1.30	1.403362	0.849993	15.000710	−0.000715
7	1.29	1.415972	0.849996	15.000430	−0.000429
8	1.28	1.429141	0.849995	15.000500	−0.000501
9	1.27	1.443070	0.850007	14.999320	0.000674
10	1.26	1.457568	0.850006	14.999440	0.000554
11	1.25	1.472833	0.850006	14.999380	0.000614
12	1.24	1.488870	0.850003	14.999740	0.000256
13	1.23	1.505880	0.850006	14.999370	0.000632
14	1.22	1.523671	0.849992	15.000770	−0.000769
15	1.21	1.542835	0.850004	14.999630	0.000370
16	1.20	1.562985	0.849997	15.000330	−0.000334
17	1.19	1.584713	0.850008	14.999170	0.000829
18	1.18	1.607829	0.850010	14.999030	0.000966
19	1.17	1.632534	0.850002	14.999810	0.000185
20	1.16	1.659419	0.850009	14.999090	0.000912
21	1.15	1.688295	0.849998	15.000190	−0.000191
22	1.14	1.720145	0.850006	14.999410	0.000590
23	1.13	1.754974	0.850003	14.999720	0.000274
24	1.12	1.793572	0.849996	15.000360	−0.000364
25	1.11	1.837116	0.849999	15.000060	−0.000060
26	1.10	1.886786	0.850005	14.999540	0.000453
27	1.09	1.944151	0.849998	15.000200	−0.000203
28	1.08	2.013125	0.850007	14.999300	0.000697
29	1.07	2.098405	0.850006	14.999360	0.000638
30	1.06	2.210543	0.849991	15.000860	−0.000858

(continued)

Table 8.2 (continued)

Parameter values					
PROB. = 0.85	RHO = 0.5		BIGH = 1.349		SMHL = 1.04
Pair No.	SMH	SMK	PROB. level	Risk level (%)	Risk error (%)
31	1.05	2.380018	0.849990	15.000960	−0.000960
32	1.04	2.762308	0.850004	14.999590	0.000405
No. of pairs		Mean COMPUT. risk errors		STD. ERR. errors	
32		0.0000771248		0.0001012811	
Z-ratio				Status of significance	
0.76149				N.S.—not significant	
Parameter values					
PROB. = 0.8	RHO = 0.5		BIGH = 1.1684		SMHL = 0.85
Pair No.	SMH	SMK	PROB. level	Risk level (%)	Risk error (%)
Table No. 8.2-109					
1	1.17	1.166900	0.800005	19.999540	0.000459
2	1.16	1.177056	0.800008	19.999200	0.000805
3	1.15	1.187485	0.799993	20.000730	−0.000727
4	1.14	1.198484	0.799999	20.000100	−0.000101
5	1.13	1.209863	0.799995	20.000500	−0.000501
6	1.12	1.221821	0.800005	19.999530	0.000471
7	1.11	1.234170	0.799998	20.000240	−0.000244
8	1.10	1.247108	0.799996	20.000390	−0.000387
9	1.09	1.260642	0.799996	20.000450	−0.000447
10	1.08	1.274778	0.799991	20.000900	−0.000900
11	1.07	1.289716	0.800001	19.999920	0.000083
12	1.06	1.305268	0.799995	20.000460	−0.000459
13	1.05	1.321635	0.799991	20.000870	−0.000870
14	1.04	1.339020	0.800003	19.999740	0.000256
15	1.03	1.357233	0.799999	20.000090	−0.000095
16	1.02	1.376477	0.799993	20.000710	−0.000715
17	1.01	1.396955	0.799993	20.000700	−0.000703
18	1.00	1.418870	0.800005	19.999470	0.000525
19	0.99	1.442034	0.799997	20.000260	−0.000256
20	0.98	1.466847	0.799990	20.000990	−0.000995
21	0.97	1.493708	0.799996	20.000360	−0.000364
22	0.96	1.522821	0.800008	19.999160	0.000840
23	0.95	1.554196	0.800001	19.999920	0.000083
24	0.94	1.588624	0.799999	20.000100	−0.000101
25	0.93	1.626506	0.799991	20.000940	−0.000936
26	0.92	1.669024	0.799998	20.000200	−0.000197
27	0.91	1.716971	0.800001	19.999880	0.000119
28	0.90	1.771921	0.799999	20.000060	−0.000060
29	0.89	1.836620	0.800003	19.999700	0.000304
30	0.88	1.914598	0.799990	20.000970	−0.000972
31	0.87	2.015241	0.800007	19.999330	0.000674

(continued)

8.2 Contents for Table 8.2 (Table No. 8.2-1 to 400)

Table 8.2 (continued)

Parameter values					
PROB. = 0.8	RHO = 0.5		BIGH = 1.1684		SMHL = 0.85
Pair No.	SMH	SMK	PROB. level	Risk level (%)	Risk error (%)
32	0.86	2.154581	0.800002	19.999800	0.000203
33	0.85	2.393568	0.800001	19.999900	0.000095
No. of pairs		Mean COMPUT. risk errors		STD. ERR. errors	
33		−0.0001504141		0.0000897397	
Z-ratio				Status of significance	
1.67612				N.S.—not significant	
Parameter values					
PROB. = 0.75	RHO = 0.5		BIGH = 1.0138		SMHL = 0.68
Pair No.	SMH	SMK	PROB. level	Risk level (%)	Risk error (%)
Table No. 8.2-110					
1	1.02	1.007833	0.750004	24.999600	0.000399
2	1.01	1.017810	0.750007	24.999330	0.000668
3	1.00	1.028084	0.749997	25.000320	−0.000316
4	0.99	1.038856	0.750005	24.999550	0.000453
5	0.98	1.049938	0.749995	25.000540	−0.000542
6	0.97	1.061531	0.749996	25.000420	−0.000417
7	0.96	1.073642	0.750004	24.999570	0.000435
8	0.95	1.086182	0.750001	24.999860	0.000143
9	0.94	1.099254	0.749998	25.000210	−0.000209
10	0.93	1.112964	0.750004	24.999610	0.000387
11	0.92	1.127222	0.750000	25.000000	0.000006
12	0.91	1.142135	0.749995	25.000480	−0.000483
13	0.90	1.157810	0.749997	25.000350	−0.000346
14	0.89	1.174352	0.750010	24.999050	0.000954
15	0.88	1.191577	0.750002	24.999830	0.000173
16	0.87	1.209688	0.749990	25.000970	−0.000966
17	0.86	1.228992	0.750000	24.999970	0.000036
18	0.85	1.249399	0.750009	24.999140	0.000858
19	0.84	1.270826	0.749993	25.000670	−0.000668
20	0.83	1.293771	0.749993	25.000660	−0.000656
21	0.82	1.318247	0.749992	25.000760	−0.000757
22	0.81	1.344658	0.750008	24.999170	0.000829
23	0.80	1.372824	0.750002	24.999840	0.000161
24	0.79	1.403344	0.750000	25.000030	−0.000030
25	0.78	1.436625	0.750003	24.999690	0.000310
26	0.77	1.473074	0.750004	24.999570	0.000435
27	0.76	1.513293	0.750001	24.999910	0.000095
28	0.75	1.558278	0.750000	25.000030	−0.000030
29	0.74	1.609218	0.749997	25.000300	−0.000298
30	0.73	1.668088	0.749999	25.000090	−0.000089
31	0.72	1.737643	0.749998	25.000190	−0.000191
32	0.71	1.822983	0.749998	25.000180	−0.000179

(continued)

Table 8.2 (continued)

Parameter values					
PROB. = 0.75	RHO = 0.5		BIGH = 1.0138		SMHL = 0.68
Pair No.	SMH	SMK	PROB. level	Risk level (%)	Risk error (%)
33	0.70	1.933897	0.749998	25.000170	−0.000167
34	0.69	2.094238	0.749994	25.000630	−0.000626
35	0.68	2.405206	0.749994	25.000580	−0.000584
No. of pairs		Mean COMPUT. risk errors		STD. ERR. errors	
35		−0.0000336104		0.0000801118	
Z-ratio				Status of significance	
0.41954				N.S.—not significant	
Parameter values					
PROB. = 0.7	RHO = 0.5		BIGH = 0.8754		SMHL = 0.53
Pair No.	SMH	SMK	PROB. level	Risk level (%)	Risk error (%)
Table No. 8.2-111					
1	0.88	0.870726	0.699992	30.000790	−0.000793
2	0.87	0.880834	0.700010	29.999020	0.000978
3	0.86	0.891174	0.700003	29.999680	0.000328
4	0.85	0.901950	0.700007	29.999310	0.000697
5	0.84	0.913073	0.700001	29.999920	0.000077
6	0.83	0.924650	0.700000	30.000040	−0.000042
7	0.82	0.936691	0.700000	29.999970	0.000030
8	0.81	0.949205	0.700000	30.000030	−0.000030
9	0.80	0.962203	0.699994	30.000560	−0.000554
10	0.79	0.975794	0.699997	30.000350	−0.000346
11	0.78	0.989988	0.700001	29.999870	0.000131
12	0.77	1.004798	0.700004	29.999600	0.000405
13	0.76	1.020237	0.700000	30.000050	−0.000048
14	0.75	1.036415	0.699997	30.000340	−0.000334
15	0.74	1.053446	0.700002	29.999770	0.000232
16	0.73	1.071245	0.699995	30.000480	−0.000477
17	0.72	1.090122	0.700007	29.999340	0.000668
18	0.71	1.109898	0.700001	29.999890	0.000113
19	0.70	1.130883	0.700005	29.999510	0.000495
20	0.69	1.152999	0.699994	30.000590	−0.000590
21	0.68	1.176753	0.700010	29.999010	0.000995
22	0.67	1.201776	0.699993	30.000690	−0.000685
23	0.66	1.228872	0.700007	29.999340	0.000668
24	0.65	1.257868	0.700007	29.999310	0.000697
25	0.64	1.289180	0.700005	29.999510	0.000495
26	0.63	1.323225	0.700003	29.999680	0.000322
27	0.62	1.360422	0.699996	30.000450	−0.000447
28	0.61	1.401583	0.699994	30.000640	−0.000638
29	0.60	1.447521	0.699991	30.000930	−0.000924
30	0.59	1.499637	0.699995	30.000500	−0.000495
31	0.58	1.559531	0.699991	30.000890	−0.000888

(continued)

8.2 Contents for Table 8.2 (Table No. 8.2-1 to 400)

Table 8.2 (continued)

Parameter values					
PROB. = 0.7	RHO = 0.5		BIGH = 0.8754		SMHL = 0.53
Pair No.	SMH	SMK	PROB. level	Risk level (%)	Risk error (%)
32	0.57	1.630367	0.699992	30.000840	−0.000840
33	0.56	1.717265	0.699998	30.000250	−0.000244
34	0.55	1.830036	0.700004	29.999590	0.000411
35	0.54	1.992558	0.700003	29.999720	0.000280
36	0.53	2.305271	0.700001	29.999910	0.000089
No. of pairs		Mean COMPUT. risk errors		STD. ERR. errors	
36		−0.0000070881		0.0000887708	
Z-ratio				Status of significance	
0.07985				N.S.—not significant	
Parameter values					
PROB. = 0.65	RHO = 0.5		BIGH = 0.7473		SMHL = 0.39
Pair No.	SMH	SMK	PROB. level	Risk level (%)	Risk error (%)
Table No. 8.2-112					
1	0.75	0.744512	0.649995	35.000470	−0.000465
2	0.74	0.754574	0.649992	35.000760	−0.000757
3	0.73	0.765010	0.649994	35.000620	−0.000614
4	0.72	0.775830	0.649998	35.000200	−0.000197
5	0.71	0.787048	0.650003	34.999670	0.000334
6	0.70	0.798675	0.650008	34.999220	0.000787
7	0.69	0.810726	0.650010	34.999050	0.000954
8	0.68	0.823214	0.650007	34.999350	0.000656
9	0.67	0.836155	0.649997	35.000330	−0.000328
10	0.66	0.849663	0.649994	35.000580	−0.000572
11	0.65	0.863755	0.649996	35.000440	−0.000441
12	0.64	0.878449	0.649997	35.000280	−0.000274
13	0.63	0.893764	0.649995	35.000500	−0.000501
14	0.62	0.909819	0.650000	34.999990	0.000012
15	0.61	0.926636	0.650007	34.999330	0.000674
16	0.60	0.944239	0.650009	34.999090	0.000918
17	0.59	0.962651	0.650001	34.999870	0.000137
18	0.58	0.981998	0.649991	35.000950	−0.000942
19	0.57	1.002601	0.650006	34.999390	0.000608
20	0.56	1.024198	0.650000	35.000040	−0.000036
21	0.55	1.047213	0.650009	34.999130	0.000870
22	0.54	1.071485	0.649997	35.000340	−0.000340
23	0.53	1.097443	0.649993	35.000670	−0.000668
24	0.52	1.125323	0.650000	34.999990	0.000018
25	0.51	1.155170	0.649995	35.000530	−0.000525
26	0.50	1.187617	0.650007	34.999280	0.000727
27	0.49	1.222716	0.650005	34.999530	0.000477
28	0.48	1.261108	0.649999	35.000100	−0.000095
29	0.47	1.303636	0.650002	34.999800	0.000209

(continued)

Table 8.2 (continued)

Parameter values					
PROB. = 0.65	RHO = 0.5		BIGH = 0.7473		SMHL = 0.39
Pair No.	SMH	SMK	PROB. level	Risk level (%)	Risk error (%)
30	0.46	1.351146	0.650006	34.999410	0.000596
31	0.45	1.404687	0.649994	35.000580	−0.000578
32	0.44	1.466877	0.650005	34.999500	0.000501
33	0.43	1.540143	0.650001	34.999910	0.000095
34	0.42	1.630441	0.650008	34.999200	0.000799
35	0.41	1.748028	0.650008	34.999250	0.000751
36	0.40	1.955517	0.650002	34.999780	0.000221
37	0.39	2.294441	0.649999	35.000100	−0.000095
No. of pairs		Mean COMPUT. risk errors		STD. ERR. errors	
37		0.0000767018		0.0000892152	
Z-ratio				Status of significance	
0.85974				N.S.—not significant	
Parameter values					
PROB. = 0.6	RHO = 0.5		BIGH = 0.6259		SMHL = 0.26
Pair No.	SMH	SMK	PROB. level	Risk level (%)	Risk error (%)
Table No. 8.2-113					
1	0.63	0.621827	0.599994	40.000570	−0.000572
2	0.62	0.631856	0.599994	40.000590	−0.000590
3	0.61	0.642214	0.599992	40.000770	−0.000775
4	0.60	0.653016	0.600008	39.999170	0.000834
5	0.59	0.664082	0.600003	39.999750	0.000250
6	0.58	0.675530	0.599995	40.000550	−0.000548
7	0.57	0.687477	0.600003	39.999740	0.000262
8	0.56	0.699848	0.600007	39.999270	0.000727
9	0.55	0.712665	0.600008	39.999210	0.000787
10	0.54	0.725953	0.600004	39.999600	0.000399
11	0.53	0.739738	0.599995	40.000490	−0.000489
12	0.52	0.754148	0.599997	40.000330	−0.000334
13	0.51	0.769115	0.599991	40.000910	−0.000906
14	0.50	0.784869	0.600008	39.999210	0.000793
15	0.49	0.801151	0.599998	40.000190	−0.000191
16	0.48	0.818296	0.600006	39.999410	0.000590
17	0.47	0.836149	0.599998	40.000200	−0.000203
18	0.46	0.854952	0.600000	39.999960	0.000036
19	0.45	0.874659	0.599995	40.000530	−0.000530
20	0.44	0.895518	0.600003	39.999710	0.000286
21	0.43	0.917493	0.600005	39.999520	0.000477
22	0.42	0.940747	0.600006	39.999380	0.000620
23	0.41	0.965450	0.600010	39.999000	0.000995
24	0.40	0.991681	0.600005	39.999460	0.000536
25	0.39	1.019723	0.600002	39.999830	0.000167
26	0.38	1.049868	0.600003	39.999730	0.000268

(continued)

8.2 Contents for Table 8.2 (Table No. 8.2-1 to 400)

Table 8.2 (continued)

Parameter values					
PROB. = 0.6	RHO = 0.5		BIGH = 0.6259		SMHL = 0.26
Pair No.	SMH	SMK	PROB. level	Risk level (%)	Risk error (%)
27	0.37	1.082418	0.600007	39.999330	0.000668
28	0.36	1.117688	0.600007	39.999350	0.000656
29	0.35	1.156201	0.600005	39.999540	0.000465
30	0.34	1.198692	0.600007	39.999290	0.000709
31	0.33	1.245716	0.599990	40.000970	−0.000972
32	0.32	1.299220	0.600002	39.999770	0.000226
33	0.31	1.360391	0.600000	40.000010	−0.000006
34	0.30	1.478100	0.599999	40.000140	−0.000143
35	0.29	1.560238	0.600005	39.999540	0.000459
36	0.28	1.667077	0.600009	39.999140	0.000864
37	0.27	1.819677	0.599995	40.000540	−0.000542
38	0.26	2.102044	0.599999	40.000140	−0.000143
No. of pairs		Mean COMPUT. risk errors		STD. ERR. errors	
38		0.0001315887		0.0000880859	
Z-ratio				Status of significance	
1.49387				N.S.—not significant	
Parameter values					
PROB. = 0.55	RHO = 0.5		BIGH = 0.5087		SMHL = 0.13
Pair No.	SMH	SMK	PROB. level	Risk level (%)	Risk error (%)
Table No. 8.2-114					
1	0.51	0.507403	0.550001	44.999900	0.000101
2	0.50	0.517551	0.550007	44.999280	0.000721
3	0.49	0.527967	0.550000	44.999990	0.000006
4	0.48	0.538725	0.549992	45.000780	−0.000775
5	0.47	0.549903	0.549996	45.000440	−0.000435
6	0.46	0.561482	0.550002	44.999780	0.000221
7	0.45	0.573397	0.549995	45.000490	−0.000495
8	0.44	0.585783	0.549996	45.000400	−0.000393
9	0.43	0.598679	0.550008	44.999240	0.000763
10	0.42	0.611933	0.549996	45.000380	−0.000381
11	0.41	0.625789	0.550001	44.999880	0.000125
12	0.40	0.640103	0.549991	45.000890	−0.000888
13	0.39	0.655129	0.550004	44.999660	0.000346
14	0.38	0.670637	0.549992	45.000800	−0.000793
15	0.37	0.686894	0.549994	45.000570	−0.000566
16	0.36	0.703880	0.549998	45.000170	−0.000167
17	0.35	0.721586	0.549994	45.000640	−0.000638
18	0.34	0.740206	0.550000	44.999960	0.000042
19	0.33	0.759754	0.550009	44.999100	0.000900
20	0.32	0.780158	0.549998	45.000240	−0.000232
21	0.31	0.801752	0.550000	44.999970	0.000030
22	0.30	0.824499	0.549997	45.000350	−0.000352

(continued)

Table 8.2 (continued)

Parameter values					
PROB. = 0.55	RHO = 0.5		BIGH = 0.5087		SMHL = 0.13
Pair No.	SMH	SMK	PROB. level	Risk level (%)	Risk error (%)
23	0.29	0.848579	0.549993	45.000690	−0.000691
24	0.28	0.874198	0.549996	45.000370	−0.000370
25	0.27	0.901494	0.550001	44.999950	0.000048
26	0.26	0.930642	0.550001	44.999940	0.000060
27	0.25	0.961860	0.549993	45.000740	−0.000739
28	0.24	0.995809	0.550009	44.999160	0.000846
29	0.23	1.032337	0.549994	45.000590	−0.000590
30	0.22	1.127424	0.549997	45.000300	−0.000304
31	0.21	1.166444	0.550005	44.999530	0.000477
32	0.20	1.210674	0.550007	44.999320	0.000685
33	0.19	1.261390	0.549999	45.000150	−0.000149
34	0.18	1.320844	0.549995	45.000470	−0.000465
35	0.17	1.392521	0.550008	44.999190	0.000817
36	0.16	1.481409	0.550007	44.999330	0.000674
37	0.15	1.598607	0.549997	45.000270	−0.000268
38	0.14	1.773396	0.550004	44.999660	0.000346
39	0.13	2.137455	0.549995	45.000520	−0.000519
No. of pairs		Mean COMPUT. risk errors		STD. ERR. errors	
39		−0.0000751019		0.0000816825	
Z-ratio				Status of significance	
0.91944				N.S.—not significant	
Parameter values					
PROB. = 0.5	RHO = 0.5		BIGH = 0.3935		SMHL = 0.01
Pair No.	SMH	SMK	PROB. level	Risk level (%)	Risk error (%)
Table No. 8.2-115					
1	0.40	0.387106	0.500001	49.999920	0.000083
2	0.39	0.397031	0.499994	50.000560	−0.000557
3	0.38	0.407284	0.499991	50.000860	−0.000858
4	0.37	0.417907	0.499997	50.000270	−0.000262
5	0.36	0.428848	0.499998	50.000190	−0.000194
6	0.35	0.440160	0.500001	49.999920	0.000083
7	0.34	0.451805	0.499994	50.000590	−0.000584
8	0.33	0.463946	0.500006	49.999370	0.000632
9	0.32	0.476460	0.500009	49.999130	0.000870
10	0.31	0.489335	0.499995	50.000520	−0.000519
11	0.30	0.502762	0.499996	50.000400	−0.000396
12	0.29	0.516751	0.500008	49.999210	0.000793
13	0.28	0.531133	0.499995	50.000540	−0.000536
14	0.27	0.546146	0.499993	50.000750	−0.000745
15	0.26	0.561855	0.500005	49.999460	0.000536
16	0.25	0.578158	0.500008	49.999190	0.000811
17	0.24	0.595078	0.499998	50.000220	−0.000218
18	0.23	0.612875	0.500007	49.999270	0.000727

(continued)

8.2 Contents for Table 8.2 (Table No. 8.2-1 to 400)

Table 8.2 (continued)

Parameter values						
PROB. = 0.5	RHO = 0.5		BIGH = 0.3935			SMHL = 0.01
Pair No.	SMH	SMK	PROB. level	Risk level (%)		Risk error (%)
19	0.22	0.631367	0.499998	50.000160		−0.000161
20	0.21	0.650820	0.500001	49.999860		0.000143
21	0.20	0.671184	0.499997	50.000270		−0.000265
22	0.19	0.692693	0.500007	49.999350		0.000656
23	0.18	0.715313	0.500009	49.999060		0.000948
24	0.17	0.739157	0.500006	49.999380		0.000620
25	0.16	0.764443	0.500007	49.999350		0.000650
26	0.15	0.851617	0.500007	49.999350		0.000650
27	0.14	0.873984	0.500005	49.999500		0.000507
28	0.13	0.898516	0.499993	50.000700		−0.000706
29	0.12	0.925703	0.499990	50.000960		−0.000957
30	0.11	0.955898	0.499992	50.000790		−0.000787
31	0.10	0.989632	0.500001	49.999920		0.000083
32	0.09	1.027304	0.499996	50.000380		−0.000376
33	0.08	1.069902	0.499993	50.000700		−0.000700
34	0.07	1.118671	0.499998	50.000240		−0.000241
35	0.06	1.175233	0.500006	49.999390		0.000614
36	0.05	1.241763	0.500001	49.999920		0.000083
37	0.04	1.322613	0.499998	50.000170		−0.000173
38	0.03	1.425461	0.500004	49.999620		0.000381
39	0.02	1.566308	0.500005	49.999510		0.000495
40	0.01	1.795025	0.500004	49.999570		0.000435
No. of pairs		Mean COMPUT. risk errors		STD. ERR. errors		
40		0.0000381615		0.0000879193		
Z-ratio				Status of significance		
0.43405				N.S.—not significant		
Parameter values						
PROB. = 0.99	RHO = 0.4		BIGH = 2.5648			SMHL = 2.33
Pair No.	SMH	SMK	PROB. level	Risk level (%)		Risk error (%)
Table No. 8.2-116						
1	2.57	2.559611	0.990002	0.999761		0.000238
2	2.56	2.569609	0.990003	0.999749		0.000250
3	.255	2.579686	0.989996	1.000398		−0.000399
4	2.54	2.591405	0.990003	0.999659		0.000340
5	2.53	2.603204	0.990003	0.999725		0.000274
6	2.52	2.615084	0.989994	1.000595		−0.000596
7	2.51	2.628609	0.989996	1.000416		−0.000417
8	2.50	2.643780	0.990006	0.999409		0.000590
9	2.49	2.659035	0.990005	0.999475		0.000525
10	2.48	2.674375	0.989994	1.000637		−0.000638
11	2.47	2.692926	0.990003	0.999749		0.000250
12	2.46	2.711565	0.989997	1.000261		−0.000262

(continued)

Table 8.2 (continued)

Parameter values					
PROB. = 0.99	RHO = 0.4		BIGH = 2.5648		SMHL = 2.33
Pair No.	SMH	SMK	PROB. level	Risk level (%)	Risk error (%)
13	2.45	2.731854	0.989993	1.000702	−0.000703
14	2.44	2.755359	0.990000	1.000029	−0.000030
15	2.43	2.780515	0.990001	0.999916	0.000083
16	2.42	2.808889	0.990005	0.999463	0.000536
17	2.41	2.838918	0.989998	1.000154	−0.000155
18	2.40	2.875291	0.990006	0.999409	0.000590
19	2.39	2.914885	0.990001	0.999880	0.000119
20	2.38	2.960824	0.989995	1.000524	−0.000525
21	2.37	3.019361	0.990001	0.999880	0.000119
22	2.36	3.087372	0.989990	1.000994	−0.000995
23	2.35	3.186733	0.990000	1.000035	−0.000036
24	2.34	3.336196	0.990003	0.999695	0.000304
25	2.33	3.673261	0.990001	0.999856	0.000143
No. of pairs		Mean COMPUT. risk errors		STD. ERR. errors	
25		−0.0000151304		0.0000852648	
Z-ratio				Status of significance	
0.17745				N.S.—not significant	
Parameter values					
PROB. = 0.98	RHO = 0.4		BIGH = 2.3093		SMHL = 2.06
Pair No.	SMH	SMK	PROB. level	Risk level (%)	Risk error (%)
Table No. 8.2-117					
1	2.31	2.308600	0.979999	2.000123	−0.000125
2	2.30	2.318638	0.979994	2.000594	−0.000596
3	2.29	2.329544	0.979998	2.000225	−0.000226
4	2.28	2.341320	0.980008	1.999199	0.000799
5	2.27	2.353187	0.980005	1.999509	0.000489
6	2.26	2.365925	0.980006	1.999402	0.000596
7	2.25	2.379538	0.980010	1.999050	0.000948
8	2.24	2.393244	0.979997	2.000284	−0.000286
9	2.23	2.408608	0.980000	1.999956	0.000042
10	2.22	2.424848	0.980001	1.999950	0.000048
11	2.21	2.441968	0.979996	2.000433	−0.000435
12	2.20	2.460749	0.979998	2.000237	−0.000238
13	2.19	2.481192	0.980003	1.999724	0.000274
14	2.18	2.502519	0.979995	2.000529	−0.000530
15	2.17	2.526292	0.979995	2.000505	−0.000507
16	2.16	2.553295	0.980008	1.999170	0.000829
17	2.15	2.581965	0.980007	1.999337	0.000662
18	2.14	2.613869	0.980005	1.999545	0.000453
19	2.13	2.649006	0.979994	2.000558	−0.000560
20	2.12	2.690503	0.979998	2.000213	−0.000215
21	2.11	2.738362	0.979998	2.000225	−0.000226

(continued)

8.2 Contents for Table 8.2 (Table No. 8.2-1 to 400)

Table 8.2 (continued)

Parameter values					
PROB. = 0.98	RHO = 0.4		BIGH = 2.3093		SMHL = 2.06
Pair No.	SMH	SMK	PROB. level	Risk level (%)	Risk error (%)
22	2.10	2.795710	0.980000	2.000034	−0.000036
23	2.09	2.867235	0.980005	1.999545	0.000453
24	2.08	2.957627	0.979994	2.000558	−0.000560
25	2.07	3.095013	0.980003	1.999664	0.000334
26	2.06	3.351269	0.980000	2.000004	−0.000006
No. of pairs		Mean COMPUT. risk errors		STD. ERR. errors	
26		0.0000509951		0.0000908965	
Z-ratio				Status of significance	
0.56102				N.S.—not significant	
Parameter values					
PROB. = 0.97	RHO = 0.4		BIGH = 2.1481		SMHL = 1.89
Pair No.	SMH	SMK	PROB. level	Risk level (%)	Risk error (%)
Table No. 8.2-118					
1	2.15	2.146201	0.970004	2.999556	0.000441
2	2.14	2.156231	0.970000	2.999997	0.000000
3	2.13	2.167135	0.970009	2.999127	0.000870
4	2.12	2.178135	0.970001	2.999854	0.000143
5	2.11	2.190012	0.970004	2.999586	0.000411
6	2.10	2.202379	0.970002	2.999830	0.000167
7	2.09	2.215627	0.970005	2.999503	0.000495
8	2.08	2.229367	0.970000	3.000021	−0.000024
9	2.07	2.243990	0.969996	3.000373	−0.000376
10	2.06	2.259499	0.969992	3.000820	−0.000823
11	2.05	2.276676	0.970005	2.999520	0.000477
12	2.04	2.293959	0.969990	3.000981	−0.000983
13	2.03	2.313695	0.970005	2.999461	0.000536
14	2.02	2.334323	0.970006	2.999407	0.000590
15	2.01	2.356626	0.970007	2.999342	0.000656
16	2.00	2.380604	0.970002	2.999771	0.000226
17	1.99	2.407041	0.970003	2.999711	0.000286
18	1.98	2.435940	0.970000	2.999997	0.000000
19	1.97	2.468082	0.969999	3.000116	−0.000119
20	1.96	2.504251	0.970000	3.000009	−0.000012
21	1.95	2.545231	0.970000	3.000021	−0.000024
22	1.94	2.592584	0.970000	2.999973	0.000024
23	1.93	2.648658	0.970002	2.999801	0.000197
24	1.92	2.715798	0.969994	3.000641	−0.000644
25	1.91	2.803381	0.969997	3.000331	−0.000334

(continued)

Table 8.2 (continued)

Parameter values					
PROB. = 0.97	RHO = 0.4		BIGH = 2.1481		SMHL = 1.89
Pair No.	SMH	SMK	PROB. level	Risk level (%)	Risk error (%)
26	1.90	2.925471	0.969998	3.000176	−0.000179
27	1.89	3.128945	0.969992	3.000778	−0.000781
No. of pairs		Mean COMPUT. risk errors		STD. ERR. errors	
27		0.0000436391		0.0000875453	
Z-ratio				Status of significance	
0.49847				N.S.—not significant	
Parameter values					
PROB. = 0.96	RHO = 0.4		BIGH = 2.0271		SMHL = 1.76
Pair No.	SMH	SMK	PROB. level	Risk level (%)	Risk error (%)
Table No. 8.2-119					
1	2.03	2.024204	0.960001	3.999901	0.000101
2	2.02	2.034225	0.959997	4.000259	−0.000256
3	2.01	2.044736	0.959994	4.000574	−0.000572
4	2.00	2.056130	0.960007	3.999305	0.000697
5	1.99	2.067626	0.960000	4.000032	−0.000030
6	1.98	2.080008	0.960004	3.999591	0.000411
7	1.97	2.092886	0.960002	3.999806	0.000197
8	1.96	2.106654	0.960006	3.999359	0.000644
9	1.95	2.120920	0.960000	3.999984	0.000018
10	1.94	2.136080	0.959995	4.000467	−0.000465
11	1.93	2.152523	0.960003	3.999746	0.000256
12	1.92	2.169861	0.960004	3.999591	0.000411
13	1.91	2.188098	0.959997	4.000330	−0.000328
14	1.90	2.208015	0.960001	3.999877	0.000125
15	1.89	2.229223	0.960000	4.000044	−0.000042
16	1.88	2.252116	0.959998	4.000223	−0.000221
17	1.87	2.277085	0.959999	4.000068	−0.000066
18	1.86	2.304525	0.960005	3.999472	0.000530
19	1.85	2.334435	0.960006	3.999436	0.000566
20	1.84	2.367600	0.960006	3.999365	0.000638
21	1.83	2.404804	0.960008	3.999168	0.000834
22	1.82	2.446828	0.960007	3.999257	0.000745
23	1.81	2.495240	0.960006	3.999412	0.000590
24	1.80	2.551602	0.959995	4.000461	−0.000459
25	1.79	2.620606	0.959990	4.000962	−0.000960

(continued)

Table 8.2 (continued)

Parameter values					
PROB. = 0.96	RHO = 0.4		BIGH = 2.0271		SMHL = 1.76
Pair No.	SMH	SMK	PROB. level	Risk level (%)	Risk error (%)
26	1.78	2.711627	0.960009	3.999120	0.000882
27	1.77	2.834044	0.959998	4.000187	−0.000185
28	1.76	3.040985	0.959994	4.000574	−0.000572
No. of pairs		Mean COMPUT. risk errors		STD. ERR. errors	
28		0.0001204425		0.0000901625	
Z-ratio				Status of significance	
1.33584				N.S.—not significant	
Parameter values					
PROB. = 0.95	RHO = 0.4		BIGH = 1.929		SMHL = 1.65
Pair No.	SMH	SMK	PROB. level	Risk level (%)	Risk error (%)
Table No. 8.2-120					
1	1.93	1.928001	0.950007	4.999262	0.000739
2	1.92	1.938042	0.950000	5.000019	−0.000018
3	1.91	1.948580	0.949993	5.000657	−0.000656
4	1.90	1.960005	0.950007	4.999280	0.000721
5	1.89	1.971539	0.949998	5.000210	−0.000209
6	1.88	1.983965	0.950004	4.999608	0.000393
7	1.87	1.996893	0.950003	4.999715	0.000286
8	1.86	2.010325	0.949993	5.000693	−0.000691
9	1.85	2.024655	0.949990	5.000973	−0.000972
10	1.84	2.040273	0.950008	4.999251	0.000751
11	1.83	2.056012	0.949990	5.000961	−0.000960
12	1.82	2.073434	0.950002	4.999829	0.000173
13	1.81	2.091761	0.950004	4.999596	0.000405
14	1.80	2.110995	0.949993	5.000657	−0.000656
15	1.79	2.131919	0.949994	5.000621	−0.000620
16	1.78	2.154535	0.949997	5.000270	−0.000268
17	1.77	2.178846	0.949996	5.000371	−0.000370
18	1.76	2.205634	0.950006	4.999381	0.000620
19	1.75	2.234121	0.949994	5.000651	−0.000650
20	1.74	2.265873	0.949991	5.000913	−0.000912
21	1.73	2.301672	0.950000	5.000019	−0.000018
22	1.72	2.341521	0.950000	4.999990	0.000012
23	1.71	2.386984	0.950001	4.999930	0.000072
24	1.70	2.439628	0.949999	5.000150	−0.000149
25	1.69	2.503361	0.950010	4.999030	0.000972
26	1.68	2.580530	0.950000	4.999995	0.000006
27	1.67	2.682855	0.949999	5.000091	−0.000089

(continued)

Table 8.2 (continued)

Parameter values					
PROB. = 0.95	RHO = 0.4		BIGH = 1.929		SMHL = 1.65
Pair No.	SMH	SMK	PROB. level	Risk level (%)	Risk error (%)
28	1.66	2.835340	0.950008	4.999251	0.000751
29	1.65	3.142675	0.950009	4.999119	0.000882
No. of pairs		Mean COMPUT. risk errors		STD. ERR. errors	
29		−0.0000150998		0.0001064235	
Z-ratio				Status of significance	
0.14188				N.S.—not significant	
Parameter values					
PROB. = 0.94	RHO = 0.4		BIGH = 1.8455		SMHL = 1.56
Pair No.	SMH	SMK	PROB. level	Risk level (%)	Risk error (%)
Table No. 8.2-121					
1	1.85	1.841011	0.939990	6.000990	−0.000989
2	1.84	1.851212	0.940002	5.999846	0.000155
3	1.83	1.861522	0.939991	6.000895	−0.000894
4	1.82	1.872529	0.939994	6.000602	−0.000602
5	1.81	1.884040	0.939996	6.000400	−0.000399
6	1.80	1.896056	0.939995	6.000507	−0.000507
7	1.79	1.908775	0.939999	6.000054	−0.000054
8	1.78	1.922004	0.939996	6.000388	−0.000387
9	1.77	1.935939	0.939993	6.000686	−0.000685
10	1.76	1.950778	0.939997	6.000299	−0.000298
11	1.75	1.966524	0.940004	5.999613	0.000387
12	1.74	1.983178	0.940010	5.999053	0.000948
13	1.73	2.000742	0.940010	5.999047	0.000954
14	1.72	2.019219	0.940000	6.000013	−0.000012
15	1.71	2.039393	0.940009	5.999124	0.000876
16	1.70	2.060484	0.939997	6.000316	−0.000316
17	1.69	2.083667	0.940004	5.999619	0.000381
18	1.68	2.108554	0.940005	5.999518	0.000483
19	1.67	2.135537	0.940004	5.999589	0.000411
20	1.66	2.165010	0.940003	5.999697	0.000304
21	1.65	2.197366	0.939999	6.000060	−0.000060
22	1.64	2.233781	0.940008	5.999172	0.000829
23	1.63	2.273866	0.939997	6.000304	−0.000304
24	1.62	2.320357	0.940004	5.999595	0.000405
25	1.61	2.373260	0.939991	6.000901	−0.000900
26	1.60	2.438043	0.940007	5.999285	0.000715
27	1.59	2.517056	0.940007	5.999309	0.000691
28	1.58	2.619676	0.939996	6.000388	−0.000387

(continued)

8.2 Contents for Table 8.2 (Table No. 8.2-1 to 400)

Table 8.2 (continued)

Parameter values					
PROB. = 0.94	RHO = 0.4		BIGH = 1.8455		SMHL = 1.56
Pair No.	SMH	SMK	PROB. level	Risk level (%)	Risk error (%)
29	1.57	2.772468	0.940000	5.999965	0.000036
30	1.56	3.076999	0.939999	6.000054	−0.000054
No. of pairs		Mean COMPUT. risk errors		STD. ERR. errors	
30		0.0000234573		0.0001018206	
Z-ratio				Status of significance	
0.23038				N.S.—not significant	
Parameter values					
PROB. = 0.93	RHO = 0.4		BIGH = 1.7726		SMHL = 1.48
Pair No.	SMH	SMK	PROB. level	Risk level (%)	Risk error (%)
Table No. 8.2-122					
1	1.78	1.765231	0.929996	7.000411	−0.000411
2	1.77	1.775204	0.930001	6.999869	0.000131
3	1.76	1.785486	0.929998	7.000226	−0.000226
4	1.75	1.796273	0.929997	7.000256	−0.000256
5	1.74	1.807569	0.929998	7.000208	−0.000209
6	1.73	1.819374	0.929997	7.000292	−0.000292
7	1.72	1.831887	0.930005	6.999481	0.000519
8	1.71	1.844914	0.930006	6.999362	0.000638
9	1.70	1.858457	0.929999	7.000137	−0.000137
10	1.69	1.872909	0.930002	6.999791	0.000209
11	1.68	1.887882	0.929990	7.000965	−0.000966
12	1.67	1.904160	0.930004	6.999642	0.000358
13	1.66	1.920963	0.929994	7.000632	−0.000632
14	1.65	1.939075	0.929998	7.000220	−0.000221
15	1.64	1.958499	0.930009	6.999106	0.000894
16	1.63	1.978847	0.930002	6.999827	0.000173
17	1.62	2.000903	0.930006	6.999409	0.000590
18	1.61	2.024278	0.929996	7.000452	−0.000453
19	1.60	2.049757	0.929994	7.000578	−0.000578
20	1.59	2.077733	0.930004	6.999594	0.000405
21	1.58	2.108209	0.930010	6.999016	0.000983
22	1.57	2.141188	0.929997	7.000309	−0.000310
23	1.56	2.178236	0.929996	7.000381	−0.000381
24	1.55	2.220137	0.930002	6.999773	0.000226
25	1.54	2.267675	0.930002	6.999791	0.000209
26	1.53	2.322417	0.929991	7.000864	−0.000864
27	1.52	2.389053	0.930004	6.999624	0.000376
28	1.51	2.470711	0.930000	7.000036	−0.000036
29	1.50	2.579115	0.930000	6.999958	0.000042

(continued)

Table 8.2 (continued)

Parameter values					
PROB. = 0.93	RHO = 0.4		BIGH = 1.7726		SMHL = 1.48
Pair No.	SMH	SMK	PROB. level	Risk level (%)	Risk error (%)
30	1.49	2.740048	0.929995	7.000548	−0.000548
31	1.48	3.085547	0.929996	7.000387	−0.000387
No. of pairs		Mean COMPUT. risk errors		STD. ERR. errors	
31		−0.0000361353		0.0000849754	
Z-ratio				Status of significance	
0.42524				N.S.—not significant	

Parameter values					
PROB. = 0.92	RHO = 0.4		BIGH = 1.7074		SMHL = 1.41
Pair No.	SMH	SMK	PROB. level	Risk level (%)	Risk error (%)
Table No. 8.2-123					
1	1.71	1.704804	0.920008	7.999200	0.000799
2	1.70	1.714832	0.920002	7.999778	0.000221
3	1.69	1.725370	0.920002	7.999778	0.000221
4	1.68	1.736419	0.920006	7.999420	0.000578
5	1.67	1.747786	0.919996	8.000428	−0.000429
6	1.66	1.759864	0.919999	8.000106	−0.000107
7	1.65	1.772461	0.919998	8.000195	−0.000197
8	1.64	1.785773	0.920004	7.999629	0.000370
9	1.63	1.799608	0.919999	8.000064	−0.000066
10	1.62	1.814359	0.920007	7.999272	0.000727
11	1.61	1.829637	0.919998	8.000172	−0.000173
12	1.60	1.845837	0.919993	8.000671	−0.000674
13	1.59	1.863156	0.919998	8.000178	−0.000179
14	1.58	1.881400	0.919996	8.000403	−0.000405
15	1.57	1.900965	0.920002	7.999808	0.000191
16	1.56	1.921852	0.920008	7.999164	0.000834
17	1.55	1.944065	0.920008	7.999248	0.000751
18	1.54	1.967996	0.920009	7.999063	0.000936
19	1.53	1.993650	0.920003	7.999688	0.000310
20	1.52	2.021420	0.919994	8.000576	−0.000578
21	1.51	2.052090	0.919999	8.000106	−0.000107
22	1.50	2.085663	0.919997	8.000267	−0.000268
23	1.49	2.122926	0.919996	8.000421	−0.000423
24	1.48	2.165052	0.920003	7.999695	0.000304
25	1.47	2.212826	0.920005	7.999504	0.000495
26	1.46	2.267816	0.919997	8.000315	−0.000316
27	1.45	2.333930	0.920002	7.999790	0.000209
28	1.44	2.415079	0.919998	8.000236	−0.000238
29	1.43	2.522986	0.920009	7.999134	0.000864

(continued)

8.2 Contents for Table 8.2 (Table No. 8.2-1 to 400)

Table 8.2 (continued)

Parameter values					
PROB. = 0.92	RHO = 0.4		BIGH = 1.7074		SMHL = 1.41
Pair No.	SMH	SMK	PROB. level	Risk level (%)	Risk error (%)
30	1.42	2.681092	0.920008	7.999170	0.000829
31	1.41	2.998777	0.919996	8.000451	−0.000453
No. of pairs		Mean COMPUT. risk errors		STD. ERR. errors	
31		0.0001257286		0.0000851271	
Z-ratio				Status of significance	
1.47695				N.S.—not significant	
Parameter values					
PROB. = 0.91	RHO = 0.4		BIGH = 1.6481		SMHL = 1.35
Pair No.	SMH	SMK	PROB. level	Risk level (%)	Risk error (%)
Table No. 8.2-124					
1	1.65	1.646202	0.910001	8.999920	0.000077
2	1.64	1.656240	0.909993	9.000712	−0.000715
3	1.63	1.666792	0.909992	9.000754	−0.000757
4	1.62	1.677859	0.909997	9.000325	−0.000328
5	1.61	1.689446	0.910003	8.999681	0.000316
6	1.60	1.701552	0.910009	8.999098	0.000900
7	1.59	1.713987	0.909996	9.000361	−0.000364
8	1.58	1.727338	0.910008	8.999229	0.000769
9	1.57	1.741023	0.909995	9.000504	−0.000507
10	1.56	1.755629	0.909998	9.000206	−0.000209
11	1.55	1.770963	0.909998	9.000224	−0.000226
12	1.54	1.787226	0.910003	8.999681	0.000316
13	1.53	1.804222	0.909996	9.000355	−0.000358
14	1.52	1.822543	0.910009	8.999121	0.000876
15	1.51	1.841799	0.910009	8.999091	0.000906
16	1.50	1.861994	0.909992	9.000832	−0.000834
17	1.49	1.883913	0.909993	9.000719	−0.000721
18	1.48	1.907559	0.910001	8.999902	0.000095
19	1.47	1.932934	0.910005	8.999550	0.000447
20	1.46	1.960043	0.909992	9.000808	−0.000811
21	1.45	1.990061	0.910001	8.999872	0.000125
22	1.44	2.022601	0.909996	9.000373	−0.000376
23	1.43	2.058839	0.910002	8.999848	0.000149
24	1.42	2.098778	0.909990	9.000987	−0.000989
25	1.41	2.144375	0.909993	9.000749	−0.000751
26	1.40	2.197198	0.910005	8.999472	0.000525

(continued)

Table 8.2 (continued)

Parameter values					
PROB. = 0.91	RHO = 0.4		BIGH = 1.6481		SMHL = 1.35
Pair No.	SMH	SMK	PROB. level	Risk level (%)	Risk error (%)
27	1.39	2.258812	0.910006	8.999366	0.000632
28	1.38	2.333129	0.909999	9.000074	−0.000077
29	1.37	2.427964	0.909996	9.000408	−0.000411
30	1.36	2.559730	0.909996	9.000373	−0.000376
31	1.35	2.780774	0.909998	9.000200	−0.000203
No. of pairs		Mean COMPUT. risk errors		STD. ERR. errors	
31		−0.0000899658		0.0000982509	
Z-ratio				Status of significance	
0.91567				N.S.—not significant	
Parameter values					
PROB. = 0.9025	RHO = 0.4		BIGH = 1.6068		SMHL = 1.30
Pair No.	SMH	SMK	PROB. level	Risk level (%)	Risk error (%)
Table No. 8.2-125					
1	1.61	1.603606	0.902491	9.750932	−0.000930
2	1.60	1.613824	0.902504	9.749603	0.000399
3	1.59	1.624168	0.902490	9.750986	−0.000983
4	1.58	1.635226	0.902500	9.749961	0.000042
5	1.57	1.646611	0.902497	9.750307	−0.000304
6	1.56	1.658520	0.902495	9.750509	−0.000507
7	1.55	1.671150	0.902507	9.749293	0.000709
8	1.54	1.684115	0.902498	9.750181	−0.000179
9	1.53	1.697807	0.902497	9.750307	−0.000304
10	1.52	1.712228	0.902499	9.750104	−0.000101
11	1.51	1.727383	0.902500	9.750014	−0.000012
12	1.50	1.743470	0.902509	9.749097	0.000906
13	1.49	1.760099	0.902494	9.750593	−0.000590
14	1.48	1.778057	0.902503	9.749669	0.000334
15	1.47	1.796956	0.902503	9.749699	0.000304
16	1.46	1.816993	0.902499	9.750069	−0.000066
17	1.45	1.838368	0.902496	9.750414	−0.000411
18	1.44	1.861280	0.902494	9.750557	−0.000554
19	1.43	1.885928	0.902495	9.750551	−0.000548
20	1.42	1.912705	0.902503	9.749681	0.000322
21	1.41	1.941615	0.902505	9.749490	0.000513
22	1.40	1.973053	0.902501	9.749889	0.000113
23	1.39	2.007805	0.902504	9.749592	0.000411
24	1.38	2.046264	0.902503	9.749735	0.000268
25	1.37	2.089217	0.902494	9.750629	−0.000626
26	1.36	2.138621	0.902499	9.750080	−0.000077
27	1.35	2.196042	0.902505	9.749460	0.000542
28	1.34	2.264220	0.902504	9.749556	0.000447
29	1.33	2.349020	0.902508	9.749211	0.000793

(continued)

8.2 Contents for Table 8.2 (Table No. 8.2-1 to 400)

Table 8.2 (continued)

Parameter values					
PROB. = 0.9025	RHO = 0.4		BIGH = 1.6068		SMHL = 1.30
Pair No.	SMH	SMK	PROB. level	Risk level (%)	Risk error (%)
30	1.32	2.460600	0.902504	9.749598	0.000405
31	1.31	2.627093	0.902499	9.750091	−0.000089
32	1.30	2.986004	0.902498	9.750241	−0.000238
No. of pairs		Mean COMPUT. risk errors		STD. ERR. errors	
32		−0.0000003612		0.0000842965	
Z-ratio				Status of significance	
0.00429				N.S.—not significant	
Parameter values					
PROB. = 0.9	RHO = 0.4		BIGH = 1.5936		SMHL = 1.29
Pair No.	SMH	SMK	PROB. level	Risk level (%)	Risk error (%)
Table No. 8.2-126					
1	1.60	1.587226	0.899991	10.000920	−0.000918
2	1.59	1.597208	0.899994	10.000560	−0.000554
3	1.58	1.607708	0.900008	9.999204	0.000799
4	1.57	1.618336	0.899993	10.000740	−0.000739
5	1.56	1.629682	0.900000	9.999954	0.000048
6	1.55	1.641356	0.899993	10.000680	−0.000679
7	1.54	1.653753	0.900002	9.999764	0.000238
8	1.53	1.666484	0.899991	10.000860	−0.000858
9	1.52	1.680139	0.900006	9.999442	0.000560
10	1.51	1.694328	0.900009	9.999121	0.000882
11	1.50	1.709056	0.899998	10.000200	−0.000197
12	1.49	1.724716	0.899999	10.000150	−0.000143
13	1.48	1.741310	0.900005	9.999519	0.000483
14	1.47	1.758647	0.899997	10.000290	−0.000292
15	1.46	1.777316	0.900009	9.999061	0.000942
16	1.45	1.796734	0.899996	10.000410	−0.000405
17	1.44	1.817685	0.899999	10.000100	−0.000101
18	1.43	1.839979	0.899997	10.000340	−0.000334
19	1.42	1.863814	0.899990	10.000960	−0.000960
20	1.41	1.889779	0.899999	10.000060	−0.000054
21	1.40	1.917878	0.900008	9.999216	0.000787
22	1.39	1.948116	0.900000	9.999979	0.000024
23	1.38	1.981668	0.900010	9.999026	0.000978
24	1.37	2.018147	0.899996	10.000440	−0.000441
25	1.36	2.059511	0.900004	9.999598	0.000405
26	1.35	2.106156	0.900007	9.999299	0.000703
27	1.34	2.159257	0.899994	10.000600	−0.000596
28	1.33	2.222725	0.900005	9.999490	0.000513
29	1.32	2.298909	0.899995	10.000490	−0.000483

(continued)

Table 8.2 (continued)

Parameter values					
PROB. = 0.9	RHO = 0.4		BIGH = 1.5936		SMHL = 1.29
Pair No.	SMH	SMK	PROB. level	Risk level (%)	Risk error (%)
30	1.31	2.397189	0.899999	10.000100	−0.000095
31	1.30	2.534758	0.900003	9.999734	0.000268
32	1.29	2.768651	0.899991	10.000890	−0.000888
No. of pairs			Mean COMPUT. risk errors	STD. ERR. errors	
32			−0.0000335953	0.0001033076	
Z-ratio				Status of significance	
0.32520				N.S.—not significant	
Parameter values					
PROB. = 0.85	RHO = 0.4		BIGH = 1.3688		SMHL = 1.04
Pair No.	SMH	SMK	PROB. level	Risk level (%)	Risk error (%)
Table No. 8.2-127					
1	1.37	1.367601	0.850006	14.999380	0.000614
2	1.36	1.377657	0.849995	15.000490	−0.000495
3	1.35	1.388253	0.850002	14.999760	0.000238
4	1.34	1.399196	0.850001	14.999890	0.000107
5	1.33	1.410587	0.850001	14.999940	0.000054
6	1.32	1.422529	0.850009	14.999090	0.000906
7	1.31	1.434732	0.849990	15.000990	−0.000995
8	1.30	1.447686	0.849996	15.000450	−0.000447
9	1.29	1.461203	0.849999	15.000080	−0.000083
10	1.28	1.475284	0.849997	15.000290	−0.000292
11	1.27	1.490130	0.850005	14.999490	0.000507
12	1.26	1.505549	0.849998	15.000180	−0.000185
13	1.25	1.521938	0.850009	14.999060	0.000936
14	1.24	1.538908	0.849995	15.000480	−0.000477
15	1.23	1.557052	0.850004	14.999640	0.000358
16	1.22	1.575983	0.849991	15.000860	−0.000864
17	1.21	1.596293	0.850001	14.999940	0.000054
18	1.20	1.617790	0.850004	14.999600	0.000399
19	1.19	1.640481	0.849992	15.000820	−0.000823
20	1.18	1.665152	0.850010	14.999030	0.000966
21	1.17	1.691223	0.850000	15.000010	−0.000012
22	1.16	1.719481	0.850000	14.999970	0.000030
23	1.15	1.749932	0.849991	15.000950	−0.000948
24	1.14	1.783364	0.849994	15.000620	−0.000620
25	1.13	1.820174	0.850001	14.999890	0.000107
26	1.12	1.860760	0.849997	15.000290	−0.000292
27	1.11	1.906299	0.849995	15.000550	−0.000548
28	1.10	1.958363	0.850001	14.999890	0.000113
29	1.09	2.018520	0.849999	15.000070	−0.000072
30	1.08	2.090296	0.850000	15.000030	−0.000030
31	1.07	2.179165	0.849997	15.000270	−0.000274

(continued)

8.2 Contents for Table 8.2 (Table No. 8.2-1 to 400)

Table 8.2 (continued)

Parameter values					
PROB. = 0.85	RHO = 0.4		BIGH = 1.3688		SMHL = 1.04
Pair No.	SMH	SMK	PROB. level	Risk level (%)	Risk error (%)
32	1.06	2.297247	0.850003	14.999690	0.000304
33	1.05	2.475018	0.850010	14.999010	0.000983
34	1.04	2.870300	0.850002	14.999780	0.000215
No. of pairs		Mean COMPUT. risk errors		STD. ERR. errors	
34		−0.0000161784		0.0000903225	
Z-ratio				Status of significance	
0.17912				N.S.—not significant	
Parameter values					
PROB. = 0.8	RHO = 0.4		BIGH = 1.1908		SMHL = 0.85
Pair No.	SMH	SMK	PROB. level	Risk level (%)	Risk error (%)
Table No. 8.2-128					
1	1.20	1.181866	0.800003	19.999750	0.000250
2	1.19	1.191698	0.799999	20.000110	−0.000113
3	1.18	1.201894	0.799997	20.000330	−0.000334
4	1.17	1.212556	0.800008	19.999200	0.000799
5	1.16	1.223492	0.800001	19.999860	0.000137
6	1.15	1.234903	0.800003	19.999710	0.000286
7	1.14	1.246793	0.800009	19.999090	0.000912
8	1.13	1.258973	0.799990	20.000980	−0.000983
9	1.12	1.271935	0.800009	19.999080	0.000918
10	1.11	1.285294	0.800009	19.999090	0.000912
11	1.10	1.299154	0.800000	20.000030	−0.000030
12	1.09	1.313715	0.800001	19.999870	0.000125
13	1.08	1.328983	0.800008	19.999210	0.000793
14	1.07	1.344769	0.799992	20.000850	−0.000852
15	1.06	1.361568	0.800003	19.999720	0.000280
16	1.05	1.378996	0.799990	20.000980	−0.000978
17	1.04	1.397450	0.799990	20.000990	−0.000989
18	1.03	1.416938	0.799993	20.000710	−0.000715
19	1.02	1.437661	0.800008	19.999170	0.000834
20	1.01	1.459433	0.800006	19.999390	0.000608
21	1.00	1.482457	0.799994	20.000570	−0.000566
22	0.99	1.507132	0.799994	20.000590	−0.000590
23	0.98	1.533662	0.800005	19.999520	0.000483
24	0.97	1.561860	0.799992	20.000810	−0.000811
25	0.96	1.592713	0.800009	19.999100	0.000900
26	0.95	1.625839	0.800000	19.999960	0.000042
27	0.94	1.662031	0.799991	20.000910	−0.000912
28	0.93	1.702079	0.799991	20.000910	−0.000912
29	0.92	1.746778	0.799996	20.000370	−0.000370
30	0.91	1.797309	0.800007	19.999350	0.000650
31	0.90	1.854857	0.799999	20.000090	−0.000095

(continued)

Table 8.2 (continued)

Parameter values					
PROB. = 0.8	RHO = 0.4		BIGH = 1.1908		SMHL = 0.85
Pair No.	SMH	SMK	PROB. level	Risk level (%)	Risk error (%)
32	0.89	1.922560	0.799999	20.000110	−0.000107
33	0.88	2.004337	0.799991	20.000880	−0.000882
34	0.87	2.109577	0.800008	19.999210	0.000793
35	0.86	2.255091	0.800003	19.999680	0.000322
36	0.85	2.504177	0.800000	19.999990	0.000012
No. of pairs		Mean COMPUT. risk errors		STD. ERR. errors	
36		−0.0000049939		0.0001078157	
Z-ratio				Status of significance	
0.04632				N.S.—not significant	
Parameter values					
PROB. = 0.75	RHO = 0.4		BIGH = 1.0387		SMHL = 0.68
Pair No.	SMH	SMK	PROB. level	Risk level (%)	Risk error (%)
Table No. 8.2-129					
1	1.04	1.037499	0.750002	24.999820	0.000179
2	1.03	1.047669	0.750009	24.999100	0.000900
3	1.02	1.058134	0.750006	24.999380	0.000626
4	1.01	1.068997	0.750008	24.999180	0.000823
5	1.00	1.080167	0.749996	25.000360	−0.000358
6	0.99	1.091847	0.750000	24.999980	0.000024
7	0.98	1.103944	0.750001	24.999920	0.000077
8	0.97	1.116465	0.749995	25.000490	−0.000489
9	0.96	1.129516	0.749995	25.000510	−0.000507
10	0.95	1.143104	0.749996	25.000360	−0.000358
11	0.94	1.157334	0.750009	24.999090	0.000912
12	0.93	1.172019	0.750001	24.999900	0.000101
13	0.92	1.187363	0.749995	25.000490	−0.000489
14	0.91	1.203473	0.749999	25.000110	−0.000107
15	0.90	1.220259	0.749993	25.000680	−0.000679
16	0.89	1.238026	0.750009	24.999140	0.000858
17	0.88	1.256489	0.750001	24.999920	0.000083
18	0.87	1.276050	0.750009	24.999100	0.000900
19	0.86	1.296524	0.750001	24.999910	0.000089
20	0.85	1.318315	0.750010	24.999050	0.000948
21	0.84	1.341238	0.750001	24.999860	0.000137
22	0.83	1.365697	0.750003	24.999660	0.000346
23	0.82	1.391705	0.750000	25.000040	−0.000042
24	0.81	1.419472	0.749990	25.000990	−0.000989
25	0.80	1.449598	0.750005	24.999530	0.000471
26	0.79	1.481904	0.750000	24.999980	0.000024
27	0.78	1.517186	0.750009	24.999140	0.000858
28	0.77	1.555463	0.749994	25.000650	−0.000644
29	0.76	1.598118	0.750004	24.999620	0.000381

(continued)

8.2 Contents for Table 8.2 (Table No. 8.2-1 to 400)

Table 8.2 (continued)

Parameter values					
PROB. = 0.75	RHO = 0.4		BIGH = 1.0387		SMHL = 0.68
Pair No.	SMH	SMK	PROB. level	Risk level (%)	Risk error (%)
30	0.75	1.645561	0.750004	24.999580	0.000417
31	0.74	1.698985	0.749992	25.000850	−0.000846
32	0.73	1.761145	0.750007	24.999350	0.000656
33	0.72	1.834015	0.749999	25.000070	−0.000072
34	0.71	1.923480	0.750000	25.000050	−0.000048
35	0.70	2.039720	0.750002	24.999770	0.000232
36	0.69	2.207369	0.749997	25.000350	−0.000346
37	0.68	2.533488	0.750003	24.999670	0.000334
No. of pairs		Mean COMPUT. risk errors		STD. ERR. errors	
37		0.0001159154		0.0000860966	
Z-ratio				Status of significance	
1.34634				N.S.—not significant	
Parameter values					
PROB. = 0.7	RHO = 0.4		BIGH = 0.9025		SMHL = 0.53
Pair No.	SMH	SMK	PROB. level	Risk level (%)	Risk error (%)
Table No. 8.2-130					
1	0.91	0.895159	0.699993	30.000680	−0.000679
2	0.90	0.905104	0.699998	30.000230	−0.000232
3	0.89	0.915371	0.700001	29.999890	0.000107
4	0.88	0.925966	0.700002	29.999830	0.000173
5	0.87	0.936897	0.699998	30.000210	−0.000203
6	0.86	0.948272	0.700006	29.999430	0.000572
7	0.85	0.960000	0.700005	29.999510	0.000489
8	0.84	0.972091	0.699994	30.000630	−0.000632
9	0.83	0.984751	0.700003	29.999700	0.000304
10	0.82	0.997792	0.699997	30.000330	−0.000328
11	0.81	1.011422	0.700004	29.999600	0.000405
12	0.80	1.025554	0.700005	29.999460	0.000542
13	0.79	1.040200	0.699997	30.000270	−0.000268
14	0.78	1.055567	0.700006	29.999440	0.000566
15	0.77	1.071472	0.699997	30.000350	−0.000346
16	0.76	1.088125	0.699992	30.000760	−0.000757
17	0.75	1.105637	0.700002	29.999820	0.000179
18	0.74	1.123926	0.700004	29.999620	0.000387
19	0.73	1.143105	0.700004	29.999560	0.000441
20	0.72	1.163290	0.700008	29.999220	0.000781
21	0.71	1.184496	0.700005	29.999540	0.000465
22	0.70	1.206939	0.700007	29.999270	0.000733
23	0.69	1.230639	0.700003	29.999660	0.000340
24	0.68	1.255811	0.700001	29.999900	0.000101
25	0.67	1.282673	0.700003	29.999660	0.000340
26	0.66	1.311248	0.699992	30.000780	−0.000775

(continued)

Table 8.2 (continued)

Parameter values					
PROB. = 0.7	RHO = 0.4		BIGH = 0.9025		SMHL = 0.53
Pair No.	SMH	SMK	PROB. level	Risk level (%)	Risk error (%)
27	0.65	1.342148	0.700000	29.999960	0.000048
28	0.64	1.375400	0.700000	29.999990	0.000012
29	0.63	1.411616	0.700009	29.999130	0.000876
30	0.62	1.451024	0.700003	29.999720	0.000280
31	0.61	1.494631	0.700006	29.999380	0.000626
32	0.60	1.543057	0.699998	30.000190	−0.000191
33	0.59	1.598096	0.700007	29.999320	0.000685
34	0.58	1.660961	0.699993	30.000740	−0.000733
35	0.57	1.735598	0.700000	30.000040	−0.000036
36	0.56	1.826740	0.700002	29.999840	0.000167
37	0.55	1.944982	0.700008	29.999220	0.000787
38	0.54	2.114594	0.699997	30.000290	−0.000292
39	0.53	2.443052	0.700008	29.999190	0.000811
No. of pairs		Mean COMPUT. risk errors		STD. ERR. errors	
39		0.0001436472		0.0000748409	
Z-ratio				Status of significance	
1.91937				N.S.—not significant	
Parameter values					
PROB. = 0.65	RHO = 0.4		BIGH = 0.7767		SMHL = 0.39
Pair No.	SMH	SMK	PROB. level	Risk level (%)	Risk error (%)
Table No. 8.2-131					
1	0.78	0.773414	0.650004	34.999590	0.000411
2	0.77	0.783458	0.650002	34.999780	0.000226
3	0.76	0.793865	0.650009	34.999110	0.000894
4	0.75	0.804546	0.650004	34.999640	0.000364
5	0.74	0.815611	0.650005	34.999550	0.000453
6	0.73	0.826973	0.649992	35.000840	−0.000834
7	0.72	0.838842	0.650001	34.999920	0.000083
8	0.71	0.851033	0.649993	35.000660	−0.000656
9	0.70	0.863757	0.650004	34.999610	0.000393
10	0.69	0.876833	0.649995	35.000500	−0.000501
11	0.68	0.890472	0.649999	35.000090	−0.000083
12	0.67	0.904592	0.649997	35.000300	−0.000298
13	0.66	0.919308	0.650002	34.999790	0.000215
14	0.65	0.934542	0.649996	35.000440	−0.000441
15	0.64	0.950411	0.649990	35.000990	−0.000989
16	0.63	0.967033	0.649996	35.000360	−0.000358
17	0.62	0.984333	0.649995	35.000460	−0.000459
18	0.61	1.002432	0.649996	35.000380	−0.000370
19	0.60	1.021454	0.650007	34.999300	0.000703
20	0.59	1.041229	0.649994	35.000560	−0.000554
21	0.58	1.062178	0.650004	34.999610	0.000393

(continued)

8.2 Contents for Table 8.2 (Table No. 8.2-1 to 400)

Table 8.2 (continued)

Parameter values					
PROB. = 0.65	RHO = 0.4		BIGH = 0.7767		SMHL = 0.39
Pair No.	SMH	SMK	PROB. level	Risk level (%)	Risk error (%)
22	0.57	1.084137	0.650001	34.999910	0.000089
23	0.56	1.107333	0.650000	34.999960	0.000048
24	0.55	1.131802	0.649991	35.000870	−0.000870
25	0.54	1.157973	0.650005	34.999540	0.000465
26	0.53	1.185693	0.650002	34.999780	0.000226
27	0.52	1.215394	0.650007	34.999280	0.000727
28	0.51	1.247126	0.649998	35.000240	−0.000238
29	0.50	1.281525	0.650002	34.999770	0.000232
30	0.49	1.318843	0.650005	34.999530	0.000477
31	0.48	1.359532	0.649998	35.000170	−0.000167
32	0.47	1.404631	0.650007	34.999280	0.000721
33	0.46	1.454800	0.650007	34.999260	0.000745
34	0.45	1.511287	0.649994	35.000630	−0.000626
35	0.44	1.576911	0.650009	34.999150	0.000852
36	0.43	1.653717	0.649991	35.000860	−0.000858
37	0.42	1.748839	0.650008	34.999230	0.000775
38	0.41	1.872153	0.650005	34.999460	0.000548
39	0.40	2.081594	0.649999	35.000070	−0.000072
40	0.39	2.437451	0.649993	34.000740	−0.000733
No. of pairs		Mean COMPUT. risk errors		STD. ERR. errors	
40		0.0000228242		0.0000852736	
Z-ratio				Status of significance	
0.26766				N.S.—not significant	
Parameter values					
PROB. = 0.6	RHO = 0.4		BIGH = 0.6575		SMHL = 0.26
Pair No.	SMH	SMK	PROB. level	Risk level (%)	Risk error (%)
Table No. 8.2-132					
1	0.66	0.655107	0.600001	39.999880	0.000119
2	0.65	0.665184	0.600001	39.999880	0.000119
3	0.64	0.675576	0.600003	39.999740	0.000256
4	0.63	0.686298	0.600005	39.999480	0.000519
5	0.62	0.697366	0.600009	39.999070	0.000930
6	0.61	0.708699	0.599995	40.000470	−0.000477
7	0.60	0.720510	0.600002	39.999810	0.000191
8	0.59	0.732722	0.600010	39.999020	0.000978
9	0.58	0.745258	0.600000	39.999990	0.000006
10	0.57	0.758237	0.599992	40.000830	−0.000834
11	0.56	0.771780	0.600002	39.999810	0.000185
12	0.55	0.785718	0.599995	40.000550	−0.000548
13	0.54	0.800274	0.600004	39.999590	0.000411
14	0.53	0.815281	0.599996	40.000430	−0.000429
15	0.52	0.830967	0.600002	39.999810	0.000191
16	0.51	0.847268	0.600005	39.999520	0.000483

(continued)

Table 8.2 (continued)

Parameter values					
PROB. = 0.6	RHO = 0.4		BIGH = 0.6575		SMHL = 0.26
Pair No.	SMH	SMK	PROB. level	Risk level (%)	Risk error (%)
17	0.50	0.864222	0.600004	39.999640	0.000358
18	0.49	0.881867	0.599997	40.000340	−0.000346
19	0.48	0.900345	0.599997	40.000310	−0.000316
20	0.47	0.919703	0.600001	39.999880	0.000119
21	0.46	0.939991	0.600006	39.999380	0.000620
22	0.45	0.961266	0.600008	39.999190	0.000805
23	0.44	0.983490	0.599990	40.000980	−0.000978
24	0.43	1.007121	0.599997	40.000270	−0.000268
25	0.42	1.032035	0.599997	40.000290	−0.000292
26	0.41	1.058507	0.600004	39.999610	0.000387
27	0.40	1.086624	0.600007	39.999280	0.000715
28	0.39	1.116485	0.599995	40.000500	−0.000501
29	0.38	1.148585	0.599992	40.000760	−0.000763
30	0.37	1.183238	0.599997	40.000330	−0.000334
31	0.36	1.220772	0.600000	40.000010	−0.000012
32	0.35	1.261722	0.600003	39.999690	0.000310
33	0.34	1.306644	0.599997	40.000340	−0.000340
34	0.33	1.356698	0.599998	40.000200	−0.000203
35	0.32	1.413065	0.599998	40.000200	−0.000203
36	0.31	1.477738	0.600002	39.999840	0.000155
37	0.30	1.591801	0.600003	39.999730	0.000268
38	0.29	1.678991	0.599991	40.000860	−0.000858
39	0.28	1.792386	0.599996	40.000420	−0.000417
40	0.27	1.954257	0.599995	40.000520	−0.000525
41	0.26	2.251776	0.600000	40.000050	−0.000048
No. of pairs		Mean COMPUT. risk errors		STD. ERR. errors	
41		−0.0000134820		0.0000753735	
Z-ratio				Status of significance	
0.17887				N.S.—not significant	
Parameter values					
PROB. = 0.55	RHO = 0.4		BIGH = 0.5425		SMHL = 0.13
Pair No.	SMH	SMK	PROB. level	Risk level (%)	Risk error (%)
Table No. 8.2-133					
1	0.55	0.535151	0.550000	45.000040	−0.000042
2	0.54	0.545109	0.550006	44.999420	0.000584
3	0.53	0.555295	0.549998	45.000190	−0.000185
4	0.52	0.565827	0.550000	45.000030	−0.000030
5	0.51	0.576680	0.550002	44.999810	0.000191
6	0.50	0.587831	0.549997	45.000320	−0.000322
7	0.49	0.599355	0.549997	45.000340	−0.000340
8	0.48	0.611282	0.550004	44.999640	0.000364
9	0.47	0.623547	0.550001	44.999880	0.000119
10	0.46	0.636183	0.549992	45.000780	−0.000781

(continued)

8.2 Contents for Table 8.2 (Table No. 8.2-1 to 400)

Table 8.2 (continued)

Parameter values						
PROB. = 0.55	RHO = 0.4		BIGH = 0.5425		SMHL = 0.13	
Pair No.	SMH	SMK	PROB. level	Risk level (%)	Risk error (%)	
11	0.45	0.649326	0.549998	45.000160	−0.000155	
12	0.44	0.662921	0.550004	44.999570	0.000435	
13	0.43	0.676914	0.549995	45.000460	−0.000459	
14	0.42	0.691452	0.549994	45.000610	−0.000608	
15	0.41	0.706589	0.550003	44.999720	0.000280	
16	0.40	0.722289	0.550010	44.999050	0.000954	
17	0.39	0.738518	0.550002	44.999780	0.000221	
18	0.38	0.755447	0.550003	44.999720	0.000286	
19	0.37	0.773059	0.550001	44.999950	0.000054	
20	0.36	0.791443	0.550001	44.999910	0.000095	
21	0.35	0.810601	0.549996	45.000430	−0.000429	
22	0.34	0.830645	0.549992	45.000850	−0.000852	
23	0.33	0.851797	0.550008	44.999170	0.000829	
24	0.32	0.873808	0.549999	45.000070	−0.000066	
25	0.31	0.897031	0.550000	45.000040	−0.000036	
26	0.30	0.921450	0.549992	45.000780	−0.000775	
27	0.29	0.947466	0.550009	44.999100	0.000900	
28	0.28	0.974823	0.550000	44.999960	0.000042	
29	0.27	1.004084	0.550010	44.999050	0.000948	
30	0.26	1.035071	0.549992	45.000770	−0.000769	
31	0.25	1.068630	0.550008	44.999240	0.000757	
32	0.24	1.104595	0.550004	44.999650	0.000352	
33	0.23	1.194044	0.549995	45.000540	−0.000542	
34	0.22	1.232090	0.549995	45.000530	−0.000530	
35	0.21	1.274699	0.550002	44.999770	0.000226	
36	0.20	1.322701	0.550003	44.999750	0.000250	
37	0.19	1.377494	0.549995	45.000490	−0.000495	
38	0.18	1.441673	0.550007	44.999270	0.000727	
39	0.17	1.517930	0.550001	44.999860	0.000143	
40	0.16	1.612458	0.549999	45.000150	−0.000149	
41	0.15	1.737039	0.550001	44.999910	0.000089	
42	0.14	1.920934	0.549996	45.000360	−0.000364	
43	0.13	2.304515	0.549997	45.000300	−0.000298	
No. of pairs		Mean COMPUT. risk errors		STD. ERR. errors		
43		0.0000140884		0.0000735699		
Z-ratio				Status of significance		
0.19150				N.S.—not significant		
Parameter values						
PROB. = 0.5	RHO = 0.4		BIGH = 0.4296		SMHL = 0.01	
Pair No.	SMH	SMK	PROB. level	Risk level (%)	Risk error (%)	
Table No. 8.2-134						
1	0.43	0.429200	0.500002	49.999800	0.000197	
2	0.42	0.439322	0.500000	49.999980	0.000018	

(continued)

Table 8.2 (continued)

Parameter values					
PROB. = 0.5	RHO = 0.4		BIGH = 0.4296		SMHL = 0.01
Pair No.	SMH	SMK	PROB. level	Risk level (%)	Risk error (%)
3	0.41	0.449746	0.500001	49.999860	0.000143
4	0.40	0.460414	0.499990	50.000990	−0.000989
5	0.39	0.471463	0.499993	50.000700	−0.000697
6	0.38	0.482842	0.499996	50.000430	−0.000432
7	0.37	0.494601	0.500005	49.999480	0.000525
8	0.36	0.506601	0.499990	50.001000	−0.000995
9	0.35	0.519100	0.499999	50.000160	−0.000152
10	0.34	0.531972	0.500001	49.999880	0.000119
11	0.33	0.545198	0.499991	50.000900	−0.000894
12	0.32	0.558964	0.499999	50.000150	−0.000143
13	0.31	0.573174	0.500001	49.999900	0.000101
14	0.30	0.587843	0.499996	50.000370	−0.000370
15	0.29	0.603099	0.500002	49.999800	0.000203
16	0.28	0.618894	0.500003	49.999660	0.000346
17	0.27	0.635299	0.500007	49.999350	0.000650
18	0.26	0.652311	0.500005	49.999550	0.000453
19	0.25	0.669962	0.499995	50.000480	−0.000474
20	0.24	0.688417	0.499997	50.000290	−0.000292
21	0.23	0.707691	0.500003	49.999700	0.000298
22	0.22	0.727758	0.500000	50.000020	−0.000018
23	0.21	0.748760	0.499998	50.000220	−0.000212
24	0.20	0.770827	0.500004	49.999610	0.000387
25	0.19	0.793906	0.499998	50.000220	−0.000221
26	0.18	0.818280	0.500003	49.999720	0.000286
27	0.17	0.905057	0.500002	49.999790	0.000215
28	0.16	0.926620	0.500001	49.999910	0.000095
29	0.15	0.950002	0.499999	50.000150	−0.000149
30	0.14	0.975484	0.500001	49.999910	0.000095
31	0.13	1.003255	0.500000	50.000050	−0.000048
32	0.12	1.033670	0.499998	50.000250	−0.000244
33	0.11	1.067235	0.500005	49.999510	0.000489
34	0.10	1.104153	0.499994	50.000570	−0.000566
35	0.09	1.145543	0.500006	49.999440	0.000560
36	0.08	1.191714	0.499995	50.000540	−0.000539
37	0.07	1.244325	0.499996	50.000370	−0.000367
38	0.06	1.304836	0.499993	50.000660	−0.000662
39	0.05	1.376271	0.500007	49.999320	0.000679

(continued)

8.2 Contents for Table 8.2 (Table No. 8.2-1 to 400)

Table 8.2 (continued)

Parameter values					
PROB. = 0.5	RHO = 0.4		BIGH = 0.4296		SMHL = 0.01
Pair No.	SMH	SMK	PROB. level	Risk level (%)	Risk error (%)
40	0.04	1.462329	0.500006	49.999370	0.000638
41	0.03	1.570598	0.499991	50.000910	−0.000903
42	0.02	1.719153	0.499995	50.000480	−0.000477
43	0.01	1.960021	0.500003	49.999670	0.000334
No. of pairs		Mean COMPUT. risk errors		STD. ERR. errors	
43		−0.0000684776		0.0000697131	
Z-ratio				Status of significance	
0.98228				N.S.—not significant	
Parameter values					
PROB. = 0.99	RHO = 0.3		BIGH = 2.569		SMHL = 2.33
Pair No.	SMH	SMK	PROB. level	Risk level (%)	Risk error (%)
Table No. 8.2-135					
1	2.57	2.568001	0.989993	1.000685	−0.000685
2	2.56	2.578032	0.989991	1.000947	−0.000948
3	2.55	2.589704	0.990003	0.999743	0.000256
4	2.54	2.601456	0.990007	0.999337	0.000662
5	2.53	2.613289	0.990003	0.999713	0.000286
6	2.52	2.625203	0.989991	1.000893	−0.000894
7	2.51	2.640325	0.990008	0.999171	0.000829
8	2.50	2.653967	0.989997	1.000303	−0.000304
9	2.49	2.669257	0.989993	1.000655	−0.000656
10	2.48	2.686194	0.989996	1.000452	−0.000453
11	2.47	2.704781	0.990001	0.999898	0.000101
12	2.46	2.723455	0.989993	1.000714	−0.000715
13	2.45	2.745342	0.989999	1.000071	−0.000072
14	2.44	2.768882	0.990002	0.999761	0.000238
15	2.43	2.794076	0.990000	0.999993	0.000006
16	2.42	2.820924	0.989990	1.000959	−0.000960
17	2.41	2.854115	0.990002	0.999832	0.000167
18	2.40	2.890525	0.990006	0.999445	0.000554
19	2.39	2.930156	0.989998	1.000214	−0.000215
20	2.38	2.979258	0.990003	0.999725	0.000274
21	2.37	3.034709	0.989993	1.000744	−0.000745
22	2.36	3.109009	0.989999	1.000112	−0.000113
23	2.35	3.205284	0.989997	1.000351	−0.000352

(continued)

Table 8.2 (continued)

Parameter values					
PROB. = 0.99	RHO = 0.3		BIGH = 2.569		SMHL = 2.33
Pair No.	SMH	SMK	PROB. level	Risk level (%)	Risk error (%)
24	2.34	3.357910	0.990003	0.999737	0.000262
25	2.33	3.682515	0.989995	1.000541	−0.000542
No. of pairs		Mean COMPUT. risk errors		STD. ERR. errors	
25		−0.0001545136		0.0001004721	
Z-ratio				Status of significance	
1.53788				N.S.—not significant	
Parameter values					
PROB. = 0.98	RHO = 0.3		BIGH = 2.3157		SMHL = 2.06
Pair No.	SMH	SMK	PROB. level	Risk level (%)	Risk error (%)
Table No. 8.2-136					
1	2.32	2.311408	0.980006	1.999390	0.000608
2	2.31	2.321414	0.980005	1.999456	0.000542
3	2.30	2.331508	0.979994	2.000600	−0.000602
4	2.29	2.342470	0.979991	2.000892	−0.000894
5	2.28	2.354303	0.979995	2.000529	−0.000530
6	2.27	2.367007	0.980003	1.999676	0.000322
7	2.26	2.379804	0.979998	2.000237	−0.000238
8	2.25	2.393475	0.979995	2.000552	−0.000554
9	2.24	2.408021	0.979992	2.000797	−0.000799
10	2.23	2.424225	0.980004	1.999587	0.000411
11	2.22	2.440526	0.979997	2.000290	−0.000292
12	2.21	2.458487	0.980000	2.000034	−0.000036
13	2.20	2.478110	0.980008	1.999205	0.000793
14	2.19	2.498615	0.980005	1.999468	0.000530
15	2.18	2.520785	0.980002	1.999772	0.000226
16	2.17	2.544620	0.979995	2.000499	−0.000501
17	2.16	2.571686	0.980001	1.999879	0.000119
18	2.15	2.600420	0.979993	2.000707	−0.000709
19	2.14	2.633951	0.980003	1.999712	0.000286
20	2.13	2.670715	0.980002	1.999784	0.000215
21	2.12	2.712278	0.979998	2.000237	−0.000238
22	2.11	2.760203	0.979991	2.000940	−0.000942
23	2.10	2.819180	0.979998	2.000248	−0.000250
24	2.09	2.890773	0.979996	2.000398	−0.000399
25	2.08	2.984358	0.979995	2.000541	−0.000542

(continued)

8.2 Contents for Table 8.2 (Table No. 8.2-1 to 400)

Table 8.2 (continued)

Parameter values					
PROB. = 0.98	RHO = 0.3		BIGH = 2.3157		SMHL = 2.06
Pair No.	SMH	SMK	PROB. level	Risk level (%)	Risk error (%)
26	2.07	3.121813	0.979997	2.000260	−0.000262
27	2.06	3.378138	0.979993	2.000737	−0.000739
No. of pairs		Mean COMPUT. risk errors		STD. ERR. errors	
27		−0.0001598682		0.0000940346	
Z-ratio				Status of significance	
1.70010				N.S.—not significant	
Parameter values					
PROB. = 0.97	RHO = 0.3		BIGH = 2.1555		SMHL = 1.89
Pair No.	SMH	SMK	PROB. level	Risk level (%)	Risk error (%)
Table No. 8.2-137					
1	2.16	2.151009	0.969995	3.000534	−0.000536
2	2.15	2.161014	0.969994	3.000605	−0.000608
3	2.14	2.171893	0.970007	2.999282	0.000715
4	2.13	2.182868	0.970005	2.999485	0.000513
5	2.12	2.194328	0.970001	2.999938	0.000060
6	2.11	2.206669	0.970005	2.999455	0.000542
7	2.10	2.219498	0.970005	2.999526	0.000471
8	2.09	2.233209	0.970009	2.999091	0.000906
9	2.08	2.247022	0.969993	3.000713	−0.000715
10	2.07	2.262500	0.970001	2.999896	0.000101
11	2.06	2.278865	0.970007	2.999294	0.000703
12	2.05	2.296116	0.970008	2.999181	0.000817
13	2.04	2.314258	0.970002	2.999765	0.000232
14	2.03	2.334071	0.970006	2.999449	0.000548
15	2.02	2.354777	0.969995	3.000510	−0.000513
16	2.01	2.377939	0.970002	2.999807	0.000191
17	2.00	2.402777	0.970003	2.999741	0.000256
18	1.99	2.430076	0.970007	2.999336	0.000662
19	1.98	2.459055	0.969992	3.000760	−0.000763
20	1.97	2.492842	0.970006	2.999419	0.000578
21	1.96	2.529875	0.970007	2.999336	0.000662
22	1.95	2.571718	0.970005	2.999497	0.000501
23	1.94	2.619938	0.970003	2.999717	0.000280
24	1.93	2.676097	0.969994	3.000653	−0.000656
25	1.92	2.746448	0.970001	2.999866	0.000131
26	1.91	2.835679	0.970002	2.999789	0.000209

(continued)

Table 8.2 (continued)

Parameter values					
PROB. = 0.97	RHO = 0.3		BIGH = 2.1555		SMHL = 1.89
Pair No.	SMH	SMK	PROB. level	Risk level (%)	Risk error (%)
27	1.90	2.957857	0.969992	3.000754	−0.000757
28	1.89	3.170795	0.970002	2.999836	0.000161
No. of pairs		Mean COMPUT. risk errors		STD. ERR. errors	
28		0.0001617547		0.0000969147	
Z-ratio				Status of significance	
1.66904				N.S.—not significant	
Parameter values					
PROB. = 0.96	RHO = 0.3		BIGH = 2.0357		SMHL = 1.76
Pair No.	SMH	SMK	PROB. level	Risk level (%)	Risk error (%)
Table No. 8.2-138					
1	2.04	2.031409	0.960001	3.999877	0.000125
2	2.03	2.041416	0.960000	3.999990	0.000012
3	2.02	2.051913	0.960000	3.999984	0.000018
4	2.01	2.062901	0.960000	3.999996	0.000006
5	2.00	2.074381	0.959998	4.000175	−0.000173
6	1.99	2.086356	0.959993	4.000693	−0.000691
7	1.98	2.099217	0.959999	4.000092	−0.000089
8	1.97	2.112576	0.959998	4.000223	−0.000221
9	1.96	2.126824	0.960003	3.999746	0.000256
10	1.95	2.141573	0.959996	4.000402	−0.000399
11	1.94	2.157215	0.959990	4.000962	−0.000960
12	1.93	2.174533	0.960009	3.999055	0.000948
13	1.92	2.191966	0.959996	4.000425	−0.000423
14	1.91	2.211079	0.959998	4.000181	−0.000179
15	1.90	2.231483	0.959999	4.000086	−0.000083
16	1.89	2.253570	0.960005	3.999549	0.000453
17	1.88	2.276951	0.959998	4.000223	−0.000221
18	1.87	2.302801	0.960004	3.999621	0.000381
19	1.86	2.330341	0.959994	4.000574	−0.000572
20	1.85	2.361134	0.959997	4.000288	−0.000286
21	1.84	2.395183	0.959999	4.000097	−0.000095
22	1.83	2.433272	0.960001	3.999907	0.000095
23	1.82	2.476183	0.959998	4.000169	−0.000167
24	1.81	2.525481	0.959994	4.000563	−0.000560
25	1.80	2.583513	0.959990	4.000992	−0.000989
26	1.79	2.655750	0.960003	3.999692	0.000310
27	1.78	2.746882	0.960005	3.999549	0.000453

(continued)

8.2 Contents for Table 8.2 (Table No. 8.2-1 to 400)

Table 8.2 (continued)

Parameter values					
PROB. = 0.96	RHO = 0.3		BIGH = 2.0357		SMHL = 1.76
Pair No.	SMH	SMK	PROB. level	Risk level (%)	Risk error (%)
28	1.77	2.872534	0.959999	4.000110	−0.000107
29	1.76	3.085837	0.960001	3.999931	0.000072
No. of pairs		Mean COMPUT. risk errors		STD. ERR. errors	
29		−0.0001029174		0.0000757393	
Z-ratio				Status of significance	
1.35884				N.S.—not significant	
Parameter values					
PROB. = 0.95	RHO = 0.3		BIGH = 1.9384		SMHL = 1.65
Pair No.	SMH	SMK	PROB. level	Risk level (%)	Risk error (%)
Table No. 8.2-139					
1	1.94	1.936801	0.949992	5.000777	−0.000775
2	1.93	1.947227	0.950008	4.999244	0.000757
3	1.92	1.957758	0.950003	4.999680	0.000322
4	1.91	1.968785	0.950000	5.000049	−0.000048
5	1.90	1.980311	0.949995	5.000508	−0.000507
6	1.89	1.992727	0.950007	4.999316	0.000685
7	1.88	2.005255	0.949994	5.000580	−0.000578
8	1.87	2.018677	0.949993	5.000657	−0.000656
9	1.86	2.032995	0.950001	4.999876	0.000125
10	1.85	2.047821	0.949997	5.000341	−0.000340
11	1.84	2.063547	0.949995	5.000538	−0.000536
12	1.83	2.080565	0.950008	4.999238	0.000763
13	1.82	2.098096	0.949999	5.000085	−0.000083
14	1.81	2.116924	0.949997	5.000276	−0.000274
15	1.80	2.137051	0.949996	5.000365	−0.000364
16	1.79	2.158478	0.949991	5.000925	−0.000924
17	1.78	2.181989	0.950001	4.999936	0.000066
18	1.77	2.207197	0.950004	4.999566	0.000435
19	1.76	2.234102	0.949994	5.000568	−0.000566
20	1.75	2.264270	0.950005	4.999471	0.000530
21	1.74	2.296922	0.950002	4.999799	0.000203
22	1.73	2.332842	0.949991	5.000896	−0.000894
23	1.72	2.374375	0.950005	4.999507	0.000495
24	1.71	2.420744	0.950001	4.999948	0.000054
25	1.70	2.474294	0.949992	5.000794	−0.000793
26	1.69	2.538935	0.949996	5.000389	−0.000387
27	1.68	2.619356	0.950005	4.999519	0.000483
28	1.67	2.723374	0.949998	5.000186	−0.000185

(continued)

Table 8.2 (continued)

Parameter values					
PROB. = 0.95	RHO = 0.3		BIGH = 1.9384		SMHL = 1.65
Pair No.	SMH	SMK	PROB. level	Risk level (%)	Risk error (%)
29	1.66	2.875990	0.949991	5.000920	−0.000918
30	1.65	3.189708	0.950003	4.999751	0.000250
No. of pairs		Mean COMPUT. risk errors		STD. ERR. errors	
30		−0.0001180557		0.0000941371	
Z-ratio				Status of significance	
1.25408				N.S.—not significant	
Parameter values					
PROB. = 0.94	RHO = 0.3		BIGH = 1.856		SMHL = 1.56
Pair No.	SMH	SMK	PROB. level	Risk level (%)	Risk error (%)
Table No. 8.2-140					
1	1.86	1.852009	0.940005	5.999523	0.000477
2	1.85	1.862019	0.940003	5.999727	0.000274
3	1.84	1.872530	0.940005	5.999512	0.000489
4	1.83	1.883541	0.940009	5.999070	0.000930
5	1.82	1.894665	0.939990	6.000972	−0.000972
6	1.81	1.906685	0.939994	6.000638	−0.000638
7	1.80	1.919211	0.939993	6.000710	−0.000709
8	1.79	1.932637	0.940008	5.999196	0.000805
9	1.78	1.946182	0.939993	6.000698	−0.000697
10	1.77	1.961022	0.940009	5.999148	0.000852
11	1.76	1.976377	0.940009	5.999070	0.000930
12	1.75	1.992248	0.939993	6.000686	−0.000685
13	1.74	2.009420	0.939995	6.000519	−0.000519
14	1.73	2.027895	0.940008	5.999208	0.000793
15	1.72	2.047284	0.940009	5.999059	0.000942
16	1.71	2.067590	0.939995	6.000495	−0.000495
17	1.70	2.089596	0.939992	6.000835	−0.000834
18	1.69	2.113305	0.939990	6.000966	−0.000966
19	1.68	2.139110	0.939997	6.000346	−0.000346
20	1.67	2.167012	0.939999	6.000102	−0.000101
21	1.66	2.197407	0.939999	6.000066	−0.000066
22	1.65	2.230687	0.939995	6.000495	−0.000495
23	1.64	2.267636	0.939991	6.000865	−0.000864
24	1.63	2.309428	0.939996	6.000429	−0.000429
25	1.62	2.356848	0.939997	6.000346	−0.000346
26	1.61	2.411462	0.939990	6.000972	−0.000972
27	1.60	2.477959	0.940008	5.999184	0.000817
28	1.59	2.558687	0.940005	5.999458	0.000542
29	1.58	2.664587	0.940003	5.999708	0.000292

(continued)

8.2 Contents for Table 8.2 (Table No. 8.2-1 to 400) 319

Table 8.2 (continued)

Parameter values					
PROB. = 0.94	RHO = 0.3		BIGH = 1.856		SMHL = 1.56
Pair No.	SMH	SMK	PROB. level	Risk level (%)	Risk error (%)
30	1.57	2.819098	0.939995	6.000531	−0.000530
31	1.56	3.133163	0.940004	5.999643	0.000358
No. of pairs		Mean COMPUT. risk errors		STD. ERR. errors	
31		−0.0000676140		0.0001179246	
Z-ratio				Status of significance	
0.57337				N.S.—not significant	
Parameter values					
PROB. = 0.93	RHO = 0.3		BIGH = 1.7838		SMHL = 1.48
Pair No.	SMH	SMK	PROB. level	Risk level (%)	Risk error (%)
Table No. 8.2-141					
1	1.79	1.777622	0.930001	6.999934	0.000066
2	1.78	1.787608	0.930003	6.999672	0.000328
3	1.77	1.797903	0.929998	7.000202	−0.000203
4	1.76	1.808703	0.929997	7.000292	−0.000292
5	1.75	1.820011	0.929999	7.000142	−0.000143
6	1.74	1.831828	0.930000	6.999994	0.000006
7	1.73	1.844156	0.929999	7.000077	−0.000077
8	1.72	1.856998	0.929994	7.000590	−0.000590
9	1.71	1.870551	0.929994	7.000595	−0.000596
10	1.70	1.885012	0.930008	6.999243	0.000757
11	1.69	1.899994	0.930008	6.999249	0.000751
12	1.68	1.915498	0.929992	7.000846	−0.000846
13	1.67	1.932308	0.930000	7.000053	−0.000054
14	1.66	1.950036	0.930004	6.999588	0.000411
15	1.65	1.968684	0.930001	6.999904	0.000095
16	1.64	1.988646	0.930004	6.999600	0.000399
17	1.63	2.009924	0.930006	6.999374	0.000626
18	1.62	2.032521	0.930001	6.999952	0.000048
19	1.61	2.056830	0.929996	7.000411	−0.000411
20	1.60	2.083245	0.929998	7.000184	−0.000185
21	1.59	2.112159	0.930009	6.999069	0.000930
22	1.58	2.143184	0.930001	6.999874	0.000125
23	1.57	2.177496	0.929999	7.000101	−0.000101
24	1.56	2.215488	0.929994	7.000625	−0.000626
25	1.55	2.258334	0.929994	7.000625	−0.000626
26	1.54	2.307602	0.930004	6.999618	0.000381
27	1.53	2.364076	0.929998	7.000208	−0.000209
28	1.52	2.432445	0.930010	6.999052	0.000948
29	1.51	2.515840	0.930000	6.999982	0.000018

(continued)

Table 8.2 (continued)

Parameter values					
PROB. = 0.93	RHO = 0.3		BIGH = 1.7838		SMHL = 1.48
Pair No.	SMH	SMK	PROB. level	Risk level (%)	Risk error (%)
30	1.50	2.627544	0.930008	6.999243	0.000757
31	1.49	2.791781	0.929996	7.000375	−0.000376
32	1.48	3.143710	0.929994	7.000619	−0.000620
No. of pairs		Mean COMPUT. risk errors		STD. ERR. errors	
32		0.0000209519		0.0000848607	
Z-ratio				Status of significance	
0.24690				N.S.—not significant	
Parameter values					
PROB. = 0.92	RHO = 0.3		BIGH = 1.7193		SMHL = 1.41
Pair No.	SMH	SMK	PROB. level	Risk level (%)	Risk error (%)
Table No. 8.2-142					
1	1.72	1.718600	0.920005	7.999456	0.000542
2	1.71	1.728651	0.919995	8.000458	−0.000459
3	1.70	1.739210	0.919993	8.000726	−0.000727
4	1.69	1.750280	0.919995	8.000499	−0.000501
5	1.68	1.761863	0.920000	8.000046	−0.000048
6	1.67	1.773962	0.920004	7.999575	0.000423
7	1.66	1.786578	0.920006	7.999378	0.000620
8	1.65	1.799714	0.920003	7.999659	0.000340
9	1.64	1.813372	0.919993	8.000690	−0.000691
10	1.63	1.827946	0.919998	8.000183	−0.000185
11	1.62	1.843242	0.920002	7.999844	0.000155
12	1.61	1.859262	0.919999	8.000064	−0.000066
13	1.60	1.876206	0.919999	8.000064	−0.000066
14	1.59	1.894271	0.920008	7.999224	0.000775
15	1.58	1.913069	0.919997	8.000338	−0.000340
16	1.57	1.933188	0.919993	8.000696	−0.000697
17	1.56	1.955023	0.920009	7.999099	0.000900
18	1.55	1.977795	0.919996	8.000356	−0.000358
19	1.54	2.002678	0.920004	7.999605	0.000393
20	1.53	2.029287	0.920001	7.999873	0.000125
21	1.52	2.058013	0.919994	8.000649	−0.000650
22	1.51	2.089642	0.919997	8.000303	−0.000304
23	1.50	2.124568	0.920006	7.999444	0.000554
24	1.49	2.162794	0.919998	8.000220	−0.000221
25	1.48	2.205886	0.919997	8.000303	−0.000304
26	1.47	2.255410	0.920009	7.999099	0.000900
27	1.46	2.312152	0.920005	7.999480	0.000519
28	1.45	2.379240	0.919995	8.000546	−0.000548
29	1.44	2.463709	0.920009	7.999057	0.000942
30	1.43	2.573377	0.920007	7.999277	0.000721

(continued)

8.2 Contents for Table 8.2 (Table No. 8.2-1 to 400)

Table 8.2 (continued)

Parameter values					
PROB. = 0.92	RHO = 0.3		BIGH = 1.7193		SMHL = 1.41
Pair No.	SMH	SMK	PROB. level	Risk level (%)	Risk error (%)
31	1.42	2.734809	0.920004	7.999587	0.000411
32	1.41	3.058947	0.919991	8.000869	−0.000870
No. of pairs		Mean COMPUT. risk errors		STD. ERR. errors	
32		0.0000390140		0.0000945807	
Z-ratio				Status of significance	
0.41249				N.S.—not significant	
Parameter values					
PROB. = 0.91	RHO = 0.3		BIGH = 1.6607		SMHL = 1.35
Pair No.	SMH	SMK	PROB. level	Risk level (%)	Risk error (%)
Table No. 8.2-143					
1	1.67	1.651647	0.910005	8.999502	0.000495
2	1.66	1.661400	0.909997	9.000254	−0.000256
3	1.65	1.671665	0.910001	8.999926	0.000072
4	1.64	1.682247	0.909996	9.000444	−0.000447
5	1.63	1.693345	0.909997	9.000349	−0.000352
6	1.62	1.704961	0.910001	8.999866	0.000131
7	1.61	1.716903	0.909992	9.000832	−0.000834
8	1.60	1.729562	0.909996	9.000373	−0.000376
9	1.59	1.742747	0.909997	9.000338	−0.000340
10	1.58	1.756459	0.909990	9.000992	−0.000995
11	1.57	1.771093	0.910002	8.999842	0.000155
12	1.56	1.786259	0.909999	9.000099	−0.000101
13	1.55	1.802353	0.910006	8.999443	0.000554
14	1.54	1.818985	0.909991	9.000932	−0.000936
15	1.53	1.836940	0.910000	8.999973	0.000024
16	1.52	1.855830	0.910003	8.999729	0.000268
17	1.51	1.875659	0.909993	9.000736	−0.000739
18	1.50	1.897210	0.910008	8.999241	0.000757
19	1.49	1.919706	0.909995	9.000486	−0.000489
20	1.48	1.944322	0.910009	8.999128	0.000870
21	1.47	1.970280	0.909997	9.000343	−0.000346
22	1.46	1.998755	0.910003	8.999693	0.000304
23	1.45	2.029360	0.909994	9.000569	−0.000572
24	1.44	2.063272	0.910001	8.999878	0.000119
25	1.43	2.100493	0.910000	8.999979	0.000018
26	1.42	2.141809	0.909993	9.000683	−0.000685
27	1.41	2.188787	0.909996	9.000421	−0.000423
28	1.40	2.242602	0.909995	9.000528	−0.000530
29	1.39	2.305993	0.909998	9.000158	−0.000161
30	1.38	2.382870	0.910002	8.999759	0.000238

(continued)

Table 8.2 (continued)

Parameter values						
PROB. = 0.91	RHO = 0.3		BIGH = 1.6607		SMHL = 1.35	
Pair No.	SMH	SMK	PROB. level	Risk level (%)	Risk error (%)	
31	1.37	2.480270	0.910000	9.000045	−0.000048	
32	1.36	2.615385	0.909999	9.000081	−0.000083	
33	1.35	2.841344	0.909997	9.000290	−0.000292	
No. of pairs			Mean COMPUT. risk errors		STD. ERR. errors	
33			−0.0001470832		0.0000785964	
Z-ratio					Status of significance	
1.87137					N.S.—not significant	
Parameter values						
PROB. = 0.9025	RHO = 0.3		BIGH = 1.62		SMHL = 1.30	
Pair No.	SMH	SMK	PROB. level	Risk level (%)	Risk error (%)	
Table No. 8.2-144						
1	1.63	1.610061	0.902494	9.750562	−0.000560	
2	1.62	1.620000	0.902507	9.749335	0.000668	
3	1.61	1.630062	0.902494	9.750586	−0.000584	
4	1.60	1.640641	0.902492	9.750796	−0.000793	
5	1.59	1.651738	0.902498	9.750234	−0.000232	
6	1.58	1.663356	0.902508	9.749180	0.000823	
7	1.57	1.675303	0.902504	9.749561	0.000441	
8	1.56	1.687776	0.902501	9.749931	0.000072	
9	1.55	1.700778	0.902494	9.750575	−0.000572	
10	1.54	1.714507	0.902498	9.750218	−0.000215	
11	1.53	1.728966	0.902507	9.749299	0.000703	
12	1.52	1.743962	0.902503	9.749693	0.000310	
13	1.51	1.759693	0.902497	9.750282	−0.000280	
14	1.50	1.776358	0.902499	9.750146	−0.000143	
15	1.49	1.793959	0.902501	9.749853	0.000149	
16	1.48	1.812501	0.902500	9.749967	0.000036	
17	1.47	1.832181	0.902501	9.749884	0.000119	
18	1.46	1.853003	0.902497	9.750319	−0.000316	
19	1.45	1.875165	0.902491	9.750879	−0.000876	
20	1.44	1.899062	0.902496	9.750355	−0.000352	
21	1.43	1.924698	0.902501	9.749937	0.000066	
22	1.42	1.952075	0.902492	9.750807	−0.000805	
23	1.41	1.981979	0.902494	9.750629	−0.000626	
24	1.40	2.014805	0.902504	9.749592	0.000411	
25	1.39	2.050557	0.902501	9.749871	0.000131	
26	1.38	2.090020	0.902492	9.750819	−0.000817	
27	1.37	2.134761	0.902498	9.750164	−0.000161	
28	1.36	2.185174	0.902490	9.750986	−0.000983	
29	1.35	2.243999	0.902491	9.750896	−0.000894	
30	1.34	2.314757	0.902506	9.749389	0.000614	
31	1.33	2.401358	0.902501	9.749884	0.000119	

(continued)

8.2 Contents for Table 8.2 (Table No. 8.2-1 to 400)

Table 8.2 (continued)

Parameter values					
PROB. = 0.9025	RHO = 0.3		BIGH = 1.62		SMHL = 1.30
Pair No.	SMH	SMK	PROB. level	Risk level (%)	Risk error (%)
32	1.32	2.516306	0.902505	9.749544	0.000459
33	1.31	2.687733	0.902505	9.749484	0.000519
34	1.30	3.056268	0.902502	9.749842	0.000161
No. of pairs		Mean COMPUT. risk errors		STD. ERR. errors	
34		−0.0000974110		0.0000869901	
Z-ratio				Status of significance	
1.11979				N.S.—not significant	
Parameter values					
PROB. = 0.9	RHO = 0.3		BIGH = 1.6069		SMHL = 1.29
Pair No.	SMH	SMK	PROB. level	Risk level (%)	Risk error (%)
Table No. 8.2-145					
1	1.61	1.603806	0.899998	10.000230	−0.000232
2	1.60	1.613830	0.899993	10.000720	−0.000715
3	1.59	1.624370	0.899999	10.000070	−0.000066
4	1.58	1.635234	0.899996	10.000370	−0.000364
5	1.57	1.646620	0.900000	10.000010	−0.000006
6	1.56	1.658335	0.899990	10.000970	−0.000972
7	1.55	1.670772	0.899999	10.000130	−0.000125
8	1.54	1.683737	0.900005	9.999460	0.000542
9	1.53	1.697040	0.899991	10.000870	−0.000870
10	1.52	1.711268	0.900001	9.999924	0.000077
11	1.51	1.726034	0.899999	10.000130	−0.000131
12	1.50	1.741535	0.899996	10.000380	−0.000376
13	1.49	1.757971	0.900003	9.999675	0.000328
14	1.48	1.775149	0.900000	9.999954	0.000048
15	1.47	1.793268	0.899996	10.000380	−0.000381
16	1.46	1.812721	0.900010	9.999026	0.000978
17	1.45	1.833121	0.900008	9.999168	0.000834
18	1.44	1.854863	0.900009	9.999108	0.000894
19	1.43	1.877949	0.900003	9.999746	0.000256
20	1.42	1.902775	0.900001	9.999895	0.000107
21	1.41	1.929343	0.899993	10.000750	−0.000745
22	1.40	1.958439	0.900001	9.999866	0.000137
23	1.39	1.989677	0.899992	10.000800	−0.000793
24	1.38	2.024232	0.899997	10.000270	−0.000262
25	1.37	2.062108	0.899992	10.000830	−0.000823
26	1.36	2.104873	0.900004	9.999627	0.000376
27	1.35	2.152530	0.899994	10.000590	−0.000584
28	1.34	2.208210	0.900009	9.999066	0.000936
29	1.33	2.272699	0.900000	9.999985	0.000018
30	1.32	2.351469	0.900000	10.000010	−0.000006
31	1.31	2.452339	0.900002	9.999776	0.000226

(continued)

Table 8.2 (continued)

Parameter values					
PROB. = 0.9	RHO = 0.3		BIGH = 1.6069		SMHL = 1.29
Pair No.	SMH	SMK	PROB. level	Risk level (%)	Risk error (%)
32	1.30	2.592501	0.899996	10.000430	−0.000423
33	1.29	2.832898	0.899991	10.000910	−0.000906
No. of pairs		Mean COMPUT. risk errors		STD. ERR. errors	
33		−0.0000888810		0.0000932181	
Z-ratio				Status of significance	
0.95347				N.S.—not significant	
Parameter values					
PROB. = 0.85	RHO = 0.3		BIGH = 1.3851		SMHL = 1.04
Pair No.	SMH	SMK	PROB. level	Risk level (%)	Risk error (%)
Table No. 8.2-146					
1	1.39	1.380120	0.849999	15.000110	−0.000107
2	1.38	1.390121	0.849999	15.000110	−0.000107
3	1.37	1.400562	0.850008	14.999180	0.000823
4	1.36	1.411249	0.850000	14.999960	0.000036
5	1.35	1.422480	0.850009	14.999060	0.000942
6	1.34	1.434062	0.850009	14.999130	0.000870
7	1.33	1.445998	0.849996	15.000390	−0.000387
8	1.32	1.458489	0.849992	15.000830	−0.000829
9	1.31	1.471536	0.849992	15.000800	−0.000805
10	1.30	1.485146	0.849993	15.000700	−0.000703
11	1.29	1.499320	0.849991	15.000890	−0.000894
12	1.28	1.514259	0.850002	14.999780	0.000221
13	1.27	1.529772	0.850002	14.999840	0.000155
14	1.26	1.546058	0.850004	14.999600	0.000399
15	1.25	1.563122	0.850004	14.999620	0.000376
16	1.24	1.580968	0.849995	15.000480	−0.000477
17	1.23	1.599992	0.850006	14.999350	0.000644
18	1.22	1.619808	0.849995	15.000460	−0.000465
19	1.21	1.641007	0.850003	14.999690	0.000310
20	1.20	1.663400	0.850003	14.999700	0.000298
21	1.19	1.687186	0.850000	15.000020	−0.000024
22	1.18	1.712568	0.849996	15.000450	−0.000453
23	1.17	1.739940	0.850002	14.999780	0.000215
24	1.16	1.769311	0.850002	14.999790	0.000209
25	1.15	1.801075	0.850000	14.999990	0.000006
26	1.14	1.835631	0.849996	15.000420	−0.000423
27	1.13	1.873570	0.849992	15.000750	−0.000757
28	1.12	1.915682	0.849993	15.000660	−0.000662
29	1.11	1.963145	0.850006	14.999390	0.000608
30	1.10	2.016749	0.850005	14.999530	0.000471
31	1.09	2.078843	0.850003	14.999740	0.000262
32	1.08	2.152953	0.850006	14.999440	0.000560

(continued)

8.2 Contents for Table 8.2 (Table No. 8.2-1 to 400)

Table 8.2 (continued)

Parameter values					
PROB. = 0.85	RHO = 0.3		BIGH = 1.3851		SMHL = 1.04
Pair No.	SMH	SMK	PROB. level	Risk level (%)	Risk error (%)
33	1.07	2.244555	0.850004	14.999620	0.000376
34	1.06	2.365375	0.849998	15.000210	−0.000215
35	1.05	2.547456	0.850001	14.999900	0.000095
36	1.04	2.950962	0.849993	15.000730	−0.000733
No. of pairs		Mean COMPUT. risk errors		STD. ERR. errors	
36		−0.0000045106		0.0000854734	
Z-ratio				Status of significance	
0.05277				N.S.—not significant	
Parameter values					
PROB. = 0.8	RHO = 0.3		BIGH = 1.2097		SMHL = 0.85
Pair No.	SMH	SMK	PROB. level	Risk level (%)	Risk error (%)
Table No. 8.2-147					
1	1.21	1.209498	0.800001	19.999880	0.000119
2	1.20	1.219674	0.800004	19.999640	0.000358
3	1.19	1.230117	0.799994	20.000640	−0.000638
4	1.18	1.241026	0.799999	20.000150	−0.000149
5	1.17	1.252310	0.800001	19.999880	0.000119
6	1.16	1.263971	0.799999	20.000070	−0.000072
7	1.15	1.276015	0.799991	20.000950	−0.000954
8	1.14	1.288642	0.799999	20.000120	−0.000125
9	1.13	1.301662	0.799994	20.000600	−0.000602
10	1.12	1.315275	0.799999	20.000100	−0.000101
11	1.11	1.329488	0.800010	19.999050	0.000948
12	1.10	1.344207	0.800009	19.999130	0.000870
13	1.09	1.359537	0.800005	19.999510	0.000489
14	1.08	1.375484	0.799993	20.000660	−0.000662
15	1.07	1.392248	0.799992	20.000820	−0.000817
16	1.06	1.409838	0.799993	20.000660	−0.000662
17	1.05	1.428260	0.799991	20.000890	−0.000894
18	1.04	1.447715	0.799998	20.000190	−0.000191
19	1.03	1.468212	0.800006	19.999450	0.000554
20	1.02	1.489758	0.800004	19.999610	0.000387
21	1.01	1.512557	0.800002	19.999820	0.000179
22	1.00	1.536811	0.800005	19.999530	0.000471
23	0.99	1.562530	0.799999	20.000140	−0.000137
24	0.98	1.590113	0.800001	19.999940	0.000060
25	0.97	1.619570	0.799992	20.000850	−0.000846
26	0.96	1.651691	0.800008	19.999210	0.000793
27	0.95	1.686096	0.799995	20.000490	−0.000495
28	0.94	1.723968	0.800001	19.999870	0.000125
29	0.93	1.765708	0.800009	19.999080	0.000918
30	0.92	1.811717	0.799996	20.000440	−0.000441

(continued)

Table 8.2 (continued)

Parameter values						
PROB. = 0.8		RHO = 0.3		BIGH = 1.2097		SMHL = 0.85
Pair No.	SMH	SMK		PROB. level	Risk level (%)	Risk error (%)
31	0.91	1.863962		0.800001	19.999860	0.000143
32	0.90	1.923627		0.799999	20.000080	−0.000083
33	0.89	1.993459		0.799994	20.000570	−0.000572
34	0.88	2.078549		0.800006	19.999400	0.000602
35	0.87	2.185945		0.799995	20.000470	−0.000471
36	0.86	2.335972		0.799997	20.000300	−0.000304
37	0.85	2.593490		0.800006	19.999370	0.000632
No. of pairs		Mean COMPUT. risk errors			STD. ERR. errors	
37		−0.0000381156			0.0000880955	
Z-ratio					Status of significance	
0.43266					N.S.—not significant	

Parameter values						
PROB. = 0.75		RHO = 0.3		BIGH = 1.06		SMHL = 0.68
Pair No.	SMH	SMK		PROB. level	Risk level (%)	Risk error (%)
Table No. 8.2-148						
1	1.06	1.060098		0.749994	25.000580	−0.000578
2	1.05	1.070291		0.750000	25.000040	−0.000036
3	1.04	1.080775		0.749998	25.000170	−0.000167
4	1.03	1.091655		0.750005	24.999480	0.000519
5	1.02	1.102838		0.750002	24.999850	0.000155
6	1.01	1.114428		0.750002	24.999830	0.000167
7	1.00	1.126432		0.750003	24.999700	0.000304
8	0.99	1.138856		0.750003	24.999690	0.000310
9	0.98	1.151706		0.749999	25.000100	−0.000101
10	0.97	1.165089		0.750003	24.999730	0.000274
11	0.96	1.178913		0.749996	25.000400	−0.000399
12	0.95	1.193284		0.749990	25.000990	−0.000989
13	0.94	1.208405		0.750008	24.999190	0.000811
14	0.93	1.223992		0.750004	24.999600	0.000405
15	0.92	1.240249		0.750001	24.999940	0.000066
16	0.91	1.257186		0.749993	25.000740	−0.000739
17	0.90	1.275007		0.749999	25.000110	−0.000107
18	0.89	1.293624		0.750000	25.000020	−0.000018
19	0.88	1.313146		0.750000	25.000010	−0.000006
20	0.87	1.333681		0.750002	24.999760	0.000244
21	0.86	1.355340		0.750009	24.999140	0.000858
22	0.85	1.378132		0.750007	24.999270	0.000727
23	0.84	1.402267		0.750008	24.999240	0.000763
24	0.83	1.427758		0.749996	25.000410	−0.000405
25	0.82	1.455009		0.749995	25.000550	−0.000548
26	0.81	1.484230		0.750001	24.999890	0.000107
27	0.80	1.515437		0.749994	25.000580	−0.000578

(continued)

8.2 Contents for Table 8.2 (Table No. 8.2-1 to 400)

Table 8.2 (continued)

Parameter values					
PROB. = 0.75	RHO = 0.3		BIGH = 1.06		SMHL = 0.68
Pair No.	SMH	SMK	PROB. level	Risk level (%)	Risk error (%)
28	0.79	1.549231	0.749997	25.000310	−0.000310
29	0.78	1.585825	0.749993	25.000680	−0.000679
30	0.77	1.626017	0.750002	24.999830	0.000167
31	0.76	1.670217	0.750002	24.999760	0.000238
32	0.75	1.719617	0.750010	24.999010	0.000989
33	0.74	1.775019	0.749996	25.000440	−0.000441
34	0.73	1.838982	0.749992	25.000820	−0.000823
35	0.72	1.914852	0.750003	24.999670	0.000328
36	0.71	2.007535	0.750009	24.999090	0.000912
37	0.70	2.127017	0.749999	25.000150	−0.000149
38	0.69	2.300280	0.750000	24.999990	0.000012
39	0.68	2.633601	0.749991	25.000930	−0.000930
No. of pairs		Mean COMPUT. risk errors		STD. ERR. errors	
39		0.0000087917		0.0000820958	
Z-ratio				Status of significance	
0.10709				N.S.—not significant	
Parameter values					
PROB. = 0.7	RHO = 0.3		BIGH = 0.9262		SMHL = 0.53
Pair No.	SMH	SMK	PROB. level	Risk level (%)	Risk error (%)
Table No. 8.2-149					
1	0.93	0.922415	0.700002	29.999800	0.000203
2	0.92	0.932442	0.700000	30.000000	0.000000
3	0.91	0.942786	0.700000	30.000010	−0.000012
4	0.90	0.953456	0.700000	29.999990	0.000012
5	0.89	0.964458	0.699999	30.000080	−0.000077
6	0.88	0.975802	0.699996	30.000440	−0.000435
7	0.87	0.987593	0.700005	29.999500	0.000507
8	0.86	0.999644	0.699991	30.000900	−0.000900
9	0.85	1.012258	0.700003	29.999750	0.000256
10	0.84	1.025250	0.700003	29.999680	0.000322
11	0.83	1.038628	0.699991	30.000880	−0.000876
12	0.82	1.052599	0.699996	30.000420	−0.000417
13	0.81	1.067077	0.699998	30.000240	−0.000238
14	0.80	1.082074	0.699993	30.000660	−0.000656
15	0.79	1.097795	0.700009	29.999090	0.000912
16	0.78	1.113963	0.699996	30.000360	−0.000358
17	0.77	1.130883	0.699995	30.000550	−0.000548
18	0.76	1.148569	0.699997	30.000260	−0.000262
19	0.75	1.167037	0.699999	30.000120	−0.000113
20	0.74	1.186400	0.700005	29.999460	0.000542
21	0.73	1.206577	0.699997	30.000310	−0.000304
22	0.72	1.227781	0.699991	30.000910	−0.000912

(continued)

Table 8.2 (continued)

Parameter values					
PROB. = 0.7	RHO = 0.3		BIGH = 0.9262		SMHL = 0.53
Pair No.	SMH	SMK	PROB. level	Risk level (%)	Risk error (%)
23	0.71	1.250226	0.700000	29.999960	0.000048
24	0.70	1.273737	0.699992	30.000840	−0.000834
25	0.69	1.298724	0.699997	30.000320	−0.000316
26	0.68	1.325210	0.700001	29.999950	0.000054
27	0.67	1.353414	0.700007	29.999350	0.000650
28	0.66	1.383361	0.699996	30.000400	−0.000393
29	0.65	1.415662	0.700002	29.999830	0.000173
30	0.64	1.450346	0.699996	30.000400	−0.000399
31	0.63	1.488027	0.699996	30.000430	−0.000429
32	0.62	1.529326	0.700005	29.999460	0.000542
33	0.61	1.574664	0.700004	29.999570	0.000435
34	0.60	1.625056	0.699998	30.000240	−0.000238
35	0.59	1.682101	0.700000	30.000010	−0.000006
36	0.58	1.747795	0.700008	29.999230	0.000775
37	0.57	1.824915	0.700004	29.999590	0.000411
38	0.56	1.919368	0.700010	29.999040	0.000960
39	0.55	2.040970	0.700000	30.000020	−0.000018
40	0.54	2.216728	0.700002	29.999760	0.000244
41	0.53	2.552951	0.699996	30.000410	−0.000411
No. of pairs		Mean COMPUT. risk errors		STD. ERR. errors	
41		−0.0000502382		0.0000745371	
Z-ratio				Status of significance	
0.67400				N.S.—not significant	
Parameter values					
PROB. = 0.65	RHO = 0.3		BIGH = 0.8026		SMHL = 0.39
Pair No.	SMH	SMK	PROB. level	Risk level (%)	Risk error (%)
Table No. 8.2-150					
1	0.81	0.795268	0.650004	34.999580	0.000423
2	0.80	0.805209	0.650006	34.999390	0.000614
3	0.79	0.815401	0.650000	35.000010	0.000000
4	0.78	0.825953	0.650005	34.999470	0.000530
5	0.77	0.836776	0.650001	34.999870	0.000131
6	0.76	0.847979	0.650007	34.999340	0.000662
7	0.75	0.859475	0.650001	34.999870	0.000131
8	0.74	0.871375	0.650003	34.999690	0.000316
9	0.73	0.883592	0.649993	35.000740	−0.000739
10	0.72	0.896336	0.650005	34.999550	0.000453
11	0.71	0.909426	0.650001	34.999860	0.000149
12	0.70	0.922973	0.650000	35.000040	−0.000036
13	0.69	0.936993	0.649997	35.000280	−0.000280
14	0.68	0.951601	0.650009	34.999150	0.000858
15	0.67	0.966619	0.649999	35.000130	−0.000131

(continued)

8.2 Contents for Table 8.2 (Table No. 8.2-1 to 400)

Table 8.2 (continued)

Parameter values					
PROB. = 0.65	RHO = 0.3		BIGH = 0.8026		SMHL = 0.39
Pair No.	SMH	SMK	PROB. level	Risk level (%)	Risk error (%)
16	0.66	0.982260	0.649997	35.000260	−0.000262
17	0.65	0.998545	0.650001	34.999860	0.000149
18	0.64	1.015495	0.650007	34.999280	0.000721
19	0.63	1.033034	0.649996	35.000380	−0.000370
20	0.62	1.051478	0.650008	34.999200	0.000805
21	0.61	1.070659	0.650009	34.999130	0.000870
22	0.60	1.090603	0.649994	35.000650	−0.000644
23	0.59	1.111730	0.650009	34.999090	0.000918
24	0.58	1.133679	0.649996	35.000390	−0.000381
25	0.57	1.156875	0.649997	35.000300	−0.000298
26	0.56	1.181352	0.650001	34.999940	0.000066
27	0.55	1.207149	0.649996	35.000410	−0.000405
28	0.54	1.234503	0.649992	35.000780	−0.000775
29	0.53	1.263651	0.649994	35.000580	−0.000572
30	0.52	1.294836	0.650003	34.999720	0.000286
31	0.51	1.328111	0.649996	35.000390	−0.000381
32	0.50	1.364114	0.650002	34.999830	0.000179
33	0.49	1.403102	0.650003	34.999690	0.000316
34	0.48	1.445529	0.649993	35.000660	−0.000656
35	0.47	1.492442	0.649996	35.000430	−0.000429
36	0.46	1.544893	0.650010	34.999020	0.000978
37	0.45	1.603747	0.650002	34.999830	0.000173
38	0.44	1.671436	0.649994	35.000630	−0.000632
39	0.43	1.751577	0.649998	35.000240	−0.000238
40	0.42	1.849745	0.650001	34.999870	0.000131
41	0.41	1.977387	0.650003	34.999730	0.000274
42	0.40	2.186978	0.649993	35.000660	−0.000656
43	0.39	2.557958	0.650003	34.999710	0.000292
No. of pairs		Mean COMPUT. risk errors		STD. ERR. errors	
43		0.0000577081		0.0000749646	
Z-ratio				Status of significance	
0.76980				N.S.—not significant	
Parameter values					
PROB. = 0.6	RHO = 0.3		BIGH = 0.6857		SMHL = 0.26
Pair No.	SMH	SMK	PROB. level	Risk level (%)	Risk error (%)
Table No. 8.2-151					
1	0.69	0.681427	0.600009	39.999080	0.000918
2	0.68	0.691448	0.600009	39.999060	0.000942
3	0.67	0.701670	0.599992	40.000760	−0.000763
4	0.66	0.712303	0.600000	39.999960	0.000036
5	0.65	0.723165	0.599992	40.000810	−0.000811
6	0.64	0.734468	0.600008	39.999200	0.000799

(continued)

Table 8.2 (continued)

Parameter values					
PROB. = 0.6	RHO = 0.3		BIGH = 0.6857		SMHL = 0.26
Pair No.	SMH	SMK	PROB. level	Risk level (%)	Risk error (%)
7	0.63	0.746032	0.600008	39.999170	0.000834
8	0.62	0.757874	0.599994	40.000580	−0.000578
9	0.61	0.770208	0.600004	39.999570	0.000429
10	0.60	0.782859	0.600000	39.999990	0.000012
11	0.59	0.795946	0.600001	39.999870	0.000125
12	0.58	0.809491	0.600007	39.999280	0.000721
13	0.57	0.823420	0.600000	39.999970	0.000024
14	0.56	0.837857	0.599998	40.000180	−0.000179
15	0.55	0.852830	0.600001	39.999900	0.000095
16	0.54	0.868368	0.600008	39.999250	0.000751
17	0.53	0.884406	0.600001	39.999870	0.000131
18	0.52	0.901076	0.599998	40.000160	−0.000161
19	0.51	0.918414	0.599998	40.000240	−0.000238
20	0.50	0.936462	0.599998	40.000200	−0.000197
21	0.49	0.955263	0.599998	40.000200	−0.000203
22	0.48	0.974863	0.599996	40.000390	−0.000393
23	0.47	0.995314	0.599990	40.000980	−0.000978
24	0.46	1.016866	0.600004	39.999560	0.000435
25	0.45	1.039385	0.600008	39.999210	0.000787
26	0.44	1.062937	0.599998	40.000220	−0.000221
27	0.43	1.087788	0.599994	40.000590	−0.000590
28	0.42	1.114115	0.600001	39.999890	0.000107
29	0.41	1.141908	0.599999	40.000100	−0.000101
30	0.40	1.171457	0.600000	39.999990	0.000012
31	0.39	1.202866	0.599992	40.000780	−0.000781
32	0.38	1.236741	0.600008	39.999200	0.000799
33	0.37	1.272916	0.599999	40.000080	−0.000077
34	0.36	1.312122	0.599996	40.000410	−0.000405
35	0.35	1.354906	0.599998	40.000210	−0.000215
36	0.34	1.402035	0.600008	39.999220	0.000775
37	0.33	1.454097	0.600000	40.000000	0.000000
38	0.32	1.512684	0.599994	40.000620	−0.000626
39	0.31	1.615160	0.599995	40.000470	−0.000477
40	0.30	1.690327	0.600004	39.999570	0.000423
41	0.29	1.781871	0.599999	40.000150	−0.000149
42	0.28	1.900322	0.600002	39.999780	0.000215

(continued)

8.2 Contents for Table 8.2 (Table No. 8.2-1 to 400) 331

Table 8.2 (continued)

Parameter values					
PROB. = 0.6	RHO = 0.3		BIGH = 0.6857		SMHL = 0.26
Pair No.	SMH	SMK	PROB. level	Risk level (%)	Risk error (%)
43	0.27	2.068765	0.599999	40.000120	−0.000119
44	0.26	2.377149	0.600001	39.999950	0.000048
No. of pairs		Mean COMPUT. risk errors		STD. ERR. errors	
44		0.0000256962		0.0000758749	
Z-ratio				Status of significance	
0.33867				N.S.—not significant	
Parameter values					
PROB. = 0.55	RHO = 0.3		BIGH = 0.5729		SMHL = 0.13
Pair No.	SMH	SMK	PROB. level	Risk level (%)	Risk error (%)
Table No. 8.2-152					
1	0.58	0.565887	0.549996	45.000450	−0.000447
2	0.57	0.575815	0.549992	45.000830	−0.000834
3	0.56	0.586097	0.550007	44.999310	0.000691
4	0.55	0.596558	0.550000	45.000010	−0.000006
5	0.54	0.607316	0.549995	45.000540	−0.000542
6	0.53	0.618394	0.549993	45.000720	−0.000721
7	0.52	0.629814	0.549997	45.000350	−0.000346
8	0.51	0.641604	0.550008	44.999200	0.000805
9	0.50	0.653694	0.550010	44.999040	0.000960
10	0.49	0.666114	0.550005	44.999540	0.000459
11	0.48	0.678897	0.549996	45.000400	−0.000393
12	0.47	0.692176	0.550006	44.999380	0.000620
13	0.46	0.705795	0.550000	44.999960	0.000042
14	0.45	0.719893	0.550001	44.999860	0.000143
15	0.44	0.734418	0.549995	45.000530	−0.000530
16	0.43	0.749520	0.550002	44.999760	0.000238
17	0.42	0.765056	0.549994	45.000590	−0.000590
18	0.41	0.781187	0.549992	45.000800	−0.000793
19	0.40	0.797977	0.550001	44.999900	0.000101
20	0.39	0.815403	0.550010	44.999020	0.000983
21	0.38	0.833448	0.550009	44.999090	0.000912
22	0.37	0.852105	0.549991	45.000920	−0.000918
23	0.36	0.871667	0.549992	45.000780	−0.000781
24	0.35	0.892052	0.549991	45.000940	−0.000942
25	0.34	0.913383	0.549994	45.000610	−0.000602
26	0.33	0.935702	0.549997	45.000290	−0.000292
27	0.32	0.959068	0.549996	45.000440	−0.000435
28	0.31	0.983658	0.550000	45.000060	−0.000054
29	0.30	1.009477	0.549995	45.000540	−0.000542
30	0.29	1.036851	0.550003	44.999730	0.000274
31	0.28	1.065748	0.550001	44.999880	0.000125
32	0.27	1.096467	0.550002	44.999810	0.000191

(continued)

Table 8.2 (continued)

Parameter values					
PROB. = 0.55	RHO = 0.3		BIGH = 0.5729		SMHL = 0.13
Pair No.	SMH	SMK	PROB. level	Risk level (%)	Risk error (%)
33	0.26	1.129159	0.549996	45.000400	−0.000399
34	0.25	1.164223	0.549995	45.000480	−0.000483
35	0.24	1.247832	0.549996	45.000360	−0.000358
36	0.23	1.284830	0.549998	45.000170	−0.000173
37	0.22	1.325671	0.550002	44.999760	0.000238
38	0.21	1.371127	0.550008	44.999180	0.000823
39	0.20	1.421929	0.549995	45.000540	−0.000536
40	0.19	1.480176	0.550003	44.999700	0.000304
41	0.18	1.547629	0.550005	44.999500	0.000501
42	0.17	1.627545	0.549993	45.000670	−0.000668
43	0.16	1.726727	0.550002	44.999810	0.000191
44	0.15	1.856842	0.550010	44.999020	0.000978
45	0.14	2.047510	0.549996	45.000400	−0.000393
46	0.13	2.443472	0.549990	45.000990	−0.000983
No. of pairs		Mean COMPUT. risk errors		STD. ERR. errors	
46		−0.0000890265		0.0000840536	
Z-ratio				Status of significance	
1.05916				N.S.—not significant	

Parameter values					
PROB. = 0.5	RHO = 0.3		BIGH = 0.4622		SMHL = 0.01
Pair No.	SMH	SMK	PROB. level	Risk level (%)	Risk error (%)
Table No. 8.2-153					
1	0.47	0.454627	0.500002	49.999850	0.000155
2	0.46	0.464508	0.499990	50.000980	−0.000980
3	0.45	0.474731	0.499998	50.000200	−0.000203
4	0.44	0.485227	0.500007	49.999300	0.000697
5	0.43	0.495932	0.500001	49.999930	0.000077
6	0.42	0.506980	0.500006	49.999400	0.000608
7	0.41	0.518312	0.500007	49.999260	0.000739
8	0.40	0.529873	0.499991	50.000900	−0.000894
9	0.39	0.541907	0.500005	49.999530	0.000471
10	0.38	0.554173	0.499995	50.000480	−0.000474
11	0.37	0.566828	0.499992	50.000810	−0.000814
12	0.36	0.579937	0.500003	49.999670	0.000328
13	0.35	0.593376	0.500002	49.999850	0.000155
14	0.34	0.607226	0.499999	50.000110	−0.000104
15	0.33	0.621579	0.500008	49.999190	0.000811
16	0.32	0.636340	0.500007	49.999300	0.000697
17	0.31	0.651527	0.499995	50.000510	−0.000507
18	0.30	0.667369	0.500007	49.999310	0.000691
19	0.29	0.683623	0.499995	50.000470	−0.000462
20	0.28	0.700557	0.500000	49.999960	0.000042

(continued)

8.2 Contents for Table 8.2 (Table No. 8.2-1 to 400)

Table 8.2 (continued)

Parameter values					
PROB. = 0.5	RHO = 0.3		BIGH = 0.4622		SMHL = 0.01
Pair No.	SMH	SMK	PROB. level	Risk level (%)	Risk error (%)
21	0.27	0.718073	0.499999	50.000120	−0.000122
22	0.26	0.736297	0.500004	49.999560	0.000441
23	0.25	0.755198	0.500005	49.999520	0.000483
24	0.24	0.774885	0.500009	49.999110	0.000888
25	0.23	0.795324	0.500003	49.999690	0.000310
26	0.22	0.816645	0.499997	50.000280	−0.000280
27	0.21	0.838959	0.499996	50.000360	−0.000364
28	0.20	0.862284	0.499992	50.000850	−0.000852
29	0.19	0.886861	0.500001	49.999880	0.000119
30	0.18	0.968076	0.500000	49.999960	0.000048
31	0.17	0.990428	0.500000	49.999970	0.000036
32	0.16	1.014476	0.500003	49.999680	0.000322
33	0.15	1.040402	0.500007	49.999270	0.000733
34	0.14	1.068302	0.499998	50.000170	−0.000167
35	0.13	1.098571	0.499991	50.000880	−0.000882
36	0.12	1.131801	0.500010	49.999010	0.000995
37	0.11	1.167859	0.500002	49.999770	0.000232
38	0.10	1.207575	0.499999	50.000130	−0.000125
39	0.09	1.251580	0.499996	50.000420	−0.000417
40	0.08	1.300825	0.499998	50.000240	−0.000241
41	0.07	1.356523	0.500002	49.999850	0.000155
42	0.06	1.420240	0.499996	50.000360	−0.000358
43	0.05	1.494833	0.499993	50.000670	−0.000668
44	0.04	1.584847	0.499997	50.000320	−0.000319
45	0.03	1.698278	0.500001	49.999870	0.000137
46	0.02	1.852459	0.500000	50.000030	−0.000024
47	0.01	2.101614	0.500002	49.999810	0.000191
No. of pairs		Mean COMPUT. risk errors		STD. ERR. errors	
47		0.0000271946		0.0000738977	
Z-ratio				Status of significance	
0.36800				N.S.—not significant	
Parameter values					
PROB. = 0.99	RHO = 0.2		BIGH = 2.5722		SMHL = 2.33
Pair No.	SMH	SMK	PROB. level	Risk level (%)	Risk error (%)
Table No. 8.2-154					
1	2.58	2.564424	0.989999	1.000088	−0.000089
2	2.57	2.574402	0.990001	0.999904	0.000095
3	2.56	2.584458	0.989996	1.000381	−0.000381
4	.255	2.596156	0.990006	0.999379	0.000620
5	2.54	2.607933	0.990008	0.999171	0.000829
6	2.53	2.619792	0.990003	0.999737	0.000262
7	2.52	2.633294	0.990008	0.999188	0.000811

(continued)

Table 8.2 (continued)

Parameter values					
PROB. = 0.99	RHO = 0.2		BIGH = 2.5722		SMHL = 2.33
Pair No.	SMH	SMK	PROB. level	Risk level (%)	Risk error (%)
8	2.51	2.646879	0.990004	0.999564	0.000435
9	2.50	2.662110	0.990009	0.999105	0.000894
10	2.49	2.677426	0.990003	0.999695	0.000304
11	2.48	2.694390	0.990003	0.999725	0.000274
12	2.47	2.713004	0.990006	0.999403	0.000596
13	2.46	2.731705	0.989996	1.000416	−0.000417
14	2.45	2.753620	0.990000	0.999987	0.000012
15	2.44	2.777188	0.990001	0.999874	0.000125
16	2.43	2.803971	0.990009	0.999087	0.000912
17	2.42	2.830847	0.989997	1.000297	−0.000298
18	2.41	2.864067	0.990005	0.999463	0.000536
19	2.40	2.900505	0.990007	0.999296	0.000703
20	2.39	2.940165	0.989997	1.000273	−0.000274
21	2.38	2.989297	0.990000	0.999963	0.000036
22	2.37	3.047901	0.990001	0.999904	0.000095
23	2.36	3.122230	0.990004	0.999606	0.000393
24	2.35	3.215409	0.989992	1.000810	−0.000811
25	2.34	3.364941	0.989994	1.000559	−0.000560
26	2.33	3.689576	0.989991	1.000893	−0.000894
No. of pairs		Mean COMPUT. risk errors		STD. ERR. errors	
26		0.0001558551		0.0000978281	
Z-ratio				Status of significance	
1.59315				N.S.—not significant	
Parameter values					
PROB. = 0.98	RHO = 0.2		BIGH = 2.3198		SMHL = 2.06
Pair No.	SMH	SMK	PROB. level	Risk level (%)	Risk error (%)
Table No. 8.2-155					
1	2.32	2.319600	0.979997	2.000260	−0.000262
2	2.31	2.329641	0.979992	2.000761	−0.000763
3	2.30	2.340552	0.979996	2.000356	−0.000358
4	2.29	2.352331	0.980008	1.999217	0.000781
5	2.28	2.364201	0.980006	1.999354	0.000644
6	2.27	2.376161	0.979992	2.000785	−0.000787
7	2.26	2.389776	0.980000	2.000034	−0.000036
8	2.25	2.404265	0.980009	1.999146	0.000852
9	2.24	2.418849	0.980001	1.999891	0.000107
10	2.23	2.435091	0.980008	1.999188	0.000811
11	2.22	2.451430	0.979996	2.000374	−0.000376
12	2.21	2.469430	0.979994	2.000570	−0.000572
13	2.20	2.489092	0.979998	2.000183	−0.000185
14	2.19	2.510418	0.980004	1.999569	0.000429
15	2.18	2.532628	0.979997	2.000320	−0.000322

(continued)

8.2 Contents for Table 8.2 (Table No. 8.2-1 to 400)

Table 8.2 (continued)

Parameter values					
PROB. = 0.98	RHO = 0.2		BIGH = 2.3198		SMHL = 2.06
Pair No.	SMH	SMK	PROB. level	Risk level (%)	Risk error (%)
16	2.17	2.558066	0.980008	1.999199	0.000799
17	2.16	2.585172	0.980009	1.999140	0.000858
18	2.15	2.613948	0.979995	2.000457	−0.000459
19	2.14	2.647519	0.980001	1.999933	0.000066
20	2.13	2.684325	0.979996	2.000415	−0.000417
21	2.12	2.727492	0.980002	1.999784	0.000215
22	2.11	2.775460	0.979991	2.000952	−0.000954
23	2.10	2.834480	0.979993	2.000678	−0.000679
24	2.09	2.909241	0.980006	1.999408	0.000590
25	2.08	3.002870	0.979998	2.000177	−0.000179
26	2.07	3.143494	0.980005	1.999497	0.000501
27	2.06	3.399864	0.979993	2.000672	−0.000674
No. of pairs		Mean COMPUT. risk errors		STD. ERR. errors	
27		−0.0000131982		0.0001083126	
Z-ratio				Status of significance	
0.12185				N.S.—not significant	
Parameter values					
PROB. = 0.97	RHO = 0.2		BIGH = 2.161		SMHL = 1.89
Pair No.	SMH	SMK	PROB. level	Risk level (%)	Risk error (%)
Table No. 8.2-156					
1	2.17	2.152037	0.970001	2.999938	0.000060
2	2.16	2.162001	0.970006	2.999401	0.000596
3	2.15	2.172056	0.969998	3.000212	−0.000215
4	2.14	2.182987	0.970004	2.999616	0.000381
5	2.13	2.194014	0.969995	3.000540	−0.000542
6	2.12	2.205918	0.969996	3.000379	−0.000381
7	2.11	2.218701	0.970006	2.999354	0.000644
8	2.10	2.231584	0.969999	3.000152	−0.000155
9	2.09	2.245350	0.969996	3.000450	−0.000453
10	2.08	2.259998	0.969996	3.000432	−0.000435
11	2.07	2.275532	0.969996	3.000367	−0.000370
12	2.06	2.291952	0.969995	3.000498	−0.000501
13	2.05	2.310041	0.970010	2.999014	0.000983
14	2.04	2.328239	0.969997	3.000349	−0.000352
15	2.03	2.348110	0.969993	3.000736	−0.000739
16	2.02	2.369654	0.969993	3.000689	−0.000691
17	2.01	2.393656	0.970010	2.999038	0.000960
18	2.00	2.418554	0.970003	2.999747	0.000250
19	1.99	2.445912	0.969999	3.000057	−0.000060
20	1.98	2.475733	0.969992	3.000772	−0.000775
21	1.97	2.509580	0.969999	3.000140	−0.000143
22	1.96	2.546675	0.969993	3.000689	−0.000691

(continued)

Table 8.2 (continued)

Parameter values					
PROB. = 0.97	RHO = 0.2		BIGH = 2.161		SMHL = 1.89
Pair No.	SMH	SMK	PROB. level	Risk level (%)	Risk error (%)
23	1.95	2.590143	0.970006	2.999365	0.000632
24	1.94	2.638425	0.969997	3.000265	−0.000268
25	1.93	2.696210	0.969998	3.000248	−0.000250
26	1.92	2.766625	0.969998	3.000170	−0.000173
27	1.91	2.857484	0.970003	2.999658	0.000340
28	1.90	2.982852	0.970001	2.999872	0.000125
29	1.89	3.195857	0.970001	2.999854	0.000143
No. of pairs		Mean COMPUT. risk errors		STD. ERR. errors	
29		−0.0000693401		0.0000895875	
Z-ratio				Status of significance	
0.77399				N.S.—not significant	
Parameter values					
PROB. = 0.96	RHO = 0.2		BIGH = 2.0419		SMHL = 1.76
Pair No.	SMH	SMK	PROB. level	Risk level (%)	Risk error (%)
Table No. 8.2-157					
1	2.05	2.034222	0.960007	3.999287	0.000715
2	2.04	2.043802	0.959993	4.000670	−0.000668
3	2.03	2.054260	0.960000	3.999972	0.000030
4	2.02	2.064819	0.959990	4.000974	−0.000972
5	2.01	2.076259	0.959997	4.000264	−0.000262
6	2.00	2.088193	0.960002	3.999794	0.000209
7	1.99	2.100622	0.960003	3.999722	0.000280
8	1.98	2.113547	0.959998	4.000187	−0.000185
9	1.97	2.127362	0.960002	3.999806	0.000197
10	1.96	2.141675	0.959996	4.000390	−0.000387
11	1.95	2.156881	0.959994	4.000616	−0.000614
12	1.94	2.173371	0.960006	3.999448	0.000554
13	1.93	2.190365	0.960000	3.999972	0.000030
14	1.92	2.208648	0.960002	3.999794	0.000209
15	1.91	2.228221	0.960006	3.999388	0.000614
16	1.90	2.249085	0.960008	3.999204	0.000799
17	1.89	2.270852	0.959992	4.000836	−0.000834
18	1.88	2.295086	0.959996	4.000378	−0.000376
19	1.87	2.321008	0.959992	4.000795	−0.000793
20	1.86	2.349401	0.959992	4.000843	−0.000840
21	1.85	2.381049	0.960002	3.999841	0.000161
22	1.84	2.415954	0.960009	3.999132	0.000870
23	1.83	2.454117	0.960000	3.999996	0.000006
24	1.82	2.497886	0.960001	3.999931	0.000072
25	1.81	2.548823	0.960009	3.999084	0.000918
26	1.80	2.606933	0.959994	4.000628	−0.000626
27	1.79	2.679248	0.959997	4.000270	−0.000268

(continued)

8.2 Contents for Table 8.2 (Table No. 8.2-1 to 400)

Table 8.2 (continued)

Parameter values					
PROB. = 0.96	RHO = 0.2		BIGH = 2.0419		SMHL = 1.76
Pair No.	SMH	SMK	PROB. level	Risk level (%)	Risk error (%)
28	1.78	2.770459	0.959992	4.000831	−0.000829
29	1.77	2.899317	0.959999	4.000097	−0.000095
30	1.76	3.112701	0.959994	4.000574	−0.000572
No. of pairs		Mean COMPUT. risk errors		STD. ERR. errors	
30		−0.0000857538		0.0000990319	
Z-ratio				Status of significance	
0.86592				N.S.—not significant	
Parameter values					
PROB. = 0.95	RHO = 0.2		BIGH = 1.9456		SMHL = 1.65
Pair No.	SMH	SMK	PROB. level	Risk level (%)	Risk error (%)
Table No. 8.2-158					
1	1.95	1.941210	0.949998	5.000180	−0.000179
2	1.94	1.951216	0.949997	5.000300	−0.000298
3	1.93	1.961717	0.949999	5.000061	−0.000060
4	1.92	1.972713	0.950003	4.999662	0.000340
5	1.91	1.984207	0.950007	4.999280	0.000721
6	1.90	1.996201	0.950009	4.999066	0.000936
7	1.89	2.008695	0.950008	4.999227	0.000775
8	1.88	2.021692	0.950001	4.999912	0.000089
9	1.87	2.035585	0.950005	4.999477	0.000525
10	1.86	2.049983	0.949999	5.000067	−0.000066
11	1.85	2.065280	0.949999	5.000103	−0.000101
12	1.84	2.081479	0.950000	4.999984	0.000018
13	1.83	2.098580	0.950000	5.000037	−0.000036
14	1.82	2.116586	0.949994	5.000621	−0.000620
15	1.81	2.135890	0.949994	5.000627	−0.000626
16	1.80	2.156884	0.950008	4.999233	0.000769
17	1.79	2.178788	0.950002	4.999817	0.000185
18	1.78	2.202388	0.949998	5.000180	−0.000179
19	1.77	2.228074	0.950001	4.999906	0.000095
20	1.76	2.255850	0.950001	4.999936	0.000066
21	1.75	2.286500	0.950009	4.999119	0.000882
22	1.74	2.319244	0.949993	5.000735	−0.000733
23	1.73	2.356819	0.950005	4.999489	0.000513
24	1.72	2.398446	0.950006	4.999447	0.000554
25	1.71	2.445691	0.950003	4.999691	0.000310
26	1.70	2.500119	0.949995	5.000520	−0.000519
27	1.69	2.566420	0.950008	4.999233	0.000769
28	1.68	2.646939	0.950003	4.999686	0.000316
29	1.67	2.752619	0.949999	5.000109	−0.000107

(continued)

Table 8.2 (continued)

Parameter values					
PROB. = 0.95	RHO = 0.2		BIGH = 1.9456		SMHL = 1.65
Pair No.	SMH	SMK	PROB. level	Risk level (%)	Risk error (%)
30	1.66	2.908461	0.949998	5.000222	−0.000221
31	1.65	3.219156	0.949992	5.000764	−0.000763
No. of pairs		Mean COMPUT. risk errors		STD. ERR. errors	
31		0.0001048669		0.0000852770	
Z-ratio				Status of significance	
1.22972				N.S.—not significant	
Parameter values					
PROB. = 0.94	RHO = 0.2		BIGH = 1.8639		SMHL = 1.56
Pair No.	SMH	SMK	PROB. level	Risk level (%)	Risk error (%)
Table No. 8.2-159					
1	1.87	1.857820	0.940002	5.999786	0.000215
2	1.86	1.867808	0.940004	5.999565	0.000435
3	1.85	1.878100	0.939999	6.000078	−0.000077
4	1.84	1.888892	0.939998	6.000197	−0.000197
5	1.83	1.900186	0.939999	6.000138	−0.000137
6	1.82	1.911984	0.939999	6.000090	−0.000089
7	1.81	1.924483	0.940009	5.999100	0.000900
8	1.80	1.937100	0.939992	6.000793	−0.000793
9	1.79	1.950616	0.939991	6.000877	−0.000876
10	1.78	1.965036	0.940002	5.999762	0.000238
11	1.77	1.979969	0.940001	5.999863	0.000137
12	1.76	1.995808	0.940006	5.999398	0.000602
13	1.75	2.012166	0.939993	6.000722	−0.000721
14	1.74	2.029825	0.939997	6.000334	−0.000334
15	1.73	2.048398	0.939994	6.000638	−0.000638
16	1.72	2.068276	0.939996	6.000358	−0.000358
17	1.71	2.089463	0.939999	6.000138	−0.000137
18	1.70	2.112352	0.940009	5.999065	0.000936
19	1.69	2.136163	0.939992	6.000853	−0.000852
20	1.68	2.162462	0.939995	6.000453	−0.000453
21	1.67	2.191250	0.940008	5.999213	0.000787
22	1.66	2.221751	0.939992	6.000793	−0.000793
23	1.65	2.256310	0.940006	5.999410	0.000590
24	1.64	2.294149	0.940006	5.999369	0.000632
25	1.63	2.336051	0.939994	6.000579	−0.000578
26	1.62	2.384364	0.939997	6.000346	−0.000346
27	1.61	2.440652	0.940004	5.999601	0.000399
28	1.60	2.507264	0.940006	5.999422	0.000578
29	1.59	2.588107	0.939990	6.000972	−0.000972
30	1.58	2.695686	0.939993	6.000680	−0.000679

(continued)

8.2 Contents for Table 8.2 (Table No. 8.2-1 to 400)

Table 8.2 (continued)

Parameter values					
PROB. = 0.94	RHO = 0.2		BIGH = 1.8639		SMHL = 1.56
Pair No.	SMH	SMK	PROB. level	Risk level (%)	Risk error (%)
31	1.57	2.853441	0.939995	6.000477	−0.000477
32	1.56	3.170751	0.940002	5.999786	0.000215
No. of pairs		Mean COMPUT. risk errors		STD. ERR. errors	
32		−0.0000861558		0.0000988571	
Z-ratio				Status of significance	
0.87152				N.S.—not significant	
Parameter values					
PROB. = 0.93	RHO = 0.2		BIGH = 1.7924		SMHL = 1.48
Pair No.	SMH	SMK	PROB. level	Risk level (%)	Risk error (%)
Table No. 8.2-160					
1	1.80	1.784832	0.929995	7.000500	−0.000501
2	1.79	1.794803	0.930001	6.999952	0.000048
3	1.78	1.805082	0.929999	7.000107	−0.000107
4	1.77	1.815865	0.930003	6.999684	0.000316
5	1.76	1.826959	0.929997	7.000268	−0.000268
6	1.75	1.838562	0.929993	7.000655	−0.000656
7	1.74	1.850870	0.930002	6.999773	0.000226
8	1.73	1.863691	0.930008	6.999195	0.000805
9	1.72	1.876832	0.929996	7.000375	−0.000376
10	1.71	1.890880	0.930001	6.999886	0.000113
11	1.70	1.905447	0.929996	7.000441	−0.000441
12	1.69	1.920926	0.930000	7.000047	−0.000048
13	1.68	1.937320	0.930009	6.999153	0.000846
14	1.67	1.954240	0.929997	7.000304	−0.000304
15	1.66	1.972469	0.930003	6.999696	0.000304
16	1.65	1.991621	0.930001	6.999946	0.000054
17	1.64	2.012087	0.930003	6.999660	0.000340
18	1.63	2.033871	0.930005	6.999546	0.000453
19	1.62	2.056974	0.929997	7.000280	−0.000280
20	1.61	2.082183	0.930006	6.999392	0.000608
21	1.60	2.109108	0.930005	6.999552	0.000447
22	1.59	2.138142	0.929997	7.000309	−0.000310
23	1.58	2.170071	0.929998	7.000214	−0.000215
24	1.57	2.205289	0.930002	6.999791	0.000209
25	1.56	2.243796	0.929990	7.000983	−0.000983
26	1.55	2.287551	0.929993	7.000715	−0.000715
27	1.54	2.337730	0.930003	6.999654	0.000346
28	1.53	2.395115	0.929996	7.000363	−0.000364
29	1.52	2.463616	0.929992	7.000816	−0.000817

(continued)

Table 8.2 (continued)

Parameter values					
PROB. = 0.93	RHO = 0.2		BIGH = 1.7924		SMHL = 1.48
Pair No.	SMH	SMK	PROB. level	Risk level (%)	Risk error (%)
30	1.51	2.549489	0.930002	6.999851	0.000149
31	1.50	2.662110	0.930001	6.999874	0.000125
32	1.49	2.831172	0.930007	6.999260	0.000739
33	1.48	3.189490	0.930001	6.999898	0.000101
No. of pairs		Mean COMPUT. risk errors		STD. ERR. errors	
33		−0.0000045580		0.0000787878	
Z-ratio				Status of significance	
0.05785				N.S.—not significant	
Parameter values					
PROB. = 0.92	RHO = 0.2		BIGH = 1.7285		SMHL = 1.41
Pair No.	SMH	SMK	PROB. level	Risk level (%)	Risk error (%)
Table No. 8.2-161					
1	1.73	1.727197	0.920004	7.999623	0.000376
2	1.72	1.737238	0.919996	8.000421	−0.000423
3	1.71	1.747786	0.919996	8.000392	−0.000393
4	1.70	1.758845	0.920003	7.999730	0.000268
5	1.69	1.770221	0.919998	8.000165	−0.000167
6	1.68	1.782111	0.919996	8.000380	−0.000381
7	1.67	1.794518	0.919994	8.000624	−0.000626
8	1.66	1.807639	0.920002	7.999766	0.000232
9	1.65	1.821281	0.920005	7.999492	0.000507
10	1.64	1.835448	0.919999	8.000058	−0.000060
11	1.63	1.850530	0.920008	7.999176	0.000823
12	1.62	1.866142	0.920002	7.999790	0.000209
13	1.61	1.882675	0.920002	7.999790	0.000209
14	1.60	1.900133	0.920004	7.999635	0.000364
15	1.59	1.918517	0.920001	7.999879	0.000119
16	1.58	1.937832	0.919990	8.000965	−0.000966
17	1.57	1.958861	0.920006	7.999378	0.000620
18	1.56	1.980825	0.920000	7.999987	0.000012
19	1.55	2.004510	0.920004	7.999611	0.000387
20	1.54	2.029916	0.920007	7.999277	0.000721
21	1.53	2.057050	0.920000	8.000004	−0.000006
22	1.52	2.086694	0.920002	7.999760	0.000238
23	1.51	2.118851	0.919999	8.000118	−0.000119
24	1.50	2.154308	0.920000	7.999981	0.000018
25	1.49	2.193457	0.919997	8.000285	−0.000286
26	1.48	2.237474	0.919998	8.000165	−0.000167
27	1.47	2.287144	0.919991	8.000928	−0.000930
28	1.46	2.345596	0.920004	7.999593	0.000405
29	1.45	2.414397	0.920003	7.999659	0.000340
30	1.44	2.499018	0.919999	8.000124	−0.000125

(continued)

Table 8.2 (continued)

Parameter values					
PROB. = 0.92	RHO = 0.2		BIGH = 1.7285		SMHL = 1.41
Pair No.	SMH	SMK	PROB. level	Risk level (%)	Risk error (%)
31	1.43	2.611184	0.920008	7.999206	0.000793
32	1.42	2.774335	0.920000	7.999993	0.000006
33	1.41	3.106444	0.919999	8.000130	−0.000131
No. of pairs		Mean COMPUT. risk errors		STD. ERR. errors	
33		0.0000548713		0.0000743814	
Z-ratio				Status of significance	
0.73770				N.S.—not significant	
Parameter values					
PROB. = 0.91	RHO = 0.2		BIGH = 1.6707		SMHL = 1.35
Pair No.	SMH	SMK	PROB. level	Risk level (%)	Risk error (%)
Table No. 8.2-162					
1	1.68	1.661452	0.909996	9.000450	−0.000453
2	1.67	1.671400	0.910005	8.999454	0.000542
3	1.66	1.681469	0.909992	9.000789	−0.000793
4	1.65	1.692246	0.910006	8.999407	0.000590
5	1.64	1.703147	0.909994	9.000624	−0.000626
6	1.63	1.714760	0.910004	8.999616	0.000381
7	1.62	1.726698	0.910001	8.999920	0.000077
8	1.61	1.739157	0.909999	9.000152	−0.000155
9	1.60	1.752336	0.910009	8.999062	0.000936
10	1.59	1.765848	0.910000	8.999956	0.000042
11	1.58	1.780083	0.909999	9.000140	−0.000143
12	1.57	1.795046	0.910000	9.000015	−0.000018
13	1.56	1.810740	0.910000	8.999956	0.000042
14	1.55	1.827361	0.910009	8.999091	0.000906
15	1.54	1.844524	0.909995	9.000456	−0.000459
16	1.53	1.863010	0.910005	8.999467	0.000530
17	1.52	1.882435	0.910007	8.999289	0.000709
18	1.51	1.902799	0.909996	9.000439	−0.000441
19	1.50	1.924888	0.910008	8.999204	0.000793
20	1.49	1.947924	0.909992	9.000767	−0.000769
21	1.48	1.973081	0.910002	8.999813	0.000185
22	1.47	1.999973	0.910004	8.999652	0.000346
23	1.46	2.028994	0.910004	8.999652	0.000346
24	1.45	2.060539	0.910004	8.999622	0.000376
25	1.44	2.095000	0.910002	8.999782	0.000215
26	1.43	2.133165	0.910006	8.999432	0.000566
27	1.42	2.175426	0.910001	8.999920	0.000077
28	1.41	2.223351	0.910003	8.999723	0.000274
29	1.40	2.278116	0.909999	9.000105	−0.000107
30	1.39	2.342460	0.909998	9.000230	−0.000232
31	1.38	2.421074	0.910009	8.999115	0.000882

(continued)

Table 8.2 (continued)

Parameter values					
PROB. = 0.91	RHO = 0.2		BIGH = 1.6707		SMHL = 1.35
Pair No.	SMH	SMK	PROB. level	Risk level (%)	Risk error (%)
32	1.37	2.520212	0.910006	8.999366	0.000632
33	1.36	2.657067	0.910000	8.999956	0.000042
34	1.35	2.886333	0.909998	9.000212	−0.000215
No. of pairs		Mean COMPUT. risk errors		STD. ERR. errors	
34		0.0001450947		0.0000798597	
Z-ratio				Status of significance	
1.81687				N.S.—not significant	
Parameter values					
PROB. = 0.9025	RHO = 0.2		BIGH = 1.6304		SMHL = 1.30
Pair No.	SMH	SMK	PROB. level	Risk level (%)	Risk error (%)
Table No. 8.2-163					
1	1.64	1.620856	0.902491	9.750914	−0.000912
2	1.63	1.630800	0.902502	9.749830	0.000173
3	1.62	1.641062	0.902507	9.749275	0.000727
4	1.61	1.651645	0.902506	9.749431	0.000572
5	1.60	1.662549	0.902496	9.750414	−0.000411
6	1.59	1.673975	0.902493	9.750664	−0.000662
7	1.58	1.685923	0.902495	9.750456	−0.000453
8	1.57	1.698397	0.902499	9.750062	−0.000060
9	1.56	1.711399	0.902503	9.749746	0.000256
10	1.55	1.724932	0.902502	9.749806	0.000197
11	1.54	1.738997	0.902495	9.750509	−0.000507
12	1.53	1.753794	0.902493	9.750676	−0.000674
13	1.52	1.769522	0.902507	9.749281	0.000721
14	1.51	1.785791	0.902503	9.749657	0.000346
15	1.50	1.802995	0.902506	9.749418	0.000584
16	1.49	1.821139	0.902509	9.749108	0.000894
17	1.48	1.840224	0.902507	9.749311	0.000691
18	1.47	1.860255	0.902494	9.750604	−0.000602
19	1.46	1.881821	0.902499	9.750098	−0.000095
20	1.45	1.904728	0.902501	9.749913	0.000089
21	1.44	1.929178	0.902501	9.749878	0.000125
22	1.43	1.955368	0.902500	9.750027	−0.000024
23	1.42	1.983693	0.902503	9.749662	0.000340
24	1.41	2.014157	0.902497	9.750301	−0.000298
25	1.40	2.047545	0.902499	9.750134	−0.000131
26	1.39	2.084252	0.902502	9.749812	0.000191
27	1.38	2.124672	0.902496	9.750408	−0.000405
28	1.37	2.169982	0.902491	9.750944	−0.000942
29	1.36	2.221749	0.902493	9.750729	−0.000727
30	1.35	2.282321	0.902508	9.749174	0.000829
31	1.34	2.353265	0.902498	9.750176	−0.000173

(continued)

Table 8.2 (continued)

Parameter values					
PROB. = 0.9025	RHO = 0.2		BIGH = 1.6304		SMHL = 1.30
Pair No.	SMH	SMK	PROB. level	Risk level (%)	Risk error (%)
32	1.33	2.441618	0.902499	9.750116	−0.000113
33	1.32	2.557540	0.902492	9.750784	−0.000781
34	1.31	2.732287	0.902502	9.749770	0.000232
35	1.30	3.107271	0.902503	9.749669	0.000334
No. of pairs		Mean COMPUT. risk errors		STD. ERR. errors	
35		−0.0000185437		0.0000860590	
Z-ratio				Status of significance	
0.21548				N.S.—not significant	
Parameter values					
PROB. = 0.9	RHO = 0.2		BIGH = 1.6175		SMHL = 1.29
Pair No.	SMH	SMK	PROB. level	Risk level (%)	Risk error (%)
Table No. 8.2-164					
1	1.62	1.615004	0.900002	9.999788	0.000215
2	1.61	1.625035	0.899996	10.000390	−0.000387
3	1.60	1.635582	0.900003	9.999728	0.000274
4	1.59	1.646452	0.900001	9.999871	0.000131
5	1.58	1.657843	0.900008	9.999191	0.000811
6	1.57	1.669562	0.900003	9.999722	0.000280
7	1.56	1.681807	0.900001	9.999901	0.000101
8	1.55	1.694580	0.900000	10.000030	−0.000024
9	1.54	1.707884	0.899996	10.000370	−0.000370
10	1.53	1.721918	0.900003	9.999657	0.000346
11	1.52	1.736488	0.900002	9.999824	0.000179
12	1.51	1.751794	0.900003	9.999693	0.000310
13	1.50	1.767837	0.900003	9.999699	0.000304
14	1.49	1.784621	0.899997	10.000260	−0.000256
15	1.48	1.802540	0.900009	9.999096	0.000906
16	1.47	1.821206	0.900004	9.999585	0.000417
17	1.46	1.840818	0.899991	10.000860	−0.000858
18	1.45	1.861966	0.900000	9.999961	0.000042
19	1.44	1.884457	0.900009	9.999055	0.000948
20	1.43	1.908100	0.899999	10.000100	−0.000101
21	1.42	1.933484	0.899993	10.000730	−0.000727
22	1.41	1.961005	0.899998	10.000220	−0.000221
23	1.40	1.990665	0.899999	10.000070	−0.000072
24	1.39	2.022859	0.899999	10.000090	−0.000089
25	1.38	2.057983	0.899995	10.000540	−0.000536
26	1.37	2.097212	0.900008	9.999216	0.000787
27	1.36	2.140160	0.899994	10.000600	−0.000602
28	1.35	2.189567	0.900008	9.999174	0.000829
29	1.34	2.245436	0.899998	10.000190	−0.000191
30	1.33	2.311678	0.900003	9.999728	0.000274

(continued)

Table 8.2 (continued)

Parameter values					
PROB. = 0.9	RHO = 0.2		BIGH = 1.6175		SMHL = 1.29
Pair No.	SMH	SMK	PROB. level	Risk level (%)	Risk error (%)
31	1.32	2.391424	0.899994	10.000590	−0.000590
32	1.31	2.494054	0.899999	10.000150	−0.000143
33	1.30	2.637542	0.900004	9.999627	0.000376
34	1.29	2.884393	0.900009	9.999144	0.000858
No. of pairs		Mean COMPUT. risk errors		STD. ERR. errors	
34		0.0000919615		0.0000811007	
Z-ratio				Status of significance	
1.13392				N.S.—not significant	
Parameter values					
PROB. = 0.85	RHO = 0.2		BIGH = 1.3985		SMHL = 1.04
Pair No.	SMH	SMK	PROB. level	Risk level (%)	Risk error (%)
Table No. 8.2-165					
1	1.40	1.397002	0.850010	14.999000	0.000995
2	1.39	1.407052	0.850001	14.999900	0.000095
3	1.38	1.417443	0.849991	15.000930	−0.000930
4	1.37	1.428374	0.850002	14.999830	0.000167
5	1.36	1.439652	0.850007	14.999330	0.000668
6	1.35	1.451281	0.850004	14.999630	0.000370
7	1.34	1.463265	0.849991	15.000920	−0.000918
8	1.33	1.475899	0.850000	15.000030	−0.000036
9	1.32	1.488895	0.849993	15.000720	−0.000721
10	1.31	1.502549	0.850001	14.999940	0.000054
11	1.30	1.516768	0.850008	14.999230	0.000769
12	1.29	1.531360	0.849991	15.000950	−0.000954
13	1.28	1.546916	0.850006	14.999410	0.000584
14	1.27	1.563049	0.850008	14.999190	0.000811
15	1.26	1.579763	0.849994	15.000620	−0.000620
16	1.25	1.597454	0.849995	15.000510	−0.000513
17	1.24	1.616127	0.850004	14.999620	0.000381
18	1.23	1.635591	0.849997	15.000350	−0.000352
19	1.22	1.656241	0.849999	15.000060	−0.000060
20	1.21	1.678084	0.850003	14.999690	0.000310
21	1.20	1.701124	0.849998	15.000200	−0.000203
22	1.19	1.725758	0.850003	14.999710	0.000292
23	1.18	1.751991	0.850004	14.999560	0.000441
24	1.17	1.780024	0.850002	14.999830	0.000173
25	1.16	1.810255	0.850003	14.999660	0.000340
26	1.15	1.842885	0.850001	14.999890	0.000113
27	1.14	1.878506	0.850005	14.999520	0.000483
28	1.13	1.917517	0.850006	14.999420	0.000578
29	1.12	1.960705	0.850008	14.999250	0.000751
30	1.11	2.008859	0.850002	14.999850	0.000149

(continued)

8.2 Contents for Table 8.2 (Table No. 8.2-1 to 400)

Table 8.2 (continued)

Parameter values					
PROB. = 0.85	RHO = 0.2		BIGH = 1.3985		SMHL = 1.04
Pair No.	SMH	SMK	PROB. level	Risk level (%)	Risk error (%)
31	1.10	2.063549	0.849996	15.000390	−0.000387
32	1.09	2.127126	0.850000	15.000050	−0.000048
33	1.08	2.202334	0.849992	15.000810	−0.000811
34	1.07	2.295821	0.849995	15.000510	−0.000513
35	1.06	2.419314	0.849998	15.000200	−0.000197
36	1.05	2.604855	0.850004	14.999620	0.000381
37	1.04	3.018078	0.850002	14.999770	0.000232
No. of pairs		Mean COMPUT. risk errors		STD. ERR. errors	
37		0.0000494091		0.0000852019	
Z-ratio				Status of significance	
0.57991				N.S.—not significant	
Parameter values					
PROB. = 0.8	RHO = 0.2		BIGH = 1.2257		SMHL = 0.85
Pair No.	SMH	SMK	PROB. level	Risk level (%)	Risk error (%)
Table No. 8.2-166					
1	1.23	1.221415	0.799992	20.000790	−0.000793
2	1.22	1.231427	0.799991	20.000950	−0.000954
3	1.21	1.241799	0.799994	20.000550	−0.000554
4	1.20	1.252536	0.800002	19.999800	0.000197
5	1.19	1.263545	0.799996	20.000360	−0.000364
6	1.18	1.274928	0.799991	20.000950	−0.000954
7	1.17	1.286786	0.799996	20.000390	−0.000387
8	1.16	1.299028	0.799996	20.000360	−0.000364
9	1.15	1.311754	0.800002	19.999760	0.000238
10	1.14	1.324874	0.799998	20.000240	−0.000244
11	1.13	1.338489	0.799993	20.000740	−0.000745
12	1.12	1.352704	0.799996	20.000360	−0.000358
13	1.11	1.367522	0.800005	19.999520	0.000477
14	1.10	1.382854	0.800001	19.999860	0.000143
15	1.09	1.398802	0.799994	20.000580	−0.000584
16	1.08	1.415568	0.800001	19.999870	0.000131
17	1.07	1.432963	0.799994	20.000590	−0.000590
18	1.06	1.451286	0.800000	20.000020	−0.000024
19	1.05	1.470449	0.800000	19.999960	0.000042
20	1.04	1.490652	0.800009	19.999120	0.000876
21	1.03	1.511708	0.799997	20.000310	−0.000310
22	1.02	1.534016	0.799995	20.000520	−0.000525
23	1.01	1.557778	0.800008	19.999180	0.000823
24	1.00	1.582809	0.800007	19.999310	0.000691
25	0.99	1.609312	0.799995	20.000470	−0.000471
26	0.98	1.637688	0.799990	20.000970	−0.000966
27	0.97	1.668336	0.800002	19.999790	0.000215

(continued)

Table 8.2 (continued)

Parameter values					
PROB. = 0.8	RHO = 0.2		BIGH = 1.2257		SMHL = 0.85
Pair No.	SMH	SMK	PROB. level	Risk level (%)	Risk error (%)
28	0.96	1.701266	0.800007	19.999260	0.000739
29	0.95	1.736880	0.800008	19.999210	0.000787
30	0.94	1.775578	0.799999	20.000110	−0.000107
31	0.93	1.818154	0.799991	20.000860	−0.000864
32	0.92	1.865791	0.800001	19.999950	0.000054
33	0.91	1.919283	0.800004	19.999640	0.000358
34	0.90	1.980204	0.799995	20.000550	−0.000548
35	0.89	2.052085	0.800005	19.999460	0.000542
36	0.88	2.138455	0.800000	20.000030	−0.000030
37	0.87	2.248702	0.800004	19.999590	0.000411
38	0.86	2.401594	0.800002	19.999750	0.000244
39	0.85	2.661208	0.799991	20.000950	−0.000954
No. of pairs		Mean COMPUT. risk errors		STD. ERR. errors	
39		−0.0001180172		0.0000865634	
Z-ratio				Status of significance	
1.36336				N.S.—not significant	
Parameter values					
PROB. = 0.75	RHO = 0.2		BIGH = 1.0785		SMHL = 0.68
Pair No.	SMH	SMK	PROB. level	Risk level (%)	Risk error (%)
Table No. 8.2-167					
1	1.08	1.076905	0.750005	24.999500	0.000501
2	1.07	1.086970	0.749998	25.000190	−0.000191
3	1.06	1.097421	0.750005	24.999520	0.000483
4	1.05	1.108164	0.750006	24.999410	0.000596
5	1.04	1.119206	0.750000	24.999980	0.000024
6	1.03	1.130651	0.750002	24.999770	0.000226
7	1.02	1.142504	0.750010	24.999030	0.000972
8	1.01	1.154673	0.750005	24.999550	0.000447
9	1.00	1.167264	0.750000	24.999990	0.000012
10	0.99	1.180380	0.750010	24.999050	0.000948
11	0.98	1.193834	0.749999	25.000100	−0.000101
12	0.97	1.207828	0.749996	25.000390	−0.000393
13	0.96	1.222369	0.749997	25.000330	−0.000328
14	0.95	1.237467	0.749998	25.000250	−0.000250
15	0.94	1.253227	0.750007	24.999260	0.000739
16	0.93	1.269462	0.749995	25.000470	−0.000465
17	0.92	1.286572	0.750009	24.999080	0.000924
18	0.91	1.304177	0.749992	25.000790	−0.000787
19	0.90	1.322773	0.750001	24.999910	0.000095
20	0.89	1.342080	0.749992	25.000820	−0.000817
21	0.88	1.362400	0.749993	25.000690	−0.000691
22	0.87	1.383843	0.750007	24.999350	0.000650

(continued)

8.2 Contents for Table 8.2 (Table No. 8.2-1 to 400)

Table 8.2 (continued)

Parameter values					
PROB. = 0.75	RHO = 0.2		BIGH = 1.0785		SMHL = 0.68
Pair No.	SMH	SMK	PROB. level	Risk level (%)	Risk error (%)
23	0.86	1.406225	0.750001	24.999950	0.000048
24	0.85	1.429949	0.750007	24.999280	0.000721
25	0.84	1.454834	0.749994	25.000570	−0.000566
26	0.83	1.481478	0.750006	24.999380	0.000626
27	0.82	1.509701	0.750007	24.999330	0.000674
28	0.81	1.539713	0.749997	25.000340	−0.000340
29	0.80	1.572116	0.750004	24.999570	0.000429
30	0.79	1.606927	0.750003	24.999730	0.000274
31	0.78	1.644749	0.750006	24.999360	0.000644
32	0.77	1.685991	0.750006	24.999450	0.000554
33	0.76	1.731258	0.749995	25.000510	−0.000507
34	0.75	1.781937	0.750000	25.000040	−0.000036
35	0.74	1.839028	0.749998	25.000200	−0.000197
36	0.73	1.904701	0.749998	25.000160	−0.000161
37	0.72	1.981909	0.749993	25.000750	−0.000745
38	0.71	2.076537	0.749996	25.000390	−0.000393
39	0.70	2.199160	0.750001	24.999870	0.000131
40	0.69	2.376366	0.750008	24.999170	0.000834
41	0.68	2.716781	0.749998	25.000170	−0.000167
No. of pairs		Mean COMPUT. risk errors		STD. ERR. errors	
41		0.0001051596		0.0000806828	
Z-ratio				Status of significance	
1.30337				N.S.—not significant	
Parameter values					
PROB. = 0.7	RHO = 0.2		BIGH = 0.947		SMHL = 0.53
Pair No.	SMH	SMK	PROB. level	Risk level (%)	Risk error (%)
Table No. 8.2-168					
1	0.95	0.943814	0.700009	29.999060	0.000948
2	0.94	0.953857	0.700007	29.999280	0.000721
3	0.93	0.964115	0.699990	30.000960	−0.000960
4	0.92	0.974792	0.699996	30.000400	−0.000393
5	0.91	0.985797	0.700004	29.999650	0.000352
6	0.90	0.997040	0.699993	30.000680	−0.000674
7	0.89	1.008725	0.700000	29.999980	0.000024
8	0.88	1.020761	0.700005	29.999540	0.000459
9	0.87	1.033159	0.700005	29.999540	0.000465
10	0.86	1.045926	0.699999	30.000130	−0.000125
11	0.85	1.059171	0.700002	29.999840	0.000167
12	0.84	1.072806	0.699994	30.000560	−0.000554
13	0.83	1.086938	0.699991	30.000900	−0.000900
14	0.82	1.101677	0.700004	29.999600	0.000399
15	0.81	1.116840	0.699999	30.000120	−0.000119

(continued)

Table 8.2 (continued)

Parameter values					
PROB. = 0.7	RHO = 0.2		BIGH = 0.947		SMHL = 0.53
Pair No.	SMH	SMK	PROB. level	Risk level (%)	Risk error (%)
16	0.80	1.132632	0.700003	29.999700	0.000304
17	0.79	1.148971	0.699998	30.000210	−0.000203
18	0.78	1.165966	0.699994	30.000600	−0.000602
19	0.77	1.183730	0.700000	29.999960	0.000042
20	0.76	1.202180	0.699998	30.000230	−0.000226
21	0.75	1.221526	0.700007	29.999290	0.000715
22	0.74	1.241591	0.699996	30.000410	−0.000411
23	0.73	1.262685	0.699995	30.000470	−0.000471
24	0.72	1.284826	0.699996	30.000390	−0.000387
25	0.71	1.308131	0.700001	29.999940	0.000060
26	0.70	1.332718	0.700008	29.999170	0.000834
27	0.69	1.358512	0.699997	30.000310	−0.000304
28	0.68	1.386024	0.700005	29.999530	0.000477
29	0.67	1.415083	0.699995	30.000530	−0.000530
30	0.66	1.446301	0.700005	29.999530	0.000477
31	0.65	1.479510	0.699994	30.000650	−0.000650
32	0.64	1.515521	0.700003	29.999710	0.000292
33	0.63	1.554365	0.699999	30.000090	−0.000083
34	0.62	1.596856	0.700003	29.999660	0.000340
35	0.61	1.643616	0.700007	29.999270	0.000727
36	0.60	1.695462	0.700001	29.999880	0.000125
37	0.59	1.753999	0.700000	30.000040	−0.000036
38	0.58	1.821222	0.699999	30.000090	−0.000083
39	0.57	1.900302	0.699998	30.000170	−0.000173
40	0.56	1.996756	0.699998	30.000250	−0.000250
41	0.55	2.121186	0.699993	30.000700	−0.000697
42	0.54	2.301381	0.700009	29.999100	0.000900
43	0.53	2.645216	0.700004	29.999570	0.000435
No. of pairs			Mean COMPUT. risk errors	STD. ERR. errors	
43			0.0000097535	0.0000752715	
Z-ratio				Status of significance	
0.12958				N.S.—not significant	
Parameter values					
PROB. = 0.65	RHO = 0.2		BIGH = 0.8255		SMHL = 0.39
Pair No.	SMH	SMK	PROB. level	Risk level (%)	Risk error (%)
Table No. 8.2-169					
1	0.83	0.821122	0.650002	34.999810	0.000191
2	0.82	0.831135	0.650001	34.999870	0.000131
3	0.81	0.841394	0.649995	35.000510	−0.000507
4	0.80	0.852008	0.650003	34.999750	0.000256
5	0.79	0.862888	0.650003	34.999670	0.000334
6	0.78	0.874045	0.649997	35.000310	−0.000304

(continued)

8.2 Contents for Table 8.2 (Table No. 8.2-1 to 400)

Table 8.2 (continued)

Parameter values					
PROB. = 0.65	RHO = 0.2		BIGH = 0.8255		SMHL = 0.39
Pair No.	SMH	SMK	PROB. level	Risk level (%)	Risk error (%)
7	0.77	0.885586	0.650002	34.999790	0.000215
8	0.76	0.897426	0.649999	35.000120	−0.000119
9	0.75	0.909674	0.650005	34.999500	0.000501
10	0.74	0.922246	0.650001	34.999880	0.000119
11	0.73	0.935251	0.650005	34.999540	0.000465
12	0.72	0.948607	0.649996	35.000380	−0.000370
13	0.71	0.962523	0.650010	34.999030	0.000972
14	0.70	0.976821	0.650008	34.999170	0.000834
15	0.69	0.991515	0.649992	35.000830	−0.000823
16	0.68	1.006820	0.649991	35.000910	−0.000906
17	0.67	1.022754	0.650003	34.999730	0.000268
18	0.66	1.039141	0.649993	35.000690	−0.000685
19	0.65	1.056197	0.649991	35.000900	−0.000900
20	0.64	1.073946	0.649992	35.000770	−0.000763
21	0.63	1.092409	0.649994	35.000630	−0.000632
22	0.62	1.111613	0.649991	35.000910	−0.000906
23	0.61	1.131780	0.650007	34.999350	0.000656
24	0.60	1.152742	0.650008	34.999250	0.000751
25	0.59	1.174629	0.650002	34.999810	0.000191
26	0.58	1.197570	0.649995	35.000460	−0.000453
27	0.57	1.221698	0.649992	35.000800	−0.000799
28	0.56	1.247149	0.649993	35.000670	−0.000662
29	0.55	1.274156	0.650010	34.999050	0.000954
30	0.54	1.302569	0.650006	34.999410	0.000590
31	0.53	1.332825	0.650009	34.999110	0.000894
32	0.52	1.364973	0.650001	34.999920	0.000083
33	0.51	1.399458	0.649997	35.000260	−0.000262
34	0.50	1.436728	0.650006	34.999390	0.000614
35	0.49	1.476847	0.649995	35.000460	−0.000459
36	0.48	1.520859	0.650003	34.999720	0.000286
37	0.47	1.569034	0.649993	35.000740	−0.000733
38	0.46	1.622819	0.649995	35.000530	−0.000525
39	0.45	1.683474	0.649995	35.000460	−0.000453
40	0.44	1.753437	0.650009	34.999140	0.000864
41	0.43	1.835548	0.650006	34.999360	0.000638
42	0.42	1.935780	0.649998	35.000230	−0.000226
43	0.41	2.066760	0.650009	34.999110	0.000894

(continued)

Table 8.2 (continued)

Parameter values					
PROB. = 0.65	RHO = 0.2		BIGH = 0.8255		SMHL = 0.39
Pair No.	SMH	SMK	PROB. level	Risk level (%)	Risk error (%)
44	0.40	2.275499	0.649998	35.000200	−0.000197
45	0.39	2.653556	0.649994	35.000570	−0.000566
No. of pairs		Mean COMPUT. risk errors		STD. ERR. errors	
45		−0.0000119209		0.0000880204	
Z-ratio				Status of significance	
0.13543				N.S.—not significant	
Parameter values					
PROB. = 0.6	RHO = 0.2		BIGH = 0.7109		SMHL = 0.26
Pair No.	SMH	SMK	PROB. level	Risk level (%)	Risk error (%)
Table No. 8.2-170					
1	0.72	0.701915	0.599999	40.000150	−0.000155
2	0.71	0.711801	0.599996	40.000370	−0.000370
3	0.70	0.721970	0.600000	40.000050	−0.000048
4	0.69	0.732433	0.600008	39.999170	0.000834
5	0.68	0.743107	0.600002	39.999770	0.000226
6	0.67	0.754101	0.600003	39.999710	0.000286
7	0.66	0.765384	0.600000	39.999990	0.000006
8	0.65	0.777018	0.600005	39.999550	0.000453
9	0.64	0.788971	0.600007	39.999350	0.000656
10	0.63	0.801212	0.599997	40.000310	−0.000310
11	0.62	0.813858	0.599996	40.000410	−0.000405
12	0.61	0.826927	0.600003	39.999680	0.000322
13	0.60	0.840345	0.600001	39.999940	0.000054
14	0.59	0.854230	0.600007	39.999310	0.000685
15	0.58	0.868511	0.600004	39.999580	0.000417
16	0.57	0.883211	0.599993	40.000660	−0.000662
17	0.56	0.898556	0.600009	39.999080	0.000918
18	0.55	0.914280	0.600000	39.999990	0.000006
19	0.54	0.930613	0.600000	39.999990	0.000012
20	0.53	0.947490	0.599993	40.000710	−0.000709
21	0.52	0.965046	0.599994	40.000600	−0.000596
22	0.51	0.983321	0.600002	39.999750	0.000244
23	0.50	1.002261	0.600002	39.999770	0.000226
24	0.49	1.022010	0.600007	39.999270	0.000727
25	0.48	1.042521	0.600002	39.999780	0.000221
26	0.47	1.063945	0.599999	40.000070	−0.000072
27	0.46	1.086344	0.599996	40.000380	−0.000381
28	0.45	1.109782	0.599990	40.000980	−0.000983
29	0.44	1.134525	0.600002	39.999810	0.000191
30	0.43	1.160455	0.600002	39.999830	0.000167
31	0.42	1.187852	0.600007	39.999260	0.000739
32	0.41	1.216713	0.600001	39.999950	0.000048

(continued)

8.2 Contents for Table 8.2 (Table No. 8.2-1 to 400)

Table 8.2 (continued)

Parameter values					
PROB. = 0.6	RHO = 0.2		BIGH = 0.7109		SMHL = 0.26
Pair No.	SMH	SMK	PROB. level	Risk level (%)	Risk error (%)
33	0.40	1.247431	0.600004	39.999620	0.000381
34	0.39	1.280022	0.599995	40.000460	−0.000465
35	0.38	1.315100	0.600008	39.999220	0.000781
36	0.37	1.352606	0.600003	39.999660	0.000340
37	0.36	1.393086	0.599994	40.000570	−0.000572
38	0.35	1.437297	0.599997	40.000280	−0.000286
39	0.34	1.485821	0.600000	40.000020	−0.000024
40	0.33	1.539652	0.600003	39.999660	0.000340
41	0.32	1.600010	0.599999	40.000060	−0.000066
42	0.31	1.697830	0.600004	39.999610	0.000387
43	0.30	1.775609	0.600004	39.999620	0.000376
44	0.29	1.870418	0.600003	39.999700	0.000298
45	0.28	1.992421	0.600003	39.999740	0.000256
46	0.27	2.165520	0.599998	40.000170	−0.000167
47	0.26	2.481261	0.599998	40.000170	−0.000173
No. of pairs		Mean COMPUT. risk errors		STD. ERR. errors	
47		0.0000865509		0.0000630077	
Z-ratio				Status of significance	
1.37366				N.S.—not significant	
Parameter values					
PROB. = 0.55	RHO = 0.2		BIGH = 0.6005		SMHL = 0.13
Pair No.	SMH	SMK	PROB. level	Risk level (%)	Risk error (%)
Table No. 8.2-171					
1	0.61	0.591001	0.550000	45.000020	−0.000018
2	0.60	0.600903	0.550003	44.999740	0.000262
3	0.59	0.610992	0.549992	45.000810	−0.000805
4	0.58	0.621432	0.550003	44.999720	0.000280
5	0.57	0.632046	0.549993	45.000670	−0.000668
6	0.56	0.643050	0.550009	44.999130	0.000876
7	0.55	0.654270	0.550008	44.999190	0.000811
8	0.54	0.665727	0.549995	45.000540	−0.000536
9	0.53	0.677546	0.549992	45.000840	−0.000840
10	0.52	0.689751	0.550001	44.999910	0.000089
11	0.51	0.702274	0.550005	44.999470	0.000530
12	0.50	0.715146	0.550008	44.999190	0.000811
13	0.49	0.728301	0.549993	45.000690	−0.000685
14	0.48	0.741973	0.550002	44.999760	0.000244
15	0.47	0.756004	0.550002	44.999840	0.000161
16	0.46	0.770437	0.549995	45.000510	−0.000507
17	0.45	0.785416	0.550005	44.999530	0.000477
18	0.44	0.800796	0.550000	45.000010	−0.000012
19	0.43	0.816730	0.550003	44.999700	0.000304

(continued)

Table 8.2 (continued)

Parameter values					
PROB. = 0.55	RHO = 0.2		BIGH = 0.6005		SMHL = 0.13
Pair No.	SMH	SMK	PROB. level	Risk level (%)	Risk error (%)
20	0.42	0.833181	0.550002	44.999770	0.000226
21	0.41	0.850215	0.550003	44.999690	0.000316
22	0.40	0.867809	0.549996	45.000400	−0.000399
23	0.39	0.886139	0.550003	44.999710	0.000292
24	0.38	0.905100	0.550000	45.000020	−0.000018
25	0.37	0.924790	0.549995	45.000530	−0.000530
26	0.36	0.945417	0.550010	44.999010	0.000989
27	0.35	0.966711	0.549997	45.000320	−0.000316
28	0.34	0.989104	0.550007	44.999290	0.000715
29	0.33	1.012454	0.550010	44.999050	0.000954
30	0.32	1.036836	0.550003	44.999730	0.000268
31	0.31	1.062444	0.549998	45.000160	−0.000155
32	0.30	1.089500	0.550008	44.999190	0.000817
33	0.29	1.117862	0.549998	45.000160	−0.000161
34	0.28	1.148013	0.550005	44.999460	0.000542
35	0.27	1.179891	0.550000	44.999990	0.000006
36	0.26	1.213875	0.549998	45.000230	−0.000226
37	0.25	1.250211	0.549993	45.000660	−0.000656
38	0.24	1.327108	0.550006	44.999380	0.000626
39	0.23	1.366067	0.549995	45.000540	−0.000542
40	0.22	1.409011	0.549994	45.000650	−0.000650
41	0.21	1.456594	0.549992	45.000850	−0.000852
42	0.20	1.510029	0.549998	45.000200	−0.000203
43	0.19	1.570549	0.549995	45.000480	−0.000483
44	0.18	1.640831	0.550008	44.999240	0.000763
45	0.17	1.723518	0.549992	45.000800	−0.000793
46	0.16	1.825622	0.549991	45.000910	−0.000912
47	0.15	1.959466	0.549999	45.000100	−0.000101
48	0.14	2.155398	0.549993	45.000730	−0.000727
49	0.13	2.561342	0.549993	45.000710	−0.000709
No. of pairs		Mean COMPUT. risk errors		STD. ERR. errors	
49		−0.0000228882		0.0000799146	
Z-ratio				Status of significance	
0.28641				N.S.—not significant	
Parameter values					
PROB. = 0.5	RHO = 0.2		BIGH = 0.4921		SMHL = 0.01
Pair No.	SMH	SMK	PROB. level	Risk level (%)	Risk error (%)
Table No. 8.2-172					
1	0.50	0.484374	0.500000	49.999990	0.000012
2	0.49	0.494307	0.500000	50.000000	0.000000
3	0.48	0.504505	0.500006	49.999380	0.000626
4	0.47	0.514897	0.500000	49.999960	0.000048

(continued)

Table 8.2 (continued)

Parameter values					
PROB. = 0.5	RHO = 0.2		BIGH = 0.4921		SMHL = 0.01
Pair No.	SMH	SMK	PROB. level	Risk level (%)	Risk error (%)
5	0.46	0.525561	0.499998	50.000160	−0.000158
6	0.45	0.536479	0.499994	50.000620	−0.000614
7	0.44	0.547732	0.500003	49.999750	0.000250
8	0.43	0.559262	0.500009	49.999130	0.000870
9	0.42	0.571011	0.499998	50.000220	−0.000218
10	0.41	0.583120	0.499998	50.000220	−0.000215
11	0.40	0.595543	0.499995	50.000480	−0.000477
12	0.39	0.608331	0.499998	50.000170	−0.000167
13	0.38	0.621449	0.499996	50.000370	−0.000367
14	0.37	0.634962	0.499999	50.000070	−0.000075
15	0.36	0.648845	0.499999	50.000150	−0.000149
16	0.35	0.663181	0.500006	49.999370	0.000638
17	0.34	0.677867	0.499999	50.000070	−0.000069
18	0.33	0.693005	0.499993	50.000720	−0.000721
19	0.32	0.708710	0.500003	49.999720	0.000286
20	0.31	0.724821	0.499996	50.000440	−0.000435
21	0.30	0.741583	0.500009	49.999070	0.000930
22	0.29	0.758773	0.500000	50.000030	−0.000024
23	0.28	0.776584	0.499994	50.000560	−0.000560
24	0.27	0.795139	0.500007	49.999350	0.000650
25	0.26	0.814206	0.499992	50.000800	−0.000796
26	0.25	0.834078	0.499990	50.000970	−0.000969
27	0.24	0.854810	0.500000	49.999990	0.000012
28	0.23	0.876316	0.500000	49.999960	0.000048
29	0.22	0.898784	0.500008	49.999160	0.000846
30	0.21	0.922098	0.499998	50.000190	−0.000191
31	0.20	0.946553	0.499998	50.000220	−0.000218
32	0.19	1.021999	0.500008	49.999240	0.000763
33	0.18	1.044759	0.499997	50.000280	−0.000280
34	0.17	1.069209	0.500008	49.999170	0.000834
35	0.16	1.095154	0.499996	50.000380	−0.000378
36	0.15	1.123105	0.499999	50.000060	−0.000060
37	0.14	1.153167	0.500003	49.999750	0.000250
38	0.13	1.185637	0.500008	49.999210	0.000793
39	0.12	1.220751	0.500006	49.999380	0.000626
40	0.11	1.258895	0.499996	50.000390	−0.000387
41	0.10	1.300917	0.500005	49.999500	0.000501
42	0.09	1.347133	0.500002	49.999760	0.000244
43	0.08	1.398508	0.499995	50.000530	−0.000530
44	0.07	1.456721	0.500009	49.999130	0.000876
45	0.06	1.522748	0.499995	50.000530	−0.000525
46	0.05	1.600264	0.500004	49.999560	0.000441

(continued)

Table 8.2 (continued)

Parameter values						
PROB. = 0.5	RHO = 0.2		BIGH = 0.4921		SMHL = 0.01	
Pair No.	SMH	SMK	PROB. level	Risk level (%)	Risk error (%)	
47	0.04	1.693097	0.500003	49.999690	0.000316	
48	0.03	1.809533	0.499997	50.000290	−0.000283	
49	0.02	1.968150	0.500004	49.999630	0.000370	
50	0.01	2.223529	0.500007	49.999310	0.000691	
No. of pairs		Mean COMPUT. risk errors		STD. ERR. errors		
50		0.0000599553		0.0000699620		
Z-ratio				Status of significance		
0.85697				N.S.—not significant		
Parameter values						
PROB. = 0.99	RHO = 0.1		BIGH = 2.574		SMHL = 2.33	
Pair No.	SMH	SMK	PROB. level	Risk level (%)	Risk error (%)	
Table No. 8.2-173						
1	2.58	2.568014	0.990001	0.999939	0.000060	
2	2.57	2.578006	0.990001	0.999886	0.000113	
3	2.56	2.588076	0.989995	1.000464	−0.000465	
4	2.55	2.599788	0.990004	0.999582	0.000417	
5	2.54	2.611580	0.990005	0.999469	0.000530	
6	2.53	2.623452	0.989999	1.000142	−0.000143	
7	2.52	2.636969	0.990003	0.999689	0.000310	
8	2.51	2.650570	0.989998	1.000154	−0.000155	
9	2.50	2.665815	0.990002	0.999785	0.000215	
10	2.49	2.681146	0.989996	1.000452	−0.000453	
11	2.48	2.698125	0.989995	1.000547	−0.000548	
12	2.47	2.716754	0.989996	1.000387	−0.000387	
13	2.46	2.737033	0.990000	1.000011	−0.000012	
14	2.45	2.758963	0.990003	0.999713	0.000286	
15	2.44	2.782546	0.990003	0.999731	0.000268	
16	2.43	2.807783	0.989998	1.000243	−0.000244	
17	2.42	2.836237	0.989996	1.000381	−0.000381	
18	2.41	2.867910	0.989994	1.000583	−0.000584	
19	2.40	2.904365	0.989996	1.000404	−0.000405	
20	2.39	2.947165	0.990003	0.999743	0.000256	
21	2.38	2.996313	0.990004	0.999594	0.000405	
22	2.37	3.051809	0.989992	1.000845	−0.000846	
23	2.36	3.126155	0.989996	1.000440	−0.000441	
24	2.35	3.225600	0.989999	1.000136	−0.000137	

(continued)

8.2 Contents for Table 8.2 (Table No. 8.2-1 to 400)

Table 8.2 (continued)

Parameter values					
PROB. = 0.99	RHO = 0.1		BIGH = 2.574		SMHL = 2.33
Pair No.	SMH	SMK	PROB. level	Risk level (%)	Risk error (%)
25	2.34	3.381399	0.990006	0.999403	0.000596
26	2.33	3.718551	0.989999	1.000071	−0.000072
No. of pairs		Mean COMPUT. risk errors		STD. ERR. errors	
26		−0.0000673312		0.0000735054	
Z-ratio				Status of significance	
0.91600				N.S.—not significant	
Parameter values					
PROB. = 0.98	RHO = 0.1		BIGH = 2.3228		SMHL = 2.06
Pair No.	SMH	SMK	PROB. level	Risk level (%)	Risk error (%)
Table No. 8.2-174					
1	2.33	2.315622	0.980004	1.999652	0.000346
2	2.32	2.325603	0.980006	1.999432	0.000566
3	2.31	2.335671	0.979998	2.000237	−0.000238
4	2.30	2.346607	0.979999	2.000123	−0.000125
5	2.29	2.358413	0.980007	1.999283	0.000715
6	2.28	2.370309	0.980003	1.999700	0.000298
7	2.27	2.383078	0.980004	1.999617	0.000381
8	2.26	2.396720	0.980008	1.999205	0.000793
9	2.25	2.410456	0.979997	2.000302	−0.000304
10	2.24	2.425848	0.980003	1.999676	0.000322
11	2.23	2.441337	0.979992	2.000809	−0.000811
12	2.22	2.458485	0.979993	2.000737	−0.000739
13	2.21	2.477295	0.980002	1.999790	0.000209
14	2.20	2.496986	0.980003	1.999706	0.000292
15	2.19	2.518340	0.980006	1.999390	0.000608
16	2.18	2.540579	0.979996	2.000421	−0.000423
17	2.17	2.566046	0.980004	1.999581	0.000417
18	2.16	2.593183	0.980003	1.999736	0.000262
19	2.15	2.623551	0.980007	1.999325	0.000674
20	2.14	2.655590	0.979991	2.000916	−0.000918
21	2.13	2.693989	0.980001	1.999939	0.000060
22	2.12	2.737187	0.980004	1.999605	0.000393
23	2.11	2.786749	0.980002	1.999766	0.000232
24	2.10	2.844237	0.979991	2.000928	−0.000930
25	2.09	2.919031	0.980002	1.999819	0.000179
26	2.08	3.012692	0.979993	2.000654	−0.000656

(continued)

Table 8.2 (continued)

Parameter values					
PROB. = 0.98	RHO = 0.1		BIGH = 2.3228		SMHL = 2.06
Pair No.	SMH	SMK	PROB. level	Risk level (%)	Risk error (%)
27	2.07	3.156473	0.980009	1.999152	0.000846
28	2.06	3.412875	0.979994	2.000630	−0.000632
No. of pairs		Mean COMPUT. risk errors		STD. ERR. errors	
28		0.0000626876		0.0000995272	
Z-ratio				Status of significance	
0.62985				N.S.—not significant	
Parameter values					
PROB. = 0.97	RHO = 0.1		BIGH = 2.1646		SMHL = 1.89
Pair No.	SMH	SMK	PROB. level	Risk level (%)	Risk error (%)
Table No. 8.2-175					
1	2.17	2.159213	0.969996	3.000426	−0.000429
2	2.16	2.169210	0.969996	3.000379	−0.000381
3	2.15	2.179689	0.969998	3.000248	−0.000250
4	2.14	2.190654	0.969999	3.000146	−0.000149
5	2.13	2.202105	0.969998	3.000206	−0.000209
6	2.12	2.214044	0.969995	3.000552	−0.000554
7	2.11	2.226863	0.970000	3.000015	−0.000018
8	2.10	2.240172	0.969999	3.000069	−0.000072
9	2.09	2.254362	0.970003	2.999663	0.000334
10	2.08	2.269047	0.969998	3.000158	−0.000161
11	2.07	2.284617	0.969994	3.000593	−0.000596
12	2.06	2.301854	0.970009	2.999067	0.000930
13	2.05	2.319200	0.969998	3.000188	−0.000191
14	2.04	2.338216	0.970000	2.999991	0.000006
15	2.03	2.358124	0.969991	3.000862	−0.000864
16	2.02	2.380488	0.970005	2.999509	0.000489
17	2.01	2.403747	0.969999	3.000057	−0.000060
18	2.00	2.429465	0.970004	2.999592	0.000405
19	1.99	2.456862	0.969996	3.000367	−0.000370
20	1.98	2.487504	0.969999	3.000128	−0.000131
21	1.97	2.521391	0.970000	2.999973	0.000024
22	1.96	2.559307	0.970002	2.999818	0.000179
23	1.95	2.602035	0.970000	2.999997	0.000000
24	1.94	2.651139	0.969997	3.000313	−0.000316
25	1.93	2.708966	0.969994	3.000641	−0.000644
26	1.92	2.780985	0.970004	2.999592	0.000405
27	1.91	2.871887	0.970005	2.999526	0.000471

(continued)

8.2 Contents for Table 8.2 (Table No. 8.2-1 to 400)

Table 8.2 (continued)

Parameter values					
PROB. = 0.97	RHO = 0.1		BIGH = 2.1646		SMHL = 1.89
Pair No.	SMH	SMK	PROB. level	Risk level (%)	Risk error (%)
28	1.90	2.997298	0.969999	3.000057	−0.000060
29	1.89	3.210345	0.969998	3.000164	−0.000167
No. of pairs		Mean COMPUT. risk errors		STD. ERR. errors	
29		−0.0000792742		0.0000698720	
Z-ratio				Status of significance	
1.13456				N.S.—not significant	
Parameter values					
PROB. = 0.96	RHO = 0.1		BIGH = 2.0464		SMHL = 1.76
Pair No.	SMH	SMK	PROB. level	Risk level (%)	Risk error (%)
Table No. 8.2-176					
1	2.05	2.042806	0.959993	4.000730	−0.000727
2	2.04	2.052820	0.959991	4.000950	−0.000948
3	2.03	2.063714	0.960009	3.999150	0.000852
4	2.02	2.074708	0.960009	3.999120	0.000882
5	2.01	2.085803	0.959991	4.000867	−0.000864
6	2.00	2.098173	0.960006	3.999430	0.000572
7	1.99	2.110648	0.959999	4.000068	−0.000066
8	1.98	2.124012	0.960004	3.999651	0.000352
9	1.97	2.137872	0.960000	4.000002	0.000000
10	1.96	2.152624	0.960002	3.999794	0.000209
11	1.95	2.167878	0.959993	4.000735	−0.000733
12	1.94	2.184417	0.959997	4.000264	−0.000262
13	1.93	2.201852	0.959999	4.000121	−0.000119
14	1.92	2.220184	0.959994	4.000616	−0.000614
15	1.91	2.240197	0.960004	3.999651	0.000352
16	1.90	2.261112	0.959999	4.000121	−0.000119
17	1.89	2.283711	0.959998	4.000169	−0.000167
18	1.88	2.307997	0.959996	4.000402	−0.000399
19	1.87	2.334753	0.960005	3.999514	0.000489
20	1.86	2.363199	0.959997	4.000264	−0.000262
21	1.85	2.394900	0.960001	3.999925	0.000077
22	1.84	2.429859	0.960002	3.999794	0.000209
23	1.83	2.468858	0.960002	3.999788	0.000215
24	1.82	2.513463	0.960010	3.999049	0.000954
25	1.81	2.563676	0.960001	3.999943	0.000060
26	1.80	2.623404	0.960000	3.999990	0.000012
27	1.79	2.695777	0.959998	4.000229	−0.000226
28	1.78	2.788607	0.960000	4.000044	−0.000042

(continued)

Table 8.2 (continued)

Parameter values					
PROB. = 0.96	RHO = 0.1		BIGH = 2.0464		SMHL = 1.76
Pair No.	SMH	SMK	PROB. level	Risk level (%)	Risk error (%)
29	1.77	2.915962	0.959992	4.000771	−0.000769
30	1.76	3.135655	0.960005	3.999519	0.000483
No. of pairs		Mean COMPUT. risk errors		STD. ERR. errors	
30		−0.0000194196		0.0000908868	
Z-ratio				Status of significance	
0.21367				N.S.—not significant	
Parameter values					
PROB. = 0.95	RHO = 0.1		BIGH = 1.9507		SMHL = 1.65
Pair No.	SMH	SMK	PROB. level	Risk level (%)	Risk error (%)
Table No. 8.2-177					
1	1.96	1.941835	0.950001	4.999876	0.000125
2	1.95	1.951791	0.950008	4.999155	0.000846
3	1.94	1.961850	0.949998	5.000222	−0.000221
4	1.93	1.972403	0.949991	5.000913	−0.000912
5	1.92	1.983844	0.950007	4.999334	0.000668
6	1.91	1.995392	0.950001	4.999882	0.000119
7	1.90	2.007440	0.949994	5.000586	−0.000584
8	1.89	2.020381	0.950003	4.999691	0.000310
9	1.88	2.033825	0.950006	4.999411	0.000590
10	1.87	2.047773	0.950001	4.999924	0.000077
11	1.86	2.062620	0.950004	4.999638	0.000364
12	1.85	2.077975	0.949994	5.000621	−0.000620
13	1.84	2.094623	0.950003	4.999721	0.000280
14	1.83	2.112173	0.950009	4.999101	0.000900
15	1.82	2.130629	0.950009	4.999108	0.000894
16	1.81	2.149994	0.949999	5.000061	−0.000060
17	1.80	2.171048	0.950004	4.999566	0.000435
18	1.79	2.193405	0.950003	4.999674	0.000328
19	1.78	2.217457	0.950004	4.999650	0.000352
20	1.77	2.243598	0.950009	4.999054	0.000948
21	1.76	2.271438	0.950000	4.999960	0.000042
22	1.75	2.301761	0.949990	5.000997	−0.000995
23	1.74	2.336132	0.950006	4.999423	0.000578
24	1.73	2.372992	0.949992	5.000812	−0.000811
25	1.72	2.415468	0.950002	4.999805	0.000197
26	1.71	2.462781	0.949993	5.000729	−0.000727
27	1.70	2.518058	0.949991	5.000878	−0.000876
28	1.69	2.584427	0.949998	5.000192	−0.000191
29	1.68	2.666580	0.950006	4.999411	0.000590
30	1.67	2.772330	0.949995	5.000460	−0.000459

(continued)

8.2 Contents for Table 8.2 (Table No. 8.2-1 to 400)

Table 8.2 (continued)

Parameter values					
PROB. = 0.95	RHO = 0.1		BIGH = 1.9507		SMHL = 1.65
Pair No.	SMH	SMK	PROB. level	Risk level (%)	Risk error (%)
31	1.66	2.929807	0.949999	5.000144	−0.000143
32	1.65	3.243699	0.949995	5.000472	−0.000471
No. of pairs		Mean COMPUT. risk errors		STD. ERR. errors	
32		0.0000476837		0.0000995811	
Z-ratio				Status of significance	
0.47884				N.S.—not significant	
Parameter values					
PROB. = 0.94	RHO = 0.1		BIGH = 1.8698		SMHL = 1.56
Pair No.	SMH	SMK	PROB. level	Risk level (%)	Risk error (%)
Table No. 8.2-178					
1	1.87	1.869600	0.940006	5.999387	0.000614
2	1.86	1.879652	0.939997	6.000316	−0.000316
3	1.85	1.890203	0.939993	6.000698	−0.000697
4	1.84	1.901254	0.939993	6.000722	−0.000721
5	1.83	1.912809	0.939994	6.000602	−0.000602
6	1.82	1.924869	0.939995	6.000513	−0.000513
7	1.81	1.937435	0.939994	6.000650	−0.000650
8	1.80	1.950900	0.940010	5.999029	0.000972
9	1.79	1.964485	0.939997	6.000316	−0.000316
10	1.78	1.978974	0.939997	6.000340	−0.000340
11	1.77	1.994368	0.940005	5.999518	0.000483
12	1.76	2.010278	0.939998	6.000197	−0.000197
13	1.75	2.027098	0.939993	6.000710	−0.000709
14	1.74	2.045220	0.940004	5.999631	0.000370
15	1.73	2.064256	0.940007	5.999321	0.000679
16	1.72	2.084209	0.939998	6.000173	−0.000173
17	1.71	2.105861	0.940006	5.999446	0.000554
18	1.70	2.128435	0.939990	6.000954	−0.000954
19	1.69	2.153104	0.939992	6.000823	−0.000823
20	1.68	2.179871	0.939999	6.000132	−0.000131
21	1.67	2.208738	0.940001	5.999917	0.000083
22	1.66	2.240100	0.939999	6.000066	−0.000066
23	1.65	2.274349	0.939992	6.000799	−0.000799
24	1.64	2.313049	0.940004	5.999589	0.000411
25	1.63	2.355816	0.940001	5.999875	0.000125
26	1.62	2.404993	0.940010	5.999011	0.000989
27	1.61	2.460585	0.939994	6.000644	−0.000644
28	1.60	2.528063	0.940001	5.999941	0.000060
29	1.59	2.611337	0.940007	5.999315	0.000685

(continued)

Table 8.2 (continued)

Parameter values					
PROB. = 0.94	RHO = 0.1		BIGH = 1.8698		SMHL = 1.56
Pair No.	SMH	SMK	PROB. level	Risk level (%)	Risk error (%)
30	1.58	2.719003	0.939999	6.000102	−0.000101
31	1.57	2.876847	0.939993	6.000710	−0.000709
32	1.56	3.191122	0.939990	6.000978	−0.000978
No. of pairs		Mean COMPUT. risk errors		STD. ERR. errors	
32		−0.0001336589		0.0001001103	
Z-ratio				Status of significance	
1.33512				N.S.—not significant	
Parameter values					
PROB. = 0.93	RHO = 0.1		BIGH = 1.799		SMHL = 1.48
Pair No.	SMH	SMK	PROB. level	Risk level (%)	Risk error (%)
Table No. 8.2-179					
1	1.80	1.798001	0.930010	6.999016	0.000983
2	1.79	1.808045	0.930002	6.999832	0.000167
3	1.78	1.818594	0.930001	6.999922	0.000077
4	1.77	1.829647	0.930005	6.999469	0.000530
5	1.76	1.841013	0.930000	7.000047	−0.000048
6	1.75	1.852888	0.929995	7.000465	−0.000465
7	1.74	1.865470	0.930004	6.999648	0.000352
8	1.73	1.878565	0.930008	6.999159	0.000840
9	1.72	1.892175	0.930008	6.999236	0.000763
10	1.71	1.906304	0.929999	7.000113	−0.000113
11	1.70	1.921344	0.930003	6.999678	0.000322
12	1.69	1.936905	0.929994	7.000643	−0.000644
13	1.68	1.953578	0.930000	6.999970	0.000030
14	1.67	1.971168	0.930007	6.999266	0.000733
15	1.66	1.989483	0.930000	7.000005	−0.000006
16	1.65	2.009111	0.930004	6.999612	0.000387
17	1.64	2.029665	0.929994	7.000613	−0.000614
18	1.63	2.051928	0.930001	6.999952	0.000048
19	1.62	2.075512	0.929998	7.000232	−0.000232
20	1.61	2.100812	0.929994	7.000578	−0.000578
21	1.60	2.128219	0.929996	7.000375	−0.000376
22	1.59	2.158128	0.930005	6.999475	0.000525
23	1.58	2.190542	0.930007	6.999278	0.000721
24	1.57	2.225855	0.929999	7.000071	−0.000072
25	1.56	2.265241	0.929999	7.000149	−0.000149
26	1.55	2.309875	0.930010	6.999040	0.000960
27	1.54	2.359371	0.929991	7.000942	−0.000942
28	1.53	2.418419	0.930006	6.999439	0.000560
29	1.52	2.487023	0.929991	7.000942	−0.000942
30	1.51	2.572999	0.929992	7.000846	−0.000846
31	1.50	2.687287	0.930001	6.999886	0.000113

(continued)

8.2 Contents for Table 8.2 (Table No. 8.2-1 to 400)

Table 8.2 (continued)

Parameter values					
PROB. = 0.93	RHO = 0.1		BIGH = 1.799		SMHL = 1.48
Pair No.	SMH	SMK	PROB. level	Risk level (%)	Risk error (%)
32	1.49	2.856455	0.930001	6.999946	0.000054
33	1.48	3.218006	0.930000	6.999964	0.000036
No. of pairs		Mean COMPUT. risk errors		STD. ERR. errors	
33		0.0000639873		0.0000918523	
Z-ratio				Status of significance	
0.69663				N.S.—not significant	
Parameter values					
PROB. = 0.92	RHO = 0.1		BIGH = 1.7358		SMHL = 1.41
Pair No.	SMH	SMK	PROB. level	Risk level (%)	Risk error (%)
Table No. 8.2-180					
1	1.74	1.731610	0.920010	7.999015	0.000983
2	1.73	1.741620	0.920008	7.999200	0.000799
3	1.72	1.751941	0.920000	7.999969	0.000030
4	1.71	1.762770	0.920001	7.999939	0.000060
5	1.70	1.773916	0.919992	8.000821	−0.000823
6	1.69	1.785771	0.920002	7.999814	0.000185
7	1.68	1.797946	0.919999	8.000141	−0.000143
8	1.67	1.810833	0.920009	7.999110	0.000888
9	1.66	1.824046	0.920002	7.999826	0.000173
10	1.65	1.837976	0.920003	7.999736	0.000262
11	1.64	1.852430	0.919995	8.000528	−0.000530
12	1.63	1.867803	0.920001	7.999879	0.000119
13	1.62	1.883706	0.919992	8.000767	−0.000769
14	1.61	1.900726	0.920001	7.999879	0.000119
15	1.60	1.918478	0.919999	8.000088	−0.000089
16	1.59	1.937157	0.919993	8.000684	−0.000685
17	1.58	1.957158	0.920000	8.000046	−0.000048
18	1.57	1.978094	0.919990	8.000958	−0.000960
19	1.56	2.000943	0.920009	7.999069	0.000930
20	1.55	2.024731	0.919998	8.000165	−0.000167
21	1.54	2.050635	0.920006	7.999450	0.000548
22	1.53	2.078266	0.920001	7.999921	0.000077
23	1.52	2.108410	0.920005	7.999492	0.000507
24	1.51	2.141068	0.920003	7.999748	0.000250
25	1.50	2.176636	0.919991	8.000935	−0.000936
26	1.49	2.216679	0.920000	7.999998	0.000000
27	1.48	2.261202	0.920000	8.000029	−0.000030
28	1.47	2.312161	0.920009	7.999087	0.000912
29	1.46	2.369949	0.919992	8.000798	−0.000799
30	1.45	2.439650	0.919997	8.000315	−0.000316
31	1.44	2.525174	0.919995	8.000523	−0.000525
32	1.43	2.638243	0.920004	7.999575	0.000423

(continued)

Table 8.2 (continued)

Parameter values					
PROB. = 0.92	RHO = 0.1		BIGH = 1.7358		SMHL = 1.41
Pair No.	SMH	SMK	PROB. level	Risk level (%)	Risk error (%)
33	1.42	2.803081	0.920001	7.999951	0.000048
34	1.41	3.136879	0.919997	8.000279	−0.000280
No. of pairs		Mean COMPUT. risk errors		STD. ERR. errors	
34		0.0000061308		0.0000917998	
Z-ratio				Status of significance	
0.06678				N.S.—not significant	
Parameter values					
PROB. = 0.91	RHO = 0.1		BIGH = 1.6784		SMHL = 1.35
Pair No.	SMH	SMK	PROB. level	Risk level (%)	Risk error (%)
Table No. 8.2-181					
1	1.68	1.676802	0.909996	9.000373	−0.000376
2	1.67	1.687038	0.910006	8.999389	0.000608
3	1.66	1.697395	0.909992	9.000760	−0.000763
4	1.65	1.708461	0.910006	8.999438	0.000560
5	1.64	1.719652	0.909993	9.000749	−0.000751
6	1.63	1.731557	0.910001	8.999896	0.000101
7	1.62	1.743788	0.909996	9.000379	−0.000381
8	1.61	1.756737	0.910007	8.999271	0.000727
9	1.60	1.770017	0.910000	8.999956	0.000042
10	1.59	1.784020	0.910004	8.999639	0.000358
11	1.58	1.798553	0.909999	9.000147	−0.000149
12	1.57	1.813816	0.909996	9.000373	−0.000376
13	1.56	1.830005	0.910007	8.999348	0.000650
14	1.55	1.846733	0.909998	9.000218	−0.000221
15	1.54	1.864394	0.909993	9.000725	−0.000727
16	1.53	1.883381	0.910010	8.999002	0.000995
17	1.52	1.902916	0.909995	9.000492	−0.000495
18	1.51	1.923979	0.910001	8.999920	0.000077
19	1.50	1.946377	0.910007	8.999289	0.000709
20	1.49	1.969919	0.909996	9.000397	−0.000399
21	1.48	1.995584	0.910008	8.999157	0.000840
22	1.47	2.022594	0.909994	9.000593	−0.000596
23	1.46	2.052126	0.909996	9.000361	−0.000364
24	1.45	2.084183	0.909998	9.000200	−0.000203
25	1.44	2.119159	0.909997	9.000312	−0.000316
26	1.43	2.157839	0.910000	8.999991	0.000006
27	1.42	2.201009	0.910006	8.999366	0.000632
28	1.41	2.249453	0.910005	8.999467	0.000530
29	1.40	2.305130	0.910008	8.999247	0.000751
30	1.39	2.370387	0.910009	8.999080	0.000918
31	1.38	2.449135	0.910007	8.999342	0.000656
32	1.37	2.548411	0.919994	9.000629	−0.000632

(continued)

8.2 Contents for Table 8.2 (Table No. 8.2-1 to 400)

Table 8.2 (continued)

Parameter values					
PROB. = 0.91	RHO = 0.1		BIGH = 1.6784		SMHL = 1.35
Pair No.	SMH	SMK	PROB. level	Risk level (%)	Risk error (%)
33	1.36	2.686967	0.909997	9.000343	−0.000346
34	1.35	2.917935	0.909995	9.000456	−0.000459
No. of pairs		Mean COMPUT. risk errors		STD. ERR. errors	
34		0.0000459807		0.0000933001	
Z-ratio				Status of significance	
0.49283				N.S.—not significant	
Parameter values					
PROB. = 0.9025	RHO = 0.1		BIGH = 1.6386		SMHL = 1.30
Pair No.	SMH	SMK	PROB. level	Risk level (%)	Risk error (%)
Table No. 8.2-182					
1	1.64	1.637201	0.902505	9.749490	0.000513
2	1.63	1.647245	0.902497	9.750319	−0.000316
3	1.62	1.657804	0.902502	9.749824	0.000179
4	1.61	1.668685	0.902499	9.750069	−0.000066
5	1.60	1.680084	0.902506	9.749412	0.000590
6	1.59	1.691810	0.902501	9.749866	0.000137
7	1.58	1.704061	0.902501	9.749878	0.000125
8	1.57	1.716838	0.902503	9.749746	0.000256
9	1.56	1.730145	0.902503	9.749722	0.000280
10	1.55	1.743983	0.902499	9.750098	−0.000095
11	1.54	1.758552	0.902504	9.749639	0.000364
12	1.53	1.773658	0.902498	9.750194	−0.000191
13	1.52	1.789501	0.902493	9.750676	−0.000674
14	1.51	1.806277	0.902499	9.750080	−0.000077
15	1.50	1.823796	0.902497	9.750319	−0.000316
16	1.49	1.842254	0.902495	9.750498	−0.000495
17	1.48	1.861852	0.902501	9.749942	0.000060
18	1.47	1.882397	0.902494	9.750611	−0.000608
19	1.46	1.904282	0.902493	9.750711	−0.000709
20	1.45	1.927707	0.902499	9.750056	−0.000054
21	1.44	1.952481	0.902493	9.750682	−0.000679
22	1.43	1.979192	0.902495	9.750521	−0.000519
23	1.42	2.008039	0.902501	9.749937	0.000066
24	1.41	2.039028	0.902496	9.750438	−0.000435
25	1.40	2.072942	0.902498	9.750241	−0.000238
26	1.39	2.110177	0.902500	9.749991	0.000012
27	1.38	2.151128	0.902493	9.750736	−0.000733
28	1.37	2.197361	0.902498	9.750247	−0.000244
29	1.36	2.250053	0.902506	9.749365	0.000638
30	1.35	2.310771	0.902507	9.749353	0.000650
31	1.34	2.382644	0.902500	9.750032	−0.000030
32	1.33	2.471929	0.902501	9.749942	0.000060

(continued)

Table 8.2 (continued)

Parameter values					
PROB. = 0.9025	RHO = 0.1		BIGH = 1.6386		SMHL = 1.30
Pair No.	SMH	SMK	PROB. level	Risk level (%)	Risk error (%)
33	1.32	2.588785	0.902492	9.750849	−0.000846
34	1.31	2.765250	0.902504	9.749627	0.000376
35	1.30	3.140391	0.902497	9.750265	−0.000262
No. of pairs			Mean COMPUT. risk errors	STD. ERR. errors	
35			−0.0000912282	0.0000679925	
Z-ratio				Status of significance	
1.34174				N.S.—not significant	
Parameter values					
PROB. = 0.9	RHO = 0.1		BIGH = 1.6258		SMHL = 1.29
Pair No.	SMH	SMK	PROB. level	Risk level (%)	Risk error (%)
Table No. 8.2-183					
1	1.63	1.621611	0.899996	10.000380	−0.000381
2	1.62	1.631621	0.899994	10.000560	−0.000560
3	1.61	1.642146	0.900007	9.999347	0.000656
4	1.60	1.652797	0.899994	10.000650	−0.000644
5	1.59	1.663969	0.899991	10.000910	−0.000906
6	1.58	1.675662	0.899996	10.000400	−0.000393
7	1.57	1.687880	0.900006	9.999389	0.000614
8	1.56	1.700430	0.900002	9.999836	0.000167
9	1.55	1.713510	0.899998	10.000250	−0.000244
10	1.54	1.727318	0.900007	9.999305	0.000697
11	1.53	1.741466	0.899995	10.000530	−0.000530
12	1.52	1.756542	0.900004	9.999561	0.000441
13	1.51	1.772160	0.900001	9.999866	0.000137
14	1.50	1.788518	0.899997	10.000340	−0.000334
15	1.49	1.805813	0.900000	9.999972	0.000030
16	1.48	1.823854	0.899993	10.000710	−0.000703
17	1.47	1.843034	0.899996	10.000410	−0.000405
18	1.46	1.863358	0.900002	9.999764	0.000238
19	1.45	1.884633	0.899993	10.000730	−0.000727
20	1.44	1.907448	0.899996	10.000440	−0.000441
21	1.43	1.931808	0.900000	9.999961	0.000042
22	1.42	1.957911	0.900007	9.999275	0.000727
23	1.41	1.985565	0.899995	10.000540	−0.000536
24	1.40	2.015752	0.899998	10.000240	−0.000238
25	1.39	2.048476	0.899998	10.000220	−0.000221
26	1.38	2.084131	0.899993	10.000690	−0.000691
27	1.37	2.123893	0.900005	9.999502	0.000501
28	1.36	2.167766	0.900002	9.999776	0.000226
29	1.35	2.217320	0.900001	9.999913	0.000089
30	1.34	2.274119	0.899998	10.000210	−0.000203
31	1.33	2.341293	0.900006	9.999365	0.000638

(continued)

8.2 Contents for Table 8.2 (Table No. 8.2-1 to 400)

Table 8.2 (continued)

Parameter values					
PROB. = 0.9	RHO = 0.1		BIGH = 1.6258		SMHL = 1.29
Pair No.	SMH	SMK	PROB. level	Risk level (%)	Risk error (%)
32	1.32	2.421975	0.899999	10.000140	−0.000137
33	1.31	2.525541	0.900001	9.999906	0.000095
34	1.30	2.670750	0.900010	9.999037	0.000966
35	1.29	2.917761	0.900003	9.999716	0.000286
No. of pairs		Mean COMPUT. risk errors		STD. ERR. errors	
35		−0.0000485116		0.0000816171	
Z-ratio				Status of significance	
0.59438				N.S.—not significant	
Parameter values					
PROB. = 0.85	RHO = 0.1		BIGH = 1.4094		SMHL = 1.04
Pair No.	SMH	SMK	PROB. level	Risk level (%)	Risk error (%)
Table No. 8.2-184					
1	1.41	1.408898	0.849998	15.000220	−0.000226
2	1.40	1.419059	0.850000	15.000040	−0.000042
3	1.39	1.429559	0.850001	14.999890	0.000113
4	1.38	1.440403	0.850001	14.999930	0.000066
5	1.37	1.451691	0.850008	14.999160	0.000834
6	1.36	1.463329	0.850010	14.999020	0.000978
7	1.35	1.475320	0.850003	14.999690	0.000304
8	1.34	1.487863	0.850009	14.999110	0.000888
9	1.33	1.500767	0.850002	14.999840	0.000155
10	1.32	1.514230	0.850001	14.999940	0.000060
11	1.31	1.528256	0.850003	14.999720	0.000274
12	1.30	1.542850	0.850004	14.999590	0.000411
13	1.29	1.558015	0.850001	14.999870	0.000125
14	1.28	1.573757	0.849990	15.000990	−0.000989
15	1.27	1.590468	0.850005	14.999480	0.000519
16	1.26	1.607765	0.850002	14.999760	0.000238
17	1.25	1.625845	0.849996	15.000450	−0.000447
18	1.24	1.644911	0.849997	15.000330	−0.000334
19	1.23	1.664966	0.849998	15.000200	−0.000203
20	1.22	1.686016	0.849992	15.000770	−0.000775
21	1.21	1.708457	0.850003	14.999740	0.000262
22	1.20	1.732098	0.850003	14.999700	0.000298
23	1.19	1.757141	0.849998	15.000200	−0.000197
24	1.18	1.783983	0.850003	14.999680	0.000316
25	1.17	1.812628	0.850003	14.999720	0.000280
26	1.16	1.843280	0.849994	15.000640	−0.000638
27	1.15	1.876530	0.849992	15.000780	−0.000781
28	1.14	1.912776	0.849995	15.000470	−0.000471
29	1.13	1.952415	0.849995	15.000520	−0.000519
30	1.12	1.996235	0.849994	15.000590	−0.000596

(continued)

Table 8.2 (continued)

Parameter values					
PROB. = 0.85	RHO = 0.1		BIGH = 1.4094		SMHL = 1.04
Pair No.	SMH	SMK	PROB. level	Risk level (%)	Risk error (%)
31	1.11	2.045416	0.850001	14.999930	0.000072
32	1.10	2.101138	0.850004	14.999630	0.000364
33	1.09	2.165361	0.849999	15.000110	−0.000107
34	1.08	2.241610	0.849993	15.000650	−0.000656
35	1.07	2.336143	0.849995	15.000520	−0.000519
36	1.06	2.461469	0.850006	14.999440	0.000560
37	1.05	2.648067	0.850001	14.999860	0.000143
38	1.04	3.066257	0.850006	14.999450	0.000548
No. of pairs		Mean COMPUT. risk errors		STD. ERR. errors	
38		0.0000079473		0.0000773434	
Z-ratio				Status of significance	
0.10275				N.S.—not significant	
Parameter values					
PROB. = 0.8	RHO = 0.1		BIGH = 1.2392		SMHL = 0.85
Pair No.	SMH	SMK	PROB. level	Risk level (%)	Risk error (%)
Table No. 8.2-185					
1	1.24	1.238401	0.800006	19.999390	0.000614
2	1.23	1.248469	0.799997	20.000340	−0.000340
3	1.22	1.258898	0.799994	20.000570	−0.000566
4	1.21	1.269691	0.799998	20.000250	−0.000250
5	1.20	1.280852	0.800004	19.999610	0.000393
6	1.19	1.292290	0.799997	20.000310	−0.000310
7	1.18	1.304202	0.800004	19.999650	0.000346
8	1.17	1.316399	0.799993	20.000720	−0.000721
9	1.16	1.329081	0.799991	20.000920	−0.000924
10	1.15	1.342350	0.800007	19.999250	0.000745
11	1.14	1.355919	0.800000	20.000030	−0.000036
12	1.13	1.370086	0.800004	19.999600	0.000399
13	1.12	1.384758	0.800004	19.999610	0.000393
14	1.11	1.400040	0.800008	19.999200	0.000799
15	1.10	1.415839	0.800000	19.999980	0.000024
16	1.09	1.432358	0.800000	19.999990	0.000006
17	1.08	1.449601	0.800002	19.999810	0.000191
18	1.07	1.467577	0.800000	19.999960	0.000036
19	1.06	1.486390	0.800000	19.999980	0.000018
20	1.05	1.506046	0.799995	20.000520	−0.000519
21	1.04	1.526749	0.799997	20.000300	−0.000304
22	1.03	1.548507	0.799998	20.000210	−0.000209
23	1.02	1.571522	0.800007	19.999330	0.000674
24	1.01	1.595608	0.799995	20.000530	−0.000530
25	1.00	1.621358	0.800003	19.999740	0.000262
26	0.99	1.648589	0.799999	20.000100	−0.000101

(continued)

8.2 Contents for Table 8.2 (Table No. 8.2-1 to 400)

Table 8.2 (continued)

Parameter values					
PROB. = 0.8	RHO = 0.1		BIGH = 1.2392		SMHL = 0.85
Pair No.	SMH	SMK	PROB. level	Risk level (%)	Risk error (%)
27	0.98	1.677697	0.800000	19.999990	0.000006
28	0.97	1.708891	0.800002	19.999810	0.000191
29	0.96	1.742374	0.799997	20.000270	−0.000274
30	0.95	1.778743	0.800001	19.999950	0.000054
31	0.94	1.818400	0.800005	19.999550	0.000447
32	0.93	1.861747	0.799995	20.000460	−0.000459
33	0.92	1.910164	0.800002	19.999830	0.000167
34	0.91	1.964444	0.800000	19.999980	0.000024
35	0.90	2.026553	0.800001	19.999920	0.000083
36	0.89	2.099239	0.800003	19.999710	0.000286
37	0.88	2.187205	0.800009	19.999080	0.000924
38	0.87	2.297888	0.799992	20.000820	−0.000823
39	0.86	2.452787	0.799998	20.000200	−0.000197
40	0.85	2.715982	0.799995	20.000490	−0.000489
No. of pairs		Mean COMPUT. risk errors		STD. ERR. errors	
40		0.0000007269		0.0000691216	
Z-ratio				Status of significance	
0.01052				N.S.—not significant	
Parameter values					
PROB. = 0.75	RHO = 0.1		BIGH = 1.0943		SMHL = 0.68
Pair No.	SMH	SMK	PROB. level	Risk level (%)	Risk error (%)
Table No. 8.2-186					
1	1.10	1.088532	0.749995	25.000510	−0.000513
2	1.09	1.098520	0.749996	25.000360	−0.000358
3	1.08	1.108790	0.749997	25.000320	−0.000322
4	1.07	1.119347	0.749995	25.000510	−0.000513
5	1.06	1.130296	0.750006	24.999390	0.000614
6	1.05	1.141446	0.749995	25.000510	−0.000513
7	1.04	1.152998	0.749994	25.000640	−0.000638
8	1.03	1.164958	0.750000	25.000000	0.000000
9	1.02	1.177235	0.749996	25.000450	−0.000447
10	1.01	1.189933	0.749994	25.000560	−0.000560
11	1.00	1.203157	0.750010	24.999040	0.000960
12	0.99	1.216620	0.749992	25.000790	−0.000787
13	0.98	1.230720	0.750000	24.999960	0.000042
14	0.97	1.245271	0.750001	24.999910	0.000089
15	0.96	1.260279	0.749991	25.000940	−0.000936
16	0.95	1.275948	0.749995	25.000540	−0.000542
17	0.94	1.292287	0.750007	24.999260	0.000745
18	0.93	1.309111	0.749998	25.000180	−0.000179
19	0.92	1.326720	0.750002	24.999810	0.000191
20	0.91	1.345027	0.750000	25.000030	−0.000030
21	0.90	1.364141	0.749999	25.000150	−0.000149

(continued)

Table 8.2 (continued)

Parameter values					
PROB. = 0.75	RHO = 0.1		BIGH = 1.0943		SMHL = 0.68
Pair No.	SMH	SMK	PROB. level	Risk level (%)	Risk error (%)
22	0.89	1.384169	0.750003	24.999700	0.000298
23	0.88	1.405123	0.750005	24.999460	0.000542
24	0.87	1.427014	0.749998	25.000200	−0.000203
25	0.86	1.450050	0.749993	25.000680	−0.000679
26	0.85	1.474439	0.750001	24.999930	0.000072
27	0.84	1.500195	0.750007	24.999260	0.000739
28	0.83	1.527332	0.750000	25.000000	0.000000
29	0.82	1.556255	0.750000	25.000020	−0.000024
30	0.81	1.587174	0.750005	24.999500	0.000501
31	0.80	1.620303	0.750010	24.999020	0.000978
32	0.79	1.655656	0.749990	25.000970	−0.000966
33	0.78	1.694231	0.749991	25.000920	−0.000924
34	0.77	1.736435	0.749999	25.000130	−0.000125
35	0.76	1.782679	0.749994	25.000580	−0.000584
36	0.75	1.834156	0.749992	25.000830	−0.000829
37	0.74	1.892451	0.750001	24.999930	0.000072
38	0.73	1.959150	0.749999	25.000130	−0.000131
39	0.72	2.037793	0.750002	24.999850	0.000155
40	0.71	2.133484	0.749996	25.000450	−0.000447
41	0.70	2.257578	0.749998	25.000180	−0.000179
42	0.69	2.437058	0.750009	24.999130	0.000876
43	0.68	2.779767	0.749992	25.000850	−0.000846
No. of pairs		Mean COMPUT. risk errors		STD. ERR. errors	
43		−0.0001261180		0.0000803628	
Z-ratio				Status of significance	
1.56936				N.S.—not significant	
Parameter values					
PROB. = 0.7	RHO = 0.1		BIGH = 0.965		SMHL = 0.53
Pair No.	SMH	SMK	PROB. level	Risk level (%)	Risk error (%)
Table No. 8.2-187					
1	0.97	0.960026	0.699999	30.000130	−0.000131
2	0.96	0.970026	0.699999	30.000140	−0.000137
3	0.95	0.980334	0.700006	29.999440	0.000560
4	0.94	0.990860	0.699999	30.000080	−0.000077
5	0.93	1.001708	0.699998	30.000220	−0.000221
6	0.92	1.012885	0.700000	29.999980	0.000018
7	0.91	1.024398	0.700005	29.999510	0.000495
8	0.90	1.036159	0.699993	30.000730	−0.000727
9	0.89	1.048371	0.699998	30.000170	−0.000173
10	0.88	1.060945	0.700002	29.999840	0.000161
11	0.87	1.073889	0.700001	29.999870	0.000137
12	0.86	1.087214	0.699996	30.000420	−0.000417
13	0.85	1.101027	0.700000	30.000040	−0.000042

(continued)

Table 8.2 (continued)

Parameter values					
PROB. = 0.7	RHO = 0.1		BIGH = 0.965		SMHL = 0.53
Pair No.	SMH	SMK	PROB. level	Risk level (%)	Risk error (%)
14	0.84	1.115242	0.699994	30.000610	−0.000608
15	0.83	1.129965	0.699993	30.000750	−0.000751
16	0.82	1.145210	0.699993	30.000750	−0.000751
17	0.81	1.160988	0.699991	30.000920	−0.000918
18	0.80	1.177410	0.699999	30.000100	−0.000101
19	0.79	1.194390	0.699998	30.000170	−0.000173
20	0.78	1.212042	0.699999	30.000090	−0.000083
21	0.77	1.230379	0.699997	30.000300	−0.000298
22	0.76	1.249514	0.700000	29.999960	0.000048
23	0.75	1.269367	0.699990	30.000970	−0.000972
24	0.74	1.290248	0.699999	30.000120	−0.000113
25	0.73	1.311978	0.699993	30.000690	−0.000685
26	0.72	1.334871	0.700001	29.999860	0.000143
27	0.71	1.358850	0.700002	29.999790	0.000209
28	0.70	1.384032	0.699996	30.000400	−0.000399
29	0.69	1.410734	0.700003	29.999680	0.000328
30	0.68	1.438979	0.700009	29.999140	0.000858
31	0.67	1.468794	0.699997	30.000340	−0.000340
32	0.66	1.500790	0.700005	29.999510	0.000489
33	0.65	1.534997	0.700009	29.999100	0.000900
34	0.64	1.571640	0.700000	29.999980	0.000018
35	0.63	1.611337	0.699995	30.000540	−0.000536
36	0.62	1.654709	0.699997	30.000320	−0.000322
37	0.61	1.702379	0.699998	30.000190	−0.000185
38	0.60	1.755361	0.700000	30.000000	0.000000
39	0.59	1.815065	0.700003	29.999700	0.000304
40	0.58	1.883294	0.699995	30.000500	−0.000495
41	0.57	1.963805	0.700000	29.999960	0.000048
42	0.56	2.061729	0.700000	30.000000	0.000006
43	0.55	2.188448	0.700008	29.999200	0.000805
44	0.54	2.369802	0.700003	29.999740	0.000262
45	0.53	2.728901	0.699997	30.000330	−0.000328
No. of pairs		Mean COMPUT. risk errors		STD. ERR. errors	
45		−0.0000912210		0.0000650977	
Z-ratio				Status of significance	
1.40129				N.S.—not significant	
Parameter values					
PROB. = 0.65	RHO = 0.1		BIGH = 0.846		SMHL = 0.39
Pair No.	SMH	SMK	PROB. level	Risk level (%)	Risk error (%)
Table No. 8.2-188					
1	0.85	0.841921	0.649999	35.000140	−0.000137
2	0.84	0.851945	0.649999	35.000150	−0.000149

(continued)

Table 8.2 (continued)

Parameter values					
PROB. = 0.65	RHO = 0.1		BIGH = 0.846		SMHL = 0.39
Pair No.	SMH	SMK	PROB. level	Risk level (%)	Risk error (%)
3	0.83	0.862211	0.649995	35.000550	−0.000548
4	0.82	0.872824	0.650007	34.999300	0.000703
5	0.81	0.883600	0.649994	35.000560	−0.000554
6	0.80	0.894743	0.649998	35.000250	−0.000244
7	0.79	0.906165	0.649995	35.000470	−0.000471
8	0.78	0.917975	0.650007	34.999320	0.000679
9	0.77	0.929989	0.649993	35.000710	−0.000709
10	0.76	0.942415	0.649991	35.000860	−0.000858
11	0.75	0.955264	0.650001	34.999900	0.000101
12	0.74	0.968453	0.650002	34.999770	0.000232
13	0.73	0.981995	0.649995	35.000550	−0.000542
14	0.72	0.996003	0.649995	35.000520	−0.000519
15	0.71	1.010492	0.650001	34.999890	0.000113
16	0.70	1.025381	0.649995	35.000480	−0.000477
17	0.69	1.040785	0.649993	35.000750	−0.000745
18	0.68	1.056820	0.650007	34.999300	0.000703
19	0.67	1.073311	0.650004	34.999580	0.000423
20	0.66	1.090375	0.649998	35.000180	−0.000179
21	0.65	1.108132	0.650002	34.999850	0.000155
22	0.64	1.126509	0.649996	35.000400	−0.000399
23	0.63	1.145627	0.649993	35.000710	−0.000703
24	0.62	1.165611	0.650002	34.999800	0.000209
25	0.61	1.186390	0.650005	34.999550	0.000453
26	0.60	1.208094	0.650009	34.999150	0.000852
27	0.59	1.230656	0.649996	35.000430	−0.000423
28	0.58	1.254500	0.650009	34.999110	0.000894
29	0.57	1.279274	0.649992	35.000850	−0.000846
30	0.56	1.305603	0.650005	34.999550	0.000453
31	0.55	1.333137	0.649991	35.000890	−0.000882
32	0.54	1.362509	0.650003	34.999670	0.000340
33	0.53	1.393571	0.650004	34.999590	0.000411
34	0.52	1.426571	0.649996	35.000410	−0.000405
35	0.51	1.461958	0.649994	35.000560	−0.000554
36	0.50	1.500181	0.650007	34.999320	0.000679
37	0.49	1.541308	0.650001	34.999920	0.000083
38	0.48	1.586191	0.650000	35.000030	−0.000030
39	0.47	1.635689	0.650009	34.999120	0.000888
40	0.46	1.690278	0.649993	35.000750	−0.000745
41	0.45	1.752197	0.649999	35.000120	−0.000119
42	0.44	1.823110	0.649998	35.000240	−0.000238
43	0.43	1.906642	0.649999	35.000120	−0.000119
44	0.42	2.008771	0.650002	34.999810	0.000191

(continued)

8.2 Contents for Table 8.2 (Table No. 8.2-1 to 400)

Table 8.2 (continued)

Parameter values					
PROB. = 0.65	RHO = 0.1		BIGH = 0.846		SMHL = 0.39
Pair No.	SMH	SMK	PROB. level	Risk level (%)	Risk error (%)
45	0.41	2.158928	0.649997	35.000280	−0.000274
46	0.40	2.348663	0.650003	34.999710	0.000292
47	0.39	2.732041	0.649995	35.000500	−0.000501
No. of pairs			Mean COMPUT. risk errors	STD. ERR. errors	
47			−0.0000732640	0.0000742318	
Z-ratio				Status of significance	
0.98696				N.S.—not significant	
Parameter values					
PROB. = 0.6	RHO = 0.1		BIGH = 0.7336		SMHL = 0.26
Pair No.	SMH	SMK	PROB. level	Risk level (%)	Risk error (%)
Table No. 8.2-189					
1	0.74	0.727255	0.599992	40.000800	−0.000805
2	0.73	0.737218	0.599991	40.000900	−0.000900
3	0.72	0.747457	0.599997	40.000280	−0.000280
4	0.71	0.757936	0.600000	39.999990	0.000012
5	0.70	0.768666	0.600000	39.999960	0.000036
6	0.69	0.779662	0.599999	40.000140	−0.000143
7	0.68	0.790937	0.599996	40.000440	−0.000447
8	0.67	0.802554	0.600002	39.999780	0.000221
9	0.66	0.814431	0.599998	40.000160	−0.000161
10	0.65	0.826683	0.600005	39.999490	0.000513
11	0.64	0.839229	0.600003	39.999680	0.000316
12	0.63	0.852088	0.599993	40.000670	−0.000674
13	0.62	0.865378	0.599996	40.000430	−0.000429
14	0.61	0.879021	0.599992	40.000830	−0.000834
15	0.60	0.893140	0.600000	39.999960	0.000042
16	0.59	0.907658	0.600004	39.999610	0.000387
17	0.58	0.922604	0.600003	39.999710	0.000286
18	0.57	0.938003	0.599999	40.000150	−0.000155
19	0.56	0.953984	0.600008	39.999220	0.000775
20	0.55	0.970383	0.599998	40.000240	−0.000244
21	0.54	0.987429	0.600002	39.999830	0.000167
22	0.53	1.005061	0.600004	39.999650	0.000352
23	0.52	1.023319	0.600003	39.999670	0.000328
24	0.51	1.042244	0.600001	39.999900	0.000101
25	0.50	1.061884	0.599997	40.000350	−0.000352
26	0.49	1.082386	0.600004	39.999650	0.000346
27	0.48	1.103607	0.599993	40.000740	−0.000739
28	0.47	1.125899	0.600004	39.999590	0.000405
29	0.46	1.149034	0.599995	40.000460	−0.000465
30	0.45	1.173274	0.599991	40.000920	−0.000924
31	0.44	1.198794	0.599999	40.000120	−0.000119
32	0.43	1.225578	0.600002	39.999760	0.000238

(continued)

Table 8.2 (continued)

Parameter values					
PROB. = 0.6	RHO = 0.1		BIGH = 0.7336		SMHL = 0.26
Pair No.	SMH	SMK	PROB. level	Risk level (%)	Risk error (%)
33	0.42	1.253815	0.600008	39.999220	0.000775
34	0.41	1.283505	0.599998	40.000250	−0.000250
35	0.40	1.315148	0.600006	39.999450	0.000548
36	0.39	1.348669	0.599999	40.000060	−0.000066
37	0.38	1.384592	0.600004	39.999640	0.000364
38	0.37	1.423064	0.600000	40.000020	−0.000024
39	0.36	1.464639	0.599999	40.000120	−0.000119
40	0.35	1.509890	0.600001	39.999860	0.000143
41	0.34	1.559411	0.599998	40.000210	−0.000215
42	0.33	1.614407	0.600003	39.999700	0.000304
43	0.32	1.701307	0.600001	39.999880	0.000119
44	0.31	1.769230	0.600003	39.999750	0.000250
45	0.30	1.848972	0.600000	39.999970	0.000030
46	0.29	1.945987	0.600000	40.000000	0.000000
47	0.28	2.070467	0.599999	40.000080	−0.000077
48	0.27	2.246340	0.599991	40.000950	−0.000954
49	0.26	2.566752	0.599994	40.000580	−0.000584
No. of pairs		Mean COMPUT. risk errors		STD. ERR. errors	
49		−0.0000580549		0.0000614467	
Z-ratio				Status of significance	
0.94480				N.S.—not significant	
Parameter values					
PROB. = 0.55	RHO = 0.1		BIGH = 0.6255		SMHL = 0.14
Pair No.	SMH	SMK	PROB. level	Risk level (%)	Risk error (%)
Table No. 8.2-190					
1	0.63	0.621032	0.550001	44.999910	0.000089
2	0.62	0.631049	0.550002	44.999800	0.000197
3	0.61	0.641296	0.550002	44.999790	0.000209
4	0.60	0.651791	0.550003	44.999690	0.000310
5	0.59	0.662550	0.550007	44.999310	0.000691
6	0.58	0.673495	0.549994	45.000620	−0.000614
7	0.57	0.684841	0.550009	44.999070	0.000930
8	0.56	0.696317	0.549992	45.000850	−0.000852
9	0.55	0.708239	0.550006	44.999360	0.000638
10	0.54	0.720338	0.549994	45.000600	−0.000596
11	0.53	0.732837	0.549999	45.000130	−0.000131
12	0.52	0.745666	0.550003	44.999740	0.000256
13	0.51	0.758857	0.550009	44.999080	0.000924
14	0.50	0.772344	0.550003	44.999740	0.000256
15	0.49	0.786263	0.550006	44.999430	0.000566
16	0.48	0.800554	0.550004	44.999620	0.000376
17	0.47	0.815260	0.550001	44.999900	0.000101
18	0.46	0.830427	0.550002	44.999810	0.000191

(continued)

8.2 Contents for Table 8.2 (Table No. 8.2-1 to 400)

Table 8.2 (continued)

Parameter values					
PROB. = 0.55	RHO = 0.1		BIGH = 0.6255		SMHL = 0.14
Pair No.	SMH	SMK	PROB. level	Risk level (%)	Risk error (%)
19	0.45	0.846007	0.549994	45.000620	−0.000620
20	0.44	0.862154	0.550000	45.000050	−0.000048
21	0.43	0.878829	0.550007	44.999340	0.000662
22	0.42	0.896001	0.550005	44.999510	0.000489
23	0.41	0.913741	0.550001	44.999920	0.000077
24	0.40	0.932129	0.550002	44.999820	0.000185
25	0.39	0.951155	0.549999	45.000090	−0.000083
26	0.38	0.970914	0.550001	44.999870	0.000137
27	0.37	0.991417	0.550002	44.999840	0.000161
28	0.36	1.012782	0.550009	44.999060	0.000936
29	0.35	1.034849	0.549993	45.000730	−0.000733
30	0.34	1.057962	0.549991	45.000960	−0.000954
31	0.33	1.082188	0.550000	45.000010	−0.000012
32	0.32	1.107422	0.549995	45.000540	−0.000536
33	0.31	1.133972	0.550001	44.999910	0.000095
34	0.30	1.161784	0.549997	45.000330	−0.000328
35	0.29	1.191130	0.549998	45.000250	−0.000244
36	0.28	1.222125	0.549997	45.000290	−0.000286
37	0.27	1.254933	0.549992	45.000850	−0.000852
38	0.26	1.289963	0.549998	45.000240	−0.000232
39	0.25	1.360507	0.550003	44.999670	0.000328
40	0.24	1.397981	0.550005	44.999520	0.000483
41	0.23	1.438586	0.549992	45.000770	−0.000763
42	0.22	1.483290	0.549997	45.000310	−0.000310
43	0.21	1.532625	0.549996	45.000360	−0.000364
44	0.20	1.587889	0.550006	44.999420	0.000578
45	0.19	1.650224	0.550001	44.999900	0.000101
46	0.18	1.722046	0.549994	45.000590	−0.000590
47	0.17	1.807328	0.550003	44.999720	0.000280
48	0.16	1.911716	0.550001	44.999920	0.000083
49	0.15	2.048174	0.550007	44.999290	0.000715
50	0.14	2.247764	0.550004	44.999610	0.000387
No. of pairs		Mean COMPUT. risk errors		STD. ERR. errors	
50		0.0000447619		0.0000691619	
Z-ratio				Status of significance	
0.64721				N.S.—not significant	
Parameter values					
PROB. = 0.5	RHO = 0.1		BIGH = 0.5196		SMHL = 0.03
Pair No.	SMH	SMK	PROB. level	Risk level (%)	Risk error (%)
Table No. 8.2-191					
1	0.52	0.519249	0.500000	49.999970	0.000030
2	0.51	0.529381	0.500007	49.999290	0.000709

(continued)

Table 8.2 (continued)

Parameter values					
PROB. = 0.5	RHO = 0.1		BIGH = 0.5196		SMHL = 0.03
Pair No.	SMH	SMK	PROB. level	Risk level (%)	Risk error (%)
3	0.50	0.539675	0.499999	50.000130	−0.000125
4	0.49	0.550207	0.499991	50.000880	−0.000882
5	0.48	0.561051	0.499999	50.000070	−0.000066
6	0.47	0.572091	0.499995	50.000520	−0.000519
7	0.46	0.583406	0.499994	50.000650	−0.000647
8	0.45	0.594984	0.499990	50.000970	−0.000966
9	0.44	0.606862	0.499991	50.000880	−0.000885
10	0.43	0.619081	0.500002	49.999790	0.000215
11	0.42	0.631589	0.500010	49.999040	0.000966
12	0.41	0.644338	0.500001	49.999900	0.000107
13	0.40	0.657480	0.500005	49.999460	0.000542
14	0.39	0.670880	0.499992	50.000790	−0.000787
15	0.38	0.684703	0.499991	50.000860	−0.000864
16	0.37	0.698925	0.499995	50.000520	−0.000519
17	0.36	0.713530	0.499996	50.000430	−0.000429
18	0.35	0.728609	0.500008	49.999240	0.000763
19	0.34	0.744071	0.500009	49.999080	0.000918
20	0.33	0.759930	0.500000	50.000040	−0.000033
21	0.32	0.776317	0.499998	50.000160	−0.000158
22	0.31	0.793182	0.499992	50.000850	−0.000849
23	0.30	0.810689	0.500002	49.999780	0.000221
24	0.29	0.828635	0.499991	50.000950	−0.000948
25	0.28	0.847334	0.500003	49.999740	0.000262
26	0.27	0.866542	0.499993	50.000730	−0.000727
27	0.26	0.886544	0.499999	50.000070	−0.000075
28	0.25	0.907279	0.500005	49.999530	0.000477
29	0.24	0.928741	0.500001	49.999930	0.000072
30	0.23	0.951089	0.500002	49.999820	0.000185
31	0.22	0.974270	0.499991	50.000880	−0.000876
32	0.21	0.998528	0.499992	50.000780	−0.000781
33	0.20	1.067888	0.500004	49.999580	0.000429
34	0.19	1.090889	0.499991	50.000920	−0.000918
35	0.18	1.115437	0.499999	50.000100	−0.000101
36	0.17	1.141460	0.500001	49.999930	0.000072
37	0.16	1.169235	0.500010	49.999040	0.000960
38	0.15	1.198720	0.499999	50.000060	−0.000063
39	0.14	1.230409	0.500000	50.000010	−0.000006
40	0.13	1.264493	0.500002	49.999830	0.000167
41	0.12	1.301231	0.499997	50.000270	−0.000268

(continued)

8.2 Contents for Table 8.2 (Table No. 8.2-1 to 400)

Table 8.2 (continued)

Parameter values					
PROB. = 0.5	RHO = 0.1		BIGH = 0.5196		SMHL = 0.03
Pair No.	SMH	SMK	PROB. level	Risk level (%)	Risk error (%)
42	0.11	1.341116	0.499995	50.000550	−0.000545
43	0.10	1.384798	0.500002	49.999780	0.000226
44	0.09	1.432631	0.499995	50.000530	−0.000525
45	0.08	1.485928	0.500001	49.999880	0.000125
46	0.07	1.545579	0.499996	50.000360	−0.000364
47	0.06	1.613786	0.500003	49.999700	0.000304
48	0.05	1.693024	0.500004	49.999580	0.000423
49	0.04	1.787857	0.500002	49.999800	0.000203
50	0.03	1.906446	0.499994	50.000650	−0.000641
No. of pairs		Mean COMPUT. risk errors		STD. ERR. errors	
50		−0.0001213714		0.0000766573	
Z-ratio				Status of significance	
1.58330				N.S.—not significant	
Parameter values					
PROB. = 0.99	RHO = 0.0		BIGH = 2.5751		SMHL = 2.34
Pair No.	SMH	SMK	PROB. level	Risk level (%)	Risk error (%)
Table No. 8.2-192					
1	2.58	2.570209	0.990003	0.999677	0.000322
2	2.57	2.580210	0.990003	0.999689	0.000310
3	2.56	2.590289	0.989997	1.000333	−0.000334
4	2.55	2.602009	0.990005	0.999522	0.000477
5	2.54	2.613810	0.990005	0.999481	0.000519
6	2.53	2.625691	0.989998	1.000220	−0.000221
7	2.52	2.639217	0.990002	0.999820	0.000179
8	2.51	2.652826	0.989996	1.000357	−0.000358
9	2.50	2.668081	0.990000	1.000041	−0.000042
10	2.49	2.684984	0.990009	0.999069	0.000930
11	2.48	2.701972	0.990007	0.999296	0.000703
12	2.47	2.719047	0.989993	1.000702	−0.000703
13	2.46	2.739335	0.989996	1.000369	−0.000370
14	2.45	2.761275	0.989999	1.000112	−0.000113
15	2.44	2.784868	0.989999	1.000142	−0.000143
16	2.43	2.810114	0.989993	1.000690	−0.000691
17	2.42	2.838578	0.989992	1.000845	−0.000846
18	2.41	2.873385	0.990009	0.999057	0.000942
19	2.40	2.906725	0.989991	1.000875	−0.000876
20	2.39	2.949535	0.989998	1.000196	−0.000197
21	2.38	2.998693	0.990000	1.000035	−0.000036
22	2.37	3.057324	0.989999	1.000083	−0.000083
23	2.36	3.128555	0.989992	1.000780	−0.000781

(continued)

Table 8.2 (continued)

Parameter values					
PROB. = 0.99	RHO = 0.0		BIGH = 2.5751		SMHL = 2.34
Pair No.	SMH	SMK	PROB. level	Risk level (%)	Risk error (%)
24	2.35	3.234261	0.990009	0.999087	0.000912
25	2.34	3.383820	0.990004	0.999582	0.000417
No. of pairs		Mean COMPUT. risk errors		STD. ERR. errors	
25		−0.0000032095		0.0001077926	
Z-ratio				Status of significance	
0.02977				N.S.—not significant	
Parameter values					
PROB. = 0.98	RHO = 0.0		BIGH = 2.3244		SMHL = 2.06
Pair No.	SMH	SMK	PROB. level	Risk level (%)	Risk error (%)
Table No. 8.2-193					
1	2.33	2.318813	0.979995	2.000469	−0.000471
2	2.32	2.328808	0.979996	2.000409	−0.000411
3	2.31	2.339671	0.980006	1.999366	0.000632
4	2.30	2.350621	0.980005	1.999462	0.000536
5	2.29	2.361660	0.979993	2.000719	−0.000721
6	2.28	2.374352	0.980006	1.999432	0.000566
7	2.27	2.387135	0.980005	1.999545	0.000453
8	2.26	2.400792	0.980007	1.999319	0.000679
9	2.25	2.414541	0.979994	2.000570	−0.000572
10	2.24	2.429949	0.979999	2.000129	−0.000131
11	2.23	2.446233	0.980001	1.999867	0.000131
12	2.22	2.463397	0.980000	1.999992	0.000006
13	2.21	2.482222	0.980008	1.999247	0.000751
14	2.20	2.501146	0.979993	2.000707	−0.000709
15	2.19	2.523298	0.980008	1.999217	0.000781
16	2.18	2.545552	0.979996	2.000421	−0.000423
17	2.17	2.571036	0.980003	1.999712	0.000286
18	2.16	2.598187	0.980000	2.000010	−0.000012
19	2.15	2.628571	0.980003	1.999718	0.000280
20	2.14	2.662189	0.980004	1.999629	0.000370
21	2.13	2.699042	0.979995	2.000511	−0.000513
22	2.12	2.742257	0.979998	2.000231	−0.000232
23	2.11	2.791835	0.979996	2.000409	−0.000411
24	2.10	2.850903	0.979995	2.000517	−0.000519
25	2.09	2.925713	0.980004	1.999575	0.000423
26	2.08	3.022516	0.980007	1.999271	0.000727

(continued)

8.2 Contents for Table 8.2 (Table No. 8.2-1 to 400)

Table 8.2 (continued)

Parameter values					
PROB. = 0.98	RHO = 0.0		BIGH = 2.3244		SMHL = 2.06
Pair No.	SMH	SMK	PROB. level	Risk level (%)	Risk error (%)
27	2.07	3.160065	0.980000	1.999968	0.000030
28	2.06	3.422735	0.979997	2.000314	−0.000316
No. of pairs		Mean COMPUT. risk errors		STD. ERR. errors	
28		0.0000417233		0.0000902462	
Z-ratio				Status of significance	
0.46233				N.S.—not significant	
Parameter values					
PROB. = 0.97	RHO = 0.0		BIGH = 2.1672		SMHL = 1.89
Pair No.	SMH	SMK	PROB. level	Risk level (%)	Risk error (%)
Table No. 8.2-194					
1	2.17	2.164404	0.970008	2.999187	0.000811
2	2.16	2.174424	0.970005	2.999473	0.000525
3	2.15	2.184928	0.970003	2.999669	0.000328
4	2.14	2.195918	0.970001	2.999908	0.000089
5	2.13	2.207393	0.969997	3.000277	−0.000280
6	2.12	2.219357	0.969991	3.000951	−0.000954
7	2.11	2.232201	0.969993	3.000701	−0.000703
8	2.10	2.245926	0.970002	2.999812	0.000185
9	2.09	2.259752	0.969991	3.000909	−0.000912
10	2.08	2.275243	0.970006	2.999365	0.000632
11	2.07	2.290840	0.969999	3.000128	−0.000131
12	2.06	2.307713	0.970000	2.999979	0.000018
13	2.05	2.325476	0.969997	3.000319	−0.000322
14	2.04	2.344519	0.969996	3.000403	−0.000405
15	2.03	2.365235	0.970003	2.999675	0.000322
16	2.02	2.386845	0.969996	3.000426	−0.000429
17	2.01	2.410913	0.970005	2.999520	0.000477
18	2.00	2.435878	0.969991	3.000897	−0.000900
19	1.99	2.464085	0.969996	3.000367	−0.000370
20	1.98	2.495536	0.970010	2.999014	0.000983
21	1.97	2.529452	0.970008	2.999175	0.000823
22	1.96	2.566617	0.969996	3.000426	−0.000429
23	1.95	2.610155	0.970003	2.999735	0.000262
24	1.94	2.660071	0.970006	2.999383	0.000614
25	1.93	2.717927	0.970000	3.000009	−0.000012
26	1.92	2.789977	0.970008	2.999222	0.000775

(continued)

Table 8.2 (continued)

Parameter values					
PROB. = 0.97	RHO = 0.0		BIGH = 2.1672		SMHL = 1.89
Pair No.	SMH	SMK	PROB. level	Risk level (%)	Risk error (%)
27	1.91	2.880909	0.970006	2.999365	0.000632
28	1.90	3.006351	0.969999	3.000057	−0.000060
29	1.89	3.222555	0.970004	2.999568	0.000429
No. of pairs		Mean COMPUT. risk errors		STD. ERR. errors	
29		0.0000665585		0.0001009764	
Z-ratio				Status of significance	
0.65915				N.S.—not significant	
Parameter values					
PROB. = 0.96	RHO = 0.0		BIGH = 2.0496		SMHL = 1.76
Pair No.	SMH	SMK	PROB. level	Risk level (%)	Risk error (%)
Table No. 8.2-195					
1	2.05	2.049200	0.960005	3.999543	0.000459
2	2.04	2.059245	0.959998	4.000247	−0.000244
3	2.03	2.069779	0.959993	4.000735	−0.000733
4	2.02	2.081196	0.960006	3.999388	0.000614
5	2.01	2.092714	0.960001	3.999895	0.000107
6	2.00	2.104726	0.959994	4.000592	−0.000590
7	1.99	2.117625	0.960000	4.000044	−0.000042
8	1.98	2.131021	0.959999	4.000092	−0.000089
9	1.97	2.144916	0.959991	4.000878	−0.000876
10	1.96	2.160092	0.960004	3.999627	0.000376
11	1.95	2.175771	0.960004	3.999591	0.000411
12	1.94	2.191954	0.959990	4.000968	−0.000966
13	1.93	2.209814	0.960001	3.999913	0.000089
14	1.92	2.228572	0.960004	3.999567	0.000435
15	1.91	2.248622	0.960009	3.999066	0.000936
16	1.90	2.269572	0.960000	3.999972	0.000030
17	1.89	2.292209	0.959996	4.000414	−0.000411
18	1.88	2.316922	0.960000	3.999990	0.000012
19	1.87	2.343324	0.959995	4.000461	−0.000459
20	1.86	2.372589	0.960003	3.999704	0.000298
21	1.85	2.404328	0.960003	3.999746	0.000256
22	1.84	2.439325	0.960000	3.999955	0.000048
23	1.83	2.478364	0.959998	4.000235	−0.000232
24	1.82	2.522226	0.959990	4.000986	−0.000983
25	1.81	2.574041	0.960003	3.999710	0.000292
26	1.80	2.633810	0.960000	4.000038	−0.000036
27	1.79	2.706224	0.959995	4.000479	−0.000477
28	1.78	2.800658	0.960008	3.999233	0.000769

(continued)

8.2 Contents for Table 8.2 (Table No. 8.2-1 to 400)

Table 8.2 (continued)

Parameter values					
PROB. = 0.96	RHO = 0.0		BIGH = 2.0496		SMHL = 1.76
Pair No.	SMH	SMK	PROB. level	Risk level (%)	Risk error (%)
29	1.77	2.929616	0.960005	3.999543	0.000459
30	1.76	3.149351	0.960010	3.999007	0.000995
No. of pairs		Mean COMPUT. risk errors		STD. ERR. errors	
30		0.0000144205		0.0000944508	
Z-ratio				Status of significance	
0.15268				N.S.—not significant	
Parameter values					
PROB. = 0.95	RHO = 0.0		BIGH = 1.9546		SMHL = 1.65
Pair No.	SMH	SMK	PROB. level	Risk level (%)	Risk error (%)
Table No. 8.2-196					
1	1.96	1.949215	0.950008	4.999203	0.000799
2	1.95	1.959211	0.950009	4.999131	0.000870
3	1.94	1.969310	0.949992	5.000806	−0.000805
4	1.93	1.980295	0.950000	4.999990	0.000012
5	1.92	1.991776	0.950010	4.999048	0.000954
6	1.91	2.003366	0.949998	5.000222	−0.000221
7	1.90	2.015847	0.950005	4.999542	0.000459
8	1.89	2.028830	0.950007	4.999292	0.000709
9	1.88	2.042316	0.950004	4.999626	0.000376
10	1.87	2.056308	0.949993	5.000735	−0.000733
11	1.86	2.071589	0.950008	4.999251	0.000751
12	1.85	2.086989	0.949992	5.000824	−0.000823
13	1.84	2.103681	0.949995	5.000514	−0.000513
14	1.83	2.121277	0.949996	5.000425	−0.000423
15	1.82	2.139779	0.949991	5.000943	−0.000942
16	1.81	2.159970	0.950005	4.999459	0.000542
17	1.80	2.181072	0.950005	4.999531	0.000471
18	1.79	2.203476	0.949998	5.000157	−0.000155
19	1.78	2.227576	0.949994	5.000627	−0.000626
20	1.77	2.253765	0.949995	5.000472	−0.000471
21	1.76	2.282435	0.950005	4.999542	0.000459
22	1.75	2.313198	0.950000	5.000019	−0.000018
23	1.74	2.347230	0.950001	4.999918	0.000083
24	1.73	2.384921	0.950001	4.999936	0.000066
25	1.72	2.427447	0.950006	4.999423	0.000578
26	1.71	2.474812	0.949993	5.000740	−0.000739
27	1.70	2.530923	0.950000	5.000014	−0.000012
28	1.69	2.596565	0.949992	5.000777	−0.000775
29	1.68	2.678771	0.949998	5.000162	−0.000161

(continued)

Table 8.2 (continued)

Parameter values					
PROB. = 0.95	RHO = 0.0		BIGH = 1.9546		SMHL = 1.65
Pair No.	SMH	SMK	PROB. level	Risk level (%)	Risk error (%)
30	1.67	2.786138	0.950000	4.999995	0.000006
31	1.66	2.945232	0.950007	4.999256	0.000745
32	1.65	3.265430	0.950009	4.999078	0.000924
No. of pairs		Mean COMPUT. risk errors		STD. ERR. errors	
32		0.0000420845		0.0001029645	
Z-ratio				Status of significance	
0.40873				N.S.—not significant	
Parameter values					
PROB. = 0.94	RHO = 0.0		BIGH = 1.8739		SMHL = 1.56
Pair No.	SMH	SMK	PROB. level	Risk level (%)	Risk error (%)
Table No. 8.2-197					
1	1.88	1.868211	0.940007	5.999267	0.000733
2	1.87	1.878199	0.940009	5.999124	0.000876
3	1.86	1.888295	0.939991	6.000871	−0.000870
4	1.85	1.899281	0.940004	5.999565	0.000435
5	1.84	1.910378	0.939996	6.000436	−0.000435
6	1.83	1.922174	0.940001	5.999935	0.000066
7	1.82	1.934475	0.940005	5.999500	0.000501
8	1.81	1.947087	0.939995	6.000459	−0.000459
9	1.80	1.960600	0.940004	5.999637	0.000364
10	1.79	1.974623	0.940005	5.999536	0.000465
11	1.78	1.989160	0.939996	6.000364	−0.000364
12	1.77	2.004602	0.939997	6.000304	−0.000304
13	1.76	2.020952	0.940002	5.999780	0.000221
14	1.75	2.038212	0.940008	5.999178	0.000823
15	1.74	2.055995	0.939993	6.000686	−0.000685
16	1.73	2.075472	0.940007	5.999333	0.000668
17	1.72	2.095477	0.939991	6.000877	−0.000876
18	1.71	2.117572	0.940008	5.999232	0.000769
19	1.70	2.140589	0.940001	5.999917	0.000083
20	1.69	2.165311	0.939995	6.000489	−0.000489
21	1.68	2.192523	0.940009	5.999089	0.000912
22	1.67	2.221445	0.940005	5.999506	0.000495
23	1.66	2.252862	0.939997	6.000316	−0.000316
24	1.65	2.287557	0.939995	6.000483	−0.000483
25	1.64	2.326315	0.940002	5.999840	0.000161
26	1.63	2.369139	0.939994	6.000602	−0.000602
27	1.62	2.418374	0.939999	6.000102	−0.000101
28	1.61	2.474806	0.939994	6.000638	−0.000638
29	1.60	2.543125	0.940008	5.999154	0.000846
30	1.59	2.625678	0.940000	5.999959	0.000042
31	1.58	2.734969	0.940005	5.999476	0.000525

(continued)

Table 8.2 (continued)

Parameter values					
PROB. = 0.94	RHO = 0.0		BIGH = 1.8739		SMHL = 1.56
Pair No.	SMH	SMK	PROB. level	Risk level (%)	Risk error (%)
32	1.57	2.892874	0.939995	6.000519	−0.000519
33	1.56	3.213461	0.940003	5.999679	0.000322
No. of pairs		Mean COMPUT. risk errors		STD. ERR. errors	
33		0.0000636367		0.0000951959	
Z-ratio				Status of significance	
0.66848				N.S.—not significant	
Parameter values					
PROB. = 0.93	RHO = 0.0		BIGH = 1.8037		SMHL = 1.48
Pair No.	SMH	SMK	PROB. level	Risk level (%)	Risk error (%)
Table No. 8.2-198					
1	1.81	1.797617	0.930003	6.999726	0.000274
2	1.80	1.807408	0.929990	7.000965	−0.000966
3	1.79	1.817896	0.930001	6.999856	0.000143
4	1.78	1.828497	0.929991	7.000923	−0.000924
5	1.77	1.839799	0.930000	7.000029	−0.000030
6	1.76	1.851415	0.929998	7.000208	−0.000209
7	1.75	1.863540	0.929997	7.000268	−0.000268
8	1.74	1.876373	0.930009	6.999130	0.000870
9	1.73	1.889524	0.930004	6.999612	0.000387
10	1.72	1.903192	0.929994	7.000619	−0.000620
11	1.71	1.917768	0.930000	7.000023	−0.000024
12	1.70	1.932866	0.929995	7.000518	−0.000519
13	1.69	1.948878	0.929999	7.000137	−0.000137
14	1.68	1.965805	0.930007	6.999308	0.000691
15	1.67	1.983260	0.929994	7.000554	−0.000554
16	1.66	2.002027	0.929999	7.000113	−0.000113
17	1.65	2.021717	0.929994	7.000566	−0.000566
18	1.64	2.042724	0.929995	7.000524	−0.000525
19	1.63	2.065051	0.929993	7.000673	−0.000674
20	1.62	2.089090	0.930000	7.000041	−0.000042
21	1.61	2.114844	0.930004	6.999583	0.000417
22	1.60	2.142318	0.929998	7.000190	−0.000191
23	1.59	2.172294	0.930000	7.000005	−0.000006
24	1.58	2.205166	0.930008	6.999171	0.000829
25	1.57	2.240937	0.930006	6.999427	0.000572
26	1.56	2.280782	0.930009	6.999088	0.000912
27	1.55	2.324706	0.929994	7.000632	−0.000632
28	1.54	2.375835	0.930006	6.999409	0.000590
29	1.53	2.434174	0.929999	7.000101	−0.000101
30	1.52	2.504413	0.930006	6.999421	0.000578
31	1.51	2.590462	0.930000	6.999999	0.000000
32	1.50	2.704826	0.930004	6.999600	0.000399

(continued)

Table 8.2 (continued)

Parameter values					
PROB. = 0.93	RHO = 0.0		BIGH = 1.8037		SMHL = 1.48
Pair No.	SMH	SMK	PROB. level	Risk level (%)	Risk error (%)
33	1.49	2.874069	0.930000	7.000029	−0.000030
34	1.48	3.235697	0.929999	7.000131	−0.000131
No. of pairs		Mean COMPUT. risk errors		STD. ERR. errors	
34		−0.0000170299		0.0000857124	
Z-ratio				Status of significance	
0.19869				N.S.—not significant	
Parameter values					
PROB. = 0.92	RHO = 0.0		BIGH = 1.7411		SMHL = 1.41
Pair No.	SMH	SMK	PROB. level	Risk level (%)	Risk error (%)
Table No. 8.2-199					
1	1.75	1.732245	0.919993	8.000708	−0.000709
2	1.74	1.742201	0.920001	7.999892	0.000107
3	1.73	1.752467	0.920004	7.999587	0.000411
4	1.72	1.763045	0.920001	7.999903	0.000095
5	1.71	1.774133	0.920005	7.999468	0.000530
6	1.70	1.785537	0.920000	7.999975	0.000024
7	1.69	1.797456	0.919999	8.000112	−0.000113
8	1.68	1.809891	0.919999	8.000100	−0.000101
9	1.67	1.822844	0.919998	8.000178	−0.000179
10	1.66	1.836318	0.919994	8.000601	−0.000602
11	1.65	1.850511	0.919998	8.000249	−0.000250
12	1.64	1.865229	0.919992	8.000791	−0.000793
13	1.63	1.880866	0.920000	7.999957	0.000042
14	1.62	1.897034	0.919993	8.000660	−0.000662
15	1.61	1.914125	0.919992	8.000815	−0.000817
16	1.60	1.932143	0.919991	8.000862	−0.000864
17	1.59	1.951481	0.920009	7.999146	0.000852
18	1.58	1.971360	0.919994	8.000601	−0.000602
19	1.57	1.992956	0.920006	7.999426	0.000572
20	1.56	2.015489	0.919995	8.000464	−0.000465
21	1.55	2.039745	0.919994	8.000565	−0.000566
22	1.54	2.065726	0.919993	8.000720	−0.000721
23	1.53	2.093826	0.919997	8.000326	−0.000328
24	1.52	2.124048	0.919993	8.000696	−0.000697
25	1.51	2.157178	0.919998	8.000225	−0.000226
26	1.50	2.193609	0.920005	7.999498	0.000501
27	1.49	2.233734	0.920006	7.999408	0.000590
28	1.48	2.278731	0.920009	7.999075	0.000924
29	1.47	2.329383	0.920002	7.999814	0.000185
30	1.46	2.388039	0.919997	8.000332	−0.000334
31	1.45	2.457828	0.919996	8.000440	−0.000441
32	1.44	2.544221	0.920001	7.999945	0.000054

(continued)

8.2 Contents for Table 8.2 (Table No. 8.2-1 to 400)

Table 8.2 (continued)

Parameter values					
PROB. = 0.92	RHO = 0.0		BIGH = 1.7411		SMHL = 1.41
Pair No.	SMH	SMK	PROB. level	Risk level (%)	Risk error (%)
33	1.43	2.657380	0.920005	7.999534	0.000465
34	1.42	2.822308	0.919998	8.000231	−0.000232
35	1.41	3.162448	0.920010	7.999003	0.000995
No. of pairs		Mean COMPUT. risk errors		STD. ERR. errors	
35		−0.0000932150		0.0000881854	
Z-ratio				Status of significance	
1.05703				N.S.—not significant	
Parameter values					
PROB. = 0.91	RHO = 0.0		BIGH = 1.6844		SMHL = 1.35
Pair No.	SMH	SMK	PROB. level	Risk level (%)	Risk error (%)
Table No. 8.2-200					
1	1.69	1.678623	0.909995	9.000522	−0.000525
2	1.68	1.688616	0.909996	9.000361	−0.000364
3	1.67	1.698924	0.909994	9.000653	−0.000656
4	1.66	1.709745	0.910002	8.999836	0.000161
5	1.65	1.720884	0.910002	8.999813	0.000185
6	1.64	1.732346	0.909993	9.000742	−0.000745
7	1.63	1.744522	0.910005	8.999532	0.000465
8	1.62	1.757024	0.910003	8.999693	0.000304
9	1.61	1.770050	0.910002	8.999848	0.000149
10	1.60	1.783603	0.909998	9.000242	−0.000244
11	1.59	1.797881	0.910003	8.999681	0.000316
12	1.58	1.812690	0.910000	8.999973	0.000024
13	1.57	1.828229	0.910000	9.000039	−0.000042
14	1.56	1.844501	0.909998	9.000200	−0.000203
15	1.55	1.861704	0.910004	8.999598	0.000399
16	1.54	1.879644	0.910000	9.000021	−0.000024
17	1.53	1.898522	0.909994	9.000647	−0.000650
18	1.52	1.918534	0.909991	9.000879	−0.000882
19	1.51	1.939880	0.909998	9.000236	−0.000238
20	1.50	1.962562	0.910004	8.999581	0.000417
21	1.49	1.986585	0.910003	8.999688	0.000310
22	1.48	2.012342	0.910005	8.999467	0.000530
23	1.47	2.039836	0.910000	9.000015	−0.000018
24	1.46	2.069852	0.910010	8.999050	0.000948
25	1.45	2.102004	0.910001	8.999861	0.000137
26	1.44	2.137467	0.910006	8.999378	0.000620
27	1.43	2.176245	0.910001	8.999920	0.000077
28	1.42	2.219514	0.909999	9.000074	−0.000077
29	1.41	2.268450	0.910003	8.999711	0.000286
30	1.40	2.324229	0.909999	9.000111	−0.000113
31	1.39	2.389590	0.909996	9.000426	−0.000429

(continued)

Table 8.2 (continued)

Parameter values					
PROB. = 0.91	RHO = 0.0		BIGH = 1.6844		SMHL = 1.35
Pair No.	SMH	SMK	PROB. level	Risk level (%)	Risk error (%)
32	1.38	2.468444	0.909990	9.000981	−0.000983
33	1.37	2.569388	0.909997	9.000254	−0.000256
34	1.36	2.708053	0.909995	9.000481	−0.000483
35	1.35	2.939131	0.909992	9.000802	−0.000805
No. of pairs		Mean COMPUT. risk errors		STD. ERR. errors	
35		−0.0000668897		0.0000761137	
Z-ratio				Status of significance	
0.87881				N.S.—not significant	
Parameter values					
PROB. = 0.9025	RHO = 0.0		BIGH = 1.645		SMHL = 1.30
Pair No.	SMH	SMK	PROB. level	Risk level (%)	Risk error (%)
Table No. 8.2-201					
1	1.65	1.639820	0.902507	9.749347	0.000656
2	1.64	1.649820	0.902507	9.749311	0.000691
3	1.63	1.660138	0.902503	9.749688	0.000316
4	1.62	1.670776	0.902494	9.750599	−0.000596
5	1.61	1.681933	0.902496	9.750414	−0.000411
6	1.60	1.693609	0.902506	9.749383	0.000620
7	1.59	1.705613	0.902505	9.749490	0.000513
8	1.58	1.718143	0.902508	9.749192	0.000811
9	1.57	1.731005	0.902496	9.750450	−0.000447
10	1.56	1.744592	0.902499	9.750134	−0.000131
11	1.55	1.758713	0.902497	9.750271	−0.000268
12	1.54	1.773565	0.902504	9.749592	0.000411
13	1.53	1.788956	0.902500	9.749991	0.000012
14	1.52	1.805084	0.902497	9.750324	−0.000322
15	1.51	1.822148	0.902504	9.749615	0.000387
16	1.50	1.839954	0.902502	9.749764	0.000238
17	1.49	1.858702	0.902501	9.749901	0.000101
18	1.48	1.878395	0.902494	9.750568	−0.000566
19	1.47	1.899427	0.902500	9.749979	0.000024
20	1.46	1.921606	0.902499	9.750122	−0.000119
21	1.45	1.945130	0.902494	9.750604	−0.000602
22	1.44	1.970590	0.902508	9.749180	0.000823
23	1.43	1.997403	0.902499	9.750134	−0.000131
24	1.42	2.026354	0.902494	9.750586	−0.000584
25	1.41	2.057838	0.902497	9.750289	−0.000286
26	1.40	2.092250	0.902506	9.749424	0.000578
27	1.39	2.129593	0.902499	9.750128	−0.000125
28	1.38	2.171044	0.902497	9.750337	−0.000334
29	1.37	2.217388	0.902493	9.750682	−0.000679
30	1.36	2.270193	0.902495	9.750486	−0.000483

(continued)

8.2 Contents for Table 8.2 (Table No. 8.2-1 to 400)

Table 8.2 (continued)

Parameter values					
PROB. = 0.9025	RHO = 0.0		BIGH = 1.645		SMHL = 1.30
Pair No.	SMH	SMK	PROB. level	Risk level (%)	Risk error (%)
31	1.35	2.331025	0.902490	9.750992	−0.000989
32	1.34	2.403796	0.902496	9.750444	−0.000441
33	1.33	2.493980	0.902504	9.749561	0.000441
34	1.32	2.612518	0.902508	9.749156	0.000846
35	1.31	2.787542	0.902499	9.750086	−0.000083
36	1.30	3.169056	0.902509	9.749115	0.000888
No. of pairs		Mean COMPUT. risk errors		STD. ERR. errors	
36		0.0000204589		0.0000842331	
Z-ratio				Status of significance	
0.24288				N.S.—not significant	
Parameter values					
PROB. = 0.9	RHO = 0.0		BIGH = 1.6322		SMHL = 1.29
Pair No.	SMH	SMK	PROB. level	Risk level (%)	Risk error (%)
Table No. 8.2-202					
1	1.64	1.624632	0.900009	9.999084	0.000918
2	1.63	1.634403	0.899996	10.000430	−0.000429
3	1.62	1.644687	0.899998	10.000160	−0.000155
4	1.61	1.655292	0.899996	10.000390	−0.000387
5	1.60	1.666415	0.900005	9.999454	0.000548
6	1.59	1.677864	0.900006	9.999371	0.000632
7	1.58	1.689640	0.899997	10.000310	−0.000310
8	1.57	1.701942	0.899993	10.000740	−0.000733
9	1.56	1.714773	0.899991	10.000860	−0.000858
10	1.55	1.728329	0.900007	9.999352	0.000650
11	1.54	1.742225	0.900002	9.999824	0.000179
12	1.53	1.756851	0.900007	9.999287	0.000715
13	1.52	1.772018	0.900003	9.999693	0.000310
14	1.51	1.787922	0.900001	9.999866	0.000137
15	1.50	1.804567	0.899998	10.000210	−0.000203
16	1.49	1.822151	0.900002	9.999770	0.000232
17	1.48	1.840677	0.900009	9.999108	0.000894
18	1.47	1.859953	0.899999	10.000120	−0.000119
19	1.46	1.880569	0.900005	9.999490	0.000513
20	1.45	1.902138	0.899995	10.000460	−0.000459
21	1.44	1.925444	0.900009	9.999096	0.000906
22	1.43	1.949709	0.899991	10.000880	−0.000876
23	1.42	1.976110	0.899998	10.000220	−0.000215
24	1.41	2.004259	0.899994	10.000570	−0.000572
25	1.40	2.034943	0.900005	9.999484	0.000519
26	1.39	2.067773	0.899995	10.000490	−0.000489
27	1.38	2.103928	0.899997	10.000310	−0.000310
28	1.37	2.143800	0.900000	10.000040	−0.000036

(continued)

Table 8.2 (continued)

Parameter values					
PROB. = 0.9	RHO = 0.0		BIGH = 1.6322		SMHL = 1.29
Pair No.	SMH	SMK	PROB. level	Risk level (%)	Risk error (%)
29	1.36	2.188176	0.900002	9.999794	0.000209
30	1.35	2.238233	0.900005	9.999550	0.000453
31	1.34	2.295148	0.899994	10.000610	−0.000608
32	1.33	2.362440	0.899996	10.000370	−0.000370
33	1.32	2.444021	0.899999	10.000130	−0.000131
34	1.31	2.547708	0.899996	10.000400	−0.000393
35	1.30	2.693039	0.900002	9.999794	0.000209
36	1.29	2.940175	0.899996	10.000380	−0.000376
No. of pairs		Mean COMPUT. risk errors		STD. ERR. errors	
36		−0.0000001611		0.0000840603	
Z-ratio				Status of significance	
0.00192				N.S.—not significant	
Parameter values					
PROB. = 0.85	RHO = 0.0		BIGH = 1.4184		SMHL = 1.04
Pair No.	SMH	SMK	PROB. level	Risk level (%)	Risk error (%)
Table No. 8.2-203					
1	1.42	1.416704	0.850002	14.999770	0.000226
2	1.41	1.426850	0.850007	14.999260	0.000739
3	1.40	1.437237	0.850001	14.999890	0.000107
4	1.39	1.447966	0.849995	15.000550	−0.000548
5	1.38	1.459138	0.949998	15.000170	−0.000173
6	1.37	1.470659	0.849998	15.000210	−0.000215
7	1.36	1.482628	0.850003	14.999700	0.000298
8	1.35	1.494953	0.850000	15.000030	−0.000036
9	1.34	1.507832	0.850009	14.999150	0.000846
10	1.33	1.521074	0.850004	14.999620	0.000381
11	1.32	1.534878	0.850005	14.999460	0.000536
12	1.31	1.549247	0.850010	14.999050	0.000954
13	1.30	1.563990	0.849992	15.000800	−0.000799
14	1.29	1.579502	0.849991	15.000890	−0.000894
15	1.28	1.595788	0.850001	14.999880	0.000119
16	1.27	1.612656	0.849998	15.000190	−0.000191
17	1.26	1.630307	0.849996	15.000370	−0.000370
18	1.25	1.648940	0.850008	14.999170	0.000829
19	1.24	1.668170	0.849992	15.000790	−0.000793
20	1.23	1.688587	0.849994	15.000620	−0.000620
21	1.22	1.710197	0.850004	14.999570	0.000429
22	1.21	1.732810	0.849999	15.000150	−0.000155
23	1.20	1.756822	0.849999	15.000140	−0.000143
24	1.19	1.782434	0.850007	14.999280	0.000715
25	1.18	1.809457	0.849998	15.000250	−0.000250
26	1.17	1.838678	0.850009	14.999100	0.000900

(continued)

8.2 Contents for Table 8.2 (Table No. 8.2-1 to 400)

Table 8.2 (continued)

Parameter values					
PROB. = 0.85	RHO = 0.0		BIGH = 1.4184		SMHL = 1.04
Pair No.	SMH	SMK	PROB. level	Risk level (%)	Risk error (%)
27	1.16	1.869712	0.849998	15.000180	−0.000179
28	1.15	1.903348	0.849995	15.000520	−0.000525
29	1.14	1.940179	0.850006	14.999370	0.000632
30	1.13	1.980015	0.849994	15.000650	−0.000650
31	1.12	2.024427	0.849999	15.000070	−0.000072
32	1.11	2.073813	0.849995	15.000490	−0.000489
33	1.10	2.130134	0.850003	14.999700	0.000298
34	1.09	2.194960	0.850002	14.999830	0.000173
35	1.08	2.272206	0.850008	14.999210	0.000787
36	1.07	2.366960	0.849999	15.000090	−0.000089
37	1.06	2.491729	0.849991	15.000950	−0.000954
38	1.05	2.680117	0.850002	14.999840	0.000161
39	1.04	3.096978	0.849998	15.000160	−0.000167
No. of pairs		Mean COMPUT. risk errors		STD. ERR. errors	
39		0.0000205636		0.0000844890	
Z-ratio				Status of significance	
0.24339				N.S.—not significant	
Parameter values					
PROB. = 0.8	RHO = 0.0		BIGH = 1.2504		SMHL = 0.85
Pair No.	SMH	SMK	PROB. level	Risk level (%)	Risk error (%)
Table No. 8.2-204					
1	1.26	1.240971	0.799999	20.000060	−0.000060
2	1.25	1.250800	0.799993	20.000710	−0.000709
3	1.24	1.260985	0.799997	20.000300	−0.000304
4	1.23	1.271529	0.800010	19.999040	0.000960
5	1.22	1.282241	0.799998	20.000180	−0.000179
6	1.21	1.293321	0.799992	20.000760	−0.000757
7	1.20	1.304870	0.800005	19.999520	0.000477
8	1.19	1.316698	0.800003	19.999660	0.000340
9	1.18	1.328906	0.800001	19.999870	0.000125
10	1.17	1.341598	0.800010	19.999020	0.000978
11	1.16	1.354583	0.799999	20.000110	−0.000113
12	1.15	1.368159	0.800007	19.999300	0.000697
13	1.14	1.382038	0.799991	20.000920	−0.000924
14	1.13	1.396617	0.800000	20.000010	−0.000012
15	1.12	1.411607	0.799992	20.000820	−0.000817
16	1.11	1.427309	0.800001	19.999950	0.000048
17	1.10	1.443629	0.800009	19.999120	0.000882
18	1.09	1.460478	0.800001	19.999930	0.000072
19	1.08	1.478154	0.800006	19.999370	0.000626
20	1.07	1.496566	0.800008	19.999180	0.000817
21	1.06	1.515722	0.800001	19.999950	0.000048

(continued)

Table 8.2 (continued)

Parameter values					
PROB. = 0.8	RHO = 0.0		BIGH = 1.2504		SMHL = 0.85
Pair No.	SMH	SMK	PROB. level	Risk level (%)	Risk error (%)
22	1.05	1.535923	0.800009	19.999150	0.000852
23	1.04	1.556881	0.799993	20.000660	−0.000662
24	1.03	1.579094	0.799997	20.000270	−0.000268
25	1.02	1.602374	0.799991	20.000940	−0.000942
26	1.01	1.627121	0.799999	20.000090	−0.000095
27	1.00	1.653344	0.800009	19.999090	0.000912
28	0.99	1.680855	0.799991	20.000890	−0.000888
29	0.98	1.710642	0.800010	19.999050	0.000954
30	0.97	1.742129	0.799998	20.000190	−0.000191
31	0.96	1.776302	0.800008	19.999200	0.000805
32	0.95	1.812977	0.799999	20.000090	−0.000089
33	0.94	1.853141	0.800004	19.999620	0.000376
34	0.93	1.897198	0.800006	19.999450	0.000554
35	0.92	1.945940	0.800001	19.999900	0.000101
36	0.91	2.000944	0.800008	19.999250	0.000751
37	0.90	2.063394	0.799998	20.000160	−0.000161
38	0.89	2.136428	0.799993	20.000730	−0.000727
39	0.88	2.225141	0.800004	19.999610	0.000393
40	0.87	2.336969	0.799997	20.000280	−0.000280
41	0.86	2.493022	0.800007	19.999320	0.000679
42	0.85	2.758161	0.800006	19.999430	0.000566
No. of pairs		Mean COMPUT. risk errors		STD. ERR. errors	
42		0.0001124171		0.0000902447	
Z-ratio				Status of significance	
1.24569				N.S.—not significant	
Parameter values					
PROB. = 0.75	RHO = 0.0		BIGH = 1.1078		SMHL = 0.68
Pair No.	SMH	SMK	PROB. level	Risk level (%)	Risk error (%)
Table No. 8.2-205					
1	1.11	1.105604	0.750001	24.999950	0.000048
2	1.10	1.115655	0.749996	25.000440	−0.000435
3	1.09	1.126086	0.750009	24.999100	0.000906
4	1.08	1.136706	0.750004	24.999630	0.000376
5	1.07	1.147619	0.749996	25.000380	−0.000381
6	1.06	1.158927	0.750002	24.999770	0.000226
7	1.05	1.170540	0.750003	24.999690	0.000310
8	1.04	1.182461	0.749997	25.000290	−0.000292
9	1.03	1.194797	0.749999	25.000100	−0.000101
10	1.02	1.207455	0.749991	25.000940	−0.000942
11	1.01	1.220637	0.750002	24.999850	0.000149
12	1.00	1.234154	0.749998	25.000230	−0.000226
13	0.99	1.248211	0.750008	24.999250	0.000751

(continued)

8.2 Contents for Table 8.2 (Table No. 8.2-1 to 400)

Table 8.2 (continued)

Parameter values					
PROB. = 0.75	RHO = 0.0		BIGH = 1.1078		SMHL = 0.68
Pair No.	SMH	SMK	PROB. level	Risk level (%)	Risk error (%)
14	0.98	1.262618	0.749998	25.000190	−0.000191
15	0.97	1.277578	0.749996	25.000390	−0.000393
16	0.96	1.293101	0.749998	25.000200	−0.000203
17	0.95	1.309194	0.750000	24.999990	0.000012
18	0.94	1.325866	0.749998	25.000210	−0.000209
19	0.93	1.343225	0.750001	24.999890	0.000107
20	0.92	1.361279	0.750004	24.999600	0.000399
21	0.91	1.380040	0.750002	24.999850	0.000155
22	0.90	1.399614	0.750000	24.999980	0.000018
23	0.89	1.420111	0.750005	24.999510	0.000489
24	0.88	1.441444	0.749997	25.000330	−0.000328
25	0.87	1.463919	0.750001	24.999880	0.000119
26	0.86	1.487548	0.750008	24.999200	0.000799
27	0.85	1.512344	0.750007	24.999350	0.000650
28	0.84	1.538516	0.750005	24.999500	0.000501
29	0.83	1.566079	0.749990	25.000960	−0.000960
30	0.82	1.595634	0.750001	24.999860	0.000143
31	0.81	1.627001	0.750001	24.999950	0.000048
32	0.80	1.660588	0.750000	25.000000	0.000000
33	0.79	1.696803	0.750005	24.999500	0.000501
34	0.78	1.735860	0.750000	25.000010	−0.000006
35	0.77	1.778558	0.750003	24.999720	0.000286
36	0.76	1.825311	0.749994	25.000620	−0.000614
37	0.75	1.877700	0.750008	24.999170	0.000834
38	0.74	1.936140	0.749993	25.000660	−0.000662
39	0.73	2.003780	0.750005	24.999520	0.000477
40	0.72	2.082988	0.750003	24.999720	0.000286
41	0.71	2.179260	0.749993	25.000750	−0.000745
42	0.70	2.304734	0.750008	24.999170	0.000829
43	0.69	2.484048	0.750000	25.000000	0.000006
44	0.68	2.829735	0.749997	25.000320	−0.000322
No. of pairs		Mean COMPUT. risk errors		STD. ERR. errors	
44		0.0000536442		0.0000693435	
Z-ratio				Status of significance	
0.77360				N.S.—not significant	
Parameter values					
PROB. = 0.7	RHO = 0.0		BIGH = 0.9808		SMHL = 0.53
Pair No.	SMH	SMK	PROB. level	Risk level (%)	Risk error (%)
Table No. 8.2-206					
1	0.99	0.971783	0.700006	29.999380	0.000626
2	0.98	0.981601	0.699990	30.000960	−0.000960
3	0.97	0.991818	0.700004	29.999600	0.000405

(continued)

Table 8.2 (continued)

Parameter values					
PROB. = 0.7	RHO = 0.0		BIGH = 0.9808		SMHL = 0.53
Pair No.	SMH	SMK	PROB. level	Risk level (%)	Risk error (%)
4	0.96	1.002246	0.700006	29.999370	0.000632
5	0.95	1.012892	0.699997	30.000330	−0.000328
6	0.94	1.023859	0.699994	30.000620	−0.000614
7	0.93	1.035156	0.699997	30.000320	−0.000322
8	0.92	1.046790	0.700004	29.999620	0.000387
9	0.91	1.058671	0.699996	30.000420	−0.000417
10	0.90	1.071002	0.700008	29.999230	0.000769
11	0.89	1.083598	0.700002	29.999800	0.000203
12	0.88	1.096564	0.699996	30.000440	−0.000435
13	0.87	1.110008	0.700004	29.999640	0.000364
14	0.86	1.123842	0.700007	29.999310	0.000697
15	0.85	1.138075	0.700004	29.999630	0.000376
16	0.84	1.152818	0.700008	29.999170	0.000829
17	0.83	1.167983	0.700002	29.999800	0.000203
18	0.82	1.183679	0.699998	30.000160	−0.000161
19	0.81	1.199920	0.699994	30.000590	−0.000590
20	0.80	1.216816	0.700001	29.999920	0.000077
21	0.79	1.234283	0.700000	30.000020	−0.000018
22	0.78	1.252433	0.700002	29.999850	0.000149
23	0.77	1.271282	0.700001	29.999880	0.000125
24	0.76	1.290943	0.700007	29.999260	0.000745
25	0.75	1.311335	0.700001	29.999890	0.000113
26	0.74	1.332672	0.700002	29.999810	0.000197
27	0.73	1.354874	0.699990	30.000980	−0.000978
28	0.72	1.378254	0.699994	30.000600	−0.000602
29	0.71	1.402736	0.699992	30.000850	−0.000846
30	0.70	1.428537	0.699995	30.000540	−0.000542
31	0.69	1.455680	0.699991	30.000910	−0.000912
32	0.68	1.484581	0.700007	29.999310	0.000697
33	0.67	1.515070	0.700006	29.999420	0.000584
34	0.66	1.547566	0.700007	29.999270	0.000727
35	0.65	1.582295	0.700006	29.999380	0.000620
36	0.64	1.619481	0.699994	30.000620	−0.000614
37	0.63	1.659941	0.700001	29.999920	0.000077
38	0.62	1.703905	0.700001	29.999920	0.000083
39	0.61	1.752192	0.700001	29.999910	0.000095
40	0.60	1.805624	0.699991	30.000870	−0.000864
41	0.59	1.866001	0.699995	30.000460	−0.000459
42	0.58	1.935128	0.699997	30.000280	−0.000274
43	0.57	2.016569	0.700010	29.999040	0.000960
44	0.56	2.115066	0.700001	29.999920	0.000083
45	0.55	2.242001	0.699994	30.000630	−0.000632

(continued)

8.2 Contents for Table 8.2 (Table No. 8.2-1 to 400)

Table 8.2 (continued)

Parameter values					
PROB. = 0.7	RHO = 0.0		BIGH = 0.9808		SMHL = 0.53
Pair No.	SMH	SMK	PROB. level	Risk level (%)	Risk error (%)
46	0.54	2.425172	0.700005	29.999510	0.000489
47	0.53	2.783783	0.700007	29.999300	0.000703
No. of pairs		Mean COMPUT. risk errors		STD. ERR. errors	
47		0.0000301749		0.0000800908	
Z-ratio				Status of significance	
0.37676				N.S.—not significant	
Parameter values					
PROB. = 0.65	RHO = 0.0		BIGH = 0.8641		SMHL = 0.39
Pair No.	SMH	SMK	PROB. level	Risk level (%)	Risk error (%)
Table No. 8.2-207					
1	0.87	0.858142	0.649990	35.000970	−0.000966
2	0.86	0.868122	0.649991	35.000920	−0.000918
3	0.85	0.878385	0.650000	35.000030	−0.000024
4	0.84	0.888891	0.650006	34.999380	0.000620
5	0.83	0.899601	0.650000	35.000030	−0.000030
6	0.82	0.910572	0.649991	35.000940	−0.000936
7	0.81	0.921911	0.649999	35.000090	−0.000083
8	0.80	0.933531	0.650004	34.999580	0.000423
9	0.79	0.945443	0.650006	34.999440	0.000566
10	0.78	0.957658	0.650003	34.999690	0.000310
11	0.77	0.970188	0.649996	35.000360	−0.000358
12	0.76	0.983142	0.650004	34.999650	0.000358
13	0.75	0.996437	0.650005	34.999490	0.000513
14	0.74	1.010086	0.650000	34.999960	0.000048
15	0.73	1.024201	0.650007	34.999310	0.000691
16	0.72	1.038700	0.650005	34.999470	0.000530
17	0.71	1.053599	0.649995	35.000500	−0.000495
18	0.70	1.069013	0.649992	35.000790	−0.000793
19	0.69	1.084961	0.649994	35.000560	−0.000554
20	0.68	1.101460	0.650000	34.999990	0.000012
21	0.67	1.118532	0.650007	34.999310	0.000691
22	0.66	1.136101	0.649997	35.000260	−0.000262
23	0.65	1.154385	0.650000	35.000020	−0.000012
24	0.64	1.173310	0.649996	35.000400	−0.000393
25	0.63	1.193001	0.649997	35.000260	−0.000262
26	0.62	1.213484	0.650000	35.000040	−0.000036
27	0.61	1.234789	0.649998	35.000170	−0.000167
28	0.60	1.257045	0.650002	34.999830	0.000173
29	0.59	1.280188	0.649991	35.000870	−0.000870
30	0.58	1.304547	0.649998	35.000240	−0.000232
31	0.57	1.330062	0.650000	35.000000	0.000006
32	0.56	1.356872	0.650001	34.999910	0.000095

(continued)

Table 8.2 (continued)

Parameter values					
PROB. = 0.65	RHO = 0.0		BIGH = 0.8641		SMHL = 0.39
Pair No.	SMH	SMK	PROB. level	Risk level (%)	Risk error (%)
33	0.55	1.385118	0.650001	34.999920	0.000083
34	0.54	1.414946	0.649998	35.000220	−0.000221
35	0.53	1.446699	0.650007	34.999320	0.000685
36	0.52	1.480432	0.650009	34.999070	0.000936
37	0.51	1.516399	0.650003	34.999680	0.000322
38	0.50	1.555055	0.649998	35.000200	−0.000197
39	0.49	1.596860	0.649995	35.000540	−0.000536
40	0.48	1.642473	0.649999	35.000130	−0.000125
41	0.47	1.692562	0.650002	34.999790	0.000215
42	0.46	1.748192	0.650006	34.999360	0.000644
43	0.45	1.810434	0.649990	35.000970	−0.000966
44	0.44	1.882129	0.649992	35.000850	−0.000846
45	0.43	1.966515	0.649996	35.000420	−0.000417
46	0.42	2.069578	0.650002	34.999800	0.000209
47	0.41	2.216454	0.650003	34.999660	0.000346
48	0.40	2.407294	0.649996	35.000400	−0.000399
49	0.39	2.795782	0.650006	34.999450	0.000554
No. of pairs		Mean COMPUT. risk errors		STD. ERR. errors	
49		−0.0000413656		0.0000712121	
Z-ratio				Status of significance	
0.58088				N.S.—not significant	
Parameter values					
PROB. = 0.6	RHO = 0.0		BIGH = 0.7541		SMHL = 0.27
Pair No.	SMH	SMK	PROB. level	Risk level (%)	Risk error (%)
Table No. 8.2-208					
1	0.76	0.748148	0.599992	40.000790	−0.000787
2	0.75	0.758125	0.599992	40.000840	−0.000840
3	0.74	0.768371	0.600000	40.000020	−0.000018
4	0.73	0.778800	0.599995	40.000490	−0.000495
5	0.72	0.789522	0.600000	39.999980	0.000018
6	0.71	0.800451	0.599994	40.000590	−0.000590
7	0.70	0.811698	0.599999	40.000110	−0.000107
8	0.69	0.823178	0.599994	40.000570	−0.000572
9	0.68	0.835005	0.600002	39.999810	0.000185
10	0.67	0.847096	0.600002	39.999830	0.000167
11	0.66	0.859468	0.599995	40.000540	−0.000536
12	0.65	0.872235	0.600002	39.999830	0.000167
13	0.64	0.885319	0.600004	39.999630	0.000370
14	0.63	0.898739	0.600001	39.999860	0.000143
15	0.62	0.912517	0.599996	40.000370	−0.000370
16	0.61	0.926772	0.600008	39.999250	0.000745
17	0.60	0.941333	0.599999	40.000120	−0.000119
18	0.59	0.956420	0.600008	39.999230	0.000763

(continued)

8.2 Contents for Table 8.2 (Table No. 8.2-1 to 400)

Table 8.2 (continued)

Parameter values					
PROB. = 0.6	RHO = 0.0		BIGH = 0.7541		SMHL = 0.27
Pair No.	SMH	SMK	PROB. level	Risk level (%)	Risk error (%)
19	0.58	0.971866	0.599999	40.000110	−0.000107
20	0.57	0.987895	0.600009	39.999120	0.000882
21	0.56	1.004343	0.600003	39.999660	0.000340
22	0.55	1.021342	0.600001	39.999950	0.000048
23	0.54	1.038926	0.600001	39.999930	0.000072
24	0.53	1.057136	0.600004	39.999570	0.000429
25	0.52	1.075914	0.599996	40.000380	−0.000381
26	0.51	1.095501	0.600007	39.999290	0.000709
27	0.50	1.115653	0.599992	40.000790	−0.000787
28	0.49	1.136716	0.599996	40.000430	−0.000435
29	0.48	1.158648	0.600002	39.999840	0.000155
30	0.47	1.181413	0.599996	40.000420	−0.000423
31	0.46	1.205177	0.599991	40.000940	−0.000936
32	0.45	1.230110	0.599997	40.000340	−0.000346
33	0.44	1.256291	0.600010	39.999030	0.000972
34	0.43	1.283613	0.600003	39.999720	0.000280
35	0.42	1.312366	0.599995	40.000460	−0.000465
36	0.41	1.342851	0.600002	39.999760	0.000232
37	0.40	1.374986	0.599994	40.000580	−0.000578
38	0.39	1.409291	0.600001	39.999920	0.000083
39	0.38	1.445904	0.600007	39.999280	0.000721
40	0.37	1.484983	0.599997	40.000300	−0.000298
41	0.36	1.527285	0.599999	40.000090	−0.000089
42	0.35	1.573198	0.599999	40.000130	−0.000131
43	0.34	1.623524	0.600001	39.999900	0.000101
44	0.33	1.700770	0.599999	40.000080	−0.000083
45	0.32	1.761066	0.599993	40.000750	−0.000751
46	0.31	1.830500	0.600005	39.999510	0.000489
47	0.30	1.911569	0.600001	39.999880	0.000119
48	0.29	2.010136	0.600005	39.999460	0.000536
49	0.28	2.135637	0.599990	40.000980	−0.000978
50	0.27	2.313982	0.599997	40.000300	−0.000298
No. of pairs		Mean COMPUT. risk errors		STD. ERR. errors	
50		−0.0000548129		0.0000689236	
Z-ratio				Status of significance	
0.79527				N.S.—not significant	
Parameter values					
PROB. = 0.55	RHO = 0.0		BIGH = 0.6484		SMHL = 0.16
Pair No.	SMH	SMK	PROB. level	Risk level (%)	Risk error (%)
Table No. 8.2-209					
1	0.65	0.646706	0.550002	44.999790	0.000209
2	0.64	0.656813	0.550010	44.999020	0.000983

(continued)

Table 8.2 (continued)

Parameter values					
PROB. = 0.55	RHO = 0.0		BIGH = 0.6484		SMHL = 0.16
Pair No.	SMH	SMK	PROB. level	Risk level (%)	Risk error (%)
3	0.63	0.667044	0.549995	45.000540	−0.000536
4	0.62	0.677613	0.550004	44.999570	0.000435
5	0.61	0.688338	0.549995	45.000480	−0.000477
6	0.60	0.699337	0.549992	45.000770	−0.000769
7	0.59	0.710627	0.549997	45.000270	−0.000268
8	0.58	0.722132	0.549991	45.000900	−0.000900
9	0.57	0.733970	0.549997	45.000260	−0.000256
10	0.56	0.746067	0.549998	45.000220	−0.000221
11	0.55	0.758448	0.549995	45.000480	−0.000483
12	0.54	0.771138	0.549993	45.000720	−0.000721
13	0.53	0.784168	0.549994	45.000630	−0.000626
14	0.52	0.797567	0.550001	44.999880	0.000125
15	0.51	0.811272	0.549999	45.000070	−0.000066
16	0.50	0.825318	0.549992	45.000780	−0.000775
17	0.49	0.839841	0.550003	44.999670	0.000328
18	0.48	0.854689	0.549999	45.000100	−0.000095
19	0.47	0.870004	0.550003	44.999690	0.000316
20	0.46	0.885739	0.550002	44.999770	0.000232
21	0.45	0.901948	0.550003	44.999740	0.000256
22	0.44	0.918591	0.549992	45.000790	−0.000787
23	0.43	0.935832	0.549995	45.000480	−0.000483
24	0.42	0.953642	0.550001	44.999910	0.000089
25	0.41	0.972002	0.550000	44.999980	0.000018
26	0.40	0.990997	0.550002	44.999850	0.000149
27	0.39	1.010623	0.549997	45.000260	−0.000256
28	0.38	1.030984	0.549998	45.000250	−0.000250
29	0.37	1.052097	0.549996	45.000360	−0.000364
30	0.36	1.074090	0.550005	44.999540	0.000459
31	0.35	1.096910	0.550005	44.999510	0.000489
32	0.34	1.120716	0.550010	44.999010	0.000995
33	0.33	1.145491	0.550006	44.999360	0.000644
34	0.32	1.171437	0.550009	44.999120	0.000882
35	0.31	1.198583	0.550007	44.999270	0.000733
36	0.30	1.226993	0.549995	45.000480	−0.000477
37	0.29	1.256958	0.549990	45.000970	−0.000972
38	0.28	1.288812	0.550009	44.999080	0.000924
39	0.27	1.322353	0.550008	44.999240	0.000757
40	0.26	1.386344	0.549996	45.000410	−0.000411
41	0.25	1.422313	0.550005	44.999530	0.000477
42	0.24	1.460938	0.549997	45.000290	−0.000286
43	0.23	1.502921	0.549995	45.000480	−0.000483
44	0.22	1.548899	0.550000	44.999970	0.000030

(continued)

8.2 Contents for Table 8.2 (Table No. 8.2-1 to 400)

Table 8.2 (continued)

Parameter values					
PROB. = 0.55	RHO = 0.0		BIGH = 0.6484		SMHL = 0.16
Pair No.	SMH	SMK	PROB. level	Risk level (%)	Risk error (%)
45	0.21	1.599470	0.549998	45.000250	−0.000250
46	0.20	1.656015	0.550007	44.999280	0.000721
47	0.19	1.719582	0.549999	45.000070	−0.000072
48	0.18	1.792708	0.549993	45.000750	−0.000745
49	0.17	1.879320	0.550001	44.999910	0.000095
50	0.16	1.985058	0.549998	45.000200	−0.000203
No. of pairs		Mean COMPUT. risk errors		STD. ERR. errors	
50		−0.0000369315		0.0000744960	
Z-ratio				Status of significance	
0.49575				N.S.—not significant	
Parameter values					
PROB. = 0.5	RHO = 0.0		BIGH = 0.545		SMHL = 0.06
Pair No.	SMH	SMK	PROB. level	Risk level (%)	Risk error (%)
Table No. 8.2-210					
1	0.55	0.539948	0.500004	49.999580	0.000429
2	0.54	0.549949	0.500005	49.999540	0.000459
3	0.53	0.560132	0.499998	50.000170	−0.000167
4	0.52	0.570567	0.500001	49.999910	0.000089
5	0.51	0.581230	0.500004	49.999580	0.000429
6	0.50	0.592097	0.500001	49.999860	0.000143
7	0.49	0.603244	0.500008	49.999190	0.000817
8	0.48	0.614603	0.500007	49.999320	0.000679
9	0.47	0.626206	0.500003	49.999700	0.000304
10	0.46	0.638089	0.500003	49.999750	0.000250
11	0.45	0.650241	0.500001	49.999940	0.000060
12	0.44	0.662654	0.499994	50.000640	−0.000638
13	0.43	0.675424	0.499999	50.000070	−0.000066
14	0.42	0.688452	0.499995	50.000530	−0.000528
15	0.41	0.701892	0.500009	49.999130	0.000870
16	0.40	0.715512	0.499991	50.000920	−0.000918
17	0.39	0.729571	0.499992	50.000850	−0.000843
18	0.38	0.744047	0.500002	49.000770	0.000232
19	0.37	0.758825	0.499997	50.000270	−0.000268
20	0.36	0.774093	0.500010	49.999040	0.000960
21	0.35	0.789658	0.499999	50.000120	−0.000116
22	0.34	0.805731	0.500001	49.999930	0.000077
23	0.33	0.822243	0.499998	50.000170	−0.000170
24	0.32	0.839238	0.499995	50.000510	−0.000507
25	0.31	0.856777	0.499997	50.000310	−0.000313
26	0.30	0.874848	0.499997	50.000350	−0.000343
27	0.29	0.893559	0.500007	49.999330	0.000668
28	0.28	0.912756	0.499996	50.000410	−0.000402

(continued)

Table 8.2 (continued)

Parameter values					
PROB. = 0.5	RHO = 0.0		BIGH = 0.545		SMHL = 0.06
Pair No.	SMH	SMK	PROB. level	Risk level (%)	Risk error (%)
29	0.27	0.932709	0.500002	49.999780	0.000226
30	0.26	0.953340	0.500006	49.999430	0.000572
31	0.25	0.974622	0.499995	50.000470	−0.000465
32	0.24	0.996782	0.499997	50.000270	−0.000259
33	0.23	1.019830	0.500004	49.999640	0.000364
34	0.22	1.043667	0.499991	50.000880	−0.000882
35	0.21	1.106688	0.499998	50.000210	−0.000209
36	0.20	1.129849	0.499993	50.000660	−0.000659
37	0.19	1.154303	0.499994	50.000590	−0.000590
38	0.18	1.180215	0.500005	49.999490	0.000513
39	0.17	1.207555	0.500005	49.999530	0.000477
40	0.16	1.236482	0.499994	50.000570	−0.000566
41	0.15	1.267468	0.500005	49.999480	0.000525
42	0.14	1.300314	0.499991	50.000900	−0.000900
43	0.13	1.335780	0.500002	49.999790	0.000215
44	0.12	1.373837	0.500002	49.999830	0.000173
45	0.11	1.415066	0.500005	49.999460	0.000542
46	0.10	1.459892	0.500004	49.999570	0.000435
47	0.09	1.509059	0.500003	49.999670	0.000334
48	0.08	1.563584	0.500009	49.999110	0.000894
49	0.07	1.624452	0.500003	49.999710	0.000292
50	0.06	1.693759	0.500003	49.999750	0.000256
No. of pairs		Mean COMPUT. risk errors		STD. ERR. errors	
50		0.0000485603		0.0000711052	
Z-ratio				Status of significance	
0.68294				N.S.—not significant	
Parameter values					
PROB. = 0.99	RHO = −0.1		BIGH = 2.5757		SMHL = 2.34
Pair No.	SMH	SMK	PROB. level	Risk level (%)	Risk error (%)
Table No. 8.2-211					
1	2.58	2.571407	0.990005	0.999540	0.000459
2	2.57	2.581413	0.990004	0.999582	0.000417
3	2.56	2.591497	0.989997	1.000285	−0.000286
4	2.55	2.603222	0.990005	0.999510	0.000489
5	2.54	2.615027	0.990005	0.999510	0.000489
6	2.53	2.626913	0.989997	1.000285	−0.000286
7	2.52	2.640444	0.990001	0.999934	0.000066
8	2.51	2.654058	0.989995	1.000518	−0.000519
9	2.50	2.669317	0.989998	1.000232	−0.000232
10	2.49	2.686225	0.990007	0.999326	0.000674
11	2.48	2.703218	0.990004	0.999606	0.000393
12	2.47	2.721861	0.990005	0.999528	0.000471

(continued)

8.2 Contents for Table 8.2 (Table No. 8.2-1 to 400)

Table 8.2 (continued)

Parameter values					
PROB. = 0.99	RHO = −0.1		BIGH = 2.5757		SMHL = 2.34
Pair No.	SMH	SMK	PROB. level	Risk level (%)	Risk error (%)
13	2.46	2.740592	0.989992	1.000798	−0.000799
14	2.45	2.762537	0.990000	0.999987	0.000012
15	2.44	2.786134	0.989999	1.000065	−0.000066
16	2.43	2.811386	0.989994	1.000643	−0.000644
17	2.42	2.839855	0.989992	1.000840	−0.000840
18	2.41	2.874668	0.990009	0.999069	0.000930
19	2.40	2.908013	0.989991	1.000905	−0.000906
20	2.39	2.950829	0.989998	1.000249	−0.000250
21	2.38	2.999992	0.989999	1.000077	−0.000077
22	2.37	3.058628	0.989999	1.000136	−0.000137
23	2.36	3.129864	0.989992	1.000834	−0.000834
24	2.35	3.235577	0.990009	0.999117	0.000882
25	2.34	3.385141	0.990004	0.999606	0.000393
No. of pairs		Mean COMPUT. risk errors		STD. ERR. errors	
25		−0.0000077945		0.0001063174	
Z-ratio				Status of significance	
0.07331				N.S.—not significant	
Parameter values					
PROB. = 0.98	RHO = −0.1		BIGH = 2.3255		SMHL = 2.06
Pair No.	SMH	SMK	PROB. level	Risk level (%)	Risk error (%)
Table No. 8.2-212					
1	2.33	2.321009	0.980000	2.000034	−0.000036
2	2.32	2.331013	0.979999	2.000070	−0.000072
3	2.31	2.341885	0.980008	1.999158	0.000840
4	2.30	2.352845	0.980006	1.999360	0.000638
5	2.29	2.363894	0.979993	2.000713	−0.000715
6	2.28	2.376596	0.980005	1.999545	0.000453
7	2.27	2.389388	0.980003	1.999742	0.000256
8	2.26	2.403055	0.980004	1.999599	0.000399
9	2.25	2.417596	0.980007	1.999283	0.000715
10	2.24	2.432232	0.979994	2.000570	−0.000572
11	2.23	2.448527	0.979996	2.000409	−0.000411
12	2.22	2.465701	0.979994	2.000600	−0.000602
13	2.21	2.484536	0.980002	1.999819	0.000179
14	2.20	2.504253	0.980000	1.999998	0.000000
15	2.19	2.525634	0.980001	1.999891	0.000107
16	2.18	2.548680	0.980001	1.999927	0.000072
17	2.17	2.573393	0.989995	2.000457	−0.000459
18	2.16	2.600556	0.989992	2.000773	−0.000775
19	2.15	2.630951	0.989995	2.000475	−0.000477
20	2.14	2.664580	0.989996	2.000392	−0.000393
21	2.13	2.703006	0.980004	1.999629	0.000370

(continued)

Table 8.2 (continued)

Parameter values					
PROB. = 0.98	RHO = −0.1		BIGH = 2.3255		SMHL = 2.06
Pair No.	SMH	SMK	PROB. level	Risk level (%)	Risk error (%)
22	2.12	2.744670	0.979991	2.000904	−0.000906
23	2.11	2.795822	0.980002	1.999807	0.000191
24	2.10	2.854901	0.980000	2.000034	−0.000036
25	2.09	2.928161	0.979999	2.000064	−0.000066
26	2.08	3.024976	0.980004	1.999635	0.000364
27	2.07	3.162536	0.979998	2.000207	−0.000209
28	2.06	3.425218	0.979996	2.000415	−0.000417
No. of pairs		Mean COMPUT. risk errors		STD. ERR. errors	
28		−0.0000538497		0.0000855209	
Z-ratio				Status of significance	
0.62967				N.S.—not significant	
Parameter values					
PROB. = 0.97	RHO = −0.1		BIGH = 2.1687		SMHL = 1.89
Pair No.	SMH	SMK	PROB. level	Risk level (%)	Risk error (%)
Table No. 8.2-213					
1	2.17	2.167401	0.970007	2.999270	0.000727
2	2.16	2.177435	0.970002	2.999765	0.000232
3	2.15	2.187953	0.969999	3.000146	−0.000149
4	2.14	2.199347	0.970008	2.999187	0.000811
5	2.13	2.210837	0.970002	2.999771	0.000226
6	2.12	2.223206	0.970007	2.999330	0.000668
7	2.11	2.236064	0.970007	2.999312	0.000685
8	2.10	2.249413	0.970001	2.999884	0.000113
9	2.09	2.263644	0.970000	2.999967	0.000030
10	2.08	2.278370	0.969991	3.000951	−0.000954
11	2.07	2.294762	0.970004	2.999616	0.000381
12	2.06	2.311261	0.969993	3.000683	−0.000685
13	2.05	2.329429	0.969999	3.000146	−0.000149
14	2.04	2.348488	0.969996	3.000391	−0.000393
15	2.03	2.369220	0.970002	2.999812	0.000185
16	2.02	2.390846	0.969993	3.000695	−0.000697
17	2.01	2.414930	0.970001	2.999914	0.000083
18	2.00	2.440692	0.970002	2.999842	0.000155
19	1.99	2.468134	0.969990	3.000957	−0.000960
20	1.98	2.499602	0.970003	2.999681	0.000316
21	1.97	2.533535	0.970001	2.999878	0.000119
22	1.96	2.571497	0.970000	3.000039	−0.000042
23	1.95	2.615052	0.970005	2.999479	0.000519
24	1.94	2.664985	0.970008	2.999216	0.000781
25	1.93	2.722859	0.970001	2.999926	0.000072
26	1.92	2.793364	0.969996	3.000432	−0.000435
27	1.91	2.884314	0.969997	3.000319	−0.000322

(continued)

Table 8.2 (continued)

Parameter values					
PROB. = 0.97	RHO = −0.1		BIGH = 2.1687		SMHL = 1.89
Pair No.	SMH	SMK	PROB. level	Risk level (%)	Risk error (%)
28	1.90	3.012899	0.970006	2.999413	0.000584
29	1.89	3.225996	0.970001	2.999908	0.000089
No. of pairs		Mean COMPUT. risk errors		STD. ERR. errors	
29		0.0000663598		0.0000892258	
Z-ratio				Status of significance	
0.74373				N.S.—not significant	
Parameter values					
PROB. = 0.96	RHO = −0.1		BIGH = 2.0516		SMHL = 1.76
Pair No.	SMH	SMK	PROB. level	Risk level (%)	Risk error (%)
Table No. 8.2-214					
1	2.06	2.043234	0.959999	4.000056	−0.000054
2	2.05	2.053201	0.960005	3.999543	0.000459
3	2.04	2.063266	0.959995	4.000550	−0.000548
4	2.03	2.074211	0.960005	3.999537	0.000465
5	2.02	2.085257	0.959997	4.000270	−0.000268
6	2.01	2.097186	0.960007	3.999341	0.000662
7	2.00	1.109219	0.959997	4.000342	−0.000340
8	1.99	2.122138	0.959999	4.000104	−0.000101
9	1.98	2.135555	0.959996	4.000425	−0.000423
10	1.97	2.149861	0.960001	3.999943	0.000060
11	1.96	2.164669	0.959995	4.000479	−0.000477
12	1.95	2.180369	0.959993	4.000664	−0.000662
13	1.94	2.196963	0.959991	4.000902	−0.000900
14	1.93	2.214846	0.959999	4.000092	−0.000089
15	1.92	2.233626	0.960000	3.999990	0.000012
16	1.91	2.253698	0.960003	3.999704	0.000298
17	1.90	2.274671	0.959992	4.000807	−0.000805
18	1.89	2.298111	0.960007	3.999257	0.000745
19	1.88	2.322457	0.959999	4.000116	−0.000113
20	1.87	2.349273	0.960002	3.999806	0.000197
21	1.86	2.378562	0.960007	3.999269	0.000733
22	1.85	2.410325	0.960005	3.999502	0.000501
23	1.84	2.445346	0.960001	3.999883	0.000119
24	1.83	2.484409	0.959997	4.000301	−0.000298
25	1.82	2.529077	0.960001	3.999919	0.000083
26	1.81	2.580136	0.960000	3.999978	0.000024
27	1.80	2.639930	0.959997	4.000330	−0.000328
28	1.79	2.713931	0.960008	3.999239	0.000763
29	1.78	2.805266	0.959993	4.000688	−0.000685

(continued)

Table 8.2 (continued)

Parameter values					
PROB. = 0.96	RHO = −0.1		BIGH = 2.0516		SMHL = 1.76
Pair No.	SMH	SMK	PROB. level	Risk level (%)	Risk error (%)
30	1.77	2.934251	0.959994	4.000569	−0.000566
31	1.76	3.154012	0.960005	3.999508	0.000495
No. of pairs		Mean COMPUT. risk errors		STD. ERR. errors	
31		−0.0000325963		0.0000837870	
Z-ratio				Status of significance	
0.38904				N.S.—not significant	
Parameter values					
PROB. = 0.95	RHO = −0.1		BIGH = 1.957		SMHL = 1.65
Pair No.	SMH	SMK	PROB. level	Risk level (%)	Risk error (%)
Table No. 8.2-215					
1	1.96	1.954005	0.950002	4.999852	0.000149
2	1.95	1.964025	0.949998	5.000198	−0.000197
3	1.94	1.974540	0.949999	5.000103	−0.000101
4	1.93	1.985549	0.950003	4.999704	0.000298
5	1.92	1.997057	0.950008	4.999185	0.000817
6	1.91	2.008672	0.949992	5.000770	−0.000769
7	1.90	2.021179	0.949995	5.000478	−0.000477
8	1.89	2.034187	0.949994	5.000615	−0.000614
9	1.88	2.048091	0.950006	4.999423	0.000578
10	1.87	2.062110	0.949991	5.000937	−0.000936
11	1.86	2.077418	0.950002	4.999829	0.000173
12	1.85	2.093236	0.950000	5.000037	−0.000036
13	1.84	2.109955	0.949999	5.000115	−0.000113
14	1.83	2.127579	0.949996	5.000401	−0.000399
15	1.82	2.146500	0.950003	4.999733	0.000268
16	1.81	2.166329	0.949999	5.000085	−0.000083
17	1.80	2.187850	0.950009	4.999054	0.000948
18	1.79	2.210283	0.950000	4.999995	0.000006
19	1.78	2.234413	0.949992	5.000758	−0.000757
20	1.77	2.260631	0.949991	5.000872	−0.000870
21	1.76	2.289332	0.949998	5.000174	−0.000173
22	1.75	2.320516	0.950002	4.999810	0.000191
23	1.74	2.354187	0.949991	5.000890	−0.000888
24	1.73	2.392691	0.950007	4.999280	0.000721
25	1.72	2.434468	0.949994	5.000562	−0.000560
26	1.71	2.483427	0.950009	4.999108	0.000894
27	1.70	2.538789	0.950001	4.999924	0.000077
28	1.69	2.605245	0.950002	4.999781	0.000221
29	1.68	2.687484	0.950006	4.999411	0.000590
30	1.67	2.793322	0.949993	5.000711	−0.000709

(continued)

8.2 Contents for Table 8.2 (Table No. 8.2-1 to 400)

Table 8.2 (continued)

Parameter values					
PROB. = 0.95	RHO = −0.1		BIGH = 1.957		SMHL = 1.65
Pair No.	SMH	SMK	PROB. level	Risk level (%)	Risk error (%)
31	1.66	2.950887	0.949995	5.000484	−0.000483
32	1.65	3.271119	0.950005	4.999519	0.000483
No. of pairs		Mean COMPUT. risk errors		STD. ERR. errors	
32		−0.0000531023		0.0000949015	
Z-ratio				Status of significance	
0.55955				N.S.—not significant	
Parameter values					
PROB. = 0.94	RHO = −0.1		BIGH = 1.8769		SMHL = 1.56
Pair No.	SMH	SMK	PROB. level	Risk level (%)	Risk error (%)
Table No. 8.2-216					
1	1.88	1.873805	0.939991	6.000871	−0.000870
2	1.87	1.884021	0.940001	5.999923	0.000077
3	1.86	1.894344	0.939991	6.000936	−0.000936
4	1.85	1.905363	0.939998	6.000161	−0.000161
5	1.84	1.916884	0.940009	5.999107	0.000894
6	1.83	1.928518	0.939996	6.000358	−0.000358
7	1.82	1.941047	0.940007	5.999279	0.000721
8	1.81	1.953694	0.939993	6.000752	−0.000751
9	1.80	1.967241	0.939996	6.000424	−0.000423
10	1.79	1.981300	0.939992	6.000793	−0.000793
11	1.78	1.996262	0.940000	6.000006	−0.000006
12	1.77	2.011740	0.939996	6.000429	−0.000429
13	1.76	2.028127	0.939997	6.000352	−0.000352
14	1.75	2.045424	0.939998	6.000185	−0.000185
15	1.74	2.063633	0.939997	6.000293	−0.000292
16	1.73	2.083149	0.940006	5.999363	0.000638
17	1.72	2.103581	0.940003	5.999673	0.000328
18	1.71	2.125324	0.940000	6.000042	−0.000042
19	1.70	2.148770	0.940004	5.999533	0.000447
20	1.69	2.173532	0.939995	6.000507	−0.000507
21	1.68	2.200783	0.940006	5.999434	0.000566
22	1.67	2.229746	0.939998	6.000173	−0.000173
23	1.66	2.261593	0.940000	6.000013	−0.000012
24	1.65	2.296721	0.940006	5.999380	0.000620
25	1.64	2.335520	0.940009	5.999070	0.000930
26	1.63	2.378386	0.939999	6.000096	−0.000095
27	1.62	2.427664	0.940002	5.999840	0.000161
28	1.61	2.484920	0.940008	5.999172	0.000829
29	1.60	2.551720	0.939997	6.000352	−0.000352
30	1.59	2.634318	0.939990	6.000978	−0.000978

(continued)

Table 8.2 (continued)

Parameter values					
PROB. = 0.94	RHO = −0.1		BIGH = 1.8769		SMHL = 1.56
Pair No.	SMH	SMK	PROB. level	Risk level (%)	Risk error (%)
31	1.58	2.743652	0.939997	6.000293	−0.000292
32	1.57	2.903166	0.939999	6.000138	−0.000137
33	1.56	3.220674	0.939998	6.000180	−0.000179
No. of pairs		Mean COMPUT. risk errors		STD. ERR. errors	
33		−0.0000620590		0.0000909885	
Z-ratio				Status of significance	
0.68205				N.S.—not significant	
Parameter values					
PROB. = 0.93	RHO = −0.1		BIGH = 1.8072		SMHL = 1.48
Pair No.	SMH	SMK	PROB. level	Risk level (%)	Risk error (%)
Table No. 8.2-217					
1	1.81	1.804404	0.930006	6.999451	0.000548
2	1.80	1.814429	0.930001	6.999910	0.000089
3	1.79	1.824956	0.930005	6.999487	0.000513
4	1.78	1.835792	0.930002	6.999779	0.000221
5	1.77	1.847135	0.930004	6.999583	0.000417
6	1.76	1.858791	0.929996	7.000417	−0.000417
7	1.75	1.871152	0.930002	6.999797	0.000203
8	1.74	1.884027	0.930007	6.999284	0.000715
9	1.73	1.897220	0.929996	7.000381	−0.000381
10	1.72	1.911321	0.930005	6.999541	0.000459
11	1.71	1.925940	0.930004	6.999564	0.000435
12	1.70	1.941081	0.929993	7.000661	−0.000662
13	1.69	1.957137	0.929992	7.000840	−0.000840
14	1.68	1.974109	0.929995	7.000536	−0.000536
15	1.67	1.992000	0.929998	7.000202	−0.000203
16	1.66	2.010812	0.929997	7.000309	−0.000310
17	1.65	2.030939	0.930007	6.999332	0.000668
18	1.64	2.051993	0.930002	6.999821	0.000179
19	1.63	2.074367	0.929995	7.000470	−0.000471
20	1.62	2.098454	0.929997	7.000304	−0.000304
21	1.61	2.124257	0.929998	7.000232	−0.000232
22	1.60	2.152170	0.930003	6.999702	0.000298
23	1.59	2.182195	0.930001	6.999922	0.000077
24	1.58	2.214727	0.929993	7.000727	−0.000727
25	1.57	2.250939	0.930001	6.999952	0.000048
26	1.56	2.290446	0.929991	7.000935	−0.000936
27	1.55	2.335203	0.929994	7.000566	−0.000566
28	1.54	2.386385	0.930005	6.999529	0.000471
29	1.53	2.444778	0.929996	7.000387	−0.000387
30	1.52	2.515071	0.930002	6.999773	0.000226
31	1.51	2.601957	0.930007	6.999338	0.000662

(continued)

8.2 Contents for Table 8.2 (Table No. 8.2-1 to 400)

Table 8.2 (continued)

Parameter values					
PROB. = 0.93	RHO = −0.1		BIGH = 1.8072		SMHL = 1.48
Pair No.	SMH	SMK	PROB. level	Risk level (%)	Risk error (%)
32	1.50	2.714814	0.929994	7.000643	−0.000644
33	1.49	2.885677	0.930003	6.999731	0.000268
34	1.48	3.244236	0.929994	7.000613	−0.000614
No. of pairs		Mean COMPUT. risk errors		STD. ERR. errors	
34		−0.0000495570		0.0000816918	
Z-ratio				Status of significance	
0.60663				N.S.—not significant	
Parameter values					
PROB. = 0.92	RHO = −0.1		BIGH = 1.7449		SMHL = 1.41
Pair No.	SMH	SMK	PROB. level	Risk level (%)	Risk error (%)
Table No. 8.2-218					
1	1.75	1.740010	0.920004	7.999582	0.000417
2	1.74	1.750009	0.920004	7.999599	0.000399
3	1.73	1.760319	0.919999	8.000076	−0.000077
4	1.72	1.771137	0.920004	7.999623	0.000376
5	1.71	1.782270	0.920000	7.999975	0.000024
6	1.70	1.793916	0.920003	7.999748	0.000250
7	1.69	1.805880	0.919993	8.000671	−0.000674
8	1.68	1.818557	0.920001	7.999951	0.000048
9	1.67	1.831753	0.920006	7.999378	0.000620
10	1.66	1.845470	0.920008	7.999200	0.000799
11	1.65	1.859711	0.920004	7.999611	0.000387
12	1.64	1.874478	0.919991	8.000887	−0.000888
13	1.63	1.890165	0.919993	8.000750	−0.000751
14	1.62	1.906773	0.920003	7.999671	0.000328
15	1.61	1.923915	0.919995	8.000506	−0.000507
16	1.60	1.942180	0.919999	8.000082	−0.000083
17	1.59	1.961375	0.919999	8.000100	−0.000101
18	1.58	1.981697	0.920000	8.000011	−0.000012
19	1.57	2.003347	0.920006	7.999432	0.000566
20	1.56	2.026324	0.920009	7.999105	0.000894
21	1.55	2.050635	0.920002	7.999784	0.000215
22	1.54	2.076671	0.919995	8.000475	−0.000477
23	1.53	2.104828	0.919995	8.000553	−0.000554
24	1.52	2.135498	0.920002	7.999802	0.000197
25	1.51	2.168685	0.920002	7.999790	0.000209
26	1.50	2.204784	0.919992	8.000755	−0.000757
27	1.49	2.244969	0.919991	8.000869	−0.000870
28	1.48	2.290025	0.919993	8.000666	−0.000668
29	1.47	2.341520	0.920005	7.999534	0.000465
30	1.46	2.400237	0.919997	8.000279	−0.000280
31	1.45	2.470089	0.919995	8.000535	−0.000536

(continued)

Table 8.2 (continued)

Parameter values					
PROB. = 0.92	RHO = −0.1		BIGH = 1.7449		SMHL = 1.41
Pair No.	SMH	SMK	PROB. level	Risk level (%)	Risk error (%)
32	1.44	2.556545	0.919999	8.000106	−0.000107
33	1.43	2.669768	0.920003	7.999701	0.000298
34	1.42	2.834762	0.919997	8.000326	−0.000328
35	1.41	3.171843	0.920002	7.999766	0.000232
No. of pairs		Mean COMPUT. risk errors		STD. ERR. errors	
35		−0.0000263254		0.0000810862	
Z-ratio				Status of significance	
0.32466				N.S.—not significant	
Parameter values					
PROB. = 0.91	RHO = −0.1		BIGH = 1.6886		SMHL = 1.35
Pair No.	SMH	SMK	PROB. level	Risk level (%)	Risk error (%)
Table No. 8.2-219					
1	1.69	1.687201	0.909994	9.000624	−0.000626
2	1.68	1.697439	0.910004	8.999604	0.000393
3	1.67	1.707798	0.909992	9.000850	−0.000852
4	1.66	1.718865	0.910008	8.999252	0.000745
5	1.65	1.730056	0.909999	9.000152	−0.000155
6	1.64	1.741765	0.909997	9.000343	−0.000346
7	1.63	1.753994	0.910000	9.000027	−0.000030
8	1.62	1.766745	0.910005	8.999496	0.000501
9	1.61	1.779826	0.909995	9.000540	−0.000542
10	1.60	1.793629	0.909997	9.000254	−0.000256
11	1.59	1.808158	0.910009	8.999091	0.000906
12	1.58	1.823024	0.909998	9.000230	−0.000232
13	1.57	1.838815	0.910003	8.999693	0.000304
14	1.56	1.855145	0.909994	9.000653	−0.000656
15	1.55	1.872406	0.909992	9.000826	−0.000829
16	1.54	1.890601	0.909993	9.000694	−0.000697
17	1.53	1.909734	0.909992	9.000808	−0.000811
18	1.52	1.930198	0.910006	8.999389	0.000608
19	1.51	1.951606	0.910005	8.999479	0.000519
20	1.50	1.974351	0.910005	8.999485	0.000513
21	1.49	1.998436	0.909998	9.000206	−0.000209
22	1.48	2.024257	0.909995	9.000522	−0.000525
23	1.47	2.052207	0.910003	8.999741	0.000256
24	1.46	2.081899	0.909990	9.000987	−0.000989
25	1.45	2.114509	0.909995	9.000534	−0.000536
26	1.44	2.150040	0.909996	9.000408	−0.000411
27	1.43	2.189277	0.910001	8.999896	0.000101
28	1.42	2.233007	0.910008	8.999157	0.000840
29	1.41	2.281623	0.909997	9.000259	−0.000262
30	1.40	2.337473	0.909992	9.000808	−0.000811

(continued)

8.2 Contents for Table 8.2 (Table No. 8.2-1 to 400)

Table 8.2 (continued)

Parameter values					
PROB. = 0.91	RHO = −0.1		BIGH = 1.6886		SMHL = 1.35
Pair No.	SMH	SMK	PROB. level	Risk level (%)	Risk error (%)
31	1.39	2.403688	0.910005	8.999550	0.000447
32	1.38	2.483397	0.910010	8.999026	0.000972
33	1.37	2.582853	0.909991	9.000903	−0.000906
34	1.36	2.721594	0.909991	9.000921	−0.000924
35	1.35	2.955875	0.910005	8.999479	0.000519
No. of pairs		Mean COMPUT. risk errors		STD. ERR. errors	
35		−0.0001105997		0.0001000822	
Z-ratio				Status of significance	
1.10509				N.S.—not significant	
Parameter values					
PROB. = 0.9025	RHO = −0.1		BIGH = 1.6495		SMHL = 1.30
Pair No.	SMH	SMK	PROB. level	Risk level (%)	Risk error (%)
Table No. 8.2-220					
1	1.65	1.649000	0.902497	9.750301	−0.000298
2	1.64	1.659250	0.902506	9.749407	0.000596
3	1.63	1.669624	0.902492	9.750807	−0.000805
4	1.62	1.680709	0.902509	9.749068	0.000936
5	1.61	1.691922	0.902501	9.749931	0.000072
6	1.60	1.703656	0.902501	9.749913	0.000089
7	1.59	1.715914	0.902507	9.749281	0.000721
8	1.58	1.728503	0.902500	9.749967	0.000036
9	1.57	1.741619	0.902495	9.750528	−0.000525
10	1.56	1.755463	0.902504	9.749574	0.000429
11	1.55	1.769840	0.902509	9.749097	0.000906
12	1.54	1.784755	0.902506	9.749359	0.000644
13	1.53	1.800208	0.902494	9.750652	−0.000650
14	1.52	1.816595	0.902496	9.750396	−0.000393
15	1.51	1.833919	0.902509	9.749150	0.000852
16	1.50	1.851791	0.902499	9.750139	−0.000137
17	1.49	1.870800	0.902502	9.749764	0.000238
18	1.48	1.890756	0.902500	9.749955	0.000048
19	1.47	1.911856	0.902499	9.750122	−0.000119
20	1.46	1.934299	0.902502	9.749806	0.000197
21	1.45	1.958089	0.902501	9.749895	0.000107
22	1.44	1.983620	0.902509	9.749138	0.000864
23	1.43	2.010505	0.902493	9.750689	−0.000685
24	1.42	2.039919	0.902502	9.749818	0.000185
25	1.41	2.071477	0.902499	9.750080	−0.000077
26	1.40	2.105964	0.902503	9.749699	0.000304
27	1.39	2.143383	0.902492	9.750796	−0.000793
28	1.38	2.185302	0.902500	9.749961	0.000042
29	1.37	2.232115	0.902506	9.749442	0.000560

(continued)

Table 8.2 (continued)

Parameter values					
PROB. = 0.9025	RHO = −0.1		BIGH = 1.6495		SMHL = 1.30
Pair No.	SMH	SMK	PROB. level	Risk level (%)	Risk error (%)
30	1.36	2.285000	0.902504	9.749639	0.000364
31	1.35	2.345913	0.902495	9.750473	−0.000471
32	1.34	2.418766	0.902498	9.750164	−0.000161
33	1.33	2.508251	0.902493	9.750671	−0.000668
34	1.32	2.626874	0.902500	9.750003	0.000000
35	1.31	2.801984	0.902494	9.750581	−0.000578
36	1.30	3.180460	0.902501	9.749949	0.000054
No. of pairs		Mean COMPUT. risk errors		STD. ERR. errors	
36		0.0000509056		0.0000822095	
Z-ratio				Status of significance	
0.61922				N.S.—not significant	
Parameter values					
PROB. = 0.9	RHO = −0.1		BIGH = 1.637		SMHL = 1.29
Pair No.	SMH	SMK	PROB. level	Risk level (%)	Risk error (%)
Table No. 8.2-221					
1	1.64	1.634005	0.900000	9.999961	0.000042
2	1.63	1.644030	0.899996	10.000410	−0.000411
3	1.62	1.654569	0.900007	9.999281	0.000721
4	1.61	1.665234	0.899994	10.000580	−0.000578
5	1.60	1.676418	0.899994	10.000640	−0.000638
6	1.59	1.688124	0.900002	9.999764	0.000238
7	1.58	1.700158	0.900001	9.999942	0.000060
8	1.57	1.712718	0.900004	9.999633	0.000370
9	1.56	1.725613	0.899992	10.000750	−0.000751
10	1.55	1.739235	0.899998	10.000220	−0.000215
11	1.54	1.753391	0.900000	10.000000	0.000000
12	1.53	1.768084	0.899996	10.000380	−0.000381
13	1.52	1.783513	0.899998	10.000160	−0.000161
14	1.51	1.799681	0.900003	9.999722	0.000280
15	1.50	1.816591	0.900005	9.999496	0.000507
16	1.49	1.834245	0.900001	9.999919	0.000083
17	1.48	1.852842	0.899999	10.000060	−0.000054
18	1.47	1.872581	0.900008	9.999222	0.000781
19	1.46	1.893271	0.900006	9.999371	0.000632
20	1.45	1.915109	0.900001	9.999877	0.000125
21	1.44	1.938294	0.899996	10.000350	−0.000352
22	1.43	1.963026	0.899994	10.000570	−0.000572
23	1.42	1.989505	0.899995	10.000530	−0.000525
24	1.41	2.018123	0.900005	9.999519	0.000483
25	1.40	2.048495	0.899992	10.000830	−0.000823
26	1.39	2.081797	0.899995	10.000530	−0.000525
27	1.38	2.118424	0.900008	9.999222	0.000781

(continued)

8.2 Contents for Table 8.2 (Table No. 8.2-1 to 400)

Table 8.2 (continued)

Parameter values					
PROB. = 0.9	RHO = −0.1		BIGH = 1.637		SMHL = 1.29
Pair No.	SMH	SMK	PROB. level	Risk level (%)	Risk error (%)
28	1.37	2.158379	0.900006	9.999418	0.000584
29	1.36	2.202449	0.899991	10.000870	−0.000870
30	1.35	2.252982	0.900004	9.999633	0.000370
31	1.34	2.309983	0.899991	10.000910	−0.000906
32	1.33	2.377363	0.899992	10.000800	−0.000793
33	1.32	2.459815	0.900008	9.999156	0.000846
34	1.31	2.562811	0.899992	10.000790	−0.000787
35	1.30	2.708235	0.900000	10.000050	−0.000048
36	1.29	2.955464	0.899996	10.000440	−0.000441
No. of pairs		Mean COMPUT. risk errors		STD. ERR. errors	
36		−0.0000790970		0.0000878753	
Z-ratio				Status of significance	
0.90010				N.S.—not significant	
Parameter values					
PROB. = 0.85	RHO = −0.1		BIGH = 1.4254		SMHL = 1.04
Pair No.	SMH	SMK	PROB. level	Risk level (%)	Risk error (%)
Table No. 8.2-222					
1	1.43	1.420815	0.850010	14.999010	0.000983
2	1.42	1.430821	0.850009	14.999100	0.000894
3	1.41	1.441066	0.849998	14.000200	−0.000197
4	1.40	1.451749	0.850002	14.999830	0.000167
5	1.39	1.462678	0.849992	15.000800	−0.000799
6	1.38	1.474052	0.849993	15.000730	−0.000727
7	1.37	1.485872	0.850001	14.999860	0.000143
8	1.36	1.498047	0.850004	14.999630	0.000364
9	1.35	1.510675	0.850009	14.999120	0.000882
10	1.34	1.523665	0.850003	14.999690	0.000310
11	1.33	1.537213	0.850007	14.999340	0.000662
12	1.32	1.551130	0.849994	15.000560	−0.000566
13	1.31	1.565810	0.850007	14.999340	0.000662
14	1.30	1.580865	0.849997	15.000330	−0.000334
15	1.29	1.596691	0.850003	14.999680	0.000316
16	1.28	1.613098	0.850001	14.999940	0.000054
17	1.27	1.630284	0.850005	14.999540	0.000459
18	1.26	1.648059	0.849991	15.000890	−0.000888
19	1.25	1.666818	0.849992	15.000800	−0.000799
20	1.24	1.686567	0.850000	14.999960	0.000036
21	1.23	1.707310	0.850008	14.999160	0.000834
22	1.22	1.729053	0.850009	14.999150	0.000846
23	1.21	1.751801	0.849993	15.000670	−0.000668
24	1.20	1.776145	0.850000	15.000030	−0.000030
25	1.19	1.801897	0.850000	14.999970	0.000024

(continued)

Table 8.2 (continued)

Parameter values					
PROB. = 0.85	RHO = −0.1		BIGH = 1.4254		SMHL = 1.04
Pair No.	SMH	SMK	PROB. level	Risk level (%)	Risk error (%)
26	1.18	1.829257	0.849997	15.000330	−0.000328
27	1.17	1.858621	0.850001	14.999860	0.000137
28	1.16	1.890193	0.850009	14.999150	0.000852
29	1.15	1.924174	0.850010	14.999010	0.000989
30	1.14	1.960766	0.849995	15.000530	−0.000530
31	1.13	2.001147	0.849998	15.000250	−0.000250
32	1.12	2.045716	0.849999	15.000060	−0.000066
33	1.11	2.095262	0.849992	15.000770	−0.000775
34	1.10	2.151746	0.849998	15.000180	−0.000179
35	1.09	2.217129	0.850008	14.999180	0.000817
36	1.08	2.293764	0.849992	15.000770	−0.000769
37	1.07	2.389470	0.850003	14.999700	0.000298
38	1.06	2.515196	0.850006	14.999400	0.000602
39	1.05	2.702201	0.849998	15.000210	−0.000215
40	1.04	3.122369	0.850007	14.999260	0.000739
No. of pairs		Mean COMPUT. risk errors		STD. ERR. errors	
40		0.0000963851		0.0000912932	
Z-ratio				Status of significance	
1.05577				N.S.—not significant	
Parameter values					
PROB. = 0.8	RHO = −0.1		BIGH = 1.2597		SMHL = 0.85
Pair No.	SMH	SMK	PROB. level	Risk level (%)	Risk error (%)
Table No. 8.2-223					
1	1.26	1.259302	0.800000	19.999990	0.000006
2	1.25	1.269475	0.800007	19.999340	0.000656
3	1.24	1.279908	0.800008	19.999230	0.000769
4	1.23	1.290605	0.800002	19.999830	0.000173
5	1.22	1.301668	0.800003	19.999710	0.000286
6	1.21	1.313004	0.799994	20.000550	−0.000554
7	1.20	1.324811	0.800004	19.999560	0.000435
8	1.19	1.336900	0.800001	19.999950	0.000054
9	1.18	1.349373	0.799996	20.000360	−0.000364
10	1.17	1.362331	0.800003	19.999710	0.000292
11	1.16	1.375586	0.799990	20.000970	−0.000966
12	1.15	1.389435	0.799997	20.000290	−0.000292
13	1.14	1.403687	0.799993	20.000730	−0.000733
14	1.13	1.418544	0.800001	19.999950	0.000054
15	1.12	1.433817	0.799992	20.000850	−0.000846
16	1.11	1.449804	0.799999	20.000060	−0.000060
17	1.10	1.466413	0.800007	19.999270	0.000727
18	1.09	1.483554	0.799999	20.000100	−0.000101
19	1.08	1.501526	0.800005	19.999540	0.000459
20	1.07	1.520141	0.799996	20.000400	−0.000399

(continued)

8.2 Contents for Table 8.2 (Table No. 8.2-1 to 400)

Table 8.2 (continued)

Parameter values					
PROB. = 0.8	RHO = −0.1		BIGH = 1.2597		SMHL = 0.85
Pair No.	SMH	SMK	PROB. level	Risk level (%)	Risk error (%)
21	1.06	1.539698	0.800000	20.000020	−0.000018
22	1.05	1.560108	0.799999	20.000140	−0.000143
23	1.04	1.581475	0.799995	20.000490	−0.000495
24	1.03	1.603906	0.799991	20.000920	−0.000924
25	1.02	1.627604	0.799995	20.000480	−0.000483
26	1.01	1.652578	0.799997	20.000330	−0.000328
27	1.00	1.679031	0.800001	19.999890	0.000107
28	0.99	1.706974	0.799994	20.000580	−0.000584
29	0.98	1.736806	0.799993	20.000720	−0.000721
30	0.97	1.768734	0.799993	20.000730	−0.000733
31	0.96	1.803157	0.800000	20.000000	0.000000
32	0.95	1.840283	0.800002	19.999830	0.000167
33	0.94	1.880710	0.800005	19.999550	0.000447
34	0.93	1.925033	0.800005	19.999470	0.000525
35	0.92	1.974049	0.800001	19.999950	0.000048
36	0.91	2.028940	0.799990	20.000980	−0.000978
37	0.90	2.092065	0.800000	20.000050	−0.000054
38	0.89	2.165783	0.800008	19.999160	0.000840
39	0.88	2.254012	0.799998	20.000210	−0.000215
40	0.87	2.366145	0.799995	20.000460	−0.000459
41	0.86	2.521729	0.799997	20.000260	−0.000256
42	0.85	2.785624	0.799993	20.000710	−0.000709
No. of pairs		Mean COMPUT. risk errors		STD. ERR. errors	
42		−0.0001248925		0.0000752682	
Z-ratio				Status of significance	
1.65930				N.S.—not significant	
Parameter values					
PROB. = 0.75	RHO = −0.1		BIGH = 1.1193		SMHL = 0.68
Pair No.	SMH	SMK	PROB. level	Risk level (%)	Risk error (%)
Table No. 8.2-224					
1	1.12	1.118503	0.750001	24.999880	0.000119
2	1.11	1.128580	0.749995	25.000550	−0.000548
3	1.10	1.139036	0.750008	24.999200	0.000805
4	1.09	1.149681	0.750004	24.999610	0.000393
5	1.08	1.160616	0.749999	25.000070	−0.000072
6	1.07	1.171848	0.749993	25.000720	−0.000715
7	1.06	1.183480	0.750001	24.999950	0.000048
8	1.05	1.195420	0.750003	24.999700	0.000298
9	1.04	1.207674	0.749999	25.000100	−0.000095
10	1.03	1.220346	0.750003	24.999660	0.000346
11	1.02	1.233345	0.749998	25.000220	−0.000221
12	1.01	1.246776	0.749996	25.000370	−0.000370
13	1.00	1.260645	0.749997	25.000350	−0.000352

(continued)

Table 8.2 (continued)

Parameter values					
PROB. = 0.75	RHO = −0.1		BIGH = 1.1193		SMHL = 0.68
Pair No.	SMH	SMK	PROB. level	Risk level (%)	Risk error (%)
14	0.99	1.274960	0.749996	25.000460	−0.000453
15	0.98	1.289729	0.749991	25.000920	−0.000918
16	0.97	1.305056	0.749994	25.000580	−0.000584
17	0.96	1.320952	0.750002	24.999830	0.000167
18	0.95	1.337326	0.749996	25.000410	−0.000405
19	0.94	1.354382	0.750001	24.999940	0.000060
20	0.93	1.372034	0.749998	25.000250	−0.000250
21	0.92	1.390485	0.750008	24.999200	0.000805
22	0.91	1.409551	0.750001	24.999910	0.000095
23	0.90	1.429536	0.750008	24.999190	0.000811
24	0.89	1.450255	0.749998	25.000160	−0.000161
25	0.88	1.472013	0.750000	25.000020	−0.000024
26	0.87	1.494725	0.749993	25.000750	−0.000751
27	0.86	1.518696	0.749999	25.000060	−0.000060
28	0.85	1.543842	0.749999	25.000110	−0.000113
29	0.84	1.570373	0.749999	25.000070	−0.000066
30	0.83	1.598499	0.750006	24.999430	0.000572
31	0.82	1.628235	0.750002	24.999830	0.000167
32	0.81	1.659988	0.750005	24.999540	0.000465
33	0.80	1.693774	0.749993	25.000710	−0.000703
34	0.79	1.730395	0.750004	24.999630	0.000376
35	0.78	1.769867	0.750005	24.999520	0.000477
36	0.77	1.812797	0.750001	24.999880	0.000125
37	0.76	1.859987	0.749999	25.000060	−0.000060
38	0.75	1.912435	0.750000	25.000000	0.000000
39	0.74	1.971532	0.750003	24.999700	0.000298
40	0.73	2.039255	0.750005	24.999510	0.000489
41	0.72	2.118560	0.749997	25.000260	−0.000262
42	0.71	2.215333	0.749996	25.000400	−0.000399
43	0.70	2.340541	0.750002	24.999830	0.000173
44	0.69	2.520385	0.750002	24.999770	0.000232
45	0.68	2.867399	0.750007	24.999350	0.000656
No. of pairs		Mean COMPUT. risk errors		STD. ERR. errors	
45		0.0000085520		0.0000631489	
Z-ratio				Status of significance	
0.13543				N.S.—not significant	
Parameter values					
PROB. = 0.7	RHO = −0.1		BIGH = 0.9946		SMHL = 0.53
Pair No.	SMH	SMK	PROB. level	Risk level (%)	Risk error (%)
Table No. 8.2-225					
1	1.00	0.989034	0.699991	30.000940	−0.000936
2	0.99	0.999026	0.699992	30.000850	−0.000846

(continued)

8.2 Contents for Table 8.2 (Table No. 8.2-1 to 400)

Table 8.2 (continued)

Parameter values					
PROB. = 0.7	RHO = −0.1		BIGH = 0.9946		SMHL = 0.53
Pair No.	SMH	SMK	PROB. level	Risk level (%)	Risk error (%)
3	0.98	1.009320	0.700003	29.999680	0.000328
4	0.97	1.019824	0.700005	29.999510	0.000495
5	0.96	1.030545	0.699996	30.000370	−0.000364
6	0.95	1.041587	0.699996	30.000370	−0.000364
7	0.94	1.052957	0.700004	29.999640	0.000364
8	0.93	1.064566	0.699998	30.000180	−0.000179
9	0.92	1.076518	0.699998	30.000160	−0.000161
10	0.91	1.088823	0.700003	29.999730	0.000268
11	0.90	1.101487	0.700010	29.999040	0.000960
12	0.89	1.114423	0.700000	30.000000	0.000006
13	0.88	1.127835	0.700008	29.999170	0.000829
14	0.87	1.141537	0.699997	30.000260	−0.000256
15	0.86	1.155735	0.700000	29.999960	0.000048
16	0.85	1.170342	0.699998	30.000180	−0.000179
17	0.84	1.185466	0.700005	29.999520	0.000483
18	0.83	1.201022	0.700002	29.999840	0.000161
19	0.82	1.217119	0.700002	29.999770	0.000232
20	0.81	1.233770	0.700003	29.999660	0.000340
21	0.80	1.250989	0.700002	29.999810	0.000197
22	0.79	1.268790	0.699995	30.000540	−0.000536
23	0.78	1.287383	0.700006	29.999450	0.000554
24	0.77	1.306588	0.700002	29.999810	0.000197
25	0.76	1.326617	0.700007	29.999330	0.000674
26	0.75	1.347390	0.700000	29.999970	0.000030
27	0.74	1.369022	0.699990	30.000970	−0.000972
28	0.73	1.391827	0.700006	29.999390	0.000608
29	0.72	1.415531	0.700002	29.999760	0.000244
30	0.71	1.440350	0.699995	30.000530	−0.000525
31	0.70	1.466504	0.699995	30.000530	−0.000530
32	0.69	1.494114	0.700000	30.000000	0.000000
33	0.68	1.523303	0.700007	29.999330	0.000674
34	0.67	1.554195	0.700008	29.999200	0.000805
35	0.66	1.586917	0.699996	30.000410	−0.000411
36	0.65	1.622086	0.700001	29.999870	0.000137
37	0.64	1.659732	0.699997	30.000340	−0.000340
38	0.63	1.700478	0.699998	30.000250	−0.000250
39	0.62	1.744749	0.699994	30.000560	−0.000554
40	0.61	1.793561	0.700006	29.999380	0.000620
41	0.60	1.847542	0.700008	29.999200	0.000805
42	0.59	1.907909	0.699993	30.000670	−0.000668
43	0.58	1.977441	0.700000	30.000020	−0.000018
44	0.57	2.058926	0.700003	29.999700	0.000298

(continued)

Table 8.2 (continued)

Parameter values					
PROB. = 0.7	RHO = −0.1		BIGH = 0.9946		SMHL = 0.53
Pair No.	SMH	SMK	PROB. level	Risk level (%)	Risk error (%)
45	0.56	2.157885	0.700002	29.999780	0.000221
46	0.55	2.285316	0.700002	29.999810	0.000197
47	0.54	2.467841	0.700000	30.000010	−0.000006
48	0.53	2.822718	0.700002	29.999790	0.000209
No. of pairs		Mean COMPUT. risk errors		STD. ERR. errors	
48		0.0000588748		0.0000682795	
Z-ratio				Status of significance	
0.86226				N.S.—not significant	
Parameter values					
PROB. = 0.65	RHO = −0.1		BIGH = 0.8800		SMHL = 0.40
Pair No.	SMH	SMK	PROB. level	Risk level (%)	Risk error (%)
Table No. 8.2-226					
1	0.89	0.870210	0.649994	35.000580	−0.000572
2	0.88	0.880098	0.649994	35.000570	−0.000566
3	0.87	0.890213	0.649994	35.000630	−0.000632
4	0.86	0.900563	0.649992	35.000760	−0.000757
5	0.85	0.911156	0.649991	35.000930	−0.000924
6	0.84	0.922100	0.650009	34.999090	0.000918
7	0.83	0.933207	0.650005	34.999470	0.000536
8	0.82	0.944586	0.650001	34.999920	0.000077
9	0.81	0.956245	0.649995	35.000500	−0.000501
10	0.80	0.968293	0.650008	34.999230	0.000775
11	0.79	0.980546	0.649999	35.000120	−0.000113
12	0.78	0.993211	0.650008	34.999250	0.000751
13	0.77	1.006105	0.649994	35.000570	−0.000560
14	0.76	1.019436	0.649997	35.000260	−0.000256
15	0.75	1.033119	0.649997	35.000290	−0.000286
16	0.74	1.047170	0.649993	35.000730	−0.000727
17	0.73	1.061701	0.650002	34.999850	0.000155
18	0.72	1.076630	0.650004	34.999550	0.000447
19	0.71	1.091973	0.650001	34.999900	0.000101
20	0.70	1.107848	0.650007	34.999320	0.000685
21	0.69	1.124174	0.650004	34.999610	0.000393
22	0.68	1.140972	0.649991	35.000900	−0.000894
23	0.67	1.158457	0.649998	35.000170	−0.000161
24	0.66	1.176458	0.649992	35.000810	−0.000805
25	0.65	1.195193	0.650000	34.999960	0.000048
26	0.64	1.214492	0.649991	35.000950	−0.000948
27	0.63	1.234675	0.650003	34.999680	0.000322
28	0.62	1.255477	0.649992	35.000810	−0.000811

(continued)

Table 8.2 (continued)

Parameter values					
PROB. = 0.65	RHO = −0.1		BIGH = 0.8800		SMHL = 0.40
Pair No.	SMH	SMK	PROB. level	Risk level (%)	Risk error (%)
29	0.61	1.277320	0.650007	34.999270	0.000739
30	0.60	1.299846	0.649991	35.000890	−0.000882
31	0.59	1.323479	0.649990	35.000960	−0.000954
32	0.58	1.348258	0.649998	35.000250	−0.000250
33	0.57	1.374221	0.650004	34.999620	0.000381
34	0.56	1.401411	0.650001	34.999930	0.000072
35	0.55	1.430070	0.650001	34.999950	0.000054
36	0.54	1.460245	0.649991	35.000910	−0.000906
37	0.53	1.492381	0.649998	35.000220	−0.000221
38	0.52	1.526535	0.650002	34.999790	0.000215
39	0.51	1.562962	0.650001	34.999860	0.000149
40	0.50	1.602119	0.650005	34.999510	0.000489
41	0.49	1.644274	0.649998	35.000210	−0.000203
42	0.48	1.690285	0.650003	34.999720	0.000286
43	0.47	1.740627	0.649998	35.000220	−0.000215
44	0.46	1.796563	0.649999	35.000070	−0.000072
45	0.45	1.859559	0.650005	34.999490	0.000513
46	0.44	1.931483	0.650000	35.000010	−0.000006
47	0.43	2.016163	0.650003	34.999700	0.000298
48	0.42	2.130526	0.649998	35.000170	−0.000167
49	0.41	2.262216	0.649996	35.000380	−0.000370
50	0.40	2.454748	0.650007	34.999280	0.000721
No. of pairs		Mean COMPUT. risk errors		STD. ERR. errors	
50		−0.0000908094		0.0000752444	
Z-ratio				Status of significance	
1.20686				N.S.—not significant	
Parameter values					
PROB. = 0.6	RHO = −0.1		BIGH = 0.7724		SMHL = 0.29
Pair No.	SMH	SMK	PROB. level	Risk level (%)	Risk error (%)
Table No. 8.2-227					
1	0.78	0.764874	0.599994	40.000570	−0.000572
2	0.77	0.774808	0.599992	40.000810	−0.000817
3	0.76	0.785002	0.599999	40.000070	−0.000072
4	0.75	0.795371	0.599995	40.000530	−0.000530
5	0.74	0.806023	0.600001	39.999870	0.000131
6	0.73	0.816872	0.599998	40.000230	−0.000226
7	0.72	0.828028	0.600007	39.999320	0.000679
8	0.71	0.839405	0.600007	39.999270	0.000733
9	0.70	0.851019	0.600000	39.999960	0.000036
10	0.69	0.862980	0.600008	39.999180	0.000823

(continued)

Table 8.2 (continued)

Parameter values					
PROB. = 0.6	RHO = −0.1		BIGH = 0.7724		SMHL = 0.29
Pair No.	SMH	SMK	PROB. level	Risk level (%)	Risk error (%)
11	0.68	0.875158	0.600000	39.999970	0.000024
12	0.67	0.887716	0.600009	39.999150	0.000846
13	0.66	0.900524	0.600003	39.999690	0.000310
14	0.65	0.913650	0.599996	40.000440	−0.000441
15	0.64	0.927210	0.600007	39.999340	0.000662
16	0.63	0.941030	0.599998	40.000210	−0.000215
17	0.62	0.955229	0.599990	40.000960	−0.000960
18	0.61	0.969930	0.600004	39.999640	0.000358
19	0.60	0.984961	0.600001	39.999900	0.000095
20	0.59	1.000447	0.600002	39.999780	0.000221
21	0.58	1.016319	0.599991	40.000890	−0.000894
22	0.57	1.032802	0.600004	39.999650	0.000346
23	0.56	1.049735	0.600005	39.999470	0.000525
24	0.55	1.067152	0.599998	40.000190	−0.000191
25	0.54	1.085189	0.600000	40.000050	−0.000048
26	0.53	1.103788	0.599994	40.000570	−0.000566
27	0.52	1.123092	0.599999	40.000120	−0.000125
28	0.51	1.143049	0.599998	40.000230	−0.000226
29	0.50	1.163711	0.599992	40.000760	−0.000763
30	0.49	1.185230	0.599997	40.000300	−0.000298
31	0.48	1.207568	0.599997	40.000280	−0.000280
32	0.47	1.230888	0.600006	39.999400	0.000602
33	0.46	1.255163	0.600009	39.999120	0.000882
34	0.45	1.280469	0.600005	39.999520	0.000477
35	0.44	1.306889	0.599992	40.000760	−0.000757
36	0.43	1.334711	0.599993	40.000730	−0.000733
37	0.42	1.364034	0.600000	40.000010	−0.000006
38	0.41	1.394969	0.600008	39.999220	0.000775
39	0.40	1.427636	0.600009	39.999120	0.000882
40	0.39	1.462169	0.599995	40.000520	−0.000519
41	0.38	1.499300	0.600010	39.999040	0.000954
42	0.37	1.538803	0.599999	40.000090	−0.000095
43	0.36	1.581444	0.599995	40.000540	−0.000536
44	0.35	1.627817	0.599998	40.000190	−0.000191
45	0.34	1.696311	0.599998	40.000160	−0.000161
46	0.33	1.750851	0.600001	39.999900	0.000101
47	0.32	1.812035	0.599998	40.000200	−0.000203
48	0.31	1.881941	0.599994	40.000620	−0.000626

(continued)

8.2 Contents for Table 8.2 (Table No. 8.2-1 to 400)

Table 8.2 (continued)

Parameter values					
PROB. = 0.6	RHO = −0.1		BIGH = 0.7724		SMHL = 0.29
Pair No.	SMH	SMK	PROB. level	Risk level (%)	Risk error (%)
49	0.30	1.964060	0.600003	39.999660	0.000340
50	0.29	2.063104	0.599999	40.000150	−0.000155
No. of pairs		Mean COMPUT. risk errors		STD. ERR. errors	
50		−0.0000079473		0.0000740104	
Z-ratio				Status of significance	
0.10738				N.S.—not significant	
Parameter values					
PROB. = 0.55	RHO = −0.1		BIGH = 0.6692		SMHL = 0.18
Pair No.	SMH	SMK	PROB. level	Risk level (%)	Risk error (%)
Table No. 8.2-228					
1	0.67	0.668303	0.550007	44.999340	0.000662
2	0.66	0.678333	0.549993	45.000720	−0.000721
3	0.65	0.688674	0.550004	44.999630	0.000370
4	0.64	0.699146	0.549994	45.000650	−0.000650
5	0.63	0.709911	0.550000	45.000040	−0.000042
6	0.62	0.720888	0.550001	44.999930	0.000072
7	0.61	0.732094	0.549999	45.000070	−0.000066
8	0.60	0.743549	0.549998	45.000220	−0.000221
9	0.59	0.755321	0.550010	44.999040	0.000960
10	0.58	0.767235	0.549994	45.000650	−0.000650
11	0.57	0.779512	0.549996	45.000360	−0.000364
12	0.56	0.792077	0.549999	45.000080	−0.000077
13	0.55	0.804956	0.550005	44.999470	0.000530
14	0.54	0.818082	0.549997	45.000280	−0.000280
15	0.53	0.831581	0.550000	45.000030	−0.000024
16	0.52	0.845388	0.549996	45.000390	−0.000387
17	0.51	0.859540	0.549991	45.000930	−0.000924
18	0.50	0.874173	0.550007	44.999250	0.000745
19	0.49	0.889033	0.549993	45.000690	−0.000685
20	0.48	0.904460	0.550010	44.999010	0.000989
21	0.47	0.920209	0.550007	44.999300	0.000703
22	0.46	0.936431	0.550009	44.999110	0.000888
23	0.45	0.953084	0.550004	44.999610	0.000387
24	0.44	0.970233	0.549999	45.000080	−0.000077
25	0.43	0.987946	0.550002	44.999850	0.000155
26	0.42	1.006199	0.550002	44.999840	0.000161
27	0.41	1.025077	0.550008	44.999220	0.000781
28	0.40	1.044473	0.549997	45.000320	−0.000316
29	0.39	1.064684	0.550010	44.999020	0.000983
30	0.38	1.085429	0.549995	45.000460	−0.000459
31	0.37	1.107026	0.549995	45.000470	−0.000471
32	0.36	1.129514	0.550006	44.999390	0.000608

(continued)

Table 8.2 (continued)

Parameter values					
PROB. = 0.55	RHO = −0.1		BIGH = 0.6692		SMHL = 0.18
Pair No.	SMH	SMK	PROB. level	Risk level (%)	Risk error (%)
33	0.35	1.152752	0.549998	45.000200	−0.000203
34	0.34	1.177005	0.549999	45.000060	−0.000060
35	0.33	1.202270	0.549998	45.000210	−0.000209
36	0.32	1.228760	0.550009	44.999060	0.000936
37	0.31	1.256326	0.550001	44.999880	0.000125
38	0.30	1.285338	0.550005	44.999530	0.000471
39	0.29	1.315719	0.549993	45.000670	−0.000668
40	0.28	1.348019	0.550010	44.999010	0.000989
41	0.27	1.405692	0.550005	44.999550	0.000447
42	0.26	1.439993	0.549998	45.000190	−0.000191
43	0.25	1.476858	0.550004	44.999640	0.000358
44	0.24	1.516340	0.549991	45.000920	−0.000918
45	0.23	1.559382	0.550002	44.999770	0.000232
46	0.22	1.606089	0.549996	45.000420	−0.000417
47	0.21	1.657513	0.549993	45.000690	−0.000685
48	0.20	1.714920	0.550005	44.999480	0.000525
49	0.19	1.779254	0.549996	45.000420	−0.000417
50	0.18	1.853557	0.550009	44.999080	0.000924
No. of pairs		Mean COMPUT. risk errors		STD. ERR. errors	
50		0.0000749149		0.0000790651	
Z-ratio				Status of significance	
0.94751				N.S.—not significant	
Parameter values					
PROB. = 0.5	RHO = −0.1		BIGH = 0.5683		SMHL = 0.08
Pair No.	SMH	SMK	PROB. level	Risk level (%)	Risk error (%)
Table No. 8.2-229					
1	0.57	0.566605	0.500002	49.999820	0.000179
2	0.56	0.576625	0.499991	50.000920	−0.000921
3	0.55	0.586916	0.499998	50.000250	−0.000247
4	0.54	0.597400	0.500001	49.999900	0.000107
5	0.53	0.608098	0.500006	49.999440	0.000560
6	0.52	0.618938	0.499992	50.000850	−0.000846
7	0.51	0.630140	0.500010	49.999010	0.000989
8	0.50	0.641438	0.499996	50.000360	−0.000358
9	0.49	0.653057	0.500002	49.999840	0.000161
10	0.48	0.664933	0.500009	49.999140	0.000858
11	0.47	0.677003	0.500001	49.999900	0.000107
12	0.46	0.689402	0.500008	49.999250	0.000751
13	0.45	0.701977	0.499992	50.000810	−0.000811
14	0.44	0.714968	0.500004	49.999580	0.000429
15	0.43	0.728181	0.500000	49.999980	0.000018
16	0.42	0.741718	0.499999	50.000070	−0.000075
17	0.41	0.755590	0.500001	49.999910	0.000089

(continued)

8.2 Contents for Table 8.2 (Table No. 8.2-1 to 400)

Table 8.2 (continued)

Parameter values					
PROB. = 0.5	RHO = −0.1		BIGH = 0.5683		SMHL = 0.08
Pair No.	SMH	SMK	PROB. level	Risk level (%)	Risk error (%)
18	0.40	0.769814	0.500006	49.999450	0.000554
19	0.39	0.784365	0.500005	49.999480	0.000519
20	0.38	0.799224	0.499993	50.000660	−0.000653
21	0.37	0.814577	0.500004	49.999630	0.000376
22	0.36	0.830230	0.499994	50.000560	−0.000563
23	0.35	0.846389	0.500001	49.999880	0.000125
24	0.34	0.862982	0.500007	49.999350	0.000656
25	0.33	0.879950	0.499996	50.000410	−0.000411
26	0.32	0.897546	0.500010	49.999030	0.000978
27	0.31	0.915552	0.500005	49.999500	0.000501
28	0.30	0.934068	0.499995	50.000540	−0.000542
29	0.29	0.953222	0.499995	50.000550	−0.000545
30	0.28	0.972976	0.499993	50.000740	−0.000742
31	0.27	0.993431	0.499998	50.000170	−0.000167
32	0.26	1.014535	0.499997	50.000300	−0.000298
33	0.25	1.036389	0.499997	50.000320	−0.000322
34	0.24	1.059064	0.500000	49.999990	0.000012
35	0.23	1.082514	0.499992	50.000810	−0.000805
36	0.22	1.139114	0.499998	50.000250	−0.000250
37	0.21	1.162340	0.500008	49.999170	0.000834
38	0.20	1.186502	0.499992	50.000850	−0.000834
39	0.19	1.212117	0.500003	49.999670	0.000334
40	0.18	1.238973	0.499999	50.000060	−0.000060
41	0.17	1.267369	0.500001	49.999940	0.000060
42	0.16	1.297434	0.500003	49.999700	0.000298
43	0.15	1.329268	0.499997	50.000350	−0.000346
44	0.14	1.363334	0.500007	49.999320	0.000679
45	0.13	1.399575	0.500000	50.000040	−0.000036
46	0.12	1.438635	0.500006	49.999420	0.000584
47	0.11	1.480570	0.499993	50.000690	−0.000691
48	0.10	1.526322	0.499997	50.000280	−0.000280
49	0.09	1.576381	0.500001	49.999910	0.000089
50	0.08	1.631582	0.499999	50.000100	−0.000101
No. of pairs		Mean COMPUT. risk errors		STD. ERR. errors	
50		−0.0000012856		0.0000734335	
Z-ratio				Status of significance	
0.01751				N.S.—not significant	
Parameter values					
PROB. = 0.99	RHO = −0.2		BIGH = 2.5758		SMHL = 2.34
Pair No.	SMH	SMK	PROB. level	Risk level (%)	Risk error (%)
Table No. 8.2-230					
1	2.58	2.571607	0.990001	0.999874	0.000125
2	2.57	2.581613	0.990001	0.999922	0.000077

(continued)

Table 8.2 (continued)

Parameter values					
PROB. = 0.99	RHO = −0.2		BIGH = 2.5758		SMHL = 2.34
Pair No.	SMH	SMK	PROB. level	Risk level (%)	Risk error (%)
3	2.56	2.591698	0.989994	1.000613	−0.000614
4	2.55	2.603423	0.990001	0.999856	0.000143
5	2.54	2.615229	0.990002	0.999844	0.000155
6	2.53	2.627116	0.989994	1.000625	−0.000626
7	2.52	2.640648	0.989997	1.000267	−0.000268
8	2.51	2.655825	0.990010	0.999016	0.000983
9	2.50	2.669523	0.989994	1.000559	−0.000560
10	2.49	2.686431	0.990004	0.999624	0.000376
11	2.48	2.703425	0.990001	0.999880	0.000119
12	2.47	2.722069	0.990002	0.999779	0.000221
13	2.46	2.742364	0.990004	0.999570	0.000429
14	2.45	2.764309	0.990006	0.999421	0.000578
15	2.44	2.787908	0.990004	0.999588	0.000411
16	2.43	2.814723	0.990009	0.999081	0.000918
17	2.42	2.841630	0.989994	1.000571	−0.000572
18	2.41	2.874881	0.990000	0.999963	0.000036
19	2.40	2.911352	0.989999	1.000094	−0.000095
20	2.39	2.954169	0.990002	0.999832	0.000167
21	2.38	3.003333	0.989998	1.000202	−0.000203
22	2.37	3.055720	0.990000	1.000035	−0.000036
23	2.36	3.130082	0.990003	0.999725	0.000274
24	2.35	3.223295	0.989991	1.000857	−0.000858
25	2.34	3.372861	0.989995	1.000476	−0.000477
No. of pairs		Mean COMPUT. risk errors		STD. ERR. errors	
25		0.0000270513		0.0000907240	
Z-ratio				Status of significance	
0.29817				N.S.—not significant	
Parameter values					
PROB. = 0.98	RHO = −0.2		BIGH = 2.3261		SMHL = 2.06
Pair No.	SMH	SMK	PROB. level	Risk level (%)	Risk error (%)
Table No. 8.2-231					
1	2.33	2.322207	0.980004	1.999617	0.000381
2	2.32	2.332216	0.980003	1.999736	0.000262
3	2.31	2.342313	0.979992	2.000851	−0.000852
4	2.30	2.354059	0.980009	1.999128	0.000870
5	2.29	2.365113	0.979995	2.000523	−0.000525
6	2.28	2.377820	0.980006	1.999402	0.000596
7	2.27	2.390618	0.980003	1.999664	0.000334
8	2.26	2.404290	0.980004	1.999569	0.000429
9	2.25	2.418837	0.980007	1.999307	0.000691
10	2.24	2.434260	0.980010	1.999033	0.000966
11	2.23	2.449779	0.979995	2.000493	−0.000495

(continued)

8.2 Contents for Table 8.2 (Table No. 8.2-1 to 400)

Table 8.2 (continued)

Parameter values					
PROB. = 0.98	RHO = −0.2		BIGH = 2.3261		SMHL = 2.06
Pair No.	SMH	SMK	PROB. level	Risk level (%)	Risk error (%)
12	2.22	2.466959	0.979993	2.000707	−0.000709
13	2.21	2.485799	0.979999	2.000070	−0.000072
14	2.20	2.505522	0.979997	2.000272	−0.000274
15	2.19	2.526909	0.979998	2.000201	−0.000203
16	2.18	2.550742	0.980010	1.999050	0.000948
17	2.17	2.574679	0.979992	2.000809	−0.000811
18	2.16	2.603410	0.980010	1.999044	0.000954
19	2.15	2.632249	0.979991	2.000868	−0.000870
20	2.14	2.665884	0.979992	2.000821	−0.000823
21	2.13	2.704317	0.979999	2.000111	−0.000113
22	2.12	2.745987	0.979991	2.000892	−0.000894
23	2.11	2.797145	0.980002	1.999778	0.000221
24	2.10	2.856231	0.980000	1.999998	0.000000
25	2.09	2.929496	0.980000	2.000016	−0.000018
26	2.08	3.026318	0.980004	1.999563	0.000435
27	2.07	3.163884	0.979999	2.000129	−0.000131
28	2.06	3.426573	0.979997	2.000326	−0.000328
No. of pairs		Mean COMPUT. risk errors		STD. ERR. errors	
28		−0.0000010277		0.0001108490	
Z-ratio				Status of significance	
0.00927				N.S.—not significant	
Parameter values					
PROB. = 0.97	RHO = −0.2		BIGH = 2.1695		SMHL = 1.89
Pair No.	SMH	SMK	PROB. level	Risk level (%)	Risk error (%)
Table No. 8.2-232					
1	2.17	2.169001	0.970004	2.999640	0.000358
2	2.16	2.179042	0.969998	3.000224	−0.000226
3	2.15	2.189958	0.970007	2.999306	0.000691
4	2.14	2.200969	0.970002	2.999842	0.000155
5	2.13	2.212858	0.970008	2.999181	0.000817
6	2.12	2.224843	0.969998	3.000158	−0.000161
7	2.11	2.237709	0.969998	3.000242	−0.000244
8	2.10	2.251457	0.970004	2.999646	0.000352
9	2.09	2.265696	0.970002	2.999830	0.000167
10	2.08	2.280820	0.970002	2.999759	0.000238
11	2.07	2.296439	0.969992	3.000778	−0.000781
12	2.06	2.313727	0.970002	2.999783	0.000215
13	2.05	2.331904	0.970006	2.999354	0.000644
14	2.04	2.350971	0.970003	2.999705	0.000292
15	2.03	2.371711	0.970008	2.999240	0.000757
16	2.02	2.393346	0.969998	3.000224	−0.000226
17	2.01	2.417438	0.970005	2.999532	0.000465

(continued)

Table 8.2 (continued)

Parameter values					
PROB. = 0.97	RHO = −0.2		BIGH = 2.1695		SMHL = 1.89
Pair No.	SMH	SMK	PROB. level	Risk level (%)	Risk error (%)
18	2.00	2.443209	0.970005	2.999550	0.000447
19	1.99	2.471441	0.970007	2.999276	0.000721
20	1.98	2.502136	0.970005	2.999538	0.000459
21	1.97	2.536078	0.970003	2.999663	0.000334
22	1.96	2.573268	0.969990	3.000993	−0.000995
23	1.95	2.616833	0.969997	3.000325	−0.000328
24	1.94	2.666774	0.970000	2.999973	0.000024
25	1.93	2.726220	0.970009	2.999067	0.000930
26	1.92	2.795172	0.969991	3.000951	−0.000954
27	1.91	2.886131	0.969993	3.000707	−0.000709
28	1.90	3.014726	0.970004	2.999640	0.000358
29	1.89	3.227833	0.970000	2.999991	0.000006
No. of pairs			Mean COMPUT. risk errors	STD. ERR. errors	
29			0.0001267592	0.0000940893	
Z-ratio				Status of significance	
1.34722				N.S.—not significant	
Parameter values					
PROB. = 0.96	RHO = −0.2		BIGH = 2.0528		SMHL = 1.76
Pair No.	SMH	SMK	PROB. level	Risk level (%)	Risk error (%)
Table No. 8.2-233					
1	2.06	2.045625	0.960002	3.999794	0.000209
2	2.05	2.055604	0.960005	3.999466	0.000536
3	2.04	2.065680	0.959993	4.000664	−0.000662
4	2.03	2.076637	0.960002	3.999817	0.000185
5	2.02	2.087695	0.959993	4.000712	−0.000709
6	2.01	2.099636	0.960001	3.999949	0.000054
7	2.00	2.112072	0.960006	3.999448	0.000554
8	1.99	2.124613	0.959990	4.000986	−0.000983
9	1.98	2.138433	0.960001	3.999877	0.000125
10	1.97	2.152752	0.960004	3.999573	0.000429
11	1.96	2.167572	0.959997	4.000264	−0.000262
12	1.95	2.183675	0.960008	3.999198	0.000805
13	1.94	2.200283	0.960004	3.999615	0.000387
14	1.93	2.217788	0.959997	4.000306	−0.000304
15	1.92	2.236972	0.960009	3.999061	0.000942
16	1.91	2.256667	0.959999	4.000134	−0.000131
17	1.90	2.278044	0.959998	4.000169	−0.000167
18	1.89	2.300717	0.959991	4.000938	−0.000936
19	1.88	2.325858	0.960003	3.999734	0.000268
20	1.87	2.352688	0.960005	3.999531	0.000471
21	1.86	2.381209	0.959991	4.000902	−0.000900
22	1.85	2.413768	0.960007	3.999353	0.000650

(continued)

8.2 Contents for Table 8.2 (Table No. 8.2-1 to 400)

Table 8.2 (continued)

Parameter values					
PROB. = 0.96	RHO = −0.2		BIGH = 2.0528		SMHL = 1.76
Pair No.	SMH	SMK	PROB. level	Risk level (%)	Risk error (%)
23	1.84	2.448804	0.960002	3.999794	0.000209
24	1.83	2.487881	0.959997	4.000259	−0.000256
25	1.82	2.532565	0.960001	3.999889	0.000113
26	1.81	2.583638	0.960001	3.999949	0.000054
27	1.80	2.644228	0.960006	3.999376	0.000626
28	1.79	2.716683	0.960000	3.999972	0.000030
29	1.78	2.809595	0.960000	4.000032	−0.000030
30	1.77	2.937033	0.959991	4.000884	−0.000882
31	1.76	3.156810	0.960004	3.999633	0.000370
No. of pairs		Mean COMPUT. risk errors		STD. ERR. errors	
31		0.0000247732		0.0000925077	
Z-ratio				Status of significance	
0.26780				N.S.—not significant	
Parameter values					
PROB. = 0.95	RHO = −0.2		BIGH = 1.9586		SMHL = 1.65
Pair No.	SMH	SMK	PROB. level	Risk level (%)	Risk error (%)
Table No. 8.2-234					
1	1.96	1.957201	0.950009	4.999066	0.000936
2	1.95	1.967238	0.950003	4.999667	0.000334
3	1.94	1.977769	0.950002	4.999841	0.000161
4	1.93	1.988796	0.950003	4.999709	0.000292
5	1.92	2.000320	0.950006	4.999423	0.000578
6	1.91	2.012343	0.950008	4.999221	0.000781
7	1.90	2.024867	0.950008	4.999203	0.000799
8	1.89	2.037893	0.950004	4.999596	0.000405
9	1.88	2.051424	0.949995	5.000532	−0.000530
10	1.87	2.065851	0.949996	5.000389	−0.000387
11	1.86	2.081177	0.950005	4.999507	0.000495
12	1.85	2.097013	0.950001	4.999936	0.000066
13	1.84	2.113751	0.949998	5.000210	−0.000209
14	1.83	2.131394	0.949993	5.000681	−0.000679
15	1.82	2.150333	0.949998	5.000168	−0.000167
16	1.81	2.170181	0.949993	5.000663	−0.000662
17	1.80	2.191722	0.950002	4.999757	0.000244
18	1.79	2.214174	0.949992	5.000836	−0.000834
19	1.78	2.238714	0.949996	5.000407	−0.000405
20	1.77	2.264952	0.949994	5.000651	−0.000650
21	1.76	2.293672	0.950000	5.000043	−0.000042
22	1.75	2.324877	0.950002	4.999775	0.000226
23	1.74	2.359350	0.950010	4.999030	0.000972
24	1.73	2.397094	0.950007	4.999352	0.000650
25	1.72	2.439674	0.950009	4.999095	0.000906

(continued)

Table 8.2 (continued)

Parameter values					
PROB. = 0.95	RHO = −0.2		BIGH = 1.9586		SMHL = 1.65
Pair No.	SMH	SMK	PROB. level	Risk level (%)	Risk error (%)
26	1.71	2.487091	0.949994	5.000615	−0.000614
27	1.70	2.543256	0.950000	5.000019	−0.000018
28	1.69	2.608952	0.949992	5.000848	−0.000846
29	1.68	2.691213	0.949998	5.000234	−0.000232
30	1.67	2.798636	0.950000	5.000037	−0.000036
31	1.66	2.954661	0.949992	5.000764	−0.000763
32	1.65	3.274917	0.950004	4.999554	0.000447
No. of pairs		Mean COMPUT. risk errors		STD. ERR. errors	
32		0.0000368465		0.0000971380	
Z-ratio				Status of significance	
0.37932				N.S.—not significant	
Parameter values					
PROB. = 0.94	RHO = −0.2		BIGH = 1.8789		SMHL = 1.56
Pair No.	SMH	SMK	PROB. level	Risk level (%)	Risk error (%)
Table No. 8.2-235					
1	1.88	1.877801	0.970007	5.999351	0.000650
2	1.87	1.887842	0.939999	6.000096	−0.000095
3	1.86	1.898383	0.939998	6.000161	−0.000161
4	1.85	1.909423	0.940003	5.999738	0.000262
5	1.84	1.920966	0.940010	5.999005	0.000995
6	1.83	1.932622	0.939994	6.000567	−0.000566
7	1.82	1.945175	0.940002	5.999816	0.000185
8	1.81	1.958235	0.940007	5.999321	0.000679
9	1.80	1.971805	0.940007	5.999327	0.000674
10	1.79	1.985887	0.940000	6.000018	−0.000018
11	1.78	2.000873	0.940005	5.999523	0.000477
12	1.77	2.016375	0.939998	6.000233	−0.000232
13	1.76	2.032785	0.939996	6.000424	−0.000423
14	1.75	2.050107	0.939995	6.000489	−0.000489
15	1.74	2.068732	0.940010	5.999011	0.000989
16	1.73	2.087881	0.939999	6.000102	−0.000101
17	1.72	2.108339	0.939994	6.000567	−0.000566
18	1.71	2.130498	0.940005	5.999506	0.000495
19	1.70	2.153970	0.940008	5.999220	0.000781
20	1.69	2.178758	0.939997	6.000334	−0.000334
21	1.68	2.206035	0.940006	5.999404	0.000596
22	1.67	2.235025	0.939997	6.000275	−0.000274
23	1.66	2.267290	0.940010	5.999023	0.000978
24	1.65	2.302055	0.940004	5.999637	0.000364
25	1.64	2.340882	0.940007	5.999351	0.000650
26	1.63	2.383775	0.939996	6.000388	−0.000387
27	1.62	2.433082	0.939999	6.000090	−0.000089

(continued)

8.2 Contents for Table 8.2 (Table No. 8.2-1 to 400)

Table 8.2 (continued)

Parameter values					
PROB. = 0.94	RHO = −0.2		BIGH = 1.8789		SMHL = 1.56
Pair No.	SMH	SMK	PROB. level	Risk level (%)	Risk error (%)
28	1.61	2.489586	0.939992	6.000757	−0.000757
29	1.60	2.557978	0.940007	5.999315	0.000685
30	1.59	2.640605	0.939999	6.000090	−0.000089
31	1.58	2.749969	0.940005	5.999530	0.000471
32	1.57	2.907951	0.939996	6.000448	−0.000447
33	1.56	3.225490	0.939998	6.000209	−0.000209
No. of pairs		Mean COMPUT. risk errors		STD. ERR. errors	
33		0.0001379672		0.0000882054	
Z-ratio				Status of significance	
1.56416				N.S.—not significant	
Parameter values					
PROB. = 0.93	RHO = −0.2		BIGH = 1.8094		SMHL = 1.48
Pair No.	SMH	SMK	PROB. level	Risk level (%)	Risk error (%)
Table No. 8.2-236					
1	1.81	1.808800	0.929997	7.000256	−0.000256
2	1.80	1.819044	0.930003	6.999684	0.000316
3	1.79	1.829596	0.930003	6.999702	0.000298
4	1.78	1.840457	0.929996	7.000428	−0.000429
5	1.77	1.852021	0.930007	6.999260	0.000739
6	1.76	1.863702	0.929995	7.000530	−0.000530
7	1.75	1.876090	0.929997	7.000315	−0.000316
8	1.74	1.888990	0.929998	7.000202	−0.000203
9	1.73	1.902600	0.930008	6.999159	0.000840
10	1.72	1.916533	0.930001	6.999928	0.000072
11	1.71	1.931375	0.930009	6.999130	0.000870
12	1.70	1.946543	0.929994	7.000595	−0.000596
13	1.69	1.962822	0.930000	7.000005	−0.000006
14	1.68	1.979821	0.929999	7.000059	−0.000060
15	1.67	1.997741	0.930000	7.000029	−0.000030
16	1.66	2.016777	0.930006	6.999439	0.000560
17	1.65	2.036542	0.929993	7.000691	−0.000691
18	1.64	2.058016	0.930005	6.999552	0.000447
19	1.63	2.080420	0.929996	7.000441	−0.000441
20	1.62	2.104537	0.929995	7.000476	−0.000477
21	1.61	2.130371	0.929994	7.000584	−0.000584
22	1.60	2.158314	0.929998	7.000214	−0.000215
23	1.59	2.188761	0.930009	6.999141	0.000858
24	1.58	2.221325	0.929999	7.000089	−0.000089
25	1.57	2.257179	0.929993	7.000679	−0.000679
26	1.56	2.297109	0.929994	7.000554	−0.000554
27	1.55	2.341899	0.929997	7.000268	−0.000268
28	1.54	2.393115	0.930007	6.999266	0.000733

(continued)

Table 8.2 (continued)

Parameter values					
PROB. = 0.93	RHO = −0.2		BIGH = 1.8094		SMHL = 1.48
Pair No.	SMH	SMK	PROB. level	Risk level (%)	Risk error (%)
29	1.53	2.451541	0.929999	7.000119	−0.000119
30	1.52	2.521087	0.929992	7.000757	−0.000757
31	1.51	2.608789	0.930010	6.999022	0.000978
32	1.50	2.721680	0.929997	7.000292	−0.000292
33	1.49	2.891016	0.929997	7.000322	−0.000322
34	1.48	3.249612	0.929993	7.000697	−0.000697
No. of pairs		Mean COMPUT. risk errors		STD. ERR. errors	
34		−0.0000543254		0.0000882809	
Z-ratio				Status of significance	
0.61537				N.S.—not significant	
Parameter values					
PROB. = 0.92	RHO = −0.2		BIGH = 1.7476		SMHL = 1.41
Pair No.	SMH	SMK	PROB. level	Risk level (%)	Risk error (%)
Table No. 8.2-237					
1	1.75	1.745203	0.920002	7.999820	0.000179
2	1.74	1.755233	0.919996	8.000363	−0.000364
3	1.73	1.765770	0.920002	7.999760	0.000238
4	1.72	1.776619	0.920002	7.999826	0.000173
5	1.71	1.787980	0.920009	7.999146	0.000852
6	1.70	1.799658	0.920006	7.999432	0.000566
7	1.69	1.811851	0.920006	7.999367	0.000632
8	1.68	1.824561	0.920008	7.999164	0.000834
9	1.67	1.837790	0.920009	7.999075	0.000924
10	1.66	1.851541	0.920007	7.999343	0.000656
11	1.65	1.865817	0.919998	8.000195	−0.000197
12	1.64	1.881009	0.920007	7.999277	0.000721
13	1.63	1.896731	0.920004	7.999599	0.000399
14	1.62	1.913180	0.919998	8.000178	−0.000179
15	1.61	1.930553	0.919998	8.000178	−0.000179
16	1.60	1.948855	0.919999	8.000106	−0.000107
17	1.59	1.968087	0.919995	8.000475	−0.000477
18	1.58	1.988447	0.919993	8.000660	−0.000662
19	1.57	2.010134	0.919996	8.000356	−0.000358
20	1.56	2.033150	0.919998	8.000231	−0.000232
21	1.55	2.057890	0.920007	7.999265	0.000733
22	1.54	2.083967	0.919998	8.000189	−0.000191
23	1.53	2.112163	0.919995	8.000475	−0.000477
24	1.52	2.142873	0.920001	7.999873	0.000125
25	1.51	2.176102	0.920000	7.999981	0.000018
26	1.50	2.212633	0.920003	7.999707	0.000292
27	1.49	2.252859	0.920000	7.999963	0.000036
28	1.48	2.297959	0.920001	7.999861	0.000137

(continued)

8.2 Contents for Table 8.2 (Table No. 8.2-1 to 400)

Table 8.2 (continued)

Parameter values					
PROB. = 0.92	RHO = −0.2		BIGH = 1.7476		SMHL = 1.41
Pair No.	SMH	SMK	PROB. level	Risk level (%)	Risk error (%)
29	1.47	2.348716	0.919993	8.000738	−0.000739
30	1.46	2.408259	0.920004	7.999587	0.000411
31	1.45	2.478154	0.920001	7.999873	0.000125
32	1.44	2.564656	0.920006	7.999450	0.000548
33	1.43	2.676362	0.919993	8.000738	−0.000739
34	1.42	2.841402	0.919992	8.000815	−0.000817
35	1.41	3.178531	0.920002	7.999796	0.000203
No. of pairs		Mean COMPUT. risk errors		STD. ERR. errors	
35		0.0000857645		0.0000809738	
Z-ratio				Status of significance	
1.05916				N.S.—not significant	
Parameter values					
PROB. = 0.91	RHO = −0.2		BIGH = 1.6916		SMHL = 1.35
Pair No.	SMH	SMK	PROB. level	Risk level (%)	Risk error (%)
Table No. 8.2-238					
1	1.70	1.683437	0.910004	8.999574	0.000423
2	1.69	1.693201	0.909993	9.000701	−0.000703
3	1.68	1.703475	0.909997	9.000338	−0.000340
4	1.67	1.714065	0.909996	9.000439	−0.000441
5	1.66	1.725169	0.910005	8.999461	0.000536
6	1.65	1.736592	0.910007	8.999276	0.000721
7	1.64	1.748339	0.909999	9.000069	−0.000072
8	1.63	1.760606	0.909997	9.000338	−0.000340
9	1.62	1.773396	0.909996	9.000379	−0.000381
10	1.61	1.786711	0.909996	9.000402	−0.000405
11	1.60	1.800553	0.909993	9.000683	−0.000685
12	1.59	1.815122	0.910000	8.999991	0.000006
13	1.58	1.830223	0.909998	9.000200	−0.000203
14	1.57	1.846056	0.909999	9.000147	−0.000149
15	1.56	1.862622	0.909998	9.000218	−0.000221
16	1.55	1.880120	0.910005	8.999538	0.000459
17	1.54	1.898358	0.910001	8.999872	0.000125
18	1.53	1.917534	0.909996	9.000402	−0.000405
19	1.52	1.938041	0.910006	8.999378	0.000620
20	1.51	1.959493	0.910002	8.999801	0.000197
21	1.50	1.982283	0.909999	9.000105	−0.000107
22	1.49	2.006805	0.910009	8.999062	0.000936
23	1.48	2.032672	0.910003	8.999676	0.000322
24	1.47	2.060668	0.910009	8.999151	0.000846
25	1.46	2.090407	0.909994	9.000617	−0.000620
26	1.45	2.123065	0.909997	9.000331	−0.000334
27	1.44	2.158644	0.909997	9.000331	−0.000334

(continued)

Table 8.2 (continued)

Parameter values					
PROB. = 0.91	RHO = −0.2		BIGH = 1.6916		SMHL = 1.35
Pair No.	SMH	SMK	PROB. level	Risk level (%)	Risk error (%)
28	1.43	2.197931	0.910001	8.999891	0.000107
29	1.42	2.241710	0.910008	8.999162	0.000834
30	1.41	2.290377	0.919997	9.000254	−0.000256
31	1.40	2.346279	0.909992	9.000760	−0.000763
32	1.39	2.412546	0.910006	8.999401	0.000596
33	1.38	2.491526	0.909999	9.000116	−0.000119
34	1.37	2.591818	0.909995	9.000522	−0.000525
35	1.36	2.730613	0.909995	9.000456	−0.000459
36	1.35	2.963386	0.910002	8.999771	0.000226
No. of pairs		Mean COMPUT. risk errors		STD. ERR. errors	
36		−0.0000244862		0.0000783251	
Z-ratio				Status of significance	
0.31262				N.S.—not significant	
Parameter values					
PROB. = 0.9025	RHO = −0.2		BIGH = 1.6528		SMHL = 1.30
Pair No.	SMH	SMK	PROB. level	Risk level (%)	Risk error (%)
Table No. 8.2-239					
1	1.66	1.645631	0.902493	9.750664	−0.000662
2	1.65	1.655605	0.902498	9.750170	−0.000167
3	1.64	1.665895	0.902500	9.749967	0.000036
4	1.63	1.676505	0.902498	9.750199	−0.000197
5	1.62	1.687436	0.902490	9.750986	−0.000983
6	1.61	1.698886	0.902493	9.750700	−0.000697
7	1.60	1.710858	0.902504	9.749556	0.000447
8	1.59	1.723159	0.902504	9.749579	0.000423
9	1.58	1.735986	0.902508	9.749222	0.000781
10	1.57	1.749147	0.902496	9.750396	−0.000393
11	1.56	1.763034	0.902500	9.750032	−0.000030
12	1.55	1.777457	0.902499	9.750134	−0.000131
13	1.54	1.792612	0.902506	9.749394	0.000608
14	1.53	1.808307	0.902503	9.749735	0.000268
15	1.52	1.824741	0.902500	9.750038	−0.000036
16	1.51	1.841917	0.902493	9.750676	−0.000674
17	1.50	1.860228	0.902506	9.749389	0.000614
18	1.49	1.879091	0.902492	9.750759	−0.000757
19	1.48	1.899291	0.902499	9.750080	−0.000077
20	1.47	1.920441	0.902494	9.750616	−0.000614
21	1.46	1.942935	0.902494	9.750629	−0.000626
22	1.45	1.966776	0.902490	9.750986	−0.000983
23	1.44	1.992359	0.902496	9.750444	−0.000441
24	1.43	2.019688	0.902498	9.750205	−0.000203
25	1.42	2.049156	0.902504	9.749561	0.000441

(continued)

Table 8.2 (continued)

Parameter values					
PROB. = 0.9025	RHO = −0.2		BIGH = 1.6528		SMHL = 1.30
Pair No.	SMH	SMK	PROB. level	Risk level (%)	Risk error (%)
26	1.41	2.080769	0.902500	9.750003	0.000000
27	1.40	2.115311	0.902503	9.749746	0.000256
28	1.39	2.152786	0.902491	9.750914	−0.000912
29	1.38	2.194761	0.902499	9.750109	−0.000107
30	1.37	2.241632	0.902504	9.749556	0.000447
31	1.36	2.294575	0.902503	9.749704	0.000298
32	1.35	2.355547	0.902496	9.750419	−0.000417
33	1.34	2.428460	0.902500	9.749967	0.000036
34	1.33	2.518007	0.902497	9.750330	−0.000328
35	1.32	2.636692	0.902505	9.749514	0.000489
36	1.31	2.811865	0.902500	9.750032	−0.000030
37	1.30	3.188843	0.902501	9.749949	0.000054
No. of pairs			Mean COMPUT. risk errors	STD. ERR. errors	
37			−0.0001123077	0.0000772144	
Z-ratio				Status of significance	
1.45449				N.S.—not significant	
Parameter values					
PROB. = 0.9	RHO = −0.2		BIGH = 1.6404		SMHL = 1.29
Pair No.	SMH	SMK	PROB. level	Risk level (%)	Risk error (%)
Table No. 8.2-240					
1	1.65	1.630856	0.899995	10.000550	−0.000548
2	1.64	1.640800	0.900004	9.999567	0.000435
3	1.63	1.650866	0.899993	10.000730	−0.000727
4	1.62	1.661448	0.899997	10.000310	−0.000310
5	1.61	1.672351	0.899996	10.000430	−0.000429
6	1.60	1.683773	0.900006	9.999371	0.000632
7	1.59	1.695523	0.900008	9.999198	0.000805
8	1.58	1.707601	0.900000	10.000050	−0.000048
9	1.57	1.720207	0.899996	10.000400	−0.000399
10	1.56	1.733342	0.899995	10.000460	−0.000453
11	1.55	1.747010	0.899995	10.000540	−0.000536
12	1.54	1.761408	0.900007	9.999305	0.000697
13	1.53	1.776149	0.899997	10.000290	−0.000286
14	1.52	1.791821	0.900009	9.999096	0.000906
15	1.51	1.807842	0.899993	10.000690	−0.000691
16	1.50	1.824801	0.899990	10.000960	−0.000954
17	1.49	1.842700	0.899996	10.000420	−0.000417
18	1.48	1.861543	0.900004	9.999651	0.000352
19	1.47	1.881334	0.900007	9.999275	0.000727
20	1.46	1.902075	0.900002	9.999818	0.000185
21	1.45	1.924161	0.900005	9.999502	0.000501
22	1.44	1.947595	0.900008	9.999186	0.000817

(continued)

Table 8.2 (continued)

Parameter values					
PROB. = 0.9	RHO = −0.2		BIGH = 1.6404		SMHL = 1.29
Pair No.	SMH	SMK	PROB. level	Risk level (%)	Risk error (%)
23	1.43	1.972382	0.900003	9.999752	0.000250
24	1.42	1.998915	0.900000	10.000010	−0.000006
25	1.41	2.027589	0.900008	9.999233	0.000769
26	1.40	2.058017	0.899993	10.000750	−0.000745
27	1.39	2.091377	0.899994	10.000590	−0.000584
28	1.38	2.128061	0.900006	9.999371	0.000632
29	1.37	2.168075	0.900004	9.999609	0.000393
30	1.36	2.212596	0.900002	9.999758	0.000244
31	1.35	2.262799	0.900003	9.999741	0.000262
32	1.34	2.319862	0.899991	10.000930	−0.000924
33	1.33	2.387304	0.899993	10.000660	−0.000662
34	1.32	2.469038	0.899997	10.000310	−0.000304
35	1.31	2.572881	0.899997	10.000350	−0.000346
36	1.30	2.716807	0.899990	10.001000	−0.000995
37	1.29	2.967227	0.900008	9.999180	0.000823
No. of pairs		Mean COMPUT. risk errors		STD. ERR. errors	
37		−0.0000246261		0.0000959173	
Z-ratio				Status of significance	
0.25674				N.S.—not significant	

Parameter values					
PROB. = 0.85	RHO = −0.2		BIGH = 1.4307		SMHL = 1.04
Pair No.	SMH	SMK	PROB. level	Risk level (%)	Risk error (%)
Table No. 8.2-241					
1	1.44	1.421460	0.849991	15.000860	−0.000858
2	1.43	1.431400	0.850001	14.999900	0.000095
3	1.42	1.441578	0.850001	14.999860	0.000143
4	1.41	1.452094	0.850005	14.999520	0.000483
5	1.40	1.462854	0.849997	15.000320	−0.000322
6	1.39	1.474056	0.850001	14.999890	0.000113
7	1.38	1.485606	0.850003	14.999690	0.000304
8	1.37	1.497507	0.850001	14.999910	0.000083
9	1.36	1.509763	0.849992	15.000760	−0.000763
10	1.35	1.522572	0.850000	15.000050	−0.000048
11	1.34	1.535742	0.849996	15.000450	−0.000453
12	1.33	1.549473	0.850001	14.999880	0.000119
13	1.32	1.563574	0.849991	15.000950	−0.000948
14	1.31	1.578341	0.849994	15.000580	−0.000584
15	1.30	1.593681	0.849997	15.000300	−0.000298
16	1.29	1.609598	0.849995	15.000470	−0.000471
17	1.28	1.626291	0.850005	14.999490	0.000513
18	1.27	1.643570	0.850002	14.999800	0.000197
19	1.26	1.661635	0.850000	14.999960	0.000042

(continued)

8.2 Contents for Table 8.2 (Table No. 8.2-1 to 400)

Table 8.2 (continued)

Parameter values					
PROB. = 0.85	RHO = −0.2		BIGH = 1.4307		SMHL = 1.04
Pair No.	SMH	SMK	PROB. level	Risk level (%)	Risk error (%)
20	1.25	1.680491	0.849995	15.000490	−0.000495
21	1.24	1.700337	0.849997	15.000260	−0.000262
22	1.23	1.721179	0.850000	14.999960	0.000036
23	1.22	1.743023	0.849996	15.000380	−0.000381
24	1.21	1.766264	0.850008	14.999160	0.000834
25	1.20	1.790518	0.849996	15.000420	−0.000423
26	1.19	1.816570	0.850008	14.999240	0.000757
27	1.18	1.844038	0.850001	14.999860	0.000143
28	1.17	1.873317	0.849991	15.000860	−0.000858
29	1.16	1.905000	0.849998	15.000200	−0.000203
30	1.15	1.939095	0.850000	15.000020	−0.000024
31	1.14	1.975997	0.849996	15.000430	−0.000429
32	1.13	2.016495	0.849999	15.000070	−0.000072
33	1.12	2.061184	0.850003	14.999740	0.000262
34	1.11	2.110853	0.849998	15.000240	−0.000244
35	1.10	2.167070	0.849992	15.000800	−0.000799
36	1.09	2.232579	0.850006	14.999360	0.000638
37	1.08	2.309342	0.849995	15.000500	−0.000501
38	1.07	2.405179	0.850010	14.999000	0.000995
39	1.06	2.529476	0.849993	15.000740	−0.000745
40	1.05	2.718180	0.850010	14.999040	0.000960
41	1.04	3.130674	0.849993	15.000730	−0.000733
No. of pairs		Mean COMPUT. risk errors		STD. ERR. errors	
41		−0.0000999087		0.0000791131	
Z-ratio				Status of significance	
1.26286				N.S.—not significant	
Parameter values					
PROB. = 0.8	RHO = −0.2		BIGH = 1.267		SMHL = 0.85
Pair No.	SMH	SMK	PROB. level	Risk level (%)	Risk error (%)
Table No. 8.2-242					
1	1.27	1.264105	0.800006	19.999430	0.000566
2	1.26	1.274136	0.800002	19.999840	0.000155
3	1.25	1.284427	0.799993	20.000670	−0.000668
4	1.24	1.295076	0.799996	20.000390	−0.000387
5	1.23	1.306089	0.800008	19.999230	0.000769
6	1.22	1.317275	0.799996	20.000440	−0.000435
7	1.21	1.328931	0.800004	19.999580	0.000423
8	1.20	1.340866	0.800001	19.999860	0.000143
9	1.19	1.353182	0.800000	19.999980	0.000018
10	1.18	1.365883	0.799999	20.000140	−0.000137
11	1.17	1.379073	0.800008	19.999180	0.000823
12	1.16	1.392561	0.799998	20.000160	−0.000155

(continued)

Table 8.2 (continued)

Parameter values						
PROB. = 0.8	RHO = −0.2		BIGH = 1.267		SMHL = 0.85	
Pair No.	SMH	SMK	PROB. level	Risk level (%)	Risk error (%)	
13	1.15	1.406645	0.800008	19.999170	0.000829	
14	1.14	1.421039	0.799994	20.000560	−0.000560	
15	1.13	1.436039	0.799993	20.000660	−0.000662	
16	1.12	1.451653	0.800001	19.999880	0.000119	
17	1.11	1.467788	0.800001	19.999870	0.000125	
18	1.10	1.484549	0.800002	19.999790	0.000215	
19	1.09	1.501941	0.799999	20.000090	−0.000089	
20	1.08	1.520070	0.799999	20.000090	−0.000095	
21	1.07	1.538942	0.799996	20.000360	−0.000358	
22	1.06	1.558759	0.800007	19.999330	0.000668	
23	1.05	1.579237	0.799991	20.000860	−0.000864	
24	1.04	1.600969	0.800005	19.999520	0.000483	
25	1.03	1.623572	0.799998	20.000180	−0.000185	
26	1.02	1.647445	0.800001	19.999900	0.000095	
27	1.01	1.672598	0.800002	19.999820	0.000179	
28	1.00	1.699234	0.800006	19.999400	0.000602	
29	0.99	1.727363	0.800000	20.000030	−0.000030	
30	0.98	1.757386	0.800000	20.000000	−0.000006	
31	0.97	1.789507	0.800002	19.999790	0.000209	
32	0.96	1.823933	0.799999	20.000120	−0.000119	
33	0.95	1.861262	0.800005	19.999530	0.000471	
34	0.94	1.901699	0.800001	19.999940	0.000060	
35	0.93	1.946038	0.799997	20.000320	−0.000316	
36	0.92	1.995269	0.800000	20.000050	−0.000054	
37	0.91	2.050381	0.799997	20.000340	−0.000340	
38	0.90	2.113341	0.799998	20.000210	−0.000215	
39	0.89	2.186897	0.800003	19.999740	0.000262	
40	0.88	2.274972	0.799992	20.000850	−0.000852	
41	0.87	2.387346	0.800000	20.000020	−0.000018	
42	0.86	2.543176	0.800010	19.999030	0.000972	
43	0.85	2.807324	0.800007	19.999290	0.000709	
No. of pairs		Mean COMPUT. risk errors		STD. ERR. errors		
43		0.0000533733		0.0000683073		
Z-ratio				Status of significance		
0.78137				N.S.—not significant		
Parameter values						
PROB. = 0.75	RHO = −0.2		BIGH = 1.1288		SMHL = 0.68	
Pair No.	SMH	SMK	PROB. level	Risk level (%)	Risk error (%)	
Table No. 8.2-243						
1	1.13	1.127504	0.750005	24.999500	0.000501	
2	1.12	1.137572	0.750000	25.000040	−0.000042	
3	1.11	1.147919	0.749998	25.000250	−0.000244	

(continued)

8.2 Contents for Table 8.2 (Table No. 8.2-1 to 400)

Table 8.2 (continued)

Parameter values					
PROB. = 0.75	RHO = −0.2		BIGH = 1.1288		SMHL = 0.68
Pair No.	SMH	SMK	PROB. level	Risk level (%)	Risk error (%)
4	1.10	1.158549	0.749998	25.000200	−0.000203
5	1.09	1.169469	0.749999	25.000080	−0.000077
6	1.08	1.180684	0.750000	24.999970	0.000030
7	1.07	1.192198	0.750000	25.000020	−0.000024
8	1.06	1.204019	0.749997	25.000320	−0.000322
9	1.05	1.216248	0.750006	24.999360	0.000644
10	1.04	1.228698	0.749993	25.000690	−0.000685
11	1.03	1.241667	0.750006	24.999440	0.000560
12	1.02	1.254869	0.749992	25.000790	−0.000793
13	1.01	1.268605	0.750000	25.000040	−0.000042
14	1.00	1.282783	0.750009	24.999100	0.000900
15	0.99	1.297314	0.750002	24.999760	0.000238
16	0.98	1.312303	0.749993	25.000670	−0.000674
17	0.97	1.327855	0.749993	25.000710	−0.000703
18	0.96	1.343980	0.749998	25.000210	−0.000209
19	0.95	1.360588	0.749990	25.000960	−0.000960
20	0.94	1.377981	0.750008	24.999230	0.000769
21	0.93	1.395877	0.750005	24.999550	0.000453
22	0.92	1.414481	0.750003	24.999700	0.000304
23	0.91	1.433802	0.749998	25.000190	−0.000191
24	0.90	1.454047	0.750008	24.999180	0.000823
25	0.89	1.475033	0.750002	24.999780	0.000221
26	0.88	1.496966	0.749997	25.000280	−0.000280
27	0.87	1.520054	0.750006	24.999360	0.000638
28	0.86	1.544115	0.749999	25.000120	−0.000119
29	0.85	1.569554	0.750006	24.999410	0.000596
30	0.84	1.596189	0.749996	25.000460	−0.000453
31	0.83	1.624425	0.749993	25.000690	−0.000685
32	0.82	1.654475	0.750000	24.999970	0.000030
33	0.81	1.686354	0.749998	25.000190	−0.000191
34	0.80	1.720471	0.749999	25.000100	−0.000101
35	0.79	1.757038	0.749993	25.000660	−0.000656
36	0.78	1.796857	0.750009	24.999120	0.000882
37	0.77	1.839947	0.750007	24.999290	0.000715
38	0.76	1.887111	0.749997	25.000310	−0.000310
39	0.75	1.939544	0.749993	25.000740	−0.000739
40	0.74	1.998830	0.750003	24.999680	0.000322
41	0.73	2.066558	0.750006	24.999450	0.000554
42	0.72	2.145879	0.750001	24.999940	0.000066
43	0.71	2.242288	0.749993	25.000710	−0.000703

(continued)

Table 8.2 (continued)

Parameter values					
PROB. = 0.75	RHO = −0.2		BIGH = 1.1288		SMHL = 0.68
Pair No.	SMH	SMK	PROB. level	Risk level (%)	Risk error (%)
44	0.70	2.367144	0.749998	25.000160	−0.000155
45	0.69	2.545869	0.749993	25.000710	−0.000709
46	0.68	2.892556	0.750009	24.999120	0.000882
No. of pairs		Mean COMPUT. risk errors		STD. ERR. errors	
46		−0.0000030436		0.0000771745	
Z-ratio				Status of significance	
0.03944				N.S.—not significant	
Parameter values					
PROB. = 0.7	RHO = −0.2		BIGH = 1.0062		SMHL = 0.53
Pair No.	SMH	SMK	PROB. level	Risk level (%)	Risk error (%)
Table No. 8.2-244					
1	1.01	1.002414	0.700000	30.000020	−0.000018
2	1.00	1.012438	0.699999	30.000130	−0.000125
3	0.99	1.022714	0.700000	29.999980	0.000018
4	0.98	1.033198	0.699993	30.000720	−0.000721
5	0.97	1.044044	0.700007	29.999310	0.000697
6	0.96	1.055014	0.699991	30.000880	−0.000882
7	0.95	1.066408	0.700005	29.999520	0.000483
8	0.94	1.078039	0.700007	29.999290	0.000709
9	0.93	1.089913	0.699998	30.000210	−0.000209
10	0.92	1.102136	0.699995	30.000470	−0.000471
11	0.91	1.114718	0.699998	30.000200	−0.000197
12	0.90	1.127666	0.700005	29.999540	0.000459
13	0.89	1.140891	0.699996	30.000410	−0.000405
14	0.88	1.154599	0.700006	29.999410	0.000596
15	0.87	1.168605	0.699998	30.000180	−0.000179
16	0.86	1.183113	0.700006	29.999450	0.000554
17	0.85	1.197940	0.699992	30.000800	−0.000799
18	0.84	1.213389	0.700006	29.999440	0.000560
19	0.83	1.229180	0.699995	30.000540	−0.000536
20	0.82	1.245618	0.700005	29.999530	0.000471
21	0.81	1.262521	0.700001	29.999880	0.000125
22	0.80	1.280001	0.699997	30.000290	−0.000292
23	0.79	1.298169	0.700003	29.999710	0.000292
24	0.78	1.316943	0.700000	29.999970	0.000030
25	0.77	1.336437	0.699999	30.000100	−0.000095
26	0.76	1.356765	0.700008	29.999230	0.000769
27	0.75	1.377847	0.700007	29.999320	0.000685
28	0.74	1.399800	0.700004	29.999600	0.000405
29	0.73	1.422643	0.699992	30.000770	−0.000763
30	0.72	1.446691	0.700000	30.000020	−0.000018
31	0.71	1.471867	0.700005	29.999540	0.000459
32	0.70	1.498293	0.700007	29.999270	0.000733

(continued)

8.2 Contents for Table 8.2 (Table No. 8.2-1 to 400)

Table 8.2 (continued)

Parameter values					
PROB. = 0.7	RHO = −0.2		BIGH = 1.0062		SMHL = 0.53
Pair No.	SMH	SMK	PROB. level	Risk level (%)	Risk error (%)
33	0.69	1.526090	0.700007	29.999270	0.000733
34	0.68	1.555286	0.699992	30.000850	−0.000846
35	0.67	1.586492	0.700003	29.999710	0.000292
36	0.66	1.619544	0.700002	29.999780	0.000221
37	0.65	1.654862	0.700004	29.999630	0.000376
38	0.64	1.692676	0.699998	30.000220	−0.000215
39	0.63	1.733607	0.700001	29.999940	0.000066
40	0.62	1.778081	0.700002	29.999810	0.000197
41	0.61	1.826922	0.700008	29.999230	0.000769
42	0.60	1.880755	0.699996	30.000370	−0.000364
43	0.59	1.941387	0.699994	30.000580	−0.000578
44	0.58	2.010817	0.699997	30.000330	−0.000328
45	0.57	2.092222	0.700001	29.999940	0.000060
46	0.56	2.190736	0.699993	30.000740	−0.000733
47	0.55	2.318139	0.700003	29.999750	0.000256
48	0.54	2.499884	0.699999	30.000100	−0.000101
49	0.53	2.850884	0.700002	29.999800	0.000203
No. of pairs		Mean COMPUT. risk errors		STD. ERR. errors	
49		0.0000468493		0.0000683786	
Z-ratio				Status of significance	
0.68514				N.S.—not significant	
Parameter values					
PROB. = 0.65	RHO = −0.2		BIGH = 0.894		SMHL = 0.41
Pair No.	SMH	SMK	PROB. level	Risk level (%)	Risk error (%)
Table No. 8.2-245					
1	0.90	0.888040	0.649993	35.000660	−0.000656
2	0.89	0.898018	0.649993	35.000680	−0.000679
3	0.88	0.908223	0.649994	35.000600	−0.000602
4	0.87	0.918662	0.649996	35.000440	−0.000441
5	0.86	0.929344	0.649998	35.000210	−0.000203
6	0.85	0.940278	0.650001	34.999890	0.000107
7	0.84	0.951471	0.650005	34.999500	0.000501
8	0.83	0.962935	0.650009	34.999060	0.000948
9	0.82	0.974580	0.649994	35.000600	−0.000596
10	0.81	0.986613	0.650000	35.000050	−0.000042
11	0.80	0.998947	0.650006	34.999440	0.000560
12	0.79	1.011496	0.649992	35.000820	−0.000817
13	0.78	1.024466	0.649998	35.000240	−0.000232
14	0.77	1.037773	0.650003	34.999690	0.000310
15	0.76	1.051431	0.650007	34.999270	0.000739
16	0.75	1.065355	0.649992	35.000810	−0.000805
17	0.74	1.079755	0.649994	35.000640	−0.000638
18	0.73	1.094551	0.649993	35.000740	−0.000739

(continued)

Table 8.2 (continued)

Parameter values					
PROB. = 0.65	RHO = −0.2		BIGH = 0.894		SMHL = 0.41
Pair No.	SMH	SMK	PROB. level	Risk level (%)	Risk error (%)
19	0.72	1.109855	0.650006	34.999380	0.000620
20	0.71	1.125489	0.649998	35.000200	−0.000191
21	0.70	1.141668	0.650002	34.999780	0.000226
22	0.69	1.158313	0.650000	34.999980	0.000024
23	0.68	1.175542	0.650007	34.999300	0.000703
24	0.67	1.193280	0.650005	34.999540	0.000465
25	0.66	1.211549	0.649992	35.000810	−0.000805
26	0.65	1.230570	0.649997	35.000280	−0.000274
27	0.64	1.250271	0.650002	34.999780	0.000221
28	0.63	1.270679	0.650004	34.999620	0.000381
29	0.62	1.291824	0.649999	35.000100	−0.000095
30	0.61	1.313836	0.649998	35.000250	−0.000250
31	0.60	1.336747	0.649995	35.000550	−0.000542
32	0.59	1.360691	0.649997	35.000260	−0.000256
33	0.58	1.385707	0.649999	35.000060	−0.000060
34	0.57	1.411933	0.650005	34.999510	0.000489
35	0.56	1.439316	0.649994	35.000620	−0.000620
36	0.55	1.468195	0.649990	35.000970	−0.000966
37	0.54	1.498816	0.650002	34.999800	0.000209
38	0.53	1.531038	0.649994	35.000590	−0.000590
39	0.52	1.565507	0.650007	34.999260	0.000745
40	0.51	1.602089	0.650002	34.999800	0.000209
41	0.50	1.641440	0.650006	34.999410	0.000596
42	0.49	1.683827	0.650004	34.999610	0.000393
43	0.48	1.729917	0.650004	34.999620	0.000381
44	0.47	1.780384	0.650000	35.000040	−0.000036
45	0.46	1.836296	0.649995	35.000500	−0.000495
46	0.45	1.899125	0.649991	35.000890	−0.000882
47	0.44	1.971131	0.649993	35.000700	−0.000697
48	0.43	2.064546	0.649995	35.000520	−0.000513
49	0.42	2.166222	0.649995	35.000530	−0.000525
50	0.41	2.297791	0.649990	35.000970	−0.000972
No. of pairs		Mean COMPUT. risk errors		STD. ERR. errors	
50		−0.0001252866		0.0000746578	
Z-ratio				Status of significance	
1.67814				N.S.—not significant	
Parameter values					
PROB. = 0.6	RHO = −0.2		BIGH = 0.7887		SMHL = 0.30
Pair No.	SMH	SMK	PROB. level	Risk level (%)	Risk error (%)
Table No. 8.2-246					
1	0.79	0.787500	0.600002	39.999780	0.000215
2	0.78	0.797595	0.600006	39.999380	0.000620

(continued)

8.2 Contents for Table 8.2 (Table No. 8.2-1 to 400)

Table 8.2 (continued)

Parameter values					
PROB. = 0.6	RHO = −0.2		BIGH = 0.7887		SMHL = 0.30
Pair No.	SMH	SMK	PROB. level	Risk level (%)	Risk error (%)
3	0.77	0.807854	0.599999	40.000140	−0.000143
4	0.76	0.818386	0.600004	39.999620	0.000381
5	0.75	0.829104	0.599999	40.000080	−0.000083
6	0.74	0.840117	0.600009	39.999110	0.000888
7	0.73	0.851290	0.600000	40.000050	−0.000054
8	0.72	0.862783	0.600006	39.999410	0.000590
9	0.71	0.874463	0.599996	40.000420	−0.000417
10	0.70	0.886491	0.600003	39.999700	0.000298
11	0.69	0.898784	0.600007	39.999320	0.000679
12	0.68	0.911358	0.600008	39.999170	0.000834
13	0.67	0.924230	0.600009	39.999120	0.000882
14	0.66	0.937418	0.600010	39.999040	0.000954
15	0.65	0.950844	0.599992	40.000830	−0.000834
16	0.64	0.964723	0.599996	40.000370	−0.000370
17	0.63	0.978979	0.600005	39.999520	0.000477
18	0.62	0.993537	0.599999	40.000120	−0.000119
19	0.61	1.008520	0.599999	40.000080	−0.000077
20	0.60	1.023953	0.600007	39.999300	0.000703
21	0.59	1.039767	0.600005	39.999490	0.000507
22	0.58	1.055992	0.599995	40.000470	−0.000471
23	0.57	1.072757	0.599997	40.000340	−0.000346
24	0.56	1.090096	0.600010	39.999020	0.000978
25	0.55	1.107851	0.600002	39.999780	0.000221
26	0.54	1.126158	0.599993	40.000730	−0.000733
27	0.53	1.145159	0.599998	40.000190	−0.000191
28	0.52	1.164800	0.600003	39.999660	0.000340
29	0.51	1.185130	0.600010	39.999050	0.000948
30	0.50	1.206106	0.600002	39.999800	0.000203
31	0.49	1.277883	0.599997	40.000260	−0.000262
32	0.48	1.250619	0.600009	39.999080	0.000918
33	0.47	1.274091	0.599996	40.000410	−0.000405
34	0.46	1.298663	0.599998	40.000170	−0.000167
35	0.45	1.324319	0.600001	39.999870	0.000131
36	0.44	1.351048	0.599991	40.000870	−0.000870
37	0.43	1.379238	0.600002	39.999830	0.000173
38	0.42	1.408799	0.600005	39.999540	0.000465
39	0.41	1.439844	0.599996	40.000390	−0.000393
40	0.40	1.472696	0.599991	40.000910	−0.000906
41	0.39	1.507688	0.600000	40.000050	−0.000048
42	0.38	1.544778	0.599990	40.000980	−0.000983
43	0.37	1.584724	0.600002	39.999830	0.000167
44	0.36	1.627518	0.599995	40.000490	−0.000495

(continued)

Table 8.2 (continued)

Parameter values					
PROB. = 0.6	RHO = −0.2		BIGH = 0.7887		SMHL = 0.30
Pair No.	SMH	SMK	PROB. level	Risk level (%)	Risk error (%)
45	0.35	1.688019	0.599997	40.000270	−0.000274
46	0.34	1.737753	0.600003	39.999660	0.000340
47	0.33	1.792803	0.600008	39.999170	0.000834
48	0.32	1.854443	0.600007	39.999350	0.000656
49	0.31	1.924571	0.599995	40.000520	−0.000519
50	0.30	2.007083	0.600009	39.999060	0.000936
No. of pairs		Mean COMPUT. risk errors		STD. ERR. errors	
50		0.0001210792		0.0000784979	
Z-ratio				Status of significance	
1.54245				N.S.—not significant	

Parameter values					
PROB. = 0.55	RHO = −0.2		BIGH = 0.688		SMHL = 0.20
Pair No.	SMH	SMK	PROB. level	Risk level (%)	Risk error (%)
Table No. 8.2-247					
1	0.69	0.685908	0.549999	45.000120	−0.000125
2	0.68	0.695997	0.550006	44.999360	0.000638
3	0.67	0.706191	0.549991	45.000930	−0.000930
4	0.66	0.716700	0.550002	44.999780	0.000221
5	0.65	0.727343	0.549994	45.000560	−0.000554
6	0.64	0.738233	0.549994	45.000610	−0.000602
7	0.63	0.749387	0.550002	44.999770	0.000226
8	0.62	0.760724	0.549999	45.000110	−0.000107
9	0.61	0.772360	0.550009	44.999080	0.000924
10	0.60	0.784121	0.549991	45.000930	−0.000924
11	0.59	0.796223	0.549991	45.000880	−0.000876
12	0.58	0.808591	0.549992	45.000850	−0.000852
13	0.57	0.821248	0.549995	45.000550	−0.000548
14	0.56	0.834222	0.550004	44.999650	0.000352
15	0.55	0.847442	0.550001	44.999910	0.000089
16	0.54	0.860938	0.549991	45.000900	−0.000900
17	0.53	0.874840	0.549999	45.000150	−0.000149
18	0.52	0.889085	0.550007	44.999320	0.000685
19	0.51	0.903614	0.550001	44.999910	0.000089
20	0.50	0.918563	0.550006	44.999430	0.000566
21	0.49	0.933879	0.550007	44.999350	0.000656
22	0.48	0.949610	0.550010	44.999040	0.000960
23	0.47	0.965708	0.550002	44.999790	0.000215
24	0.46	0.982329	0.550009	44.999090	0.000906
25	0.45	0.999336	0.550001	44.999890	0.000113
26	0.44	1.016895	0.550004	44.999620	0.000376

(continued)

8.2 Contents for Table 8.2 (Table No. 8.2-1 to 400)

Table 8.2 (continued)

Parameter values					
PROB. = 0.55	RHO = −0.2		BIGH = 0.688		SMHL = 0.20
Pair No.	SMH	SMK	PROB. level	Risk level (%)	Risk error (%)
27	0.43	1.034979	0.550007	44.999300	0.000703
28	0.42	1.053571	0.550003	44.999710	0.000292
29	0.41	1.072759	0.550001	44.999920	0.000077
30	0.40	1.092539	0.549994	45.000580	−0.000578
31	0.39	1.113116	0.550009	44.999060	0.000936
32	0.38	1.134216	0.549996	45.000440	−0.000435
33	0.37	1.156163	0.549996	45.000360	−0.000364
34	0.36	1.178906	0.549995	45.000520	−0.000519
35	0.35	1.202606	0.550005	44.999500	0.000501
36	0.34	1.227148	0.550001	44.999920	0.000077
37	0.33	1.252734	0.549999	45.000080	−0.000077
38	0.32	1.279394	0.549993	45.000750	−0.000745
39	0.31	1.307383	0.550000	45.000060	−0.000054
40	0.30	1.336601	0.549992	45.000770	−0.000769
41	0.29	1.387883	0.550003	44.999700	0.000304
42	0.28	1.418832	0.549995	45.000530	−0.000530
43	0.27	1.451756	0.549997	45.000280	−0.000280
44	0.26	1.486762	0.549995	45.000480	−0.000477
45	0.25	1.524232	0.549998	45.000240	−0.000232
46	0.24	1.564451	0.549997	45.000340	−0.000340
47	0.23	1.608014	0.550003	44.999670	0.000334
48	0.22	1.655271	0.549998	45.000220	−0.000221
49	0.21	1.707140	0.549992	45.000810	−0.000811
50	0.20	1.764763	0.549991	45.000930	−0.000930
No. of pairs		Mean COMPUT. risk errors		STD. ERR. errors	
50		−0.0000723437		0.0000786311	
Z-ratio				Status of significance	
0.92004				N.S.—not significant	
Parameter values					
PROB. = 0.5	RHO = −0.2		BIGH = 0.5898		SMHL = 0.10
Pair No.	SMH	SMK	PROB. level	Risk level (%)	Risk error (%)
Table No. 8.2-248					
1	0.59	0.589502	0.500004	49.999560	0.000441
2	0.58	0.599570	0.499999	50.000070	−0.000066
3	0.57	0.609848	0.500000	50.000030	−0.000030
4	0.56	0.620307	0.499997	50.000340	−0.000331
5	0.55	0.631015	0.500006	49.999380	0.000620
6	0.54	0.641849	0.499996	50.000410	−0.000408
7	0.53	0.652929	0.499994	50.000590	−0.000590

(continued)

Table 8.2 (continued)

Parameter values					
PROB. = 0.5	RHO = −0.2		BIGH = 0.5898		SMHL = 0.10
Pair No.	SMH	SMK	PROB. level	Risk level (%)	Risk error (%)
8	0.52	0.664282	0.500005	49.999480	0.000519
9	0.51	0.675788	0.499999	50.000110	−0.000107
10	0.50	0.687525	0.499992	50.000770	−0.000760
11	0.49	0.699575	0.500003	49.999720	0.000280
12	0.48	0.711826	0.500002	49.999800	0.000203
13	0.47	0.724316	0.499997	50.000310	−0.000304
14	0.46	0.737085	0.499995	50.000490	−0.000492
15	0.45	0.750180	0.500004	49.999620	0.000381
16	0.44	0.763549	0.500010	49.999030	0.000972
17	0.43	0.777150	0.500001	49.999870	0.000131
18	0.42	0.791138	0.500010	49.999050	0.000954
19	0.41	0.805382	0.500004	49.999630	0.000376
20	0.40	0.819953	0.499996	50.000360	−0.000364
21	0.39	0.834927	0.500000	49.999990	0.000012
22	0.38	0.850295	0.500009	49.999090	0.000906
23	0.37	0.865954	0.500000	49.999970	0.000036
24	0.36	0.882011	0.499992	50.000840	−0.000837
25	0.35	0.898584	0.500001	49.999860	0.000143
26	0.34	0.915512	0.499995	50.000480	−0.000477
27	0.33	0.932941	0.499997	50.000340	−0.000334
28	0.32	0.950844	0.499995	50.000460	−0.000459
29	0.31	0.969309	0.500003	49.999700	0.000298
30	0.30	0.988257	0.500000	49.999990	0.000006
31	0.29	1.007831	0.500006	49.999410	0.000596
32	0.28	1.027917	0.499996	50.000400	−0.000396
33	0.27	1.048736	0.500000	50.000030	−0.000021
34	0.26	1.070164	0.499991	50.000910	−0.000912
35	0.25	1.092431	0.499998	50.000220	−0.000221
36	0.24	1.115447	0.499998	50.000180	−0.000179
37	0.23	1.165575	0.499997	50.000300	−0.000301
38	0.22	1.188620	0.500006	49.999410	0.000590
39	0.21	1.212548	0.499994	50.000610	−0.000602
40	0.20	1.237755	0.500004	49.999590	0.000417
41	0.19	1.264064	0.499998	50.000190	−0.000185
42	0.18	1.291755	0.499998	50.000190	−0.000194
43	0.17	1.320865	0.499992	50.000840	−0.000840
44	0.16	1.351686	0.499994	50.000570	−0.000566
45	0.15	1.384324	0.499997	50.000320	−0.000319
46	0.14	1.418919	0.499991	50.000940	−0.000945

(continued)

Table 8.2 (continued)

Parameter values					
PROB. = 0.5	RHO = −0.2		BIGH = 0.5898		SMHL = 0.10
Pair No.	SMH	SMK	PROB. level	Risk level (%)	Risk error (%)
47	0.13	1.455951	0.499997	50.000300	−0.000298
48	0.12	1.495541	0.499996	50.000430	−0.000429
49	0.11	1.538176	0.499994	50.000610	−0.000608
50	0.10	1.584493	0.500001	49.999880	0.000125
No. of pairs		Mean COMPUT. risk errors		STD. ERR. errors	
50		−0.0000895823		0.0000678346	
Z-ratio				Status of significance	
1.32060				N.S.—not significant	
Parameter values					
PROB. = 0.99	RHO = −0.3		BIGH = 2.5758		SMHL = 2.34
Pair No.	SMH	SMK	PROB. level	Risk level (%)	Risk error (%)
Table No. 8.2-249					
1	2.58	2.571607	0.989999	1.000088	−0.000089
2	2.57	2.581613	0.989999	1.000130	−0.000131
3	2.56	2.591698	0.989992	1.000822	−0.000823
4	2.55	2.603423	0.989999	1.000065	−0.000066
5	2.54	2.615229	0.989999	1.000059	−0.000060
6	2.53	2.627116	0.989992	1.000834	−0.000834
7	2.52	2.640648	0.989995	1.000464	−0.000465
8	2.51	2.655825	0.990008	0.999182	0.000817
9	2.50	2.669523	0.989993	1.000732	−0.000733
10	2.49	2.686431	0.990002	0.999796	0.000203
11	2.48	2.703425	0.990000	1.000029	−0.000030
12	2.47	2.722069	0.990001	0.999916	0.000083
13	2.46	2.742364	0.990003	0.999683	0.000316
14	2.45	2.764309	0.990005	0.999504	0.000495
15	2.44	2.787908	0.990004	0.999641	0.000358
16	2.43	2.814723	0.990009	0.999093	0.000906
17	2.42	2.841630	0.989995	1.000536	−0.000536
18	2.41	2.874881	0.990002	0.999826	0.000173
19	2.40	2.911352	0.990001	0.999868	0.000131
20	2.39	2.954169	0.990006	0.999451	0.000548
21	2.38	3.000208	0.989991	1.000923	−0.000924
22	2.37	3.061970	0.990000	0.999963	0.000036
23	2.36	3.136332	0.989997	1.000333	−0.000334

(continued)

Table 8.2 (continued)

Parameter values					
PROB. = 0.99	RHO = −0.3		BIGH = 2.5758		SMHL = 2.34
Pair No.	SMH	SMK	PROB. level	Risk level (%)	Risk error (%)
24	2.35	3.242046	0.989999	1.000124	−0.000125
25	2.34	3.435361	0.989995	1.000536	−0.000536
No. of pairs		Mean COMPUT. risk errors		STD. ERR. errors	
25		−0.0000623556		0.0000958580	
Z-ratio				Status of significance	
0.65050				N.S.—not significant	
Parameter values					
PROB. = 0.98	RHO = −0.3		BIGH = 2.3261		SMHL = 2.06
Pair No.	SMH	SMK	PROB. level	Risk level (%)	Risk error (%)
Table No. 8.2-250					
1	2.33	2.322207	0.979992	2.000851	−0.000852
2	2.32	2.332216	0.979990	2.000964	−0.000969
3	2.31	2.343094	0.979999	2.000099	−0.000101
4	2.30	2.354059	0.979997	2.000350	−0.000352
5	2.29	2.365894	0.980001	1.999855	0.000143
6	2.28	2.377820	0.979994	2.000612	−0.000614
7	2.27	2.391399	0.980009	1.999074	0.000924
8	2.26	2.405071	0.980010	1.999027	0.000972
9	2.25	2.418837	0.979995	2.000463	−0.000465
10	2.24	2.434260	0.979998	2.000165	−0.000167
11	2.23	2.450560	0.979999	2.000070	−0.000072
12	2.22	2.467740	0.979997	2.000320	−0.000322
13	2.21	2.486581	0.980003	1.999700	0.000298
14	2.20	2.506303	0.980001	1.999939	0.000060
15	2.19	2.527690	0.980001	1.999885	0.000113
16	2.18	2.550742	0.980000	1.999956	0.000042
17	2.17	2.576242	0.960006	1.999408	0.000590
18	2.16	2.603410	0.980002	1.999837	0.000161
19	2.15	2.633812	0.980004	1.999652	0.000346
20	2.14	2.667446	0.980003	1.999664	0.000334
21	2.13	2.705879	0.980010	1.999039	0.000960
22	2.12	2.749112	0.980010	1.999015	0.000983
23	2.11	2.798707	0.980005	1.999480	0.000519
24	2.10	2.857793	0.980001	1.999927	0.000072
25	2.09	2.932621	0.980006	1.999414	0.000584
26	2.08	3.029443	0.980002	1.999843	0.000155

(continued)

8.2 Contents for Table 8.2 (Table No. 8.2-1 to 400)

Table 8.2 (continued)

Parameter values					
PROB. = 0.98	RHO = −0.3		BIGH = 2.3261		SMHL = 2.06
Pair No.	SMH	SMK	PROB. level	Risk level (%)	Risk error (%)
27	2.07	3.165432	0.980000	1.999998	0.000000
28	2.06	3.401573	0.979990	2.000994	−0.000995
No. of pairs		Mean COMPUT. risk errors		STD. ERR. errors	
28		0.0000809801		0.0001015903	
Z-ratio				Status of significance	
0.79712				N.S.—not significant	
Parameter values					
PROB. = 0.97	RHO = −0.3		BIGH = 2.1698		SMHL = 1.89
Pair No.	SMH	SMK	PROB. level	Risk level (%)	Risk error (%)
Table No. 8.2-251					
1	2.17	2.169600	0.969995	3.000516	−0.000519
2	2.16	2.180035	0.970003	2.999693	0.000304
3	2.15	2.190564	0.969998	3.000248	−0.000250
4	2.14	2.201577	0.969992	3.000826	−0.000829
5	2.13	2.213468	0.969998	3.000188	−0.000191
6	2.12	2.225847	0.970001	2.999878	0.000119
7	2.11	2.238717	0.970000	3.000021	−0.000024
8	2.10	2.252077	0.969993	3.000713	−0.000715
9	2.09	2.266709	0.970003	2.999705	0.000292
10	2.08	2.281446	0.969992	3.000832	−0.000834
11	2.07	2.297849	0.970004	2.999628	0.000370
12	2.06	2.314359	0.969992	3.000844	−0.000846
13	2.05	2.332538	0.969996	3.000414	−0.000417
14	2.04	2.351609	0.969992	3.000760	−0.000763
15	2.03	2.372352	0.969997	3.000260	−0.000262
16	2.02	2.394771	0.970006	2.999449	0.000548
17	2.01	2.418085	0.969995	3.000492	−0.000495
18	2.00	2.443859	0.969995	3.000456	−0.000459
19	1.99	2.472095	0.969999	3.000128	−0.000131
20	1.98	2.502794	0.969997	3.000319	−0.000322
21	1.97	2.536739	0.969995	3.000546	−0.000548
22	1.96	2.575494	0.970005	2.999544	0.000453
23	1.95	2.619062	0.970009	2.999061	0.000936
24	1.94	2.667445	0.969993	3.000659	−0.000662
25	1.93	2.726894	0.970003	2.999735	0.000262
26	1.92	2.798974	0.970008	2.999175	0.000823
27	1.91	2.889938	0.970003	2.999723	0.000274

(continued)

Table 8.2 (continued)

Parameter values					
PROB. = 0.97	RHO = −0.3		BIGH = 2.1698		SMHL = 1.89
Pair No.	SMH	SMK	PROB. level	Risk level (%)	Risk error (%)
28	1.90	3.015411	0.970010	2.999038	0.000960
29	1.89	3.228521	0.970007	2.999342	0.000656
No. of pairs		Mean COMPUT. risk errors		STD. ERR. errors	
29		−0.0000756979		0.0001007346	
Z-ratio				Status of significance	
0.75146				N.S.—not significant	
Parameter values					
PROB. = 0.96	RHO = −0.3		BIGH = 2.0533		SMHL = 1.76
Pair No.	SMH	SMK	PROB. level	Risk level (%)	Risk error (%)
Table No. 8.2-252					
1	2.06	2.046622	0.959991	4.000950	−0.000948
2	2.05	2.056605	0.959993	4.000682	−0.000679
3	2.04	2.067077	0.959999	4.000145	−0.000143
4	2.03	2.078039	0.960006	3.999394	0.000608
5	2.02	2.089102	0.959996	4.000390	−0.000387
6	2.01	2.101049	0.960003	3.999734	0.000268
7	2.00	2.113489	0.960007	3.999311	0.000691
8	1.99	2.126426	0.960007	3.999329	0.000674
9	1.98	2.139861	0.960001	3.999925	0.000077
10	1.97	2.154184	0.960003	3.999716	0.000286
11	1.96	2.169010	0.959995	4.000455	−0.000453
12	1.95	2.184728	0.959991	4.000878	−0.000876
13	1.94	2.201733	0.960001	3.999925	0.000077
14	1.93	2.219633	0.960007	3.999353	0.000650
15	1.92	2.238042	0.959993	4.000712	−0.000709
16	1.91	2.258132	0.959994	4.000569	−0.000566
17	1.90	2.279906	0.960005	3.999478	0.000525
18	1.89	2.302584	0.959997	4.000318	−0.000316
19	1.88	2.327731	0.960008	3.999168	0.000834
20	1.87	2.353786	0.959991	4.000950	−0.000948
21	1.86	2.383095	0.959996	4.000437	−0.000435
22	1.85	2.414878	0.959994	4.000652	−0.000650
23	1.84	2.450701	0.960005	3.999466	0.000536
24	1.83	2.489784	0.960000	3.999955	0.000048
25	1.82	2.533693	0.959991	4.000890	−0.000888
26	1.81	2.585554	0.960003	3.999746	0.000256
27	1.80	2.645369	0.959998	4.000169	−0.000167
28	1.79	2.717830	0.959993	4.000712	−0.000709
29	1.78	2.810749	0.959999	4.000097	−0.000095

(continued)

8.2 Contents for Table 8.2 (Table No. 8.2-1 to 400)

Table 8.2 (continued)

Parameter values					
PROB. = 0.96	RHO = −0.3		BIGH = 2.0533		SMHL = 1.76
Pair No.	SMH	SMK	PROB. level	Risk level (%)	Risk error (%)
30	1.77	2.938193	0.959992	4.000813	−0.000811
31	1.76	3.157977	0.960006	3.999436	0.000566
No. of pairs		Mean COMPUT. risk errors		STD. ERR. errors	
31		−0.0001151115		0.0001005799	
Z-ratio				Status of significance	
1.14448				N.S.—not significant	
Parameter values					
PROB. = 0.95	RHO = −0.3		BIGH = 1.9594		SMHL = 1.65
Pair No.	SMH	SMK	PROB. level	Risk level (%)	Risk error (%)
Table No. 8.2-253					
1	1.96	1.958800	0.950003	4.999697	0.000304
2	1.95	1.968845	0.949996	5.000443	−0.000441
3	1.94	1.979385	0.949993	5.000729	−0.000727
4	1.93	1.990420	0.949993	5.000705	−0.000703
5	1.92	2.001952	0.949995	5.000544	−0.000542
6	1.91	2.013984	0.949996	5.000443	−0.000441
7	1.90	2.026517	0.949995	5.000532	−0.000530
8	1.89	2.039942	0.950009	4.999060	0.000942
9	1.88	2.053482	0.949999	5.000144	−0.000143
10	1.87	2.067918	0.949999	5.000120	−0.000119
11	1.86	2.083252	0.950007	4.999346	0.000656
12	1.85	2.099097	0.950001	4.999900	0.000101
13	1.84	2.115845	0.949997	5.000270	−0.000268
14	1.83	2.133887	0.950008	4.999221	0.000781
15	1.82	2.152446	0.949996	5.000377	−0.000376
16	1.81	2.172694	0.950005	4.999465	0.000536
17	1.80	2.193854	0.949999	5.000079	−0.000077
18	1.79	2.216707	0.950002	4.999829	0.000173
19	1.78	2.241256	0.950005	4.999519	0.000483
20	1.77	2.267504	0.950002	4.999841	0.000161
21	1.76	2.296234	0.950007	4.999328	0.000674
22	1.75	2.327450	0.950009	4.999119	0.000882
23	1.74	2.361152	0.949997	5.000329	−0.000328
24	1.73	2.398906	0.949995	5.000538	−0.000536
25	1.72	2.441496	0.949999	5.000139	−0.000137
26	1.71	2.489705	0.949999	5.000127	−0.000125
27	1.70	2.545881	0.950004	4.999584	0.000417
28	1.69	2.612369	0.950008	4.999227	0.000775
29	1.68	2.693079	0.949995	5.000496	−0.000495

(continued)

Table 8.2 (continued)

Parameter values					
PROB. = 0.95	RHO = −0.3		BIGH = 1.9594		SMHL = 1.65
Pair No.	SMH	SMK	PROB. level	Risk level (%)	Risk error (%)
30	1.67	2.800513	0.949999	5.000127	−0.000125
31	1.66	2.956549	0.949993	5.000705	−0.000703
32	1.65	3.276816	0.950006	4.999418	0.000584
No. of pairs		Mean COMPUT. risk errors		STD. ERR. errors	
32		0.0000196876		0.0000890135	
Z-ratio				Status of significance	
0.22118				N.S.—not significant	
Parameter values					
PROB. = 0.94	RHO = −0.3		BIGH = 1.88		SMHL = 1.56
Pair No.	SMH	SMK	PROB. level	Risk level (%)	Risk error (%)
Table No. 8.2-254					
1	1.89	1.870053	0.939997	6.000334	−0.000334
2	1.88	1.880000	0.940006	5.999387	0.000614
3	1.87	1.890054	0.939997	6.000328	−0.000328
4	1.86	1.900606	0.939994	6.000585	−0.000584
5	1.85	1.911658	0.939997	6.000352	−0.000352
6	1.84	1.923213	0.940002	5.999798	0.000203
7	1.83	1.935273	0.940009	5.999142	0.000858
8	1.82	1.947447	0.939991	6.000901	−0.000900
9	1.81	1.960519	0.939994	6.000555	−0.000554
10	1.80	1.974102	0.939993	6.000692	−0.000691
11	1.79	1.988588	0.940006	5.999357	0.000644
12	1.78	2.003587	0.940010	5.999029	0.000972
13	1.77	2.019102	0.940001	5.999881	0.000119
14	1.76	2.035525	0.939998	6.000209	−0.000209
15	1.75	2.052860	0.939996	6.000418	−0.000417
16	1.74	2.071108	0.939992	6.000853	−0.000852
17	1.73	2.090662	0.939998	6.000215	−0.000215
18	1.72	2.111134	0.939992	6.000763	−0.000763
19	1.71	2.133307	0.940002	5.999780	0.000221
20	1.70	2.156793	0.940005	5.999530	0.000471
21	1.69	2.181986	0.940007	5.999261	0.000739
22	1.68	2.208887	0.940002	5.999768	0.000232
23	1.67	2.238282	0.940006	5.999363	0.000638
24	1.66	2.269781	0.939995	6.000543	−0.000542
25	1.65	2.304951	0.940001	5.999941	0.000060
26	1.64	2.343403	0.939994	6.000596	−0.000596
27	1.63	2.387093	0.940003	5.999685	0.000316
28	1.62	2.436415	0.940006	5.999422	0.000578
29	1.61	2.492935	0.940000	6.000018	−0.000018
30	1.60	2.560562	0.940002	5.999762	0.000238
31	1.59	2.643205	0.939997	6.000340	−0.000340

(continued)

8.2 Contents for Table 8.2 (Table No. 8.2-1 to 400)

Table 8.2 (continued)

Parameter values					
PROB. = 0.94	RHO = −0.3		BIGH = 1.88		SMHL = 1.56
Pair No.	SMH	SMK	PROB. level	Risk level (%)	Risk error (%)
32	1.58	2.752586	0.940004	5.999583	0.000417
33	1.57	2.910584	0.939996	6.000358	−0.000358
34	1.56	3.228140	0.940000	6.000048	−0.000048
No. of pairs		Mean COMPUT. risk errors		STD. ERR. errors	
34		−0.0000223092		0.0000879385	
Z-ratio				Status of significance	
0.25369				N.S.—not significant	
Parameter values					
PROB. = 0.93	RHO = −0.3		BIGH = 1.8108		SMHL = 1.48
Pair No.	SMH	SMK	PROB. level	Risk level (%)	Risk error (%)
Table No. 8.2-255					
1	1.82	1.801646	0.929994	7.000584	−0.000584
2	1.81	1.811600	0.930002	6.999761	0.000238
3	1.80	1.821665	0.929991	7.000894	−0.000894
4	1.79	1.832428	0.930003	6.999713	0.000286
5	1.78	1.843305	0.929993	7.000697	−0.000697
6	1.77	1.854884	0.930002	6.999767	0.000232
7	1.76	1.866777	0.930001	6.999886	0.000113
8	1.75	1.879181	0.930001	6.999916	0.000083
9	1.74	1.892098	0.930000	7.000029	−0.000030
10	1.73	1.905530	0.929996	7.000452	−0.000453
11	1.72	1.919674	0.929998	7.000190	−0.000191
12	1.71	1.934339	0.929992	7.000757	−0.000757
13	1.70	1.949915	0.930000	7.000023	−0.000024
14	1.69	1.966016	0.929993	7.000745	−0.000745
15	1.68	1.983033	0.929991	7.000935	−0.000936
16	1.67	2.001166	0.930000	6.999982	0.000018
17	1.66	2.020221	0.930005	6.999499	0.000501
18	1.65	2.040005	0.929991	7.000864	−0.000864
19	1.64	2.061497	0.930002	6.999797	0.000203
20	1.63	2.084310	0.930010	6.999016	0.000983
21	1.62	2.108447	0.930008	6.999171	0.000829
22	1.61	2.134300	0.930006	6.999409	0.000590
23	1.60	2.161872	0.929994	7.000619	−0.000620
24	1.59	2.191949	0.929991	7.000935	−0.000936
25	1.58	2.225314	0.930008	6.999177	0.000823
26	1.57	2.261189	0.930002	6.999785	0.000215
27	1.56	2.301139	0.930003	6.999689	0.000310
28	1.55	2.345950	0.930006	6.999392	0.000608
29	1.54	2.396405	0.929999	7.000119	−0.000119
30	1.53	2.455634	0.930008	6.999225	0.000775
31	1.52	2.525203	0.930001	6.999874	0.000125

(continued)

Table 8.2 (continued)

Parameter values				
PROB. = 0.93	RHO = −0.3	BIGH = 1.8108		SMHL = 1.48
Pair No.	SMH	SMK	PROB. level	Risk level (%)	Risk error (%)

Pair No.	SMH	SMK	PROB. level	Risk level (%)	Risk error (%)
32	1.51	2.612145	0.930008	6.999183	0.000817
33	1.50	2.725059	0.929998	7.000238	−0.000238
34	1.49	2.894418	0.929999	7.000125	−0.000125
35	1.48	3.253037	0.929995	7.000494	−0.000495
No. of pairs		Mean COMPUT. risk errors		STD. ERR. errors	
35		−0.0000266565		0.0000941302	
Z-ratio				Status of significance	
0.28319				N.S.—not significant	

Parameter values				
PROB. = 0.92	RHO = −0.3	BIGH = 1.7492		SMHL = 1.41
Pair No.	SMH	SMK	PROB. level	Risk level (%)	Risk error (%)

Table No. 8.2-256

Pair No.	SMH	SMK	PROB. level	Risk level (%)	Risk error (%)
1	1.75	1.748400	0.919995	8.000493	−0.000495
2	1.74	1.758644	0.920003	7.999707	0.000292
3	1.73	1.769199	0.920006	7.999439	0.000560
4	1.72	1.780068	0.920002	7.999820	0.000179
5	1.71	1.791447	0.920005	7.999468	0.000530
6	1.70	1.803144	0.920000	8.000051	−0.000054
7	1.69	1.815357	0.919997	8.000267	−0.000268
8	1.68	1.828086	0.919997	8.000332	−0.000334
9	1.67	1.841531	0.920009	7.999075	0.000924
10	1.66	1.855302	0.920004	7.999599	0.000399
11	1.65	1.869598	0.919993	8.000696	−0.000697
12	1.64	1.884812	0.920000	7.999993	0.000006
13	1.63	1.900554	0.919995	8.000493	−0.000495
14	1.62	1.917220	0.920000	8.000029	−0.000030
15	1.61	1.934810	0.920010	7.999045	0.000954
16	1.60	1.952938	0.919997	8.000338	−0.000340
17	1.59	1.972387	0.920003	7.999736	0.000262
18	1.58	1.992769	0.919999	8.000106	−0.000107
19	1.57	2.014479	0.920001	7.999915	0.000083
20	1.56	2.037519	0.920001	7.999909	0.000089
21	1.55	2.062281	0.920010	7.999033	0.000966
22	1.54	2.088381	0.920000	8.000016	−0.000018
23	1.53	2.116601	0.919997	8.000326	−0.000328
24	1.52	2.147336	0.920003	7.999730	0.000268
25	1.51	2.180588	0.920002	7.999826	0.000173
26	1.50	2.217144	0.920005	7.999540	0.000459
27	1.49	2.257396	0.920003	7.999736	0.000262
28	1.48	2.302521	0.920005	7.999540	0.000459
29	1.47	2.353303	0.919997	8.000315	−0.000316
30	1.46	2.412872	0.920009	7.999075	0.000924

(continued)

Table 8.2 (continued)

Parameter values					
PROB. = 0.92	RHO = −0.3		BIGH = 1.7492		SMHL = 1.41
Pair No.	SMH	SMK	PROB. level	Risk level (%)	Risk error (%)
31	1.45	2.482013	0.919993	8.000666	−0.000668
32	1.44	2.568541	0.920001	7.999897	0.000101
33	1.43	2.680275	0.919991	8.000869	−0.000870
34	1.42	2.845343	0.919993	8.000690	−0.000691
35	1.41	3.182499	0.920004	7.999582	0.000417
No. of pairs		Mean COMPUT. risk errors		STD. ERR. errors	
35		0.0000721878		0.0000810247	
Z-ratio				Status of significance	
0.89094				N.S.—not significant	
Parameter values					
PROB. = 0.91	RHO = −0.3		BIGH = 1.6935		SMHL = 1.35
Pair No.	SMH	SMK	PROB. level	Risk level (%)	Risk error (%)
Table No. 8.2-257					
1	1.70	1.687220	0.910002	8.999824	0.000173
2	1.69	1.697203	0.910005	8.999544	0.000453
3	1.68	1.707499	0.910004	8.999574	0.000423
4	1.67	1.718112	0.909999	9.000099	−0.000101
5	1.66	1.729239	0.910005	8.999502	0.000495
6	1.65	1.740686	0.910003	8.999711	0.000286
7	1.64	1.752652	0.910008	8.999216	0.000781
8	1.63	1.764942	0.910001	8.999861	0.000137
9	1.62	1.777757	0.909998	9.000212	−0.000215
10	1.61	1.791096	0.909994	9.000588	−0.000590
11	1.60	1.805159	0.910004	8.999634	0.000364
12	1.59	1.819753	0.910007	8.999300	0.000697
13	1.58	1.834880	0.910002	8.999771	0.000226
14	1.57	1.850738	0.910000	9.000004	−0.000006
15	1.56	1.867331	0.909997	9.000325	−0.000328
16	1.55	1.884856	0.910001	8.999866	0.000131
17	1.54	1.903315	0.910009	8.999121	0.000876
18	1.53	1.922519	0.910001	8.999861	0.000137
19	1.52	1.943054	0.910010	8.999014	0.000983
20	1.51	1.964534	0.910004	8.999568	0.000429
21	1.50	1.987352	0.910000	8.999986	0.000012
22	1.49	2.011902	0.910010	8.999014	0.000983
23	1.48	2.037603	0.909993	9.000665	−0.000668
24	1.47	2.065824	0.910008	8.999193	0.000805
25	1.46	2.095594	0.909993	9.000659	−0.000662
26	1.45	2.128282	0.909997	9.000338	−0.000340
27	1.44	2.163892	0.909997	9.000278	−0.000280
28	1.43	2.203210	0.910003	8.999723	0.000274
29	1.42	2.246630	0.909999	9.000122	−0.000125

(continued)

Table 8.2 (continued)

Parameter values					
PROB. = 0.91	RHO = −0.3		BIGH = 1.6935		SMHL = 1.35
Pair No.	SMH	SMK	PROB. level	Risk level (%)	Risk error (%)
30	1.41	2.295720	0.910001	8.999861	0.000137
31	1.40	2.351654	0.909998	9.000224	−0.000226
32	1.39	2.417174	0.909996	9.000385	−0.000387
33	1.38	2.496968	0.910007	8.999329	0.000668
34	1.37	2.596513	0.909993	9.000701	−0.000703
35	1.36	2.735343	0.909997	9.000312	−0.000316
36	1.35	2.968150	0.910005	8.999467	0.000530
No. of pairs		Mean COMPUT. risk errors		STD. ERR. errors	
36		0.0001366074		0.0000781756	
Z-ratio				Status of significance	
1.74744				N.S.—not significant	
Parameter values					
PROB. = 0.9025	RHO = −0.3		BIGH = 1.655		SMHL = 1.30
Pair No.	SMH	SMK	PROB. level	Risk level (%)	Risk error (%)
Table No. 8.2-258					
1	1.66	1.650015	0.902506	9.749401	0.000602
2	1.65	1.660015	0.902506	9.749407	0.000596
3	1.64	1.670332	0.902504	9.749651	0.000352
4	1.63	1.680969	0.902497	9.750337	−0.000334
5	1.62	1.692123	0.902503	9.749722	0.000280
6	1.61	1.703601	0.902501	9.749866	0.000137
7	1.60	1.715406	0.902491	9.750920	−0.000918
8	1.59	1.727931	0.902504	9.749567	0.000435
9	1.58	1.740787	0.902504	9.749556	0.000447
10	1.57	1.754172	0.902505	9.749460	0.000542
11	1.56	1.768090	0.902506	9.749442	0.000560
12	1.55	1.782543	0.902501	9.749884	0.000119
13	1.54	1.797728	0.902506	9.749449	0.000554
14	1.53	1.813454	0.902499	9.750074	−0.000072
15	1.52	1.830115	0.902508	9.749174	0.000829
16	1.51	1.847322	0.902499	9.750091	−0.000089
17	1.50	1.865470	0.902496	9.750384	−0.000381
18	1.49	1.884756	0.902507	9.749329	0.000674
19	1.48	1.904794	0.902499	9.750109	−0.000107
20	1.47	1.926173	0.902504	9.749579	0.000423
21	1.46	1.948701	0.902503	9.749746	0.000256
22	1.45	1.972576	0.902498	9.750223	−0.000221
23	1.44	1.998194	0.902502	9.749776	0.000226
24	1.43	2.025558	0.902504	9.749598	0.000405
25	1.42	2.054672	0.902492	9.750825	−0.000823
26	1.41	2.086711	0.902506	9.749412	0.000590
27	1.40	2.121290	0.902509	9.749115	0.000888

(continued)

Table 8.2 (continued)

Parameter values					
PROB. = 0.9025	RHO = −0.3		BIGH = 1.655		SMHL = 1.30
Pair No.	SMH	SMK	PROB. level	Risk level (%)	Risk error (%)
28	1.39	2.158802	0.902498	9.750229	−0.000226
29	1.38	2.200425	0.902493	9.750689	−0.000685
30	1.37	2.247335	0.902501	9.749901	0.000101
31	1.36	2.299926	0.902491	9.750872	−0.000870
32	1.35	2.361719	0.902507	9.749311	0.000691
33	1.34	2.433891	0.902497	9.750342	−0.000340
34	1.33	2.523479	0.902497	9.750337	−0.000334
35	1.32	2.642205	0.902508	9.749198	0.000805
36	1.31	2.815858	0.902493	9.750706	−0.000703
37	1.30	3.194441	0.902504	9.749592	0.000411
No. of pairs		Mean COMPUT. risk errors		STD. ERR. errors	
37		0.0001268952		0.0000823599	
Z-ratio				Status of significance	
1.54074				N.S.—not significant	
Parameter values					
PROB. = 0.9	RHO = −0.3		BIGH = 1.6426		SMHL = 1.29
Pair No.	SMH	SMK	PROB. level	Risk level (%)	Risk error (%)
Table No. 8.2-259					
1	1.65	1.635233	0.899993	10.000720	−0.000721
2	1.64	1.645204	0.899998	10.000190	−0.000191
3	1.63	1.655493	0.900001	9.999895	0.000107
4	1.62	1.666102	0.899997	10.000310	−0.000310
5	1.61	1.677032	0.899995	10.000540	−0.000536
6	1.60	1.688483	0.900001	9.999906	0.000095
7	1.59	1.700260	0.899999	10.000150	−0.000143
8	1.58	1.712563	0.900004	9.999633	0.000370
9	1.57	1.725198	0.899996	10.000390	−0.000387
10	1.56	1.738558	0.900009	9.999132	0.000870
11	1.55	1.752256	0.900004	9.999633	0.000370
12	1.54	1.766489	0.899996	10.000370	−0.000364
13	1.53	1.781456	0.899999	10.000080	−0.000077
14	1.52	1.796964	0.899993	10.000720	−0.000715
15	1.51	1.813407	0.900004	9.999603	0.000399
16	1.50	1.830397	0.899998	10.000160	−0.000161
17	1.49	1.848329	0.900001	9.999906	0.000095
18	1.48	1.867205	0.900007	9.999352	0.000650
19	1.47	1.887028	0.900008	9.999186	0.000817
20	1.46	1.907803	0.900001	9.999871	0.000131
21	1.45	1.929923	0.900003	9.999668	0.000334
22	1.44	1.953392	0.900005	9.999454	0.000548
23	1.43	1.978214	0.899999	10.000080	−0.000077
24	1.42	2.004782	0.899996	10.000360	−0.000358

(continued)

Table 8.2 (continued)

Parameter values					
PROB. = 0.9	RHO = −0.3		BIGH = 1.6426		SMHL = 1.29
Pair No.	SMH	SMK	PROB. level	Risk level (%)	Risk error (%)
25	1.41	2.033493	0.900004	9.999572	0.000429
26	1.40	2.064348	0.900008	9.999216	0.000787
27	1.39	2.097354	0.899992	10.000800	−0.000799
28	1.38	2.134076	0.900005	9.999454	0.000548
29	1.37	2.174128	0.900005	9.999538	0.000465
30	1.36	2.218297	0.899992	10.000830	−0.000823
31	1.35	2.268931	0.900007	9.999341	0.000662
32	1.34	2.326033	0.899997	10.000330	−0.000328
33	1.33	2.393516	0.900001	9.999884	0.000119
34	1.32	2.475291	0.900007	9.999347	0.000656
35	1.31	2.578394	0.899996	10.000410	−0.000405
36	1.30	2.722362	0.899993	10.000720	−0.000715
37	1.29	2.969701	0.899998	10.000230	−0.000232
No. of pairs		Mean COMPUT. risk errors		STD. ERR. errors	
37		0.0000291749		0.0000799171	
Z-ratio				Status of significance	
0.36506				N.S.—not significant	
Parameter values					
PROB. = 0.85	RHO = −0.3		BIGH = 1.4346		SMHL = 1.04
Pair No.	SMH	SMK	PROB. level	Risk level (%)	Risk error (%)
Table No. 8.2-260					
1	1.44	1.429123	0.849997	15.000330	−0.000328
2	1.43	1.439117	0.849998	15.000240	−0.000238
3	1.42	1.449350	0.849990	15.000990	−0.000989
4	1.41	1.460020	0.849999	15.000100	−0.000101
5	1.40	1.471031	0.850009	14.999120	0.000882
6	1.39	1.482193	0.849993	15.000700	−0.000703
7	1.38	1.493899	0.850000	14.999960	0.000036
8	1.37	1.505957	0.850003	14.999670	0.000328
9	1.36	1.518370	0.850000	14.999970	0.000024
10	1.35	1.531240	0.850001	14.999910	0.000083
11	1.34	1.544472	0.849991	15.000920	−0.000918
12	1.33	1.558364	0.850002	14.999780	0.000215
13	1.32	1.572626	0.849997	15.000280	−0.000286
14	1.31	1.587457	0.849996	15.000420	−0.000417
15	1.30	1.602862	0.849994	15.000590	−0.000596
16	1.29	1.619041	0.850009	14.999130	0.000864
17	1.28	1.635607	0.849995	15.000530	−0.000530
18	1.27	1.653150	0.850008	14.999220	0.000781
19	1.26	1.671285	0.850003	14.999680	0.000316
20	1.25	1.690211	0.849995	15.000470	−0.000471
21	1.24	1.710130	0.849996	15.000420	−0.000417

(continued)

8.2 Contents for Table 8.2 (Table No. 8.2-1 to 400)

Table 8.2 (continued)

Parameter values					
PROB. = 0.85	RHO = −0.3		BIGH = 1.4346		SMHL = 1.04
Pair No.	SMH	SMK	PROB. level	Risk level (%)	Risk error (%)
22	1.23	1.731046	0.849998	15.000240	−0.000238
23	1.22	1.752964	0.849993	15.000720	−0.000721
24	1.21	1.776281	0.850005	14.999530	0.000471
25	1.20	1.800611	0.849992	15.000760	−0.000763
26	1.19	1.826742	0.850005	14.999470	0.000530
27	1.18	1.854289	0.850001	14.999940	0.000060
28	1.17	1.883649	0.849993	15.000750	−0.000751
29	1.16	1.915415	0.850002	14.999840	0.000161
30	1.15	1.949398	0.849995	15.000500	−0.000501
31	1.14	1.986580	0.850006	14.999450	0.000548
32	1.13	2.026775	0.849993	15.000660	−0.000662
33	1.12	2.071553	0.850002	14.999790	0.000209
34	1.11	2.121310	0.850002	14.999750	0.000244
35	1.10	2.177619	0.850002	14.999770	0.000226
36	1.09	2.242831	0.850010	14.999030	0.000966
37	1.08	2.319688	0.850005	14.999530	0.000471
38	1.07	2.414060	0.849992	15.000780	−0.000781
39	1.06	2.539237	0.849999	15.000120	−0.000119
40	1.05	2.725697	0.849998	15.000190	−0.000191
41	1.04	3.141419	0.850002	14.999760	0.000238
No. of pairs		Mean COMPUT. risk errors		STD. ERR. errors	
41		−0.0000730866		0.0000807283	
Z-ratio				Status of significance	
0.90534				N.S.—not significant	
Parameter values					
PROB. = 0.8	RHO = −0.3		BIGH = 1.273		SMHL = 0.85
Pair No.	SMH	SMK	PROB. level	Risk level (%)	Risk error (%)
Table No. 8.2-261					
1	1.28	1.265452	0.799999	20.000060	−0.000060
2	1.27	1.275421	0.800004	19.999560	0.000435
3	1.26	1.285548	0.799991	20.000940	−0.000942
4	1.25	1.296130	0.800005	19.999460	0.000542
5	1.24	1.306878	0.799999	20.000090	−0.000095
6	1.23	1.317992	0.800002	19.999800	0.000203
7	1.22	1.329377	0.799997	20.000290	−0.000292
8	1.21	1.341136	0.799998	20.000180	−0.000185
9	1.20	1.353273	0.800003	19.999690	0.000310
10	1.19	1.365695	0.799995	20.000480	−0.000477
11	1.18	1.378603	0.800002	19.999820	0.000179
12	1.17	1.391903	0.800006	19.999450	0.000554
13	1.16	1.405601	0.800005	19.999540	0.000459
14	1.15	1.419702	0.799996	20.000360	−0.000364

(continued)

Table 8.2 (continued)

Parameter values					
PROB. = 0.8	RHO = −0.3		BIGH = 1.273		SMHL = 0.85
Pair No.	SMH	SMK	PROB. level	Risk level (%)	Risk error (%)
15	1.14	1.434407	0.800005	19.999510	0.000489
16	1.13	1.449526	0.800000	19.999980	0.000024
17	1.12	1.465260	0.800005	19.999480	0.000519
18	1.11	1.481420	0.799991	20.000940	−0.000936
19	1.10	1.498403	0.800002	19.999770	0.000232
20	1.09	1.515923	0.799998	20.000160	−0.000161
21	1.08	1.534278	0.800010	19.999050	0.000948
22	1.07	1.553185	0.799996	20.000390	−0.000387
23	1.06	1.573137	0.800008	19.999230	0.000769
24	1.05	1.593751	0.799994	20.000580	−0.000584
25	1.04	1.615427	0.799990	20.000960	−0.000960
26	1.03	1.638368	0.800007	19.999310	0.000685
27	1.02	1.662191	0.799995	20.000490	−0.000489
28	1.01	1.687492	0.800001	19.999870	0.000125
29	1.00	1.714083	0.799994	20.000570	−0.000566
30	0.99	1.742366	0.799995	20.000490	−0.000489
31	0.98	1.772546	0.800003	19.999720	0.000280
32	0.97	1.804632	0.799999	20.000150	−0.000149
33	0.96	1.839222	0.800004	19.999580	0.000417
34	0.95	1.876523	0.800007	19.999330	0.000674
35	0.94	1.916935	0.800001	19.999880	0.000119
36	0.93	1.961254	0.799998	20.000240	−0.000244
37	0.92	2.010272	0.799992	20.000780	−0.000775
38	0.91	2.065566	0.800002	19.999760	0.000238
39	0.90	2.128321	0.800000	19.999960	0.000042
40	0.89	2.201287	0.799992	20.000820	−0.000823
41	0.88	2.289947	0.800007	19.999340	0.000662
42	0.87	2.401738	0.800008	19.999170	0.000829
43	0.86	2.556210	0.800006	19.999450	0.000548
44	0.85	2.819003	0.800004	19.999610	0.000393
No. of pairs		Mean COMPUT. risk errors		STD. ERR. errors	
44		0.0000377496		0.0000773684	
Z-ratio				Status of significance	
0.48792				N.S.—not significant	
Parameter values					
PROB. = 0.75	RHO = −0.3		BIGH = 1.1364		SMHL = 0.68
Pair No.	SMH	SMK	PROB. level	Risk level (%)	Risk error (%)
Table No. 8.2-262					
1	1.14	1.132714	0.749996	25.000420	−0.000417
2	1.13	1.142739	0.749994	25.000600	−0.000596
3	1.12	1.153040	0.749997	25.000270	−0.000268
4	1.11	1.163623	0.750004	24.999600	0.000399

(continued)

8.2 Contents for Table 8.2 (Table No. 8.2-1 to 400)

Table 8.2 (continued)

Parameter values					
PROB. = 0.75	RHO = −0.3		BIGH = 1.1364		SMHL = 0.68
Pair No.	SMH	SMK	PROB. level	Risk level (%)	Risk error (%)
5	1.10	1.174395	0.749995	25.000500	−0.000501
6	1.09	1.185556	0.750006	24.999420	0.000584
7	1.08	1.196917	0.749999	25.000110	−0.000107
8	1.07	1.208678	0.750009	24.999110	0.000888
9	1.06	1.220650	0.749999	25.000080	−0.000077
10	1.05	1.233034	0.750003	24.999690	0.000310
11	1.04	1.245739	0.750002	24.999850	0.000155
12	1.03	1.258869	0.750010	24.999020	0.000978
13	1.02	1.272333	0.750009	24.999100	0.000906
14	1.01	1.286138	0.749998	25.000230	−0.000226
15	1.00	1.300389	0.749990	25.000990	−0.000989
16	0.99	1.315192	0.749998	25.000190	−0.000185
17	0.98	1.330455	0.750004	24.999610	0.000393
18	0.97	1.346189	0.750004	24.999560	0.000441
19	0.96	1.362401	0.749997	25.000320	−0.000322
20	0.95	1.379295	0.750006	24.999410	0.000590
21	0.94	1.396686	0.750000	25.000050	−0.000048
22	0.93	1.414779	0.750001	24.999870	0.000131
23	0.92	1.433584	0.750006	24.999430	0.000566
24	0.91	1.453110	0.750007	24.999270	0.000733
25	0.90	1.473371	0.750001	24.999920	0.000077
26	0.89	1.494571	0.750004	24.999630	0.000370
27	0.88	1.516724	0.750008	24.999180	0.000823
28	0.87	1.539842	0.750006	24.999390	0.000608
29	0.86	1.564036	0.750000	25.000040	−0.000036
30	0.85	1.589612	0.750010	24.999050	0.000954
31	0.84	1.616293	0.749993	25.000680	−0.000679
32	0.83	1.644776	0.750006	24.999420	0.000578
33	0.82	1.674688	0.749993	25.000710	−0.000703
34	0.81	1.706827	0.750008	24.999220	0.000787
35	0.80	1.740818	0.749994	25.000560	−0.000560
36	0.79	1.777658	0.750007	24.999260	0.000739
37	0.78	1.817170	0.750000	24.999980	0.000018
38	0.77	1.860352	0.750006	24.999360	0.000644
39	0.76	1.907420	0.749994	25.000630	−0.000632
40	0.75	1.959763	0.749990	25.001000	−0.000995
41	0.74	2.018969	0.750004	24.999600	0.000405
42	0.73	2.086235	0.749996	25.000460	−0.000453
43	0.72	2.165492	0.750000	24.999970	0.000036
44	0.71	2.261848	0.750003	24.999720	0.000286

(continued)

Table 8.2 (continued)

Parameter values					
PROB. = 0.75	RHO = −0.3		BIGH = 1.1364		SMHL = 0.68
Pair No.	SMH	SMK	PROB. level	Risk level (%)	Risk error (%)
45	0.70	2.385488	0.749993	25.000700	−0.000697
46	0.69	2.563787	0.749997	25.000300	−0.000298
47	0.68	2.905373	0.749993	25.000660	−0.000656
No. of pairs		Mean COMPUT. risk errors		STD. ERR. errors	
47		0.0000823289		0.000804599	
Z-ratio				Status of significance	
1.02323				N.S.—not significant	
Parameter values					
PROB. = 0.7	RHO = −0.3		BIGH = 1.0159		SMHL = 0.53
Pair No.	SMH	SMK	PROB. level	Risk level (%)	Risk error (%)
Table No. 8.2-263					
1	1.02	1.011817	0.699992	30.000780	−0.000775
2	1.01	1.021835	0.699992	30.000840	−0.000834
3	1.00	1.032151	0.700005	29.999500	0.000501
4	0.99	1.042575	0.699990	30.000990	−0.000983
5	0.98	1.053408	0.700009	29.999100	0.000900
6	0.97	1.064363	0.699999	30.000100	−0.000095
7	0.96	1.075641	0.700001	29.999950	0.000054
8	0.95	1.087153	0.699993	30.000730	−0.000727
9	0.94	1.099003	0.699995	30.000530	−0.000530
10	0.93	1.111199	0.700005	29.999500	0.000507
11	0.92	1.123652	0.700004	29.999590	0.000411
12	0.91	1.136468	0.700009	29.999070	0.000930
13	0.90	1.149557	0.700002	29.999840	0.000161
14	0.89	1.163028	0.699998	30.000230	−0.000232
15	0.88	1..176889	0.699997	30.000340	−0.000334
16	0.87	1.191150	0.699997	30.000350	−0.000352
17	0.86	1.205823	0.699996	30.000420	−0.000423
18	0.85	1.221016	0.700009	29.999090	0.000912
19	0.84	1.236544	0.700001	29.999890	0.000113
20	0.83	1.252617	0.700003	29.999710	0.000292
21	0.82	1.269148	0.699996	30.000420	−0.000423
22	0.81	1.286248	0.699993	30.000740	−0.000733
23	0.80	1.303933	0.699990	30.000960	−0.000954
24	0.79	1.322314	0.700000	29.999970	0.000030
25	0.78	1.341308	0.700003	29.999710	0.000292
26	0.77	1.361031	0.700009	29.999140	0.000864
27	0.76	1.381401	0.699999	30.000070	−0.000066
28	0.75	1.402633	0.699996	30.000360	−0.000358
29	0.74	1.424744	0.699994	30.000620	−0.000620
30	0.73	1.447950	0.700009	29.999140	0.000864
31	0.72	1.472078	0.700008	29.999180	0.000817
32	0.71	1.497345	0.700008	29.999220	0.000781

(continued)

8.2 Contents for Table 8.2 (Table No. 8.2-1 to 400)

Table 8.2 (continued)

Parameter values					
PROB. = 0.7	RHO = −0.3		BIGH = 1.0159		SMHL = 0.53
Pair No.	SMH	SMK	PROB. level	Risk level (%)	Risk error (%)
33	0.70	1.523775	0.699997	30.000260	−0.000262
34	0.69	1.551783	0.700008	29.999160	0.000840
35	0.68	1.581201	0.700005	29.999480	0.000525
36	0.67	1.612252	0.699993	30.000670	−0.000668
37	0.66	1.645356	0.699992	30.000790	−0.000793
38	0.65	1.680742	0.699996	30.000400	−0.000399
39	0.64	1.718637	0.699996	30.000440	−0.000435
40	0.63	1.759663	0.700006	29.999390	0.000614
41	0.62	1.804054	0.700004	29.999620	0.000381
42	0.61	1.852631	0.699998	30.000230	−0.000232
43	0.60	1.906415	0.699992	30.000850	−0.000846
44	0.59	1.967210	0.700008	29.999240	0.000763
45	0.58	2.036041	0.699993	30.000670	−0.000668
46	0.57	2.117258	0.700004	29.999580	0.000423
47	0.56	2.215606	0.700006	29.999400	0.000602
48	0.55	2.342083	0.700008	29.999160	0.000840
49	0.54	2.525269	0.699996	30.000380	−0.000381
50	0.53	2.869142	0.700001	29.999920	0.000083
No. of pairs		Mean COMPUT. risk errors		STD. ERR. errors	
50		0.0000073629		0.0000841370	
Z-ratio				Status of significance	
0.08751				N.S.—not significant	
Parameter values					
PROB. = 0.65	RHO = −0.3		BIGH = 0.9062		SMHL = 0.42
Pair No.	SMH	SMK	PROB. level	Risk level (%)	Risk error (%)
Table No. 8.2-264					
1	0.91	0.901928	0.649996	35.000430	−0.000429
2	0.90	0.911954	0.649997	35.000270	−0.000268
3	0.89	0.922207	0.650001	34.999930	0.000072
4	0.88	0.932692	0.650007	34.999350	0.000656
5	0.87	0.943320	0.649993	35.000750	−0.000745
6	0.86	0.954296	0.650003	34.999690	0.000310
7	0.85	0.965432	0.649994	35.000580	−0.000572
8	0.84	0.976934	0.650009	34.999060	0.000948
9	0.83	0.988614	0.650006	34.999440	0.000566
10	0.82	1.000583	0.650005	34.999540	0.000465
11	0.81	1.012849	0.650006	34.999420	0.000584
12	0.80	1.025424	0.650009	34.999070	0.000936
13	0.79	1.038222	0.649995	35.000500	−0.000495
14	0.78	1.051451	0.650003	34.999700	0.000298
15	0.77	1.064929	0.649993	35.000710	−0.000703
16	0.76	1.078864	0.650004	34.999640	0.000370
17	0.75	1.093075	0.649996	35.000360	−0.000358

(continued)

Table 8.2 (continued)

Parameter values					
PROB. = 0.65	RHO = −0.3		BIGH = 0.9062		SMHL = 0.42
Pair No.	SMH	SMK	PROB. level	Risk level (%)	Risk error (%)
18	0.74	1.107774	0.650009	34.999150	0.000852
19	0.73	1.122781	0.650002	34.999800	0.000209
20	0.72	1.138209	0.649995	35.000460	−0.000459
21	0.71	1.154176	0.650005	34.999500	0.000507
22	0.70	1.170503	0.649995	35.000500	−0.000501
23	0.69	1.187408	0.649999	35.000130	−0.000131
24	0.68	1.204812	0.649998	35.000230	−0.000226
25	0.67	1.222739	0.649991	35.000910	−0.000906
26	0.66	1.241310	0.649992	35.000760	−0.000751
27	0.65	1.260550	0.650000	35.000030	−0.000030
28	0.64	1.280388	0.649995	35.000520	−0.000513
29	0.63	1.300950	0.649991	35.000930	−0.000930
30	0.62	1.322365	0.649998	35.000230	−0.000226
31	0.61	1.344566	0.649998	35.000240	−0.000238
32	0.60	1.367687	0.650000	35.000040	−0.000036
33	0.59	1.391763	0.649999	35.000150	−0.000149
34	0.58	1.416933	0.650001	34.999950	0.000054
35	0.57	1.443237	0.649999	35.000150	−0.000143
36	0.56	1.470917	0.650007	34.999290	0.000715
37	0.55	1.499923	0.650005	34.999550	0.000453
38	0.54	1.530503	0.650001	34.999890	0.000113
39	0.53	1.562906	0.650003	34.999690	0.000310
40	0.52	1.597195	0.649994	35.000650	−0.000644
41	0.51	1.634020	0.650008	34.999250	0.000751
42	0.50	1.673255	0.650001	34.999930	0.000072
43	0.49	1.715757	0.650010	34.999010	0.000989
44	0.48	1.761610	0.649997	35.000350	−0.000346
45	0.47	1.812073	0.650000	35.000040	−0.000036
46	0.46	1.868025	0.650007	34.999310	0.000691
47	0.45	1.930743	0.650008	34.999180	0.000817
48	0.44	2.008545	0.649993	35.000730	−0.000727
49	0.43	2.091793	0.649994	35.000640	−0.000638
50	0.42	2.193513	0.650003	34.999680	0.000322
No. of pairs		Mean COMPUT. risk errors		STD. ERR. errors	
50		0.0000168295		0.0000759706	
Z-ratio				Status of significance	
0.22153				N.S.—not significant	
Parameter values					
PROB. = 0.6	RHO = −0.3		BIGH = 0.8032		SMHL = 0.32
Pair No.	SMH	SMK	PROB. level	Risk level (%)	Risk error (%)
Table No. 8.2-265					
1	0.81	0.796164	0.599995	40.000480	−0.000483
2	0.80	0.806120	0.599994	40.000630	−0.000632

(continued)

8.2 Contents for Table 8.2 (Table No. 8.2-1 to 400)

Table 8.2 (continued)

Parameter values					
PROB. = 0.6	RHO = −0.3		BIGH = 0.8032		SMHL = 0.32
Pair No.	SMH	SMK	PROB. level	Risk level (%)	Risk error (%)
3	0.79	0.816328	0.600005	39.999500	0.000501
4	0.78	0.826700	0.600006	39.999380	0.000614
5	0.77	0.837246	0.599998	40.000220	−0.000221
6	0.76	0.848074	0.600005	39.999520	0.000483
7	0.75	0.859099	0.600005	39.999530	0.000471
8	0.74	0.870333	0.599999	40.000130	−0.000137
9	0.73	0.881836	0.599999	40.000080	−0.000083
10	0.72	0.893573	0.599996	40.000400	−0.000399
11	0.71	0.905607	0.600001	39.999860	0.000143
12	0.70	0.917904	0.600005	39.999470	0.000530
13	0.69	0.930479	0.600009	39.999090	0.000912
14	0.68	0.943252	0.599993	40.000710	−0.000715
15	0.67	0.956436	0.600000	39.999970	0.000024
16	0.66	0.969853	0.599991	40.000910	−0.000906
17	0.65	0.983719	0.600007	39.999300	0.000703
18	0.64	0.997860	0.600009	39.999080	0.000918
19	0.63	1.012297	0.600000	40.000040	−0.000036
20	0.62	1.027153	0.599999	40.000060	−0.000060
21	0.61	1.042356	0.599991	40.000910	−0.000912
22	0.60	1.058029	0.599995	40.000530	−0.000530
23	0.59	1.074105	0.599993	40.000670	−0.000668
24	0.58	1.090711	0.600007	39.999350	0.000656
25	0.57	1.107686	0.600000	39.999970	0.000024
26	0.56	1.125162	0.599994	40.000570	−0.000572
27	0.55	1.143276	0.600007	39.999300	0.000697
28	0.54	1.161873	0.600006	39.999410	0.000590
29	0.53	1.181094	0.600010	39.999050	0.000948
30	0.52	1.200889	0.600003	39.999680	0.000322
31	0.51	1.221406	0.600004	39.999570	0.000429
32	0.50	1.242604	0.599999	40.000140	−0.000143
33	0.49	1.264638	0.600002	39.999780	0.000215
34	0.48	1.287477	0.600001	39.999920	0.000083
35	0.47	1.311288	0.600009	39.999090	0.000906
36	0.46	1.335952	0.600000	39.999990	0.000006
37	0.45	1.361747	0.600000	40.000010	−0.000012
38	0.44	1.388665	0.599995	40.000530	−0.000530
39	0.43	1.416903	0.599995	40.000520	−0.000525
40	0.42	1.446570	0.599997	40.000350	−0.000352
41	0.41	1.477784	0.599996	40.000450	−0.000453
42	0.40	1.510871	0.600006	39.999370	0.000632
43	0.39	1.545780	0.600000	39.999970	0.000024
44	0.38	1.583061	0.600005	39.999460	0.000536

(continued)

Table 8.2 (continued)

Parameter values					
PROB. = 0.6	RHO = −0.3		BIGH = 0.8032		SMHL = 0.32
Pair No.	SMH	SMK	PROB. level	Risk level (%)	Risk error (%)
45	0.37	1.622890	0.600006	39.999430	0.000572
46	0.36	1.676011	0.599999	40.000110	−0.000113
47	0.35	1.721547	0.599996	40.000370	−0.000370
48	0.34	1.771463	0.600001	39.999940	0.000060
49	0.33	1.826422	0.599991	40.000950	−0.000948
50	0.32	1.888295	0.600000	40.000040	−0.000036
No. of pairs		Mean COMPUT. risk errors		STD. ERR. errors	
50		0.0000424245		0.0000736967	
Z-ratio				Status of significance	
0.57566				N.S.—not significant	
Parameter values					
PROB. = 0.55	RHO = −0.3		BIGH = 0.7048		SMHL = 0.22
Pair No.	SMH	SMK	PROB. level	Risk level (%)	Risk error (%)
Table No. 8.2-266					
1	0.71	0.699638	0.550001	44.999950	0.000054
2	0.70	0.709633	0.550000	44.999990	0.000012
3	0.69	0.719820	0.550001	44.999930	0.000066
4	0.68	0.730212	0.550004	44.999580	0.000417
5	0.67	0.740773	0.550000	44.999970	0.000030
6	0.66	0.751567	0.550003	44.999670	0.000328
7	0.65	0.762560	0.550003	44.999660	0.000340
8	0.64	0.773720	0.549991	45.000910	−0.000906
9	0.63	0.785161	0.549993	45.000750	−0.000751
10	0.62	0.796901	0.550010	44.999010	0.000995
11	0.61	0.808766	0.550000	45.000010	−0.000006
12	0.60	0.820923	0.550000	44.999980	0.000024
13	0.59	0.833344	0.550002	44.999780	0.000221
14	0.58	0.846004	0.549998	45.000170	−0.000173
15	0.57	0.858979	0.550003	44.999710	0.000292
16	0.56	0.872197	0.549998	45.000190	−0.000191
17	0.55	0.885786	0.550009	44.999060	0.000936
18	0.54	0.899582	0.549998	45.000160	−0.000161
19	0.53	0.913716	0.549991	45.000860	−0.000864
20	0.52	0.928224	0.549993	45.000750	−0.000745
21	0.51	0.943146	0.550007	44.999320	0.000679
22	0.50	0.958330	0.550000	45.000040	−0.000042
23	0.49	0.973917	0.549997	45.000280	−0.000274
24	0.48	0.989959	0.550005	44.999460	0.000542
25	0.47	1.006314	0.549993	45.000700	−0.000703
26	0.46	1.023235	0.550005	44.999540	0.000465
27	0.45	1.040494	0.549993	45.000750	−0.000751
28	0.44	1.058355	0.550002	44.999840	0.000161

(continued)

8.2 Contents for Table 8.2 (Table No. 8.2-1 to 400)

Table 8.2 (continued)

Parameter values					
PROB. = 0.55	RHO = −0.3		BIGH = 0.7048		SMHL = 0.22
Pair No.	SMH	SMK	PROB. level	Risk level (%)	Risk error (%)
29	0.43	1.076700	0.550005	44.999550	0.000447
30	0.42	1.095514	0.549994	45.000570	−0.000572
31	0.41	1.114986	0.549998	45.000220	−0.000221
32	0.40	1.135021	0.549993	45.000700	−0.000703
33	0.39	1.155731	0.549992	45.000840	−0.000840
34	0.38	1.177139	0.549991	45.000940	−0.000942
35	0.37	1.199384	0.550003	44.999710	0.000292
36	0.36	1.222321	0.549999	45.000130	−0.000131
37	0.35	1.246218	0.550008	44.999180	0.000823
38	0.34	1.270969	0.550006	44.999360	0.000644
39	0.33	1.296687	0.549999	45.000150	−0.000149
40	0.32	1.323609	0.550004	44.999570	0.000435
41	0.31	1.351614	0.549994	45.000580	−0.000578
42	0.30	1.396628	0.549999	45.000100	−0.000095
43	0.29	1.426479	0.550000	44.999980	0.000018
44	0.28	1.457967	0.550000	45.000040	−0.000042
45	0.27	1.491349	0.540004	44.999640	0.000364
46	0.26	1.526758	0.550001	44.999930	0.000072
47	0.25	1.564508	0.549994	45.000590	−0.000584
48	0.24	1.605111	0.549998	45.000220	−0.000215
49	0.23	1.648811	0.549993	45.000750	−0.000745
50	0.22	1.696591	0.550010	44.999010	0.000995
No. of pairs		Mean COMPUT. risk errors		STD. ERR. errors	
50		−0.0000340097		0.0000734170	
Z-ratio				Status of significance	
0.46324				N.S.—not significant	
Parameter values					
PROB. = 0.5	RHO = −0.3		BIGH = 0.6093		SMHL = 0.12
Pair No.	SMH	SMK	PROB. level	Risk level (%)	Risk error (%)
Table No. 8.2-267					
1	0.61	0.608601	0.500003	49.999740	0.000262
2	0.60	0.618647	0.499995	50.000510	−0.000513
3	0.59	0.628938	0.500005	49.999530	0.000477
4	0.58	0.639348	0.499998	50.000220	−0.000218
5	0.57	0.649942	0.499990	50.000990	−0.000989
6	0.56	0.660792	0.499998	50.000190	−0.000188
7	0.55	0.671820	0.500001	49.999930	0.000072
8	0.54	0.683001	0.499991	50.000940	−0.000936
9	0.53	0.694508	0.500009	49.999130	0.000876
10	0.52	0.706123	0.500000	50.000000	−0.000003

(continued)

Table 8.2 (continued)

Parameter values					
PROB. = 0.5	RHO = −0.3		BIGH = 0.6093		SMHL = 0.12
Pair No.	SMH	SMK	PROB. level	Risk level (%)	Risk error (%)
11	0.51	0.717973	0.499994	50.000610	−0.000605
12	0.50	0.730091	0.499996	50.000360	−0.000355
13	0.49	0.742411	0.499991	50.000940	−0.000942
14	0.48	0.755071	0.500006	49.999410	0.000596
15	0.47	0.767913	0.500004	49.999580	0.000429
16	0.46	0.780983	0.499994	50.000570	−0.000569
17	0.45	0.794426	0.500006	49.999370	0.000638
18	0.44	0.808097	0.500006	49.999400	0.000608
19	0.43	0.822055	0.500004	49.999640	0.000358
20	0.42	0.836313	0.500000	50.000040	−0.000042
21	0.41	0.850889	0.499996	50.000430	−0.000423
22	0.40	0.865811	0.499995	50.000520	−0.000519
23	0.39	0.881113	0.500000	49.999960	0.000048
24	0.38	0.896788	0.500008	49.999160	0.000846
25	0.37	0.912743	0.499997	50.000330	−0.000328
26	0.36	0.929189	0.500004	49.999610	0.000393
27	0.35	0.945957	0.499993	50.000670	−0.000665
28	0.34	0.963288	0.500007	49.999310	0.000691
29	0.33	0.980945	0.499996	50.000370	−0.000370
30	0.32	0.999207	0.500008	49.999200	0.000805
31	0.31	1.017881	0.500003	49.999740	0.000262
32	0.30	1.037096	0.499999	50.000150	−0.000149
33	0.29	1.056917	0.500001	49.999910	0.000089
34	0.28	1.077344	0.500004	49.999610	0.000387
35	0.27	1.098423	0.500009	49.999130	0.000870
36	0.26	1.120057	0.499994	50.000650	−0.000650
37	0.25	1.164476	0.499992	50.000800	−0.000802
38	0.24	1.186409	0.500005	49.999510	0.000489
39	0.23	1.209035	0.499992	50.000850	−0.000852
40	0.22	1.232794	0.500008	49.999210	0.000793
41	0.21	1.257389	0.499999	50.000080	−0.000077
42	0.20	1.283183	0.500005	49.999530	0.000471
43	0.19	1.310078	0.499998	50.000240	−0.000238
44	0.18	1.338356	0.500000	50.000010	−0.000015
45	0.17	1.367979	0.499990	50.000980	−0.000972

(continued)

8.2 Contents for Table 8.2 (Table No. 8.2-1 to 400)

Table 8.2 (continued)

Parameter values					
PROB. = 0.5	RHO = −0.3		BIGH = 0.6093		SMHL = 0.12
Pair No.	SMH	SMK	PROB. level	Risk level (%)	Risk error (%)
46	0.16	1.399388	0.500003	49.999720	0.000286
47	0.15	1.432394	0.499994	50.000570	−0.000566
48	0.14	1.467576	0.500004	49.999560	0.000447
49	0.13	1.504955	0.500007	49.999290	0.000715
50	0.12	1.544886	0.500006	49.999400	0.000602
No. of pairs			Mean COMPUT. risk errors	STD. ERR. errors	
50			0.0000103432	0.0000789997	
Z-ratio				Status of significance	
0.13093				N.S.—not significant	
Parameter values					
PROB. = 0.99	RHO = −0.4		BIGH = 2.5759		SMHL = 2.34
Pair No.	SMH	SMK	PROB. level	Risk level (%)	Risk error (%)
Table No. 8.2-268					
1	2.58	2.571807	0.990002	0.999850	0.000149
2	2.57	2.581814	0.990001	0.999904	0.000095
3	2.56	2.591899	0.989994	1.000595	−0.000596
4	2.55	2.603626	0.990002	0.999832	0.000167
5	2.54	2.615433	0.990002	0.999844	0.000155
6	2.53	2.627320	0.989994	1.000631	−0.000632
7	2.52	2.640853	0.989997	1.000261	−0.000262
8	2.51	2.654468	0.989992	1.000822	−0.000823
9	2.50	2.669729	0.989995	1.000536	−0.000536
10	2.49	2.686639	0.990004	0.999606	0.000393
11	2.48	2.703634	0.990002	0.999850	0.000149
12	2.47	2.722278	0.990003	0.999737	0.000262
13	2.46	2.741011	0.989991	1.000953	−0.000954
14	2.45	2.764520	0.990007	0.999320	0.000679
15	2.44	2.788119	0.990005	0.999457	0.000542
16	2.43	2.813373	0.989999	1.000094	−0.000095
17	2.42	2.841843	0.989997	1.000321	−0.000322
18	2.41	2.875095	0.990004	0.999612	0.000387
19	2.40	2.911567	0.990004	0.999624	0.000376
20	2.39	2.951260	0.989993	1.000738	−0.000739
21	2.38	3.000425	0.989995	1.000541	−0.000542
22	2.37	3.062188	0.990006	0.999415	0.000584
23	2.36	3.136551	0.990005	0.999486	0.000513

(continued)

Table 8.2 (continued)

Parameter values					
PROB. = 0.99	RHO = −0.4		BIGH = 2.5759		SMHL = 2.34
Pair No.	SMH	SMK	PROB. level	Risk level (%)	Risk error (%)
24	2.35	3.236015	0.990002	0.999755	0.000244
25	2.34	3.385581	0.989992	1.000798	−0.000799
No. of pairs		Mean COMPUT. risk errors		STD. ERR. errors	
25		−0.0000616679		0.0000978924	
Z-ratio				Status of significance	
0.62996				N.S.—not significant	
Parameter values					
PROB. = 0.98	RHO = −0.4		BIGH = 2.3262		SMHL = 2.06
Pair No.	SMH	SMK	PROB. level	Risk level (%)	Risk error (%)
Table No. 8.2-269					
1	2.33	2.322406	0.979993	2.000737	−0.000739
2	2.32	2.332417	0.979991	2.000874	−0.000876
3	2.31	2.343295	0.980000	2.000010	−0.000012
4	2.30	2.354261	0.979997	2.000278	−0.000280
5	2.29	2.366097	0.980002	1.999784	0.000215
6	2.28	2.378023	0.979995	2.000552	−0.000554
7	2.27	2.391604	0.980010	1.999033	0.000966
8	2.26	2.404496	0.979993	2.000701	−0.000703
9	2.25	2.419043	0.979996	2.000421	−0.000423
10	2.24	2.434467	0.979999	2.000135	−0.000137
11	2.23	2.450769	0.980000	2.000034	−0.000036
12	2.22	2.467949	0.979997	2.000284	−0.000286
13	2.21	2.486791	0.980003	1.999670	0.000328
14	2.20	2.506514	0.980001	1.999903	0.000095
15	2.19	2.527902	0.980001	1.999867	0.000131
16	2.18	2.550955	0.980001	1.999950	0.000048
17	2.17	2.576456	0.980006	1.999378	0.000620
18	2.16	2.603625	0.980002	1.999807	0.000191
19	2.15	2.634028	0.980004	1.999617	0.000381
20	2.14	2.667663	0.980004	1.999605	0.000393
21	2.13	2.704535	0.979994	2.000564	−0.000566
22	2.12	2.747768	0.979997	2.000344	−0.000346
23	2.11	2.798927	0.980007	1.999325	0.000674
24	2.10	2.858015	0.980003	1.999658	0.000340
25	2.09	2.929718	0.979993	2.000678	−0.000679
26	2.08	3.026541	0.979997	2.000284	−0.000286

(continued)

8.2 Contents for Table 8.2 (Table No. 8.2-1 to 400)

Table 8.2 (continued)

Parameter values					
PROB. = 0.98	RHO = −0.4		BIGH = 2.3262		SMHL = 2.06
Pair No.	SMH	SMK	PROB. level	Risk level (%)	Risk error (%)
27	2.07	3.170359	0.980005	1.999545	0.000453
28	2.06	3.476798	0.980002	1.999819	0.000179
No. of pairs		Mean COMPUT. risk errors		STD. ERR. errors	
28		−0.0000314466		0.0000864760	
Z-ratio				Status of significance	
0.36365				N.S.—not significant	
Parameter values					
PROB. = 0.97	RHO = −0.4		BIGH = 2.1701		SMHL = 1.88
Pair No.	SMH	SMK	PROB. level	Risk level (%)	Risk error (%)
Table No. 8.2-270					
1	2.18	2.160245	0.969999	3.000111	−0.000113
2	2.17	2.170200	0.970005	2.999497	0.000501
3	2.16	2.180247	0.969999	3.000152	−0.000155
4	2.15	2.191169	0.970007	2.999306	0.000691
5	2.14	2.202186	0.970001	2.999926	0.000072
6	2.13	2.214080	0.970007	2.999318	0.000679
7	2.12	2.226071	0.969996	3.000379	−0.000381
8	2.11	2.238943	0.969995	3.000486	−0.000489
9	2.10	2.252696	0.970001	2.999938	0.000060
10	2.09	2.266942	0.969998	3.000182	−0.000185
11	2.08	2.282071	0.969999	3.000146	−0.000149
12	2.07	2.298087	0.969999	3.000075	−0.000077
13	2.06	2.314991	0.969998	3.000224	−0.000226
14	2.05	2.333173	0.970002	2.999830	0.000167
15	2.04	2.352247	0.969998	3.000182	−0.000185
16	2.03	2.372994	0.970003	2.999729	0.000268
17	2.02	2.394635	0.969993	3.000707	−0.000709
18	2.01	2.418733	0.970000	3.000003	−0.000006
19	2.00	2.444510	0.970000	2.999997	0.000000
20	1.99	2.472749	0.970003	2.999675	0.000322
21	1.98	2.503451	0.970001	2.999896	0.000101
22	1.97	2.537399	0.969999	3.000128	−0.000131
23	1.96	2.576158	0.970009	2.999151	0.000846
24	1.95	2.618167	0.969993	3.000695	−0.000697
25	1.94	2.668116	0.969997	3.000283	−0.000286
26	1.93	2.727569	0.970007	2.999336	0.000662
27	1.92	2.798090	0.970000	2.999967	0.000030
28	1.91	2.890619	0.970009	2.999085	0.000912
29	1.90	3.016096	0.969998	3.000194	−0.000197

(continued)

Table 8.2 (continued)

Parameter values					
PROB. = 0.97	RHO = −0.4		BIGH = 2.1701		SMHL = 1.88
Pair No.	SMH	SMK	PROB. level	Risk level (%)	Risk error (%)
30	1.89	3.241710	0.970005	2.999455	0.000542
31	1.88	3.654963	0.969999	3.000134	−0.000137
No. of pairs		Mean COMPUT. risk errors		STD. ERR. errors	
31		0.0000540167		0.0000731994	
Z-ratio				Status of significance	
0.73794				N.S.—not significant	
Parameter values					
PROB. = 0.96	RHO = −0.4		BIGH = 2.0537		SMHL = 1.76
Pair No.	SMH	SMK	PROB. level	Risk level (%)	Risk error (%)
Table No. 8.2-271					
1	2.06	2.047419	0.960003	3.999686	0.000316
2	2.05	2.057407	0.960005	3.999502	0.000501
3	2.04	2.067492	0.959992	4.000836	−0.000834
4	2.03	2.078458	0.959999	4.000116	−0.000113
5	2.02	2.089915	0.960006	3.999376	0.000626
6	2.01	2.101475	0.959995	4.000461	−0.000459
7	2.00	2.113920	0.959999	4.000062	−0.000060
8	1.99	2.126861	0.959999	4.000092	−0.000089
9	1.98	2.140300	0.959993	4.000688	−0.000685
10	1.97	2.154628	0.959995	4.000461	−0.000459
11	1.96	2.169848	0.960003	3.999734	0.000268
12	1.95	2.185571	0.959998	4.000193	−0.000191
13	1.94	2.202189	0.959993	4.000664	−0.000662
14	1.93	2.220094	0.959999	4.000074	−0.000072
15	1.92	2.238897	0.959999	4.000151	−0.000149
16	1.91	2.258993	0.960000	4.000026	−0.000024
17	1.90	2.280380	0.959999	4.000121	−0.000119
18	1.89	2.303454	0.960002	3.999823	0.000179
19	1.88	2.327824	0.959992	4.000783	−0.000781
20	1.87	2.354664	0.959995	4.000515	−0.000513
21	1.86	2.383978	0.960000	4.000014	−0.000012
22	1.85	2.415766	0.959998	4.000253	−0.000250
23	1.84	2.451594	0.960009	3.999061	0.000942
24	1.83	2.490682	0.960004	3.999579	0.000423
25	1.82	2.534596	0.959995	4.000503	−0.000501
26	1.81	2.586462	0.960007	3.999353	0.000650
27	1.80	2.646282	0.960002	3.999758	0.000244
28	1.79	2.718748	0.959997	4.000264	−0.000262

(continued)

8.2 Contents for Table 8.2 (Table No. 8.2-1 to 400)

Table 8.2 (continued)

Parameter values					
PROB. = 0.96	RHO = −0.4		BIGH = 2.0537		SMHL = 1.76
Pair No.	SMH	SMK	PROB. level	Risk level (%)	Risk error (%)
29	1.78	2.813234	0.960009	3.999114	0.000888
30	1.77	2.942247	0.960003	3.999692	0.000310
31	1.76	3.165160	0.960001	3.999871	0.000131
No. of pairs		Mean COMPUT. risk errors		STD. ERR. errors	
31		−0.0000236556		0.0000819418	
Z-ratio				Status of significance	
0.28869				N.S.—not significant	
Parameter values					
PROB. = 0.95	RHO = −0.4		BIGH = 1.9597		SMHL = 1.65
Pair No.	SMH	SMK	PROB. level	Risk level (%)	Risk error (%)
Table No. 8.2-272					
1	1.96	1.959400	0.949991	5.000901	−0.000900
2	1.95	1.969839	0.950006	4.999453	0.000548
3	1.94	1.980381	0.950001	4.999858	0.000143
4	1.93	1.991419	0.950001	4.999912	0.000089
5	1.92	2.002955	0.950002	4.999841	0.000161
6	1.91	2.014990	0.950002	4.999810	0.000191
7	1.90	2.027525	0.950000	4.999990	0.000012
8	1.89	2.040564	0.949995	5.000526	−0.000525
9	1.88	2.054497	0.950003	4.999721	0.000280
10	1.87	2.068937	0.950002	4.999769	0.000232
11	1.86	2.084276	0.950009	4.999066	0.000936
12	1.85	2.100124	0.950004	4.999650	0.000352
13	1.84	2.116874	0.949999	5.000073	−0.000072
14	1.83	2.134530	0.949994	5.000651	−0.000650
15	1.82	2.153483	0.949998	5.000246	−0.000244
16	1.81	2.173344	0.949992	5.000836	−0.000834
17	1.80	2.194897	0.950000	4.999990	0.000012
18	1.79	2.217754	0.950002	4.999763	0.000238
19	1.78	2.241916	0.949993	5.000723	−0.000721
20	1.77	2.268169	0.949990	5.000973	−0.000972
21	1.76	2.296903	0.949996	5.000371	−0.000370
22	1.75	2.328121	0.949999	5.000067	−0.000066
23	1.74	2.362609	0.950007	4.999256	0.000745
24	1.73	2.400367	0.950005	4.999537	0.000465
25	1.72	2.442179	0.949992	5.000782	−0.000781
26	1.71	2.491174	0.950008	4.999244	0.000757
27	1.70	2.546572	0.950000	4.999960	0.000042
28	1.69	2.613065	0.950003	4.999691	0.000310
29	1.68	2.695341	0.950008	4.999244	0.000757

(continued)

Table 8.2 (continued)

Parameter values					
PROB. = 0.95	RHO = −0.4		BIGH = 1.9597		SMHL = 1.65
Pair No.	SMH	SMK	PROB. level	Risk level (%)	Risk error (%)
30	1.67	2.802779	0.950007	4.999328	0.000674
31	1.66	2.960383	0.950000	4.999972	0.000030
32	1.65	3.252528	0.949990	5.000967	−0.000966
No. of pairs		Mean COMPUT. risk errors		STD. ERR. errors	
32		−0.0000037930		0.0000946738	
Z-ratio				Status of significance	
0.04006				N.S.—not significant	
Parameter values					
PROB. = 0.94	RHO = −0.4		BIGH = 1.8805		SMHL = 1.56
Pair No.	SMH	SMK	PROB. level	Risk level (%)	Risk error (%)
Table No. 8.2-273					
1	1.89	1.871048	0.939991	6.000942	−0.000942
2	1.88	1.881000	0.939999	6.000108	−0.000107
3	1.87	1.891254	0.940002	5.999833	0.000167
4	1.86	1.901812	0.939998	6.000203	−0.000203
5	1.85	1.912870	0.939999	6.000060	−0.000060
6	1.84	1.924235	0.939992	6.000829	−0.000829
7	1.83	1.936300	0.939998	6.000245	−0.000244
8	1.82	1.948870	0.940003	5.999732	0.000268
9	1.81	1.961949	0.940005	5.999500	0.000501
10	1.80	1.975537	0.940003	5.999744	0.000256
11	1.79	1.989638	0.939994	6.000644	−0.000644
12	1.78	2.004643	0.939997	6.000328	−0.000328
13	1.77	2.020555	0.940008	5.999196	0.000805
14	1.76	2.036984	0.940004	5.999607	0.000393
15	1.75	2.054325	0.940001	5.999863	0.000137
16	1.74	2.072579	0.939996	6.000358	−0.000358
17	1.73	2.092139	0.940002	5.999768	0.000232
18	1.72	2.112617	0.939996	6.000376	−0.000376
19	1.71	2.134406	0.939990	6.000996	−0.000995
20	1.70	2.158290	0.940008	5.999178	0.000823
21	1.69	2.183098	0.939997	6.000352	−0.000352
22	1.68	2.210397	0.940006	5.999446	0.000554
23	1.67	2.239408	0.939997	6.000299	−0.000298
24	1.66	2.271695	0.940010	5.999011	0.000989
25	1.65	2.306481	0.940004	5.999577	0.000423
26	1.64	2.345330	0.940008	5.999213	0.000787
27	1.63	2.388247	0.939998	6.000161	−0.000161
28	1.62	2.437576	0.940002	5.999774	0.000226
29	1.61	2.494103	0.939997	6.000346	−0.000346
30	1.60	2.561737	0.940000	5.999995	0.000006
31	1.59	2.644387	0.939995	6.000495	−0.000495

(continued)

8.2 Contents for Table 8.2 (Table No. 8.2-1 to 400)

Table 8.2 (continued)

Parameter values					
PROB. = 0.94	RHO = −0.4		BIGH = 1.8805		SMHL = 1.56
Pair No.	SMH	SMK	PROB. level	Risk level (%)	Risk error (%)
32	1.58	2.753776	0.940002	5.999762	0.000238
33	1.57	2.914907	0.940008	5.999172	0.000829
34	1.56	3.223095	0.940003	5.999720	0.000280
No. of pairs		Mean COMPUT. risk errors		STD. ERR. errors	
34		0.0000337192		0.0000867667	
Z-ratio				Status of significance	
0.38862				N.S.—not significant	
Parameter values					
PROB. = 0.93	RHO = −0.4		BIGH = 1.8115		SMHL = 1.48
Pair No.	SMH	SMK	PROB. level	Risk level (%)	Risk error (%)
Table No. 8.2-274					
1	1.82	1.803040	0.929992	7.000757	−0.000757
2	1.81	1.813001	0.929999	7.000065	−0.000066
3	1.80	1.823269	0.930001	6.999869	0.000131
4	1.79	1.833844	0.929997	7.000274	−0.000274
5	1.78	1.844925	0.930000	6.999982	0.000018
6	1.77	1.856317	0.929995	7.000548	−0.000548
7	1.76	1.868413	0.930006	6.999433	0.000566
8	1.75	1.880630	0.929991	7.000905	−0.000906
9	1.74	1.893750	0.930002	6.999809	0.000191
10	1.73	1.907386	0.930009	6.999076	0.000924
11	1.72	1.921344	0.929999	7.000142	−0.000143
12	1.71	1.936212	0.930004	6.999606	0.000393
13	1.70	1.951797	0.930010	6.999004	0.000995
14	1.69	1.967907	0.930002	6.999832	0.000167
15	1.68	1.984933	0.929999	7.000083	−0.000083
16	1.67	2.002880	0.929997	7.000268	−0.000268
17	1.66	2.021749	0.929992	7.000804	−0.000805
18	1.65	2.041932	0.929998	7.000238	−0.000238
19	1.64	2.063434	0.930008	6.999219	0.000781
20	1.63	2.085866	0.929998	7.000208	−0.000209
21	1.62	2.110012	0.929997	7.000322	−0.000322
22	1.61	2.135875	0.929995	7.000476	−0.000477
23	1.60	2.163848	0.929999	7.000095	−0.000095
24	1.59	2.193934	0.929996	7.000399	−0.000399
25	1.58	2.226919	0.930001	6.999916	0.000083
26	1.57	2.262804	0.929996	7.000393	−0.000393
27	1.56	2.302764	0.929998	7.000172	−0.000173
28	1.55	2.347586	0.930003	6.999726	0.000274
29	1.54	2.398052	0.929996	7.000363	−0.000364
30	1.53	2.457291	0.930007	6.999332	0.000668
31	1.52	2.526871	0.930001	6.999869	0.000131

(continued)

Table 8.2 (continued)

Parameter values					
PROB. = 0.93	RHO = −0.4		BIGH = 1.8115		SMHL = 1.48
Pair No.	SMH	SMK	PROB. level	Risk level (%)	Risk error (%)
32	1.51	2.613824	0.930009	6.999123	0.000876
33	1.50	2.726750	0.929998	7.000214	−0.000215
34	1.49	2.896119	0.930008	6.999171	0.000829
35	1.48	3.254750	0.930005	6.999469	0.000530
No. of pairs		Mean COMPUT. risk errors		STD. ERR. errors	
35		0.0000228484		0.0000833645	
Z-ratio				Status of significance	
0.27408				N.S.—not significant	
Parameter values					
PROB. = 0.92	RHO = −0.4		BIGH = 1.7501		SMHL = 1.41
Pair No.	SMH	SMK	PROB. level	Risk level (%)	Risk error (%)
Table No. 8.2-275					
1	1.76	1.740451	0.920002	7.999790	0.000209
2	1.75	1.750200	0.919995	8.000499	−0.000501
3	1.74	1.760454	0.920001	7.999909	0.000089
4	1.73	1.771020	0.920002	7.999814	0.000185
5	1.72	1.781899	0.919996	8.000356	−0.000358
6	1.71	1.793289	0.919999	8.000136	−0.000137
7	1.70	1.805192	0.920007	7.999349	0.000650
8	1.69	1.817415	0.920003	7.999742	0.000256
9	1.68	1.830156	0.920001	7.999939	0.000060
10	1.67	1.843417	0.919998	8.000231	−0.000232
11	1.66	1.857200	0.919991	8.000862	−0.000864
12	1.65	1.871898	0.920007	7.999337	0.000662
13	1.64	1.886927	0.919999	8.000082	−0.000083
14	1.63	1.902877	0.920006	7.999415	0.000584
15	1.62	1.919554	0.920010	7.999039	0.000960
16	1.61	1.936766	0.919995	8.000546	−0.000548
17	1.60	1.955297	0.920005	7.999528	0.000471
18	1.59	1.974758	0.920010	7.999003	0.000995
19	1.58	1.995153	0.920006	7.999415	0.000584
20	1.57	2.016875	0.920007	7.999283	0.000715
21	1.56	2.039928	0.920007	7.999307	0.000691
22	1.55	2.064313	0.919997	8.000273	−0.000274
23	1.54	2.090817	0.920006	7.999415	0.000584
24	1.53	2.119050	0.920003	7.999718	0.000280
25	1.52	2.149408	0.919994	8.000642	−0.000644
26	1.51	2.183065	0.920009	7.999134	0.000864
27	1.50	2.219243	0.919999	8.000118	−0.000119
28	1.49	2.259510	0.919999	8.000147	−0.000149
29	1.48	2.304649	0.920002	7.999808	0.000191
30	1.47	2.355446	0.919996	8.000410	−0.000411

(continued)

Table 8.2 (continued)

Parameter values					
PROB. = 0.92	RHO = −0.4		BIGH = 1.7501		SMHL = 1.41
Pair No.	SMH	SMK	PROB. level	Risk level (%)	Risk error (%)
31	1.46	2.415029	0.920010	7.999015	0.000983
32	1.45	2.484185	0.919995	8.000481	−0.000483
33	1.44	2.570728	0.920004	7.999623	0.000376
34	1.43	2.682477	0.919994	8.000601	−0.000602
35	1.42	2.847561	0.920001	7.999897	0.000101
36	1.41	3.178483	0.919996	8.000440	−0.000441
No. of pairs		Mean COMPUT. risk errors		STD. ERR. errors	
36		0.0001254920		0.0000846751	
Z-ratio				Status of significance	
1.48204				N.S.—not significant	
Parameter values					
PROB. = 0.91	RHO = −0.4		BIGH = 1.6947		SMHL = 1.35
Pair No.	SMH	SMK	PROB. level	Risk level (%)	Risk error (%)
Table No. 8.2-276					
1	1.70	1.689417	0.910004	8.999628	0.000370
2	1.69	1.699413	0.910004	8.999586	0.000411
3	1.68	1.709724	0.910002	8.999819	0.000179
4	1.67	1.720351	0.909995	9.000546	−0.000548
5	1.66	1.731493	0.909999	9.000147	−0.000149
6	1.65	1.742955	0.909995	9.000522	−0.000525
7	1.64	1.754935	0.909998	9.000182	−0.000185
8	1.63	1.767437	0.910007	8.999342	0.000656
9	1.62	1.780266	0.910001	8.999884	0.000113
10	1.61	1.793621	0.909996	9.000391	−0.000393
11	1.60	1.807700	0.910004	8.999598	0.000399
12	1.59	1.822310	0.910006	8.999389	0.000608
13	1.58	1.837453	0.910000	8.999956	0.000042
14	1.57	1.853328	0.909997	9.000283	−0.000286
15	1.56	1.869937	0.909993	9.000683	−0.000685
16	1.55	1.887479	0.909997	9.000265	−0.000268
17	1.54	1.905956	0.910005	8.999550	0.000447
18	1.53	1.925176	0.909997	9.000325	−0.000328
19	1.52	1.945729	0.910005	8.999467	0.000530
20	1.51	1.967226	0.910000	9.000015	−0.000018
21	1.50	1.990063	0.909996	9.000397	−0.000399
22	1.49	2.014631	0.910006	8.999353	0.000644
23	1.48	2.040350	0.909991	9.000927	−0.000930
24	1.47	2.068590	0.910006	8.999378	0.000620
25	1.46	2.098378	0.909993	9.000736	−0.000739
26	1.45	2.131086	0.909997	9.000296	−0.000298
27	1.44	2.166715	0.909999	9.000087	−0.000089
28	1.43	2.206053	0.910006	8.999414	0.000584

(continued)

Table 8.2 (continued)

Parameter values					
PROB. = 0.91	RHO = −0.4		BIGH = 1.6947		SMHL = 1.35
Pair No.	SMH	SMK	PROB. level	Risk level (%)	Risk error (%)
29	1.42	2.249103	0.919991	9.000910	−0.000912
30	1.41	2.298603	0.910007	8.999276	0.000721
31	1.40	2.354558	0.910005	8.999509	0.000489
32	1.39	2.420098	0.910004	8.999574	0.000423
33	1.38	2.499133	0.910006	8.999378	0.000620
34	1.37	2.599480	0.910001	8.999866	0.000131
35	1.36	2.736770	0.909993	9.000731	−0.000733
36	1.35	2.968037	0.909999	9.000134	−0.000137
No. of pairs		Mean COMPUT. risk errors		STD. ERR. errors	
36		0.0000098267		0.0000814738	
Z-ratio				Status of significance	
0.12061				N.S.—not significant	
Parameter values					
PROB. = 0.9025	RHO = −0.4		BIGH = 1.6562		SMHL = 1.30
Pair No.	SMH	SMK	PROB. level	Risk level (%)	Risk error (%)
Table No. 8.2-277					
1	1.66	1.652409	0.902496	9.750426	−0.000423
2	1.65	1.662426	0.902493	9.750682	−0.000679
3	1.64	1.672951	0.902507	9.749288	0.000715
4	1.63	1.683603	0.902498	9.750218	−0.000215
5	1.62	1.694772	0.902501	9.749860	0.000143
6	1.61	1.706265	0.902498	9.750229	−0.000226
7	1.60	1.718280	0.902503	9.749735	0.000268
8	1.59	1.730625	0.902496	9.750335	−0.000352
9	1.58	1.743497	0.902495	9.750539	−0.000536
10	1.57	1.756899	0.902494	9.750593	−0.000590
11	1.56	1.770832	0.902493	9.750729	−0.000727
12	1.55	1.785497	0.902503	9.749699	0.000304
13	1.54	1.800699	0.902506	9.749436	0.000566
14	1.53	1.816442	0.902498	9.750176	−0.000173
15	1.52	1.833120	0.902506	9.749394	0.000608
16	1.51	1.850344	0.902496	9.750402	−0.000399
17	1.50	1.868509	0.902493	9.750742	−0.000739
18	1.49	1.887814	0.902503	9.749717	0.000286
19	1.48	1.908065	0.902507	9.749269	0.000733
20	1.47	1.929267	0.902500	9.750009	−0.000006
21	1.46	1.951813	0.902499	9.750128	−0.000125
22	1.45	1.975707	0.902494	9.750568	−0.000566
23	1.44	2.001344	0.902499	9.750056	−0.000054
24	1.43	2.028728	0.902502	9.749776	0.000226
25	1.42	2.058251	0.902510	9.749042	0.000960
26	1.41	2.089920	0.902507	9.749353	0.000650

(continued)

8.2 Contents for Table 8.2 (Table No. 8.2-1 to 400)

Table 8.2 (continued)

Parameter values					
PROB. = 0.9025	RHO = −0.4		BIGH = 1.6562		SMHL = 1.30
Pair No.	SMH	SMK	PROB. level	Risk level (%)	Risk error (%)
27	1.40	2.124128	0.902495	9.750551	−0.000548
28	1.39	2.162052	0.902501	9.749871	0.000131
29	1.38	2.203695	0.902498	9.750187	−0.000185
30	1.37	2.250626	0.902508	9.749240	0.000763
31	1.36	2.303238	0.902499	9.750062	−0.000060
32	1.35	2.364663	0.902507	9.749311	0.000691
33	1.34	2.436857	0.902499	9.750091	−0.000089
34	1.33	2.526467	0.902501	9.749901	0.000101
35	1.32	2.643653	0.902497	9.750348	−0.000346
36	1.31	2.818891	0.902501	9.749949	0.000054
37	1.30	3.197497	0.902509	9.749055	0.000948
No. of pairs		Mean COMPUT. risk errors		STD. ERR. errors	
37		0.0000291749		0.0000796681	
Z-ratio				Status of significance	
0.36621				N.S.—not significant	
Parameter values					
PROB. = 0.9	RHO = −0.4		BIGH = 1.6439		SMHL = 1.29
Pair No.	SMH	SMK	PROB. level	Risk level (%)	Risk error (%)
Table No. 8.2-278					
1	1.65	1.637823	0.899993	10.000750	−0.000745
2	1.64	1.647809	0.899995	10.000530	−0.000525
3	1.63	1.658114	0.899995	10.000490	−0.000489
4	1.62	1.668739	0.899992	10.000800	−0.000793
5	1.61	1.679881	0.900003	9.999728	0.000274
6	1.60	1.691348	0.900006	9.999365	0.000638
7	1.59	1.703143	0.900002	9.999848	0.000155
8	1.58	1.715463	0.900004	9.999572	0.000429
9	1.57	1.728115	0.899995	10.000540	−0.000536
10	1.56	1.741492	0.900005	9.999472	0.000530
11	1.55	1.755207	0.899999	10.000130	−0.000131
12	1.54	1.769654	0.900006	9.999407	0.000596
13	1.53	1.784639	0.900007	9.999299	0.000703
14	1.52	1.800165	0.899999	10.000090	−0.000089
15	1.51	1.816627	0.900009	9.999108	0.000894
16	1.50	1.833636	0.900002	9.999776	0.000226
17	1.49	1.851587	0.900004	9.999591	0.000411
18	1.48	1.870482	0.900009	9.999132	0.000870
19	1.47	1.890130	0.899997	10.000280	−0.000280
20	1.46	1.911120	0.900003	9.999710	0.000292
21	1.45	1.933260	0.900005	9.999508	0.000495
22	1.44	1.956750	0.900007	9.999264	0.000739
23	1.43	1.981592	0.900002	9.999836	0.000167

(continued)

Table 8.2 (continued)

Parameter values					
PROB. = 0.9	RHO = −0.4		BIGH = 1.6439		SMHL = 1.29
Pair No.	SMH	SMK	PROB. level	Risk level (%)	Risk error (%)
24	1.42	2.008181	0.899999	10.000080	−0.000077
25	1.41	2.036913	0.900008	9.999180	0.000823
26	1.40	2.067400	0.899995	10.000550	−0.000548
27	1.39	2.100818	0.899998	10.000200	−0.000197
28	1.38	2.137172	0.899997	10.000300	−0.000298
29	1.37	2.177247	0.899999	10.000130	−0.000125
30	1.36	2.221830	0.900002	9.999824	0.000179
31	1.35	2.272095	0.900006	9.999371	0.000632
32	1.34	2.329221	0.899999	10.000110	−0.000107
33	1.33	2.396728	0.900006	9.999424	0.000578
34	1.32	2.477746	0.899999	10.000150	−0.000149
35	1.31	2.581656	0.900005	9.999514	0.000489
36	1.30	2.725649	0.900002	9.999788	0.000215
37	1.29	2.969888	0.899990	10.000990	−0.000983
No. of pairs		Mean COMPUT. risk errors		STD. ERR. errors	
37		0.0001121509		0.0000812927	
Z-ratio				Status of significance	
1.37959				N.S.—not significant	
Parameter values					
PROB. = 0.85	RHO = −0.4		BIGH = 1.4371		SMHL = 1.04
Pair No.	SMH	SMK	PROB. level	Risk level (%)	Risk error (%)
Table No. 8.2-279					
1	1.44	1.434206	0.849996	15.000420	−0.000417
2	1.43	1.444235	0.849991	15.000870	−0.000876
3	1.42	1.454601	0.849992	15.000830	−0.000834
4	1.41	1.465307	0.849995	15.000470	−0.000471
5	1.40	1.476355	0.850000	14.999960	0.000036
6	1.39	1.487749	0.850005	14.999490	0.000513
7	1.38	1.499492	0.850008	14.999200	0.000799
8	1.37	1.511588	0.850007	14.999350	0.000650
9	1.36	1.524040	0.850000	15.000030	−0.000030
10	1.35	1.537046	0.850008	14.999180	0.000823
11	1.34	1.550416	0.850006	14.999380	0.000620
12	1.33	1.564153	0.849992	15.000840	−0.000840
13	1.32	1.578651	0.850006	14.999380	0.000614
14	1.31	1.593524	0.850002	14.999770	0.000232
15	1.30	1.608971	0.849998	15.000160	−0.000167
16	1.29	1.624997	0.849991	15.000920	−0.000918
17	1.28	1.641802	0.849996	15.000400	−0.000405
18	1.27	1.659389	0.850008	14.999190	0.000811
19	1.26	1.677569	0.850003	14.999700	0.000298
20	1.25	1.696541	0.849995	15.000480	−0.000483

(continued)

8.2 Contents for Table 8.2 (Table No. 8.2-1 to 400)

Table 8.2 (continued)

Parameter values					
PROB. = 0.85	RHO = −0.4		BIGH = 1.4371		SMHL = 1.04
Pair No.	SMH	SMK	PROB. level	Risk level (%)	Risk error (%)
21	1.24	1.716506	0.849996	15.000380	−0.000381
22	1.23	1.737469	0.849999	15.000140	−0.000143
23	1.22	1.759435	0.849995	15.000520	−0.000525
24	1.21	1.782605	0.949993	15.000710	−0.000709
25	1.20	1.807180	0.849998	15.000160	−0.000161
26	1.19	1.833166	0.849999	15.000060	−0.000060
27	1.18	1.860764	0.849998	15.000190	−0.000191
28	1.17	1.890176	0.849994	15.000630	−0.000632
29	1.16	1.921799	0.849995	15.000520	−0.000519
30	1.15	1.956031	0.850004	14.999560	0.000441
31	1.14	1.992878	0.849998	15.000160	−0.000161
32	1.13	2.033129	0.849992	15.000780	−0.000781
33	1.12	2.077963	0.850007	14.999290	0.000703
34	1.11	2.127388	0.849997	15.000270	−0.000274
35	1.10	2.183755	0.850005	14.999550	0.000447
36	1.09	2.248636	0.850006	14.999370	0.000632
37	1.08	2.325555	0.850009	14.999100	0.000900
38	1.07	2.419989	0.850003	14.999710	0.000292
39	1.06	2.544448	0.850002	14.999840	0.000155
40	1.05	2.729410	0.849991	15.000870	−0.000876
41	1.04	3.142072	0.849995	15.000480	−0.000477
No. of pairs		Mean COMPUT. risk errors		STD. ERR. errors	
41		−0.0000563406		0.0000864007	
Z-ratio				Status of significance	
0.65208				N.S.—not significant	
Parameter values					
PROB. = 0.8	RHO = −0.4		BIGH = 1.2768		SMHL = 0.85
Pair No.	SMH	SMK	PROB. level	Risk level (%)	Risk error (%)
Table No. 8.2-280					
1	1.28	1.273608	0.800009	19.999090	0.000906
2	1.27	1.283637	0.800006	19.999420	0.000578
3	1.26	1.293922	0.800000	19.999980	0.000018
4	1.25	1.304565	0.800007	19.999250	0.000745
5	1.24	1.315473	0.800010	19.999030	0.000972
6	1.23	1.326553	0.799990	20.000980	−0.000983
7	1.22	1.338198	0.800010	19.999020	0.000983
8	1.21	1.350022	0.800005	19.999520	0.000483
9	1.20	1.362226	0.800004	19.999570	0.000429
10	1.19	1.374814	0.800006	19.999410	0.000590
11	1.18	1.387791	0.800008	19.999220	0.000781
12	1.17	1.401161	0.800008	19.999230	0.000769
13	1.16	1.414931	0.800003	19.999690	0.000310

(continued)

Table 8.2 (continued)

Parameter values					
PROB. = 0.8	RHO = −0.4		BIGH = 1.2768		SMHL = 0.85
Pair No.	SMH	SMK	PROB. level	Risk level (%)	Risk error (%)
14	1.15	1.429104	0.799992	20.000790	−0.000793
15	1.14	1.443883	0.799998	20.000170	−0.000173
16	1.13	1.459077	0.799992	20.000790	−0.000787
17	1.12	1.474888	0.799996	20.000390	−0.000393
18	1.11	1.491321	0.800006	19.999400	0.000602
19	1.10	1.508188	0.799993	20.000690	−0.000685
20	1.09	1.525886	0.800002	19.999800	0.000197
21	1.08	1.544227	0.800003	19.999680	0.000322
22	1.07	1.563217	0.799992	20.000820	−0.000817
23	1.06	1.583254	0.800006	19.999360	0.000638
24	1.05	1.603956	0.799996	20.000390	−0.000387
25	1.04	1.625720	0.799997	20.000350	−0.000346
26	1.03	1.648556	0.799998	20.000160	−0.000155
27	1.02	1.672472	0.799993	20.000730	−0.000733
28	1.01	1.697866	0.800005	19.999470	0.000530
29	1.00	1.724554	0.800005	19.999470	0.000530
30	0.99	1.752739	0.799997	20.000280	−0.000286
31	0.98	1.782824	0.799998	20.000200	−0.000197
32	0.97	1.815012	0.800003	19.999670	0.000328
33	0.96	1.849315	0.799991	20.000860	−0.000858
34	0.95	1.886722	0.800006	19.999430	0.000566
35	0.94	1.927048	0.800000	20.000040	−0.000042
36	0.93	1.971282	0.799997	20.000280	−0.000280
37	0.92	2.020413	0.800005	19.999550	0.000453
38	0.91	2.075042	0.799991	20.000860	−0.000864
39	0.90	2.137916	0.800004	19.999590	0.000411
40	0.89	2.210611	0.799995	20.000460	−0.000459
41	0.88	2.298614	0.800002	19.999830	0.000167
42	0.87	2.409751	0.800001	19.999900	0.000095
43	0.86	2.563570	0.800001	19.999870	0.000131
44	0.85	2.824152	0.799994	20.000550	−0.000554
No. of pairs		Mean COMPUT. risk errors		STD. ERR. errors	
44		0.0000609292		0.0000853316	
Z-ratio				Status of significance	
0.71403				N.S.—not significant	
Parameter values					
PROB. = 0.75	RHO = −0.4		BIGH = 1.1423		SMHL = 0.68
Pair No.	SMH	SMK	PROB. level	Risk level (%)	Risk error (%)
Table No. 8.2-281					
1	1.15	1.134456	0.750003	24.999710	0.000292
2	1.14	1.144409	0.750007	24.999300	0.000703
3	1.13	1.154539	0.749998	25.000220	−0.000221

(continued)

8.2 Contents for Table 8.2 (Table No. 8.2-1 to 400)

Table 8.2 (continued)

Parameter values					
PROB. = 0.75	RHO = −0.4		BIGH = 1.1423		SMHL = 0.68
Pair No.	SMH	SMK	PROB. level	Risk level (%)	Risk error (%)
4	1.12	1.164946	0.749994	25.000580	−0.000578
5	1.11	1.175638	0.749995	25.000500	−0.000501
6	1.10	1.186617	0.749999	25.000100	−0.000095
7	1.09	1.197891	0.750005	24.999490	0.000507
8	1.08	1.209366	0.749994	25.000590	−0.000590
9	1.07	1.221243	0.750001	24.999920	0.000083
10	1.06	1.233431	0.750006	24.999420	0.000584
11	1.05	1.245838	0.749990	25.000960	−0.000960
12	1.04	1.258764	0.750005	24.999530	0.000471
13	1.03	1.271922	0.749996	25.000400	−0.000399
14	1.02	1.285514	0.749996	25.000430	−0.000429
15	1.01	1.299547	0.750002	24.999810	0.000191
16	1.00	1.313931	0.749996	25.000390	−0.000393
17	0.99	1.328772	0.749992	25.000820	−0.000823
18	0.98	1.344174	0.750001	24.999860	0.000137
19	0.97	1.360049	0.750007	24.999350	0.000656
20	0.96	1.376405	0.750004	24.999570	0.000429
21	0.95	1.393252	0.749992	25.000820	−0.000823
22	0.94	1.410794	0.749993	25.000690	−0.000685
23	0.93	1.429040	0.750004	24.999650	0.000352
24	0.92	1.447806	0.749991	25.000870	−0.000864
25	0.91	1.467494	0.750004	24.999640	0.000358
26	0.90	1.487918	0.750008	24.999170	0.000834
27	0.89	1.509091	0.750000	25.000020	−0.000024
28	0.88	1.531220	0.749995	25.000530	−0.000525
29	0.87	1.554514	0.750008	24.999250	0.000751
30	0.86	1.578790	0.750006	24.999450	0.000548
31	0.85	1.604257	0.750001	24.999930	0.000072
32	0.84	1.631126	0.750002	24.999810	0.000191
33	0.83	1.659412	0.749995	25.000510	−0.000507
34	0.82	1.689521	0.750002	24.999830	0.000167
35	0.81	1.721471	0.750003	24.999740	0.000262
36	0.80	1.755670	0.750010	24.999010	0.000995
37	0.79	1.792137	0.749999	25.000080	−0.000077
38	0.78	1.831672	0.750001	24.999910	0.000089
39	0.77	1.874687	0.750004	24.999560	0.000441
40	0.76	1.921594	0.749992	25.000820	−0.000817
41	0.75	1.973783	0.749991	25.000890	−0.000888
42	0.74	2.032450	0.749991	25.000860	−0.000858
43	0.73	2.099964	0.750009	24.999130	0.000876
44	0.72	2.178696	0.750008	24.999240	0.000763
45	0.71	2.274533	0.750010	24.999050	0.000948

(continued)

Table 8.2 (continued)

Parameter values					
PROB. = 0.75	RHO = −0.4		BIGH = 1.1423		SMHL = 0.68
Pair No.	SMH	SMK	PROB. level	Risk level (%)	Risk error (%)
46	0.70	2.397663	0.750003	24.999730	0.000274
47	0.69	2.574678	0.750001	24.999940	0.000060
48	0.68	2.915769	0.750000	25.000050	−0.000048
No. of pairs		Mean COMPUT. risk errors		STD. ERR. errors	
48		0.0000189762		0.0000805011	
Z-ratio				Status of significance	
0.23573				N.S.—not significant	
Parameter values					
PROB. = 0.7	RHO = −0.4		BIGH = 1.0237		SMHL = 0.54
Pair No.	SMH	SMK	PROB. level	Risk level (%)	Risk error (%)
Table No. 8.2-282					
1	1.03	1.017439	0.700001	29.999880	0.000125
2	1.02	1.027414	0.700002	29.999800	0.000203
3	1.01	1.037586	0.699997	30.000320	−0.000316
4	1.00	1.048059	0.700007	29.999300	0.000703
5	0.99	1.058694	0.700000	30.000000	0.000000
6	0.98	1.069642	0.700007	29.999310	0.000697
7	0.97	1.080764	0.699996	30.000370	−0.000364
8	0.96	1.092213	0.699998	30.000160	−0.000161
9	0.95	1.103899	0.699992	30.000790	−0.000793
10	0.94	1.115927	0.699997	30.000350	−0.000346
11	0.93	1.128208	0.699992	30.000850	−0.000846
12	0.92	1.140847	0.699995	30.000470	−0.000471
13	0.91	1.153852	0.700007	29.999320	0.000679
14	0.90	1.167136	0.700006	29.999420	0.000584
15	0.89	1.180708	0.699992	30.000770	−0.000769
16	0.88	1.194772	0.700001	29.999940	0.000066
17	0.87	1.209143	0.699994	30.000630	−0.000626
18	0.86	1.224029	0.700005	29.999510	0.000495
19	0.85	1.239244	0.699998	30.000220	−0.000215
20	0.84	1.254995	0.700005	29.999530	0.000471
21	0.83	1.271198	0.700006	29.999390	0.000608
22	0.82	1.287865	0.700000	29.999980	0.000018
23	0.81	1.305108	0.700000	30.000000	0.000006
24	0.80	1.322940	0.700002	29.999760	0.000244
25	0.79	1.341377	0.700004	29.999630	0.000376
26	0.78	1.360435	0.700001	29.999950	0.000054
27	0.77	1.380227	0.700003	29.999740	0.000262
28	0.76	1.400772	0.700005	29.999480	0.000519
29	0.75	1.422087	0.700003	29.999750	0.000250
30	0.74	1.444289	0.700002	29.999780	0.000226
31	0.73	1.467399	0.699997	30.000270	−0.000268

(continued)

8.2 Contents for Table 8.2 (Table No. 8.2-1 to 400)

Table 8.2 (continued)

Parameter values					
PROB. = 0.7	RHO = −0.4		BIGH = 1.0237		SMHL = 0.54
Pair No.	SMH	SMK	PROB. level	Risk level (%)	Risk error (%)
32	0.72	1.491635	0.700004	29.999570	0.000429
33	0.71	1.516822	0.699990	30.000960	−0.000954
34	0.70	1.543377	0.699992	30.000810	−0.000805
35	0.69	1.571323	0.699996	30.000420	−0.000417
36	0.68	1.600885	0.700009	29.999130	0.000876
37	0.67	1.631895	0.699995	30.000480	−0.000477
38	0.66	1.664969	0.699996	30.000450	−0.000447
39	0.65	1.700334	0.700004	29.999620	0.000387
40	0.64	1.738025	0.699994	30.000560	−0.000554
41	0.63	1.778860	0.700001	29.999920	0.000077
42	0.62	1.823072	0.699998	30.000190	−0.000191
43	0.61	1.871485	0.699996	30.000450	−0.000447
44	0.60	1.925117	0.699996	30.000390	−0.000387
45	0.59	1.985189	0.699992	30.000830	−0.000829
46	0.58	2.054095	0.700001	29.999870	0.000137
47	0.57	2.134622	0.700005	29.999510	0.000495
48	0.56	2.232296	0.700007	29.999310	0.000697
49	0.55	2.359290	0.699996	30.000400	−0.000399
50	0.54	2.538324	0.700000	29.999980	0.000024
No. of pairs		Mean COMPUT. risk errors		STD. ERR. errors	
50		−0.0000268805		0.0000683154	
Z-ratio				Status of significance	
0.39348				N.S.—not significant	
Parameter values					
PROB. = 0.65	RHO = −0.4		BIGH = 0.9159		SMHL = 0.43
Pair No.	SMH	SMK	PROB. level	Risk level (%)	Risk error (%)
Table No. 8.2-283					
1	0.92	0.911818	0.650004	34.999580	0.000423
2	0.91	0.921838	0.650005	34.999540	0.000459
3	0.90	0.932081	0.650008	34.999170	0.000834
4	0.89	0.942456	0.649992	35.000760	−0.000757
5	0.88	0.953167	0.650004	34.999650	0.000352
6	0.87	0.964026	0.649996	35.000430	−0.000429
7	0.86	0.975140	0.649992	35.000780	−0.000775
8	0.85	0.986519	0.649993	35.000730	−0.000721
9	0.84	0.998170	0.649997	35.000260	−0.000256
10	0.83	1.010104	0.650006	34.999380	0.000620
11	0.82	1.022234	0.649998	35.000170	−0.000167
12	0.81	1.034669	0.649995	35.000550	−0.000542

(continued)

Table 8.2 (continued)

Parameter values					
PROB. = 0.65	RHO = −0.4		BIGH = 0.9159		SMHL = 0.43
Pair No.	SMH	SMK	PROB. level	Risk level (%)	Risk error (%)
13	0.80	1.047419	0.649995	35.000470	−0.000465
14	0.79	1.060497	0.650000	34.999980	0.000024
15	0.78	1.073915	0.650009	34.999110	0.000894
16	0.77	1.087590	0.650002	34.999820	0.000185
17	0.76	1.101631	0.649999	35.000130	−0.000131
18	0.75	1.116055	0.649999	35.000130	−0.000131
19	0.74	1.130877	0.650002	34.999820	0.000185
20	0.73	1.146113	0.650007	34.999280	0.000721
21	0.72	1.161683	0.649997	35.000350	−0.000346
22	0.71	1.177800	0.650005	34.999490	0.000513
23	0.70	1.194288	0.649996	35.000360	−0.000352
24	0.69	1.211363	0.650004	34.999560	0.000441
25	0.68	1.228851	0.649994	35.000620	−0.000620
26	0.67	1.246970	0.649997	35.000340	−0.000340
27	0.66	1.265648	0.649995	35.000520	−0.000513
28	0.65	1.285007	0.650002	34.999800	0.000209
29	0.64	1.304977	0.650000	34.999970	0.000036
30	0.63	1.325684	0.650002	34.999760	0.000238
31	0.62	1.347161	0.650005	34.999530	0.000477
32	0.61	1.369439	0.650004	34.999640	0.000370
33	0.60	1.392652	0.650008	34.999160	0.000846
34	0.59	1.416739	0.650001	34.999920	0.000077
35	0.58	1.441937	0.650001	34.999930	0.000072
36	0.57	1.468288	0.650001	34.999940	0.000066
37	0.56	1.495838	0.649993	35.000740	−0.000739
38	0.55	1.524929	0.650001	34.999940	0.000060
39	0.54	1.555420	0.649991	35.000930	−0.000930
40	0.53	1.587856	0.650001	34.999870	0.000137
41	0.52	1.622200	0.650004	34.999590	0.000417
42	0.51	1.658714	0.649999	35.000130	−0.000131
43	0.50	1.697861	0.649996	35.000360	−0.000352
44	0.49	1.740109	0.650000	35.000040	−0.000036
45	0.48	1.785931	0.650001	34.999930	0.000072
46	0.47	1.836006	0.649996	35.000420	−0.000417
47	0.46	1.891604	0.650003	34.999750	0.000250
48	0.45	1.957910	0.649995	35.000550	−0.000542

(continued)

8.2 Contents for Table 8.2 (Table No. 8.2-1 to 400)

Table 8.2 (continued)

Parameter values					
PROB. = 0.65	RHO = −0.4		BIGH = 0.9159		SMHL = 0.43
Pair No.	SMH	SMK	PROB. level	Risk level (%)	Risk error (%)
49	0.44	2.028793	0.650006	34.999410	0.000596
50	0.43	2.111412	0.649995	35.000540	−0.000536
No. of pairs			Mean COMPUT. risk errors	STD. ERR. errors	
50			−0.0000128559	0.0000659138	
Z-ratio				Status of significance	
0.19504				N.S.—not significant	
Parameter values					
PROB. = 0.6	RHO = −0.4		BIGH = 0.8153		SMHL = 0.33
Pair No.	SMH	SMK	PROB. level	Risk level (%)	Risk error (%)
Table No. 8.2-284					
1	0.82	0.810725	0.600006	39.999410	0.000584
2	0.81	0.820732	0.600006	39.999380	0.000614
3	0.80	0.830893	0.599996	40.000380	−0.000381
4	0.79	0.841313	0.600002	39.999840	0.000161
5	0.78	0.851905	0.599999	40.000110	−0.000107
6	0.77	0.862728	0.600001	39.999880	0.000119
7	0.76	0.873745	0.599998	40.000230	−0.000232
8	0.75	0.885016	0.600001	39.999860	0.000137
9	0.74	0.896504	0.600001	39.999860	0.000143
10	0.73	0.908223	0.599999	40.000070	−0.000072
11	0.72	0.920187	0.599996	40.000370	−0.000370
12	0.71	0.932408	0.599994	40.000610	−0.000608
13	0.70	0.944904	0.599993	40.000690	−0.000691
14	0.69	0.957690	0.599995	40.000460	−0.000465
15	0.68	0.970782	0.600002	39.999790	0.000209
16	0.67	0.984103	0.599993	40.000650	−0.000656
17	0.66	0.997767	0.599992	40.000780	−0.000781
18	0.65	1.011797	0.600000	40.000000	0.000000
19	0.64	1.026115	0.599998	40.000250	−0.000250
20	0.63	1.040844	0.600007	39.999310	0.000691
21	0.62	1.055908	0.600010	39.999020	0.000983
22	0.61	1.071336	0.600008	39.999180	0.000823
23	0.60	1.087153	0.600004	39.999560	0.000441
24	0.59	1.103391	0.600000	40.000000	0.000000
25	0.58	1.120082	0.599997	40.000280	−0.000286
26	0.57	1.137258	0.599997	40.000270	−0.000268
27	0.56	1.154958	0.600003	39.999720	0.000280
28	0.55	1.173121	0.599998	40.000230	−0.000226
29	0.54	1.191889	0.600001	39.999860	0.000143
30	0.53	1.211208	0.599999	40.000120	−0.000125
31	0.52	1.231225	0.600008	39.999190	0.000811
32	0.51	1.251798	0.599999	40.000080	−0.000077

(continued)

Table 8.2 (continued)

Parameter values					
PROB. = 0.6	RHO = −0.4		BIGH = 0.8153		SMHL = 0.33
Pair No.	SMH	SMK	PROB. level	Risk level (%)	Risk error (%)
33	0.50	1.273177	0.600005	39.999500	0.000501
34	0.49	1.295230	0.599997	40.000330	−0.000328
35	0.48	1.318219	0.600005	39.999500	0.000501
36	0.47	1.342019	0.600002	39.999820	0.000179
37	0.46	1.366807	0.600002	39.999810	0.000191
38	0.45	1.392571	0.599992	40.000770	−0.000769
39	0.44	1.419599	0.599998	40.000170	−0.000173
40	0.43	1.447799	0.599993	40.000740	−0.000739
41	0.42	1.477573	0.600009	39.999120	0.000882
42	0.41	1.508752	0.600008	39.999230	0.000763
43	0.40	1.541667	0.600007	39.999350	0.000656
44	0.39	1.576464	0.599999	40.000150	−0.000155
45	0.38	1.621121	0.599992	40.000800	−0.000805
46	0.37	1.660194	0.599992	40.000840	−0.000846
47	0.36	1.702285	0.599994	40.000610	−0.000608
48	0.35	1.748009	0.600005	39.999540	0.000465
49	0.34	1.797813	0.600003	39.999700	0.000304
50	0.33	1.852759	0.599999	40.000110	−0.000107
No. of pairs		Mean COMPUT. risk errors		STD. ERR. errors	
50		0.0000088823		0.0000696412	
Z-ratio				Status of significance	
0.12754				N.S.—not significant	
Parameter values					
PROB. = 0.55	RHO = −0.4		BIGH = 0.7196		SMHL = 0.23
Pair No.	SMH	SMK	PROB. level	Risk level (%)	Risk error (%)
Table No. 8.2-285					
1	0.72	0.719249	0.549999	45.000110	−0.000113
2	0.71	0.729330	0.549998	45.000170	−0.000167
3	0.70	0.739602	0.550001	44.999950	0.000054
4	0.69	0.750079	0.550007	44.999270	0.000733
5	0.68	0.760725	0.550008	44.999200	0.000799
6	0.67	0.771505	0.549993	45.000750	−0.000745
7	0.66	0.782580	0.550001	44.999930	0.000066
8	0.65	0.793820	0.549998	45.000250	−0.000244
9	0.64	0.805292	0.549998	45.000190	−0.000185
10	0.63	0.816963	0.549993	45.000680	−0.000679
11	0.62	0.828950	0.550009	44.999070	0.000930
12	0.61	0.841080	0.550002	44.999760	0.000238

(continued)

8.2 Contents for Table 8.2 (Table No. 8.2-1 to 400)

Table 8.2 (continued)

Parameter values					
PROB. = 0.55	RHO = −0.4		BIGH = 0.7196		SMHL = 0.23
Pair No.	SMH	SMK	PROB. level	Risk level (%)	Risk error (%)
13	0.60	0.853470	0.549999	45.000110	−0.000113
14	0.59	0.866144	0.550003	44.999730	0.000274
15	0.58	0.879031	0.549995	45.000540	−0.000536
16	0.57	0.892252	0.550001	44.999900	0.000101
17	0.56	0.905740	0.550004	44.999640	0.000364
18	0.55	0.919526	0.550007	44.999320	0.000685
19	0.54	0.933543	0.549994	45.000610	−0.000602
20	0.53	0.947925	0.549992	45.000820	−0.000817
21	0.52	0.962710	0.550005	44.999510	0.000489
22	0.51	0.977743	0.549997	45.000260	−0.000256
23	0.50	0.993167	0.549997	45.000350	−0.000352
24	0.49	1.009030	0.550008	44.999220	0.000781
25	0.48	1.025187	0.549999	45.000090	−0.000083
26	0.47	1.041792	0.549998	45.000240	−0.000238
27	0.46	1.058810	0.549992	45.000810	−0.000805
28	0.45	1.076306	0.549991	45.000940	−0.000942
29	0.44	1.094353	0.550002	44.999810	0.000191
30	0.43	1.112835	0.550000	45.000010	−0.000012
31	0.42	1.131840	0.549995	45.000470	−0.000465
32	0.41	1.151461	0.549999	45.000070	−0.000072
33	0.40	1.171708	0.550006	44.999360	0.000638
34	0.39	1.192499	0.549997	45.000290	−0.000286
35	0.38	1.214061	0.550002	44.999840	0.000161
36	0.37	1.236340	0.550003	44.999700	0.000298
37	0.36	1.259395	0.550003	44.999740	0.000262
38	0.35	1.283305	0.550003	44.999730	0.000274
39	0.34	1.308167	0.550008	44.999220	0.000781
40	0.33	1.334006	0.550010	44.999020	0.000978
41	0.32	1.372691	0.550006	44.999400	0.000602
42	0.31	1.399988	0.549999	45.000120	−0.000125
43	0.30	1.428617	0.549994	45.000620	−0.000614
44	0.29	1.458840	0.550006	44.999430	0.000566
45	0.28	1.490580	0.550005	44.999530	0.000471
46	0.27	1.524114	0.550001	44.999870	0.000137
47	0.26	1.559792	0.550008	44.999240	0.000757
48	0.25	1.597660	0.549999	45.000110	−0.000113

(continued)

Table 8.2 (continued)

Parameter values					
PROB. = 0.55	RHO = −0.4		BIGH = 0.7196		SMHL = 0.23
Pair No.	SMH	SMK	PROB. level	Risk level (%)	Risk error (%)
49	0.24	1.638261	0.549994	45.000560	−0.000554
50	0.23	1.682069	0.549996	45.000410	−0.000411
No. of pairs		Mean COMPUT. risk errors		STD. ERR. errors	
50		0.0000411389		0.0000707694	
Z-ratio				Status of significance	
0.58131				N.S.—not significant	
Parameter values					
PROB. = 0.5	RHO = −0.4		BIGH = 0.6269		SMHL = 0.14
Pair No.	SMH	SMK	PROB. level	Risk level (%)	Risk error (%)
Table No. 8.2-286					
1	0.63	0.623767	0.499997	50.000300	−0.000301
2	0.62	0.633779	0.499992	50.000800	−0.000799
3	0.61	0.643975	0.499991	50.000880	−0.000885
4	0.60	0.654371	0.499998	50.000230	−0.000226
5	0.59	0.664936	0.500002	49.999810	0.000191
6	0.58	0.675639	0.499995	50.000500	−0.000501
7	0.57	0.686550	0.499994	50.000640	−0.000638
8	0.56	0.697691	0.500002	49.999820	0.000185
9	0.55	0.708986	0.499999	50.000070	−0.000072
10	0.54	0.720460	0.499991	50.000910	−0.000912
11	0.53	0.732239	0.500006	49.999370	0.000638
12	0.52	0.744155	0.500002	49.999780	0.000226
13	0.51	0.756337	0.500009	49.999090	0.000906
14	0.50	0.768722	0.500009	49.999070	0.000930
15	0.49	0.781296	0.499999	50.000160	−0.000152
16	0.48	0.794148	0.499996	50.000420	−0.000417
17	0.47	0.807272	0.499998	50.000240	−0.000238
18	0.46	0.820664	0.500001	49.999900	0.000107
19	0.45	0.834279	0.499993	50.000660	−0.000662
20	0.44	0.848268	0.500007	49.999300	0.000697
21	0.43	0.862497	0.500009	49.999120	0.000882
22	0.42	0.876933	0.499990	50.001000	−0.000998
23	0.41	0.891846	0.500006	49.999380	0.000620
24	0.40	0.907020	0.500008	49.999180	0.000823
25	0.39	0.922447	0.499992	50.000840	−0.000840
26	0.38	0.938321	0.499994	50.000590	−0.000590
27	0.37	0.954554	0.499994	50.000610	−0.000608
28	0.36	0.971168	0.499992	50.000790	−0.000790
29	0.35	0.988199	0.499992	50.000780	−0.000775
30	0.34	1.005697	0.500000	50.000010	−0.000006

(continued)

8.2 Contents for Table 8.2 (Table No. 8.2-1 to 400)

Table 8.2 (continued)

Parameter values					
PROB. = 0.5	RHO = −0.4		BIGH = 0.6269		SMHL = 0.14
Pair No.	SMH	SMK	PROB. level	Risk level (%)	Risk error (%)
31	0.33	1.023634	0.500005	49.999480	0.000525
32	0.32	1.042002	0.500003	49.999720	0.000280
33	0.31	1.060916	0.500009	49.999110	0.000894
34	0.30	1.080323	0.500009	49.999120	0.000882
35	0.29	1.100203	0.499994	50.000590	−0.000587
36	0.28	1.120770	0.499996	50.000370	−0.000370
37	0.27	1.159180	0.500000	49.999970	0.000036
38	0.26	1.180007	0.499999	50.000160	−0.000152
39	0.25	1.201602	0.500003	49.999740	0.000262
40	0.24	1.223939	0.500000	50.000010	−0.000012
41	0.23	1.247185	0.500007	49.999260	0.000739
42	0.22	1.271241	0.500000	50.000010	−0.000015
43	0.21	1.296345	0.500002	49.999820	0.000179
44	0.20	1.322437	0.499993	50.000750	−0.000751
45	0.19	1.349882	0.500008	49.999180	0.000823
46	0.18	1.378467	0.500005	49.999480	0.000525
47	0.17	1.408462	0.500003	49.999700	0.000298
48	0.16	1.440057	0.500005	49.999460	0.000542
49	0.15	1.473340	0.500002	49.999790	0.000209
50	0.14	1.508531	0.499994	50.000560	−0.000560
No. of pairs		Mean COMPUT. risk errors		STD. ERR. errors	
50		−0.0000089991		0.0000823227	
Z-ratio				Status of significance	
0.10932				N.S.—not significant	
Parameter values					
PROB. = 0.99	RHO = −0.5		BIGH = 2.5761		SMHL = 2.34
Pair No.	SMH	SMK	PROB. level	Risk level (%)	Risk error (%)
Table No. 8.2-287					
1	2.58	2.572206	0.990007	0.999272	0.000727
2	2.57	2.582215	0.990007	0.999337	0.000662
3	2.56	2.592301	0.990000	1.000053	−0.000054
4	2.55	2.604030	0.990007	0.999302	0.000697
5	2.54	2.615838	0.990007	0.999314	0.000685
6	2.53	2.627728	0.989999	1.000100	−0.000101
7	2.52	2.641262	0.990002	0.999767	0.000232
8	2.51	2.654879	0.989996	1.000363	−0.000364
9	2.50	2.670142	0.989999	1.000077	−0.000077
10	2.49	2.687053	0.990008	0.999159	0.000840
11	2.48	2.704049	0.990006	0.999421	0.000578
12	2.47	2.721133	0.989991	1.000863	−0.000864
13	2.46	2.741429	0.989994	1.000559	−0.000560
14	2.45	2.763378	0.989997	1.000321	−0.000322

(continued)

Table 8.2 (continued)

Parameter values					
PROB. = 0.99	RHO = −0.5		BIGH = 2.5761		SMHL = 2.34
Pair No.	SMH	SMK	PROB. level	Risk level (%)	Risk error (%)
15	2.44	2.788542	0.990009	0.999093	0.000906
16	2.43	2.812234	0.989991	1.000935	−0.000936
17	2.42	2.842269	0.990000	1.000011	−0.000012
18	2.41	2.875523	0.990007	0.999343	0.000656
19	2.40	2.911996	0.990007	0.999332	0.000668
20	2.39	2.951691	0.989996	1.000452	−0.000453
21	2.38	3.000858	0.989997	1.000255	−0.000256
22	2.37	3.062623	0.990009	0.999141	0.000858
23	2.36	3.136988	0.990009	0.999129	0.000870
24	2.35	3.236453	0.990007	0.999254	0.000745
25	2.34	3.386022	0.990002	0.999796	0.000203
No. of pairs		Mean COMPUT. risk errors		STD. ERR. errors	
25		0.0002049483		0.0001121421	
Z-ratio				Status of significance	
1.82758				N.S.—not significant	
Parameter values					
PROB. = 0.98	RHO = −0.5		BIGH = 2.3264		SMHL = 2.06
Pair No.	SMH	SMK	PROB. level	Risk level (%)	Risk error (%)
Table No. 8.2-288					
1	2.33	2.322806	0.980002	1.999772	0.000226
2	2.32	2.332818	0.980001	1.999909	0.000089
3	2.31	2.343698	0.980009	1.999074	0.000924
4	2.30	2.354666	0.980006	1.999366	0.000632
5	2.29	2.365722	0.979992	2.000791	−0.000793
6	2.28	2.378431	0.980003	1.999700	0.000298
7	2.27	2.391232	0.980000	1.999968	0.000030
8	2.26	2.404907	0.980001	1.999891	0.000107
9	2.25	2.419457	0.980004	1.999623	0.000376
10	2.24	2.434883	0.980006	1.999372	0.000626
11	2.23	2.450405	0.979992	2.000845	−0.000846
12	2.22	2.468368	0.980004	1.999581	0.000417
13	2.21	2.486431	0.979996	2.000398	−0.000399
14	2.20	2.506937	0.980008	1.999253	0.000745
15	2.19	2.527546	0.979995	2.000517	−0.000519
16	2.18	2.551382	0.980006	1.999354	0.000644
17	2.17	2.576104	0.980001	1.999945	0.000054
18	2.16	2.604056	0.980007	1.999277	0.000721
19	2.15	2.634461	0.980009	1.999116	0.000882
20	2.14	2.666536	0.979991	2.000928	−0.000930
21	2.13	2.704972	0.979999	2.000129	−0.000131
22	2.12	2.748207	0.980001	1.999933	0.000066
23	2.11	2.797806	0.979998	2.000189	−0.000191

(continued)

8.2 Contents for Table 8.2 (Table No. 8.2-1 to 400)

Table 8.2 (continued)

Parameter values					
PROB. = 0.98	RHO = −0.5		BIGH = 2.3264		SMHL = 2.06
Pair No.	SMH	SMK	PROB. level	Risk level (%)	Risk error (%)
24	2.10	2.858458	0.980007	1.999307	0.000691
25	2.09	2.930164	0.979997	2.000308	−0.000310
26	2.08	3.026989	0.980002	1.999849	0.000149
27	2.07	3.164558	0.979996	2.000415	−0.000417
28	2.06	3.427250	0.979991	2.000904	−0.000906
No. of pairs		Mean COMPUT. risk errors		STD. ERR. errors	
28		0.0000770750		0.0001023453	
Z-ratio				Status of significance	
0.75309				N.S.—not significant	
Parameter values					
PROB. = 0.97	RHO = −0.5		BIGH = 2.1702		SMHL = 1.89
Pair No.	SMH	SMK	PROB. level	Risk level (%)	Risk error (%)
Table No. 8.2-289					
1	2.18	2.160444	0.970003	2.999723	0.000274
2	2.17	2.170400	0.970009	2.999109	0.000888
3	2.16	2.180448	0.970002	2.999789	0.000209
4	2.15	2.190981	0.969997	3.000349	−0.000352
5	2.14	2.202389	0.970004	2.999568	0.000429
6	2.13	2.213893	0.969997	3.000319	−0.000322
7	2.12	2.226276	0.970000	3.000033	−0.000036
8	2.11	2.239149	0.969998	3.000170	−0.000173
9	2.10	2.252903	0.970004	2.999640	0.000358
10	2.09	2.267150	0.970001	2.999884	0.000113
11	2.08	2.282280	0.970001	2.999860	0.000137
12	2.07	2.297907	0.969991	3.000915	−0.000918
13	2.06	2.315201	0.970000	2.999962	0.000036
14	2.05	2.333385	0.970004	2.999574	0.000423
15	2.04	2.352460	0.970001	2.999950	0.000048
16	2.03	2.373208	0.970005	2.999497	0.000501
17	2.02	2.394850	0.969995	3.000480	−0.000483
18	2.01	2.418949	0.970002	2.999795	0.000203
19	2.00	2.444728	0.970002	2.999795	0.000203
20	1.99	2.472967	0.970005	2.999485	0.000513
21	1.98	2.503670	0.970003	2.999711	0.000286
22	1.97	2.537620	0.970001	2.999938	0.000060
23	1.96	2.575599	0.969999	3.000105	−0.000107
24	1.95	2.618390	0.969995	3.000534	−0.000536
25	1.94	2.668340	0.969999	3.000116	−0.000119

(continued)

Table 8.2 (continued)

Parameter values					
PROB. = 0.97	RHO = −0.5		BIGH = 2.1702		SMHL = 1.89
Pair No.	SMH	SMK	PROB. level	Risk level (%)	Risk error (%)
26	1.93	2.727794	0.970009	2.999151	0.000846
27	1.92	2.798316	0.970002	2.999759	0.000238
28	1.91	2.887721	0.969993	3.000730	−0.000733
29	1.90	3.016325	0.970003	2.999687	0.000310
30	1.89	3.229440	0.969997	3.000283	−0.000286
No. of pairs		Mean COMPUT. risk errors		STD. ERR. errors	
30		0.0000647960		0.0000742040	
Z-ratio				Status of significance	
0.87321				N.S.—not significant	
Parameter values					
PROB. = 0.96	RHO = −0.5		BIGH = 2.0538		SMHL = 1.76
Pair No.	SMH	SMK	PROB. level	Risk level (%)	Risk error (%)
Table No. 8.2-290					
1	2.06	2.047619	0.960004	3.999591	0.000411
2	2.05	2.057607	0.960006	3.999412	0.000590
3	2.04	2.067694	0.959992	4.000759	−0.000757
4	2.03	2.078660	0.959999	4.000062	−0.000060
5	2.02	2.090119	0.960007	3.999329	0.000674
6	2.01	2.101680	0.959996	4.000431	−0.000429
7	2.00	2.114125	0.960000	4.000050	−0.000048
8	1.99	2.127068	0.959999	4.000068	−0.000066
9	1.98	2.140507	0.959993	4.000688	−0.000685
10	1.97	2.154837	0.959995	4.000473	−0.000471
11	1.96	2.170058	0.960003	3.999728	0.000274
12	1.95	2.185782	0.959998	4.000205	−0.000203
13	1.94	2.202401	0.959993	4.000682	−0.000679
14	1.93	2.220307	0.959999	4.000068	−0.000066
15	1.92	2.239112	0.959998	4.000169	−0.000167
16	1.91	2.259208	0.960000	4.000044	−0.000042
17	1.90	2.280597	0.959999	4.000134	−0.000131
18	1.89	2.303671	0.960002	3.999835	0.000167
19	1.88	2.328042	0.959992	4.000777	−0.000775
20	1.87	2.354884	0.959995	4.000509	−0.000507
21	1.86	2.384119	0.960000	3.999996	0.000006
22	1.85	2.415989	0.959998	4.000217	−0.000215
23	1.84	2.451818	0.960010	3.999031	0.000972
24	1.83	2.490907	0.960005	3.999525	0.000477
25	1.82	2.534822	0.959996	4.000437	−0.000435
26	1.81	2.586689	0.960007	3.999269	0.000733
27	1.80	2.646510	0.960004	3.999651	0.000352
28	1.79	2.718977	0.959999	4.000139	−0.000137
29	1.78	2.811903	0.959999	4.000086	−0.000083

(continued)

8.2 Contents for Table 8.2 (Table No. 8.2-1 to 400)

Table 8.2 (continued)

Parameter values					
PROB. = 0.96	RHO = −0.5		BIGH = 2.0538		SMHL = 1.76
Pair No.	SMH	SMK	PROB. level	Risk level (%)	Risk error (%)
30	1.77	2.939354	0.959991	4.000914	−0.000912
31	1.76	3.159144	0.960001	3.999913	0.000089
No. of pairs		Mean COMPUT. risk errors		STD. ERR. errors	
31		−0.0000663102		0.0000824060	
Z-ratio				Status of significance	
0.80468				N.S.—not significant	
Parameter values					
PROB. = 0.95	RHO = −0.5		BIGH = 1.9599		SMHL = 1.65
Pair No.	SMH	SMK	PROB. level	Risk level (%)	Risk error (%)
Table No. 8.2-291					
1	1.96	1.959800	0.949997	5.000294	−0.000292
2	1.95	1.970046	0.950000	5.000007	−0.000006
3	1.94	1.980785	0.950007	4.999304	0.000697
4	1.93	1.991826	0.950006	4.999399	0.000602
5	1.92	2.003363	0.950007	4.999352	0.000650
6	1.91	2.015400	0.950007	4.999352	0.000650
7	1.90	2.027938	0.950005	4.999542	0.000459
8	1.89	2.040979	0.949999	5.000115	−0.000113
9	1.88	2.054915	0.950007	4.999334	0.000668
10	1.87	2.069356	0.950006	4.999405	0.000596
11	1.86	2.084306	0.949995	5.000478	−0.000477
12	1.85	2.100547	0.950007	4.999322	0.000679
13	1.84	2.117300	0.950003	4.999751	0.000250
14	1.83	2.134958	0.949997	5.000353	−0.000352
15	1.82	2.153913	0.950001	4.999948	0.000054
16	1.81	2.173777	0.949994	5.000556	−0.000554
17	1.80	2.195332	0.950003	4.999704	0.000298
18	1.79	2.217801	0.949992	5.000824	−0.000823
19	1.78	2.242357	0.949996	5.000443	−0.000441
20	1.77	2.268611	0.949993	5.000699	−0.000697
21	1.76	2.297348	0.949999	5.000091	−0.000089
22	1.75	2.328569	0.950002	4.999799	0.000203
23	1.74	2.362278	0.949991	5.000901	−0.000900
24	1.73	2.400820	0.950007	4.999268	0.000733
25	1.72	2.442635	0.949995	5.000502	−0.000501
26	1.71	2.490851	0.949996	5.000359	−0.000358
27	1.70	2.547033	0.950003	4.999674	0.000328
28	1.69	2.613528	0.950006	4.999423	0.000578
29	1.68	2.694245	0.949994	5.000592	−0.000590
30	1.67	2.801686	0.949998	5.000174	−0.000173

(continued)

Table 8.2 (continued)

Parameter values					
PROB. = 0.95	RHO = −0.5		BIGH = 1.9599		SMHL = 1.65
Pair No.	SMH	SMK	PROB. level	Risk level (%)	Risk error (%)
31	1.66	2.957730	0.949992	5.000842	−0.000840
32	1.65	3.278004	0.949995	5.000502	−0.000501
No. of pairs		Mean COMPUT. risk errors		STD. ERR. errors	
32		−0.0000079473		0.0000922388	
Z-ratio				Status of significance	
0.08616				N.S.—not significant	
Parameter values					
PROB. = 0.94	RHO = −0.5		BIGH = 1.8807		SMHL = 1.56
Pair No.	SMH	SMK	PROB. level	Risk level (%)	Risk error (%)
Table No. 8.2-292					
1	1.89	1.871641	0.940002	5.999804	0.000197
2	1.88	1.881400	0.939996	6.000358	−0.000358
3	1.87	1.891656	0.939999	6.000126	−0.000125
4	1.86	1.902412	0.940008	5.999232	0.000769
5	1.85	1.913472	0.940008	5.999184	0.000817
6	1.84	1.924839	0.940000	5.999983	0.000018
7	1.83	1.936711	0.939994	6.000644	−0.000644
8	1.82	1.949284	0.939998	6.000161	−0.000161
9	1.81	1.962364	0.940001	5.999935	0.000066
10	1.80	1.975955	0.939998	6.000203	−0.000203
11	1.79	1.990254	0.940000	6.000013	−0.000012
12	1.78	2.005065	0.939992	6.000799	−0.000799
13	1.77	2.020979	0.940003	5.999661	0.000340
14	1.76	2.037412	0.939999	6.000054	−0.000054
15	1.75	2.054755	0.939997	6.000310	−0.000310
16	1.74	2.073011	0.939992	6.000805	−0.000805
17	1.73	2.092574	0.939998	6.000185	−0.000185
18	1.72	2.113445	0.940009	5.999094	0.000906
19	1.71	2.135237	0.940002	5.999774	0.000226
20	1.70	2.158732	0.940005	5.999506	0.000495
21	1.69	2.183543	0.939993	6.000656	−0.000656
22	1.68	2.210845	0.940003	5.999703	0.000298
23	1.67	2.239858	0.939995	6.000519	−0.000519
24	1.66	2.272148	0.940008	5.999184	0.000817
25	1.65	2.306937	0.940003	5.999697	0.000304
26	1.64	2.345789	0.940007	5.999285	0.000715
27	1.63	2.388708	0.939998	6.000203	−0.000203
28	1.62	2.438040	0.940003	5.999744	0.000256
29	1.61	2.494570	0.949997	6.000269	−0.000268
30	1.60	2.562207	0.940002	5.999851	0.000149
31	1.59	2.644861	0.939997	6.000304	−0.000304
32	1.58	2.754252	0.940005	5.999482	0.000519

(continued)

8.2 Contents for Table 8.2 (Table No. 8.2-1 to 400)

Table 8.2 (continued)

Parameter values					
PROB. = 0.94	RHO = −0.5		BIGH = 1.8807		SMHL = 1.56
Pair No.	SMH	SMK	PROB. level	Risk level (%)	Risk error (%)
33	1.57	2.912261	0.939997	6.000334	−0.000334
34	1.56	3.236077	0.940002	5.999774	0.000226
No. of pairs		Mean COMPUT. risk errors		STD. ERR. errors	
34		0.0000337192		0.0000784338	
Z-ratio				Status of significance	
0.42991				N.S.—not significant	
Parameter values					
PROB. = 0.93	RHO = −0.5		BIGH = 1.8118		SMHL = 1.48
Pair No.	SMH	SMK	PROB. level	Risk level (%)	Risk error (%)
Table No. 8.2-293					
1	1.82	1.803637	0.929992	7.000768	−0.000769
2	1.81	1.813602	0.929999	7.000119	−0.000119
3	1.80	1.823873	0.930000	7.000005	−0.000006
4	1.79	1.834647	0.930010	6.999016	0.000983
5	1.78	1.845535	0.929998	7.000190	−0.000191
6	1.77	1.856931	0.929992	7.000816	−0.000817
7	1.76	1.869031	0.930003	6.999749	0.000250
8	1.75	1.881446	0.930001	6.999928	0.000072
9	1.74	1.894375	0.929998	7.000184	−0.000185
10	1.73	1.908015	0.930005	6.999469	0.000530
11	1.72	1.922171	0.930007	6.999350	0.000650
12	1.71	1.936848	0.930000	7.000041	−0.000042
13	1.70	1.952437	0.930006	6.999451	0.000548
14	1.69	1.968550	0.929997	7.000262	−0.000262
15	1.68	1.985581	0.929995	7.000530	−0.000530
16	1.67	2.003531	0.929993	7.000697	−0.000697
17	1.66	2.022794	0.930008	6.999213	0.000787
18	1.65	2.042591	0.929994	7.000613	−0.000614
19	1.64	2.064097	0.930004	6.999570	0.000429
20	1.63	2.086533	0.929995	7.000542	−0.000542
21	1.62	2.110683	0.929994	7.000595	−0.000596
22	1.61	2.136550	0.929993	7.000703	−0.000703
23	1.60	2.164528	0.929997	7.000280	−0.000280
24	1.59	2.195009	0.930009	6.999135	0.000864
25	1.58	2.227607	0.930000	6.999982	0.000018
26	1.57	2.263496	0.929996	7.000405	−0.000405
27	1.56	2.303461	0.929999	7.000107	−0.000107
28	1.55	2.348287	0.930004	6.999618	0.000381
29	1.54	2.398758	0.929998	7.000184	−0.000185
30	1.53	2.458002	0.930009	6.999111	0.000888
31	1.52	2.526805	0.929991	7.000876	−0.000876
32	1.51	2.612981	0.929992	7.000816	−0.000817

(continued)

Table 8.2 (continued)

Parameter values					
PROB. = 0.93	RHO = −0.5		BIGH = 1.8118		SMHL = 1.48
Pair No.	SMH	SMK	PROB. level	Risk level (%)	Risk error (%)
33	1.50	2.727474	0.930002	6.999809	0.000191
34	1.49	2.896849	0.930001	6.999856	0.000143
35	1.48	3.267985	0.930003	6.999660	0.000340
No. of pairs		Mean COMPUT. risk errors		STD. ERR. errors	
35		−0.0000463592		0.0000892824	
Z-ratio				Status of significance	
0.51924				N.S.—not significant	
Parameter values					
PROB. = 0.92	RHO = −0.5		BIGH = 1.7505		SMHL = 1.41
Pair No.	SMH	SMK	PROB. level	Risk level (%)	Risk error (%)
Table No. 8.2-294					
1	1.76	1.741247	0.920002	7.999778	0.000221
2	1.75	1.751000	0.919994	8.000565	−0.000566
3	1.74	1.761259	0.920000	8.000051	−0.000054
4	1.73	1.771829	0.920000	8.000029	−0.000030
5	1.72	1.782713	0.919993	8.000671	−0.000674
6	1.71	1.794107	0.919995	8.000499	−0.000501
7	1.70	1.806016	0.920002	7.999760	0.000238
8	1.69	1.818244	0.919998	8.000195	−0.000197
9	1.68	1.830990	0.919996	8.000440	−0.000441
10	1.67	1.844255	0.919992	8.000767	−0.000769
11	1.66	1.858239	0.920000	8.000035	−0.000036
12	1.65	1.872746	0.920001	7.999933	0.000066
13	1.64	1.887976	0.920006	7.999373	0.000626
14	1.63	1.903736	0.920000	8.000011	−0.000012
15	1.62	1.920419	0.920004	7.999635	0.000364
16	1.61	1.937831	0.920001	7.999951	0.000048
17	1.60	1.956172	0.919999	8.000094	−0.000095
18	1.59	1.975639	0.920005	7.999545	0.000453
19	1.58	1.996039	0.920001	7.999927	0.000072
20	1.57	2.017767	0.920003	7.999730	0.000268
21	1.56	2.040825	0.920003	7.999725	0.000274
22	1.55	2.065217	0.919994	8.000613	−0.000614
23	1.54	2.091725	0.920003	7.999707	0.000292
24	1.53	2.119965	0.920001	7.999915	0.000083
25	1.52	2.150329	0.919992	8.000767	−0.000769
26	1.51	2.183992	0.920008	7.999206	0.000793
27	1.50	2.220177	0.919999	8.000070	−0.000072
28	1.49	2.260450	0.920000	8.000029	−0.000030
29	1.48	2.305595	0.920004	7.999605	0.000393
30	1.47	2.356398	0.919999	8.000136	−0.000137
31	1.46	2.415207	0.919997	8.000345	−0.000346

(continued)

8.2 Contents for Table 8.2 (Table No. 8.2-1 to 400)

Table 8.2 (continued)

Parameter values					
PROB. = 0.92	RHO = −0.5		BIGH = 1.7505		SMHL = 1.41
Pair No.	SMH	SMK	PROB. level	Risk level (%)	Risk error (%)
32	1.45	2.485150	0.919999	8.000070	−0.000072
33	1.44	2.571701	0.920009	7.999134	0.000864
34	1.43	2.683456	0.919999	8.000082	−0.000083
35	1.42	2.848546	0.919999	8.000106	−0.000107
36	1.41	3.185726	0.919994	8.000649	−0.000650
No. of pairs		Mean COMPUT. risk errors		STD. ERR. errors	
36		−0.0000323798		0.0000671089	
Z-ratio				Status of significance	
0.48250				N.S.—not significant	
Parameter values					
PROB. = 0.91	RHO = −0.5		BIGH = 1.6952		SMHL = 1.34
Pair No.	SMH	SMK	PROB. level	Risk level (%)	Risk error (%)
Table No. 8.2-295					
1	1.70	1.690414	0.910001	8.999931	0.000066
2	1.69	1.700416	0.910000	8.999968	0.000030
3	1.68	1.710733	0.909997	9.000338	−0.000340
4	1.67	1.721561	0.910006	8.999366	0.000632
5	1.66	1.732709	0.910009	8.999073	0.000924
6	1.65	1.744177	0.910004	8.999562	0.000435
7	1.64	1.756164	0.910007	8.999311	0.000685
8	1.63	1.768477	0.909998	9.000200	−0.000203
9	1.62	1.781313	0.909992	9.000814	−0.000817
10	1.61	1.794870	0.910002	8.999801	0.000197
11	1.60	1.808955	0.910009	8.999068	0.000930
12	1.59	1.823376	0.909996	9.000385	−0.000387
13	1.58	1.838721	0.910005	8.999538	0.000459
14	1.57	1.854603	0.910001	8.999891	0.000107
15	1.56	1.871219	0.909997	9.000325	−0.000328
16	1.55	1.888767	0.910001	8.999926	0.000072
17	1.54	1.907056	0.909995	9.000492	−0.000495
18	1.53	1.926479	0.910000	8.999997	0.000000
19	1.52	1.946844	0.909997	9.000296	−0.000298
20	1.51	1.968349	0.909992	9.000778	−0.000781
21	1.50	1.991388	0.910000	9.000004	−0.000006
22	1.49	2.015769	0.910000	8.999956	0.000042
23	1.48	2.041691	0.909995	9.000474	−0.000477
24	1.47	2.069743	0.910002	8.999759	0.000238
25	1.46	2.099930	0.910007	8.999311	0.000685
26	1.45	2.132254	0.909995	9.000456	−0.000459
27	1.44	2.167893	0.909999	9.000134	−0.000137
28	1.43	2.207238	0.910007	8.999336	0.000662
29	1.42	2.250297	0.909993	9.000712	−0.000715

(continued)

Table 8.2 (continued)

Parameter values						
PROB. = 0.91		RHO = −0.5		BIGH = 1.6952		SMHL = 1.34
Pair No.	SMH	SMK		PROB. level	Risk level (%)	Risk error (%)
30	1.41	2.299415		0.909999	9.000063	−0.000066
31	1.40	2.355769		0.910009	8.999086	0.000912
32	1.39	2.420536		0.909993	9.000701	−0.000703
33	1.38	2.499580		0.909994	9.000606	−0.000608
34	1.37	2.600717		0.910008	8.999241	0.000757
35	1.36	2.738016		0.909995	9.000492	−0.000495
36	1.35	2.972417		0.910010	8.999014	0.000983
37	1.34	3.744552		0.909991	9.000874	−0.000876
No. of pairs		Mean COMPUT. risk errors			STD. ERR. errors	
36		0.0000169148			0.0000918966	
Z-ratio					Status of significance	
0.18406					N.S.—not significant	
Parameter values						
PROB. = 0.9025		RHO = −0.5		BIGH = 1.6568		SMHL = 1.30
Pair No.	SMH	SMK		PROB. level	Risk level (%)	Risk error (%)
Table No. 8.2-296						
1	1.66	1.653606		0.902494	9.750634	−0.000632
2	1.65	1.663823		0.902509	9.749079	0.000924
3	1.64	1.674163		0.902503	9.749735	0.000268
4	1.63	1.684822		0.902492	9.750789	−0.000787
5	1.62	1.695998		0.902495	9.750509	−0.000507
6	1.61	1.707695		0.902508	9.749174	0.000829
7	1.60	1.719523		0.902494	9.750568	−0.000566
8	1.59	1.732071		0.902505	9.749526	0.000477
9	1.58	1.744950		0.902502	9.749812	0.000191
10	1.57	1.758360		0.902500	9.749955	0.000048
11	1.56	1.772302		0.902498	9.750158	−0.000155
12	1.55	1.786974		0.902508	9.749211	0.000793
13	1.54	1.801990		0.902495	9.750528	−0.000525
14	1.53	1.817937		0.902502	9.749788	0.000215
15	1.52	1.834428		0.902495	9.750486	−0.000483
16	1.51	1.851856		0.902500	9.750038	−0.000036
17	1.50	1.870225		0.902510	9.749049	0.000954
18	1.49	1.889148		0.902493	9.750676	−0.000674
19	1.48	1.909408		0.902498	9.750164	−0.000161
20	1.47	1.930619		0.902492	9.750825	−0.000823
21	1.46	1.953174		0.902491	9.750867	−0.000864
22	1.45	1.977274		0.902499	9.750091	−0.000089
23	1.44	2.002725		0.902494	9.750575	−0.000572
24	1.43	2.030118		0.902498	9.750164	−0.000161
25	1.42	2.059651		0.902507	9.749299	0.000703
26	1.41	2.091330		0.902505	9.749472	0.000530

(continued)

8.2 Contents for Table 8.2 (Table No. 8.2-1 to 400)

Table 8.2 (continued)

Parameter values					
PROB. = 0.9025	RHO = −0.5		BIGH = 1.6568		SMHL = 1.30
Pair No.	SMH	SMK	PROB. level	Risk level (%)	Risk error (%)
27	1.40	2.125548	0.902495	9.750515	−0.000513
28	1.39	2.163482	0.902503	9.749699	0.000304
29	1.38	2.205135	0.902501	9.749878	0.000125
30	1.37	2.251686	0.902500	9.750027	−0.000024
31	1.36	2.304309	0.902494	9.750581	−0.000578
32	1.35	2.365354	0.902495	9.750521	−0.000519
33	1.34	2.437559	0.902491	9.750920	−0.000918
34	1.33	2.527961	0.902509	9.749090	0.000912
35	1.32	2.645159	0.902502	9.749788	0.000215
36	1.31	2.820408	0.902503	9.749699	0.000304
37	1.30	3.205277	0.902502	9.749824	0.000179
No. of pairs		Mean COMPUT. risk errors		STD. ERR. errors	
37		−0.0000425075		0.0000899739	
Z-ratio				Status of significance	
0.47244				N.S.—not significant	
Parameter values					
PROB. = 0.9	RHO = −0.5		BIGH = 1.6447		SMHL = 1.29
Pair No.	SMH	SMK	PROB. level	Risk level (%)	Risk error (%)
Table No. 8.2-297					
1	1.65	1.639222	0.900005	9.999519	0.000483
2	1.64	1.649218	0.900006	9.999424	0.000578
3	1.63	1.659533	0.900005	9.999514	0.000489
4	1.62	1.670167	0.900000	9.999961	0.000042
5	1.61	1.681125	0.899991	10.000890	−0.000888
6	1.60	1.692602	0.899994	10.000600	−0.000596
7	1.59	1.704602	0.900007	9.999317	0.000685
8	1.58	1.716737	0.899991	10.000890	−0.000888
9	1.57	1.729595	0.899999	10.000150	−0.000149
10	1.56	1.742983	0.900008	9.999174	0.000829
11	1.55	1.756709	0.900001	9.999884	0.000119
12	1.54	1.770971	0.899992	10.000810	−0.000805
13	1.53	1.785968	0.899993	10.000680	−0.000679
14	1.52	1.801701	0.900001	9.999942	0.000060
15	1.51	1.817978	0.899996	10.000450	−0.000447
16	1.50	1.835195	0.900004	9.999627	0.000376
17	1.49	1.852962	0.899992	10.000830	−0.000829
18	1.48	1.871869	0.899997	10.000290	−0.000286
19	1.47	1.891725	0.899999	10.000070	−0.000072
20	1.46	1.912532	0.899993	10.000690	−0.000685
21	1.45	1.934684	0.899996	10.000380	−0.000376
22	1.44	1.958186	0.900000	10.000010	−0.000006
23	1.43	1.983041	0.899995	10.000460	−0.000453

(continued)

Table 8.2 (continued)

Parameter values					
PROB. = 0.9	RHO = −0.5		BIGH = 1.6447		SMHL = 1.29
Pair No.	SMH	SMK	PROB. level	Risk level (%)	Risk error (%)
24	1.42	2.009644	0.899995	10.000540	−0.000536
25	1.41	2.038389	0.900005	9.999502	0.000501
26	1.40	2.068888	0.899993	10.000710	−0.000709
27	1.39	2.102320	0.899998	10.000190	−0.000191
28	1.38	2.138688	0.899999	10.000130	−0.000131
29	1.37	2.178777	0.900002	9.999800	0.000203
30	1.36	2.223373	0.900007	9.999347	0.000656
31	1.35	2.273263	0.900001	9.999919	0.000083
32	1.34	2.330403	0.899996	10.000380	−0.000381
33	1.33	2.397926	0.900005	9.999460	0.000542
34	1.32	2.478958	0.900000	9.999979	0.000024
35	1.31	2.582101	0.899996	10.000430	−0.000429
36	1.30	2.726110	0.899996	10.000390	−0.000387
37	1.29	2.975053	0.900004	9.999572	0.000429
No. of pairs		Mean COMPUT. risk errors		STD. ERR. errors	
37		−0.0001007005		0.0000805517	
Z-ratio				Status of significance	
1.25014				N.S.—not significant	
Parameter values					
PROB. = 0.85	RHO = −0.5		BIGH = 1.4386		SMHL = 1.04
Pair No.	SMH	SMK	PROB. level	Risk level (%)	Risk error (%)
Table No. 8.2-298					
1	1.44	1.437201	0.850000	15.000020	−0.000024
2	1.43	1.447252	0.849992	15.000820	−0.000823
3	1.42	1.457737	0.850003	14.999720	0.000280
4	1.41	1.468366	0.849990	15.000980	−0.000978
5	1.40	1.479436	0.849993	15.000700	−0.000703
6	1.39	1.490852	0.849996	15.000450	−0.000447
7	1.38	1.502618	0.849996	15.000380	−0.000381
8	1.37	1.514736	0.849993	15.000680	−0.000685
9	1.36	1.527407	0.850009	14.999130	0.000864
10	1.35	1.540241	0.849992	15.000810	−0.000811
11	1.34	1.553732	0.850001	14.999940	0.000060
12	1.33	1.567591	0.849996	15.000370	−0.000370
13	1.32	1.582016	0.849999	15.000080	−0.000083
14	1.31	1.597012	0.850006	14.999440	0.000554
15	1.30	1.612484	0.850001	14.999870	0.000131
16	1.29	1.628536	0.849994	15.000620	−0.000620
17	1.28	1.645367	0.849999	15.000080	−0.000083
18	1.27	1.662785	0.849993	15.000740	−0.000739
19	1.26	1.681187	0.850008	14.999220	0.000781
20	1.25	1.700187	0.850001	14.999900	0.000095

(continued)

8.2 Contents for Table 8.2 (Table No. 8.2-1 to 400)

Table 8.2 (continued)

Parameter values					
PROB. = 0.85	RHO = −0.5		BIGH = 1.4386		SMHL = 1.04
Pair No.	SMH	SMK	PROB. level	Risk level (%)	Risk error (%)
21	1.24	1.720179	0.850003	14.999670	0.000328
22	1.23	1.741171	0.850008	14.999250	0.000745
23	1.22	1.763165	0.850006	14.999440	0.000560
24	1.21	1.786169	0.849990	15.000960	−0.000966
25	1.20	1.810774	0.849999	15.000130	−0.000131
26	1.19	1.836790	0.850003	14.999690	0.000304
27	1.18	1.864419	0.850005	14.999500	0.000501
28	1.17	1.893863	0.850004	14.999580	0.000417
29	1.16	1.925518	0.850009	14.999100	0.000894
30	1.15	1.959391	0.849999	15.000060	−0.000060
31	1.14	1.996271	0.849999	15.000130	−0.000131
32	1.13	2.036555	0.849998	15.000240	−0.000238
33	1.12	2.081033	0.850000	15.000050	−0.000048
34	1.11	2.130492	0.849996	15.000420	−0.000417
35	1.10	2.186895	0.850009	14.999090	0.000906
36	1.09	2.251031	0.849991	15.000860	−0.000864
37	1.08	2.327986	0.850002	14.999800	0.000197
38	1.07	2.422458	0.850003	14.999750	0.000250
39	1.06	2.546173	0.849994	15.000630	−0.000632
40	1.05	2.731955	0.849996	15.000420	−0.000417
41	1.04	3.139969	0.849996	15.000390	−0.000393
No. of pairs		Mean COMPUT. risk errors		STD. ERR. errors	
41		−0.0000756411		0.0000838474	
Z-ratio				Status of significance	
0.90213				N.S.—not significant	
Parameter values					
PROB. = 0.8	RHO = −0.5		BIGH = 1.2794		SMHL = 0.85
Pair No.	SMH	SMK	PROB. level	Risk level (%)	Risk error (%)
Table No. 8.2-299					
1	1.28	1.278800	0.799992	20.000770	−0.000769
2	1.27	1.288967	0.800001	19.999870	0.000125
3	1.26	1.299294	0.799991	20.000890	−0.000894
4	1.25	1.309980	0.799994	20.000570	−0.000572
5	1.24	1.321028	0.800009	19.999110	0.000888
6	1.23	1.332249	0.800002	19.999830	0.000167
7	1.22	1.343840	0.800003	19.999730	0.000268
8	1.21	1.355710	0.799995	20.000500	−0.000501
9	1.20	1.367960	0.799992	20.000790	−0.000793
10	1.19	1.380594	0.799992	20.000790	−0.000787
11	1.18	1.393618	0.799993	20.000710	−0.000709
12	1.17	1.407037	0.799992	20.000800	−0.000805
13	1.16	1.420953	0.800001	19.999890	0.000107

(continued)

Table 8.2 (continued)

Parameter values					
PROB. = 0.8	RHO = −0.5		BIGH = 1.2794		SMHL = 0.85
Pair No.	SMH	SMK	PROB. level	Risk level (%)	Risk error (%)
14	1.15	1.435274	0.800004	19.999650	0.000346
15	1.14	1.450006	0.799997	20.000340	−0.000340
16	1.13	1.465349	0.800004	19.999560	0.000441
17	1.12	1.481212	0.800010	19.999040	0.000960
18	1.11	1.497504	0.799996	20.000390	−0.000387
19	1.10	1.514523	0.799998	20.000180	−0.000185
20	1.09	1.532179	0.799998	20.000180	−0.000185
21	1.08	1.550576	0.800003	19.999700	0.000298
22	1.07	1.569623	0.799996	20.000420	−0.000417
23	1.06	1.589524	0.799993	20.000670	−0.000674
24	1.05	1.610481	0.800010	19.999000	0.000995
25	1.04	1.632111	0.799995	20.000460	−0.000459
26	1.03	1.655009	0.800004	19.999620	0.000381
27	1.02	1.678988	0.800005	19.999510	0.000489
28	1.01	1.704251	0.800007	19.999330	0.000674
29	1.00	1.730809	0.799998	20.000230	−0.000226
30	0.99	1.759062	0.799999	20.000130	−0.000131
31	0.98	1.789019	0.799993	20.000670	−0.000668
32	0.97	1.821082	0.799994	20.000600	−0.000602
33	0.96	1.855457	0.799993	20.000710	−0.000709
34	0.95	1.892546	0.799993	20.000750	−0.000751
35	0.94	1.932946	0.799999	20.000140	−0.000137
36	0.93	1.977256	0.800008	19.999210	0.000793
37	0.92	2.026075	0.800006	19.999360	0.000644
38	0.91	2.080783	0.800005	19.999520	0.000483
39	0.90	2.142956	0.799997	20.000270	−0.000274
40	0.89	2.215735	0.800001	19.999870	0.000131
41	0.88	2.303041	0.799996	20.000360	−0.000364
42	0.87	2.414265	0.800008	19.999180	0.000823
43	0.86	2.567392	0.800005	19.999540	0.000465
44	0.85	2.828846	0.800009	19.999060	0.000942
No. of pairs		Mean COMPUT. risk errors		STD. ERR. errors	
44		−0.0000426504		0.0000863631	
Z-ratio				Status of significance	
0.49385				N.S.—not significant	
Parameter values					
PROB. = 0.75	RHO = −0.5		BIGH = 1.1463		SMHL = 0.68
Pair No.	SMH	SMK	PROB. level	Risk level (%)	Risk error (%)
Table No. 8.2-300					
1	1.15	1.142612	0.750002	24.999800	0.000203
2	1.14	1.152635	0.750001	24.999950	0.000054
3	1.13	1.162933	0.750006	24.999450	0.000548

(continued)

Table 8.2 (continued)

Parameter values					
PROB. = 0.75	RHO = −0.5		BIGH = 1.1463		SMHL = 0.68
Pair No.	SMH	SMK	PROB. level	Risk level (%)	Risk error (%)
4	1.12	1.173413	0.749997	25.000280	−0.000280
5	1.11	1.184178	0.749994	25.000600	−0.000602
6	1.10	1.195232	0.749995	25.000540	−0.000536
7	1.09	1.206582	0.749998	25.000200	−0.000203
8	1.08	1.218232	0.750003	24.999720	0.000280
9	1.07	1.230189	0.750008	24.999200	0.000799
10	1.06	1.242360	0.749994	25.000560	−0.000560
11	1.05	1.254948	0.749996	25.000380	−0.000381
12	1.04	1.267860	0.749994	25.000600	−0.000596
13	1.03	1.281200	0.750003	24.999670	0.000328
14	1.02	1.294879	0.750005	24.999510	0.000495
15	1.01	1.308904	0.749997	25.000260	−0.000256
16	1.00	1.323379	0.749995	25.000500	−0.000501
17	0.99	1.338311	0.749995	25.000510	−0.000513
18	0.98	1.353711	0.749995	25.000540	−0.000542
19	0.97	1.369682	0.750006	24.999430	0.000566
20	0.96	1.386039	0.749996	25.000430	−0.000429
21	0.95	1.403083	0.750005	24.999510	0.000495
22	0.94	1.420630	0.750001	24.999940	0.000060
23	0.93	1.438883	0.750007	24.999340	0.000662
24	0.92	1.457757	0.750004	24.999570	0.000429
25	0.91	1.477358	0.750002	24.999800	0.000197
26	0.90	1.497699	0.749994	25.000590	−0.000590
27	0.89	1.518986	0.749999	25.000150	−0.000149
28	0.88	1.541232	0.750007	24.999270	0.000727
29	0.87	1.564352	0.750001	24.999940	0.000066
30	0.86	1.588653	0.750004	24.999630	0.000370
31	0.85	1.614050	0.749995	25.000540	−0.000536
32	0.84	1.640852	0.749993	25.000660	−0.000662
33	0.83	1.669269	0.750005	24.999540	0.000465
34	0.82	1.699123	0.749994	25.000640	−0.000638
35	0.81	1.731015	0.749997	25.000260	−0.000262
36	0.80	1.765160	0.750010	24.999050	0.000948
37	0.79	1.801576	0.750004	24.999570	0.000429
38	0.78	1.840869	0.749999	25.000110	−0.000107
39	0.77	1.883646	0.749999	25.000130	−0.000131
40	0.76	1.930514	0.749997	25.000340	−0.000340
41	0.75	1.982473	0.749996	25.000390	−0.000393
42	0.74	2.040914	0.749999	25.000070	−0.000072
43	0.73	2.107816	0.750005	24.999550	0.000447
44	0.72	2.185941	0.749997	25.000260	−0.000262
45	0.71	2.281177	0.749999	25.000110	−0.000107

(continued)

Table 8.2 (continued)

Parameter values					
PROB. = 0.75	RHO = −0.5		BIGH = 1.1463		SMHL = 0.68
Pair No.	SMH	SMK	PROB. level	Risk level (%)	Risk error (%)
46	0.70	2.403709	0.749997	25.000350	−0.000352
47	0.69	2.579352	0.749990	25.000980	−0.000978
48	0.68	2.919856	0.750004	24.999610	0.000387
No. of pairs		Mean COMPUT. risk errors		STD. ERR. errors	
48		−0.0000413583		0.0000669867	
Z-ratio				Status of significance	
0.61741				N.S.—not significant	
Parameter values					
PROB. = 0.7	RHO = −0.5		BIGH = 1.0296		SMHL = 0.54
Pair No.	SMH	SMK	PROB. level	Risk level (%)	Risk error (%)
Table No. 8.2-301					
1	1.03	1.029103	0.700006	29.999410	0.000596
2	1.02	1.039193	0.700004	29.999600	0.000399
3	1.01	1.049483	0.699997	30.000310	−0.000304
4	1.00	1.060076	0.700006	29.999390	0.000614
5	0.99	1.070882	0.700009	29.999060	0.000948
6	0.98	1.081906	0.700007	29.999320	0.000685
7	0.97	1.093155	0.699998	30.000230	−0.000232
8	0.96	1.104734	0.700002	29.999760	0.000244
9	0.95	1.116553	0.700000	30.000040	−0.000036
10	0.94	1.128717	0.700009	29.999140	0.000864
11	0.93	1.141136	0.700009	29.999110	0.000894
12	0.92	1.153819	0.700000	29.999980	0.000024
13	0.91	1.166872	0.700001	29.999950	0.000054
14	0.90	1.180206	0.699990	30.000990	−0.000983
15	0.89	1.194027	0.700005	29.999520	0.000483
16	0.88	1.208148	0.700007	29.999340	0.000662
17	0.87	1.222580	0.699994	30.000600	−0.000596
18	0.86	1.237530	0.700002	29.999840	0.000161
19	0.85	1.252812	0.699992	30.000770	−0.000763
20	0.84	1.268636	0.699999	30.000150	−0.000149
21	0.83	1.284915	0.700001	29.999890	0.000107
22	0.82	1.301663	0.699998	30.000230	−0.000232
23	0.81	1.318990	0.700002	29.999820	0.000179
24	0.80	1.336814	0.699995	30.000540	−0.000542
25	0.79	1.355345	0.700004	29.999630	0.000370
26	0.78	1.374502	0.700009	29.999060	0.000948
27	0.77	1.394300	0.700008	29.999230	0.000775
28	0.76	1.414759	0.699995	30.000480	−0.000477

(continued)

8.2 Contents for Table 8.2 (Table No. 8.2-1 to 400)

Table 8.2 (continued)

Parameter values					
PROB. = 0.7	RHO = −0.5		BIGH = 1.0296		SMHL = 0.54
Pair No.	SMH	SMK	PROB. level	Risk level (%)	Risk error (%)
29	0.75	1.436091	0.699994	30.000630	−0.000632
30	0.74	1.458316	0.699997	30.000340	−0.000334
31	0.73	1.481456	0.699997	30.000320	−0.000322
32	0.72	1.505628	0.699999	30.000120	−0.000119
33	0.71	1.530955	0.700006	29.999440	0.000566
34	0.70	1.557363	0.699996	30.000450	−0.000447
35	0.69	1.585170	0.699991	30.000870	−0.000870
36	0.68	1.614599	0.700000	30.000040	−0.000042
37	0.67	1.645679	0.700003	29.999680	0.000328
38	0.66	1.678636	0.700004	29.999620	0.000387
39	0.65	1.713698	0.699998	30.000170	−0.000167
40	0.64	1.751290	0.699995	30.000470	−0.000471
41	0.63	1.791840	0.699996	30.000440	−0.000435
42	0.62	1.835971	0.700005	29.999510	0.000495
43	0.61	1.883923	0.699990	30.001000	−0.000995
44	0.60	1.937496	0.700008	29.999250	0.000757
45	0.59	1.997129	0.700000	30.000050	−0.000048
46	0.58	2.065216	0.699993	30.000750	−0.000745
47	0.57	2.145329	0.700004	29.999640	0.000364
48	0.56	2.242992	0.699995	30.000510	−0.000507
49	0.55	2.367253	0.699997	30.000330	−0.000328
50	0.54	2.545133	0.699998	30.000190	−0.000185
No. of pairs		Mean COMPUT. risk errors		STD. ERR. errors	
50		0.0000184658		0.0000750897	
Z-ratio				Status of significance	
0.24392				N.S.—not significant	
Parameter values					
PROB. = 0.65	RHO = −0.5		BIGH = 0.9238		SMHL = 0.44
Pair No.	SMH	SMK	PROB. level	Risk level (%)	Risk error (%)
Table No. 8.2-302					
1	0.93	0.917544	0.650008	34.999220	0.000787
2	0.92	0.927518	0.650008	34.999240	0.000763
3	0.91	0.937663	0.650000	35.000010	0.000000
4	0.90	0.948034	0.649997	35.000260	−0.000262
5	0.89	0.958639	0.650000	35.000050	−0.000042
6	0.88	0.969487	0.650007	34.999300	0.000703
7	0.87	0.980536	0.650009	34.999150	0.000852
8	0.86	0.991747	0.649993	35.000720	−0.000715
9	0.85	1.003324	0.650006	34.999450	0.000554

(continued)

Table 8.2 (continued)

Parameter values					
PROB. = 0.65	RHO = −0.5		BIGH = 0.9238		SMHL = 0.44
Pair No.	SMH	SMK	PROB. level	Risk level (%)	Risk error (%)
10	0.84	1.015081	0.650002	34.999820	0.000185
11	0.83	1.027126	0.650004	34.999570	0.000429
12	0.82	1.039372	0.649992	35.000850	−0.000846
13	0.81	1.052026	0.650006	34.999390	0.000608
14	0.80	1.064902	0.650006	34.999420	0.000584
15	0.79	1.078015	0.649991	35.000880	−0.000876
16	0.78	1.091572	0.650003	34.999660	0.000346
17	0.77	1.105390	0.650001	34.999880	0.000119
18	0.76	1.119583	0.650005	34.999500	0.000507
19	0.75	1.134066	0.649995	35.000490	−0.000489
20	0.74	1.148955	0.649991	35.000920	−0.000918
21	0.73	1.164265	0.649992	35.000840	−0.000834
22	0.72	1.180013	0.649997	35.000330	−0.000322
23	0.71	1.196219	0.650005	34.999460	0.000548
24	0.70	1.212804	0.649999	35.000060	−0.000060
25	0.69	1.229887	0.649995	35.000460	−0.000453
26	0.68	1.247587	0.650010	34.999040	0.000960
27	0.67	1.265733	0.650006	34.999360	0.000644
28	0.66	1.284446	0.650002	34.999810	0.000191
29	0.65	1.303753	0.649994	35.000610	−0.000608
30	0.64	1.323779	0.649996	35.000420	−0.000417
31	0.63	1.344456	0.649990	35.000990	−0.000983
32	0.62	1.366012	0.650003	34.999730	0.000274
33	0.61	1.388284	0.650001	34.999930	0.000072
34	0.60	1.411406	0.649995	35.000500	−0.000501
35	0.59	1.435513	0.649994	35.000610	−0.000608
36	0.58	1.460647	0.649991	35.000860	−0.000858
37	0.57	1.487047	0.650005	34.999500	0.000501
38	0.56	1.514564	0.650002	34.999820	0.000185
39	0.55	1.543444	0.649996	35.000360	−0.000352
40	0.54	1.573936	0.650000	35.000040	−0.000036
41	0.53	1.606098	0.649996	35.000360	−0.000358
42	0.52	1.640189	0.649992	35.000850	−0.000846
43	0.51	1.676665	0.650002	34.999830	0.000179
44	0.50	1.715601	0.650001	34.999870	0.000131
45	0.49	1.757464	0.649995	35.000460	−0.000459
46	0.48	1.802929	0.649994	35.000600	−0.000596
47	0.47	1.855405	0.650002	34.999840	0.000161
48	0.46	1.910308	0.650009	34.999100	0.000900

(continued)

8.2 Contents for Table 8.2 (Table No. 8.2-1 to 400)

Table 8.2 (continued)

Parameter values					
PROB. = 0.65	RHO = −0.5		BIGH = 0.9238		SMHL = 0.44
Pair No.	SMH	SMK	PROB. level	Risk level (%)	Risk error (%)
49	0.45	1.971457	0.649990	35.000960	−0.000960
50	0.44	2.041902	0.650000	34.999960	0.000042
No. of pairs		Mean COMPUT. risk errors		STD. ERR. errors	
50		−0.0000426582		0.0000801002	
Z-ratio				Status of significance	
0.53256				N.S.—not significant	
Parameter values					
PROB. = 0.6	RHO = −0.5		BIGH = 0.8255		SMHL = 0.34
Pair No.	SMH	SMK	PROB. level	Risk level (%)	Risk error (%)
Table No. 8.2-303					
1	0.83	0.820927	0.600002	39.999780	0.000215
2	0.82	0.830939	0.600003	39.999700	0.000298
3	0.81	0.841101	0.599995	40.000520	−0.000525
4	0.80	0.851520	0.600003	39.999680	0.000322
5	0.79	0.862107	0.600005	39.999550	0.000447
6	0.78	0.872873	0.600000	40.000020	−0.000024
7	0.77	0.883828	0.599991	40.000950	−0.000954
8	0.76	0.895083	0.600002	39.999820	0.000179
9	0.75	0.906501	0.599999	40.000130	−0.000131
10	0.74	0.918144	0.599995	40.000520	−0.000525
11	0.73	0.930075	0.600003	39.999690	0.000310
12	0.72	0.942211	0.600002	39.999840	0.000155
13	0.71	0.954613	0.600003	39.999680	0.000316
14	0.70	0.967250	0.599998	40.000170	−0.000167
15	0.69	0.980187	0.600000	40.000020	−0.000018
16	0.68	0.993441	0.600009	39.999120	0.000882
17	0.67	1.006933	0.600005	39.999460	0.000542
18	0.66	1.020684	0.599992	40.000810	−0.000811
19	0.65	1.034811	0.599991	40.000880	−0.000882
20	0.64	1.049336	0.600005	39.999550	0.000447
21	0.63	1.064089	0.599993	40.000730	−0.000727
22	0.62	1.079289	0.599999	40.000110	−0.000107
23	0.61	1.094866	0.600005	39.999550	0.000447
24	0.60	1.110750	0.599992	40.000770	−0.000775
25	0.59	1.127168	0.600003	39.999690	0.000310
26	0.58	1.143957	0.600001	39.999920	0.000083
27	0.57	1.161249	0.600007	39.999350	0.000650
28	0.56	1.178984	0.600004	39.999590	0.000405
29	0.55	1.197203	0.599996	40.000360	−0.000358
30	0.54	1.216046	0.600003	39.999720	0.000280

(continued)

Table 8.2 (continued)

Parameter values					
PROB. = 0.6	RHO = −0.5		BIGH = 0.8255		SMHL = 0.34
Pair No.	SMH	SMK	PROB. level	Risk level (%)	Risk error (%)
31	0.53	1.235364	0.599992	40.000850	−0.000852
32	0.52	1.255403	0.599998	40.000190	−0.000191
33	0.51	1.276020	0.599992	40.000800	−0.000805
34	0.50	1.297470	0.600007	39.999330	0.000668
35	0.49	1.319523	0.599998	40.000230	−0.000226
36	0.48	1.342441	0.599998	40.000250	−0.000250
37	0.47	1.366299	0.600007	39.999330	0.000668
38	0.46	1.390983	0.599998	40.000180	−0.000179
39	0.45	1.416774	0.600000	39.999960	0.000036
40	0.44	1.443671	0.600000	40.000050	−0.000048
41	0.43	1.471876	0.600007	39.999280	0.000715
42	0.42	1.501406	0.600009	39.999110	0.000888
43	0.41	1.532385	0.600002	39.999800	0.000203
44	0.40	1.570030	0.599998	40.000200	−0.000203
45	0.39	1.604338	0.599998	40.000190	−0.000191
46	0.38	1.640945	0.600009	39.999150	0.000846
47	0.37	1.679841	0.599997	40.000350	−0.000352
48	0.36	1.721821	0.599998	40.000160	−0.000161
49	0.35	1.767311	0.600003	39.999680	0.000322
50	0.34	1.816763	0.599993	40.000730	−0.000727
No. of pairs		Mean COMPUT. risk errors		STD. ERR. errors	
50		0.0000087654		0.0000700254	
Z-ratio				Status of significance	
0.12517				N.S.—not significant	
Parameter values					
PROB. = 0.55	RHO = −0.5		BIGH = 0.7323		SMHL = 0.25
Pair No.	SMH	SMK	PROB. level	Risk level (%)	Risk error (%)
Table No. 8.2-304					
1	0.74	0.724582	0.549998	45.000170	−0.000173
2	0.73	0.734510	0.549991	45.000920	−0.000918
3	0.72	0.744664	0.549999	45.000100	−0.000095
4	0.71	0.754959	0.549998	45.000200	−0.000203
5	0.70	0.765456	0.550003	44.999720	0.000280
6	0.69	0.776119	0.550003	44.999730	0.000268
7	0.68	0.786962	0.550000	44.999980	0.000024
8	0.67	0.798000	0.549998	45.000240	−0.000232
9	0.66	0.809249	0.549998	45.000230	−0.000226
10	0.65	0.820723	0.550003	44.999730	0.000274
11	0.64	0.832345	0.549991	45.000890	−0.000888
12	0.63	0.844278	0.550003	44.999750	0.000250

(continued)

8.2 Contents for Table 8.2 (Table No. 8.2-1 to 400)

Table 8.2 (continued)

Parameter values						
PROB. = 0.55	RHO = −0.5		BIGH = 0.7323			SMHL = 0.25
Pair No.	SMH	SMK	PROB. level	Risk level (%)		Risk error (%)
13	0.62	0.856347	0.549992	45.000830		−0.000834
14	0.61	0.868768	0.550010	44.999030		0.000966
15	0.60	0.881321	0.550001	44.999920		0.000077
16	0.59	0.894174	0.550004	44.999590		0.000411
17	0.58	0.907258	0.550001	44.999940		0.000060
18	0.57	0.920598	0.549995	45.000540		−0.000536
19	0.56	0.934273	0.550002	44.999820		0.000185
20	0.55	0.948168	0.549993	45.000660		−0.000662
21	0.54	0.962416	0.549997	45.000320		−0.000322
22	0.53	0.976954	0.549996	45.000450		−0.000453
23	0.52	0.991822	0.549995	45.000480		−0.000477
24	0.51	1.007063	0.550002	44.999790		0.000215
25	0.50	1.022624	0.550002	44.999790		0.000209
26	0.49	1.038555	0.550002	44.999820		0.000179
27	0.48	1.054910	0.550009	44.999090		0.000912
28	0.47	1.071552	0.549993	45.000730		−0.000733
29	0.46	1.088738	0.550000	45.000000		0.000000
30	0.45	1.106344	0.550001	44.999870		0.000137
31	0.44	1.124443	0.550007	44.999310		0.000691
32	0.43	1.143021	0.550009	44.999130		0.000876
33	0.42	1.162070	0.550001	44.999930		0.000066
34	0.41	1.181689	0.549995	45.000530		−0.000530
35	0.40	1.201985	0.550004	44.999620		0.000376
36	0.39	1.222884	0.550008	44.999240		0.000757
37	0.38	1.244421	0.550006	44.999450		0.000554
38	0.37	1.266644	0.549998	45.000180		−0.000179
39	0.36	1.289716	0.550003	44.999710		0.000292
40	0.35	1.313624	0.550008	44.999220		0.000781
41	0.34	1.346384	0.549997	45.000280		−0.000274
42	0.33	1.371718	0.550007	44.999330		0.000674
43	0.32	1.397989	0.549998	45.000240		−0.000238
44	0.31	1.425582	0.550009	44.999110		0.000894
45	0.30	1.454339	0.550004	44.999640		0.000358
46	0.29	1.484532	0.550000	44.999990		0.000006
47	0.28	1.516395	0.550006	44.999360		0.000638
48	0.27	1.549926	0.550000	44.999990		0.000006

(continued)

Table 8.2 (continued)

Parameter values					
PROB. = 0.55	RHO = −0.5		BIGH = 0.7323		SMHL = 0.25
Pair No.	SMH	SMK	PROB. level	Risk level (%)	Risk error (%)
49	0.26	1.585594	0.550007	44.999260	0.000739
50	0.25	1.623370	0.549996	45.000400	−0.000393
No. of pairs		Mean COMPUT. risk errors		STD. ERR. errors	
50		0.0000742136		0.0000695583	
Z-ratio				Status of significance	
1.06693				N.S.—not significant	
Parameter values					
PROB. = 0.5	RHO = −0.5		BIGH = 0.6424		SMHL = 0.16
Pair No.	SMH	SMK	PROB. level	Risk level (%)	Risk error (%)
Table No. 8.2-305					
1	0.65	0.634742	0.500003	49.999690	0.000310
2	0.64	0.644711	0.500006	49.999410	0.000596
3	0.63	0.654800	0.499998	50.000190	−0.000194
4	0.62	0.665072	0.499997	50.000350	−0.000346
5	0.61	0.675544	0.500004	49.999580	0.000429
6	0.60	0.686136	0.499998	50.000180	−0.000182
7	0.59	0.696915	0.499996	50.000450	−0.000444
8	0.58	0.707900	0.500001	49.999950	0.000054
9	0.57	0.719016	0.499991	50.000900	−0.000900
10	0.56	0.730382	0.499998	50.000200	−0.000203
11	0.55	0.741925	0.500001	49.999940	0.000060
12	0.54	0.753671	0.500004	49.999610	0.000393
13	0.53	0.765600	0.500001	49.999860	0.000143
14	0.52	0.777742	0.499999	50.000070	−0.000066
15	0.51	0.790129	0.500004	49.999600	0.000405
16	0.50	0.802699	0.499998	50.000160	−0.000155
17	0.49	0.815539	0.500002	49.999790	0.000209
18	0.48	0.828593	0.500000	50.000050	−0.000051
19	0.47	0.841905	0.499999	50.000070	−0.000066
20	0.46	0.855475	0.500000	50.000040	−0.000033
21	0.45	0.869308	0.499999	50.000080	−0.000077
22	0.44	0.883411	0.499998	50.000200	−0.000203
23	0.43	0.897801	0.499997	50.000320	−0.000325
24	0.42	0.912546	0.500009	49.999120	0.000882
25	0.41	0.927527	0.500004	49.999560	0.000441
26	0.40	0.942827	0.500000	49.999990	0.000006
27	0.39	0.958441	0.499991	50.000860	−0.000861
28	0.38	0.974470	0.499997	50.000300	−0.000298
29	0.37	0.990833	0.499996	50.000370	−0.000373
30	0.36	1.007557	0.499992	50.000850	−0.000846
31	0.35	1.024782	0.500007	49.999290	0.000709
32	0.34	1.042273	0.499993	50.000740	−0.000739

(continued)

8.2 Contents for Table 8.2 (Table No. 8.2-1 to 400)

Table 8.2 (continued)

Parameter values					
PROB. = 0.5	RHO = −0.5		BIGH = 0.6424		SMHL = 0.16
Pair No.	SMH	SMK	PROB. level	Risk level (%)	Risk error (%)
33	0.33	1.060303	0.499998	50.000230	−0.000229
34	0.32	1.078777	0.499999	50.000100	−0.000101
35	0.31	1.097721	0.499997	50.000290	−0.000292
36	0.30	1.117193	0.499997	50.000270	−0.000259
37	0.29	1.149490	0.499996	50.000410	−0.000411
38	0.28	1.169453	0.500006	49.999420	0.000584
39	0.27	1.189958	0.500001	49.999870	0.000131
40	0.26	1.211146	0.500000	49.999990	0.000006
41	0.25	1.233033	0.499997	50.000310	−0.000304
42	0.24	1.255719	0.500000	50.000030	−0.000024
43	0.23	1.279210	0.500000	49.999980	0.000024
44	0.22	1.303539	0.499994	50.000610	−0.000602
45	0.21	1.328899	0.499999	50.000140	−0.000137
46	0.20	1.355280	0.500000	49.999960	0.000048
47	0.19	1.382727	0.499993	50.000670	−0.000665
48	0.18	1.411596	0.500009	49.999060	0.000942
49	0.17	1.441575	0.499993	50.000720	−0.000721
50	0.16	1.473275	0.500001	49.999870	0.000131
No. of pairs		Mean COMPUT. risk errors		STD. ERR. errors	
50		−0.0000706490		0.0000593810	
Z-ratio				Status of significance	
1.18976				N.S.—not significant	
Parameter values					
PROB. = 0.99	RHO = −0.6		BIGH = 2.5761		SMHL = 2.34
Pair No.	SMH	SMK	PROB. level	Risk level (%)	Risk error (%)
Table No. 8.2-306					
1	2.58	2.572206	0.990007	0.999260	0.000739
2	2.57	2.582215	0.990007	0.999337	0.000662
3	2.56	2.592301	0.990000	1.000047	−0.000048
4	2.55	2.604030	0.990007	0.999284	0.000715
5	2.54	2.615838	0.990007	0.999308	0.000691
6	2.53	2.627728	0.989999	1.000100	−0.000101
7	2.52	2.641262	0.990002	0.999761	0.000238
8	2.51	2.654879	0.989997	1.000351	−0.000352
9	2.50	2.670142	0.989999	1.000071	−0.000072
10	2.49	2.687053	0.990008	0.999159	0.000840
11	2.48	2.704049	0.990006	0.999427	0.000572
12	2.47	2.721133	0.989991	1.000857	−0.000858
13	2.46	2.741429	0.989994	1.000553	−0.000554
14	2.45	2.763378	0.989997	1.000321	−0.000322
15	2.44	2.788542	0.990009	0.999093	0.000906
16	2.43	2.812234	0.989991	1.000923	−0.000924

(continued)

Table 8.2 (continued)

Parameter values					
PROB. = 0.99	RHO = −0.6		BIGH = 2.5761		SMHL = 2.34
Pair No.	SMH	SMK	PROB. level	Risk level (%)	Risk error (%)
17	2.42	2.841488	0.989995	1.000547	−0.000548
18	2.41	2.875523	0.990007	0.999314	0.000685
19	2.40	2.911996	0.990007	0.999320	0.000679
20	2.39	2.951691	0.989996	1.000440	−0.000441
21	2.38	3.000858	0.989998	1.000243	−0.000244
22	2.37	3.062623	0.990009	0.999117	0.000882
23	2.36	3.136988	0.990009	0.999093	0.000906
24	2.35	3.236453	0.990008	0.999188	0.000811
25	2.34	3.386022	0.990003	0.999659	0.000340
No. of pairs		Mean COMPUT. risk errors		STD. ERR. errors	
25		0.0002001341		0.0001166464	
Z-ratio				Status of significance	
1.71573				N.S.—not significant	
Parameter values					
PROB. = 0.98	RHO = −0.6		BIGH = 2.3265		SMHL = 2.06
Pair No.	SMH	SMK	PROB. level	Risk level (%)	Risk error (%)
Table No. 8.2-307					
1	2.33	2.323005	0.980007	1.999265	0.000733
2	2.32	2.333018	0.980006	1.999408	0.000590
3	2.31	2.343118	0.979994	2.000582	−0.000584
4	2.30	2.354087	0.979992	2.000821	−0.000823
5	2.29	2.365925	0.979997	2.000320	−0.000322
6	2.28	2.378636	0.980008	1.999223	0.000775
7	2.27	2.391437	0.980005	1.999521	0.000477
8	2.26	2.405113	0.980006	1.999450	0.000548
9	2.25	2.419663	0.980008	1.999199	0.000799
10	2.24	2.434309	0.979994	2.000558	−0.000560
11	2.23	2.450613	0.979996	2.000427	−0.000429
12	2.22	2.467796	0.979993	2.000672	−0.000674
13	2.21	2.486641	0.980000	2.000040	−0.000042
14	2.20	2.506367	0.979997	2.000254	−0.000256
15	2.19	2.527758	0.979998	2.000177	−0.000179
16	2.18	2.551595	0.980010	1.999033	0.000966
17	2.17	2.575536	0.979992	2.000761	−0.000763
18	2.16	2.603490	0.980000	2.000040	−0.000042
19	2.15	2.633114	0.979992	2.000785	−0.000787
20	2.14	2.666753	0.979993	2.000690	−0.000691
21	2.13	2.705190	0.980001	1.999909	0.000089
22	2.12	2.748427	0.980003	1.999736	0.000262
23	2.11	2.798026	0.980000	2.000004	−0.000006
24	2.10	2.858679	0.980009	1.999140	0.000858
25	2.09	2.930386	0.979998	2.000165	−0.000167

(continued)

8.2 Contents for Table 8.2 (Table No. 8.2-1 to 400)

Table 8.2 (continued)

Parameter values					
PROB. = 0.98	RHO = −0.6		BIGH = 2.3265		SMHL = 2.06
Pair No.	SMH	SMK	PROB. level	Risk level (%)	Risk error (%)
26	2.08	3.027212	0.980003	1.999694	0.000304
27	2.07	3.164783	0.979998	2.000225	−0.000226
28	2.06	3.427476	0.979996	2.000415	−0.000417
No. of pairs		Mean COMPUT. risk errors		STD. ERR. errors	
28		−0.0000195257		0.0001028293	
Z-ratio				Status of significance	
0.18988				N.S.—not significant	
Parameter values					
PROB. = 0.97	RHO = −0.6		BIGH = 2.1702		SMHL = 1.89
Pair No.	SMH	SMK	PROB. level	Risk level (%)	Risk error (%)
Table No. 8.2-308					
1	2.18	2.160444	0.970002	2.999783	0.000215
2	2.17	2.170400	0.970008	2.999163	0.000834
3	2.16	2.180448	0.970001	2.999854	0.000143
4	2.15	2.190981	0.969996	3.000403	−0.000405
5	2.14	2.202389	0.970004	2.999634	0.000364
6	2.13	2.214284	0.970010	2.999038	0.000960
7	2.12	2.226276	0.969999	3.000081	−0.000083
8	2.11	2.239149	0.969998	3.000236	−0.000238
9	2.10	2.252903	0.970003	2.999687	0.000310
10	2.09	2.267150	0.970001	2.999944	0.000054
11	2.08	2.282280	0.970001	2.999938	0.000060
12	2.07	2.297907	0.969990	3.000975	−0.000978
13	2.06	2.315201	0.970000	3.000021	−0.000024
14	2.05	2.333385	0.970004	2.999628	0.000370
15	2.04	2.352460	0.970000	3.000009	−0.000012
16	2.03	2.373208	0.970005	2.999544	0.000453
17	2.02	2.394850	0.969995	3.000534	−0.000536
18	2.01	2.418949	0.970002	2.999830	0.000167
19	2.00	2.444728	0.970002	2.999824	0.000173
20	1.99	2.472967	0.970005	2.999514	0.000483
21	1.98	2.503670	0.970003	2.999741	0.000256
22	1.97	2.537620	0.970000	2.999962	0.000036
23	1.96	2.576380	0.970010	2.999002	0.000995
24	1.95	2.618390	0.969995	3.000546	−0.000548
25	1.94	2.668340	0.969999	3.000128	−0.000131
26	1.93	2.727794	0.970009	2.999151	0.000846
27	1.92	2.796754	0.969990	3.000975	−0.000978
28	1.91	2.887721	0.969993	3.000695	−0.000697

(continued)

Table 8.2 (continued)

Parameter values				
PROB. = 0.97	RHO = −0.6	BIGH = 2.1702		SMHL = 1.89
Pair No.	SMH	SMK	PROB. level	Risk level (%)	Risk error (%)
29	1.90	3.016325	0.970004	2.999598	0.000399
30	1.89	3.229440	0.970001	2.999944	0.000054
No. of pairs		Mean COMPUT. risk errors		STD. ERR. errors	
30		0.0000819083		0.0000898711	
Z-ratio				Status of significance	
0.91140				N.S.—not significant	
Parameter values				
PROB. = 0.96	RHO = −0.6	BIGH = 2.0538		SMHL = 1.76
Pair No.	SMH	SMK	PROB. level	Risk level (%)	Risk error (%)
Table No. 8.2-309					
1	2.06	2.047619	0.960002	3.999770	0.000232
2	2.05	2.057607	0.960004	3.999597	0.000405
3	2.04	2.067694	0.959991	4.000944	−0.000942
4	2.03	2.078660	0.959998	4.000247	−0.000244
5	2.02	2.090119	0.960005	3.999508	0.000495
6	2.01	2.101680	0.959994	4.000616	−0.000614
7	2.00	2.114125	0.959998	4.000235	−0.000232
8	1.99	2.127068	0.959998	4.000247	−0.000244
9	1.98	2.140507	0.959991	4.000867	−0.000864
10	1.97	2.155227	0.960009	3.999102	0.000900
11	1.96	2.170058	0.960001	3.999895	0.000107
12	1.95	2.185782	0.959996	4.000366	−0.000364
13	1.94	2.202401	0.959992	4.000836	−0.000834
14	1.93	2.220307	0.959998	4.000229	−0.000226
15	1.92	2.239112	0.959997	4.000301	−0.000298
16	1.91	2.259208	0.959998	4.000175	−0.000173
17	1.90	2.280987	0.960009	3.999102	0.000900
18	1.89	2.303671	0.960000	3.999955	0.000048
19	1.88	2.328042	0.959991	4.000884	−0.000882
20	1.87	2.354884	0.959994	4.000604	−0.000602
21	1.86	2.384199	0.959999	4.000086	−0.000083
22	1.85	2.415989	0.959997	4.000306	−0.000304
23	1.84	2.451818	0.960009	3.999102	0.000900
24	1.83	2.490907	0.960004	3.999585	0.000417
25	1.82	2.534822	0.959995	4.000479	−0.000477
26	1.81	2.586689	0.960007	3.999311	0.000691
27	1.80	2.646510	0.960003	3.999668	0.000334
28	1.79	2.718977	0.959999	4.000139	−0.000137
29	1.78	2.811903	0.959999	4.000068	−0.000066

(continued)

8.2 Contents for Table 8.2 (Table No. 8.2-1 to 400)

Table 8.2 (continued)

Parameter values					
PROB. = 0.96	RHO = −0.6		BIGH = 2.0538		SMHL = 1.76
Pair No.	SMH	SMK	PROB. level	Risk level (%)	Risk error (%)
30	1.77	2.939354	0.959992	4.000819	−0.000817
31	1.76	3.159144	0.960005	3.999549	0.000453
No. of pairs		Mean COMPUT. risk errors		STD. ERR. errors	
31		−0.0000787899		0.0000953406	
Z-ratio				Status of significance	
0.82640				N.S.—not significant	
Parameter values					
PROB. = 0.95	RHO = −0.6		BIGH = 1.96		SMHL = 1.65
Pair No.	SMH	SMK	PROB. level	Risk level (%)	Risk error (%)
Table No. 8.2-310					
1	1.97	1.950051	0.949996	5.000377	−0.000376
2	1.96	1.960000	0.950004	4.999554	0.000447
3	1.95	1.970051	0.949996	5.000371	−0.000370
4	1.94	1.980597	0.949992	5.000782	−0.000781
5	1.93	1.991638	0.949992	5.000842	−0.000840
6	1.92	2.003177	0.949992	5.000764	−0.000763
7	1.91	2.015215	0.949993	5.000729	−0.000727
8	1.90	2.027754	0.949991	5.000883	−0.000882
9	1.89	2.041186	0.950005	4.999465	0.000536
10	1.88	2.054732	0.949994	5.000603	−0.000602
11	1.87	2.069175	0.949994	5.000639	−0.000638
12	1.86	2.084517	0.950001	4.999888	0.000113
13	1.85	2.100369	0.949995	5.000472	−0.000471
14	1.84	2.117513	0.950008	4.999197	0.000805
15	1.83	2.135172	0.950002	4.999805	0.000197
16	1.82	2.153738	0.949990	5.000961	−0.000960
17	1.81	2.173993	0.950000	5.000043	−0.000042
18	1.80	2.195160	0.949994	5.000621	−0.000620
19	1.79	2.218020	0.949997	5.000341	−0.000340
20	1.78	2.242577	0.950000	4.999990	0.000012
21	1.77	2.268833	0.949998	5.000246	−0.000244
22	1.76	2.297571	0.950003	4.999680	0.000322
23	1.75	2.328793	0.950006	4.999399	0.000602
24	1.74	2.362503	0.949995	5.000532	−0.000530
25	1.73	2.400265	0.949993	5.000675	−0.000674
26	1.72	2.442863	0.949998	5.000162	−0.000161
27	1.71	2.491081	0.950000	5.000049	−0.000048
28	1.70	2.547264	0.950006	4.999399	0.000602
29	1.69	2.613761	0.950008	4.999155	0.000846

(continued)

Table 8.2 (continued)

Parameter values					
PROB. = 0.95	RHO = −0.6		BIGH = 1.96		SMHL = 1.65
Pair No.	SMH	SMK	PROB. level	Risk level (%)	Risk error (%)
30	1.68	2.694479	0.949996	5.000377	−0.000376
31	1.67	2.801921	0.950000	4.999966	0.000036
32	1.66	2.957966	0.949994	5.000573	−0.000572
33	1.65	3.278241	0.950005	4.999471	0.000530
No. of pairs		Mean COMPUT. risk errors		STD. ERR. errors	
33		−0.0001754831		0.0000903273	
Z-ratio				Status of significance	
1.94275				N.S.—not significant	
Parameter values					
PROB. = 0.94	RHO = −0.6		BIGH = 1.8808		SMHL = 1.56
Pair No.	SMH	SMK	PROB. level	Risk level (%)	Risk error (%)
Table No. 8.2-311					
1	1.89	1.871645	0.939994	6.000579	−0.000578
2	1.88	1.881600	0.940002	5.999804	0.000197
3	1.87	1.891662	0.939991	6.000889	−0.000888
4	1.86	1.902418	0.940000	6.000006	−0.000006
5	1.85	1.913480	0.940001	5.999911	0.000089
6	1.84	1.924848	0.939993	6.000722	−0.000721
7	1.83	1.936916	0.939998	6.000161	−0.000161
8	1.82	1.949490	0.940003	5.999697	0.000304
9	1.81	1.962572	0.940005	5.999482	0.000519
10	1.80	1.976164	0.940002	5.999756	0.000244
11	1.79	1.990268	0.939993	6.000686	−0.000685
12	1.78	2.005277	0.939996	6.000382	−0.000381
13	1.77	2.021192	0.940007	5.999261	0.000739
14	1.76	2.037625	0.940003	5.999673	0.000328
15	1.75	2.054970	0.940001	5.999935	0.000066
16	1.74	2.073227	0.939996	6.000424	−0.000423
17	1.73	2.092792	0.940002	5.999840	0.000161
18	1.72	2.113273	0.939996	6.000424	−0.000423
19	1.71	2.135457	0.940006	5.999417	0.000584
20	1.70	2.158954	0.940008	5.999178	0.000823
21	1.69	2.183766	0.939997	6.000323	−0.000322
22	1.68	2.211069	0.940006	5.999375	0.000626
23	1.67	2.240084	0.939998	6.000209	−0.000209
24	1.66	2.271984	0.940000	6.000048	−0.000048
25	1.65	2.307165	0.940006	5.999387	0.000614
26	1.64	2.345237	0.939990	6.000990	−0.000989
27	1.63	2.388939	0.940001	5.999911	0.000089
28	1.62	2.438273	0.940005	5.999476	0.000525
29	1.61	2.494804	0.940000	6.000006	−0.000006
30	1.60	2.562442	0.940004	5.999613	0.000387

(continued)

Table 8.2 (continued)

Parameter values					
PROB. = 0.94	RHO = −0.6		BIGH = 1.8808		SMHL = 1.56
Pair No.	SMH	SMK	PROB. level	Risk level (%)	Risk error (%)
31	1.59	2.645097	0.939999	6.000078	−0.000077
32	1.58	2.754491	0.940007	5.999274	0.000727
33	1.57	2.912501	0.939999	6.000072	−0.000072
34	1.56	3.230068	0.940000	6.000006	−0.000006
No. of pairs		Mean COMPUT. risk errors		STD. ERR. errors	
34		0.0000292914		0.0000801900	
Z-ratio				Status of significance	
0.36528				N.S.—not significant	
Parameter values					
PROB. = 0.93	RHO = −0.6		BIGH = 1.8119		SMHL = 1.48
Pair No.	SMH	SMK	PROB. level	Risk level (%)	Risk error (%)
Table No. 8.2-312					
1	1.82	1.803836	0.929994	7.000584	−0.000584
2	1.81	1.813802	0.930000	6.999975	0.000024
3	1.80	1.824074	0.930001	6.999869	0.000131
4	1.79	1.834654	0.929997	7.000333	−0.000334
5	1.78	1.845739	0.929999	7.000095	−0.000095
6	1.77	1.857136	0.929993	7.000727	−0.000727
7	1.76	1.869237	0.930003	6.999660	0.000340
8	1.75	1.881653	0.930001	6.999862	0.000137
9	1.74	1.894584	0.929999	7.000125	−0.000125
10	1.73	1.908224	0.930006	6.999433	0.000566
11	1.72	1.922382	0.930007	6.999302	0.000697
12	1.71	1.937060	0.930000	7.000005	−0.000006
13	1.70	1.952650	0.930006	6.999409	0.000590
14	1.69	1.968764	0.929998	7.000220	−0.000221
15	1.68	1.985796	0.929995	7.000482	−0.000483
16	1.67	2.003748	0.929993	7.000661	−0.000662
17	1.66	2.023012	0.930008	6.999159	0.000840
18	1.65	2.042811	0.929995	7.000548	−0.000548
19	1.64	2.064318	0.930005	6.999499	0.000501
20	1.63	2.086755	0.929996	7.000441	−0.000441
21	1.62	2.110906	0.929995	7.000506	−0.000507
22	1.61	2.136775	0.929994	7.000602	−0.000602
23	1.60	2.164754	0.929999	7.000149	−0.000149
24	1.59	2.194846	0.929996	7.000375	−0.000376
25	1.58	2.227836	0.930002	6.999845	0.000155
26	1.57	2.263727	0.929998	7.000232	−0.000232
27	1.56	2.303694	0.930001	6.999928	0.000072
28	1.55	2.348521	0.930006	6.999433	0.000566
29	1.54	2.398993	0.930000	6.999994	0.000006
30	1.53	2.457458	0.929996	7.000435	−0.000435

(continued)

Table 8.2 (continued)

Parameter values					
PROB. = 0.93	RHO = −0.6		BIGH = 1.8119		SMHL = 1.48
Pair No.	SMH	SMK	PROB. level	Risk level (%)	Risk error (%)
31	1.52	2.527043	0.929993	7.000679	−0.000679
32	1.51	2.613222	0.929994	7.000632	−0.000632
33	1.50	2.727716	0.930004	6.999606	0.000393
34	1.49	2.897092	0.930004	6.999559	0.000441
35	1.48	3.255729	0.929996	7.000405	−0.000405
No. of pairs		Mean COMPUT. risk errors		STD. ERR. errors	
35		−0.0000773205		0.0000746408	
Z-ratio				Status of significance	
1.03590				N.S.—not significant	
Parameter values					
PROB. = 0.92	RHO = −0.6		BIGH = 1.7507		SMHL = 1.41
Pair No.	SMH	SMK	PROB. level	Risk level (%)	Risk error (%)
Table No. 8.2-313					
1	1.76	1.741449	0.919998	8.000220	−0.000221
2	1.75	1.751400	0.920006	7.999373	0.000626
3	1.74	1.761466	0.919995	8.000528	−0.000530
4	1.73	1.772038	0.919995	8.000528	−0.000530
5	1.72	1.783120	0.920004	7.999558	0.000441
6	1.71	1.794322	0.919990	8.000983	−0.000983
7	1.70	1.806232	0.919997	8.000255	−0.000256
8	1.69	1.818658	0.920008	7.999182	0.000817
9	1.68	1.831406	0.920005	7.999462	0.000536
10	1.67	1.844675	0.920002	7.999826	0.000173
11	1.66	1.858465	0.919995	8.000506	−0.000507
12	1.65	1.873171	0.920010	7.999027	0.000972
13	1.64	1.888208	0.920002	7.999814	0.000185
14	1.63	1.904166	0.920008	7.999170	0.000829
15	1.62	1.920656	0.920000	8.000040	−0.000042
16	1.61	1.938071	0.919997	8.000290	−0.000292
17	1.60	1.956609	0.920007	7.999277	0.000721
18	1.59	1.975689	0.919990	8.000958	−0.000960
19	1.58	1.996483	0.920009	7.999146	0.000852
20	1.57	2.018018	0.920000	7.999981	0.000018
21	1.56	2.040884	0.919991	8.000875	−0.000876
22	1.55	2.065668	0.920002	7.999850	0.000149
23	1.54	2.091790	0.919993	8.000678	−0.000679
24	1.53	2.120423	0.920008	7.999170	0.000829
25	1.52	2.150789	0.920000	8.000029	−0.000030
26	1.51	2.184065	0.920001	7.999897	0.000101
27	1.50	2.220644	0.920006	7.999361	0.000638
28	1.49	2.260920	0.920007	7.999331	0.000668
29	1.48	2.305678	0.920000	8.000016	−0.000018

(continued)

Table 8.2 (continued)

Parameter values					
PROB. = 0.92	RHO = −0.6		BIGH = 1.7507		SMHL = 1.41
Pair No.	SMH	SMK	PROB. level	Risk level (%)	Risk error (%)
30	1.47	2.356875	0.920005	7.999492	0.000507
31	1.46	2.415687	0.920003	7.999730	0.000268
32	1.45	2.485633	0.920005	7.999492	0.000507
33	1.44	2.570624	0.919991	8.000935	−0.000936
34	1.43	2.683946	0.920004	7.999635	0.000364
35	1.42	2.849040	0.920003	7.999671	0.000328
36	1.41	3.186222	0.920008	7.999248	0.000751
No. of pairs		Mean COMPUT. risk errors		STD. ERR. errors	
36		0.0001193704		0.0000946408	
Z-ratio				Status of significance	
1.26130				N.S.—not significant	
Parameter values					
PROB. = 0.91	RHO = −0.6		BIGH = 1.6954		SMHL = 1.35
Pair No.	SMH	SMK	PROB. level	Risk level (%)	Risk error (%)
Table No. 8.2-314					
1	1.70	1.690812	0.910005	8.999491	0.000507
2	1.69	1.700817	0.910004	8.999581	0.000417
3	1.68	1.711136	0.910000	8.999968	0.000030
4	1.67	1.721968	0.910010	8.999044	0.000954
5	1.66	1.732922	0.909995	9.000522	−0.000525
6	1.65	1.744588	0.910007	8.999300	0.000697
7	1.64	1.756578	0.910009	8.999086	0.000912
8	1.63	1.768893	0.910000	9.000004	−0.000006
9	1.62	1.781731	0.909994	9.000611	−0.000614
10	1.61	1.795291	0.910004	8.999622	0.000376
11	1.60	1.809183	0.909996	9.000408	−0.000411
12	1.59	1.823802	0.909998	9.000230	−0.000232
13	1.58	1.839150	0.910006	8.999389	0.000608
14	1.57	1.855035	0.910003	8.999741	0.000256
15	1.56	1.871653	0.909998	9.000164	−0.000167
16	1.55	1.889205	0.910002	8.999759	0.000238
17	1.54	1.907497	0.909997	9.000319	−0.000322
18	1.53	1.926922	0.910002	8.999806	0.000191
19	1.52	1.947290	0.909999	9.000074	−0.000077
20	1.51	1.968798	0.909995	9.000546	−0.000548
21	1.50	1.991644	0.909992	9.000814	−0.000817
22	1.49	2.016224	0.910003	8.999663	0.000334
23	1.48	2.042149	0.909998	9.000164	−0.000167
24	1.47	2.070205	0.910006	8.999414	0.000584
25	1.46	2.100004	0.909994	9.000629	−0.000632
26	1.45	2.132722	0.910000	9.000045	−0.000048
27	1.44	2.168363	0.910003	8.999699	0.000298

(continued)

Table 8.2 (continued)

Parameter values					
PROB. = 0.91	RHO = −0.6		BIGH = 1.6954		SMHL = 1.35
Pair No.	SMH	SMK	PROB. level	Risk level (%)	Risk error (%)
28	1.43	2.207322	0.909998	9.000230	−0.000232
29	1.42	2.250774	0.909998	9.000224	−0.000226
30	1.41	2.299505	0.909993	9.000665	−0.000668
31	1.40	2.355472	0.909995	9.000528	−0.000530
32	1.39	2.421024	0.909998	9.000218	−0.000221
33	1.38	2.500072	0.909999	9.000122	−0.000125
34	1.37	2.599650	0.909991	9.000916	−0.000918
35	1.36	2.738515	0.909999	9.000087	−0.000089
36	1.35	2.969795	0.910001	8.999884	0.000113
No. of pairs		Mean COMPUT. risk errors		STD. ERR. errors	
36		−0.0000286747		0.0000775435	
Z-ratio				Status of significance	
0.36979				N.S.—not significant	
Parameter values					
PROB. = 0.9025	RHO = −0.6		BIGH = 1.657		SMHL = 1.30
Pair No.	SMH	SMK	PROB. level	Risk level (%)	Risk error (%)
Table No. 8.2-315					
1	1.66	1.654201	0.902509	9.749068	0.000936
2	1.65	1.664225	0.902505	9.749532	0.000471
3	1.64	1.674567	0.902498	9.750223	−0.000221
4	1.63	1.685424	0.902506	9.749431	0.000572
5	1.62	1.696603	0.902508	9.749216	0.000787
6	1.61	1.708106	0.902503	9.749746	0.000256
7	1.60	1.720132	0.902506	9.749394	0.000608
8	1.59	1.732487	0.902499	9.750128	−0.000125
9	1.58	1.745370	0.902496	9.750419	−0.000417
10	1.57	1.758782	0.902494	9.750581	−0.000578
11	1.56	1.772922	0.902508	9.749180	0.000823
12	1.55	1.787402	0.902502	9.749830	0.000173
13	1.54	1.802615	0.902504	9.749592	0.000411
14	1.53	1.818370	0.902496	9.750361	−0.000358
15	1.52	1.835059	0.902504	9.749574	0.000429
16	1.51	1.852295	0.902494	9.750562	−0.000560
17	1.50	1.870667	0.902505	9.749544	0.000459
18	1.49	1.889787	0.902502	9.749818	0.000185
19	1.48	1.909856	0.902494	9.750568	−0.000566
20	1.47	1.931265	0.902500	9.749961	0.000042
21	1.46	1.953824	0.902500	9.749991	0.000012
22	1.45	1.977926	0.902508	9.749222	0.000781
23	1.44	2.003185	0.902493	9.750724	−0.000721
24	1.43	2.030581	0.902497	9.750265	−0.000262
25	1.42	2.060118	0.902507	9.749317	0.000685

(continued)

8.2 Contents for Table 8.2 (Table No. 8.2-1 to 400)

Table 8.2 (continued)

Parameter values					
PROB. = 0.9025	RHO = −0.6		BIGH = 1.657		SMHL = 1.30
Pair No.	SMH	SMK	PROB. level	Risk level (%)	Risk error (%)
26	1.41	2.091800	0.902506	9.749436	0.000566
27	1.40	2.126021	0.902496	9.750408	−0.000405
28	1.39	2.163959	0.902505	9.749538	0.000465
29	1.38	2.205225	0.902490	9.750992	−0.000989
30	1.37	2.251779	0.902490	9.750962	−0.000960
31	1.36	2.304796	0.902498	9.750223	−0.000221
32	1.35	2.365845	0.902499	9.750139	−0.000137
33	1.34	2.438053	0.902495	9.750521	−0.000519
34	1.33	2.527678	0.902500	9.749961	0.000042
35	1.32	2.645661	0.902506	9.749389	0.000614
36	1.31	2.820914	0.902508	9.749251	0.000751
37	1.30	3.199536	0.902508	9.749180	0.000823
No. of pairs		Mean COMPUT. risk errors		STD. ERR. errors	
37		0.0001013279		0.0000882724	
Z-ratio				Status of significance	
1.14790				N.S.—not significant	
Parameter values					
PROB. = 0.9	RHO = −0.6		BIGH = 1.6448		SMHL = 1.29
Pair No.	SMH	SMK	PROB. level	Risk level (%)	Risk error (%)
Table No. 8.2-316					
1	1.65	1.639616	0.899997	10.000290	−0.000292
2	1.64	1.649614	0.899998	10.000230	−0.000232
3	1.63	1.659930	0.899996	10.000380	−0.000381
4	1.62	1.670566	0.899991	10.000870	−0.000870
5	1.61	1.681719	0.900001	9.999942	0.000060
6	1.60	1.693198	0.900003	9.999728	0.000274
7	1.59	1.705004	0.899997	10.000340	−0.000334
8	1.58	1.717336	0.899998	10.000160	−0.000155
9	1.57	1.730195	0.900005	9.999490	0.000513
10	1.56	1.743389	0.899998	10.000230	−0.000232
11	1.55	1.757117	0.899990	10.000960	−0.000954
12	1.54	1.771575	0.899997	10.000280	−0.000274
13	1.53	1.786573	0.899998	10.000190	−0.000185
14	1.52	1.802112	0.899990	10.000980	−0.000978
15	1.51	1.818587	0.900000	9.999979	0.000024
16	1.50	1.835609	0.899994	10.000620	−0.000614
17	1.49	1.853573	0.899996	10.000380	−0.000381
18	1.48	1.872482	0.900002	9.999848	0.000155
19	1.47	1.892339	0.900004	9.999615	0.000387
20	1.46	1.913147	0.899998	10.000230	−0.000232
21	1.45	1.935301	0.900001	9.999906	0.000095
22	1.44	1.958805	0.900005	9.999508	0.000495

(continued)

Table 8.2 (continued)

Parameter values					
PROB. = 0.9	RHO = −0.6		BIGH = 1.6448		SMHL = 1.29
Pair No.	SMH	SMK	PROB. level	Risk level (%)	Risk error (%)
23	1.43	1.983662	0.900001	9.999937	0.000066
24	1.42	2.010265	0.900000	9.999991	0.000012
25	1.41	2.038622	0.899991	10.000880	−0.000876
26	1.40	2.069514	0.899999	10.000120	−0.000113
27	1.39	2.102947	0.900004	9.999572	0.000429
28	1.38	2.139317	0.900005	9.999490	0.000513
29	1.37	2.179407	0.900009	9.999126	0.000876
30	1.36	2.223615	0.900000	9.999972	0.000030
31	1.35	2.273507	0.899996	10.000400	−0.000393
32	1.34	2.330649	0.899993	10.000720	−0.000715
33	1.33	2.398173	0.900003	9.999657	0.000346
34	1.32	2.479207	0.900000	10.000050	−0.000048
35	1.31	2.582352	0.899996	10.000410	−0.000405
36	1.30	2.726363	0.899998	10.000240	−0.000238
37	1.29	2.972182	0.899995	10.000480	−0.000477
No. of pairs		Mean COMPUT. risk errors		STD. ERR. errors	
37		−0.0001344242		0.0000710765	
Z-ratio				Status of significance	
1.89126				N.S.—not significant	
Parameter values					
PROB. = 0.85	RHO = −0.6		BIGH = 1.4393		SMHL = 1.04
Pair No.	SMH	SMK	PROB. level	Risk level (%)	Risk error (%)
Table No. 8.2-317					
1	1.44	1.438600	0.850002	14.999760	0.000238
2	1.43	1.448660	0.849993	15.000690	−0.000691
3	1.42	1.459155	0.850003	14.999720	0.000274
4	1.41	1.469892	0.850002	14.999760	0.000238
5	1.40	1.480875	0.849990	15.000970	−0.000972
6	1.39	1.492301	0.849992	15.000760	−0.000763
7	1.38	1.504078	0.849992	15.000770	−0.000775
8	1.37	1.516304	0.850001	14.999900	0.000095
9	1.36	1.528888	0.850004	14.999630	0.000370
10	1.35	1.541831	0.849999	15.000150	−0.000155
11	1.34	1.555333	0.850007	14.999300	0.000697
12	1.33	1.569105	0.849991	15.000890	−0.000894
13	1.32	1.583640	0.850005	14.999490	0.000513
14	1.31	1.598549	0.850001	14.999900	0.000095
15	1.30	1.614034	0.849997	15.000280	−0.000280
16	1.29	1.630098	0.849990	15.000960	−0.000960
17	1.28	1.646941	0.849997	15.000330	−0.000328
18	1.27	1.664372	0.849991	15.000880	−0.000882
19	1.26	1.682786	0.850008	14.999240	0.000763

(continued)

8.2 Contents for Table 8.2 (Table No. 8.2-1 to 400)

Table 8.2 (continued)

Parameter values					
PROB. = 0.85	RHO = −0.6		BIGH = 1.4393		SMHL = 1.04
Pair No.	SMH	SMK	PROB. level	Risk level (%)	Risk error (%)
20	1.25	1.701798	0.850002	14.999780	0.000215
21	1.24	1.721804	0.850006	14.999400	0.000596
22	1.23	1.742613	0.849995	15.000510	−0.000513
23	1.22	1.764621	0.849995	15.000460	−0.000465
24	1.21	1.787834	0.849998	15.000180	−0.000179
25	1.20	1.812257	0.849994	15.000640	−0.000638
26	1.19	1.838288	0.850001	14.999940	0.000060
27	1.18	1.865736	0.849992	15.000820	−0.000823
28	1.17	1.895194	0.849995	15.000540	−0.000542
29	1.16	1.926864	0.850003	14.999710	0.000286
30	1.15	1.960752	0.849997	15.000320	−0.000322
31	1.14	1.997647	0.849999	15.000070	−0.000072
32	1.13	2.037947	0.850002	14.999840	0.000155
33	1.12	2.082441	0.850006	14.999370	0.000626
34	1.11	2.131917	0.850006	14.999430	0.000572
35	1.10	2.187945	0.850007	14.999340	0.000662
36	1.09	2.252098	0.849993	15.000740	−0.000745
37	1.08	2.329070	0.850006	14.999370	0.000632
38	1.07	2.423559	0.850009	14.999100	0.000894
39	1.06	2.547292	0.850001	14.999880	0.000119
40	1.05	2.732311	0.849996	15.000360	−0.000364
41	1.04	3.148156	0.850002	14.999790	0.000209
No. of pairs		Mean COMPUT. risk errors		STD. ERR. errors	
41		−0.0000726609		0.0000847700	
Z-ratio				Status of significance	
0.85715				N.S.—not significant	
Parameter values					
PROB. = 0.8	RHO = −0.6		BIGH = 1.2809		SMHL = 0.85
Pair No.	SMH	SMK	PROB. level	Risk level (%)	Risk error (%)
Table No. 8.2-318					
1	1.29	1.271864	0.800004	19.999600	0.000399
2	1.28	1.281703	0.799994	20.000570	−0.000566
3	1.27	1.291894	0.800001	19.999920	0.000077
4	1.26	1.302342	0.800005	19.999480	0.000519
5	1.25	1.313052	0.800006	19.999360	0.000638
6	1.24	1.324028	0.800003	19.999680	0.000322
7	1.23	1.335274	0.799995	20.000540	−0.000536
8	1.22	1.346891	0.799995	20.000510	−0.000513
9	1.21	1.358884	0.800002	19.999810	0.000191
10	1.20	1.371160	0.799999	20.000150	−0.000149
11	1.19	1.383822	0.799998	20.000170	−0.000167
12	1.18	1.396873	0.799999	20.000090	−0.000089

(continued)

Table 8.2 (continued)

Parameter values					
PROB. = 0.8	RHO = −0.6		BIGH = 1.2809		SMHL = 0.85
Pair No.	SMH	SMK	PROB. level	Risk level (%)	Risk error (%)
13	1.17	1.410320	0.799999	20.000140	−0.000143
14	1.16	1.424166	0.799995	20.000540	−0.000542
15	1.15	1.438516	0.799998	20.000160	−0.000155
16	1.14	1.453277	0.799993	20.000680	−0.000679
17	1.13	1.468650	0.800003	19.999700	0.000304
18	1.12	1.484446	0.799998	20.000230	−0.000232
19	1.11	1.500866	0.800000	20.000010	−0.000012
20	1.10	1.517917	0.800005	19.999520	0.000477
21	1.09	1.535507	0.799996	20.000380	−0.000381
22	1.08	1.553937	0.800005	19.999490	0.000513
23	1.07	1.573017	0.800002	19.999800	0.000197
24	1.06	1.592952	0.800004	19.999590	0.000405
25	1.05	1.613748	0.800004	19.999580	0.000423
26	1.04	1.635413	0.799995	20.000460	−0.000465
27	1.03	1.658152	0.799990	20.000970	−0.000972
28	1.02	1.682167	0.799998	20.000210	−0.000209
29	1.01	1.707468	0.800006	19.999360	0.000644
30	1.00	1.734064	0.800005	19.999550	0.000447
31	0.99	1.762160	0.799996	20.000400	−0.000399
32	0.98	1.792157	0.799998	20.000180	−0.000179
33	0.97	1.824260	0.800006	19.999360	0.000644
34	0.96	1.858481	0.799999	20.000080	−0.000077
35	0.95	1.895417	0.799993	20.000660	−0.000662
36	0.94	1.935665	0.799996	20.000390	−0.000393
37	0.93	1.979823	0.800003	19.999680	0.000322
38	0.92	2.028297	0.799991	20.000920	−0.000924
39	0.91	2.083049	0.800000	20.000040	−0.000042
40	0.90	2.145270	0.800001	19.999910	0.000089
41	0.89	2.217707	0.799999	20.000090	−0.000089
42	0.88	2.305061	0.800002	19.999790	0.000209
43	0.87	2.415554	0.800002	19.999790	0.000209
44	0.86	2.568733	0.800005	19.999470	0.000530
45	0.85	2.828677	0.799999	20.000080	−0.000083
No. of pairs		Mean COMPUT. risk errors		STD. ERR. errors	
45		−0.0000239714		0.0000630199	
Z-ratio				Status of significance	
0.38038				N.S.—not significant	
Parameter values					
PROB. = 0.75	RHO = −0.6		BIGH = 1.1488		SMHL = 0.68
Pair No.	SMH	SMK	PROB. level	Risk level (%)	Risk error (%)
Table No. 8.2-319					
1	1.15	1.147601	0.749994	25.000630	−0.000632
2	1.14	1.157766	0.750009	24.999110	0.000888

(continued)

8.2 Contents for Table 8.2 (Table No. 8.2-1 to 400)

Table 8.2 (continued)

Parameter values					
PROB. = 0.75	RHO = −0.6		BIGH = 1.1488		SMHL = 0.68
Pair No.	SMH	SMK	PROB. level	Risk level (%)	Risk error (%)
3	1.13	1.168011	0.749992	25.000820	−0.000823
4	1.12	1.178634	0.750001	24.999920	0.000077
5	1.11	1.189445	0.749996	25.000440	−0.000435
6	1.10	1.200546	0.749995	25.000500	−0.000495
7	1.09	1.211944	0.749998	25.000230	−0.000232
8	1.08	1.223643	0.750002	24.999760	0.000238
9	1.07	1.235649	0.750008	24.999220	0.000787
10	1.06	1.247871	0.749995	25.000500	−0.000495
11	1.05	1.260510	0.749998	25.000170	−0.000167
12	1.04	1.273474	0.749998	25.000200	−0.000203
13	1.03	1.286771	0.749993	25.000710	−0.000703
14	1.02	1.300505	0.749998	25.000200	−0.000197
15	1.01	1.314585	0.749994	25.000570	−0.000572
16	1.00	1.329116	0.749996	25.000370	−0.000364
17	0.99	1.344107	0.750001	24.999880	0.000119
18	0.98	1.359565	0.750007	24.999350	0.000650
19	0.97	1.375402	0.749994	25.000610	−0.000608
20	0.96	1.391918	0.750006	24.999430	0.000572
21	0.95	1.408928	0.750008	24.999170	0.000834
22	0.94	1.426441	0.749998	25.000180	−0.000179
23	0.93	1.444663	0.750000	25.000020	−0.000018
24	0.92	1.463603	0.750007	24.999290	0.000715
25	0.91	1.483175	0.750003	24.999750	0.000250
26	0.90	1.503488	0.749993	25.000680	−0.000679
27	0.89	1.524848	0.750009	24.999060	0.000936
28	0.88	1.546972	0.750007	24.999320	0.000685
29	0.87	1.570069	0.750002	24.999800	0.000203
30	0.86	1.594348	0.750008	24.999230	0.000769
31	0.85	1.619629	0.749992	25.000790	−0.000793
32	0.84	1.646511	0.750006	24.999440	0.000560
33	0.83	1.674620	0.749994	25.000620	−0.000620
34	0.82	1.704557	0.749999	25.000090	−0.000083
35	0.81	1.736341	0.750002	24.999800	0.000203
36	0.80	1.770184	0.749998	25.000180	−0.000179
37	0.79	1.806496	0.749996	25.000450	−0.000447
38	0.78	1.845882	0.750008	24.999190	0.000811
39	0.77	1.888363	0.749998	25.000160	−0.000155
40	0.76	1.934939	0.749990	25.000990	−0.000989
41	0.75	1.986998	0.750008	24.999250	0.000751
42	0.74	2.045152	0.750007	24.999290	0.000715
43	0.73	2.111380	0.749995	25.000550	−0.000548
44	0.72	2.189223	0.749990	25.000960	−0.000960

(continued)

Table 8.2 (continued)

Parameter values					
PROB. = 0.75	RHO = −0.6		BIGH = 1.1488		SMHL = 0.68
Pair No.	SMH	SMK	PROB. level	Risk level (%)	Risk error (%)
45	0.71	2.284571	0.750008	24.999240	0.000757
46	0.70	2.406438	0.749995	25.000480	−0.000477
47	0.69	2.582199	0.750000	25.000040	−0.000042
48	0.68	2.922044	0.750004	24.999630	0.000370
No. of pairs		Mean COMPUT. risk errors		STD. ERR. errors	
48		−0.0000041358		0.0000817501	
Z-ratio				Status of significance	
0.05059				N.S.—not significant	
Parameter values					
PROB. = 0.7	RHO = −0.6		BIGH = 1.0335		SMHL = 0.55
Pair No.	SMH	SMK	PROB. level	Risk level (%)	Risk error (%)
Table No. 8.2-320					
1	1.04	1.026943	0.699990	30.000990	−0.000989
2	1.03	1.036914	0.699991	30.000880	−0.000882
3	1.02	1.047179	0.700010	29.999010	0.000995
4	1.01	1.057547	0.700002	29.999810	0.000197
5	1.00	1.068171	0.700000	29.999960	0.000042
6	0.99	1.079009	0.699994	30.000600	−0.000602
7	0.98	1.090116	0.699993	30.000710	−0.000709
8	0.97	1.101548	0.700007	29.999280	0.000721
9	0.96	1.113115	0.699994	30.000560	−0.000554
10	0.95	1.125023	0.699996	30.000420	−0.000423
11	0.94	1.137277	0.700010	29.999010	0.000995
12	0.93	1.149690	0.699996	30.000370	−0.000364
13	0.92	1.162467	0.699994	30.000570	−0.000566
14	0.91	1.175616	0.700003	29.999750	0.000250
15	0.90	1.189048	0.700001	29.999950	0.000054
16	0.89	1.202872	0.700006	29.999360	0.000638
17	0.88	1.216998	0.700000	29.999970	0.000030
18	0.87	1.231535	0.699999	30.000060	−0.000054
19	0.86	1.246495	0.700002	29.999790	0.000209
20	0.85	1.261888	0.700006	29.999380	0.000620
21	0.84	1.277628	0.699993	30.000680	−0.000674
22	0.83	1.293925	0.699995	30.000500	−0.000495
23	0.82	1.310791	0.700009	29.999140	0.000864
24	0.81	1.328044	0.699999	30.000140	−0.000137
25	0.80	1.345895	0.699995	30.000500	−0.000495
26	0.79	1.364358	0.699994	30.000590	−0.000590
27	0.78	1.383449	0.699992	30.000810	−0.000811
28	0.77	1.403285	0.699999	30.000150	−0.000143
29	0.76	1.423783	0.699995	30.000460	−0.000453
30	0.75	1.445061	0.699991	30.000890	−0.000888

(continued)

Table 8.2 (continued)

Parameter values						
PROB. = 0.7	RHO = −0.6		BIGH = 1.0335		SMHL = 0.55	
Pair No.	SMH	SMK	PROB. level	Risk level (%)	Risk error (%)
31	0.74	1.467236	0.699993	30.000670	−0.000668
32	0.73	1.490329	0.699995	30.000530	−0.000530
33	0.72	1.514459	0.700000	30.000050	−0.000048
34	0.71	1.539651	0.699999	30.000060	−0.000060
35	0.70	1.566122	0.700007	29.999300	0.000703
36	0.69	1.593803	0.700000	30.000020	−0.000018
37	0.68	1.623111	0.700007	29.999260	0.000745
38	0.67	1.653977	0.700003	29.999700	0.000304
39	0.66	1.686726	0.699998	30.000160	−0.000161
40	0.65	1.721780	0.700008	29.999240	0.000763
41	0.64	1.759174	0.700004	29.999620	0.000381
42	0.63	1.799337	0.699991	30.000940	−0.000936
43	0.62	1.843089	0.699991	30.000910	−0.000912
44	0.61	1.891253	0.700009	29.999060	0.000948
45	0.60	1.944070	0.699999	30.000100	−0.000095
46	0.59	2.003344	0.699992	30.000790	−0.000793
47	0.58	2.071666	0.699994	30.000630	−0.000626
48	0.57	2.151241	0.700006	29.999390	0.000608
49	0.56	2.247203	0.699997	30.000350	−0.000352
50	0.55	2.370945	0.699998	30.000190	−0.000185
No. of pairs		Mean COMPUT. risk errors		STD. ERR. errors	
50		−0.0001008604		0.0000816709	
Z-ratio				Status of significance	
1.23496				N.S.—not significant	
Parameter values					
PROB. = 0.65	RHO = −0.6		BIGH = 0.9294		SMHL = 0.44	
Pair No.	SMH	SMK	PROB. level	Risk level (%)	Risk error (%)
Table No. 8.2-321					
1	0.93	0.928800	0.649992	35.000830	−0.000823
2	0.92	0.938896	0.649994	35.000580	−0.000578
3	0.91	0.949214	0.650002	34.999780	0.000226
4	0.90	0.959663	0.649992	35.000760	−0.000751
5	0.89	0.970398	0.650001	34.999910	0.000089
6	0.88	0.981280	0.649992	35.000770	−0.000763
7	0.87	0.992465	0.650002	34.999770	0.000232
8	0.86	1.003814	0.649996	35.000380	−0.000370
9	0.85	1.015436	0.649998	35.000230	−0.000226
10	0.84	1.027338	0.640006	34.999380	0.000620
11	0.83	1.039435	0.650000	34.999970	0.000036
12	0.82	1.051833	0.650003	34.999750	0.000250

(continued)

Table 8.2 (continued)

Parameter values					
PROB. = 0.65	RHO = −0.6		BIGH = 0.9294		SMHL = 0.44
Pair No.	SMH	SMK	PROB. level	Risk level (%)	Risk error (%)
13	0.81	1.064447	0.649991	35.000920	−0.000912
14	0.80	1.077484	0.650009	34.999140	0.000864
15	0.79	1.090664	0.649993	35.000750	−0.000745
16	0.78	1.104291	0.650005	34.999490	0.000513
17	0.77	1.118184	0.650005	34.999490	0.000513
18	0.76	1.132359	0.649993	35.000670	−0.000668
19	0.75	1.147025	0.650009	34.999110	0.000894
20	0.74	1.161905	0.649993	35.000700	−0.000691
21	0.73	1.177309	0.650004	34.999630	0.000376
22	0.72	1.193060	0.650002	34.999800	0.000203
23	0.71	1.209273	0.650007	34.999350	0.000656
24	0.70	1.225872	0.649998	35.000170	−0.000167
25	0.69	1.242974	0.649995	35.000500	−0.000501
26	0.68	1.260603	0.649995	35.000470	−0.000471
27	0.67	1.278781	0.649998	35.000230	−0.000226
28	0.66	1.297533	0.650001	34.999920	0.000083
29	0.65	1.316887	0.650003	34.999690	0.000310
30	0.64	1.336870	0.650002	34.999780	0.000221
31	0.63	1.357512	0.649996	35.000360	−0.000358
32	0.62	1.378942	0.649997	35.000260	−0.000262
33	0.61	1.401196	0.650001	34.999860	0.000149
34	0.60	1.424210	0.649991	35.000920	−0.000918
35	0.59	1.448318	0.650001	34.999860	0.000143
36	0.58	1.473365	0.650000	34.999970	0.000036
37	0.57	1.499590	0.650007	34.999340	0.000668
38	0.56	1.526944	0.649999	35.000060	−0.000054
39	0.55	1.555672	0.649994	35.000570	−0.000560
40	0.54	1.585928	0.649990	35.000960	−0.000954
41	0.53	1.618062	0.650005	34.999490	−0.000513
42	0.52	1.651943	0.650000	34.999970	0.000036
43	0.51	1.688030	0.649997	35.000340	−0.000334
44	0.50	1.728154	0.650004	34.999650	0.000352
45	0.49	1.769660	0.650004	34.999620	0.000387
46	0.48	1.814588	0.649997	35.000280	−0.000274
47	0.47	1.864009	0.650009	34.999120	0.000888

(continued)

8.2 Contents for Table 8.2 (Table No. 8.2-1 to 400)

Table 8.2 (continued)

Parameter values					
PROB. = 0.65	RHO = −0.6		BIGH = 0.9294		SMHL = 0.44
Pair No.	SMH	SMK	PROB. level	Risk level (%)	Risk error (%)
48	0.46	1.918415	0.650003	34.999720	0.000286
49	0.45	1.979480	0.649999	35.000060	−0.000054
50	0.44	2.049472	0.650007	34.999310	0.000697
No. of pairs		Mean COMPUT. risk errors		STD. ERR. errors	
50		−0.0000278155		0.0000725008	
Z-ratio				Status of significance	
0.38366				N.S.—not significant	
Parameter values					
PROB. = 0.6	RHO = −0.6		BIGH = 0.8332		SMHL = 0.35
Pair No.	SMH	SMK	PROB. level	Risk level (%)	Risk error (%)
Table No. 8.2-322					
1	0.84	0.826455	0.599993	40.000750	−0.000751
2	0.83	0.836461	0.600003	39.999730	0.000268
3	0.82	0.846612	0.600004	39.999640	0.000358
4	0.81	0.856967	0.600010	39.999020	0.000978
5	0.80	0.867485	0.600009	39.999080	0.000918
6	0.79	0.878176	0.600003	39.999660	0.000340
7	0.78	0.889052	0.599994	40.000640	−0.000644
8	0.77	0.900220	0.600006	39.999380	0.000620
9	0.76	0.911497	0.599992	40.000760	−0.000757
10	0.75	0.923090	0.600003	39.999680	0.000322
11	0.74	0.934818	0.599991	40.000870	−0.000876
12	0.73	0.946888	0.600006	39.999400	0.000596
13	0.72	0.959119	0.600002	39.999840	0.000161
14	0.71	0.971625	0.600003	39.999700	0.000298
15	0.70	0.984373	0.600001	39.999950	0.000048
16	0.69	0.997428	0.600007	39.999300	0.000697
17	0.68	1.010661	0.599991	40.000920	−0.000918
18	0.67	1.024336	0.600009	39.999070	0.000930
19	0.66	1.038180	0.599998	40.000170	−0.000173
20	0.65	1.052409	0.600003	39.999670	0.000334
21	0.64	1.066948	0.600005	39.999510	0.000495
22	0.63	1.081823	0.600005	39.999460	0.000536
23	0.62	1.097057	0.600007	39.999320	0.000679
24	0.61	1.112581	0.599992	40.000830	−0.000834
25	0.60	1.128619	0.600002	39.999840	0.000161
26	0.59	1.145007	0.600000	40.000010	−0.000012
27	0.58	1.161876	0.600008	39.999210	0.000793
28	0.57	1.179066	0.599991	40.000880	−0.000882
29	0.56	1.196909	0.600007	39.999300	0.000703
30	0.55	1.215151	0.600004	39.999640	0.000358
31	0.54	1.233936	0.600002	39.999840	0.000161

(continued)

Table 8.2 (continued)

Parameter values					
PROB. = 0.6	RHO = −0.6		BIGH = 0.8332		SMHL = 0.35
Pair No.	SMH	SMK	PROB. level	Risk level (%)	Risk error (%)
32	0.53	1.253310	0.600003	39.999660	0.000340
33	0.52	1.273226	0.599995	40.000500	−0.000501
34	0.51	1.293836	0.599995	40.000470	−0.000477
35	0.50	1.315201	0.600006	39.999440	0.000560
36	0.49	1.337189	0.599998	40.000230	−0.000232
37	0.48	1.360065	0.600004	39.999600	0.000399
38	0.47	1.383708	0.599997	40.000330	−0.000334
39	0.46	1.408396	0.600006	39.999420	0.000578
40	0.45	1.433926	0.599991	40.000910	−0.000906
41	0.44	1.460784	0.600006	39.999400	0.000596
42	0.43	1.491520	0.600000	39.999980	0.000018
43	0.42	1.520682	0.600005	39.999520	0.000477
44	0.41	1.551231	0.599996	40.000370	−0.000370
45	0.40	1.583601	0.600004	39.999620	0.000381
46	0.39	1.617750	0.599999	40.000150	−0.000149
47	0.38	1.654047	0.599993	40.000710	−0.000715
48	0.37	1.692876	0.599992	40.000790	−0.000793
49	0.36	1.734643	0.599995	40.000470	−0.000477
50	0.35	1.779974	0.600010	39.999050	0.000948
No. of pairs		Mean COMPUT. risk errors		STD. ERR. errors	
50		0.0000833296		0.0000808903	
Z-ratio				Status of significance	
1.03016				N.S.—not significant	
Parameter values					
PROB. = 0.55	RHO = −0.6		BIGH = 0.7425		SMHL = 0.26
Pair No.	SMH	SMK	PROB. level	Risk level (%)	Risk error (%)
Table No. 8.2-323					
1	0.75	0.735026	0.550004	44.999640	0.000358
2	0.74	0.745008	0.550010	44.999010	0.000995
3	0.73	0.755116	0.550006	44.999450	0.000554
4	0.72	0.765410	0.550006	44.999370	0.000632
5	0.71	0.775853	0.550001	44.999910	0.000089
6	0.70	0.786506	0.550004	44.999560	0.000441
7	0.69	0.797334	0.550006	44.999380	0.000626
8	0.68	0.808303	0.549996	45.000410	−0.000405
9	0.67	0.819525	0.550002	44.999800	0.000197
10	0.66	0.830918	0.550001	44.999880	0.000125

(continued)

Table 8.2 (continued)

Parameter values					
PROB. = 0.55	RHO = −0.6		BIGH = 0.7425		SMHL = 0.26
Pair No.	SMH	SMK	PROB. level	Risk level (%)	Risk error (%)
11	0.65	0.842499	0.549997	45.000340	−0.000340
12	0.64	0.854287	0.549992	45.000840	−0.000840
13	0.63	0.866349	0.550001	44.999870	0.000137
14	0.62	0.878559	0.549993	45.000730	−0.000733
15	0.61	0.891085	0.550006	44.999450	0.000554
16	0.60	0.903804	0.550007	44.999250	0.000745
17	0.59	0.916741	0.550003	44.999740	0.000262
18	0.58	0.929922	0.549996	45.000430	−0.000423
19	0.57	0.943375	0.549991	45.000880	−0.000876
20	0.56	0.957131	0.549994	45.000600	−0.000596
21	0.55	0.971222	0.550009	44.999130	0.000876
22	0.54	0.985488	0.549996	45.000380	−0.000376
23	0.53	1.000161	0.550007	44.999330	0.000674
24	0.52	1.015087	0.550003	44.999720	0.000286
25	0.51	1.030308	0.549991	45.000880	−0.000876
26	0.50	1.045971	0.550000	44.999960	0.000042
27	0.49	1.061931	0.549996	45.000450	−0.000447
28	0.48	1.078339	0.550005	44.999470	0.000530
29	0.47	1.095062	0.549998	45.000180	−0.000179
30	0.46	1.112261	0.550003	44.999670	0.000328
31	0.45	1.129812	0.549992	45.000840	−0.000840
32	0.44	1.147890	0.549993	45.000720	−0.000721
33	0.43	1.166482	0.549999	45.000100	−0.000101
34	0.42	1.185582	0.550004	44.999620	0.000376
35	0.41	1.205195	0.550003	44.999730	0.000268
36	0.40	1.225335	0.549992	45.000770	−0.000769
37	0.39	1.246222	0.550005	44.999540	0.000459
38	0.38	1.267699	0.550005	44.999480	0.000519
39	0.37	1.294801	0.550004	44.999640	0.000358
40	0.36	1.317440	0.550001	44.999930	0.000066
41	0.35	1.340882	0.549996	45.000450	−0.000447
42	0.34	1.365237	0.549995	45.000470	−0.000471
43	0.33	1.390545	0.549994	45.000610	−0.000602
44	0.32	1.416970	0.550003	44.999730	0.000268
45	0.31	1.444421	0.549995	45.000530	−0.000530
46	0.30	1.473232	0.550000	44.999960	0.000042
47	0.29	1.503397	0.550001	44.999910	0.000089

(continued)

Table 8.2 (continued)

Parameter values					
PROB. = 0.55	RHO = −0.6		BIGH = 0.7425		SMHL = 0.26
Pair No.	SMH	SMK	PROB. level	Risk level (%)	Risk error (%)
48	0.28	1.535159	0.550006	44.999370	0.000632
49	0.27	1.568629	0.550008	44.999230	0.000769
50	0.26	1.603998	0.550002	44.999840	0.000161
No. of pairs		Mean COMPUT. risk errors		STD. ERR. errors	
50		0.0000369315		0.0000736449	
Z-ratio				Status of significance	
0.50148				N.S.—not significant	
Parameter values					
PROB. = 0.5	RHO = −0.6		BIGH = 0.6554		SMHL = 0.17
Pair No.	SMH	SMK	PROB. level	Risk level (%)	Risk error (%)
Table No. 8.2-324					
1	0.66	0.650734	0.499993	50.000660	−0.000656
2	0.65	0.660747	0.499996	50.000450	−0.000447
3	0.64	0.670926	0.500002	49.999820	0.000185
4	0.63	0.681238	0.500002	49.999830	0.000167
5	0.62	0.691698	0.499999	50.000150	−0.000143
6	0.61	0.702323	0.499996	50.000380	−0.000381
7	0.60	0.713132	0.499998	50.000160	−0.000155
8	0.59	0.724143	0.500010	49.999050	0.000954
9	0.58	0.735231	0.499993	50.000680	−0.000679
10	0.57	0.746564	0.499995	50.000480	−0.000477
11	0.56	0.758068	0.499993	50.000670	−0.000674
12	0.55	0.769768	0.499993	50.000680	−0.000682
13	0.54	0.781692	0.500001	49.999930	0.000072
14	0.53	0.793771	0.499996	50.000400	−0.000393
15	0.52	0.806085	0.499999	50.000090	−0.000089
16	0.51	0.818620	0.500004	49.999640	0.000364
17	0.50	0.831364	0.500006	49.999440	0.000560
18	0.49	0.844306	0.500001	49.999950	0.000054
19	0.48	0.857491	0.499998	50.000230	−0.000224
20	0.47	0.870965	0.500007	49.999350	0.000656
21	0.46	0.884632	0.500001	49.999880	0.000125
22	0.45	0.898596	0.500005	49.999480	0.000525
23	0.44	0.912771	0.499996	50.000430	−0.000423
24	0.43	0.927271	0.499997	50.000290	−0.000292
25	0.42	0.942071	0.500001	49.999880	0.000125
26	0.41	0.957153	0.500001	49.999930	0.000077
27	0.40	0.972505	0.499991	50.000870	−0.000870
28	0.39	0.988224	0.499991	50.000900	−0.000897
29	0.38	1.004317	0.499998	50.000170	−0.000167
30	0.37	1.020708	0.499993	50.000680	−0.000677
31	0.36	1.037527	0.500000	49.999990	0.000012

(continued)

8.2 Contents for Table 8.2 (Table No. 8.2-1 to 400)

Table 8.2 (continued)

Parameter values					
PROB. = 0.5	RHO = −0.6		BIGH = 0.6554		SMHL = 0.17
Pair No.	SMH	SMK	PROB. level	Risk level (%)	Risk error (%)
32	0.35	1.054723	0.500005	49.999540	0.000465
33	0.34	1.072265	0.499997	50.000340	−0.000334
34	0.33	1.090334	0.500008	49.999160	0.000846
35	0.32	1.116754	0.499994	50.000630	−0.000626
36	0.31	1.135348	0.500008	49.999220	0.000781
37	0.30	1.154290	0.499992	50.000830	−0.000823
38	0.29	1.173878	0.499997	50.000340	−0.000340
39	0.28	1.194063	0.500007	49.999350	0.000650
40	0.27	1.214748	0.499998	50.000240	−0.000241
41	0.26	1.236094	0.499992	50.000830	−0.000826
42	0.25	1.258232	0.500003	49.999660	0.000340
43	0.24	1.280996	0.499996	50.000400	−0.000390
44	0.23	1.304613	0.499999	50.000150	−0.000146
45	0.22	1.329055	0.499996	50.000400	−0.000399
46	0.21	1.354452	0.499998	50.000240	−0.000241
47	0.20	1.380847	0.499998	50.000230	−0.000224
48	0.19	1.408339	0.499998	50.000250	−0.000244
49	0.18	1.437063	0.500001	49.999940	0.000060
50	0.17	1.467086	0.500000	50.000030	−0.000021
No. of pairs		Mean COMPUT. risk errors		STD. ERR. errors	
50		−0.0001209039		0.0000656646	
Z-ratio				Status of significance	
1.84123				N.S.—not significant	
Parameter values					
PROB. = 0.99	RHO = −0.7		BIGH = 2.5761		SMHL = 2.34
Pair No.	SMH	SMK	PROB. level	Risk level (%)	Risk error (%)
Table No. 8.2-325					
1	2.58	2.572206	0.990007	0.999266	0.000733
2	2.57	2.582215	0.990007	0.999332	0.000668
3	2.56	2.592301	0.990000	1.000047	−0.000048
4	2.55	2.604030	0.990007	0.999290	0.000709
5	2.54	2.615838	0.990007	0.999308	0.000691
6	2.53	2.627728	0.989999	1.000094	−0.000095
7	2.52	2.641262	0.990002	0.999761	0.000238
8	2.51	2.654879	0.989996	1.000357	−0.000358
9	2.50	2.670142	0.989999	1.000071	−0.000072
10	2.49	2.687053	0.990009	0.999153	0.000846
11	2.48	2.704049	0.990006	0.999415	0.000584
12	2.47	2.721133	0.989991	1.000857	−0.000858
13	2.46	2.741429	0.989994	1.000559	−0.000560
14	2.45	2.763378	0.989997	1.000327	−0.000328
15	2.44	2.788542	0.990009	0.999093	0.000906

(continued)

Table 8.2 (continued)

Parameter values					
PROB. = 0.99	RHO = −0.7		BIGH = 2.5761		SMHL = 2.34
Pair No.	SMH	SMK	PROB. level	Risk level (%)	Risk error (%)
16	2.43	2.812234	0.989991	1.000929	−0.000930
17	2.42	2.841488	0.989995	1.000547	−0.000548
18	2.41	2.875523	0.990007	0.999308	0.000691
19	2.40	2.911996	0.990007	0.999320	0.000679
20	2.39	2.951691	0.989996	1.000440	−0.000441
21	2.38	3.000858	0.989998	1.000243	−0.000244
22	2.37	3.062623	0.990009	0.999117	0.000882
23	2.36	3.136988	0.990009	0.999093	0.000906
24	2.35	3.236453	0.990008	0.999188	0.000811
25	2.34	3.386022	0.990003	0.999653	0.000346
No. of pairs		Mean COMPUT. risk errors		STD. ERR. errors	
25		0.0002003633		0.0001169691	
Z-ratio				Status of significance	
1.71296				N.S.—not significant	
Parameter values					
PROB. = 0.98	RHO = −0.7		BIGH = 2.3265		SMHL = 2.06
Pair No.	SMH	SMK	PROB. level	Risk level (%)	Risk error (%)
Table No. 8.2-326					
1	2.33	2.323005	0.980007	1.999259	0.000739
2	2.32	2.333018	0.980006	1.999408	0.000590
3	2.31	2.343118	0.979994	2.000570	−0.000572
4	2.30	2.354087	0.979992	2.000827	−0.000829
5	2.29	2.365925	0.979997	2.000320	−0.000322
6	2.28	2.378636	0.980008	1.999223	0.000775
7	2.27	2.391437	0.980005	1.999515	0.000483
8	2.26	2.405113	0.980006	1.999444	0.000554
9	2.25	2.419663	0.980008	1.999199	0.000799
10	2.24	2.434309	0.979994	2.000558	−0.000560
11	2.23	2.450613	0.979996	2.000433	−0.000435
12	2.22	2.467796	0.979993	2.000672	−0.000674
13	2.21	2.486641	0.980000	2.000034	−0.000036
14	2.20	2.506367	0.979998	2.000248	−0.000250
15	2.19	2.527758	0.979998	2.000177	−0.000179
16	2.18	2.551595	0.980010	1.999021	0.000978
17	2.17	2.575536	0.979992	2.000767	−0.000769
18	2.16	2.603490	0.980000	2.000034	−0.000036
19	2.15	2.633114	0.979992	2.000791	−0.000793
20	2.14	2.666753	0.979993	2.000690	−0.000691
21	2.13	2.705190	0.980001	1.999903	0.000095
22	2.12	2.748427	0.980003	1.999736	0.000262
23	2.11	2.798026	0.980000	1.999998	0.000000
24	2.10	2.858679	0.980009	1.999140	0.000858

(continued)

8.2 Contents for Table 8.2 (Table No. 8.2-1 to 400)

Table 8.2 (continued)

Parameter values					
PROB. = 0.98	RHO = −0.7		BIGH = 2.3265		SMHL = 2.06
Pair No.	SMH	SMK	PROB. level	Risk level (%)	Risk error (%)
25	2.09	2.930386	0.979998	2.000159	−0.000161
26	2.08	3.027212	0.980003	1.999682	0.000316
27	2.07	3.164783	0.979998	2.000219	−0.000221
28	2.06	3.427476	0.979996	2.000380	−0.000381
No. of pairs		Mean COMPUT. risk errors		STD. ERR. errors	
28		−0.0000158261		0.0001030632	
Z-ratio				Status of significance	
0.15356				N.S.—not significant	
Parameter values					
PROB. = 0.97	RHO = −0.7		BIGH = 2.1701		SMHL = 1.89
Pair No.	SMH	SMK	PROB. level	Risk level (%)	Risk error (%)
Table No. 8.2-327					
1	2.18	2.160245	0.969995	3.000546	−0.000548
2	2.17	2.170200	0.970001	2.999926	0.000072
3	2.16	2.180247	0.969994	3.000593	−0.000596
4	2.15	2.191169	0.970003	2.999735	0.000262
5	2.14	2.202186	0.969996	3.000355	−0.000358
6	2.13	2.214080	0.970003	2.999741	0.000256
7	2.12	2.226071	0.969992	3.000784	−0.000787
8	2.11	2.238943	0.969991	3.000903	−0.000906
9	2.10	2.252696	0.969996	3.000355	−0.000358
10	2.09	2.267332	0.970006	2.999389	0.000608
11	2.08	2.282071	0.969995	3.000546	−0.000548
12	2.07	2.298478	0.970007	2.999348	0.000650
13	2.06	2.314991	0.969994	3.000593	−0.000596
14	2.05	2.333173	0.969998	3.000182	−0.000185
15	2.04	2.352247	0.969995	3.000534	−0.000536
16	2.03	2.372994	0.969999	3.000057	−0.000060
17	2.02	2.395416	0.970008	2.999246	0.000751
18	2.01	2.418733	0.969997	3.000301	−0.000304
19	2.00	2.444510	0.969997	3.000260	−0.000262
20	1.99	2.472749	0.970001	2.999932	0.000066
21	1.98	2.503451	0.969999	3.000116	−0.000119
22	1.97	2.537399	0.969997	3.000319	−0.000322
23	1.96	2.576158	0.970007	2.999306	0.000691
24	1.95	2.618167	0.969992	3.000832	−0.000834
25	1.94	2.668116	0.969996	3.000379	−0.000381
26	1.93	2.727569	0.970006	2.999359	0.000638
27	1.92	2.798090	0.970001	2.999926	0.000072
28	1.91	2.887494	0.969992	3.000832	−0.000834

(continued)

Table 8.2 (continued)

Parameter values					
PROB. = 0.97	RHO = −0.7		BIGH = 2.1701		SMHL = 1.89
Pair No.	SMH	SMK	PROB. level	Risk level (%)	Risk error (%)
29	1.90	3.016096	0.970003	2.999681	0.000316
30	1.89	3.229210	0.970000	2.999973	0.000024
No. of pairs		Mean COMPUT. risk errors		STD. ERR. errors	
30		−0.0001332452		0.0000873121	
Z-ratio				Status of significance	
1.52608				N.S.—not significant	
Parameter values					
PROB. = .96	RHO = −0.7		BIGH = 2.0538		SMHL = 1.76
Pair No.	SMH	SMK	PROB. level	Risk level (%)	Risk error (%)
Table No. 8.2-328					
1	2.06	2.047619	0.960002	3.999794	0.000209
2	2.05	2.057607	0.960004	3.999615	0.000387
3	2.04	2.067694	0.959990	4.000968	−0.000966
4	2.03	2.078660	0.959997	4.000259	−0.000256
5	2.02	2.090119	0.960005	3.999531	0.000471
6	2.01	2.101680	0.959994	4.000628	−0.000626
7	2.00	2.114125	0.959998	4.000240	−0.000238
8	1.99	2.127068	0.959998	4.000253	−0.000250
9	1.98	2.140507	0.959991	4.000878	−0.000876
10	1.97	2.155227	0.960009	3.999126	0.000876
11	1.96	2.170058	0.960001	3.999919	0.000083
12	1.95	2.185782	0.959996	4.000384	−0.000381
13	1.94	2.202401	0.959992	4.000849	−0.000846
14	1.93	2.220307	0.959998	4.000235	−0.000232
15	1.92	2.239112	0.959997	4.000318	−0.000316
16	1.91	2.259208	0.959998	4.000187	−0.000185
17	1.90	2.280987	0.960009	3.999114	0.000888
18	1.89	2.303671	0.960000	3.999967	0.000036
19	1.88	2.328042	0.959991	4.000902	−0.000900
20	1.87	2.354884	0.959994	4.000616	−0.000614
21	1.86	2.384199	0.959999	4.000097	−0.000095
22	1.85	2.415989	0.959997	4.000312	−0.000310
23	1.84	2.451818	0.960009	3.999108	0.000894
24	1.83	2.490907	0.960004	3.999585	0.000417
25	1.82	2.534822	0.959995	4.000497	−0.000495
26	1.81	2.586689	0.960007	3.999305	0.000697
27	1.80	2.646510	0.960003	3.999668	0.000334
28	1.79	2.718977	0.959999	4.000134	−0.000131

(continued)

8.2 Contents for Table 8.2 (Table No. 8.2-1 to 400)

Table 8.2 (continued)

Parameter values					
PROB. = .96	RHO = −0.7		BIGH = 2.0538		SMHL = 1.76
Pair No.	SMH	SMK	PROB. level	Risk level (%)	Risk error (%)
29	1.78	2.811903	0.959999	4.000056	−0.000054
30	1.77	2.939354	0.959992	4.000807	−0.000805
31	1.76	3.159144	0.960005	3.999508	0.000495
No. of pairs		Mean COMPUT. risk errors		STD. ERR. errors	
31		−0.0000871718		0.0000956939	
Z-ratio				Status of significance	
0.91094				N.S.—not significant	
Parameter values					
PROB. = 0.95	RHO = −0.7		BIGH = 1.96		SMHL = 1.65
Pair No.	SMH	SMK	PROB. level	Risk level (%)	Risk error (%)
Table No. 8.2-329					
1	1.97	1.950051	0.949996	5.000425	−0.000423
2	1.96	1.960000	0.950004	4.999590	0.000411
3	1.95	1.970051	0.949996	5.000425	−0.000423
4	1.94	1.980597	0.949992	5.000818	−0.000817
5	1.93	1.991638	0.949991	5.000878	−0.000876
6	1.92	2.003177	0.949992	5.000806	−0.000805
7	1.91	2.015215	0.949992	5.000770	−0.000769
8	1.90	2.027754	0.949991	5.000925	−0.000924
9	1.89	2.041186	0.950005	4.999519	0.000483
10	1.88	2.054732	0.949994	5.000639	−0.000638
11	1.87	2.069175	0.949993	5.000675	−0.000674
12	1.86	2.084517	0.950001	4.999942	0.000060
13	1.85	2.100369	0.949995	5.000496	−0.000495
14	1.84	2.117513	0.950008	4.999238	0.000763
15	1.83	2.135172	0.950002	4.999852	0.000149
16	1.82	2.154129	0.950005	4.999483	0.000519
17	1.81	2.173993	0.949999	5.000085	−0.000083
18	1.80	2.195160	0.949994	5.000651	−0.000650
19	1.79	2.218020	0.949996	5.000383	−0.000381
20	1.78	2.242577	0.950000	5.000031	−0.000030
21	1.77	2.268833	0.949997	5.000287	−0.000286
22	1.76	2.297571	0.950003	4.999697	0.000304
23	1.75	2.328793	0.950006	4.999429	0.000572
24	1.74	2.362503	0.949994	5.000562	−0.000560
25	1.73	2.400265	0.949993	5.000693	−0.000691
26	1.72	2.442863	0.949998	5.000174	−0.000173
27	1.71	2.491081	0.949999	5.000061	−0.000060
28	1.70	2.547264	0.950006	4.999405	0.000596
29	1.69	2.613761	0.950008	4.999167	0.000834

(continued)

Table 8.2 (continued)

Parameter values					
PROB. = 0.95	RHO = −0.7		BIGH = 1.96		SMHL = 1.65
Pair No.	SMH	SMK	PROB. level	Risk level (%)	Risk error (%)
30	1.68	2.694479	0.949996	5.000365	−0.000364
31	1.67	2.801921	0.950000	4.999954	0.000048
32	1.66	2.957966	0.949995	5.000544	−0.000542
33	1.65	3.278241	0.950006	4.999364	0.000638
No. of pairs		Mean COMPUT. risk errors		STD. ERR. errors	
33		−0.0001554980		0.0000907315	
Z-ratio				Status of significance	
1.71383				N.S.—not significant	
Parameter values					
PROB. = 0.94	RHO = −0.7		BIGH = 1.8808		SMHL = 1.56
Pair No.	SMH	SMK	PROB. level	Risk level (%)	Risk error (%)
Table No. 8.2-330					
1	1.89	1.871645	0.939993	6.000680	−0.000679
2	1.88	1.881600	0.940001	5.999911	0.000089
3	1.87	1.891662	0.939990	6.000990	−0.000989
4	1.86	1.902418	0.939999	6.000102	−0.000101
5	1.85	1.913480	0.940000	6.000018	−0.000018
6	1.84	1.924848	0.939992	6.000817	−0.000817
7	1.83	1.936916	0.939997	6.000275	−0.000274
8	1.82	1.949490	0.940002	5.999798	0.000203
9	1.81	1.962572	0.940004	5.999583	0.000417
10	1.80	1.976164	0.940002	5.999851	0.000149
11	1.79	1.990268	0.939992	6.000781	−0.000781
12	1.78	2.005277	0.939995	6.000477	−0.000477
13	1.77	2.021192	0.940007	5.999345	0.000656
14	1.76	2.037625	0.940002	5.999768	0.000232
15	1.75	2.054970	0.940000	6.000018	−0.000018
16	1.74	2.073227	0.939995	6.000513	−0.000513
17	1.73	2.092792	0.940001	5.999911	0.000089
18	1.72	2.113273	0.939995	6.000495	−0.000495
19	1.71	2.135457	0.940005	5.999482	0.000519
20	1.70	2.158954	0.940008	5.999226	0.000775
21	1.69	2.183766	0.939996	6.000376	−0.000376
22	1.68	2.211069	0.940006	5.999417	0.000584
23	1.67	2.240084	0.939998	6.000245	−0.000244
24	1.66	2.271984	0.939999	6.000090	−0.000089
25	1.65	2.307165	0.940006	5.999434	0.000566
26	1.64	2.346018	0.940010	5.999029	0.000972
27	1.63	2.388939	0.940001	5.999941	0.000060

(continued)

8.2 Contents for Table 8.2 (Table No. 8.2-1 to 400)

Table 8.2 (continued)

Parameter values						
PROB. = 0.94	RHO = −0.7		BIGH = 1.8808			SMHL = 1.56
Pair No.	SMH	SMK	PROB. level	Risk level (%)		Risk error (%)
28	1.62	2.438273	0.940005	5.999506		0.000495
29	1.61	2.494804	0.940000	6.000030		−0.000030
30	1.60	2.562442	0.940004	5.999619		0.000381
31	1.59	2.645097	0.939999	6.000078		−0.000077
32	1.58	2.754491	0.940007	5.999261		0.000739
33	1.57	2.912501	0.940000	6.000042		−0.000042
34	1.56	3.230068	0.940001	5.999881		0.000119
No. of pairs		Mean COMPUT. risk errors		STD. ERR. errors		
34		0.0000292914		0.000814372		
Z-ratio				Status of significance		
0.35968				N.S.—not significant		
Parameter values						
PROB. = 0.93	RHO = −0.7		BIGH = 1.8119			SMHL = 1.48
Pair No.	SMH	SMK	PROB. level	Risk level (%)		Risk error (%)
Table No. 8.2-331						
1	1.82	1.803836	0.929992	7.000792		−0.000793
2	1.81	1.813802	0.929998	7.000184		−0.000185
3	1.80	1.824074	0.929999	7.000071		−0.000072
4	1.79	1.834849	0.930009	6.999100		0.000900
5	1.78	1.845739	0.929997	7.000309		−0.000310
6	1.77	1.857136	0.929991	7.000935		−0.000936
7	1.76	1.869237	0.930001	6.999869		0.000131
8	1.75	1.881653	0.929999	7.000077		−0.000077
9	1.74	1.894584	0.929997	7.000327		−0.000328
10	1.73	1.908224	0.930004	6.999630		0.000370
11	1.72	1.922382	0.930005	6.999499		0.000501
12	1.71	1.937060	0.929998	7.000196		−0.000197
13	1.70	1.952650	0.930004	6.999600		0.000399
14	1.69	1.968764	0.929996	7.000405		−0.000405
15	1.68	1.985796	0.929993	7.000655		−0.000656
16	1.67	2.003748	0.929992	7.000828		−0.000829
17	1.66	2.023012	0.930007	6.999314		0.000685
18	1.65	2.042811	0.929993	7.000697		−0.000697
19	1.64	2.064318	0.930004	6.999636		0.000364
20	1.63	2.086755	0.929994	7.000572		−0.000572
21	1.62	2.110906	0.929994	7.000625		−0.000626
22	1.61	2.136775	0.929993	7.000709		−0.000709
23	1.60	2.164754	0.929997	7.000256		−0.000256
24	1.59	2.195237	0.930009	6.999088		0.000912

(continued)

Table 8.2 (continued)

Parameter values					
PROB. = 0.93	RHO = −0.7		BIGH = 1.8119		SMHL = 1.48
Pair No.	SMH	SMK	PROB. level	Risk level (%)	Risk error (%)
25	1.58	2.227836	0.930001	6.999916	0.000083
26	1.57	2.263727	0.929997	7.000298	−0.000298
27	1.56	2.303694	0.930000	6.999988	0.000012
28	1.55	2.348521	0.930005	6.999481	0.000519
29	1.54	2.398993	0.930000	7.000036	−0.000036
30	1.53	2.457458	0.929995	7.000458	−0.000459
31	1.52	2.527043	0.929993	7.000685	−0.000685
32	1.51	2.613222	0.929994	7.000643	−0.000644
33	1.50	2.727716	0.930004	6.999600	0.000399
34	1.49	2.897092	0.930005	6.999522	0.000477
35	1.48	3.255729	0.929998	7.000214	−0.000215
No. of pairs		Mean COMPUT. risk errors		STD. ERR. errors	
35		−0.0001175536		0.0000847523	
Z-ratio				Status of significance	
1.38703				N.S.—not significant	
Parameter values					
PROB. = 0.92	RHO = −0.7		BIGH = 1.7507		SMHL = 1.41
Pair No.	SMH	SMK	PROB. level	Risk level (%)	Risk error (%)
Table No. 8.2-332					
1	1.76	1.741449	0.919994	8.000618	−0.000620
2	1.75	1.751400	0.920002	7.999754	0.000244
3	1.74	1.761466	0.919991	8.000928	−0.000930
4	1.73	1.772038	0.919991	8.000916	−0.000918
5	1.72	1.783120	0.920001	7.999933	0.000066
6	1.71	1.794517	0.920002	7.999814	0.000185
7	1.70	1.806428	0.920009	7.999110	0.000888
8	1.69	1.818658	0.920005	7.999552	0.000447
9	1.68	1.831406	0.920002	7.999838	0.000161
10	1.67	1.844675	0.919998	8.000195	−0.000197
11	1.66	1.858465	0.919991	8.000856	−0.000858
12	1.65	1.873171	0.920006	7.999385	0.000614
13	1.64	1.888208	0.919998	8.000154	−0.000155
14	1.63	1.904166	0.920005	7.999498	0.000501
15	1.62	1.920851	0.920009	7.999122	0.000876
16	1.61	1.938071	0.919994	8.000606	−0.000608
17	1.60	1.956609	0.920004	7.999575	0.000423
18	1.59	1.976079	0.920010	7.999033	0.000966
19	1.58	1.996483	0.920006	7.999402	0.000596
20	1.57	2.018213	0.920008	7.999206	0.000793

(continued)

Table 8.2 (continued)

Parameter values					
PROB. = 0.92	RHO = −0.7		BIGH = 1.7507		SMHL = 1.41
Pair No.	SMH	SMK	PROB. level	Risk level (%)	Risk error (%)
21	1.56	2.041274	0.920008	7.999176	0.000823
22	1.55	2.065668	0.919999	8.000076	−0.000077
23	1.54	2.091790	0.919991	8.000875	−0.000876
24	1.53	2.120423	0.920007	7.999349	0.000650
25	1.52	2.150789	0.919998	8.000189	−0.000191
26	1.51	2.184065	0.920000	8.000040	−0.000042
27	1.50	2.220644	0.920005	7.999480	0.000519
28	1.49	2.260920	0.920006	7.999444	0.000554
29	1.48	2.306068	0.920010	7.999003	0.000995
30	1.47	2.356875	0.920004	7.999563	0.000435
31	1.46	2.415687	0.920002	7.999772	0.000226
32	1.45	2.485633	0.920005	7.999534	0.000465
33	1.44	2.570624	0.919991	8.000946	−0.000948
34	1.43	2.683946	0.920004	7.999635	0.000364
35	1.42	2.849040	0.920004	7.999641	0.000358
36	1.41	3.186222	0.920009	7.999057	0.000942
No. of pairs		Mean COMPUT. risk errors		STD. ERR. errors	
36		0.0001802638		0.000974165	
Z-ratio				Status of significance	
1.85044				N.S.—not significant	
Parameter values					
PROB. = 0.91	RHO = −0.7		BIGH = 1.6954		SMHL = 1.35
Pair No.	SMH	SMK	PROB. level	Risk level (%)	Risk error (%)
Table No. 8.2-333					
1	1.70	1.690812	0.909999	9.000147	−0.000149
2	1.69	1.700817	0.909998	9.000242	−0.000244
3	1.68	1.711136	0.909994	9.000617	−0.000620
4	1.67	1.721968	0.910003	8.999693	0.000304
5	1.66	1.733117	0.910006	8.999432	0.000566
6	1.65	1.744588	0.910001	8.999944	0.000054
7	1.64	1.756578	0.910003	8.999723	0.000274
8	1.63	1.768893	0.909994	9.000641	−0.000644
9	1.62	1.781926	0.910004	8.999639	0.000358
10	1.61	1.795291	0.909998	9.000224	−0.000226
11	1.60	1.809379	0.910005	8.999496	0.000501
12	1.59	1.823802	0.909992	9.000814	−0.000817
13	1.58	1.839150	0.910000	8.999956	0.000042
14	1.57	1.855035	0.909997	9.000296	−0.000298
15	1.56	1.871849	0.910007	8.999348	0.000650

(continued)

Table 8.2 (continued)

Parameter values					
PROB. = 0.91	RHO = −0.7		BIGH = 1.6954		SMHL = 1.35
Pair No.	SMH	SMK	PROB. level	Risk level (%)	Risk error (%)
16	1.55	1.889205	0.909997	9.000272	−0.000274
17	1.54	1.907497	0.909992	9.000814	−0.000817
18	1.53	1.927118	0.910009	8.999056	0.000942
19	1.52	1.947290	0.909995	9.000522	−0.000525
20	1.51	1.968798	0.909990	9.000974	−0.000978
21	1.50	1.992035	0.910009	8.999073	0.000924
22	1.49	2.016224	0.910000	9.000021	−0.000024
23	1.48	2.042149	0.909995	9.000504	−0.000507
24	1.47	2.070205	0.910003	8.999718	0.000280
25	1.46	2.100004	0.909991	9.000903	−0.000906
26	1.45	2.132722	0.909997	9.000301	−0.000304
27	1.44	2.168363	0.910001	8.999909	0.000089
28	1.43	2.207713	0.910010	8.999044	0.000954
29	1.42	2.250774	0.909996	9.000361	−0.000364
30	1.41	2.299505	0.909992	9.000784	−0.000787
31	1.40	2.355472	0.909994	9.000617	−0.000620
32	1.39	2.421024	0.909997	9.000290	−0.000292
33	1.38	2.500072	0.909998	9.000182	−0.000185
34	1.37	2.599650	0.909991	9.000932	−0.000936
35	1.36	2.738515	0.909999	9.000081	−0.000083
36	1.35	2.969795	0.910002	8.999806	0.000191
No. of pairs		Mean COMPUT. risk errors		STD. ERR. errors	
36		−0.0001208202		0.0000887619	
Z-ratio				Status of significance	
1.36117				N.S.—not significant	
Parameter values					
PROB. = 0.9025	RHO = −0.7		BIGH = 1.6571		SMHL = 1.30
Pair No.	SMH	SMK	PROB. level	Risk level (%)	Risk error (%)
Table No. 8.2-334					
1	1.66	1.654205	0.902501	9.749949	0.000054
2	1.65	1.664231	0.902496	9.750396	−0.000393
3	1.64	1.674769	0.902508	9.749168	0.000834
4	1.63	1.685432	0.902497	9.750277	−0.000274
5	1.62	1.696612	0.902500	9.750051	−0.000048
6	1.61	1.708117	0.902494	9.750568	−0.000566
7	1.60	1.720144	0.902498	9.750170	−0.000167
8	1.59	1.732500	0.902491	9.750896	−0.000894
9	1.58	1.745579	0.902505	9.749496	0.000507
10	1.57	1.758993	0.902504	9.749639	0.000364
11	1.56	1.772939	0.902501	9.749878	0.000125
12	1.55	1.787420	0.902495	9.750504	−0.000501
13	1.54	1.802635	0.902498	9.750234	−0.000232

(continued)

8.2 Contents for Table 8.2 (Table No. 8.2-1 to 400)

Table 8.2 (continued)

Parameter values					
PROB. = 0.9025	RHO = −0.7		BIGH = 1.6571		SMHL = 1.30
Pair No.	SMH	SMK	PROB. level	Risk level (%)	Risk error (%)
14	1.53	1.818586	0.902505	9.749508	0.000495
15	1.52	1.835082	0.902498	9.750176	−0.000173
16	1.51	1.852514	0.902503	9.749722	0.000280
17	1.50	1.870692	0.902499	9.750069	−0.000066
18	1.49	1.889815	0.902497	9.750324	−0.000322
19	1.48	1.910080	0.902502	9.749776	0.000226
20	1.47	1.931295	0.902496	9.750390	−0.000387
21	1.46	1.953855	0.902496	9.750378	−0.000376
22	1.45	1.977764	0.902493	9.750671	−0.000668
23	1.44	2.003415	0.902500	9.749979	0.000024
24	1.43	2.030813	0.902505	9.749544	0.000459
25	1.42	2.059961	0.902495	9.750486	−0.000483
26	1.41	2.091644	0.902495	9.750486	−0.000483
27	1.40	2.126258	0.902503	9.749741	0.000262
28	1.39	2.163806	0.902496	9.750390	−0.000387
29	1.38	2.205465	0.902496	9.750396	−0.000393
30	1.37	2.252021	0.902496	9.750378	−0.000376
31	1.36	2.305040	0.902503	9.749699	0.000304
32	1.35	2.366090	0.902504	9.749651	0.000352
33	1.34	2.438301	0.902499	9.750098	−0.000095
34	1.33	2.527146	0.902491	9.750885	−0.000882
35	1.32	2.645912	0.902509	9.749102	0.000900
36	1.31	2.821167	0.902510	9.749037	0.000966
37	1.30	3.193541	0.902497	9.750324	−0.000322
No. of pairs		Mean COMPUT. risk errors		STD. ERR. errors	
37		−0.0000614869		0.0000751260	
Z-ratio				Status of significance	
0.81845				N.S.—not significant	
Parameter values					
PROB. = 0.9	RHO = −0.7		BIGH = 1.6448		SMHL = 1.29
Pair No.	SMH	SMK	PROB. level	Risk level (%)	Risk error (%)
Table No. 8.2-335					
1	1.65	1.639812	0.900007	9.999293	0.000709
2	1.64	1.649809	0.900007	9.999264	0.000739
3	1.63	1.660125	0.900005	9.999460	0.000542
4	1.62	1.670761	0.900000	9.999985	0.000018
5	1.61	1.681915	0.900009	9.999079	0.000924
6	1.60	1.693198	0.899993	10.000750	−0.000745
7	1.59	1.705200	0.900005	9.999538	0.000465
8	1.58	1.717531	0.900006	9.999376	0.000626
9	1.57	1.730195	0.899995	10.000470	−0.000465
10	1.56	1.743584	0.900005	9.999502	0.000501

(continued)

Table 8.2 (continued)

Parameter values					
PROB. = 0.9	RHO = −0.7		BIGH = 1.6448		SMHL = 1.29
Pair No.	SMH	SMK	PROB. level	Risk level (%)	Risk error (%)
11	1.55	1.757312	0.899998	10.000250	−0.000250
12	1.54	1.771771	0.900004	9.999579	0.000423
13	1.53	1.786768	0.900005	9.999514	0.000489
14	1.52	1.802308	0.899997	10.000350	−0.000346
15	1.51	1.818587	0.899992	10.000840	−0.000834
16	1.50	1.835804	0.900000	9.999996	0.000006
17	1.49	1.853768	0.900002	9.999794	0.000209
18	1.48	1.872482	0.899994	10.000610	−0.000608
19	1.47	1.892339	0.899997	10.000340	−0.000334
20	1.46	1.913147	0.899991	10.000910	−0.000912
21	1.45	1.935301	0.899995	10.000540	−0.000542
22	1.44	1.958805	0.899999	10.000120	−0.000119
23	1.43	1.983662	0.899995	10.000500	−0.000501
24	1.42	2.010265	0.899995	10.000510	−0.000507
25	1.41	2.039012	0.900006	9.999400	0.000602
26	1.40	2.069514	0.899995	10.000540	−0.000536
27	1.39	2.102947	0.900001	9.999942	0.000060
28	1.38	2.139317	0.900002	9.999811	0.000191
29	1.37	2.179407	0.900006	9.999400	0.000602
30	1.36	2.223615	0.899998	10.000210	−0.000203
31	1.35	2.273507	0.899994	10.000580	−0.000578
32	1.34	2.330649	0.899992	10.000850	−0.000846
33	1.33	2.398173	0.900002	9.999770	0.000232
34	1.32	2.479207	0.899999	10.000110	−0.000107
35	1.31	2.582352	0.899996	10.000440	−0.000441
36	1.30	2.726363	0.899998	10.000240	−0.000238
37	1.29	2.972182	0.899996	10.000380	−0.000376
No. of pairs		Mean COMPUT. risk errors		STD. ERR. errors	
37		−0.0000566244		0.0000833852	
Z-ratio				Status of significance	
0.67907				N.S.—not significant	
Parameter values					
PROB. = 0.85	RHO = −0.7		BIGH = 1.4395		SMHL = 1.04
Pair No.	SMH	SMK	PROB. level	Risk level (%)	Risk error (%)
Table No. 8.2-336					
1	1.44	1.439000	0.849999	15.000110	−0.000113
2	1.43	1.449161	0.850003	14.999690	0.000304
3	1.42	1.459561	0.849999	15.000130	−0.000131
4	1.41	1.470301	0.849998	15.000200	−0.000197

(continued)

8.2 Contents for Table 8.2 (Table No. 8.2-1 to 400)

Table 8.2 (continued)

Parameter values						
PROB. = 0.85	RHO = −0.7		BIGH = 1.4395		SMHL = 1.04	
Pair No.	SMH	SMK	PROB. level	Risk level (%)	Risk error (%)	
5	1.40	1.481384	0.849999	15.000100	−0.000101	
6	1.39	1.492814	0.850000	14.999960	0.000036	
7	1.38	1.504593	0.850000	15.000000	−0.000006	
8	1.37	1.516823	0.850009	14.999150	0.000846	
9	1.36	1.529311	0.849999	15.000110	−0.000113	
10	1.35	1.542355	0.850006	14.999430	0.000572	
11	1.34	1.555763	0.850002	14.999770	0.000232	
12	1.33	1.569734	0.850009	14.999060	0.000942	
13	1.32	1.584076	0.850001	14.999880	0.000119	
14	1.31	1.598989	0.849997	15.000270	−0.000274	
15	1.30	1.614477	0.849994	15.000600	−0.000602	
16	1.29	1.630740	0.850008	14.999160	0.000834	
17	1.28	1.647391	0.849995	15.000550	−0.000548	
18	1.27	1.665021	0.850009	14.999090	0.000906	
19	1.26	1.683243	0.850007	14.999310	0.000685	
20	1.25	1.702259	0.850002	14.999790	0.000209	
21	1.24	1.722269	0.850007	14.999330	0.000668	
22	1.23	1.743081	0.849996	15.000360	−0.000358	
23	1.22	1.765093	0.849998	15.000220	−0.000226	
24	1.21	1.788310	0.850002	14.999840	0.000155	
25	1.20	1.812737	0.849998	15.000220	−0.000226	
26	1.19	1.838576	0.849991	15.000860	−0.000858	
27	1.18	1.866224	0.849998	15.000220	−0.000221	
28	1.17	1.895686	0.850001	14.999860	0.000143	
29	1.16	1.927165	0.849998	15.000160	−0.000167	
30	1.15	1.961253	0.850005	14.999470	0.000530	
31	1.14	1.998153	0.850009	14.999150	0.000846	
32	1.13	2.038066	0.849992	15.000830	−0.000834	
33	1.12	2.082564	0.849999	15.000120	−0.000119	
34	1.11	2.132045	0.850000	15.000010	−0.000012	
35	1.10	2.188078	0.850003	14.999730	0.000268	
36	1.09	2.252626	0.850003	14.999740	0.000262	
37	1.08	2.328822	0.849995	15.000490	−0.000495	
38	1.07	2.423316	0.850001	14.999920	0.000077	
39	1.06	2.547054	0.849996	15.000410	−0.000411	

(continued)

Table 8.2 (continued)

Parameter values					
PROB. = 0.85	RHO = −0.7		BIGH = 1.4395		SMHL = 1.04
Pair No.	SMH	SMK	PROB. level	Risk level (%)	Risk error (%)
40	1.05	2.732860	0.850002	14.999840	0.000161
41	1.04	3.142460	0.849992	15.000790	−0.000793
No. of pairs		Mean COMPUT. risk errors		STD. ERR. errors	
41		0.0000473999		0.0000735532	
Z-ratio				Status of significance	
0.64443				N.S.—not significant	
Parameter values					
PROB. = 0.8	RHO = −0.7		BIGH = 1.2815		SMHL = 0.85
Pair No.	SMH	SMK	PROB. level	Risk level (%)	Risk error (%)
Table No. 8.2-337					
1	1.29	1.272958	0.799999	20.000090	−0.000095
2	1.28	1.282904	0.800006	19.999450	0.000548
3	1.27	1.293006	0.799994	20.000590	−0.000590
4	1.26	1.303464	0.799998	20.000200	−0.000197
5	1.25	1.314184	0.799999	20.000140	−0.000137
6	1.24	1.325170	0.799995	20.000480	−0.000483
7	1.23	1.336523	0.800002	19.999770	0.000232
8	1.22	1.348151	0.800002	19.999780	0.000221
9	1.21	1.360155	0.800009	19.999080	0.000924
10	1.20	1.372442	0.800006	19.999390	0.000614
11	1.19	1.385114	0.800006	19.999370	0.000626
12	1.18	1.398176	0.800007	19.999270	0.000733
13	1.17	1.411634	0.800007	19.999250	0.000745
14	1.16	1.425492	0.800004	19.999580	0.000417
15	1.15	1.439755	0.799995	20.000480	−0.000483
16	1.14	1.454626	0.800005	19.999520	0.000483
17	1.13	1.469913	0.800003	19.999720	0.000280
18	1.12	1.485721	0.799999	20.000090	−0.000089
19	1.11	1.502154	0.800003	19.999710	0.000292
20	1.10	1.519119	0.799998	20.000210	−0.000209
21	1.09	1.536722	0.799992	20.000830	−0.000829
22	1.08	1.555165	0.800003	19.999690	0.000310
23	1.07	1.574259	0.800003	19.999720	0.000280
24	1.06	1.594207	0.800008	19.999230	0.000769
25	1.05	1.614919	0.800000	19.999970	0.000030
26	1.04	1.636696	0.800005	19.999500	0.000501
27	1.03	1.659449	0.800003	19.999720	0.000280
28	1.02	1.683283	0.799995	20.000540	−0.000542
29	1.01	1.708599	0.800007	19.999280	0.000721

(continued)

8.2 Contents for Table 8.2 (Table No. 8.2-1 to 400)

Table 8.2 (continued)

Parameter values					
PROB. = 0.8	RHO = −0.7		BIGH = 1.2815		SMHL = 0.85
Pair No.	SMH	SMK	PROB. level	Risk level (%)	Risk error (%)
30	1.00	1.735211	0.800009	19.999110	0.000894
31	0.99	1.763322	0.800004	19.999590	0.000405
32	0.98	1.793140	0.799995	20.000550	−0.000548
33	0.97	1.825064	0.799992	20.000760	−0.000763
34	0.96	1.859497	0.800004	19.999620	0.000381
35	0.95	1.896449	0.800002	19.999800	0.000197
36	0.94	1.936519	0.799996	20.000380	−0.000381
37	0.93	1.980695	0.800008	19.999240	0.000757
38	0.92	2.029186	0.799999	20.000140	−0.000143
39	0.91	2.083567	0.799993	20.000740	−0.000745
40	0.90	2.145806	0.799998	20.000170	−0.000173
41	0.89	2.218262	0.800000	19.999960	0.000042
42	0.88	2.305246	0.799995	20.000490	−0.000495
43	0.87	2.415758	0.799999	20.000090	−0.000089
44	0.86	2.568958	0.800005	19.999460	0.000536
45	0.85	2.828923	0.800002	19.999790	0.000209
No. of pairs		Mean COMPUT. risk errors		STD. ERR. errors	
45		0.0001181727		0.0000712759	
Z-ratio				Status of significance	
1.65796				N.S.—not significant	
Parameter values					
PROB. = 0.75	RHO = −0.7		BIGH = 1.15		SMHL = 0.68
Pair No.	SMH	SMK	PROB. level	Risk level (%)	Risk error (%)
Table No. 8.2-338					
1	1.16	1.140086	0.749991	25.000870	−0.000870
2	1.15	1.150000	0.749997	25.000290	−0.000292
3	1.14	1.160088	0.749991	25.000870	−0.000870
4	1.13	1.170452	0.749993	25.000710	−0.000709
5	1.12	1.181097	0.750001	24.999880	0.000125
6	1.11	1.191930	0.749996	25.000420	−0.000417
7	1.10	1.203054	0.749995	25.000500	−0.000495
8	1.09	1.214475	0.749998	25.000200	−0.000203
9	1.08	1.226197	0.750004	24.999650	0.000352
10	1.07	1.238130	0.749992	25.000820	−0.000823
11	1.06	1.250474	0.749998	25.000190	−0.000185
12	1.05	1.263137	0.750003	24.999690	0.000310
13	1.04	1.276127	0.750005	24.999520	0.000477
14	1.03	1.289449	0.750002	24.999800	0.000203
15	1.02	1.303209	0.750010	24.999040	0.000960
16	1.01	1.317218	0.749993	25.000750	−0.000745
17	1.00	1.331777	0.749998	25.000170	−0.000167
18	0.99	1.346796	0.750007	24.999290	0.000715

(continued)

Table 8.2 (continued)

Parameter values					
PROB. = 0.75	RHO = −0.7		BIGH = 1.15		SMHL = 0.68
Pair No.	SMH	SMK	PROB. level	Risk level (%)	Risk error (%)
19	0.98	1.362185	0.750001	24.999870	0.000131
20	0.97	1.378050	0.749994	25.000650	−0.000644
21	0.96	1.394498	0.749996	25.000410	−0.000411
22	0.95	1.411539	0.750004	24.999560	0.000441
23	0.94	1.429083	0.750000	24.999960	0.000042
24	0.93	1.447238	0.749995	25.000530	−0.000530
25	0.92	1.466211	0.750009	24.999080	0.000924
26	0.91	1.485718	0.749999	25.000140	−0.000143
27	0.90	1.506163	0.750010	24.999050	0.000948
28	0.89	1.527361	0.750009	24.999110	0.000894
29	0.88	1.549325	0.749991	25.000870	−0.000864
30	0.87	1.572458	0.749996	25.000440	−0.000441
31	0.86	1.596677	0.750000	25.000040	−0.000036
32	0.85	1.622093	0.750004	24.999600	0.000399
33	0.84	1.648818	0.750007	24.999320	0.000679
34	0.83	1.676967	0.750005	24.999550	0.000453
35	0.82	1.706750	0.750001	24.999880	0.000125
36	0.81	1.738380	0.749997	25.000320	−0.000322
37	0.80	1.772265	0.750004	24.999650	0.000352
38	0.79	1.808425	0.749996	25.000440	−0.000441
39	0.78	1.847465	0.749990	25.000990	−0.000989
40	0.77	1.890188	0.750005	24.999540	0.000459
41	0.76	1.936615	0.749994	25.000620	−0.000620
42	0.75	1.988333	0.749999	25.000110	−0.000113
43	0.74	2.046537	0.750008	24.999150	0.000846
44	0.73	2.112815	0.750003	24.999750	0.000250
45	0.72	2.190711	0.750005	24.999490	0.000513
46	0.71	2.285331	0.750004	24.999630	0.000370
47	0.70	2.407253	0.750001	24.999890	0.000107
48	0.69	2.582290	0.749996	25.000400	−0.000399
49	0.68	2.919851	0.749996	25.000380	−0.000381
No. of pairs		Mean COMPUT. risk errors		STD. ERR. errors	
49		−0.0000207424		0.0000774849	
Z-ratio				Status of significance	
0.26770				N.S.—not significant	
Parameter values					
PROB. = 0.7	RHO = −0.7		BIGH = 1.0357		SMHL = 0.55
Pair No.	SMH	SMK	PROB. level	Risk level (%)	Risk error (%)
Table No. 8.2-339					
1	1.04	1.031222	0.699993	30.000670	−0.000668
2	1.03	1.041236	0.699994	30.000630	−0.000626
3	1.02	1.051446	0.699990	30.000990	−0.000983

(continued)

Table 8.2 (continued)

Parameter values					
PROB. = 0.7	RHO = −0.7		BIGH = 1.0357		SMHL = 0.55
Pair No.	SMH	SMK	PROB. level	Risk level (%)	Risk error (%)
4	1.01	1.061956	0.700005	29.999520	0.000483
5	1.00	1.072577	0.699993	30.000670	−0.000662
6	0.99	1.083510	0.699999	30.000080	−0.000077
7	0.98	1.094663	0.700000	29.999970	0.000030
8	0.97	1.106045	0.699996	30.000360	−0.000358
9	0.96	1.117760	0.700008	29.999230	0.000775
10	0.95	1.129619	0.699992	30.000770	−0.000769
11	0.94	1.141827	0.699991	30.000890	−0.000888
12	0.93	1.154390	0.700003	29.999710	0.000292
13	0.92	1.167220	0.700007	29.999350	0.000650
14	0.91	1.180325	0.700002	29.999810	0.000197
15	0.90	1.193814	0.700007	29.999280	0.000721
16	0.89	1.207596	0.700002	29.999770	0.000232
17	0.88	1.221780	0.700005	29.999510	0.000495
18	0.87	1.236279	0.699996	30.000450	−0.000447
19	0.86	1.251202	0.699991	30.000940	−0.000936
20	0.85	1.266657	0.700006	29.999410	0.000596
21	0.84	1.282462	0.700005	29.999550	0.000453
22	0.83	1.298726	0.700002	29.999820	0.000179
23	0.82	1.315463	0.699996	30.000430	−0.000429
24	0.81	1.332785	0.700000	30.000020	−0.000018
25	0.80	1.350609	0.699995	30.000490	−0.000489
26	0.79	1.369144	0.700009	29.999070	0.000936
27	0.78	1.388114	0.699993	30.000700	−0.000697
28	0.77	1.407927	0.700002	29.999780	0.000226
29	0.76	1.428405	0.700002	29.999780	0.000221
30	0.75	1.449666	0.700002	29.999790	0.000209
31	0.74	1.471825	0.700009	29.999060	0.000948
32	0.73	1.494807	0.700004	29.999590	0.000411
33	0.72	1.518731	0.699992	30.000830	−0.000829
34	0.71	1.543914	0.700000	29.999960	0.000042
35	0.70	1.570282	0.700006	29.999380	0.000626
36	0.69	1.597959	0.700009	29.999080	0.000918
37	0.68	1.627071	0.700007	29.999320	0.000685
38	0.67	1.657842	0.700004	29.999560	0.000441
39	0.66	1.690498	0.700003	29.999700	0.000304
40	0.65	1.725268	0.699999	30.000100	−0.000101
41	0.64	1.762576	0.700001	29.999880	0.000119
42	0.63	1.802657	0.699994	30.000560	−0.000554
43	0.62	1.846525	0.700004	29.999600	0.000405
44	0.61	1.894028	0.699992	30.000770	−0.000763
45	0.60	1.946774	0.699993	30.000690	−0.000685

(continued)

Table 8.2 (continued)

Parameter values					
PROB. = 0.7	RHO = −0.7		BIGH = 1.0357		SMHL = 0.55
Pair No.	SMH	SMK	PROB. level	Risk level (%)	Risk error (%)
46	0.59	2.005982	0.699997	30.000350	−0.000352
47	0.58	2.073656	0.700002	29.999770	0.000232
48	0.57	2.152587	0.699996	30.000380	−0.000381
49	0.56	2.248692	0.700001	29.999920	0.000077
50	0.55	2.372190	0.700002	29.999780	0.000226
No. of pairs		Mean COMPUT. risk errors		STD. ERR. errors	
50		0.0000081810		0.0000772671	
Z-ratio				Status of significance	
0.10588				N.S.—not significant	
Parameter values					
PROB. = 0.65	RHO = −0.7		BIGH = 0.9329		SMHL = 0.45
Pair No.	SMH	SMK	PROB. level	Risk level (%)	Risk error (%)
Table No. 8.2-340					
1	0.94	0.925854	0.650005	34.999510	0.000489
2	0.93	0.935809	0.650004	34.999650	0.000352
3	0.92	0.945981	0.650009	34.999150	0.000858
4	0.91	0.956279	0.649996	35.000420	−0.000417
5	0.90	0.966807	0.649991	35.000930	−0.000930
6	0.89	0.977575	0.649993	35.000730	−0.000727
7	0.88	0.988589	0.650002	34.999770	0.000232
8	0.87	0.999762	0.649996	35.000380	−0.000376
9	0.86	1.011198	0.649999	35.000150	−0.000149
10	0.85	1.022909	0.650009	34.999060	0.000942
11	0.84	1.034805	0.650006	34.999420	0.000584
12	0.83	1.046946	0.650000	34.999970	0.000036
13	0.82	1.059391	0.650004	34.999590	0.000411
14	0.81	1.072104	0.650006	34.999390	0.000608
15	0.80	1.085046	0.649996	35.000390	−0.000381
16	0.79	1.098328	0.649996	35.000410	−0.000405
17	0.78	1.111963	0.650005	34.999480	0.000525
18	0.77	1.125868	0.650003	34.999670	0.000334
19	0.76	1.140056	0.649991	35.000920	−0.000912
20	0.75	1.154739	0.650008	34.999230	0.000775
21	0.74	1.169639	0.649994	35.000580	−0.000572
22	0.73	1.184968	0.649990	35.000990	−0.000989
23	0.72	1.200745	0.649995	35.000550	−0.000548
24	0.71	1.216989	0.650007	34.999340	0.000668
25	0.70	1.233621	0.650007	34.999270	0.000733
26	0.69	1.250663	0.649997	35.000330	−0.000322
27	0.68	1.268333	0.650009	34.999080	0.000924
28	0.67	1.286459	0.650009	34.999150	0.000852
29	0.66	1.305065	0.649994	35.000580	−0.000578

(continued)

8.2 Contents for Table 8.2 (Table No. 8.2-1 to 400)

Table 8.2 (continued)

Parameter values					
PROB. = 0.65	RHO = −0.7		BIGH = 0.9329		SMHL = 0.45
Pair No.	SMH	SMK	PROB. level	Risk level (%)	Risk error (%)
30	0.65	1.324376	0.649998	35.000240	−0.000232
31	0.64	1.344320	0.650000	35.000010	0.000000
32	0.63	1.364928	0.649999	35.000090	−0.000083
33	0.62	1.386330	0.650007	34.999280	0.000727
34	0.61	1.408365	0.649992	35.000840	−0.000834
35	0.60	1.431363	0.649992	35.000810	−0.000811
36	0.59	1.455362	0.650002	34.999780	0.000226
37	0.58	1.480306	0.650003	34.999740	0.000262
38	0.57	1.506338	0.650001	34.999950	0.000054
39	0.56	1.533603	0.650000	35.000010	0.000000
40	0.55	1.562250	0.650003	34.999700	0.000298
41	0.54	1.592920	0.650007	34.999350	0.000656
42	0.53	1.624696	0.650007	34.999340	0.000662
43	0.52	1.658228	0.649991	35.000900	−0.000894
44	0.51	1.694169	0.649998	35.000160	−0.000155
45	0.50	1.732596	0.650002	34.999810	0.000191
46	0.49	1.773978	0.650006	34.999420	0.000584
47	0.48	1.818793	0.650005	34.999540	0.000465
48	0.47	1.867721	0.659996	35.000380	−0.000370
49	0.46	1.922038	0.650001	34.999920	0.000083
50	0.45	1.982832	0.649998	35.000220	−0.000215
No. of pairs		Mean COMPUT. risk errors		STD. ERR. errors	
50		0.0000515405		0.0000789203	
Z-ratio				Status of significance	
0.65307				N.S.—not significant	
Parameter values					
PROB. = 0.6	RHO = −0.7		BIGH = 0.8385		SMHL = 0.35
Pair No.	SMH	SMK	PROB. level	Risk level (%)	Risk error (%)
Table No. 8.2-341					
1	0.84	0.836905	0.600002	39.999820	0.000179
2	0.83	0.846989	0.600009	39.999150	0.000852
3	0.82	0.857222	0.600007	39.999260	0.000739
4	0.81	0.867612	0.600000	39.999990	0.000012
5	0.80	0.878218	0.600001	39.999900	0.000095
6	0.79	0.889001	0.599998	40.000220	−0.000221
7	0.78	0.900020	0.600005	39.999470	0.000530
8	0.77	0.911190	0.599999	40.000110	−0.000113
9	0.76	0.922569	0.599993	40.000700	−0.000703
10	0.75	0.934220	0.600002	39.999840	0.000155

(continued)

Table 8.2 (continued)

Parameter values					
PROB. = 0.6	RHO = −0.7		BIGH = 0.8385		SMHL = 0.35
Pair No.	SMH	SMK	PROB. level	Risk level (%)	Risk error (%)
11	0.74	0.946058	0.600001	39.999880	0.000119
12	0.73	0.958146	0.600005	39.999460	0.000542
13	0.72	0.970448	0.600004	39.999560	0.000441
14	0.71	0.982981	0.600000	40.000010	−0.000006
15	0.70	0.995809	0.600005	39.999480	0.000519
16	0.69	1.008852	0.599999	40.000100	−0.000101
17	0.68	1.022226	0.600006	39.999370	0.000626
18	0.67	1.035802	0.599995	40.000520	−0.000525
19	0.66	1.049749	0.600001	39.999940	0.000060
20	0.65	1.063989	0.600004	39.999640	0.000358
21	0.64	1.078546	0.600006	39.999450	0.000548
22	0.63	1.093445	0.600009	39.999120	0.000882
23	0.62	1.108613	0.599995	40.000460	−0.000459
24	0.61	1.124273	0.600009	39.999090	0.000906
25	0.60	1.140163	0.599991	40.000930	−0.000930
26	0.59	1.156606	0.600004	39.999640	0.000358
27	0.58	1.173343	0.599991	40.000910	−0.000906
28	0.57	1.190606	0.599994	40.000570	−0.000572
29	0.56	1.208335	0.599997	40.000300	−0.000298
30	0.55	1.226573	0.600002	39.999790	0.000209
31	0.54	1.245265	0.599994	40.000570	−0.000572
32	0.53	1.264558	0.599994	40.000580	−0.000584
33	0.52	1.284502	0.600004	39.999580	0.000417
34	0.51	1.304959	0.599994	40.000630	−0.000638
35	0.50	1.326183	0.599998	40.000190	−0.000191
36	0.49	1.348142	0.600004	39.999650	0.000346
37	0.48	1.370808	0.599997	40.000290	−0.000292
38	0.47	1.395528	0.599994	40.000620	−0.000620
39	0.46	1.419943	0.599997	40.000310	−0.000316
40	0.45	1.445412	0.600007	39.999300	0.000703
41	0.44	1.471829	0.599999	40.000130	−0.000131
42	0.43	1.499527	0.599998	40.000180	−0.000185
43	0.42	1.528496	0.599992	40.000810	−0.000811
44	0.41	1.559071	0.600001	39.999900	0.000101
45	0.40	1.591298	0.600008	39.999180	0.000823
46	0.39	1.625234	0.599998	40.000240	−0.000238
47	0.38	1.661347	0.599993	40.000740	−0.000739
48	0.37	1.700025	0.599998	40.000200	−0.000203

(continued)

Table 8.2 (continued)

Parameter values					
PROB. = 0.6	RHO = −0.7		BIGH = 0.8385		SMHL = 0.35
Pair No.	SMH	SMK	PROB. level	Risk level (%)	Risk error (%)
49	0.36	1.741481	0.599995	40.000470	−0.000477
50	0.35	1.786343	0.599995	40.000540	−0.000536
No. of pairs		Mean COMPUT. risk errors		STD. ERR. errors	
50		−0.0000165958		0.0000718095	
Z-ratio				Status of significance	
0.23111				N.S.—not significant	
Parameter values					
PROB. = 0.55	RHO = −0.7		BIGH = 0.7499		SMHL = 0.26
Pair No.	SMH	SMK	PROB. level	Risk level (%)	Risk error (%)
Table No. 8.2-342					
1	0.75	0.749800	0.549993	45.000730	−0.000727
2	0.74	0.759932	0.550009	44.999150	0.000852
3	0.73	0.770147	0.550002	44.999840	0.000161
4	0.72	0.780553	0.550002	44.999760	0.000244
5	0.71	0.791114	0.549999	45.000070	−0.000066
6	0.70	0.801892	0.550008	44.999180	0.000823
7	0.69	0.812754	0.549991	45.000890	−0.000888
8	0.68	0.823860	0.549991	45.000880	−0.000876
9	0.67	0.835178	0.549998	45.000220	−0.000221
10	0.66	0.846674	0.550001	44.999950	0.000048
11	0.65	0.858367	0.550002	44.999780	0.000221
12	0.64	0.870273	0.550007	44.999320	0.000679
13	0.63	0.882365	0.550004	44.999570	0.000429
14	0.62	0.894662	0.549999	45.000090	−0.000089
15	0.61	0.907237	0.550007	44.999270	0.000733
16	0.60	0.919965	0.549996	45.000440	−0.000435
17	0.59	0.933018	0.550006	44.999420	0.000578
18	0.58	0.946229	0.549993	45.000660	−0.000656
19	0.57	0.959772	0.550000	45.000030	−0.000024
20	0.56	0.973532	0.549994	45.000590	−0.000584
21	0.55	0.987591	0.549994	45.000590	−0.000590
22	0.54	1.001935	0.549994	45.000620	−0.000620
23	0.53	1.016604	0.550000	45.000060	−0.000054
24	0.52	1.031539	0.549995	45.000470	−0.000471
25	0.51	1.046836	0.549999	45.000070	−0.000066
26	0.50	1.062492	0.550008	44.999180	0.000823
27	0.49	1.078414	0.549998	45.000220	−0.000221
28	0.48	1.094804	0.550009	44.999140	0.000864
29	0.47	1.111528	0.550008	44.999210	0.000793

(continued)

Table 8.2 (continued)

Parameter values					
PROB. = 0.55	RHO = −0.7		BIGH = 0.7499		SMHL = 0.26
Pair No.	SMH	SMK	PROB. level	Risk level (%)	Risk error (%)
30	0.46	1.128652	0.550006	44.999390	0.000614
31	0.45	1.146150	0.549994	45.000630	−0.000632
32	0.44	1.164200	0.550001	44.999870	0.000137
33	0.43	1.182692	0.550002	44.999760	0.000244
34	0.42	1.201623	0.549991	45.000880	−0.000876
35	0.41	1.223537	0.549998	45.000210	−0.000209
36	0.40	1.243569	0.550010	44.999030	0.000972
37	0.39	1.264090	0.550001	44.999930	0.000072
38	0.38	1.285336	0.550006	44.999420	0.000578
39	0.37	1.307168	0.549993	45.000670	−0.000674
40	0.36	1.329855	0.550000	45.000040	−0.000036
41	0.35	1.353294	0.549998	45.000190	−0.000191
42	0.34	1.377601	0.549997	45.000310	−0.000310
43	0.33	1.402917	0.550006	44.999410	0.000590
44	0.32	1.429216	0.550009	44.999120	0.000882
45	0.31	1.456608	0.550007	44.999280	0.000721
46	0.30	1.485239	0.550007	44.999350	0.000650
47	0.29	1.515112	0.549990	45.000990	−0.000983
48	0.28	1.546671	0.549995	45.000540	−0.000536
49	0.27	1.580040	0.550010	44.999010	0.000989
50	0.26	1.615029	0.549990	45.000990	−0.000983
No. of pairs		Mean COMPUT. risk errors		STD. ERR. errors	
50		0.0000329579		0.0000839639	
Z-ratio				Status of significance	
0.39252				N.S.—not significant	
Parameter values					
PROB. = 0.5	RHO = −0.7		BIGH = 0.6655		SMHL = 0.18
Pair No.	SMH	SMK	PROB. level	Risk level (%)	Risk error (%)
Table No. 8.2-343					
1	0.67	0.660981	0.499999	50.000120	−0.000119
2	0.66	0.670997	0.500001	49.999870	0.000131
3	0.65	0.681174	0.500009	49.999110	0.000894
4	0.64	0.691430	0.499996	50.000430	−0.000432
5	0.63	0.701926	0.500009	49.999140	0.000864
6	0.62	0.712484	0.499994	50.000560	−0.000560
7	0.61	0.723266	0.499999	50.000060	−0.000057
8	0.60	0.734195	0.500000	50.000030	−0.000030
9	0.59	0.745290	0.500000	50.000030	−0.000024
10	0.58	0.756572	0.500004	49.999560	0.000447
11	0.57	0.768016	0.500005	49.999480	0.000519

(continued)

Table 8.2 (continued)

Parameter values					
PROB. = 0.5	RHO = −0.7		BIGH = 0.6655		SMHL = 0.18
Pair No.	SMH	SMK	PROB. level	Risk level (%)	Risk error (%)
12	0.56	0.779645	0.500007	49.999260	0.000745
13	0.55	0.791388	0.499990	50.000970	−0.000969
14	0.54	0.803419	0.500000	49.999990	0.000012
15	0.53	0.815622	0.500004	49.999640	0.000364
16	0.52	0.827981	0.499995	50.000510	−0.000516
17	0.51	0.840580	0.499995	50.000500	−0.000501
18	0.50	0.853407	0.499999	50.000120	−0.000119
19	0.49	0.866455	0.500003	49.999740	0.000268
20	0.48	0.879719	0.500004	49.999620	0.000381
21	0.47	0.893198	0.500000	50.000010	−0.000015
22	0.46	0.906945	0.500002	49.999800	0.000203
23	0.45	0.920919	0.499998	50.000170	−0.000173
24	0.44	0.935181	0.500002	49.999850	0.000149
25	0.43	0.949703	0.500002	49.999820	0.000185
26	0.42	0.964461	0.499992	50.000830	−0.000826
27	0.41	0.979585	0.499999	50.000090	−0.000083
28	0.40	0.995018	0.500008	49.999220	0.000781
29	0.39	1.010713	0.500004	49.999600	0.000405
30	0.38	1.026730	0.499999	50.000130	−0.000131
31	0.37	1.043093	0.499994	50.000620	−0.000614
32	0.36	1.064136	0.500001	49.999870	0.000131
33	0.35	1.081024	0.499994	50.000640	−0.000629
34	0.34	1.098418	0.500009	49.999090	0.000912
35	0.33	1.116114	0.500000	49.999970	0.000036
36	0.32	1.134323	0.500007	49.999300	0.000703
37	0.31	1.152895	0.499993	50.000660	−0.000656
38	0.30	1.172100	0.500008	49.999230	0.000775
39	0.29	1.191659	0.499991	50.000940	−0.000936
40	0.28	1.211924	0.500003	49.999720	0.000286
41	0.27	1.232714	0.500004	49.999650	0.000352
42	0.26	1.254105	0.500000	50.000040	−0.000036
43	0.25	1.276148	0.499994	50.000650	−0.000644
44	0.24	1.298986	0.500001	49.999930	0.000077
45	0.23	1.322579	0.500005	49.999500	0.000501
46	0.22	1.346923	0.499997	50.000340	−0.000340

(continued)

Table 8.2 (continued)

Parameter values					
PROB. = 0.5	RHO = −0.7		BIGH = 0.6655		SMHL = 0.18
Pair No.	SMH	SMK	PROB. level	Risk level (%)	Risk error (%)
47	0.21	1.372278	0.500005	49.999470	0.000530
48	0.20	1.398529	0.500001	49.999870	0.000137
49	0.19	1.425919	0.500007	49.999270	0.000727
50	0.18	1.454442	0.500008	49.999240	0.000763
No. of pairs		Mean COMPUT. risk errors		STD. ERR. errors	
50		0.0000759083		0.0000704783	
Z-ratio				Status of significance	
1.07705				N.S.—not significant	
Parameter values					
PROB. = 0.99	RHO = −0.8		BIGH = 2.5758		SMHL = 2.33
Pair No.	SMH	SMK	PROB. level	Risk level (%)	Risk error (%)
Table No. 8.2-344					
1	2.58	2.571607	0.989999	1.000142	−0.000143
2	2.57	2.581613	0.989998	1.000190	−0.000191
3	2.56	2.591698	0.989991	1.000881	−0.000882
4	2.55	2.603423	0.989999	1.000118	−0.000119
5	2.54	2.615229	0.989999	1.000106	−0.000107
6	2.53	2.627116	0.989991	1.000881	−0.000882
7	2.52	2.640648	0.989995	1.000512	−0.000513
8	2.51	2.655825	0.990008	0.999236	0.000763
9	2.50	2.669523	0.989992	1.000774	−0.000775
10	2.49	2.686431	0.990002	0.999826	0.000173
11	2.48	2.703425	0.989999	1.000059	−0.000060
12	2.47	2.722069	0.990001	0.999934	0.000066
13	2.46	2.742364	0.990003	0.999689	0.000310
14	2.45	2.764309	0.990005	0.999504	0.000495
15	2.44	2.787908	0.990004	0.999618	0.000381
16	2.43	2.814723	0.990010	0.999045	0.000954
17	2.42	2.841630	0.989996	1.000446	−0.000447
18	2.41	2.874881	0.990003	0.999719	0.000280
19	2.40	2.911352	0.990003	0.999689	0.000310
20	2.39	2.951044	0.989992	1.000762	−0.000763
21	2.38	3.000208	0.989995	1.000518	−0.000519
22	2.37	3.061970	0.990007	0.999332	0.000668
23	2.36	3.136332	0.990007	0.999260	0.000739

(continued)

8.2 Contents for Table 8.2 (Table No. 8.2-1 to 400)

Table 8.2 (continued)

Parameter values					
PROB. = 0.99	RHO = −0.8		BIGH = 2.5758		SMHL = 2.33
Pair No.	SMH	SMK	PROB. level	Risk level (%)	Risk error (%)
24	2.35	3.235795	0.990007	0.999278	0.000721
25	2.34	3.385361	0.990004	0.999576	0.000423
26	2.33	3.697529	0.990002	0.999832	0.000167
No. of pairs		Mean COMPUT. risk errors		STD. ERR. errors	
26		0.0000388534		0.0001028189	
Z-ratio				Status of significance	
0.37788				N.S.—not significant	
Parameter values					
PROB. = 0.98	RHO = −0.8		BIGH = 2.3265		SMHL = 2.06
Pair No.	SMH	SMK	PROB. level	Risk level (%)	Risk error (%)
Table No. 8.2-345					
1	2.33	2.323005	0.980007	1.999259	0.000739
2	2.32	2.333018	0.980006	1.999408	0.000590
3	2.31	2.343118	0.979994	2.000570	−0.000572
4	2.30	2.354087	0.979992	2.000821	−0.000823
5	2.29	2.365925	0.979997	2.000320	−0.000322
6	2.28	2.378636	0.980008	1.999223	0.000775
7	2.27	2.391437	0.980005	1.999515	0.000483
8	2.26	2.405113	0.980006	1.999444	0.000554
9	2.25	2.419663	0.980008	1.999199	0.000799
10	2.24	2.434309	0.979994	2.000558	−0.000560
11	2.23	2.450613	0.979996	2.000433	−0.000435
12	2.22	2.467796	0.979993	2.000672	−0.000674
13	2.21	2.486641	0.980000	2.000034	−0.000036
14	2.20	2.506367	0.979998	2.000248	−0.000250
15	2.19	2.527758	0.979998	2.000177	−0.000179
16	2.18	2.551595	0.980010	1.999027	0.000972
17	2.17	2.575536	0.979992	2.000767	−0.000769
18	2.16	2.603490	0.980000	2.000034	−0.000036
19	2.15	2.633114	0.979992	2.000791	−0.000793
20	2.14	2.666753	0.979993	2.000690	−0.000691
21	2.13	2.705190	0.980001	1.999903	0.000095
22	2.12	2.748427	0.980003	1.999736	0.000262
23	2.11	2.798026	0.980000	1.999998	0.000000
24	2.10	2.858679	0.980009	1.999140	0.000858
25	2.09	2.930386	0.979998	2.000159	−0.000161
26	2.08	3.027212	0.980003	1.999682	0.000316

(continued)

Table 8.2 (continued)

Parameter values					
PROB. = 0.98	RHO = −0.8		BIGH = 2.3265		SMHL = 2.06
Pair No.	SMH	SMK	PROB. level	Risk level (%)	Risk error (%)
27	2.07	3.164783	0.979998	2.000219	−0.000221
28	2.06	3.427476	0.979997	2.000350	−0.000352
No. of pairs		Mean COMPUT. risk errors		STD. ERR. errors	
28		−0.0000147984		0.0001028096	
Z-ratio				Status of significance	
0.14394				N.S.—not significant	
Parameter values					
PROB. = 0.97	RHO = −0.8		BIGH = 2.1701		SMHL = 1.89
Pair No.	SMH	SMK	PROB. level	Risk level (%)	Risk error (%)
Table No. 8.2-346					
1	2.18	2.160245	0.969995	3.000552	−0.000554
2	2.17	2.170200	0.970001	2.999926	0.000072
3	2.16	2.180247	0.969994	3.000593	−0.000596
4	2.15	2.191169	0.970003	2.999735	0.000262
5	2.14	2.202186	0.969996	3.000355	−0.000358
6	2.13	2.214080	0.970003	2.999735	0.000262
7	2.12	2.226071	0.969992	3.000790	−0.000793
8	2.11	2.238943	0.969991	3.000903	−0.000906
9	2.10	2.252696	0.969996	3.000355	−0.000358
10	2.09	2.267332	0.970006	2.999389	0.000608
11	2.08	2.282071	0.969995	3.000540	−0.000542
12	2.07	2.298478	0.970006	2.999354	0.000644
13	2.06	2.314991	0.969994	3.000593	−0.000596
14	2.05	2.333173	0.969998	3.000188	−0.000191
15	2.04	2.352247	0.969995	3.000534	−0.000536
16	2.03	2.372994	0.969999	3.000057	−0.000060
17	2.02	2.395416	0.970008	2.999246	0.000751
18	2.01	2.418733	0.969997	3.000295	−0.000298
19	2.00	2.444510	0.969997	3.000260	−0.000262
20	1.99	2.472749	0.970001	2.999932	0.000066
21	1.98	2.503451	0.969999	3.000116	−0.000119
22	1.97	2.537399	0.969997	3.000319	−0.000322
23	1.96	2.576158	0.970007	2.999306	0.000691
24	1.95	2.618167	0.969992	3.000832	−0.000834
25	1.94	2.668116	0.969996	3.000379	−0.000381
26	1.93	2.727569	0.970006	2.999365	0.000632
27	1.92	2.798090	0.970001	2.999926	0.000072

(continued)

8.2 Contents for Table 8.2 (Table No. 8.2-1 to 400)

Table 8.2 (continued)

Parameter values					
PROB. = 0.97	RHO = −0.8		BIGH = 2.1701		SMHL = 1.89
Pair No.	SMH	SMK	PROB. level	Risk level (%)	Risk error (%)
28	1.91	2.887494	0.969992	3.000832	−0.000834
29	1.90	3.016096	0.970003	2.999681	0.000316
30	1.89	3.229210	0.970000	2.999973	0.000024
No. of pairs		Mean COMPUT. risk errors		STD. ERR. errors	
30		−0.0001336298		0.0000872676	
Z-ratio				Status of significance	
1.53126				N.S.—not significant	
Parameter values					
PROB. = 0.96	RHO = −0.8		BIGH = 2.0538		SMHL = 1.76
Pair No.	SMH	SMK	PROB. level	Risk level (%)	Risk error (%)
Table No. 8.2-347					
1	2.06	2.047619	0.960002	3.999794	0.000209
2	2.05	2.057607	0.960004	3.999615	0.000387
3	2.04	2.067694	0.959990	4.000962	−0.000960
4	2.03	2.078660	0.959997	4.000259	−0.000256
5	2.02	2.090119	0.960005	3.999519	0.000483
6	2.01	2.101680	0.959994	4.000628	−0.000626
7	2.00	2.114125	0.959998	4.000247	−0.000244
8	1.99	2.127068	0.959997	4.000264	−0.000262
9	1.98	2.140507	0.959991	4.000873	−0.000870
10	1.97	2.155227	0.960009	3.999114	0.000888
11	1.96	2.170058	0.960001	3.999913	0.000089
12	1.95	2.185782	0.959996	4.000378	−0.000376
13	1.94	2.202401	0.959991	4.000855	−0.000852
14	1.93	2.220307	0.959998	4.000235	−0.000232
15	1.92	2.239112	0.959997	4.000318	−0.000316
16	1.91	2.259208	0.959998	4.000187	−0.000185
17	1.90	2.280987	0.960009	3.999114	0.000888
18	1.89	2.303671	0.960000	3.999961	0.000042
19	1.88	2.328042	0.959991	4.000897	−0.000894
20	1.87	2.354884	0.959994	4.000622	−0.000620
21	1.86	2.384199	0.959999	4.000092	−0.000089
22	1.85	2.415989	0.959997	4.000306	−0.000304
23	1.84	2.451818	0.960009	3.999102	0.000900
24	1.83	2.490907	0.960004	3.999591	0.000411
25	1.82	2.534822	0.959995	4.000485	−0.000483
26	1.81	2.586689	0.960007	3.999305	0.000697
27	1.80	2.646510	0.960003	3.999674	0.000328
28	1.79	2.718977	0.959999	4.000139	−0.000137
29	1.78	2.811903	0.959999	4.000056	−0.000054

(continued)

Table 8.2 (continued)

Parameter values					
PROB. = 0.96	RHO = −0.8		BIGH = 2.0538		SMHL = 1.76
Pair No.	SMH	SMK	PROB. level	Risk level (%)	Risk error (%)
30	1.77	2.939354	0.959992	4.000807	−0.000805
31	1.76	3.159144	0.960005	3.999508	0.000495
No. of pairs		Mean COMPUT. risk errors		STD. ERR. errors	
31		−0.0000858679		0.0000957859	
Z-ratio				Status of significance	
0.89646				N.S.—not significant	
Parameter values					
PROB. = 0.95	RHO = −0.8		BIGH = 1.9599		SMHL = 1.65
Pair No.	SMH	SMK	PROB. level	Risk level (%)	Risk error (%)
Table No. 8.2-348					
1	1.96	1.959800	0.949992	5.000758	−0.000757
2	1.95	1.970241	0.950007	4.999346	0.000656
3	1.94	1.980785	0.950002	4.999757	0.000244
4	1.93	1.991826	0.950002	4.999852	0.000149
5	1.92	2.003363	0.950002	4.999805	0.000197
6	1.91	2.015400	0.950002	4.999805	0.000197
7	1.90	2.027938	0.950000	4.999990	0.000012
8	1.89	2.040979	0.949995	5.000544	−0.000542
9	1.88	2.054915	0.950002	4.999757	0.000244
10	1.87	2.069356	0.950002	4.999829	0.000173
11	1.86	2.084697	0.950009	4.999113	0.000888
12	1.85	2.100547	0.950003	4.999709	0.000292
13	1.84	2.117300	0.949999	5.000139	−0.000137
14	1.83	2.134958	0.949993	5.000729	−0.000727
15	1.82	2.153913	0.949997	5.000305	−0.000304
16	1.81	2.173777	0.949991	5.000890	−0.000888
17	1.80	2.195332	0.950000	5.000031	−0.000030
18	1.79	2.218192	0.950002	4.999793	0.000209
19	1.78	2.242357	0.949993	5.000729	−0.000727
20	1.77	2.268611	0.949990	5.000961	−0.000960
21	1.76	2.297348	0.949997	5.000335	−0.000334
22	1.75	2.328569	0.950000	5.000019	−0.000018
23	1.74	2.363059	0.950008	4.999185	0.000817
24	1.73	2.400820	0.950006	4.999429	0.000572
25	1.72	2.442635	0.949994	5.000639	−0.000638
26	1.71	2.491633	0.950009	4.999078	0.000924
27	1.70	2.547033	0.950002	4.999763	0.000238
28	1.69	2.613528	0.950005	4.999471	0.000530

(continued)

8.2 Contents for Table 8.2 (Table No. 8.2-1 to 400)

Table 8.2 (continued)

Parameter values					
PROB. = 0.95	RHO = −0.8		BIGH = 1.9599		SMHL = 1.65
Pair No.	SMH	SMK	PROB. level	Risk level (%)	Risk error (%)
29	1.68	2.694245	0.949994	5.000603	−0.000602
30	1.67	2.801686	0.949999	5.000139	−0.000137
31	1.66	2.957730	0.949993	5.000675	−0.000674
32	1.65	3.278004	0.950006	4.999411	0.000590
No. of pairs		Mean COMPUT. risk errors		STD. ERR. errors	
32		−0.0000164364		0.0000934485	
Z-ratio				Status of significance	
0.17589				N.S.—not significant	
Parameter values					
PROB. = 0.94	RHO = −0.8		BIGH = 1.8808		SMHL = 1.56
Pair No.	SMH	SMK	PROB. level	Risk level (%)	Risk error (%)
Table No. 8.2-349					
1	1.89	1.871645	0.939993	6.000686	−0.000685
2	1.88	1.881600	0.940001	5.999911	0.000089
3	1.87	1.891662	0.939990	6.000990	−0.000989
4	1.86	1.902418	0.939999	6.000108	−0.000107
5	1.85	1.913480	0.940000	6.000018	−0.000018
6	1.84	1.924848	0.939992	6.000823	−0.000823
7	1.83	1.936916	0.939997	6.000281	−0.000280
8	1.82	1.949490	0.940002	5.999804	0.000197
9	1.81	1.962572	0.940004	5.999589	0.000411
10	1.80	1.976164	0.940002	5.999851	0.000149
11	1.79	1.990268	0.939992	6.000781	−0.000781
12	1.78	2.005277	0.939995	6.000477	−0.000477
13	1.77	2.021192	0.940007	5.999351	0.000650
14	1.76	2.037625	0.940002	5.999768	0.000232
15	1.75	2.054970	0.940000	6.000018	−0.000018
16	1.74	2.073227	0.939995	6.000501	−0.000501
17	1.73	2.092792	0.940001	5.999911	0.000089
18	1.72	2.113273	0.939995	6.000495	−0.000495
19	1.71	2.135457	0.940005	5.999482	0.000519
20	1.70	2.158954	0.940008	5.999237	0.000763
21	1.69	2.183766	0.939996	6.000376	−0.000376
22	1.68	2.211069	0.940006	5.999422	0.000578
23	1.67	2.240084	0.939998	6.000251	−0.000250
24	1.66	2.271984	0.939999	6.000090	−0.000089
25	1.65	2.307165	0.940006	5.999428	0.000572
26	1.64	2.346018	0.940010	5.999023	0.000978
27	1.63	2.388939	0.940001	5.999935	0.000066
28	1.62	2.438273	0.940005	5.999494	0.000507
29	1.61	2.494804	0.940000	6.000018	−0.000018
30	1.60	2.562442	0.940004	5.999613	0.000387

(continued)

Table 8.2 (continued)

Parameter values					
PROB. = 0.94	RHO = −0.8		BIGH = 1.8808		SMHL = 1.56
Pair No.	SMH	SMK	PROB. level	Risk level (%)	Risk error (%)
31	1.59	2.645097	0.939999	6.000072	−0.000072
32	1.58	2.754491	0.940007	5.999261	0.000739
33	1.57	2.912501	0.940000	6.000042	−0.000042
34	1.56	3.230068	0.940001	5.999875	0.000125
No. of pairs			Mean COMPUT. risk errors	STD. ERR. errors	
34			0.0000294617	0.0000814799	
Z-ratio				Status of significance	
0.36158				N.S.—not significant	
Parameter values					
PROB. = 0.93	RHO = −0.8		BIGH = 1.8119		SMHL = 1.48
Pair No.	SMH	SMK	PROB. level	Risk level (%)	Risk error (%)
Table No. 8.2-350					
1	1.82	1.803836	0.929992	7.000804	−0.000805
2	1.81	1.813802	0.929998	7.000196	−0.000197
3	1.80	1.824074	0.929999	7.000089	−0.000089
4	1.79	1.834849	0.930009	6.999106	0.000894
5	1.78	1.845739	0.929997	7.000315	−0.000316
6	1.77	1.857136	0.929991	7.000947	−0.000948
7	1.76	1.869237	0.930001	6.999874	0.000125
8	1.75	1.881653	0.929999	7.000077	−0.000077
9	1.74	1.894584	0.929997	7.000327	−0.000328
10	1.73	1.908224	0.930004	6.999636	0.000364
11	1.72	1.922382	0.930005	6.999505	0.000495
12	1.71	1.937060	0.929998	7.000214	−0.000215
13	1.70	1.952650	0.930004	6.999600	0.000399
14	1.69	1.968764	0.929996	7.000411	−0.000411
15	1.68	1.985796	0.929993	7.000661	−0.000662
16	1.67	2.003748	0.929992	7.000828	−0.000829
17	1.66	2.023012	0.930007	6.999314	0.000685
18	1.65	2.042811	0.929993	7.000709	−0.000709
19	1.64	2.064318	0.930004	6.999648	0.000352
20	1.63	2.086755	0.929994	7.000578	−0.000578
21	1.62	2.110906	0.929994	7.000625	−0.000626
22	1.61	2.136775	0.929993	7.000709	−0.000709
23	1.60	2.164754	0.929997	7.000256	−0.000256
24	1.59	2.195237	0.930009	6.999082	0.000918
25	1.58	2.227836	0.930001	6.999922	0.000077
26	1.57	2.263727	0.929997	7.000304	−0.000304
27	1.56	2.303694	0.930000	6.999994	0.000006

(continued)

8.2 Contents for Table 8.2 (Table No. 8.2-1 to 400)

Table 8.2 (continued)

Parameter values					
PROB. = 0.93	RHO = −0.8		BIGH = 1.8119		SMHL = 1.48
Pair No.	SMH	SMK	PROB. level	Risk level (%)	Risk error (%)
28	1.55	2.348521	0.930005	6.999481	0.000519
29	1.54	2.398993	0.930000	7.000023	−0.000024
30	1.53	2.457458	0.929995	7.000458	−0.000459
31	1.52	2.527043	0.929993	7.000685	−0.000685
32	1.51	2.613222	0.929994	7.000632	−0.000632
33	1.50	2.727716	0.930004	6.999588	0.000411
34	1.49	2.897092	0.930005	6.999529	0.000471
35	1.48	3.255729	0.929998	7.000208	−0.000209
No. of pairs		Mean COMPUT. risk errors		STD. ERR. errors	
35		−0.0001208650		0.0000849201	
Z-ratio				Status of significance	
1.42328				N.S.—not significant	
Parameter values					
PROB. = 0.92	RHO = −0.8		BIGH = 1.7507		SMHL = 1.41
Pair No.	SMH	SMK	PROB. level	Risk level (%)	Risk error (%)
Table No. 8.2-351					
1	1.76	1.741449	0.919994	8.000642	−0.000644
2	1.75	1.751400	0.920002	7.999772	0.000226
3	1.74	1.761466	0.919991	8.000952	−0.000954
4	1.73	1.772038	0.919991	8.000935	−0.000936
5	1.72	1.783120	0.920000	7.999963	0.000036
6	1.71	1.794517	0.920002	7.999844	0.000155
7	1.70	1.806428	0.920009	7.999140	0.000858
8	1.69	1.818658	0.920004	7.999575	0.000423
9	1.68	1.831406	0.920001	7.999855	0.000143
10	1.67	1.844675	0.919998	8.000207	−0.000209
11	1.66	1.858465	0.919991	8.000875	−0.000876
12	1.65	1.873171	0.920006	7.999391	0.000608
13	1.64	1.888208	0.919998	8.000165	−0.000167
14	1.63	1.904166	0.920005	7.999516	0.000483
15	1.62	1.920851	0.920009	7.999134	0.000864
16	1.61	1.938071	0.919994	8.000618	−0.000620
17	1.60	1.956609	0.920004	7.999593	0.000405
18	1.59	1.976079	0.920010	7.999033	0.000966
19	1.58	1.996483	0.920006	7.999420	0.000578
20	1.57	2.018213	0.920008	7.999218	0.000781
21	1.56	2.041274	0.920008	7.999176	0.000823
22	1.55	2.065668	0.919999	8.000070	−0.000072
23	1.54	2.091790	0.919991	8.000887	−0.000888
24	1.53	2.120423	0.920007	7.999349	0.000650
25	1.52	2.150789	0.919998	8.000195	−0.000197
26	1.51	2.184065	0.920000	8.000040	−0.000042

(continued)

Table 8.2 (continued)

Parameter values					
PROB. = 0.92	RHO = −0.8		BIGH = 1.7507		SMHL = 1.41
Pair No.	SMH	SMK	PROB. level	Risk level (%)	Risk error (%)
27	1.50	2.220644	0.920005	7.999486	0.000513
28	1.49	2.260920	0.920006	7.999444	0.000554
29	1.48	2.306068	0.920010	7.999003	0.000995
30	1.47	2.356875	0.920004	7.999563	0.000435
31	1.46	2.415687	0.920002	7.999778	0.000221
32	1.45	2.485633	0.920005	7.999528	0.000471
33	1.44	2.570624	0.919991	8.000946	−0.000948
34	1.43	2.683946	0.920004	7.999635	0.000364
35	1.42	2.849040	0.920004	7.999641	0.000358
36	1.41	3.186222	0.920010	7.999051	0.000948
No. of pairs		Mean COMPUT. risk errors		STD. ERR. errors	
36		0.0001704371		0.0000978798	
Z-ratio				Status of significance	
1.74129				N.S.—not significant	
Parameter values					
PROB. = 0.91	RHO = −0.8		BIGH = 1.6954		SMHL = 1.35
Pair No.	SMH	SMK	PROB. level	Risk level (%)	Risk error (%)
Table No. 8.2-352					
1	1.70	1.690812	0.909998	9.000182	−0.000185
2	1.69	1.700817	0.909997	9.000278	−0.000280
3	1.68	1.711136	0.909993	9.000665	−0.000668
4	1.67	1.721968	0.910003	8.999735	0.000262
5	1.66	1.733117	0.910005	8.999472	0.000525
6	1.65	1.744588	0.910000	8.999979	0.000018
7	1.64	1.756578	0.910002	8.999759	0.000238
8	1.63	1.768893	0.909993	9.000671	−0.000674
9	1.62	1.781926	0.910003	8.999681	0.000316
10	1.61	1.795291	0.909997	9.000265	−0.000268
11	1.60	1.809379	0.910005	8.999550	0.000447
12	1.59	1.823802	0.909991	9.000856	−0.000858
13	1.58	1.839150	0.910000	8.999991	0.000006
14	1.57	1.855035	0.909997	9.000325	−0.000328
15	1.56	1.871849	0.910006	8.999384	0.000614
16	1.55	1.889205	0.909997	9.000307	−0.000310
17	1.54	1.907497	0.909992	9.000844	−0.000846
18	1.53	1.927118	0.910009	8.999073	0.000924
19	1.52	1.947290	0.909995	9.000551	−0.000554
20	1.51	1.968798	0.909990	9.000992	−0.000995
21	1.50	1.992035	0.910009	8.999086	0.000912
22	1.49	2.016224	0.910000	9.000045	−0.000048
23	1.48	2.042149	0.909995	9.000510	−0.000513
24	1.47	2.070205	0.910003	8.999735	0.000262

(continued)

Table 8.2 (continued)

Parameter values					
PROB. = 0.91	RHO = −0.8		BIGH = 1.6954		SMHL = 1.35
Pair No.	SMH	SMK	PROB. level	Risk level (%)	Risk error (%)
25	1.46	2.100004	0.909991	9.000921	−0.000924
26	1.45	2.132722	0.909997	9.000301	−0.000304
27	1.44	2.168363	0.910001	8.999914	0.000083
28	1.43	2.207713	0.910009	8.999062	0.000936
29	1.42	2.250774	0.909996	9.000379	−0.000381
30	1.41	2.299505	0.909992	9.000789	−0.000793
31	1.40	2.355472	0.909994	9.000617	−0.000620
32	1.39	2.421024	0.909997	9.000290	−0.000292
33	1.38	2.500072	0.909998	9.000170	−0.000173
34	1.37	2.599650	0.909991	9.000932	−0.000936
35	1.36	2.738515	0.909999	9.000074	−0.000077
36	1.35	2.969795	0.910002	8.999801	0.000197
No. of pairs		Mean COMPUT. risk errors		STD. ERR. errors	
36		−0.0001428901		0.0000883711	
Z-ratio				Status of significance	
1.61693				N.S.—not significant	

Parameter values					
PROB. = 0.9025	RHO = −0.8		BIGH = 1.6571		SMHL = 1.30
Pair No.	SMH	SMK	PROB. level	Risk level (%)	Risk error (%)
Table No. 8.2-353					
1	1.66	1.654205	0.902500	9.750027	−0.000024
2	1.65	1.664231	0.902495	9.750468	−0.000465
3	1.64	1.674769	0.902508	9.749233	0.000769
4	1.63	1.685432	0.902497	9.750348	−0.000346
5	1.62	1.696612	0.902499	9.750122	−0.000119
6	1.61	1.708117	0.902494	9.750634	−0.000632
7	1.60	1.720144	0.902498	9.750241	−0.000238
8	1.59	1.732500	0.902490	9.750968	−0.000966
9	1.58	1.745579	0.902505	9.749544	0.000459
10	1.57	1.758993	0.902503	9.749699	0.000304
11	1.56	1.772939	0.902501	9.749949	0.000054
12	1.55	1.787420	0.902494	9.750568	−0.000566
13	1.54	1.802635	0.902497	9.750294	−0.000292
14	1.53	1.818586	0.902504	9.749561	0.000441
15	1.52	1.835082	0.902498	9.750223	−0.000221
16	1.51	1.852514	0.902502	9.749770	0.000232
17	1.50	1.870692	0.902499	9.750116	−0.000113
18	1.49	1.889815	0.902496	9.750366	−0.000364
19	1.48	1.910080	0.902502	9.749812	0.000191
20	1.47	1.931295	0.902496	9.750426	−0.000423
21	1.46	1.953855	0.902496	9.750419	−0.000417
22	1.45	1.977764	0.902493	9.750694	−0.000691

(continued)

Table 8.2 (continued)

Parameter values					
PROB. = 0.9025	RHO = −0.8		BIGH = 1.6571		SMHL = 1.30
Pair No.	SMH	SMK	PROB. level	Risk level (%)	Risk error (%)
23	1.44	2.003415	0.902500	9.750003	0.000000
24	1.43	2.030813	0.902504	9.749574	0.000429
25	1.42	2.059961	0.902495	9.750498	−0.000495
26	1.41	2.091644	0.902495	9.750509	−0.000507
27	1.40	2.126258	0.902503	9.749746	0.000256
28	1.39	2.163806	0.902496	9.750402	−0.000399
29	1.38	2.205465	0.902496	9.750402	−0.000399
30	1.37	2.252021	0.902496	9.750396	−0.000393
31	1.36	2.305040	0.902503	9.749710	0.000292
32	1.35	2.366090	0.902504	9.749651	0.000352
33	1.34	2.438301	0.902499	9.750098	−0.000095
34	1.33	2.527146	0.902491	9.750879	−0.000876
35	1.32	2.645912	0.902509	9.749090	0.000912
36	1.31	2.821167	0.902510	9.749031	0.000972
37	1.30	3.193541	0.902497	9.750319	−0.000316
No. of pairs		Mean COMPUT. risk errors		STD. ERR. errors	
37		−0.0000972497		0.0000758007	
Z-ratio				Status of significance	
1.28297				N.S.—not significant	
Parameter values					
PROB. = 0.9	RHO = −0.8		BIGH = 1.645		SMHL = 1.29
Pair No.	SMH	SMK	PROB. level	Risk level (%)	Risk error (%)
Table No. 8.2-354					
1	1.65	1.639820	0.900007	9.999275	0.000727
2	1.64	1.649820	0.900008	9.999251	0.000751
3	1.63	1.660138	0.900006	9.999418	0.000584
4	1.62	1.670776	0.900001	9.999913	0.000089
5	1.61	1.681737	0.899991	10.000900	−0.000894
6	1.60	1.693219	0.899994	10.000630	−0.000632
7	1.59	1.705223	0.900006	9.999407	0.000596
8	1.58	1.717362	0.899990	10.001000	−0.000995
9	1.57	1.730223	0.899997	10.000290	−0.000292
10	1.56	1.743616	0.900007	9.999299	0.000703
11	1.55	1.757346	0.900000	10.000010	−0.000012
12	1.54	1.771612	0.899990	10.000960	−0.000954
13	1.53	1.786612	0.899992	10.000840	−0.000834
14	1.52	1.802350	0.899999	10.000080	−0.000077
15	1.51	1.818632	0.899994	10.000560	−0.000554
16	1.50	1.835852	0.900003	9.999704	0.000298
17	1.49	1.853624	0.899991	10.000890	−0.000882
18	1.48	1.872536	0.899997	10.000280	−0.000274
19	1.47	1.892396	0.900000	10.000000	0.000000

(continued)

8.2 Contents for Table 8.2 (Table No. 8.2-1 to 400)

Table 8.2 (continued)

Parameter values					
PROB. = 0.9	RHO = −0.8		BIGH = 1.645		SMHL = 1.29
Pair No.	SMH	SMK	PROB. level	Risk level (%)	Risk error (%)
20	1.46	1.913207	0.899994	10.000590	−0.000584
21	1.45	1.935365	0.899998	10.000210	−0.000209
22	1.44	1.958871	0.900002	9.999782	0.000221
23	1.43	1.983731	0.899999	10.000150	−0.000149
24	1.42	2.010338	0.899998	10.000160	−0.000155
25	1.41	2.038698	0.899990	10.001000	−0.000995
26	1.40	2.069593	0.899998	10.000180	−0.000173
27	1.39	2.103030	0.900004	9.999603	0.000399
28	1.38	2.139403	0.900005	9.999478	0.000525
29	1.37	2.179106	0.899995	10.000530	−0.000525
30	1.36	2.223708	0.900001	9.999901	0.000101
31	1.35	2.273994	0.900009	9.999121	0.000882
32	1.34	2.331140	0.900004	9.999561	0.000441
33	1.33	2.397886	0.899996	10.000410	−0.000405
34	1.32	2.479705	0.900008	9.999198	0.000805
35	1.31	2.582854	0.900003	9.999728	0.000274
36	1.30	2.726869	0.900003	9.999741	0.000262
37	1.29	2.972693	0.899999	10.000130	−0.000131
No. of pairs		Mean COMPUT. risk errors		STD. ERR. errors	
37		−0.0000544285		0.0000903659	
Z-ratio				Status of significance	
0.60231				N.S.—not significant	
Parameter values					
PROB. = 0.85	RHO = −0.8		BIGH = 1.4396		SMHL = 1.04
Pair No.	SMH	SMK	PROB. level	Risk level (%)	Risk error (%)
Table No. 8.2-355					
1	1.44	1.439005	0.849992	15.000820	−0.000823
2	1.43	1.449264	0.850010	14.999030	0.000972
3	1.42	1.459666	0.850005	14.999470	0.000525
4	1.41	1.470407	0.850005	14.999550	0.000447
5	1.40	1.481492	0.850006	14.999430	0.000566
6	1.39	1.492923	0.850007	14.999300	0.000697
7	1.38	1.504704	0.850007	14.999330	0.000668
8	1.37	1.516837	0.850003	14.999720	0.000274
9	1.36	1.529327	0.849994	15.000650	−0.000650
10	1.35	1.542373	0.850001	14.999940	0.000054
11	1.34	1.555783	0.849998	15.000250	−0.000250
12	1.33	1.569755	0.850005	14.999510	0.000489
13	1.32	1.584099	0.849997	15.000300	−0.000298
14	1.31	1.599013	0.849993	15.000670	−0.000674
15	1.30	1.614503	0.849991	15.000950	−0.000954
16	1.29	1.630767	0.850005	14.999490	0.000507

(continued)

Table 8.2 (continued)

Parameter values				
PROB. = 0.85	RHO = −0.8	BIGH = 1.4396		SMHL = 1.04
Pair No.	SMH	SMK	PROB. level	Risk level (%)	Risk error (%)
---	---	---	---	---	---
17	1.28	1.647420	0.849992	15.000830	−0.000834
18	1.27	1.665052	0.850007	14.999350	0.000650
19	1.26	1.683276	0.850004	14.999560	0.000441
20	1.25	1.702294	0.850000	14.999980	0.000018
21	1.24	1.722305	0.850005	14.999470	0.000530
22	1.23	1.743120	0.849995	15.000480	−0.000483
23	1.22	1.765134	0.849997	15.000300	−0.000298
24	1.21	1.788353	0.850001	14.999900	0.000095
25	1.20	1.812782	0.849998	15.000220	−0.000226
26	1.19	1.838818	0.850006	14.999400	0.000596
27	1.18	1.866273	0.849998	15.000180	−0.000179
28	1.17	1.895737	0.850002	14.999770	0.000226
29	1.16	1.927218	0.849999	15.000060	−0.000060
30	1.15	1.961113	0.849995	15.000480	−0.000483
31	1.14	1.998015	0.849999	15.000080	−0.000077
32	1.13	2.038321	0.850003	14.999680	0.000316
33	1.12	2.082431	0.849992	15.000840	−0.000846
34	1.11	2.131913	0.849994	15.000630	−0.000632
35	1.10	2.187949	0.849997	15.000250	−0.000256
36	1.09	2.252500	0.849998	15.000190	−0.000191
37	1.08	2.329089	0.850002	14.999820	0.000179
38	1.07	2.423585	0.850006	14.999370	0.000626
39	1.06	2.547326	0.850000	14.999990	0.000012
40	1.05	2.733134	0.850004	14.999580	0.000423
41	1.04	3.142737	0.849993	15.000680	−0.000679
No. of pairs		Mean COMPUT. risk errors		STD. ERR. errors	
41		0.0000099341		0.0000795404	
Z-ratio				Status of significance	
0.12489				N.S.—not significant	

Parameter values				
PROB. = 0.8	RHO = −0.8	BIGH = 1.2815		SMHL = 0.85
Pair No.	SMH	SMK	PROB. level	Risk level (%)	Risk error (%)
---	---	---	---	---	---
Table No. 8.2-356					
1	1.29	1.273154	0.799994	20.000570	−0.000572
2	1.28	1.283099	0.800000	19.999980	0.000018
3	1.27	1.293299	0.800005	19.999460	0.000536
4	1.26	1.303757	0.800009	19.999130	0.000864
5	1.25	1.314380	0.799992	20.000760	−0.000763
6	1.24	1.325463	0.800005	19.999500	0.000495

(continued)

8.2 Contents for Table 8.2 (Table No. 8.2-1 to 400)

Table 8.2 (continued)

Parameter values					
PROB. = 0.8	RHO = −0.8		BIGH = 1.2815		SMHL = 0.85
Pair No.	SMH	SMK	PROB. level	Risk level (%)	Risk error (%)
7	1.23	1.336719	0.799996	20.000440	−0.000441
8	1.22	1.348346	0.799995	20.000460	−0.000459
9	1.21	1.360350	0.800002	19.999760	0.000238
10	1.20	1.372637	0.799999	20.000090	−0.000089
11	1.19	1.385309	0.799999	20.000090	−0.000089
12	1.18	1.398371	0.800000	19.999960	0.000042
13	1.17	1.411829	0.800001	19.999950	0.000054
14	1.16	1.425687	0.799997	20.000270	−0.000268
15	1.15	1.440048	0.800002	19.999750	0.000244
16	1.14	1.454821	0.799999	20.000140	−0.000143
17	1.13	1.470108	0.799997	20.000330	−0.000328
18	1.12	1.486014	0.800006	19.999360	0.000638
19	1.11	1.502349	0.799998	20.000200	−0.000203
20	1.10	1.519314	0.799993	20.000680	−0.000679
21	1.09	1.537015	0.800000	20.000030	−0.000030
22	1.08	1.555360	0.800000	20.000040	−0.000042
23	1.07	1.574454	0.800000	20.000000	−0.000006
24	1.06	1.594402	0.800005	19.999460	0.000542
25	1.05	1.615212	0.800009	19.999080	0.000924
26	1.04	1.636891	0.800004	19.999590	0.000405
27	1.03	1.659644	0.800003	19.999750	0.000250
28	1.02	1.683479	0.799995	20.000490	−0.000495
29	1.01	1.708795	0.800008	19.999170	0.000829
30	1.00	1.735211	0.799993	20.000650	−0.000656
31	0.99	1.763518	0.800007	19.999340	0.000662
32	0.98	1.793335	0.799998	20.000240	−0.000238
33	0.97	1.825259	0.799996	20.000410	−0.000411
34	0.96	1.859497	0.799994	20.000580	−0.000584
35	0.95	1.896644	0.800007	19.999320	0.000679
36	0.94	1.936714	0.800001	19.999880	0.000119
37	0.93	1.980695	0.800002	19.999830	0.000173
38	0.92	2.029186	0.799994	20.000590	−0.000590
39	0.91	2.083958	0.800007	19.999310	0.000691
40	0.90	2.145806	0.799996	20.000440	−0.000435
41	0.89	2.218262	0.799999	20.000140	−0.000137
42	0.88	2.305246	0.799994	20.000600	−0.000596
43	0.87	2.415758	0.799999	20.000150	−0.000149

(continued)

Table 8.2 (continued)

Parameter values					
PROB. = 0.8	RHO = −0.8		BIGH = 1.2815		SMHL = 0.85
Pair No.	SMH	SMK	PROB. level	Risk level (%)	Risk error (%)
44	0.86	2.568958	0.800005	19.999470	0.000530
45	0.85	2.828923	0.800002	19.999770	0.000226
No. of pairs		Mean COMPUT. risk errors		STD. ERR. errors	
45		0.0000164561		0.0000688844	
Z-ratio				Status of significance	
0.23889				N.S.—not significant	
Parameter values					
PROB. = 0.75	RHO = −0.8		BIGH = 1.1504		SMHL = 0.68
Pair No.	SMH	SMK	PROB. level	Risk level (%)	Risk error (%)
Table No. 8.2-357					
1	1.16	1.140782	0.750004	24.999590	0.000411
2	1.15	1.150703	0.750010	24.999050	0.000954
3	1.14	1.160797	0.750004	24.999640	0.000364
4	1.13	1.171169	0.750005	24.999480	0.000525
5	1.12	1.181723	0.749994	25.000580	−0.000584
6	1.11	1.192661	0.750008	24.999180	0.000823
7	1.10	1.203793	0.750008	24.999220	0.000781
8	1.09	1.215123	0.749992	25.000760	−0.000757
9	1.08	1.226854	0.749999	25.000130	−0.000131
10	1.07	1.238892	0.750006	24.999410	0.000596
11	1.06	1.251146	0.749995	25.000480	−0.000477
12	1.05	1.263818	0.750001	24.999860	0.000143
13	1.04	1.276816	0.750004	24.999570	0.000435
14	1.03	1.290148	0.750003	24.999700	0.000304
15	1.02	1.303818	0.749996	25.000440	−0.000441
16	1.01	1.317934	0.749997	25.000320	−0.000316
17	1.00	1.332502	0.750004	24.999580	0.000423
18	0.99	1.347433	0.749999	25.000070	−0.000072
19	0.98	1.362831	0.749996	25.000430	−0.000429
20	0.97	1.378804	0.750005	24.999460	0.000542
21	0.96	1.395164	0.749995	25.000480	−0.000477
22	0.95	1.412214	0.750006	24.999380	0.000620
23	0.94	1.429769	0.750005	24.999510	0.000489
24	0.93	1.447935	0.750002	24.999800	0.000203
25	0.92	1.466820	0.750006	24.999410	0.000596
26	0.91	1.486339	0.749998	25.000160	−0.000155
27	0.90	1.506697	0.750000	25.000010	−0.000012
28	0.89	1.527809	0.749991	25.000910	−0.000906
29	0.88	1.549980	0.750001	24.999910	0.000095
30	0.87	1.573125	0.750008	24.999180	0.000823
31	0.86	1.597259	0.750004	24.999570	0.000429
32	0.85	1.622590	0.750002	24.999850	0.000155

(continued)

Table 8.2 (continued)

Parameter values					
PROB. = 0.75	RHO = −0.8		BIGH = 1.1504		SMHL = 0.68
Pair No.	SMH	SMK	PROB. level	Risk level (%)	Risk error (%)
33	0.84	1.649328	0.750008	24.999200	0.000799
34	0.83	1.677294	0.749990	25.000970	−0.000972
35	0.82	1.707286	0.750009	24.999080	0.000918
36	0.81	1.738930	0.750008	24.999230	0.000775
37	0.80	1.772634	0.750001	24.999920	0.000083
38	0.79	1.808809	0.749997	25.000350	−0.000352
39	0.78	1.847864	0.749994	25.000570	−0.000566
40	0.77	1.890407	0.749999	25.000110	−0.000113
41	0.76	1.937045	0.750003	24.999680	0.000316
42	0.75	1.988583	0.749999	25.000070	−0.000066
43	0.74	2.046608	0.750002	24.999800	0.000197
44	0.73	2.112708	0.749992	25.000780	−0.000775
45	0.72	2.190426	0.749991	25.000910	−0.000906
46	0.71	2.285065	0.749994	25.000650	−0.000644
47	0.70	2.407005	0.749995	25.000550	−0.000548
48	0.69	2.582061	0.749992	25.000760	−0.000763
49	0.68	2.921204	0.750004	24.999560	0.000441
No. of pairs		Mean COMPUT. risk errors		STD. ERR. errors	
49		0.0000555515		0.0000780409	
Z-ratio				Status of significance	
0.71183				N.S.—not significant	
Parameter values					
PROB. = 0.7	RHO = −0.8		BIGH = 1.0363		SMHL = 0.55
Pair No.	SMH	SMK	PROB. level	Risk level (%)	Risk error (%)
Table No. 8.2-358					
1	1.04	1.032711	0.699997	30.000290	−0.000292
2	1.03	1.042736	0.699996	30.000360	−0.000358
3	1.02	1.052958	0.699992	30.000770	−0.000769
4	1.01	1.063480	0.700007	29.999340	0.000668
5	1.00	1.074113	0.699995	30.000510	−0.000513
6	0.99	1.085058	0.700001	29.999890	0.000113
7	0.98	1.096225	0.700003	29.999750	0.000256
8	0.97	1.107620	0.699999	30.000070	−0.000066
9	0.96	1.119250	0.699991	30.000930	−0.000924
10	0.95	1.131221	0.699998	30.000250	−0.000244
11	0.94	1.143442	0.699998	30.000220	−0.000215
12	0.93	1.155922	0.699991	30.000860	−0.000852
13	0.92	1.168767	0.699998	30.000230	−0.000226
14	0.91	1.181887	0.699996	30.000410	−0.000411
15	0.90	1.195390	0.700004	29.999580	0.000423
16	0.89	1.209188	0.700003	29.999750	0.000250
17	0.88	1.223291	0.699990	30.000960	−0.000960

(continued)

Table 8.2 (continued)

Parameter values					
PROB. = 0.7	RHO = −0.8		BIGH = 1.0363		SMHL = 0.55
Pair No.	SMH	SMK	PROB. level	Risk level (%)	Risk error (%)
18	0.87	1.237904	0.700003	29.999700	0.000304
19	0.86	1.252843	0.700002	29.999780	0.000226
20	0.85	1.268218	0.700005	29.999550	0.000453
21	0.84	1.283942	0.699991	30.000900	−0.000900
22	0.83	1.300224	0.699994	30.000630	−0.000626
23	0.82	1.317077	0.700010	29.999020	0.000978
24	0.81	1.334320	0.700003	29.999680	0.000328
25	0.80	1.352163	0.700005	29.999530	0.000471
26	0.79	1.370522	0.699995	30.000520	−0.000519
27	0.78	1.389610	0.700000	29.999970	0.000036
28	0.77	1.409346	0.700002	29.999830	0.000173
29	0.76	1.429846	0.700009	29.999100	0.000906
30	0.75	1.451030	0.700003	29.999730	0.000268
31	0.74	1.473115	0.700004	29.999550	0.000447
32	0.73	1.496120	0.700007	29.999320	0.000685
33	0.72	1.520067	0.700003	29.999750	0.000256
34	0.71	1.545177	0.700007	29.999290	0.000709
35	0.70	1.571472	0.700007	29.999340	0.000668
36	0.69	1.598980	0.699997	30.000350	−0.000352
37	0.68	1.628118	0.700003	29.999680	0.000328
38	0.67	1.658916	0.700010	29.999050	0.000954
39	0.66	1.691405	0.699998	30.000250	−0.000250
40	0.65	1.726204	0.700002	29.999790	0.000209
41	0.64	1.763347	0.699996	30.000430	−0.000429
42	0.63	1.803459	0.699997	30.000300	−0.000298
43	0.62	1.846968	0.699997	30.000290	−0.000286
44	0.61	1.894504	0.699992	30.000850	−0.000846
45	0.60	1.947283	0.699998	30.000230	−0.000226
46	0.59	2.006526	0.700005	29.999460	0.000542
47	0.58	2.073847	0.699996	30.000360	−0.000358
48	0.57	2.152815	0.699995	30.000500	−0.000501
49	0.56	2.248958	0.700003	29.999700	0.000304
50	0.55	2.372107	0.699997	30.000290	−0.000292
No. of pairs		Mean COMPUT. risk errors		STD. ERR. errors	
50		−0.0000148427		0.0000729907	
Z-ratio				Status of significance	
0.20335				N.S.—not significant	
Parameter values					
PROB. = 0.65	RHO = −0.8		BIGH = 0.9343		SMHL = 0.45
Pair No.	SMH	SMK	PROB. level	Risk level (%)	Risk error (%)
Table No. 8.2-359					
1	0.94	0.928732	0.649996	35.000440	−0.000435
2	0.93	0.938718	0.649995	35.000520	−0.000513

(continued)

8.2 Contents for Table 8.2 (Table No. 8.2-1 to 400)

Table 8.2 (continued)

Parameter values					
PROB. = 0.65	RHO = −0.8		BIGH = 0.9343		SMHL = 0.45
Pair No.	SMH	SMK	PROB. level	Risk level (%)	Risk error (%)
3	0.92	0.948920	0.650001	34.999880	0.000119
4	0.91	0.959298	0.650002	34.999780	0.000226
5	0.90	0.969858	0.649999	35.000120	−0.000113
6	0.89	0.980610	0.649991	35.000870	−0.000870
7	0.88	0.991658	0.650004	34.999580	0.000423
8	0.87	1.002864	0.650002	34.999810	0.000191
9	0.86	1.014336	0.650008	34.999160	0.000846
10	0.85	1.025984	0.650001	34.999890	0.000107
11	0.84	1.037917	0.650003	34.999680	0.000322
12	0.83	1.050047	0.649993	35.000730	−0.000727
13	0.82	1.062481	0.649992	35.000790	−0.000787
14	0.81	1.075233	0.650002	34.999840	0.000167
15	0.80	1.088216	0.650000	35.000030	−0.000030
16	0.79	1.101539	0.650008	34.999190	0.000811
17	0.78	1.115120	0.650006	34.999430	0.000572
18	0.77	1.128970	0.649993	35.000700	−0.000697
19	0.76	1.143203	0.649991	35.000870	−0.000870
20	0.75	1.157834	0.650000	35.000020	−0.000018
21	0.74	1.172781	0.649998	35.000170	−0.000167
22	0.73	1.188159	0.650007	34.999340	0.000668
23	0.72	1.203888	0.650005	34.999480	0.000525
24	0.71	1.219987	0.649994	35.000590	−0.000584
25	0.70	1.236672	0.650010	34.999020	0.000978
26	0.69	1.253670	0.649996	35.000360	−0.000358
27	0.68	1.271200	0.649990	35.000990	−0.000983
28	0.67	1.289383	0.650006	34.999380	0.000620
29	0.66	1.307952	0.649992	35.000780	−0.000775
30	0.65	1.327225	0.649997	35.000290	−0.000286
31	0.64	1.347135	0.650002	34.999810	0.000191
32	0.63	1.367710	0.650004	34.999570	0.000429
33	0.62	1.389082	0.650002	34.999800	0.000203
34	0.61	1.411088	0.649992	35.000760	−0.000757
35	0.60	1.434059	0.650000	35.000050	−0.000042
36	0.59	1.457839	0.649990	35.000970	−0.000972
37	0.58	1.482762	0.650000	35.000050	−0.000048
38	0.57	1.508775	0.650007	34.999350	0.000650
39	0.56	1.535927	0.650003	34.999670	S.000334
40	0.55	1.564464	0.650005	34.999530	0.000477
41	0.54	1.594441	0.650000	35.000050	−0.000048
42	0.53	1.626113	0.649996	35.000420	−0.000417
43	0.52	1.659739	0.649997	35.000260	−0.000256
44	0.51	1.695584	0.650003	34.999730	0.000268

(continued)

Table 8.2 (continued)

Parameter values					
PROB. = 0.65	RHO = −0.8		BIGH = 0.9343		SMHL = 0.45
Pair No.	SMH	SMK	PROB. level	Risk level (%)	Risk error (%)
45	0.50	1.733918	0.650005	34.999500	0.000501
46	0.49	1.775211	0.650009	34.999120	0.000882
47	0.48	1.819746	0.649993	35.000670	−0.000662
48	0.47	1.868791	0.650001	34.999890	0.000107
49	0.46	1.923033	0.650007	34.999280	0.000727
50	0.45	1.983563	0.649995	35.000520	−0.000513
No. of pairs		Mean COMPUT. risk errors		STD. ERR. errors	
50		−0.0000114534		0.0000768679	
Z-ratio				Status of significance	
0.14900				N.S.—not significant	
Parameter values					
PROB. = 0.6	RHO = −0.8		BIGH = 0.841		SMHL = 0.36
Pair No.	SMH	SMK	PROB. level	Risk level (%)	Risk error (%)
Table No. 8.2-360					
1	0.85	0.832193	0.600006	39.999390	0.000608
2	0.84	0.842099	0.599998	40.000180	−0.000179
3	0.83	0.852243	0.600010	39.999030	0.000966
4	0.82	0.862489	0.600001	39.999930	0.000072
5	0.81	0.872942	0.600000	40.000020	−0.000024
6	0.80	0.883613	0.600008	39.999190	0.000805
7	0.79	0.894414	0.600000	39.999970	0.000030
8	0.78	0.905403	0.599991	40.000860	−0.000858
9	0.77	0.916691	0.600008	39.999170	0.000829
10	0.76	0.928094	0.600001	39.999880	0.000119
11	0.75	0.939721	0.599997	40.000260	−0.000262
12	0.74	0.951586	0.599998	40.000170	−0.000167
13	0.73	0.963702	0.600005	39.999460	0.000542
14	0.72	0.975987	0.599996	40.000370	−0.000370
15	0.71	0.988553	0.599997	40.000300	−0.000298
16	0.70	1.001417	0.600009	39.999120	0.000882
17	0.69	1.014449	0.599999	40.000110	−0.000107
18	0.68	1.027814	0.600004	39.999600	0.000393
19	0.67	1.041385	0.599992	40.000820	−0.000823
20	0.66	1.055329	0.599999	40.000130	−0.000131
21	0.65	1.069570	0.600004	39.999590	0.000405
22	0.64	1.084130	0.600010	39.999000	0.000995
23	0.63	1.098937	0.599998	40.000210	−0.000215
24	0.62	1.114115	0.599992	40.000800	−0.000799
25	0.61	1.129692	0.599994	40.000590	−0.000590
26	0.60	1.145696	0.600007	39.999350	0.000650
27	0.59	1.161965	0.599992	40.000800	−0.000805
28	0.58	1.178727	0.599993	40.000680	−0.000679

(continued)

8.2 Contents for Table 8.2 (Table No. 8.2-1 to 400)

Table 8.2 (continued)

Parameter values					
PROB. = 0.6	RHO = −0.8		BIGH = 0.841		SMHL = 0.36
Pair No.	SMH	SMK	PROB. level	Risk level (%)	Risk error (%)
29	0.57	1.195922	0.599993	40.000700	−0.000697
30	0.56	1.213587	0.599994	40.000570	−0.000572
31	0.55	1.232059	0.600002	39.999770	0.000226
32	0.54	1.250697	0.600000	40.000050	−0.000054
33	0.53	1.269941	0.600006	39.999360	0.000644
34	0.52	1.289647	0.599991	40.000860	−0.000864
35	0.51	1.310067	0.599992	40.000820	−0.000823
36	0.50	1.331163	0.599993	40.000700	−0.000697
37	0.49	1.353000	0.599998	40.000240	−0.000238
38	0.48	1.375650	0.600008	39.999230	0.000763
39	0.47	1.398993	0.599995	40.000460	−0.000459
40	0.46	1.423308	0.599993	40.000700	−0.000703
41	0.45	1.448687	0.600001	39.999940	0.000060
42	0.44	1.475034	0.599992	40.000800	−0.000799
43	0.43	1.502651	0.599994	40.000620	−0.000620
44	0.42	1.531657	0.600004	39.999570	0.000423
45	0.41	1.561988	0.599996	40.000360	−0.000358
46	0.40	1.593982	0.599991	40.000910	−0.000912
47	0.39	1.627992	0.600002	39.999760	0.000232
48	0.38	1.663998	0.600000	40.000010	−0.000012
49	0.37	1.702583	0.600010	39.999000	0.000995
50	0.36	1.743769	0.599997	40.000340	−0.000346
No. of pairs		Mean COMPUT. risk errors		STD. ERR. errors	
50		−0.0000749149		0.0000815954	
Z-ratio				Status of significance	
0.91813				N.S.—not significant	
Parameter values					
PROB. = 0.55	RHO = −0.8		BIGH = 0.7542		SMHL = 0.27
Pair No.	SMH	SMK	PROB. level	Risk level (%)	Risk error (%)
Table No. 8.2-361					
1	0.76	0.748347	0.549997	45.000320	−0.000316
2	0.75	0.758326	0.549995	45.000540	−0.000542
3	0.74	0.768477	0.549997	45.000260	−0.000262
4	0.73	0.778812	0.550008	44.999250	0.000751
5	0.72	0.789243	0.549999	45.000120	−0.000125
6	0.71	0.799882	0.550002	44.999800	0.000197
7	0.70	0.810643	0.549992	45.000840	−0.000840
8	0.69	0.821639	0.549998	45.000170	−0.000173
9	0.68	0.832786	0.549997	45.000320	−0.000316

(continued)

Table 8.2 (continued)

Parameter values					
PROB. = 0.55	RHO = −0.8		BIGH = 0.7542		SMHL = 0.27
Pair No.	SMH	SMK	PROB. level	Risk level (%)	Risk error (%)
10	0.67	0.844099	0.549991	45.000930	−0.000930
11	0.66	0.855692	0.550010	44.999040	0.000960
12	0.65	0.867389	0.550003	44.999690	0.000316
13	0.64	0.879305	0.550002	44.999800	0.000203
14	0.63	0.891459	0.550009	44.999080	0.000924
15	0.62	0.903776	0.550003	44.999670	0.000328
16	0.61	0.916326	0.550001	44.999930	0.000072
17	0.60	0.929084	0.549994	45.000650	−0.000650
18	0.59	0.942125	0.549999	45.000140	−0.000137
19	0.58	0.955427	0.550008	44.999180	0.000823
20	0.57	0.968921	0.550003	44.999690	0.000316
21	0.56	0.982689	0.550001	44.999890	0.000113
22	0.55	0.996714	0.549996	45.000390	−0.0000387
23	0.54	1.011080	0.550006	44.999450	0.000554
24	0.53	1.025682	0.550001	44.999880	0.000125
25	0.52	1.040559	0.549991	45.000910	−0.000912
26	0.51	1.055855	0.550004	44.999640	0.000364
27	0.50	1.071424	0.550003	44.999740	0.000262
28	0.49	1.088000	0.549995	45.000480	−0.000483
29	0.48	1.104274	0.550000	44.999960	0.000042
30	0.47	1.120894	0.549998	45.000170	−0.000173
31	0.46	1.137926	0.549999	45.000110	−0.000113
32	0.45	1.155347	0.549993	45.000700	−0.000703
33	0.44	1.173235	0.549992	45.000780	−0.000775
34	0.43	1.191678	0.550009	44.999130	0.000870
35	0.42	1.210479	0.549998	45.000180	−0.000179
36	0.41	1.229839	0.549995	45.000540	−0.000542
37	0.40	1.249777	0.549995	45.000510	−0.000507
38	0.39	1.270322	0.549997	45.000260	−0.000262
39	0.38	1.291516	0.550002	44.999780	0.000221
40	0.37	1.313320	0.549995	45.000510	−0.000507
41	0.36	1.335907	0.549997	45.000350	−0.000352
42	0.35	1.359273	0.549997	45.000320	−0.000316
43	0.34	1.383538	0.550004	44.999620	0.000376
44	0.33	1.408649	0.549998	45.000160	−0.000155
45	0.32	1.434779	0.549995	45.000510	−0.000513
46	0.31	1.462041	0.549995	45.000510	−0.000513
47	0.30	1.490587	0.550003	44.999700	0.000304
48	0.29	1.520421	0.550003	44.999730	0.000268

(continued)

8.2 Contents for Table 8.2 (Table No. 8.2-1 to 400)

Table 8.2 (continued)

Parameter values					
PROB. = 0.55	RHO = −0.8		BIGH = 0.7542		SMHL = 0.27
Pair No.	SMH	SMK	PROB. level	Risk level (%)	Risk error (%)
49	0.28	1.551801	0.550007	44.999340	0.000662
50	0.27	1.584853	0.550008	44.999220	0.000781
No. of pairs			Mean COMPUT. risk errors	STD. ERR. errors	
50			−0.0000363471	0.0000705460	
Z-ratio				Status of significance	
0.51523				N.S.—not significant	
Parameter values					
PROB. = 0.5	RHO = −0.8		BIGH = 0.672		SMHL = 0.19
Pair No.	SMH	SMK	PROB. level	Risk level (%)	Risk error (%)
Table No. 8.2-362					
1	0.68	0.664045	0.500004	49.999580	0.000423
2	0.67	0.674006	0.500005	49.999540	0.000465
3	0.66	0.684072	0.499995	50.000510	−0.000513
4	0.65	0.694354	0.500008	49.999170	0.000834
5	0.64	0.704721	0.500003	49.999670	0.000328
6	0.63	0.715237	0.499999	50.000120	−0.000122
7	0.62	0.725920	0.499999	50.000110	−0.000107
8	0.61	0.736786	0.500008	49.999240	0.000763
9	0.60	0.747757	0.500000	49.999970	0.000036
10	0.59	0.758902	0.499997	50.000320	−0.000319
11	0.58	0.770243	0.500002	49.999820	0.000185
12	0.57	0.781706	0.499993	50.000750	−0.000748
13	0.56	0.793412	0.500003	49.999700	0.000304
14	0.55	0.805241	0.499998	50.000240	−0.000241
15	0.54	0.817272	0.499997	50.000310	−0.000313
16	0.53	0.829486	0.499994	50.000590	−0.000581
17	0.52	0.841917	0.499997	50.000270	−0.000259
18	0.51	0.854550	0.500001	49.999920	0.000083
19	0.50	0.867377	0.500000	49.999960	0.000048
20	0.49	0.880389	0.499992	50.000750	−0.000754
21	0.48	0.893680	0.500000	49.999990	0.000006
22	0.47	0.907203	0.500008	49.999180	0.000823
23	0.46	0.920864	0.499991	50.000880	−0.000876
24	0.45	0.934867	0.499999	50.000100	−0.000092
25	0.44	0.949080	0.499996	50.000380	−0.000378
26	0.43	0.965039	0.499999	50.000140	−0.000137
27	0.42	0.979740	0.499999	50.000150	−0.000149
28	0.41	0.994734	0.500000	49.999970	0.000030
29	0.40	1.010014	0.500000	50.000010	−0.000006
30	0.39	1.025583	0.499995	50.000460	−0.000459
31	0.38	1.041503	0.499998	50.000210	−0.000209
32	0.37	1.057801	0.500010	49.999010	0.000989

(continued)

Table 8.2 (continued)

Parameter values					
PROB. = 0.5	RHO = −0.8		BIGH = 0.672		SMHL = 0.19
Pair No.	SMH	SMK	PROB. level	Risk level (%)	Risk error (%)
33	0.36	1.074321	0.499993	50.000720	−0.000721
34	0.35	1.091313	0.499999	50.000100	−0.000101
35	0.34	1.108656	0.499997	50.000270	−0.000265
36	0.33	1.126443	0.500004	49.999620	0.000381
37	0.32	1.144597	0.499998	50.000230	−0.000226
38	0.31	1.163264	0.500004	49.999600	0.000405
39	0.30	1.182329	0.499995	50.000540	−0.000536
40	0.29	1.202008	0.500007	49.999350	0.000650
41	0.28	1.222173	0.500008	49.999190	0.000811
42	0.27	1.242843	0.499998	50.000220	−0.000221
43	0.26	1.264202	0.500002	49.999800	0.000203
44	0.25	1.286216	0.500007	49.999350	0.000656
45	0.24	1.308843	0.499996	50.000370	−0.000373
46	0.23	1.332350	0.500007	49.999260	0.000745
47	0.22	1.356557	0.500000	49.999990	0.000012
48	0.21	1.381646	0.499993	50.000690	−0.000691
49	0.20	1.407720	0.499991	50.000910	−0.000912
50	0.19	1.434956	0.500005	49.999550	0.000453
No. of pairs		Mean COMPUT. risk errors		STD. ERR. errors	
50		−0.0000133234		0.0000683641	
Z-ratio				Status of significance	
0.19489				N.S.—not significant	
Parameter values					
PROB. = 0.99	RHO = −0.9		BIGH = 2.4534		SMHL = 1.97
Pair No.	SMH	SMK	PROB. level	Risk level (%)	Risk error (%)
Table No. 8.2-363					
1	2.46	2.446818	0.990000	0.999957	0.000042
2	2.45	2.456804	0.990002	0.999844	0.000155
3	2.44	2.466873	0.989997	1.000345	−0.000346
4	2.43	2.477416	0.990004	0.999647	0.000352
5	2.42	2.488042	0.990004	0.999576	0.000423
6	2.41	2.498753	0.989999	1.000130	−0.000131
7	2.40	2.509941	0.990004	0.999570	0.000429
8	2.39	2.521216	0.990003	0.999671	0.000328
9	2.38	2.532579	0.989996	1.000428	−0.000429
10	2.37	2.544422	0.989999	1.000136	−0.000137
11	2.36	2.556355	0.989995	1.000506	−0.000507
12	2.35	2.568771	0.990002	0.999850	0.000149
13	2.34	2.581279	0.990002	0.999850	0.000149

(continued)

Table 8.2 (continued)

Parameter values					
PROB. = 0.99	RHO = −0.9		BIGH = 2.4534		SMHL = 1.97
Pair No.	SMH	SMK	PROB. level	Risk level (%)	Risk error (%)
14	2.33	2.593882	0.989995	1.000500	−0.000501
15	2.32	2.606970	0.989999	1.000088	−0.000089
16	2.31	2.620155	0.989997	1.000297	−0.000298
17	2.30	2.633828	0.990006	0.999391	0.000608
18	2.29	2.647209	0.989992	1.000798	−0.000799
19	2.28	2.661471	0.990007	0.999290	0.000709
20	2.27	2.675445	0.990000	1.000005	−0.000006
21	2.26	2.689912	0.990005	0.999457	0.000542
22	2.25	2.704484	0.990007	0.999326	0.000674
23	2.24	2.719161	0.990004	0.999576	0.000423
24	2.23	2.733945	0.989998	1.000172	−0.000173
25	2.22	2.749229	0.990008	0.999159	0.000840
26	2.21	2.764231	0.989997	1.000273	−0.000274
27	2.20	2.909424	0.990002	0.999755	0.000244
28	2.19	2.928167	0.989991	1.000941	−0.000942
29	2.18	2.947806	0.990009	0.999129	0.000870
30	2.17	2.966780	0.989998	1.000226	−0.000226
31	2.16	2.986262	0.990004	0.999635	0.000364
32	2.15	3.005473	0.989996	1.000357	−0.000358
33	2.14	3.025196	0.980009	0.999141	0.000858
34	2.13	3.044651	0.990009	0.999141	0.000858
35	2.12	3.063841	0.989996	1.000434	−0.000435
36	2.11	3.083546	0.990006	0.999403	0.000596
37	2.10	3.102990	0.990004	0.999564	0.000435
38	2.09	3.122173	0.989990	1.000965	−0.000966
39	2.08	3.141878	0.990003	0.999671	0.000328
40	2.07	3.161327	0.990005	0.999498	0.000501
41	2.06	3.180521	0.989995	1.000536	−0.000536
42	2.05	3.199852	0.989993	1.000673	−0.000674
43	2.04	3.219323	0.990003	0.999743	0.000256
44	2.03	3.238546	0.990000	1.000017	−0.000018
45	2.02	3.257912	0.990009	0.999111	0.000888
46	2.01	3.277033	0.990006	0.999415	0.000584
47	2.00	3.295912	0.989990	1.000994	−0.000995
48	1.99	3.315138	0.990000	1.000011	−0.000012

(continued)

Table 8.2 (continued)

Parameter values					
PROB. = 0.99	RHO = −0.9		BIGH = 2.4534		SMHL = 1.97
Pair No.	SMH	SMK	PROB. level	Risk level (%)	Risk error (%)
49	1.98	3.334125	0.989997	1.000279	−0.000280
50	1.97	3.353072	0.989995	1.000464	−0.000465
No. of pairs		Mean COMPUT. risk errors		STD. ERR. errors	
50		0.0000590203		0.0000725336	
Z-ratio				Status of significance	
0.81370				N.S.—not significant	
Parameter values					
PROB. = 0.98	RHO = −0.9		BIGH = 2.3155		SMHL = 1.83
Pair No.	SMH	SMK	PROB. level	Risk level (%)	Risk error (%)
Table No. 8.2-364					
1	2.32	2.311009	0.980008	1.999205	0.000793
2	2.31	1.321013	0.980007	1.999265	0.000733
3	2.30	2.331105	0.979997	2.000314	−0.000316
4	2.29	2.342065	0.980000	2.000004	−0.000006
5	2.28	2.353115	0.979992	2.000797	−0.000799
6	2.27	2.365037	0.979995	2.000475	−0.000477
7	2.26	2.377832	0.980009	1.999134	0.000864
8	2.25	2.390720	0.980009	1.999146	0.000852
9	2.24	2.403701	0.979995	2.000487	−0.000489
10	2.23	2.417950	0.980000	2.000034	−0.000036
11	2.22	2.433077	0.980010	1.999033	0.000966
12	2.21	2.447912	0.979994	2.000570	−0.000572
13	2.20	2.464408	0.980003	1.999718	0.000280
14	2.19	2.481004	0.979995	2.000493	−0.000495
15	2.18	2.499266	0.980008	1.999164	0.000834
16	2.17	2.517631	0.980005	1.999515	0.000483
17	2.16	2.536882	0.980003	1.999748	0.000250
18	2.15	2.557021	0.980002	1.999831	0.000167
19	2.14	2.578049	0.980002	1.999778	0.000221
20	2.13	2.599968	0.980005	1.999480	0.000519
21	2.12	2.621997	0.979993	2.000666	−0.000668
22	2.11	2.645702	0.980004	1.999599	0.000399
23	2.10	2.731239	0.979995	2.000475	−0.000477
24	2.09	2.763767	0.980001	1.999921	0.000077
25	2.08	2.796414	0.979993	2.000654	−0.000656
26	2.07	2.830741	0.980004	1.999593	0.000405
27	2.06	2.864408	0.979999	2.000099	−0.000101
28	2.05	2.898198	0.979996	2.000368	−0.000370

(continued)

8.2 Contents for Table 8.2 (Table No. 8.2-1 to 400)

Table 8.2 (continued)

Parameter values					
PROB. = 0.98	RHO = −0.9		BIGH = 2.3155		SMHL = 1.83
Pair No.	SMH	SMK	PROB. level	Risk level (%)	Risk error (%)
29	2.04	2.932112	0.980002	1.999837	0.000161
30	2.03	2.965371	0.980004	1.999557	0.000441
31	2.02	2.997977	0.980008	1.999253	0.000745
32	2.01	3.029151	0.979994	2.000558	−0.000560
33	2.00	3.060457	0.980003	1.999724	0.000274
34	1.99	3.090335	0.979995	2.000523	−0.000525
35	1.98	3.119957	0.980001	1.999927	0.000072
36	1.97	3.148546	0.979998	2.000159	−0.000161
37	1.96	3.176495	0.979998	2.000219	−0.000221
38	1.95	3.203804	0.979998	2.000213	−0.000215
39	1.94	3.230477	0.979997	2.000302	−0.000304
40	1.93	3.256906	0.980009	1.999104	0.000894
41	1.92	3.282312	0.980001	1.999855	0.000143
42	1.91	3.307479	0.980004	1.999569	0.000429
43	1.90	3.332019	0.979999	2.000094	−0.000095
44	1.89	3.356325	0.980002	1.999754	0.000244
45	1.88	3.380007	0.979994	2.000636	−0.000638
46	1.87	3.403461	0.979991	2.000874	−0.000876
47	1.86	3.426688	0.979995	2.000511	−0.000513
48	1.85	3.449691	0.980004	1.999605	0.000393
49	1.84	3.472083	0.979993	2.000701	−0.000703
50	1.83	3.494451	0.979998	2.000243	−0.000244
No. of pairs		Mean COMPUT. risk errors		STD. ERR. errors	
50		0.0000220888		0.0000719392	
Z-ratio				Status of significance	
0.30705				N.S.—not significant	
Parameter values					
PROB. = 0.97	RHO = −0.9		BIGH = 2.1694		SMHL = 1.69
Pair No.	SMH	SMK	PROB. level	Risk level (%)	Risk error (%)
Table No. 8.2-365					
1	2.17	2.168800	0.970000	3.000021	−0.000024
2	2.16	2.178841	0.969994	3.000581	−0.000584
3	2.15	2.189756	0.970004	2.999580	0.000417
4	2.14	2.200766	0.970000	3.000045	−0.000048
5	2.13	2.212654	0.970008	2.999234	0.000763
6	2.12	2.224638	0.969999	3.000075	−0.000077
7	2.11	2.237503	0.970001	2.999944	0.000054
8	2.10	2.251250	0.970009	2.999085	0.000912
9	2.09	2.265097	0.969998	3.000158	−0.000161

(continued)

Table 8.2 (continued)

Parameter values					
PROB. = 0.97	RHO = −0.9		BIGH = 2.1694		SMHL = 1.69
Pair No.	SMH	SMK	PROB. level	Risk level (%)	Risk error (%)
10	2.08	2.279830	0.969992	3.000850	−0.000852
11	2.07	2.296229	0.970009	2.999061	0.000936
12	2.06	2.312734	0.970003	2.999681	0.000316
13	2.05	2.330129	0.969994	3.000611	−0.000614
14	2.04	2.349195	0.970000	3.000009	−0.000012
15	2.03	2.369154	0.969997	3.000295	−0.000298
16	2.02	2.390787	0.970002	2.999848	0.000149
17	2.01	2.413316	0.969991	3.000867	−0.000870
18	2.00	2.438304	0.969998	3.000200	−0.000203
19	1.99	2.464973	0.969999	3.000087	−0.000089
20	1.98	2.494105	0.970006	2.999413	0.000584
21	1.97	2.532732	0.970007	2.999300	0.000697
22	1.96	2.568359	0.969996	3.000373	−0.000376
23	1.95	2.608797	0.969999	3.000140	−0.000143
24	1.94	2.654050	0.970002	2.999830	0.000167
25	1.93	2.704120	0.969998	3.000236	−0.000238
26	1.92	2.760570	0.970000	2.999973	0.000024
27	1.91	2.821841	0.969995	3.000480	−0.000483
28	1.90	2.887935	0.970002	2.999818	0.000179
29	1.89	2.954166	0.970004	2.999586	0.000411
30	1.88	3.017410	0.970004	2.999586	0.000411
31	1.87	3.076110	0.970007	2.999282	0.000715
32	1.86	3.128703	0.969996	3.000450	−0.000453
33	1.85	3.177536	0.970000	3.000027	−0.000030
34	1.84	3.221831	0.969995	3.000498	−0.000501
35	1.83	3.263152	0.970000	3.000003	−0.000006
36	1.82	3.301500	0.970001	2.999926	0.000072
37	1.81	3.337662	0.970006	2.999425	0.000572
38	1.80	3.371639	0.970003	2.999711	0.000286
39	1.79	3.404215	0.970008	2.999157	0.000840
40	1.78	3.435001	0.969997	3.000265	−0.000268
41	1.77	3.464782	0.969993	3.000713	−0.000715
42	1.76	3.493952	0.970008	2.999187	0.000811
43	1.75	3.521732	0.970000	3.000009	−0.000012
44	1.74	3.548907	0.970001	2.999920	0.000077
45	1.73	3.575479	0.970007	2.999288	0.000709
46	1.72	3.601061	0.969992	3.000814	−0.000817
47	1.71	3.626437	0.969998	3.000224	−0.000226

(continued)

8.2 Contents for Table 8.2 (Table No. 8.2-1 to 400)

Table 8.2 (continued)

Parameter values					
PROB. = 0.97	RHO = −0.9		BIGH = 2.1694		SMHL = 1.69
Pair No.	SMH	SMK	PROB. level	Risk level (%)	Risk error (%)
48	1.70	3.651220	0.969998	3.000218	−0.000221
49	1.69	3.675609	0.970002	2.999765	0.000232
No. of pairs		Mean COMPUT. risk errors		STD. ERR. errors	
49		0.0000402927		0.0000676706	
Z-ratio				Status of significance	
0.59543				N.S.—not significant	
Parameter values					
PROB. = 0.96	RHO = −0.9		BIGH = 2.0537		SMHL = 1.63
Pair No.	SMH	SMK	PROB. level	Risk level (%)	Risk error (%)
Table No. 8.2-366					
1	2.06	2.047419	0.959998	4.000199	−0.000197
2	2.05	2.057407	0.960000	4.000008	−0.000006
3	2.04	2.067883	0.960005	3.999484	0.000519
4	2.03	2.078458	0.959994	4.000604	−0.000602
5	2.02	2.089915	0.960002	3.999841	0.000161
6	2.01	2.101475	0.959991	4.000920	−0.000918
7	2.00	2.113920	0.959995	4.000497	−0.000495
8	1.99	2.126861	0.959995	4.000485	−0.000483
9	1.98	2.140690	0.960005	3.999472	0.000530
10	1.97	2.155018	0.960008	3.999251	0.000751
11	1.96	2.169848	0.960000	3.999984	0.000018
12	1.95	2.185571	0.959996	4.000378	−0.000376
13	1.94	2.202189	0.959992	4.000771	−0.000769
14	1.93	2.220094	0.959999	4.000080	−0.000077
15	1.92	2.238897	0.960000	4.000050	−0.000048
16	1.91	2.258993	0.960002	3.999788	0.000215
17	1.90	2.279990	0.959991	4.000860	−0.000858
18	1.89	2.303454	0.960008	3.999204	0.000799
19	1.88	2.327824	0.960001	3.999901	0.000101
20	1.87	2.354664	0.960007	3.999305	0.000697
21	1.86	2.383978	0.960002	3.999764	0.000238
22	1.85	2.415766	0.960002	3.999764	0.000238
23	1.84	2.450813	0.960002	3.999806	0.000197
24	1.83	2.489901	0.960002	3.999758	0.000244
25	1.82	2.533815	0.960001	3.999937	0.000066
26	1.81	2.584118	0.960001	3.999889	0.000113
27	1.80	2.642376	0.959998	4.000163	−0.000161
28	1.79	2.712497	0.960007	3.999323	0.000679
29	1.78	2.796047	0.960004	3.999609	0.000393
30	1.77	2.895371	0.959993	4.000747	−0.000745

(continued)

Table 8.2 (continued)

Parameter values					
PROB. = 0.96	RHO = −0.9		BIGH = 2.0537		SMHL = 1.63
Pair No.	SMH	SMK	PROB. level	Risk level (%)	Risk error (%)
31	1.76	3.008911	0.960004	3.999573	0.000429
32	1.75	3.114792	0.960005	3.999549	0.000453
33	1.74	3.202080	0.959992	4.000807	−0.000805
34	1.73	3.275466	0.960008	3.999233	0.000769
35	1.72	3.335734	0.960002	3.999776	0.000226
36	1.71	3.388356	0.960006	3.999430	0.000572
37	1.70	3.434895	0.960003	3.999722	0.000280
38	1.69	3.476920	0.959992	4.000765	−0.000763
39	1.68	3.515993	0.959992	4.000789	−0.000787
40	1.67	3.552901	0.960009	3.999055	0.000948
41	1.66	3.586866	0.959993	4.000700	−0.000697
42	1.65	3.619843	0.960009	3.999090	0.000912
43	1.64	3.650858	0.960000	3.999984	0.000018
44	1.63	3.680894	0.960004	3.999645	0.000358
No. of pairs		Mean COMPUT. risk errors		STD. ERR. errors	
44		0.0000475513		0.0000793774	
Z-ratio				Status of significance	
0.59905				N.S.—not significant	

Parameter values					
PROB. = 0.95	RHO = −0.9		BIGH = 1.96		SMHL = 1.57
Pair No.	SMH	SMK	PROB. level	Risk level (%)	Risk error (%)
Table No. 8.2-367					
1	1.97	1.950051	0.949997	5.000329	−0.000328
2	1.96	1.960000	0.950005	4.999507	0.000495
3	1.95	1.970051	0.949997	5.000329	−0.000328
4	1.94	1.980597	0.949993	5.000740	−0.000739
5	1.93	1.991638	0.949992	5.000806	−0.000805
6	1.92	2.003177	0.949993	5.000716	−0.000715
7	1.91	2.015215	0.949993	5.000681	−0.000679
8	1.90	2.027754	0.949992	5.000836	−0.000834
9	1.89	2.041186	0.950006	4.999423	0.000578
10	1.88	2.054732	0.949995	5.000538	−0.000536
11	1.87	2.069175	0.949994	5.000573	−0.000572
12	1.86	2.084517	0.950002	4.999829	0.000173
13	1.85	2.100369	0.949996	5.000383	−0.000381
14	1.84	2.117513	0.950009	4.999101	0.000900
15	1.83	2.135172	0.950003	4.999715	0.000286
16	1.82	2.153738	0.949992	5.000848	−0.000846
17	1.81	2.173993	0.950001	4.999906	0.000095
18	1.80	2.195160	0.949995	5.000467	−0.000465
19	1.79	2.218020	0.949998	5.000162	−0.000161
20	1.78	2.242577	0.950002	4.999763	0.000238

(continued)

Table 8.2 (continued)

Parameter values					
PROB. = 0.95	RHO = −0.9		BIGH = 1.96		SMHL = 1.57
Pair No.	SMH	SMK	PROB. level	Risk level (%)	Risk error (%)
21	1.77	2.268833	0.949998	5.000162	−0.000161
22	1.76	2.297571	0.950004	4.999561	0.000441
23	1.75	2.328793	0.950007	4.999262	0.000739
24	1.74	2.362503	0.949997	5.000353	−0.000352
25	1.73	2.400265	0.949996	5.000425	−0.000423
26	1.72	2.442863	0.950002	4.999823	0.000179
27	1.71	2.490299	0.949990	5.000961	−0.000960
28	1.70	2.546483	0.950001	4.999871	0.000131
29	1.69	2.612198	0.950000	4.999984	0.000018
30	1.68	2.692916	0.950002	4.999763	0.000238
31	1.67	2.795671	0.949999	5.000061	−0.000060
32	1.66	2.932966	0.949997	5.000353	−0.000352
33	1.65	3.103242	0.949999	5.000097	−0.000095
34	1.64	3.251813	0.950009	4.999125	0.000876
35	1.63	3.355246	0.949996	5.000377	−0.000376
36	1.62	3.433857	0.950001	4.999900	0.000101
37	1.61	3.497023	0.950002	4.999799	0.000203
38	1.60	3.550608	0.949999	5.000067	−0.000066
39	1.59	3.598131	0.950005	4.999477	0.000525
40	1.58	3.640766	0.950001	4.999900	0.000101
41	1.57	3.680081	0.950000	5.000031	−0.000030
No. of pairs		Mean COMPUT. risk errors		STD. ERR. errors	
41		−0.0000939483		0.0000741860	
Z-ratio				Status of significance	
1.26639				N.S.—not significant	
Parameter values					
PROB. = 0.94	RHO = −0.9		BIGH = 1.8808		SMHL = 1.51
Pair No.	SMH	SMK	PROB. level	Risk level (%)	Risk error (%)
Table No. 8.2-368					
1	1.89	1.871645	0.939993	6.000668	−0.000668
2	1.88	1.881600	0.940001	5.999887	0.000113
3	1.87	1.891662	0.939990	6.000966	−0.000966
4	1.86	1.902418	0.939999	6.000090	−0.000089
5	1.85	1.913480	0.940000	6.000000	0.000000
6	1.84	1.924848	0.939992	6.000805	−0.000805
7	1.83	1.936916	0.939997	6.000263	−0.000262
8	1.82	1.949490	0.940002	5.999780	0.000221
9	1.81	1.962572	0.940004	5.999565	0.000435
10	1.80	1.976164	0.940002	5.999833	0.000167
11	1.79	1.990268	0.939992	6.000757	−0.000757
12	1.78	2.005277	0.939995	6.000453	−0.000453
13	1.77	2.021192	0.940007	5.999327	0.000674

(continued)

Table 8.2 (continued)

Parameter values					
PROB. = 0.94	RHO = −0.9		BIGH = 1.8808		SMHL = 1.51
Pair No.	SMH	SMK	PROB. level	Risk level (%)	Risk error (%)
14	1.76	2.037625	0.940003	5.999744	0.000256
15	1.75	2.054970	0.940000	5.999995	0.000006
16	1.74	2.073227	0.939995	6.000477	−0.000477
17	1.73	2.092792	0.940001	5.999887	0.000113
18	1.72	2.113273	0.939995	6.000459	−0.000459
19	1.71	2.135457	0.940006	5.999446	0.000554
20	1.70	2.158954	0.940008	5.999184	0.000817
21	1.69	2.183766	0.939997	6.000352	−0.000352
22	1.68	2.211069	0.940006	5.999398	0.000602
23	1.67	2.240084	0.939998	6.000209	−0.000209
24	1.66	2.271984	0.939999	6.000054	−0.000054
25	1.65	2.307165	0.940006	5.999375	0.000626
26	1.64	2.345237	0.939990	6.000960	−0.000960
27	1.63	2.388939	0.940002	5.999851	0.000149
28	1.62	2.438273	0.940006	5.999380	0.000620
29	1.61	2.494804	0.940002	5.999846	0.000155
30	1.60	2.562442	0.940007	5.999303	0.000697
31	1.59	2.645097	0.940005	5.999470	0.000530
32	1.58	2.751366	0.939994	6.000644	−0.000644
33	1.57	2.903126	0.939992	6.000763	−0.000763
34	1.56	3.130068	0.939999	6.000108	−0.000107
35	1.55	3.338448	0.940002	5.999768	0.000232
36	1.54	3.464204	0.940008	5.999244	0.000757
37	1.53	3.549530	0.939994	6.000644	−0.000644
38	1.52	3.617084	0.940003	5.999697	0.000304
39	1.51	3.673122	0.940003	5.999720	0.000280
No. of pairs			Mean COMPUT. risk errors	STD. ERR. errors	
39			−0.0000089407	0.0000815076	
Z-ratio				Status of significance	
0.10969				N.S.—not significant	
Parameter values					
PROB. = 0.93	RHO = −0.9		BIGH = 1.8119		SMHL = 1.45
Pair No.	SMH	SMK	PROB. level	Risk level (%)	Risk error (%)
Table No. 8.2-369					
1	1.82	1.803836	0.929992	7.000804	−0.000805
2	1.81	1.813802	0.929998	7.000196	−0.000197
3	1.80	1.824074	0.929999	7.000089	−0.000089
4	1.79	1.834849	0.930009	6.999111	0.000888
5	1.78	1.845739	0.929997	7.000315	−0.000316
6	1.77	1.857136	0.929991	7.000947	−0.000948
7	1.76	1.869237	0.930001	6.999880	0.000119
8	1.75	1.881653	0.929999	7.000077	−0.000077

(continued)

8.2 Contents for Table 8.2 (Table No. 8.2-1 to 400)

Table 8.2 (continued)

Parameter values					
PROB. = 0.93	RHO = −0.9		BIGH = 1.8119		SMHL = 1.45
Pair No.	SMH	SMK	PROB. level	Risk level (%)	Risk error (%)
9	1.74	1.894584	0.929997	7.000327	−0.000328
10	1.73	1.908224	0.930004	6.999636	0.000364
11	1.72	1.922382	0.930005	6.999505	0.000495
12	1.71	1.937060	0.929998	7.000208	−0.000209
13	1.70	1.952650	0.930004	6.999600	0.000399
14	1.69	1.968764	0.929996	7.000411	−0.000411
15	1.68	1.985796	0.929993	7.000661	−0.000662
16	1.67	2.003748	0.929992	7.000816	−0.000817
17	1.66	2.023012	0.930007	6.999314	0.000685
18	1.65	2.042811	0.929993	7.000709	−0.000709
19	1.64	2.064318	0.930004	6.999624	0.000376
20	1.63	2.086755	0.929994	7.000578	−0.000578
21	1.62	2.110906	0.929994	7.000619	−0.000620
22	1.61	2.136775	0.929993	7.000703	−0.000703
23	1.60	2.164754	0.929998	7.000244	−0.000244
24	1.59	2.195237	0.930009	6.999076	0.000924
25	1.58	2.227836	0.930001	6.999910	0.000089
26	1.57	2.263727	0.929997	7.000292	−0.000292
27	1.56	2.303694	0.930000	6.999982	0.000018
28	1.55	2.348521	0.930005	6.999457	0.000542
29	1.54	2.398993	0.930000	6.999994	0.000006
30	1.53	2.457458	0.929996	7.000411	−0.000411
31	1.52	2.527043	0.929994	7.000590	−0.000590
32	1.51	2.613222	0.929996	7.000435	−0.000435
33	1.50	2.726153	0.929994	7.000602	−0.000602
34	1.49	2.893967	0.930007	6.999350	0.000650
35	1.48	3.174480	0.929995	7.000542	−0.000542
36	1.47	3.439569	0.930005	6.999481	0.000519
37	1.46	3.573616	0.930005	6.999517	0.000483
38	1.45	3.661780	0.930001	6.999880	0.000119
No. of pairs		Mean COMPUT. risk errors		STD. ERR. errors	
38		−0.0001002581		0.0000821442	
Z-ratio				Status of significance	
1.22051				N.S.—not significant	
Parameter values					
PROB. = 0.92	RHO = −0.9		BIGH = 1.7507		SMHL = 1.39
Pair No.	SMH	SMK	PROB. level	Risk level (%)	Risk error (%)
Table No. 8.3-370					
1	1.76	1.741449	0.919994	8.000636	−0.000638
2	1.75	1.751400	0.920002	7.999778	0.000221

(continued)

Table 8.2 (continued)

Parameter values					
PROB. = 0.92	RHO = −0.9		BIGH = 1.7507		SMHL = 1.39
Pair No.	SMH	SMK	PROB. level	Risk level (%)	Risk error (%)
3	1.74	1.761466	0.919991	8.000946	−0.000948
4	1.73	1.772038	0.919991	8.000935	−0.000936
5	1.72	1.783120	0.920000	7.999957	0.000042
6	1.71	1.794517	0.920002	7.999838	0.000161
7	1.70	1.806428	0.920009	7.999129	0.000870
8	1.69	1.818658	0.920004	7.999575	0.000423
9	1.68	1.831406	0.920002	7.999850	0.000149
10	1.67	1.844675	0.919998	8.000207	−0.000209
11	1.66	1.858465	0.919991	8.000875	−0.000876
12	1.65	1.873171	0.920006	7.999397	0.000602
13	1.64	1.888208	0.919998	8.000160	−0.000161
14	1.63	1.904166	0.920005	7.999516	0.000483
15	1.62	1.920851	0.920009	7.999134	0.000864
16	1.61	1.938071	0.919994	8.000613	−0.000614
17	1.60	1.956609	0.920004	7.999587	0.000411
18	1.59	1.976079	0.920010	7.999039	0.000960
19	1.58	1.996483	0.920006	7.999420	0.000578
20	1.57	2.018213	0.920008	7.999218	0.000781
21	1.56	2.041274	0.920008	7.999188	0.000811
22	1.55	2.065668	0.919999	8.000076	−0.000077
23	1.54	2.091790	0.919991	8.000887	−0.000888
24	1.53	2.120423	0.920006	7.999355	0.000644
25	1.52	2.150789	0.919998	8.000195	−0.000197
26	1.51	2.184065	0.920000	8.000040	−0.000042
27	1.50	2.220644	0.920005	7.999486	0.000513
28	1.49	2.260920	0.920006	7.999432	0.000566
29	1.48	2.305678	0.919999	8.000094	−0.000095
30	1.47	2.356875	0.920005	7.999552	0.000447
31	1.46	2.415687	0.920002	7.999766	0.000232
32	1.45	2.485633	0.920005	7.999498	0.000501
33	1.44	2.570624	0.919991	8.000887	−0.000888
34	1.43	2.683946	0.920005	7.999468	0.000530
35	1.42	2.847477	0.920000	8.000035	−0.000036
36	1.41	3.148722	0.919992	8.000803	−0.000805

(continued)

8.2 Contents for Table 8.2 (Table No. 8.2-1 to 400)

Table 8.2 (continued)

Parameter values					
PROB. = 0.92	RHO = −0.9		BIGH = 1.7507		SMHL = 1.39
Pair No.	SMH	SMK	PROB. level	Risk level (%)	Risk error (%)
37	1.40	3.504873	0.920008	7.999194	0.000805
38	1.39	3.656561	0.920007	7.999295	0.000703
No. of pairs		Mean COMPUT. risk errors		STD. ERR. errors	
38		0.0001253226		0.0000938682	
Z-ratio				Status of significance	
1.33509				N.S.—not significant	
Parameter values					
PROB. = 0.91	RHO = −0.9		BIGH = 1.6954		SMHL = 1.33
Pair No.	SMH	SMK	PROB. level	Risk level (%)	Risk error (%)
Table No. 8.2-371					
1	1.70	1.690812	0.909998	9.000182	−0.000185
2	1.69	1.700817	0.909997	9.000278	−0.000280
3	1.68	1.711136	0.909993	9.000659	−0.000662
4	1.67	1.721968	0.910003	8.999735	0.000262
5	1.66	1.733117	0.910005	8.999467	0.000530
6	1.65	1.744588	0.910000	8.999979	0.000018
7	1.64	1.756578	0.910002	8.999759	0.000238
8	1.63	1.768893	0.909993	9.000671	−0.000674
9	1.62	1.781926	0.910003	8.999676	0.000322
10	1.61	1.795291	0.909997	9.000259	−0.000262
11	1.60	1.809379	0.910005	8.999538	0.000459
12	1.59	1.823802	0.909992	9.000850	−0.000852
13	1.58	1.839150	0.910000	8.999991	0.000006
14	1.57	1.855035	0.909996	9.000391	−0.000393
15	1.56	1.871849	0.910006	8.999371	0.000626
16	1.55	1.889205	0.909997	9.000301	−0.000304
17	1.54	1.907497	0.909992	9.000838	−0.000840
18	1.53	1.927118	0.910009	8.999073	0.000924
19	1.52	1.947290	0.909995	9.000546	−0.000548
20	1.51	1.968798	0.909990	9.000992	−0.000995
21	1.50	1.992035	0.910009	8.999091	0.000906
22	1.49	2.016224	0.910000	9.000039	−0.000042
23	1.48	2.042149	0.909995	9.000516	−0.000519
24	1.47	2.070205	0.910003	8.999729	0.000268
25	1.46	2.100004	0.909991	9.000921	−0.000924
26	1.45	2.132722	0.909997	9.000307	−0.000310
27	1.44	2.168363	0.910001	8.999920	0.000077
28	1.43	2.207713	0.910009	8.999062	0.000936
29	1.42	2.250774	0.909996	9.000379	−0.000381
30	1.41	2.299505	0.909992	9.000789	−0.000793
31	1.40	2.355472	0.909994	9.000624	−0.000626
32	1.39	2.421024	0.909997	9.000283	−0.000286

(continued)

Table 8.2 (continued)

Parameter values					
PROB. = 0.91	RHO = −0.9		BIGH = 1.6954		SMHL = 1.33
Pair No.	SMH	SMK	PROB. level	Risk level (%)	Risk error (%)
33	1.38	2.500072	0.909998	9.000158	−0.000161
34	1.37	2.599650	0.909991	9.000898	−0.000900
35	1.36	2.738515	0.910001	8.999944	0.000054
36	1.35	2.966670	0.909997	9.000319	−0.000322
37	1.34	3.441934	0.910000	9.000039	−0.000042
38	1.33	3.683062	0.910005	8.999461	0.000536
No. of pairs		Mean COMPUT. risk errors		STD. ERR. errors	
38		−0.0001317416		0.0000853998	
Z-ratio				Status of significance	
1.54264				N.S.—not significant	
Parameter values					
PROB. = 0.9025	RHO = −0.9		BIGH = 1.6571		SMHL = 1.29
Pair No.	SMH	SMK	PROB. level	Risk level (%)	Risk error (%)
Table No. 8.2-372					
1	1.66	1.654205	0.902500	9.750021	−0.000018
2	1.65	1.664231	0.902495	9.750462	−0.000459
3	1.64	1.674769	0.902508	9.749240	0.000763
4	1.63	1.685432	0.902497	9.750342	−0.000340
5	1.62	1.696612	0.902499	9.750116	−0.000113
6	1.61	1.708117	0.902494	9.750629	−0.000626
7	1.60	1.720144	0.902498	9.750234	−0.000232
8	1.59	1.732500	0.902490	9.750962	−0.000960
9	1.58	1.745579	0.902505	9.749550	0.000453
10	1.57	1.758993	0.902503	9.749699	0.000304
11	1.56	1.772939	0.902501	9.749937	0.000066
12	1.55	1.787420	0.902494	9.750557	−0.000554
13	1.54	1.802635	0.902497	9.750289	−0.000286
14	1.53	1.818586	0.902504	9.749561	0.000441
15	1.52	1.835082	0.902498	9.750223	−0.000221
16	1.51	1.852514	0.902502	9.749770	0.000232
17	1.50	1.870692	0.902499	9.750116	−0.000113
18	1.49	1.889815	0.902496	9.750366	−0.000364
19	1.48	1.910080	0.902502	9.749812	0.000191
20	1.47	1.931295	0.902496	9.750426	−0.000423
21	1.46	1.953855	0.902496	9.750419	−0.000417
22	1.45	1.977764	0.902493	9.750700	−0.000697
23	1.44	2.003415	0.902500	9.750009	−0.000006
24	1.43	2.030813	0.902504	9.749567	0.000435
25	1.42	2.059961	0.902495	9.750498	−0.000495
26	1.41	2.091644	0.902495	9.750509	−0.000507
27	1.40	2.126258	0.902503	9.749752	0.000250
28	1.39	2.163806	0.902496	9.750402	−0.000399

(continued)

8.2 Contents for Table 8.2 (Table No. 8.2-1 to 400)

Table 8.2 (continued)

Parameter values					
PROB. = 0.9025	RHO = −0.9		BIGH = 1.6571		SMHL = 1.29
Pair No.	SMH	SMK	PROB. level	Risk level (%)	Risk error (%)
29	1.38	2.205465	0.902496	9.750402	−0.000399
30	1.37	2.252021	0.902496	9.750390	−0.000387
31	1.36	2.305040	0.902503	9.749710	0.000292
32	1.35	2.366090	0.902504	9.749651	0.000352
33	1.34	2.438301	0.902499	9.750098	−0.000095
34	1.33	2.527146	0.902491	9.750867	−0.000864
35	1.32	2.645912	0.902509	9.749055	0.000948
36	1.31	2.819605	0.902500	9.750032	−0.000030
37	1.30	3.181041	0.902505	9.749478	0.000525
38	1.29	3.650540	0.902501	9.749926	0.000077
No. of pairs		Mean COMPUT. risk errors		STD. ERR. errors	
38		−0.0000942976		0.0000702704	
Z-ratio				Status of significance	
1.34193				N.S.—not significant	
Parameter values					
PROB. = 0.9	RHO = −0.9		BIGH = 1.645		SMHL = 1.28
Pair No.	SMH	SMK	PROB. level	Risk level (%)	Risk error (%)
Table No. 8.2-373					
1	1.65	1.639820	0.900007	9.999275	0.000727
2	1.64	1.649820	0.900008	9.999251	0.000751
3	1.63	1.660138	0.900006	9.999418	0.000584
4	1.62	1.670776	0.900001	9.999906	0.000095
5	1.61	1.681737	0.899991	10.000890	−0.000888
6	1.60	1.693219	0.899994	10.000630	−0.000626
7	1.59	1.705223	0.900006	9.999394	0.000608
8	1.58	1.717362	0.899990	10.001000	−0.000995
9	1.57	1.730223	0.899997	10.000290	−0.000286
10	1.56	1.743616	0.900007	9.999299	0.000703
11	1.55	1.757346	0.900000	10.000020	−0.000018
12	1.54	1.771612	0.899990	10.000960	−0.000960
13	1.53	1.786612	0.899992	10.000840	−0.000834
14	1.52	1.802350	0.899999	10.000070	−0.000072
15	1.51	1.818632	0.899995	10.000550	−0.000548
16	1.50	1.835852	0.900003	9.999704	0.000298
17	1.49	1.853624	0.899991	10.000870	−0.000870
18	1.48	1.872536	0.899997	10.000280	−0.000274
19	1.47	1.892396	0.900000	10.000000	0.000000
20	1.46	1.913207	0.899994	10.000580	−0.000578
21	1.45	1.935365	0.899998	10.000210	−0.000203
22	1.44	1.958871	0.900002	9.999776	0.000226
23	1.43	1.983731	0.899999	10.000150	−0.000149
24	1.42	2.010338	0.899998	10.000160	−0.000155

(continued)

Table 8.2 (continued)

Parameter values					
PROB. = 0.9	RHO = −0.9		BIGH = 1.645		SMHL = 1.28
Pair No.	SMH	SMK	PROB. level	Risk level (%)	Risk error (%)
25	1.41	2.038698	0.899990	10.000990	−0.000989
26	1.40	2.069593	0.899998	10.000180	−0.000179
27	1.39	2.103030	0.900004	9.999609	0.000393
28	1.38	2.139403	0.900005	9.999478	0.000525
29	1.37	2.179106	0.899995	10.000530	−0.000525
30	1.36	2.223708	0.900001	9.999895	0.000107
31	1.35	2.273994	0.900009	9.999121	0.000882
32	1.34	2.331140	0.900004	9.999567	0.000435
33	1.33	2.397886	0.899996	10.000410	−0.000405
34	1.32	2.479705	0.900008	9.999198	0.000805
35	1.31	2.582854	0.900003	9.999716	0.000286
36	1.30	2.726869	0.900003	9.999681	0.000322
37	1.29	2.972693	0.900005	9.999550	0.000453
38	1.28	3.532830	0.900009	9.999073	0.000930
No. of pairs		Mean COMPUT. risk errors		STD. ERR. errors	
38		−0.0000108511		0.0000923360	
Z-ratio				Status of significance	
0.11752				N.S.—not significant	
Parameter values					
PROB. = 0.85	RHO = −0.9		BIGH = 1.4395		SMHL = 1.04
Pair No.	SMH	SMK	PROB. level	Risk level (%)	Risk error (%)
Table No. 8.2-374					
1	1.44	1.439000	0.849991	15.000900	−0.000906
2	1.43	1.449258	0.850009	14.999130	0.000870
3	1.42	1.459658	0.850004	14.999590	0.000405
4	1.41	1.470398	0.850003	14.999680	0.000322
5	1.40	1.481482	0.850004	14.999580	0.000417
6	1.39	1.492911	0.850005	14.999460	0.000536
7	1.38	1.504690	0.850005	14.999510	0.000489
8	1.37	1.516823	0.850001	14.999930	0.000072
9	1.36	1.529311	0.849991	15.000860	−0.000864
10	1.35	1.542355	0.849998	15.000160	−0.000167
11	1.34	1.555763	0.849995	15.000500	−0.000501
12	1.33	1.569734	0.850002	14.999770	0.000226
13	1.32	1.584076	0.849994	15.000570	−0.000572
14	1.31	1.598989	0.849991	15.000940	−0.000942
15	1.30	1.614672	0.850009	14.999130	0.000870
16	1.29	1.630740	0.850002	14.999800	0.000197
17	1.28	1.647586	0.850009	14.999130	0.000870
18	1.27	1.665021	0.850003	14.999660	0.000334
19	1.26	1.683243	0.850001	14.999870	0.000131
20	1.25	1.702259	0.849997	15.000310	−0.000310

(continued)

8.2 Contents for Table 8.2 (Table No. 8.2-1 to 400)

Table 8.2 (continued)

Parameter values					
PROB. = 0.85	RHO = −0.9		BIGH = 1.4395		SMHL = 1.04
Pair No.	SMH	SMK	PROB. level	Risk level (%)	Risk error (%)
21	1.24	1.722269	0.850002	14.999820	0.000179
22	1.23	1.743277	0.850009	14.999120	0.000876
23	1.22	1.765289	0.850010	14.999010	0.000989
24	1.21	1.788310	0.849998	15.000240	−0.000238
25	1.20	1.812737	0.849994	15.000580	−0.000584
26	1.19	1.838772	0.850003	14.999750	0.000250
27	1.18	1.866224	0.849995	15.000510	−0.000513
28	1.17	1.895686	0.849999	15.000120	−0.000119
29	1.16	1.927360	0.850008	14.999180	0.000823
30	1.15	1.961253	0.850003	14.999660	0.000334
31	1.14	1.998153	0.850007	14.999310	0.000685
32	1.13	2.038457	0.850010	14.999010	0.000989
33	1.12	2.082564	0.849998	15.000240	−0.000238
34	1.11	2.132045	0.849999	15.000090	−0.000095
35	1.10	2.188078	0.850002	14.999800	0.000203
36	1.09	2.252626	0.850002	14.999780	0.000215
37	1.08	2.328822	0.849995	15.000520	−0.000525
38	1.07	2.423316	0.850001	14.999940	0.000060
39	1.06	2.547054	0.849996	15.000410	−0.000411
40	1.05	2.732860	0.850002	14.999830	0.000173
41	1.04	3.142460	0.849995	15.000530	−0.000536
No. of pairs		Mean COMPUT. risk errors		STD. ERR. errors	
41		0.0000950836		0.0000831749	
Z-ratio				Status of significance	
1.14318				N.S.—not significant	
Parameter values					
PROB. = 0.8	RHO = −0.9		BIGH = 1.2815		SMHL = 0.85
Pair No.	SMH	SMK	PROB. level	Risk level (%)	Risk error (%)
Table No. 8.2-375					
1	1.29	1.273154	0.799993	20.000720	−0.000721
2	1.28	1.283099	0.799999	20.000120	−0.000125
3	1.27	1.293299	0.800004	19.999610	0.000387
4	1.26	1.303757	0.800007	19.999280	0.000721
5	1.25	1.314380	0.799991	20.000910	−0.000912
6	1.24	1.325463	0.800004	19.999650	0.000352
7	1.23	1.336719	0.799994	20.000580	−0.000584
8	1.22	1.348444	0.800010	19.999030	0.000972
9	1.21	1.360350	0.800001	19.999910	0.000089
10	1.20	1.372637	0.799998	20.000230	−0.000232
11	1.19	1.385309	0.799998	20.000220	−0.000221
12	1.18	1.398371	0.799999	20.000090	−0.000089
13	1.17	1.411829	0.799999	20.000070	−0.000072

(continued)

Table 8.2 (continued)

Parameter values					
PROB. = 0.8	RHO = −0.9		BIGH = 1.2815		SMHL = 0.85
Pair No.	SMH	SMK	PROB. level	Risk level (%)	Risk error (%)
14	1.16	1.425687	0.799996	20.000380	−0.000381
15	1.15	1.440048	0.800001	19.999870	0.000125
16	1.14	1.454821	0.799997	20.000260	−0.000256
17	1.13	1.470108	0.799996	20.000440	−0.000435
18	1.12	1.486014	0.800005	19.999460	0.000542
19	1.11	1.502349	0.799997	20.000300	−0.000304
20	1.10	1.519314	0.799992	20.000770	−0.000769
21	1.09	1.537015	0.799999	20.000130	−0.000131
22	1.08	1.555360	0.799999	20.000120	−0.000125
23	1.07	1.574454	0.799999	20.000090	−0.000089
24	1.06	1.594402	0.800005	19.999540	0.000465
25	1.05	1.615212	0.800009	19.999140	0.000858
26	1.04	1.636891	0.800004	19.999650	0.000352
27	1.03	1.659644	0.800002	19.999800	0.000197
28	1.02	1.683479	0.799995	20.000540	−0.000542
29	1.01	1.708795	0.800008	19.999220	0.000781
30	1.00	1.735211	0.799993	20.000700	−0.000697
31	0.99	1.763518	0.800006	19.999370	0.000626
32	0.98	1.793335	0.799997	20.000270	−0.000268
33	0.97	1.825259	0.799996	20.000440	−0.000435
34	0.96	1.859497	0.799994	20.000600	−0.000602
35	0.95	1.896644	0.800007	19.999340	0.000662
36	0.94	1.936714	0.800001	19.999890	0.000113
37	0.93	1.980695	0.800002	19.999830	0.000173
38	0.92	2.029186	0.799994	20.000600	−0.000596
39	0.91	2.083958	0.800007	19.999310	0.000685
40	0.90	2.145806	0.799996	20.000440	−0.000435
41	0.89	2.218262	0.799999	20.000140	−0.000137
42	0.88	2.305246	0.799994	20.000600	−0.000596
43	0.87	2.415758	0.799999	20.000140	−0.000143
44	0.86	2.568958	0.800005	19.999470	0.000530
45	0.85	2.828923	0.800002	19.999770	0.000232
No. of pairs		Mean COMPUT. risk errors		STD. ERR. errors	
45		−0.0000225461		0.0000719883	
Z-ratio				Status of significance	
0.31319				N.S.—not significant	
Parameter values					
PROB. = 0.75	RHO = −0.9		BIGH = 1.1504		SMHL = 0.68
Pair No.	SMH	SMK	PROB. level	Risk level (%)	Risk error (%)
Table No. 8.2-376					
1	1.16	1.140782	0.749995	25.000480	−0.000477
2	1.15	1.150703	0.750001	24.999940	0.000060

(continued)

8.2 Contents for Table 8.2 (Table No. 8.2-1 to 400)

Table 8.2 (continued)

Parameter values					
PROB. = 0.75	RHO = −0.9		BIGH = 1.1504		SMHL = 0.68
Pair No.	SMH	SMK	PROB. level	Risk level (%)	Risk error (%)
3	1.14	1.160797	0.749995	25.000540	−0.000536
4	1.13	1.171169	0.749996	25.000370	−0.000370
5	1.12	1.181821	0.750005	24.999540	0.000465
6	1.11	1.192661	0.749999	25.000060	−0.000060
7	1.10	1.203793	0.749999	25.000090	−0.000083
8	1.09	1.215221	0.750002	24.999760	0.000244
9	1.08	1.226952	0.750009	24.999140	0.000858
10	1.07	1.238892	0.749998	25.000250	−0.000250
11	1.06	1.251244	0.750005	24.999510	0.000495
12	1.05	1.263818	0.749993	25.000670	−0.000668
13	1.04	1.276816	0.749996	25.000370	−0.000364
14	1.03	1.290148	0.749995	25.000480	−0.000477
15	1.02	1.303916	0.750005	24.999540	0.000465
16	1.01	1.318032	0.750006	24.999410	0.000590
17	1.00	1.332502	0.749997	25.000290	−0.000286
18	0.99	1.347530	0.750008	24.999180	0.000823
19	0.98	1.362929	0.750005	24.999550	0.000453
20	0.97	1.378804	0.749999	25.000100	−0.000095
21	0.96	1.395262	0.750004	24.999610	0.000387
22	0.95	1.412214	0.750001	24.999950	0.000048
23	0.94	1.429769	0.749999	25.000070	−0.000066
24	0.93	1.447935	0.749997	25.000320	−0.000316
25	0.92	1.466820	0.750001	24.999880	0.000119
26	0.91	1.486339	0.749994	25.000600	−0.000596
27	0.90	1.506795	0.750008	24.999170	0.000829
28	0.89	1.527907	0.749999	25.000070	−0.000072
29	0.88	1.549980	0.749997	25.000270	−0.000268
30	0.87	1.573125	0.750005	24.999500	0.000501
31	0.86	1.597259	0.750001	24.999860	0.000143
32	0.85	1.622590	0.749999	25.000110	−0.000107
33	0.84	1.649328	0.750006	24.999440	0.000560
34	0.83	1.677490	0.750007	24.999260	0.000739
35	0.82	1.707286	0.750007	24.999260	0.000745
36	0.81	1.738930	0.750006	24.999360	0.000644
37	0.80	1.772634	0.750000	25.000000	0.000000
38	0.79	1.808809	0.749996	25.000420	−0.000417
39	0.78	1.847864	0.749994	25.000620	−0.000620
40	0.77	1.890407	0.749998	25.000170	−0.000167
41	0.76	1.937045	0.750003	24.999720	0.000286
42	0.75	1.988778	0.750010	24.999010	0.000989
43	0.74	2.046608	0.750002	24.999820	0.000179
44	0.73	2.112708	0.749992	25.000790	−0.000787

(continued)

Table 8.2 (continued)

Parameter values					
PROB. = 0.75	RHO = −0.9		BIGH = 1.1504		SMHL = 0.68
Pair No.	SMH	SMK	PROB. level	Risk level (%)	Risk error (%)
45	0.72	2.190426	0.749991	25.000910	−0.000906
46	0.71	2.285065	0.749994	25.000650	−0.000644
47	0.70	2.407005	0.749995	25.000550	−0.000548
48	0.69	2.582061	0.749992	25.000760	−0.000757
49	0.68	2.921204	0.750004	24.999560	0.000441
No. of pairs			Mean COMPUT. risk errors	STD. ERR. errors	
49			0.0000225306	0.0000711594	
Z-ratio				Status of significance	
0.31662				N.S.—not significant	
Parameter values					
PROB. = 0.7	RHO = −0.9		BIGH = 1.037		SMHL = .55
Pair No.	SMH	SMK	PROB. level	Risk level (%)	Risk error (%)
Table No. 8.2-377					
1	1.04	1.032886	0.700001	29.999870	0.000137
2	1.03	1.042876	0.699992	30.000780	−0.000775
3	1.02	1.053111	0.699991	30.000910	−0.000906
4	1.01	1.063648	0.700008	29.999170	0.000829
5	1.00	1.074295	0.699999	30.000060	−0.000054
6	0.99	1.085255	0.700009	29.999150	0.000852
7	0.98	1.096339	0.699992	30.000830	−0.000823
8	0.97	1.107749	0.699992	30.000820	−0.000817
9	0.96	1.119492	0.700007	29.999270	0.000727
10	0.95	1.131381	0.699997	30.000340	−0.000340
11	0.94	1.143619	0.700000	29.999970	0.000036
12	0.93	1.156115	0.699997	30.000260	−0.000256
13	0.92	1.168977	0.700007	29.999290	0.000709
14	0.91	1.182115	0.700009	29.999140	0.000864
15	0.90	1.195538	0.700001	29.999860	0.000143
16	0.89	1.209354	0.700004	29.999640	0.000364
17	0.88	1.223475	0.699995	30.000490	−0.000483
18	0.87	1.238009	0.699993	30.000660	−0.000656
19	0.86	1.252968	0.699997	30.000320	−0.000322
20	0.85	1.268362	0.700003	29.999680	0.000328
21	0.84	1.284107	0.699994	30.000630	−0.000632
22	0.83	1.300410	0.700000	29.999960	0.000042
23	0.82	1.317187	0.700004	29.999600	0.000399
24	0.81	1.334452	0.700002	29.999810	0.000197
25	0.80	1.352219	0.699992	30.000810	−0.000805
26	0.79	1.370699	0.700002	29.999830	0.000173

(continued)

Table 8.2 (continued)

Parameter values					
PROB. = 0.7	RHO = −0.9		BIGH = 1.037		SMHL = .55
Pair No.	SMH	SMK	PROB. level	Risk level (%)	Risk error (%)
27	0.78	1.389713	0.699997	30.000340	−0.000340
28	0.77	1.409473	0.700003	29.999750	0.000250
29	0.76	1.429900	0.700000	30.000010	−0.000006
30	0.75	1.451110	0.699998	30.000200	−0.000197
31	0.74	1.473123	0.699991	30.000890	−0.000888
32	0.73	1.496154	0.699998	30.000170	−0.000173
33	0.72	1.520130	0.699999	30.000150	−0.000149
34	0.71	1.545268	0.700007	29.999270	0.000733
35	0.70	1.571397	0.699991	30.000880	−0.000882
36	0.69	1.599130	0.700007	29.999310	0.000697
37	0.68	1.628104	0.699996	30.000370	−0.000364
38	0.67	1.658934	0.700007	29.999340	0.000662
39	0.66	1.691456	0.699998	30.000170	−0.000173
40	0.65	1.726288	0.700006	29.999390	0.000614
41	0.64	1.763466	0.700003	29.999720	0.000286
42	0.63	1.803418	0.699992	30.000850	−0.000846
43	0.62	1.846965	0.699995	30.000500	−0.000495
44	0.61	1.894539	0.699993	30.000750	−0.000745
45	0.60	1.947359	0.700001	29.999880	0.000125
46	0.59	2.006252	0.699990	30.001000	−0.000995
47	0.58	2.074005	0.700003	29.999670	0.000334
48	0.57	2.153017	0.700003	29.999740	0.000262
49	0.56	2.248816	0.699998	30.000160	−0.000161
50	0.55	2.372402	0.700004	29.999600	0.000399
No. of pairs		Mean COMPUT. risk errors		STD. ERR. errors	
50		−0.0000612409		0.0000768968	
Z-ratio				Status of significance	
0.79640				N.S.—not significant	
Parameter values					
PROB. = 0.65	RHO = −0.9		BIGH = 0.9345		SMHL = 0.45
Pair No.	SMH	SMK	PROB. level	Risk level (%)	Risk error (%)
Table No. 8.2-378					
1	0.94	0.929227	0.650006	34.999380	0.000620
2	0.93	0.939217	0.650005	34.999460	0.000542
3	0.92	0.949375	0.649999	35.000080	−0.000077
4	0.91	0.959757	0.650001	34.999950	0.000054
5	0.90	0.970322	0.649998	35.000240	−0.000238
6	0.89	0.981127	0.650003	34.999730	0.000268
7	0.88	0.992082	0.649992	35.000780	−0.000775

(continued)

Table 8.2 (continued)

Parameter values					
PROB. = 0.65	RHO = −0.9		BIGH = 0.9345		SMHL = 0.45
Pair No.	SMH	SMK	PROB. level	Risk level (%)	Risk error (%)
8	0.87	1.003294	0.649991	35.000920	−0.000918
9	0.86	1.014770	0.649998	35.000180	−0.000173
10	0.85	1.026424	0.649992	35.000790	−0.000793
11	0.84	1.038362	0.649995	35.000460	−0.000459
12	0.83	1.050594	0.650009	34.999130	0.000870
13	0.82	1.063035	0.650009	34.999090	0.000918
14	0.81	1.075695	0.649998	35.000180	−0.000173
15	0.80	1.088683	0.649998	35.000200	−0.000197
16	0.79	1.102012	0.650008	34.999180	0.000823
17	0.78	1.115599	0.650008	34.999210	0.000793
18	0.77	1.129455	0.649997	35.000260	−0.000256
19	0.76	1.143695	0.649998	35.000220	−0.000215
20	0.75	1.158332	0.650009	34.999130	0.000870
21	0.74	1.173286	0.650010	34.999040	0.000960
22	0.73	1.188573	0.650001	34.999880	0.000119
23	0.72	1.204309	0.650003	34.999730	0.000274
24	0.71	1.220416	0.649995	35.000550	−0.000542
25	0.70	1.237010	0.649995	35.000500	−0.000501
26	0.69	1.254114	0.650003	34.999710	0.000292
27	0.68	1.271652	0.650000	35.000040	−0.000036
28	0.67	1.289746	0.650002	34.999830	0.000179
29	0.66	1.308323	0.649991	35.000870	−0.000864
30	0.65	1.327605	0.650000	35.000030	−0.000030
31	0.64	1.347523	0.650008	34.999210	0.000793
32	0.63	1.368010	0.649999	35.000150	−0.000143
33	0.62	1.389294	0.650000	35.000050	−0.000042
34	0.61	1.411310	0.649993	35.000740	−0.000733
35	0.60	1.434292	0.650003	34.999750	0.000250
36	0.59	1.458082	0.649996	35.000410	−0.000405
37	0.58	1.483016	0.650008	34.999230	0.000775
38	0.57	1.508845	0.649992	35.000780	−0.000781
39	0.56	1.536008	0.649992	35.000780	−0.000775
40	0.55	1.564557	0.649997	35.000310	−0.000304
41	0.54	1.594547	0.649995	35.000520	−0.000513
42	0.53	1.626232	0.649994	35.000610	−0.000608
43	0.52	1.659872	0.649998	35.000170	−0.000167
44	0.51	1.695731	0.650006	34.999410	0.000596
45	0.50	1.733884	0.649993	35.000680	−0.000679
46	0.49	1.775192	0.650000	34.999990	0.000012

(continued)

8.2 Contents for Table 8.2 (Table No. 8.2-1 to 400)

Table 8.2 (continued)

Parameter values					
PROB. = 0.65	RHO = −0.9		BIGH = 0.9345		SMHL = 0.45
Pair No.	SMH	SMK	PROB. level	Risk level (%)	Risk error (%)
47	0.48	1.819939	0.650002	34.999780	0.000226
48	0.47	1.868805	0.649998	35.000240	−0.000238
49	0.46	1.923064	0.650006	34.999390	0.000608
50	0.45	1.983612	0.649995	35.000460	−0.000459
No. of pairs		Mean COMPUT. risk errors		STD. ERR. errors	
50		−0.0000245431		0.0000762995	
Z-ratio				Status of significance	
0.32167				N.S.—not significant	
Parameter values					
PROB. = 0.6	RHO = −0.9		BIGH = 0.8416		SMHL = 0.36
Pair No.	SMH	SMK	PROB. level	Risk level (%)	Risk error (%)
Table No. 8.2-379					
1	0.85	0.833283	0.599999	40.000150	−0.000155
2	0.84	0.843203	0.599992	40.000800	−0.000805
3	0.83	0.853362	0.600005	39.999510	0.000489
4	0.82	0.863622	0.599998	40.000210	−0.000215
5	0.81	0.874091	0.599999	40.000090	−0.000089
6	0.80	0.884777	0.600010	39.999010	0.000989
7	0.79	0.895594	0.600005	39.999510	0.000489
8	0.78	0.906600	0.599999	40.000110	−0.000107
9	0.77	0.917807	0.599994	40.000630	−0.000638
10	0.76	0.929227	0.599990	40.000970	−0.000972
11	0.75	0.940872	0.599991	40.000940	−0.000942
12	0.74	0.952755	0.599995	40.000520	−0.000519
13	0.73	0.964890	0.600007	39.999330	0.000674
14	0.72	0.977194	0.600003	39.999740	0.000262
15	0.71	0.989780	0.600009	39.999120	0.000882
16	0.70	1.002566	0.600003	39.999700	0.000298
17	0.69	1.015619	0.599999	40.000080	−0.000083
18	0.68	1.028957	0.600000	40.000030	−0.000030
19	0.67	1.042599	0.600006	39.999410	0.000584
20	0.66	1.056468	0.599998	40.000240	−0.000244
21	0.65	1.070683	0.600000	40.000050	−0.000048
22	0.64	1.085219	0.600003	39.999710	0.000286
23	0.63	1.100052	0.599999	40.000060	−0.000060
24	0.62	1.115255	0.600002	39.999780	0.000215
25	0.61	1.130761	0.599993	40.000740	−0.000739
26	0.60	1.146695	0.599994	40.000560	−0.000560
27	0.59	1.163090	0.600009	39.999060	0.000942
28	0.58	1.179784	0.600001	39.999930	0.000072
29	0.57	1.196911	0.599992	40.000830	−0.000834
30	0.56	1.214609	0.600004	39.999630	0.000370

(continued)

Table 8.2 (continued)

Parameter values					
PROB. = 0.6	RHO = −0.9		BIGH = 0.8416		SMHL = 0.36
Pair No.	SMH	SMK	PROB. level	Risk level (%)	Risk error (%)
31	0.55	1.232722	0.600002	39.999850	0.000149
32	0.54	1.251394	0.600007	39.999280	0.000715
33	0.53	1.270576	0.600005	39.999470	0.000525
34	0.52	1.290319	0.599999	40.000090	−0.000095
35	0.51	1.310679	0.599992	40.000810	−0.000811
36	0.50	1.331814	0.600003	39.999740	0.000256
37	0.49	1.353595	0.600001	39.999920	0.000083
38	0.48	1.376190	0.600006	39.999440	0.000554
39	0.47	1.399578	0.600003	39.999690	0.000310
40	0.46	1.423843	0.599996	40.000360	−0.000358
41	0.45	1.449173	0.600001	39.999950	0.000048
42	0.44	1.475570	0.600002	39.999780	0.000215
43	0.43	1.503143	0.600001	39.999860	0.000143
44	0.42	1.532108	0.600010	39.999040	0.000954
45	0.41	1.562399	0.600000	39.999970	0.000030
46	0.40	1.594357	0.599994	40.000600	−0.000596
47	0.39	1.628237	0.599995	40.000550	−0.000548
48	0.38	1.664311	0.600002	39.999760	0.000238
49	0.37	1.702773	0.600004	39.999640	0.000358
50	0.36	1.744034	0.600000	40.000010	−0.000012
No. of pairs			Mean COMPUT. risk errors	STD. ERR. errors	
50			0.0000327241	0.0000715699	
Z-ratio				Status of significance	
0.45723				N.S.—not significant	
Parameter values					
PROB. = 0.55	RHO = −0.9		BIGH = 0.7553		SMHL = 0.27
Pair No.	SMH	SMK	PROB. level	Risk level (%)	Risk error (%)
Table No. 8.2-380					
1	0.76	0.750776	0.549996	45.000400	−0.000399
2	0.75	0.760784	0.549997	45.000320	−0.000322
3	0.74	0.770965	0.550003	44.999670	0.000328
4	0.73	0.781282	0.550004	44.999650	0.000352
5	0.72	0.791745	0.550000	45.000020	−0.000018
6	0.71	0.802416	0.550009	44.999130	0.000870
7	0.70	0.813211	0.550005	44.999540	0.000459
8	0.69	0.824143	0.549991	45.000950	−0.000948
9	0.68	0.835325	0.549997	45.000320	−0.000316
10	0.67	0.846675	0.549999	45.000090	−0.000089
11	0.66	0.858208	0.549994	45.000640	−0.000644

(continued)

8.2 Contents for Table 8.2 (Table No. 8.2-1 to 400)

Table 8.2 (continued)

Parameter values					
PROB. = 0.55	RHO = −0.9		BIGH = 0.7553		SMHL = 0.27
Pair No.	SMH	SMK	PROB. level	Risk level (%)	Risk error (%)
12	0.65	0.869944	0.549997	45.000270	−0.000268
13	0.64	0.881899	0.550007	44.999290	0.000709
14	0.63	0.893997	0.550000	45.000010	−0.000012
15	0.62	0.906356	0.550006	44.999360	0.000644
16	0.61	0.918901	0.550004	44.999570	0.000435
17	0.60	0.931656	0.549999	45.000150	−0.000149
18	0.59	0.944646	0.549994	45.000650	−0.000650
19	0.58	0.957899	0.549994	45.000580	−0.000578
20	0.57	0.971444	0.550006	44.999430	0.000572
21	0.56	0.985214	0.550009	44.999110	0.000888
22	0.55	0.999195	0.549998	45.000190	−0.000185
23	0.54	1.013472	0.549992	45.000830	−0.000834
24	0.53	1.028131	0.550007	44.999250	0.000745
25	0.52	1.042971	0.549995	45.000550	−0.000548
26	0.51	1.058232	0.550006	44.999380	0.000626
27	0.50	1.073768	0.550006	44.999430	0.000566
28	0.49	1.089631	0.550001	44.999860	0.000143
29	0.48	1.105878	0.550003	44.999730	0.000268
30	0.47	1.122474	0.549998	45.000230	−0.000226
31	0.46	1.139486	0.549997	45.000300	−0.000304
32	0.45	1.156888	0.549991	45.000910	−0.000912
33	0.44	1.174763	0.549991	45.000860	−0.000864
34	0.43	1.193099	0.549991	45.000880	−0.000876
35	0.42	1.211991	0.550004	44.999650	0.000352
36	0.41	1.231350	0.550005	44.999510	0.000489
37	0.40	1.251194	0.549993	45.000670	−0.000668
38	0.39	1.271748	0.550004	44.999650	0.000352
39	0.38	1.292858	0.550000	44.999990	0.000012
40	0.37	1.314683	0.550003	44.999690	0.000316
41	0.36	1.337199	0.550000	45.000030	−0.000024
42	0.35	1.360502	0.549997	45.000290	−0.000286
43	0.34	1.384711	0.550003	44.999690	0.000316
44	0.33	1.409774	0.549998	45.000160	−0.000161
45	0.32	1.435867	0.549997	45.000290	−0.000292
46	0.31	1.463101	0.550001	44.999920	0.000077
47	0.30	1.491532	0.550002	44.999850	0.000155
48	0.29	1.521264	0.549996	45.000390	−0.000387

(continued)

Table 8.2 (continued)

Parameter values					
PROB. = 0.55	RHO = −0.9		BIGH = 0.7553		SMHL = 0.27
Pair No.	SMH	SMK	PROB. level	Risk level (%)	Risk error (%)
49	0.28	1.552653	0.550009	44.999110	0.000888
50	0.27	1.585534	0.549998	45.000210	−0.000209
No. of pairs		Mean COMPUT. risk errors		STD. ERR. errors	
50		−0.0000119209		0.0000718600	
Z-ratio				Status of significance	
0.16589				N.S.—not significant	
Parameter values					
PROB. = 0.5	RHO = −0.9		BIGH = 0.6744		SMHL = 0.19
Pair No.	SMH	SMK	PROB. level	Risk level (%)	Risk error (%)
Table No. 8.2-381					
1	0.68	0.668797	0.500001	49.999880	0.000119
2	0.67	0.678780	0.499998	50.000160	−0.000161
3	0.66	0.688919	0.500002	49.999830	0.000167
4	0.65	0.699179	0.499999	50.000070	−0.000066
5	0.64	0.709575	0.499995	50.000500	−0.000495
6	0.63	0.720171	0.500008	49.999210	0.000799
7	0.62	0.730839	0.499997	50.000300	−0.000295
8	0.61	0.741693	0.499997	50.000340	−0.000340
9	0.60	0.752703	0.499997	50.000350	−0.000352
10	0.59	0.763989	0.500003	49.999660	0.000340
11	0.58	0.775278	0.499991	50.000930	−0.000930
12	0.57	0.786789	0.499994	50.000590	−0.000581
13	0.56	0.798498	0.500005	49.999460	0.000536
14	0.55	0.810335	0.500002	49.999770	0.000238
15	0.54	0.822329	0.499992	50.000780	−0.000781
16	0.53	0.834607	0.500010	49.999010	0.000989
17	0.52	0.847008	0.500008	49.999210	0.000799
18	0.51	0.859568	0.499995	50.000460	−0.000453
19	0.50	0.872423	0.500008	49.999190	0.000817
20	0.49	0.885421	0.500002	49.999790	0.000209
21	0.48	0.898655	0.500001	49.999860	0.000143
22	0.47	0.912125	0.500004	49.999600	0.000405
23	0.46	0.925838	0.500009	49.999090	0.000906
24	0.45	0.939704	0.499992	50.000850	−0.000846
25	0.44	0.953885	0.499991	50.000860	−0.000858
26	0.43	0.968354	0.499998	50.000170	−0.000167
27	0.42	0.983088	0.500004	49.999570	0.000435
28	0.41	0.998074	0.500003	49.999660	0.000346
29	0.40	1.013307	0.499992	50.000800	−0.000793

(continued)

8.2 Contents for Table 8.2 (Table No. 8.2-1 to 400)

Table 8.2 (continued)

Parameter values					
PROB. = 0.5	RHO = −0.9		BIGH = 0.6744		SMHL = 0.19
Pair No.	SMH	SMK	PROB. level	Risk level (%)	Risk error (%)
30	0.39	1.028888	0.499992	50.000850	−0.000846
31	0.38	1.044831	0.500002	49.999830	0.000167
32	0.37	1.061065	0.500003	49.999740	0.000262
33	0.36	1.077633	0.500000	49.999970	0.000030
34	0.35	1.094588	0.500004	49.999630	0.000376
35	0.34	1.111910	0.500004	49.999630	0.000370
36	0.33	1.129594	0.499996	50.000450	−0.000441
37	0.32	1.147761	0.500000	49.999970	0.000036
38	0.31	1.166363	0.500003	49.999740	0.000262
39	0.30	1.185385	0.499995	50.000540	−0.000536
40	0.29	1.204948	0.499995	50.000540	−0.000536
41	0.28	1.225119	0.500009	49.999060	0.000942
42	0.27	1.245729	0.500000	49.999980	0.000018
43	0.26	1.266963	0.499995	50.000510	−0.000513
44	0.25	1.288887	0.499998	50.000240	−0.000232
45	0.24	1.311467	0.499994	50.000620	−0.000620
46	0.23	1.334876	0.500004	49.999630	0.000370
47	0.22	1.359038	0.500004	49.999590	0.000417
48	0.21	1.384045	0.500000	50.000010	−0.000012
49	0.20	1.410010	0.499997	50.000340	−0.000334
50	0.19	1.437119	0.500008	49.999210	0.000799
No. of pairs		Mean COMPUT. risk errors		STD. ERR. errors	
50		0.0000021037		0.0000743314	
Z-ratio				Status of significance	
0.02830				N.S.—not significant	
Parameter values					
PROB. = 0.99	RHO = −0.95		BIGH = 2.5758		SMHL = 2.34
Pair No.	SMH	SMK	PROB. level	Risk level (%)	Risk error (%)
Table No. 8.2-382					
1	2.58	2.571607	0.989999	1.000142	−0.000143
2	2.57	2.581613	0.989998	1.000190	−0.000191
3	2.56	2.591698	0.989991	1.000881	−0.000882
4	2.55	2.603423	0.989999	1.000118	−0.000119
5	2.54	2.615229	0.989999	1.000106	−0.000107
6	2.53	2.627116	0.989991	1.000881	−0.000882
7	2.52	2.640648	0.989995	1.000512	−0.000513
8	2.51	2.655825	0.990008	0.999236	0.000763
9	2.50	2.669523	0.989992	1.000774	−0.000775
10	2.49	2.686431	0.990002	0.999832	0.000167
11	2.48	2.703425	0.989999	1.000059	−0.000060
12	2.47	2.722069	0.990001	0.999928	0.000072
13	2.46	2.742364	0.990003	0.999689	0.000310

(continued)

Table 8.2 (continued)

Parameter values					
PROB. = 0.99	RHO = −0.95		BIGH = 2.5758		SMHL = 2.34
Pair No.	SMH	SMK	PROB. level	Risk level (%)	Risk error (%)
14	2.45	2.764309	0.990005	0.999498	0.000501
15	2.44	2.787908	0.990004	0.999618	0.000381
16	2.43	2.814723	0.990010	0.999045	0.000954
17	2.42	2.841630	0.989996	1.000452	−0.000453
18	2.41	2.874881	0.990003	0.999713	0.000286
19	2.40	2.911352	0.990003	0.999689	0.000310
20	2.39	2.951044	0.989992	1.000762	−0.000763
21	2.38	3.000208	0.989995	1.000524	−0.000525
22	2.37	3.061970	0.990007	0.999343	0.000656
23	2.36	3.136332	0.990007	0.999284	0.000715
24	2.35	3.235795	0.990007	0.999332	0.000668
25	2.34	3.385361	0.990003	0.999725	0.000274
No. of pairs		Mean COMPUT. risk errors		STD. ERR. errors	
25		0.0000247589		0.0001053138	
Z-ratio				Status of significance	
0.23510				N.S.—not significant	
Parameter values					
PROB. = 0.98	RHO = −0.95		BIGH = 2.3265		SMHL = 2.07
Pair No.	SMH	SMK	PROB. level	Risk level (%)	Risk error (%)
Table No. 8.2-383					
1	2.33	2.323005	0.980007	1.999259	0.000739
2	2.32	2.333018	0.980006	1.999402	0.000596
3	2.31	2.343118	0.979994	2.000570	−0.000572
4	2.30	2.354087	0.979992	2.000821	−0.000823
5	2.29	2.365925	0.979997	2.000320	−0.000322
6	2.28	2.378636	0.980008	1.999223	0.000775
7	2.27	2.391437	0.980005	1.999521	0.000477
8	2.26	2.405113	0.980006	1.999450	0.000548
9	2.25	2.419663	0.980008	1.999199	0.000799
10	2.24	2.434309	0.979994	2.000558	−0.000560
11	2.23	2.450613	0.979996	2.000427	−0.000429
12	2.22	2.467796	0.979993	2.000666	−0.000668
13	2.21	2.486641	0.980000	2.000034	−0.000036
14	2.20	2.506367	0.979998	2.000248	−0.000250
15	2.19	2.527758	0.979998	2.000177	−0.000179
16	2.18	2.551595	0.980010	1.999021	0.000978
17	2.17	2.575536	0.979992	2.000761	−0.000763
18	2.16	2.603490	0.980000	2.000034	−0.000036
19	2.15	2.633114	0.979992	2.000785	−0.000787
20	2.14	2.666753	0.979993	2.000690	−0.000691
21	2.13	2.705190	0.980001	1.999903	0.000095
22	2.12	2.748427	0.980003	1.999736	0.000262

(continued)

8.2 Contents for Table 8.2 (Table No. 8.2-1 to 400)

Table 8.2 (continued)

Parameter values					
PROB. = 0.98	RHO = −0.95		BIGH = 2.3265		SMHL = 2.07
Pair No.	SMH	SMK	PROB. level	Risk level (%)	Risk error (%)
23	2.11	2.798026	0.980000	1.999998	0.000000
24	2.10	2.858679	0.980009	1.999140	0.000858
25	2.09	2.930386	0.979998	2.000165	−0.000167
26	2.08	3.027212	0.980003	1.999688	0.000310
27	2.07	3.164783	0.979998	2.000213	−0.000215
No. of pairs			Mean COMPUT. risk errors	STD. ERR. errors	
27			−0.0000021287	0.0001056323	
Z-ratio				Status of significance	
0.02015				N.S.—not significant	
Parameter values					
PROB. = 0.97	RHO = −0.95		BIGH = 2.1702		SMHL = 1.89
Pair No.	SMH	SMK	PROB. level	Risk level (%)	Risk error (%)
Table No. 8.2-384					
1	2.18	2.160444	0.970002	2.999783	0.000215
2	2.17	2.170400	0.970008	2.999175	0.000823
3	2.16	2.180448	0.970001	2.999854	0.000143
4	2.15	2.190981	0.969996	3.000414	−0.000417
5	2.14	2.202389	0.970004	2.999640	0.000358
6	2.13	2.214284	0.970010	2.999044	0.000954
7	2.12	2.226276	0.969999	3.000093	−0.000095
8	2.11	2.239149	0.969998	3.000236	−0.000238
9	2.10	2.252903	0.970003	2.999699	0.000298
10	2.09	2.267150	0.970001	2.999950	0.000048
11	2.08	2.282280	0.970001	2.999926	0.000072
12	2.07	2.297907	0.969990	3.000975	−0.000978
13	2.06	2.315201	0.970000	3.000021	−0.000024
14	2.05	2.333385	0.970004	2.999628	0.000370
15	2.04	2.352460	0.970000	2.999997	0.000000
16	2.03	2.373208	0.970005	2.999544	0.000453
17	2.02	2.394850	0.969995	3.000522	−0.000525
18	2.01	2.418949	0.970002	2.999830	0.000167
19	2.00	2.444728	0.970002	2.999830	0.000167
20	1.99	2.472967	0.970005	2.999520	0.000477
21	1.98	2.503670	0.970003	2.999735	0.000262
22	1.97	2.537620	0.970000	2.999962	0.000036
23	1.96	2.575599	0.969999	3.000116	−0.000119
24	1.95	2.618390	0.969995	3.000546	−0.000548
25	1.94	2.668340	0.969999	3.000116	−0.000119
26	1.93	2.727794	0.970009	2.999151	0.000846
27	1.92	2.796754	0.969990	3.000975	−0.000978

(continued)

Table 8.2 (continued)

Parameter values					
PROB. = 0.97	RHO = −0.95		BIGH = 2.1702		SMHL = 1.89
Pair No.	SMH	SMK	PROB. level	Risk level (%)	Risk error (%)
28	1.91	2.887721	0.969993	3.000689	−0.000691
29	1.90	3.016325	0.970004	2.999580	0.000417
30	1.89	3.092027	0.969627	3.037298	−0.037301
No. of pairs		Mean COMPUT. risk errors		STD. ERR. errors	
30		−0.00011590220		0.00012076760	
Z-ratio				Status of significance	
0.95971				N.S.—not significant	
Parameter values					
PROB. = 0.96	RHO = −0.95		BIGH = 2.0538		SMHL = 1.78
Pair No.	SMH	SMK	PROB. level	Risk level (%)	Risk error (%)
Table No. 8.2-385					
1	2.06	2.047619	0.960002	3.999794	0.000209
2	2.05	2.057607	0.960004	3.999615	0.000387
3	2.04	2.067694	0.959990	4.000962	−0.000960
4	2.03	2.078660	0.959997	4.000259	−0.000256
5	2.02	2.090119	0.960005	3.999519	0.000483
6	2.01	2.101680	0.959994	4.000628	−0.000626
7	2.00	2.114125	0.959998	4.000247	−0.000244
8	1.99	2.127068	0.959997	4.000259	−0.000256
9	1.98	2.140507	0.959991	4.000878	−0.000876
10	1.97	2.155227	0.960009	3.999114	0.000888
11	1.96	2.170058	0.960001	3.999913	0.000089
12	1.95	2.185782	0.959996	4.000378	−0.000376
13	1.94	2.202401	0.959991	4.000855	−0.000852
14	1.93	2.220307	0.959998	4.000235	−0.000232
15	1.92	2.239112	0.959997	4.000318	−0.000316
16	1.91	2.259208	0.959998	4.000187	−0.000185
17	1.90	2.280987	0.960009	3.999114	0.000888
18	1.89	2.303671	0.960000	3.999961	0.000042
19	1.88	2.328042	0.959991	4.000902	−0.000900
20	1.87	2.354884	0.959994	4.000616	−0.000614
21	1.86	2.384199	0.959999	4.000092	−0.000089
22	1.85	2.415989	0.959997	4.000306	−0.000304
23	1.84	2.451818	0.960009	3.999102	0.000900
24	1.83	2.490907	0.960004	3.999591	0.000411
25	1.82	2.534822	0.959995	4.000485	−0.000483
26	1.81	2.586689	0.960007	3.999305	0.000697
27	1.80	2.646510	0.960003	3.999674	0.000328

(continued)

8.2 Contents for Table 8.2 (Table No. 8.2-1 to 400)

Table 8.2 (continued)

Parameter values					
PROB. = 0.96	RHO = −0.95		BIGH = 2.0538		SMHL = 1.78
Pair No.	SMH	SMK	PROB. level	Risk level (%)	Risk error (%)
28	1.79	2.718977	0.959999	4.000139	−0.000137
29	1.78	2.811903	0.959999	4.000056	−0.000054
No. of pairs		Mean COMPUT. risk errors		STD. ERR. errors	
29		−0.0000812610		0.0000974256	
Z-ratio				Status of significance	
0.83408				N.S.—not significant	
Parameter values					
PROB. = 0.95	RHO = −0.95		BIGH = 1.9599		SMHL = 1.68
Pair No.	SMH	SMK	PROB. level	Risk level (%)	Risk error (%)
Table No. 8.2-386					
1	1.96	1.959800	0.949992	5.000758	−0.000757
2	1.95	1.970241	0.950007	4.999352	0.000650
3	1.94	1.980785	0.950002	4.999757	0.000244
4	1.93	1.991826	0.950002	4.999852	0.000149
5	1.92	2.003363	0.950002	4.999805	0.000197
6	1.91	2.015400	0.950002	4.999805	0.000197
7	1.90	2.027938	0.950000	4.999995	0.000006
8	1.89	2.040979	0.949995	5.000544	−0.000542
9	1.88	2.054915	0.950002	4.999757	0.000244
10	1.87	2.069356	0.950002	4.999829	0.000173
11	1.86	2.084697	0.950009	4.999113	0.000888
12	1.85	2.100547	0.950003	4.999709	0.000292
13	1.84	2.117300	0.949999	5.000139	−0.000137
14	1.83	2.134958	0.949993	5.000723	−0.000721
15	1.82	2.153913	0.949997	5.000305	−0.000304
16	1.81	2.173777	0.949991	5.000890	−0.000888
17	1.80	2.195332	0.950000	5.000031	−0.000030
18	1.79	2.218192	0.950002	4.999793	0.000209
19	1.78	2.242357	0.949993	5.000735	−0.000733
20	1.77	2.268611	0.949990	5.000961	−0.000960
21	1.76	2.297348	0.949997	5.000329	−0.000328
22	1.75	2.328569	0.950000	5.000019	−0.000018
23	1.74	2.363059	0.950008	4.999185	0.000817
24	1.73	2.400820	0.950006	4.999423	0.000578
25	1.72	2.442635	0.949994	5.000639	−0.000638
26	1.71	2.491633	0.950009	4.999078	0.000924
27	1.70	2.547033	0.950002	4.999757	0.000244

(continued)

Table 8.2 (continued)

Parameter values				
PROB. = 0.95	RHO = −0.95	BIGH = 1.9599		SMHL = 1.68
Pair No.	SMH	SMK	PROB. level	Risk level (%)	Risk error (%)
28	1.69	2.613528	0.950005	4.999471	0.000530
29	1.68	2.694245	0.949994	5.000603	−0.000602
No. of pairs		Mean COMPUT. risk errors		STD. ERR. errors	
29		−0.0000105302		0.0000982874	
Z-ratio				Status of significance	
0.10714				N.S.—not significant	
Parameter values				
PROB. = 0.94	RHO = −0.95	BIGH = 1.8808		SMHL = 1.59
Pair No.	SMH	SMK	PROB. level	Risk level (%)	Risk error (%)
Table No. 8.2-387					
1	1.89	1.871645	0.939993	6.000686	−0.000685
2	1.88	1.881600	0.940001	5.999911	0.000089
3	1.87	1.891662	0.939990	6.000996	−0.000995
4	1.86	1.902418	0.939999	6.000114	−0.000113
5	1.85	1.913480	0.940000	6.000018	−0.000018
6	1.84	1.924848	0.939992	6.000829	−0.000829
7	1.83	1.936916	0.939997	6.000281	−0.000280
8	1.82	1.949490	0.940002	5.999804	0.000197
9	1.81	1.962572	0.940004	5.999589	0.000411
10	1.80	1.976164	0.940002	5.999851	0.000149
11	1.79	1.990268	0.939992	6.000781	−0.000781
12	1.78	2.005277	0.939995	6.000471	−0.000471
13	1.77	2.021192	0.940007	5.999351	0.000650
14	1.76	2.037625	0.940002	5.999768	0.000232
15	1.75	2.054970	0.940000	6.000018	−0.000018
16	1.74	2.073227	0.939995	6.000507	−0.000507
17	1.73	2.092792	0.940001	5.999911	0.000089
18	1.72	2.113273	0.939995	6.000495	−0.000495
19	1.71	2.135457	0.940005	5.999482	0.000519
20	1.70	2.158954	0.940008	5.999232	0.000769
21	1.69	2.183766	0.939996	6.000376	−0.000376
22	1.68	2.211069	0.940006	5.999422	0.000578
23	1.67	2.240084	0.939998	6.000245	−0.000244
24	1.66	2.271984	0.939999	6.000090	−0.000089
25	1.65	2.307165	0.940006	5.999422	0.000578
26	1.64	2.346018	0.940010	5.999017	0.000983
27	1.63	2.388939	0.940001	5.999935	0.000066
28	1.62	2.438273	0.940005	5.999494	0.000507
29	1.61	2.494804	0.940000	6.000018	−0.000018

(continued)

8.2 Contents for Table 8.2 (Table No. 8.2-1 to 400)

Table 8.2 (continued)

Parameter values					
PROB. = 0.94	RHO = −0.95	BIGH = 1.8808		SMHL = 1.59	
Pair No.	SMH	SMK	PROB. level	Risk level (%)	Risk error (%)
30	1.60	2.562442	0.940004	5.999613	0.000387
31	1.59	2.601229	0.939438	6.056202	−0.056201
No. of pairs		Mean COMPUT. risk errors		STD. ERR. errors	
31		−0.00017473470		0.00017587000	
Z-ratio				Status of significance	
0.99354				N.S.—not significant	
Parameter values					
PROB. = 0.93	RHO = −0.95	BIGH = 1.8119		SMHL = 1.53	
Pair No.	SMH	SMK	PROB. level	Risk level (%)	Risk error (%)
Table No. 8.2-388					
1	1.82	1.803836	0.929992	7.000804	−0.000805
2	1.81	1.813802	0.929998	7.000196	−0.000197
3	1.80	1.824074	0.929999	7.000089	−0.000089
4	1.79	1.834849	0.930009	6.999111	0.000888
5	1.78	1.845739	0.929997	7.000315	−0.000316
6	1.77	1.857136	0.929991	7.000947	−0.000948
7	1.76	1.869237	0.930001	6.999874	0.000125
8	1.75	1.881653	0.929999	7.000077	−0.000077
9	1.74	1.894584	0.929997	7.000327	−0.000328
10	1.73	1.908224	0.930004	6.999636	0.000364
11	1.72	1.922382	0.930005	6.999505	0.000495
12	1.71	1.937060	0.929998	7.000208	−0.000209
13	1.70	1.952650	0.930004	6.999600	0.000399
14	1.69	1.968764	0.929996	7.000411	−0.000411
15	1.68	1.985796	0.929993	7.000661	−0.000662
16	1.67	2.003748	0.929992	7.000828	−0.000829
17	1.66	2.023012	0.930007	6.999314	0.000685
18	1.65	2.042811	0.929993	7.000709	−0.000709
19	1.64	2.064318	0.930004	6.999636	0.000364
20	1.63	2.086755	0.929994	7.000578	−0.000578
21	1.62	2.110906	0.929994	7.000625	−0.000626
22	1.61	2.136775	0.929993	7.000709	−0.000709
23	1.60	2.164754	0.929997	7.000256	−0.000256
24	1.59	2.195237	0.930009	6.999088	0.000912
25	1.58	2.227836	0.930001	6.999922	0.000077
26	1.57	2.263727	0.929997	7.000304	−0.000304
27	1.56	2.303694	0.930000	6.999994	0.000006
28	1.55	2.348521	0.930005	6.999481	0.000519

(continued)

Table 8.2 (continued)

Parameter values					
PROB. = 0.93	RHO = −0.95		BIGH = 1.8119		SMHL = 1.53
Pair No.	SMH	SMK	PROB. level	Risk level (%)	Risk error (%)
29	1.54	2.398993	0.930000	7.000029	−0.000030
30	1.53	2.457458	0.929995	7.000458	−0.000459
No. of pairs		Mean COMPUT. risk errors		STD. ERR. errors	
30		−0.0001195938		0.0000918758	
Z-ratio				Status of significance	
1.30169				N.S.—not significant	
Parameter values					
PROB. = 0.92	RHO = −0.95		BIGH = 1.7507		SMHL = 1.46
Pair No.	SMH	SMK	PROB. level	Risk level (%)	Risk error (%)
Table No. 8.2-389					
1	1.76	1.741449	0.919994	8.000636	−0.000638
2	1.75	1.751400	0.920002	7.999778	0.000221
3	1.74	1.761466	0.919991	8.000946	−0.000948
4	1.73	1.772038	0.919991	8.000935	−0.000936
5	1.72	1.783120	0.920000	7.999957	0.000042
6	1.71	1.794517	0.920002	7.999838	0.000161
7	1.70	1.806428	0.920009	7.999134	0.000864
8	1.69	1.818658	0.920004	7.999575	0.000423
9	1.68	1.831406	0.920002	7.999850	0.000149
10	1.67	1.844675	0.919998	8.000207	−0.000209
11	1.66	1.858465	0.919991	8.000875	−0.000876
12	1.65	1.873171	0.920006	7.999397	0.000602
13	1.64	1.888208	0.919998	8.000160	−0.000161
14	1.63	1.904166	0.920005	7.999516	0.000483
15	1.62	1.920851	0.920009	7.999134	0.000864
16	1.61	1.938071	0.919994	8.000613	−0.000614
17	1.60	1.956609	0.920004	7.999587	0.000411
18	1.59	1.976079	0.920010	7.999039	0.000960
19	1.58	1.996483	0.920006	7.999420	0.000578
20	1.57	2.018213	0.920008	7.999218	0.000781
21	1.56	2.041274	0.920008	7.999182	0.000817
22	1.55	2.065668	0.919999	8.000076	−0.000077
23	1.54	2.091790	0.919991	8.000887	−0.000888
24	1.53	2.120423	0.920007	7.999349	0.000650
25	1.52	2.150789	0.919998	8.000195	−0.000197
26	1.51	2.184065	0.920000	8.000040	−0.000042
27	1.50	2.220644	0.920005	7.999492	0.000507
28	1.49	2.260920	0.920006	7.999444	0.000554

(continued)

8.2 Contents for Table 8.2 (Table No. 8.2-1 to 400)

Table 8.2 (continued)

Parameter values					
PROB. = 0.92	RHO = −0.95		BIGH = 1.7507		SMHL = 1.46
Pair No.	SMH	SMK	PROB. level	Risk level (%)	Risk error (%)
29	1.48	2.306068	0.920010	7.999003	0.000995
30	1.47	2.356875	0.920004	7.999563	0.000435
31	1.46	2.388550	0.919398	8.060253	−0.060254
No. of pairs		Mean COMPUT. risk errors		STD. ERR. errors	
31		−0.00017294660		0.00018907500	
Z-ratio				Status of significance	
0.91470				N.S.—not significant	
Parameter values					
PROB. = 0.91	RHO = −0.95		BIGH = 1.6954		SMHL = 1.41
Pair No.	SMH	SMK	PROB. level	Risk level (%)	Risk error (%)
Table No. 8.2-390					
1	1.70	1.690812	0.909998	9.000182	−0.000185
2	1.69	1.700817	0.909997	9.000278	−0.000280
3	1.68	1.711136	0.909993	9.000659	−0.000662
4	1.67	1.721968	0.910003	8.999729	0.000268
5	1.66	1.733117	0.910005	8.999467	0.000530
6	1.65	1.744588	0.910000	8.999979	0.000018
7	1.64	1.756578	0.910003	8.999753	0.000244
8	1.63	1.768893	0.909993	9.000671	−0.000674
9	1.62	1.781926	0.910003	8.999681	0.000316
10	1.61	1.795291	0.909997	9.000265	−0.000268
11	1.60	1.809379	0.910005	8.999538	0.000459
12	1.59	1.823802	0.909992	9.000850	−0.000852
13	1.58	1.839150	0.910000	8.999991	0.000006
14	1.57	1.855035	0.909997	9.000325	−0.000328
15	1.56	1.871849	0.900006	8.999371	0.000626
16	1.55	1.889205	0.909997	9.000301	−0.000304
17	1.54	1.907497	0.909992	9.000838	−0.000840
18	1.53	1.927118	0.910009	8.999073	0.000924
19	1.52	1.947290	0.909995	9.000540	−0.000542
20	1.51	1.968798	0.909990	9.000992	−0.000995
21	1.50	1.992035	0.910009	8.999086	0.000912
22	1.49	2.016224	0.910000	9.000039	−0.000042
23	1.48	2.042149	0.909995	9.000516	−0.000519
24	1.47	2.070205	0.910003	8.999729	0.000268
25	1.46	2.100004	0.909991	9.000921	−0.000924
26	1.45	2.132722	0.909997	9.000301	−0.000304
27	1.44	2.168363	0.910001	8.999920	0.000077

(continued)

Table 8.2 (continued)

Parameter values					
PROB. = 0.91	RHO = −0.95		BIGH = 1.6954		SMHL = 1.41
Pair No.	SMH	SMK	PROB. level	Risk level (%)	Risk error (%)
28	1.43	2.207713	0.910009	8.999062	0.000936
29	1.42	2.250774	0.909996	9.000373	−0.000376
30	1.41	2.299505	0.909992	9.000789	−0.000793
No. of pairs		Mean COMPUT. risk errors		STD. ERR. errors	
30		−0.0001065193		0.0001004418	
Z-ratio				Status of significance	
1.06051				N.S.—not significant	
Parameter values					
PROB. = 0.9025	RHO = −0.95		BIGH = 1.6571		SMHL = 1.37
Pair No.	SMH	SMK	PROB. level	Risk level (%)	Risk error (%)
Table No. 8.2-391					
1	1.66	1.654205	0.902500	9.750021	−0.000018
2	1.65	1.664231	0.902495	9.750462	−0.000459
3	1.64	1.674769	0.902508	9.749233	0.000769
4	1.63	1.685432	0.902497	9.750342	−0.000340
5	1.62	1.696612	0.902499	9.750116	−0.000113
6	1.61	1.708117	0.902494	9.750629	−0.000626
7	1.60	1.720144	0.902498	9.750234	−0.000232
8	1.59	1.732500	0.902490	9.750962	−0.000960
9	1.58	1.745579	0.902505	9.749544	0.000459
10	1.57	1.758993	0.902503	9.749699	0.000304
11	1.56	1.772939	0.902501	9.749937	0.000066
12	1.55	1.787420	0.902494	9.750557	−0.000554
13	1.54	1.802635	0.902497	9.750294	−0.000292
14	1.53	1.818586	0.902504	9.749556	0.000447
15	1.52	1.835082	0.902498	9.750223	−0.000221
16	1.51	1.852514	0.902502	9.749770	0.000232
17	1.50	1.870692	0.902499	9.750116	−0.000113
18	1.49	1.889815	0.902496	9.750366	−0.000364
19	1.48	1.910080	0.902502	9.749818	0.000185
20	1.47	1.931295	0.902496	9.750426	−0.000423
21	1.46	1.953855	0.902496	9.750414	−0.000411
22	1.45	1.977764	0.902493	9.750700	−0.000697
23	1.44	2.003415	0.902500	9.750009	−0.000006
24	1.43	2.030813	0.902504	9.749567	0.000435
25	1.42	2.059961	0.902495	9.750498	−0.000495
26	1.41	2.091644	0.902495	9.750509	−0.000507
27	1.40	2.126258	0.902503	9.749746	0.000256

(continued)

8.2 Contents for Table 8.2 (Table No. 8.2-1 to 400)

Table 8.2 (continued)

Parameter values					
PROB. = 0.9025	RHO = −0.95		BIGH = 1.6571		SMHL = 1.37
Pair No.	SMH	SMK	PROB. level	Risk level (%)	Risk error (%)
28	1.39	2.163806	0.902496	9.750402	−0.000399
29	1.38	2.205465	0.902496	9.750402	−0.000399
30	1.37	2.241310	0.902154	9.784651	−0.034648
No. of pairs		Mean COMPUT. risk errors		STD. ERR. errors	
30		−0.00012620800		0.00011151480	
Z-ratio				Status of significance	
1.13176				N.S.—not significant	
Parameter values					
PROB. = 0.9	RHO = −0.95		BIGH = 1.645		SMHL = 1.36
Pair No.	SMH	SMK	PROB. level	Risk level (%)	Risk error (%)
Table No. 8.2-392					
1	1.65	1.639820	0.900007	9.999275	0.000727
2	1.64	1.649820	0.900008	9.999251	0.000751
3	1.63	1.660138	0.900006	9.999418	0.000584
4	1.62	1.670776	0.900001	9.999906	0.000095
5	1.61	1.681737	0.899991	10.000900	−0.000894
6	1.60	1.693219	0.899994	10.000630	−0.000632
7	1.59	1.705223	0.900006	9.999394	0.000608
8	1.58	1.717362	0.899990	10.000990	−0.000989
9	1.57	1.730223	0.899997	10.000290	−0.000286
10	1.56	1.743616	0.900007	9.999299	0.000703
11	1.55	1.757346	0.900000	10.000010	−0.000012
12	1.54	1.771612	0.899990	10.000970	−0.000966
13	1.53	1.786612	0.899992	10.000840	−0.000834
14	1.52	1.802350	0.899999	10.000070	−0.000072
15	1.51	1.818632	0.899995	10.000550	−0.000548
16	1.50	1.835852	0.900003	9.999704	0.000298
17	1.49	1.853624	0.899991	10.000870	−0.000870
18	1.48	1.872536	0.899997	10.000280	−0.000274
19	1.47	1.892396	0.900000	10.000000	0.000000
20	1.46	1.913207	0.899994	10.000590	−0.000584
21	1.45	1.935365	0.899998	10.000210	−0.000203
22	1.44	1.958871	0.900002	9.999776	0.000226
23	1.43	1.983731	0.899998	10.000160	−0.000155
24	1.42	2.010338	0.899998	10.000160	−0.000155
25	1.41	2.038698	0.899990	10.000990	−0.000989
26	1.40	2.069593	0.899998	10.000180	−0.000179
27	1.39	2.103030	0.900004	9.999609	0.000393

(continued)

Table 8.2 (continued)

Parameter values					
PROB. = 0.9	RHO = −0.95		BIGH = 1.645		SMHL = 1.36
Pair No.	SMH	SMK	PROB. level	Risk level (%)	Risk error (%)
28	1.38	2.139403	0.900005	9.999478	0.000525
29	1.37	2.179106	0.899995	10.000530	−0.000525
30	1.36	2.223708	0.900001	9.999895	0.000107
No. of pairs		Mean COMPUT. risk errors		STD. ERR. errors	
30		−0.0001338221		0.0000987920	
Z-ratio				Status of significance	
1.35458				N.S.—not significant	
Parameter values					
PROB. = 0.85	RHO = −0.95		BIGH = 1.4395		SMHL = 0.95
Pair No.	SMH	SMK	PROB. level	Risk level (%)	Risk error (%)
Table No. 8.2-393					
1	1.44	1.439000	0.849991	15.000900	−0.000906
2	1.43	1.449258	0.850009	14.999130	0.000870
3	1.42	1.459658	0.850004	14.999590	0.000405
4	1.41	1.470398	0.850003	14.999680	0.000322
5	1.40	1.481482	0.850004	14.999580	0.000417
6	1.39	1.492911	0.850005	14.999460	0.000536
7	1.38	1.504690	0.850005	14.999510	0.000489
8	1.37	1.516823	0.850001	14.999930	0.000072
9	1.36	1.529311	0.849991	15.000870	−0.000870
10	1.35	1.542355	0.849998	15.000160	−0.000167
11	1.34	1.555763	0.849995	15.000500	−0.000501
12	1.33	1.569734	0.850002	14.999770	0.000226
13	1.32	1.584076	0.849994	15.000570	−0.000572
14	1.31	1.598989	0.849991	15.000940	−0.000942
15	1.30	1.614672	0.850009	14.999130	0.000870
16	1.29	1.630740	0.850002	14.999800	0.000203
17	1.28	1.647586	0.850009	14.999130	0.000870
18	1.27	1.665021	0.850003	14.999660	0.000334
19	1.26	1.683243	0.850001	14.999870	0.000131
20	1.25	1.702259	0.849997	15.000310	−0.000310
21	1.24	1.722269	0.850002	14.999820	0.000179
22	1.23	1.743277	0.850009	14.999130	0.000870
23	1.22	1.765289	0.850010	14.999010	0.000989
24	1.21	1.788310	0.849998	15.000240	−0.000238
25	1.20	1.812737	0.849994	15.000580	−0.000584
26	1.19	1.838772	0.850003	14.999750	0.000250
27	1.18	1.866224	0.849995	15.000510	−0.000513

(continued)

8.2 Contents for Table 8.2 (Table No. 8.2-1 to 400)

Table 8.2 (continued)

Parameter values					
PROB. = 0.85	RHO = −0.95	BIGH = 1.4395		SMHL = 0.95	
Pair No.	SMH	SMK	PROB. level	Risk level (%)	Risk error (%)
28	1.17	1.895686	0.849999	15.000120	−0.000119
29	1.16	1.926579	0.850001	14.999890	0.000107
30	1.15	1.960277	0.850001	14.999860	0.000137
31	1.14	1.996590	0.849997	15.000310	−0.000310
32	1.13	2.036113	0.849998	15.000210	−0.000209
33	1.12	2.079439	0.850008	14.999210	0.000793
34	1.11	2.126576	0.850003	14.999660	0.000334
35	1.10	2.178313	0.850002	14.999810	0.000191
36	1.09	2.234657	0.850002	14.999830	0.000173
37	1.08	2.294447	0.849991	15.000890	−0.000894
38	1.07	2.356129	0.849995	15.000520	−0.000525
39	1.06	2.416195	0.849998	15.000250	−0.000250
40	1.05	2.471923	0.849991	15.000860	−0.000858
41	1.04	2.522929	0.850009	14.999150	0.000852
42	1.03	2.568445	0.849997	15.000250	−0.000256
43	1.02	2.610044	0.850010	14.999020	0.000978
44	1.01	2.647736	0.849992	15.000810	−0.000811
45	1.00	2.682901	0.850001	14.999890	0.000107
46	0.99	2.715551	0.849991	15.000860	−0.000858
47	0.98	2.746479	0.850003	14.999750	0.000250
48	0.97	2.775699	0.849999	15.000070	−0.000072
49	0.96	2.803616	0.849999	15.000090	−0.000095
50	0.95	2.830438	0.850005	14.999540	0.000459
No. of pairs		Mean COMPUT. risk errors		STD. ERR. errors	
50		0.0000305036		0.0000776683	
Z-ratio				Status of significance	
0.39274				N.S.—not significant	
Parameter values					
PROB. = 0.8	RHO = −0.95	BIGH = 1.2815		SMHL = 0.8	
Pair No.	SMH	SMK	PROB. level	Risk level (%)	Risk error (%)
Table No. 8.2-394					
1	1.29	1.273154	0.799993	20.000720	−0.000721
2	1.28	1.283099	0.799999	20.000120	−0.000125
3	1.27	1.293299	0.800004	19.999610	0.000387
4	1.26	1.303757	0.800007	19.999280	0.000721
5	1.25	1.314380	0.799991	20.000910	−0.000912
6	1.24	1.325463	0.800004	19.999650	0.000352
7	1.23	1.336719	0.799994	20.000580	−0.000578
8	1.22	1.348444	0.800010	19.999030	0.000972
9	1.21	1.360350	0.800001	19.999910	0.000089
10	1.20	1.372637	0.799998	20.000230	−0.000232
11	1.19	1.385309	0.799998	20.000220	−0.000221

(continued)

Table 8.2 (continued)

Parameter values					
PROB. = 0.8	RHO = −0.95		BIGH = 1.2815		SMHL = 0.8
Pair No.	SMH	SMK	PROB. level	Risk level (%)	Risk error (%)
12	1.18	1.398371	0.799999	20.000090	−0.000089
13	1.17	1.411829	0.799999	20.000080	−0.000077
14	1.16	1.425687	0.799996	20.000390	−0.000387
15	1.15	1.440048	0.800001	19.999870	0.000125
16	1.14	1.454821	0.799997	20.000260	−0.000256
17	1.13	1.470108	0.799996	20.000440	−0.000435
18	1.12	1.486014	0.800005	19.999460	0.000542
19	1.11	1.502349	0.799997	20.000300	−0.000304
20	1.10	1.519314	0.799992	20.000770	−0.000769
21	1.09	1.537015	0.799999	20.000120	−0.000125
22	1.08	1.555360	0.799999	20.000120	−0.000125
23	1.07	1.574454	0.799999	20.000090	−0.000089
24	1.06	1.594402	0.800005	19.999530	0.000471
25	1.05	1.615212	0.800009	19.999150	0.000852
26	1.04	1.636891	0.800004	19.999650	0.000352
27	1.03	1.659644	0.800002	19.999800	0.000197
28	1.02	1.683479	0.799995	20.000520	−0.000519
29	1.01	1.708795	0.800008	19.999200	0.000805
30	1.00	1.735211	0.799993	20.000660	−0.000662
31	0.99	1.763518	0.800007	19.999340	0.000662
32	0.98	1.793335	0.799998	20.000220	−0.000221
33	0.97	1.825259	0.799996	20.000360	−0.000364
34	0.96	1.859497	0.799995	20.000500	−0.000501
35	0.95	1.896644	0.800008	19.999180	0.000817
36	0.94	1.936519	0.799992	20.000850	−0.000852
37	0.93	1.980695	0.800005	19.999460	0.000536
38	0.92	2.029186	0.800000	19.999980	0.000018
39	0.91	2.083567	0.800000	19.999990	0.000012
40	0.90	2.145416	0.800002	19.999830	0.000173
41	0.89	2.217090	0.800005	19.999540	0.000465
42	0.88	2.301340	0.799994	20.000560	−0.000560
43	0.87	2.402477	0.800009	19.999120	0.000882
44	0.86	2.514270	0.799993	20.000730	−0.000733
45	0.85	2.619548	0.800010	19.999000	0.000995
46	0.84	2.702705	0.800006	19.999370	0.000626
47	0.83	2.768057	0.800006	19.999360	0.000638
48	0.82	2.821483	0.799997	20.000290	−0.000292

(continued)

Table 8.2 (continued)

Parameter values					
PROB. = 0.8	RHO = −0.95		BIGH = 1.2815		SMHL = 0.8
Pair No.	SMH	SMK	PROB. level	Risk level (%)	Risk error (%)
49	0.81	2.867301	0.799996	20.000440	−0.000435
50	0.80	2.907879	0.800004	19.999650	0.000346
No. of pairs		Mean COMPUT. risk errors		STD. ERR. errors	
50		0.0000283999		0.0000741898	
Z-ratio				Status of significance	
0.38280				N.S.—not significant	

Parameter values					
PROB. = 0.75	RHO = −0.95		BIGH = 1.1504		SMHL = 0.67
Pair No.	SMH	SMK	PROB. level	Risk level (%)	Risk error (%)
Table No. 8.2-395					
1	1.16	1.140782	0.749995	25.000480	−0.000477
2	1.15	1.150703	0.750001	24.999930	0.000072
3	1.14	1.160797	0.749995	25.000530	−0.000525
4	1.13	1.171169	0.749996	25.000370	−0.000370
5	1.12	1.181821	0.750005	24.999520	0.000477
6	1.11	1.192661	0.750000	25.000050	−0.000048
7	1.10	1.203793	0.749999	25.000090	−0.000083
8	1.09	1.215221	0.750003	24.999750	0.000250
9	1.08	1.226952	0.750009	24.999140	0.000858
10	1.07	1.238892	0.749998	25.000240	−0.000238
11	1.06	1.251244	0.750005	24.999500	0.000501
12	1.05	1.263818	0.749993	25.000670	−0.000668
13	1.04	1.276816	0.749996	25.000360	−0.000358
14	1.03	1.290148	0.749995	25.000480	−0.000477
15	1.02	1.303916	0.750005	24.999540	0.000465
16	1.01	1.318032	0.750006	24.999410	0.000596
17	1.00	1.332502	0.749997	25.000290	−0.000286
18	0.99	1.347530	0.750008	24.999180	0.000823
19	0.98	1.362929	0.750005	24.999550	0.000453
20	0.97	1.378804	0.749999	25.000100	−0.000095
21	0.96	1.395262	0.750004	24.999610	0.000393
22	0.95	1.412214	0.750001	24.999950	0.000048
23	0.94	1.429769	0.749999	25.000060	−0.000060
24	0.93	1.447935	0.749997	25.000310	−0.000310
25	0.92	1.466820	0.750001	24.999880	0.000119
26	0.91	1.486339	0.749994	25.000600	−0.000596
27	0.90	1.506795	0.750008	24.999170	0.000834
28	0.89	1.527907	0.749999	25.000070	−0.000072
29	0.88	1.549980	0.749997	25.000260	−0.000262

(continued)

Table 8.2 (continued)

Parameter values					
PROB. = 0.75	RHO = −0.95		BIGH = 1.1504		SMHL = 0.67
Pair No.	SMH	SMK	PROB. level	Risk level (%)	Risk error (%)
30	0.87	1.573125	0.750005	24.999510	0.000495
31	0.86	1.597259	0.750001	24.999860	0.000143
32	0.85	1.622590	0.749999	25.000110	−0.000107
33	0.84	1.649328	0.750006	24.999430	0.000572
34	0.83	1.677490	0.750007	24.999260	0.000739
35	0.82	1.707286	0.750008	24.999250	0.000751
36	0.81	1.738930	0.750006	24.999360	0.000644
37	0.80	1.772634	0.750000	25.000000	0.000000
38	0.79	1.808809	0.749996	25.000420	−0.000417
39	0.78	1.847864	0.749994	25.000620	−0.000620
40	0.77	1.890407	0.749998	25.000160	−0.000155
41	0.76	1.937045	0.750003	24.999700	0.000304
42	0.75	1.988583	0.749999	25.000070	−0.000072
43	0.74	2.046608	0.750002	24.999760	0.000238
44	0.73	2.112708	0.749993	25.000660	−0.000662
45	0.72	2.190426	0.749994	25.000590	−0.000590
46	0.71	2.285065	0.750004	24.999650	0.000352
47	0.70	2.405443	0.750001	24.999950	0.000054
48	0.69	2.566436	0.749996	25.000410	−0.000411
49	0.68	2.745423	0.750003	24.999660	0.000346
50	0.67	2.868221	0.750000	25.000020	−0.000024
No. of pairs		Mean COMPUT. risk errors		STD. ERR. errors	
50		0.0000499043		0.0000615784	
Z-ratio				Status of significance	
0.81042				N.S.—not significant	
Parameter values					
PROB. = 0.7	RHO = −0.95		BIGH = 1.0364		SMHL = 0.55
Pair No.	SMH	SMK	PROB. level	Risk level (%)	Risk error (%)
Table No. 8.2-396					
1	1.04	1.032910	0.700007	29.999310	0.000697
2	1.03	1.042937	0.700006	29.999360	0.000638
3	1.02	1.053161	0.700002	29.999780	0.000221
4	1.01	1.063588	0.699995	30.000540	−0.000542
5	1.00	1.074320	0.700005	29.999500	0.000507
6	0.99	1.085219	0.700001	29.999950	0.000054
7	0.98	1.096339	0.699992	30.000830	−0.000829
8	0.97	1.107833	0.700010	29.999020	0.000983
9	0.96	1.119466	0.700002	29.999840	0.000161
10	0.95	1.131439	0.700009	29.999140	0.000864

(continued)

8.2 Contents for Table 8.2 (Table No. 8.2-1 to 400)

Table 8.2 (continued)

Parameter values					
PROB. = 0.7	RHO = −0.95		BIGH = 1.0364		SMHL = 0.55
Pair No.	SMH	SMK	PROB. level	Risk level (%)	Risk error (%)
11	0.94	1.143663	0.700009	29.999070	0.000936
12	0.93	1.156145	0.700003	29.999660	0.000340
13	0.92	1.168894	0.699991	30.000950	−0.000948
14	0.91	1.182115	0.700009	29.999140	0.000864
15	0.90	1.195523	0.699999	30.000150	−0.000149
16	0.89	1.209323	0.699998	30.000230	−0.000232
17	0.88	1.223526	0.700005	29.999520	0.000483
18	0.87	1.238044	0.700000	30.000000	0.000006
19	0.86	1.252986	0.700000	29.999980	0.000018
20	0.85	1.268364	0.700003	29.999660	0.000340
21	0.84	1.284189	0.700008	29.999210	0.000793
22	0.83	1.300376	0.699995	30.000540	−0.000542
23	0.82	1.317135	0.699995	30.000470	−0.000471
24	0.81	1.334478	0.700006	29.999370	0.000632
25	0.80	1.352226	0.699993	30.000690	−0.000685
26	0.79	1.370687	0.700000	30.000020	−0.000018
27	0.78	1.389778	0.700007	29.999350	0.000650
28	0.77	1.409420	0.699995	30.000540	−0.000542
29	0.76	1.429923	0.700003	29.999680	0.000328
30	0.75	1.451111	0.699998	30.000190	−0.000185
31	0.74	1.473199	0.700002	29.999850	0.000149
32	0.73	1.496208	0.700005	29.999470	0.000530
33	0.72	1.520160	0.700002	29.999780	0.000221
34	0.71	1.545273	0.700008	29.999200	0.000805
35	0.70	1.571475	0.700000	29.999970	0.000030
36	0.69	1.599085	0.700002	29.999810	0.000197
37	0.68	1.628228	0.700010	29.999050	0.000954
38	0.67	1.658835	0.699997	30.000340	−0.000340
39	0.66	1.691524	0.700005	29.999520	0.000483
40	0.65	1.726327	0.700010	29.999030	0.000972
41	0.64	1.763475	0.700004	29.999640	0.000364
42	0.63	1.803592	0.700005	29.999480	0.000519
43	0.62	1.846911	0.699991	30.000880	−0.000876
44	0.61	1.894648	0.700000	30.000030	−0.000030
45	0.60	1.947238	0.699994	30.000600	−0.000602
46	0.59	2.006487	0.700003	29.999730	0.000268
47	0.58	2.073813	0.699995	30.000550	−0.000548
48	0.57	2.153178	0.700009	29.999080	0.000918

(continued)

Table 8.2 (continued)

Parameter values					
PROB. = 0.7	RHO = −0.95		BIGH = 1.0364		SMHL = 0.55
Pair No.	SMH	SMK	PROB. level	Risk level (%)	Risk error (%)
49	0.56	2.248547	0.699990	30.000960	−0.000960
50	0.55	2.371702	0.699991	30.000940	−0.000936
No. of pairs		Mean COMPUT. risk errors		STD. ERR. errors	
50		0.0001271566		0.0000812055	
Z-ratio				Status of significance	
1.56586				N.S.—not significant	
Parameter values					
PROB. = 0.65	RHO = −0.95		BIGH = 0.9346		SMHL = 0.45
Pair No.	SMH	SMK	PROB. level	Risk level (%)	Risk error (%)
Table No. 8.2-397					
1	0.94	0.929231	0.650007	34.999350	0.000656
2	0.93	0.939223	0.650006	34.999370	0.000632
3	0.92	0.949383	0.650001	34.999930	0.000072
4	0.91	0.959767	0.650003	34.999740	0.000262
5	0.90	0.970335	0.650000	34.999990	0.000012
6	0.89	0.981142	0.650006	34.999440	0.000560
7	0.88	0.992099	0.649996	35.000420	−0.000417
8	0.87	1.003313	0.649995	35.000520	−0.000513
9	0.86	1.014792	0.650003	34.999700	0.000298
10	0.85	1.026448	0.649997	35.000260	−0.000262
11	0.84	1.038389	0.650001	34.999870	0.000131
12	0.83	1.050527	0.649993	35.000740	−0.000739
13	0.82	1.062970	0.649994	35.000590	−0.000584
14	0.81	1.075730	0.650006	34.999410	0.000596
15	0.80	1.088721	0.650006	34.999370	0.000632
16	0.79	1.101956	0.649996	35.000420	−0.000417
17	0.78	1.115545	0.649996	35.000380	−0.000370
18	0.77	1.129503	0.650007	34.999280	0.000727
19	0.76	1.143745	0.650008	34.999190	0.000811
20	0.75	1.158288	0.650000	35.000050	−0.000048
21	0.74	1.173245	0.650001	34.999870	0.000131
22	0.73	1.188536	0.649994	35.000630	−0.000632
23	0.72	1.204276	0.649996	35.000410	−0.000405
24	0.71	1.220483	0.650007	34.999280	0.000727
25	0.70	1.237082	0.650008	34.999190	0.000811
26	0.69	1.254092	0.649999	35.000130	−0.000131
27	0.68	1.271634	0.649996	35.000380	−0.000370
28	0.67	1.289732	0.649999	35.000070	−0.000072
29	0.66	1.308411	0.650006	34.999370	0.000632
30	0.65	1.327600	0.649999	35.000120	−0.000119
31	0.64	1.347425	0.649992	35.000800	−0.000799
32	0.63	1.368014	0.649999	35.000110	−0.000107

(continued)

8.2 Contents for Table 8.2 (Table No. 8.2-1 to 400)

Table 8.2 (continued)

Parameter values					
PROB. = 0.65	RHO = −0.95		BIGH = 0.9346		SMHL = 0.45
Pair No.	SMH	SMK	PROB. level	Risk level (%)	Risk error (%)
33	0.62	1.389302	0.650001	34.999920	0.000083
34	0.61	1.411421	0.650009	34.999100	0.000906
35	0.60	1.434310	0.650005	34.999490	0.000513
36	0.59	1.458106	0.649999	35.000090	−0.000083
37	0.58	1.482947	0.649999	35.000150	−0.000143
38	0.57	1.508978	0.650009	34.999100	0.000906
39	0.56	1.536147	0.650009	34.999090	0.000918
40	0.55	1.564506	0.649991	35.000900	−0.000894
41	0.54	1.594600	0.650001	34.999920	0.000077
42	0.53	1.626292	0.650000	34.999970	0.000036
43	0.52	1.659841	0.649995	35.000490	−0.000489
44	0.51	1.695707	0.650004	34.999640	0.000370
45	0.50	1.734062	0.650009	34.999090	0.000918
46	0.49	1.775183	0.649999	35.000060	−0.000060
47	0.48	1.819938	0.650002	34.999780	0.000226
48	0.47	1.868812	0.649998	35.000180	−0.000179
49	0.46	1.923080	0.650007	34.999300	0.000703
50	0.45	1.983637	0.649997	35.000330	−0.000322
No. of pairs		Mean COMPUT. risk errors		STD. ERR. errors	
50		0.0001017954		0.0000711813	
Z-ratio				Status of significance	
1.43009				N.S.—not significant	
Parameter values					
PROB. = 0.6	RHO = −0.95		BIGH = 0.8416		SMHL = 0.36
Pair No.	SMH	SMK	PROB. level	Risk level (%)	Risk error (%)
Table No. 8.2-398					
1	0.85	0.833283	0.599995	40.000510	−0.000513
2	0.84	0.843252	0.600002	39.999800	0.000203
3	0.83	0.853362	0.600001	39.999870	0.000131
4	0.82	0.863671	0.600008	39.999220	0.000775
5	0.81	0.874140	0.600009	39.999120	0.000882
6	0.80	0.884777	0.600006	39.999370	0.000632
7	0.79	0.895594	0.600001	39.999870	0.000131
8	0.78	0.906600	0.599995	40.000460	−0.000465
9	0.77	0.917807	0.599990	40.000990	−0.000989
10	0.76	0.929276	0.600000	40.000040	−0.000036
11	0.75	0.940920	0.600000	40.000010	−0.000012

(continued)

Table 8.2 (continued)

Parameter values					
PROB. = 0.6	RHO = −0.95		BIGH = 0.8416		SMHL = 0.36
Pair No.	SMH	SMK	PROB. level	Risk level (%)	Risk error (%)
12	0.74	0.952755	0.599993	40.000700	−0.000703
13	0.73	0.964890	0.600005	39.999510	0.000489
14	0.72	0.977194	0.600001	39.999920	0.000083
15	0.71	0.989780	0.600007	39.999300	0.000703
16	0.70	1.002566	0.600001	39.999880	0.000119
17	0.69	1.015668	0.600009	39.999080	0.000918
18	0.68	1.029006	0.600009	39.999060	0.000942
19	0.67	1.042599	0.600004	39.999580	0.000417
20	0.66	1.056468	0.599996	40.000410	−0.000405
21	0.65	1.070732	0.600009	39.999100	0.000894
22	0.64	1.085219	0.600001	39.999860	0.000143
23	0.63	1.100052	0.599998	40.000200	−0.000203
24	0.62	1.115255	0.600001	39.999930	0.000072
25	0.61	1.130761	0.599991	40.000870	−0.000870
26	0.60	1.146695	0.599993	40.000690	−0.000691
27	0.59	1.163090	0.600008	39.999180	0.000823
28	0.58	1.179784	0.600000	40.000040	−0.000042
29	0.57	1.196911	0.599991	40.000940	−0.000942
30	0.56	1.214609	0.600003	39.999730	0.000274
31	0.55	1.232722	0.600001	39.999930	0.000066
32	0.54	1.251394	0.600006	39.999370	0.000632
33	0.53	1.270576	0.600005	39.999550	0.000447
34	0.52	1.290319	0.599998	40.000170	−0.000167
35	0.51	1.310679	0.599991	40.000870	−0.000870
36	0.50	1.331814	0.600002	39.999800	0.000203
37	0.49	1.353595	0.600000	39.999960	0.000036
38	0.48	1.376190	0.600005	39.999490	0.000513
39	0.47	1.399578	0.600003	39.999730	0.000274
40	0.46	1.423843	0.599996	40.000390	−0.000393
41	0.45	1.449173	0.600000	39.999970	0.000024
42	0.44	1.475570	0.600002	39.999810	0.000191
43	0.43	1.503143	0.600001	39.999880	0.000119
44	0.42	1.532108	0.600009	39.999060	0.000942
45	0.41	1.562399	0.600000	39.999980	0.000018
46	0.40	1.594357	0.599994	40.000610	−0.000608
47	0.39	1.628237	0.599994	40.000560	−0.000560
48	0.38	1.664311	0.600002	39.999760	0.000232

(continued)

8.2 Contents for Table 8.2 (Table No. 8.2-1 to 400)

Table 8.2 (continued)

Parameter values					
PROB. = 0.6	RHO = −0.95		BIGH = 0.8416		SMHL = 0.36
Pair No.	SMH	SMK	PROB. level	Risk level (%)	Risk error (%)
49	0.37	1.702773	0.600004	39.999640	0.000358
50	0.36	1.744034	0.600000	40.000010	−0.000006
No. of pairs		Mean COMPUT. risk errors		STD. ERR. errors	
50		0.0000825115		0.0000734359	
Z-ratio				Status of significance	
1.12359				N.S.—not significant	
Parameter values					
PROB. = 0.55	RHO = −0.95		BIGH = 0.7554		SMHL = 0.27
Pair No.	SMH	SMK	PROB. level	Risk level (%)	Risk error (%)
Table No. 8.2-399					
1	0.76	0.750828	0.549995	45.000540	−0.000542
2	0.75	0.760839	0.549996	45.000400	−0.000399
3	0.74	0.771023	0.550003	44.999670	0.000328
4	0.73	0.781342	0.550004	44.999580	0.000423
5	0.72	0.791808	0.550001	44.999890	0.000113
6	0.71	0.802434	0.549997	45.000320	−0.000316
7	0.70	0.813231	0.549994	45.000640	−0.000638
8	0.69	0.824264	0.550008	44.999170	0.000829
9	0.68	0.835401	0.550002	44.999830	0.000173
10	0.67	0.846705	0.549991	45.000860	−0.000864
11	0.66	0.858242	0.549994	45.000630	−0.000626
12	0.65	0.869981	0.549998	45.000160	−0.000161
13	0.64	0.881940	0.550009	44.999090	0.000912
14	0.63	0.894042	0.550003	44.999710	0.000292
15	0.62	0.906356	0.549997	45.000260	−0.000256
16	0.61	0.918954	0.550009	44.999080	0.000924
17	0.60	0.931713	0.550005	44.999550	0.000447
18	0.59	0.944707	0.550000	44.999960	0.000042
19	0.58	0.957964	0.550002	44.999760	0.000238
20	0.57	0.971416	0.549991	45.000940	−0.000942
21	0.56	0.985191	0.549995	45.000470	−0.000471
22	0.55	0.999226	0.549998	45.000220	−0.000221
23	0.54	1.013556	0.550004	44.999580	0.000423
24	0.53	1.028123	0.549998	45.000190	−0.000185
25	0.52	1.043066	0.550010	44.999050	0.000948
26	0.51	1.058236	0.550000	44.999960	0.000036
27	0.50	1.073777	0.550001	44.999880	0.000125

(continued)

Table 8.2 (continued)

Parameter values					
PROB. = 0.55	RHO = −0.95		BIGH = 0.7554		SMHL = 0.27
Pair No.	SMH	SMK	PROB. level	Risk level (%)	Risk error (%)
28	0.49	1.089646	0.549999	45.000140	−0.000143
29	0.48	1.105900	0.550001	44.999870	0.000137
30	0.47	1.122502	0.549998	45.000190	−0.000185
31	0.46	1.139521	0.549999	45.000100	−0.000101
32	0.45	1.156931	0.549995	45.000540	−0.000542
33	0.44	1.174813	0.549997	45.000320	−0.000316
34	0.43	1.193157	0.549998	45.000170	−0.000173
35	0.42	1.211960	0.549994	45.000650	−0.000650
36	0.41	1.231328	0.549997	45.000300	−0.000298
37	0.40	1.251279	0.550005	44.999460	0.000536
38	0.39	1.271744	0.550000	45.000010	−0.000012
39	0.38	1.292865	0.549999	45.000140	−0.000143
40	0.37	1.314701	0.550004	44.999650	0.000352
41	0.36	1.337228	0.550002	44.999760	0.000244
42	0.35	1.360543	0.550002	44.999830	0.000173
43	0.34	1.384764	0.550010	44.999030	0.000972
44	0.33	1.409841	0.550007	44.999320	0.000685
45	0.32	1.435948	0.550007	44.999270	0.000733
46	0.31	1.463100	0.550000	45.000060	−0.000054
47	0.30	1.491548	0.550003	44.999740	0.000256
48	0.29	1.521297	0.550000	45.000060	−0.000054
49	0.28	1.552606	0.550003	44.999710	0.000292
50	0.27	1.585508	0.549995	45.000540	−0.000542
No. of pairs		Mean COMPUT. risk errors		STD. ERR. errors	
50		0.0000352953		0.0000668608	
Z-ratio				Status of significance	
0.52789				N.S.—not significant	
Parameter values					
PROB. = 0.5	RHO = −0.95		BIGH = 0.6745		SMHL = 0.19
Pair No.	SMH	SMK	PROB. level	Risk level (%)	Risk error (%)
Table No. 8.2-400					
1	0.68	0.668996	0.499999	50.000130	−0.000131
2	0.67	0.678981	0.499996	50.000370	−0.000370
3	0.66	0.689123	0.500000	49.999980	0.000024
4	0.65	0.699386	0.499999	50.000150	−0.000143
5	0.64	0.709785	0.499995	50.000480	−0.000480
6	0.63	0.720385	0.500009	49.999120	0.000882
7	0.62	0.731056	0.499999	50.000120	−0.000122

(continued)

8.2 Contents for Table 8.2 (Table No. 8.2-1 to 400)

Table 8.2 (continued)

Parameter values					
PROB. = 0.5	RHO = −0.95		BIGH = 0.6745		SMHL = 0.19
Pair No.	SMH	SMK	PROB. level	Risk level (%)	Risk error (%)
8	0.61	0.741914	0.499999	50.000060	−0.000060
9	0.60	0.752928	0.500000	49.999970	0.000036
10	0.59	0.764071	0.499992	50.000790	−0.000784
11	0.58	0.775412	0.499995	50.000530	−0.000525
12	0.57	0.786928	0.499999	50.000100	−0.000095
13	0.56	0.798593	0.499997	50.000310	−0.000307
14	0.55	0.810483	0.500009	49.999090	0.000906
15	0.54	0.822481	0.500000	50.000010	−0.000012
16	0.53	0.834666	0.499991	50.000880	−0.000882
17	0.52	0.847072	0.499991	50.000920	−0.000918
18	0.51	0.859735	0.500007	49.999320	0.000679
19	0.50	0.872498	0.499994	50.000590	−0.000584
20	0.49	0.885550	0.500003	49.999700	0.000304
21	0.48	0.898789	0.500004	49.999600	0.000405
22	0.47	0.912217	0.499996	50.000450	−0.000441
23	0.46	0.925936	0.500002	49.999770	0.000238
24	0.45	0.939857	0.500000	50.000040	−0.000042
25	0.44	0.954046	0.500001	49.999870	0.000131
26	0.43	0.968473	0.499998	50.000190	−0.000185
27	0.42	0.983214	0.500007	49.999350	0.000656
28	0.41	0.998208	0.500008	49.999210	0.000799
29	0.40	1.013449	0.499999	50.000090	−0.000089
30	0.39	1.029038	0.500001	49.999900	0.000101
31	0.38	1.044893	0.499991	50.000880	−0.000876
32	0.37	1.061137	0.499995	50.000500	−0.000495
33	0.36	1.077715	0.499996	50.000410	−0.000402
34	0.35	1.094681	0.500002	49.999760	0.000244
35	0.34	1.112014	0.500006	49.999430	0.000572
36	0.33	1.129710	0.500001	49.999920	0.000083
37	0.32	1.147890	0.500009	49.999110	0.000894
38	0.31	1.166408	0.499995	50.000510	−0.000513
39	0.30	1.185444	0.499991	50.000940	−0.000939
40	0.29	1.205022	0.499995	50.000540	−0.000539
41	0.28	1.225113	0.499995	50.000500	−0.000495
42	0.27	1.245740	0.499990	50.000980	−0.000978
43	0.26	1.267091	0.500007	49.999310	0.000691
44	0.25	1.288939	0.499997	50.000320	−0.000322
45	0.24	1.311541	0.499997	50.000260	−0.000256

(continued)

Table 8.2 (continued)

Parameter values					
PROB. = 0.5	RHO = −0.95		BIGH = 0.6745		SMHL = 0.19
Pair No.	SMH	SMK	PROB. level	Risk level (%)	Risk error (%)
46	0.23	1.334877	0.499996	50.000410	−0.000408
47	0.22	1.359065	0.500002	49.999830	0.000167
48	0.21	1.384101	0.500003	49.999750	0.000256
49	0.20	1.410099	0.500005	49.999550	0.000453
50	0.19	1.437049	0.499993	50.000660	−0.000662
No. of pairs		Mean COMPUT. risk errors		STD. ERR. errors	
50		−0.0000888226		0.0000720535	
Z-ratio				Status of significance	
1.23273				N.S.—not significant	

Chapter 9
Tables Generated for Software Testing

9.1 The Table of COMP-T to Test Its Equivalence to OWEN's

Details of tables of this chapter are available at Sect. 5.4 of Chap. 5 (Table 9.1).

Table 9.1 The table of COMP-T to test its equivalence to OWEN's-T

Sl. No.	H	A	OWEN's-T	COMP-T	DIFF-T
1	0.00	0.25	0.038990	0.038990	0.000000
2	0.08	0.25	0.038862	0.038862	−0.000000
3	0.25	0.75	0.098755	0.098755	0.000000
4	0.30	0.25	0.037240	0.037240	0.000000
5	0.42	0.50	0.067098	0.067098	0.000000
6	0.50	0.50	0.064489	0.064489	0.000000
7	0.56	1.00	0.102473	0.102473	0.000000
8	0.60	0.75	0.083069	0.083069	0.000000
9	0.75	0.75	0.073866	0.073866	0.000000
10	0.88	1.00	0.076773	0.076773	−0.000000
11	0.95	0.50	0.045384	0.045384	0.000000
12	1.00	0.50	0.043065	0.043065	0.000000
13	1.15	0.75	0.047646	0.047646	0.000000
14	1.25	0.50	0.031828	0.031828	0.000000
15	1.40	1.00	0.037118	0.037118	0.000000
16	1.50	0.75	0.028029	0.028029	−0.000000
17	1.65	1.00	0.023512	0.023512	−0.000000
18	1.70	0.50	0.015617	0.015617	0.000000
19	1.75	0.25	0.008175	0.008175	0.000000

(continued)

© Springer India 2015
N.C. Das, *Decision Processes by Using Bivariate Normal Quantile Pairs*,
DOI 10.1007/978-81-322-2364-1_9

Table 9.1 (continued)

Sl. No.	H	A	OWEN's-T	COMP-T	DIFF-T
20	1.90	0.75	0.012983	0.012983	0.000000
21	2.00	0.25	0.005068	0.005068	−0.000000
22	2.24	1.00	0.006194	0.006194	−0.000000
23	2.50	0.50	0.002600	0.002600	−0.000000
24	2.80	0.25	0.000716	0.000716	−0.000000
25	3.00	0.75	0.000665	0.000665	0.000000
26	0.00	0.05	0.007951	0.007951	−0.000000
27	0.75	0.05	0.006000	0.006000	−0.000000
28	0.00	0.10	0.015863	0.015863	0.000000
29	1.25	0.10	0.007244	0.007244	0.000000
30	2.00	0.15	0.003160	0.003160	0.000000
31	1.25	0.19	0.013555	0.013555	−0.000000
32	0.75	0.24	0.028148	0.028148	0.000000
33	2.50	0.25	0.001609	0.001609	−0.000000
34	1.75	0.30	0.009599	0.009599	−0.000000
35	2.00	0.35	0.006717	0.006717	−0.000000
36	0.50	0.42	0.055458	0.055458	0.000000
37	1.00	0.50	0.043065	0.043065	0.000000
38	2.75	0.50	0.001295	0.001295	−0.000000
39	2.25	0.55	0.005140	0.005140	0.000000
40	1.50	0.60	0.024831	0.024831	0.000000
41	0.50	0.69	0.083347	0.083347	0.000000
42	2.00	0.85	0.010796	0.010796	−0.000000
43	1.50	0.92	0.030408	0.030408	0.000000
44	2.74	1.00	0.001581	0.001581	−0.000000
45	3.00	0.35	0.000505	0.000505	0.000000
46	3.20	0.80	0.000341	0.000342	−0.000001
47	3.45	0.10	0.000040	0.000040	−0.000000
48	3.60	0.40	0.000070	0.000070	0.000000
49	3.70	1.00	0.000054	0.000054	0.000000
50	4.00	0.25	0.000011	0.000011	−0.000000

9.2 Tables for Testing BIVNOR COMP. PROB. with JNT. PROB Under Zero Correlation

See Table 9.2.

9.2 Tables for Testing BIVNOR COMP. PROB. with JNT. PROB ...

Table 9.2 Testing BIVNOR COMP. PROB. with JNT. PROB under zero correlation

Parameter values						
PROB = 0.99	RHO = 0		BIGH = 2.5751		SMHL = 2.34	
Pair No.	SMH	SMK	PROB. COMP.	Risk level%	Risk% error	
1	2.58	2.570209	0.9900032	0.9996772	0.0003219	
	XH	XK	PROB-JNT.	PROB-XH	PROB-XK	DIFF-PROBS
	2.58	2.570209	0.9900032	0.9950599	0.9949182	0.0000000
2	2.57	2.580210	0.9900031	0.9996891	0.0003099	
	XH	XK	PROB-JNT.	PROB-XH	PROB-XK	DIFF-PROBS
	2.57	2.580210	0.9900031	0.9949151	0.9950629	0.0000000
3	2.56	2.590289	0.9899967	1.0003330	−0.0003338	
	XH	XK	PROB-JNT.	PROB-XH	PROB-XK	DIFF-PROBS
	2.56	2.590289	0.9899967	0.9947664	0.9952052	0.0000000
4	2.55	2.602009	0.9900048	0.9995222	0.0004768	
	XH	XK	PROB-JNT.	PROB-XH	PROB-XK	DIFF-PROBS
	2.55	2.602009	0.9900048	0.9946138	0.995366	0.0000000
5	2.54	2.613810	0.9900052	0.9994805	0.0005186	
	XH	XK	PROB-JNT.	PROB-XH	PROB-XK	DIFF-PROBS
	2.54	2.613810	0.9900051	0.9944574	0.995523	0.0000001
No. of pairs	Mean PROB. DIFF.			STD. ERR. Mean PROB. DIFF.		
5	0.000000			0.000000		
Z-RAT. for mean	PROB. DIFF.	0.9128711		N.S.—Not Significant		
6	2.53	2.625691	0.9899978	1.0002200	−0.0002205	
	XH	XK	PROB-JNT.	PROB-XH	PROB-XK	DIFF-PROBS
	2.53	2.625691	0.9899978	0.9942969	0.9956762	0.0000000
7	2.52	2.639217	0.9900018	0.9998202	0.0001788	
	XH	XK	PROB-JNT.	PROB-XH	PROB-XK	DIFF-PROBS
	2.52	2.639217	0.9900018	0.9941322	0.9958451	0.0000000
8	2.51	2.652826	0.9899964	1.0003570	−0.0003576	
	XH	XK	PROB-JNT.	PROB-XH	PROB-XK	DIFF-PROBS
	2.51	2.652826	0.9899964	0.993964	0.9960089	0.0000000
9	2.50	2.668081	0.9899996	1.0000410	−0.0000417	
	XH	XK	PROB-JNT.	PROB-XH	PROB-XK	DIFF-PROBS
	2.50	2.668081	0.9899996	0.9937902	0.9961857	0.0000000
10	2.49	2.684984	0.9900093	0.9990692	0.0009298	
	XH	XK	PROB-JNT.	PROB-XH	PROB-XK	DIFF-PROBS
	2.49	2.684984	0.9900093	0.9936128	0.9963764	0.0000000
11	2.48	2.701972	0.9900071	0.9992957	0.0007033	
	XH	XK	PROB-JNT.	PROB-XH	PROB-XK	DIFF-PROBS
	2.48	2.701972	0.9900071	0.9934309	0.9965536	0.0000000
12	2.47	2.719047	0.9899930	1.0007020	−0.0007033	
	XH	XK	PROB-JNT.	PROB-XH	PROB-XK	DIFF-PROBS
	2.47	2.719047	0.9899930	0.9932444	0.9967264	0.0000000

(continued)

Table 9.2 (continued)

Parameter values						
PROB = 0.99	RHO = 0		BIGH = 2.5751		SMHL = 2.34	
Pair No.	SMH	SMK	PROB. COMP.	Risk level%	Risk% error	
13	2.46	2.739335	0.9899963	1.0003690	−0.0003695	
	XH	XK	PROB-JNT.	PROB-XH	PROB-XK	DIFF-PROBS
	2.46	2.739335	0.9899963	0.9930531	0.9969218	0.0000000
14	2.45	2.761275	0.9899989	1.0001120	−0.0001132	
	XH	XK	PROB-JNT.	PROB-XH	PROB-XK	DIFF-PROBS
	2.45	2.761275	0.9899989	0.9928572	0.9971211	0.0000000
15	2.44	2.784868	0.9899986	1.0001420	−0.0001431	
	XH	XK	PROB-JNT.	PROB-XH	PROB-XK	DIFF-PROBS
	2.44	2.784868	0.9899986	0.9926564	0.9973226	0.0000000
16	2.43	2.810114	0.9899931	1.0006900	−0.0006914	
	XH	XK	PROB-JNT.	PROB-XH	PROB-XK	DIFF-PROBS
	2.43	2.810114	0.9899931	0.994506	0.9975238	0.0000000
17	2.42	2.838578	0.9899916	1.0008450	−0.0008464	
	XH	XK	PROB-JNT.	PROB-XH	PROB-XK	DIFF-PROBS
	2.42	2.838578	0.9899916	0.9922397	0.9977342	0.0000000
18	2.41	2.873385	0.9900094	0.9990573	0.0009418	
	XH	XK	PROB-JNT.	PROB-XH	PROB-XK	DIFF-PROBS
	2.41	2.873385	0.9900094	0.9920237	0.9979695	0.0000000
19	2.40	2.906725	0.9899912	1.0008750	−0.0008762	
	XH	XK	PROB-JNT.	PROB-XH	PROB-XK	DIFF-PROBS
	2.40	2.906725	0.9899912	0.9918024	0.9981738	0.0000000
20	2.39	2.949535	0.9899981	1.0001960	−0.0001967	
	XH	XK	PROB-JNT.	PROB-XH	PROB-XK	DIFF-PROBS
	2.39	2.949535	0.9899981	0.9915758	0.9984088	0.0000000
21	2.38	2.998693	0.9899996	1.0000350	−0.0000358	
	XH	XK	PROB-JNT.	PROB-XH	PROB-XK	DIFF-PROBS
	2.38	2.998693	0.9899996	0.9913437	0.9986442	0.0000000
22	2.37	3.057324	0.9899992	1.0000830	−0.0000834	
	XH	XK	PROB-JNT.	PROB-XH	PROB-XK	DIFF-PROBS
	2.37	3.057324	0.9899992	0.9911059	0.9988834	0.0000000
23	2.36	3.128555	0.9899922	1.0007800	−0.0007808	
	XH	XK	PROB-JNT.	PROB-XH	PROB-XK	DIFF-PROBS
	2.36	3.128555	0.9899922	0.9908625	0.9991217	0.0000000
24	2.35	3.234261	0.9900091	0.9990871	0.0009120	
	XH	XK	PROB-JNT.	PROB-XH	PROB-XK	DIFF-PROBS
	2.35	3.234261	0.9900091	0.9906133	0.9993901	0.0000000

(continued)

9.2 Tables for Testing BIVNOR COMP. PROB. with JNT. PROB … 625

Table 9.2 (continued)

Parameter values						
PROB = 0.99	RHO = 0		BIGH = 2.5751		SMHL = 2.34	
Pair No.	SMH	SMK	PROB. COMP.	Risk level%	Risk% error	
25	2.34	3.383820	0.9900042	0.9995818	0.0004172	
	XH	XK	PROB-JNT.	PROB-XH	PROB-XK	DIFF-PROBS
	2.34	3.383820	0.9900042	0.9903581	0.9996426	0.0000000
Parameter values						
PROB = 0.95	RHO = 0		BIGH = 1.9546		SMHL = 1.65	
Pair No.	SMH	SMK	PROB. COMP.	Risk level%	Risk% error	
1	1.96	1.949215	0.9500080	4.9992030	0.0007987	
	XH	XK	PROB-JNT.	PROB-XH	PROB-XK	DIFF-PROBS
	1.96	1.949215	0.9500080	0.9750021	0.9743651	0.0000000
2	1.95	1.959211	0.9500087	4.9991310	0.0008702	
	XH	XK	PROB-JNT.	PROB-XH	PROB-XK	DIFF-PROBS
	1.95	1.959211	0.9500087	0.9744119	0.9749559	0.0000000
3	1.94	1.969310	0.9499919	5.0008060	−0.0008047	
	XH	XK	PROB-JNT.	PROB-XH	PROB-XK	DIFF-PROBS
	1.94	1.969310	0.9499919	0.9738101	0.9755413	0.0000000
4	1.93	1.980295	0.9500001	4.9999900	0.0000119	
	XH	XK	PROB-JNT.	PROB-XH	PROB-XK	DIFF-PROBS
	1.93	1.980295	0.9500001	0.9731966	0.9761647	0.0000000
5	1.92	1.991776	0.9500095	4.9990480	0.0009537	
	XH	XK	PROB-JNT.	PROB-XH	PROB-XK	DIFF-PROBS
	1.92	1.991776	0.9500096	0.9725711	0.9768022	−0.0000001
6	1.91	2.003366	0.9499978	5.0002220	−0.0002205	
	XH	XK	PROB-JNT.	PROB-XH	PROB-XK	DIFF-PROBS
	1.91	2.003366	0.9499978	0.9719334	0.977431	0.0000000
7	1.90	2.015847	0.9500046	4.9995420	0.0004590	
	XH	XK	PROB-JNT.	PROB-XH	PROB-XK	DIFF-PROBS
	1.90	2.015847	0.9500046	0.9712835	0.9780919	0.0000000
8	1.89	2.028830	0.9500071	4.9992920	0.0007093	
	XH	XK	PROB-JNT.	PROB-XH	PROB-XK	DIFF-PROBS
	1.89	2.028830	0.9500071	0.970621	0.9787621	0.0000000
9	1.88	2.042316	0.9500038	4.9996260	0.0003755	
	XH	XK	PROB-JNT.	PROB-XH	PROB-XK	DIFF-PROBS
	1.88	2.042316	0.9500038	0.9699459	0.9794399	0.0000000
10	1.87	2.056308	0.9499927	5.0007350	−0.0007331	
	XH	XK	PROB-JNT.	PROB-XH	PROB-XK	DIFF-PROBS
	1.87	2.056308	0.9499927	0.9692581	0.9801235	0.0000000
11	1.86	2.071589	0.9500075	4.9992510	0.0007510	
	XH	XK	PROB-JNT.	PROB-XH	PROB-XK	DIFF-PROBS
	1.86	2.071589	0.9500075	0.9685572	0.9808481	0.0000000

(continued)

Table 9.2 (continued)

Parameter values						
PROB = 0.95	RHO = 0		BIGH = 1.9546		SMHL = 1.65	
Pair No.	SMH	SMK	PROB. COMP.	Risk level%	Risk% error	
12	1.85	2.086989	0.9499918	5.0008240	−0.0008225	
	XH	XK	PROB-JNT.	PROB-XH	PROB-XK	DIFF-PROBS
	1.85	2.086989	0.9499918	0.9678432	0.9815554	0.0000000
13	1.84	2.103681	0.9499949	5.0005140	−0.0005126	
	XH	XK	PROB-JNT.	PROB-XH	PROB-XK	DIFF-PROBS
	1.84	2.103681	0.9499949	0.9671159	0.9822968	0.0000000
14	1.83	2.121277	0.9499958	5.0004250	−0.0004232	
	XH	XK	PROB-JNT.	PROB-XH	PROB-XK	DIFF-PROBS
	1.83	2.121277	0.9499958	0.9663751	0.9830508	0.0000000
15	1.82	2.139779	0.9499906	5.0009430	−0.0009418	
	XH	XK	PROB-JNT.	PROB-XH	PROB-XK	DIFF-PROBS
	1.82	2.139779	0.9499906	0.9656204	0.9838137	0.0000000
16	1.81	2.159970	0.9500054	4.9994590	0.0005424	
	XH	XK	PROB-JNT.	PROB-XH	PROB-XK	DIFF-PROBS
	1.81	2.159970	0.9500054	0.9648521	0.9846124	0.0000000
17	1.80	2.181072	0.9500047	4.9995310	0.0004709	
	XH	XK	PROB-JNT.	PROB-XH	PROB-XK	DIFF-PROBS
	1.80	2.181072	0.9500047	0.9640696	0.9854109	0.0000000
18	1.79	2.203476	0.9499984	5.0001570	−0.0001550	
	XH	XK	PROB-JNT.	PROB-XH	PROB-XK	DIFF-PROBS
	1.79	2.203476	0.9499984	0.9632729	0.9862194	0.0000000
19	1.78	2.227576	0.9499938	5.0006270	−0.0006258	
	XH	XK	PROB-JNT.	PROB-XH	PROB-XK	DIFF-PROBS
	1.78	2.227576	0.9499938	0.9624619	0.9870455	0.0000000
20	1.77	2.253765	0.9499952	5.0004720	−0.0004709	
	XH	XK	PROB-JNT.	PROB-XH	PROB-XK	DIFF-PROBS
	1.77	2.253765	0.9499952	0.9616364	0.9878945	0.0000000
21	1.76	2.282435	0.9500046	4.9995420	0.0004590	
	XH	XK	PROB-JNT.	PROB-XH	PROB-XK	DIFF-PROBS
	1.76	2.282435	0.9500046	0.9607961	0.9887681	0.0000000
22	1.75	2.313198	0.9499998	5.0000190	−0.0000179	
	XH	XK	PROB-JNT.	PROB-XH	PROB-XK	DIFF-PROBS
	1.75	2.313198	0.9499998	0.9599408	0.9896442	0.0000000
23	1.74	2.347230	0.9500008	4.9999180	0.0000834	
	XH	XK	PROB-JNT.	PROB-XH	PROB-XK	DIFF-PROBS
	1.74	2.347230	0.9500008	0.9590705	0.9905432	0.0000000
24	1.73	2.384921	0.950001	4.9999360	0.0000656	
	XH	XK	PROB-JNT.	PROB-XH	PROB-XK	DIFF-PROBS
	1.73	2.384921	0.9500007	0.9581849	0.9914586	−0.0000001

(continued)

9.2 Tables for Testing BIVNOR COMP. PROB. with JNT. PROB …

Table 9.2 (continued)

Parameter values							
PROB = 0.95	RHO = 0		BIGH = 1.9546		SMHL = 1.65		
Pair No.	SMH	SMK	PROB. COMP.	Risk level%	Risk% error		
25	1.72	2.427447	0.9500058	4.9994230	0.0005782		
	XH	XK	PROB-JNT.	PROB-XH	PROB-XK	DIFF-PROBS	
	1.72	2.427447	0.9500058	0.9572839	0.9923972	0.0000000	
26	1.71	2.474812	0.9499926	5.0007400	−0.0007391		
	XH	XK	PROB-JNT.	PROB-XH	PROB-XK	DIFF-PROBS	
	1.71	2.474812	0.9499926	0.9563671	0.9933346	0.0000000	
27	1.70	2.530923	0.9499999	5.0000140	−0.0000119		
	XH	XK	PROB-JNT.	PROB-XH	PROB-XK	DIFF-PROBS	
	1.70	2.530923	0.9499999	0.9554346	0.9943118	0.0000000	
28	1.69	2.596565	0.9499922	5.0007770	−0.0007749		
	XH	XK	PROB-JNT.	PROB-XH	PROB-XK	DIFF-PROBS	
	1.69	2.596565	0.9499922	0.954486	0.9952919	0.0000000	
29	1.68	2.678771	0.9499984	5.0001620	−0.0001609		
	XH	XK	PROB-JNT.	PROB-XH	PROB-XK	DIFF-PROBS	
	1.68	2.678771	0.9499984	0.9535214	0.9963054	−0.0000001	
30	1.67	2.786138	0.9500001	4.9999950	0.0000060		
	XH	XK	PROB-JNT.	PROB-XH	PROB-XK	DIFF-PROBS	
	1.67	2.786138	0.9500001	0.9525404	0.9973331	0.0000000	
31	1.66	2.945232	0.9500074	4.9992560	0.0007451		
	XH	XK	PROB-JNT.	PROB-XH	PROB-XK	DIFF-PROBS	
	1.66	2.945232	0.9500074	0.9515429	0.9983864	0.0000000	
32	1.65	3.265430	0.9500092	4.9990780	0.0009239		
	XH	XK	PROB-JNT.	PROB-XH	PROB-XK	DIFF-PROBS	
	1.65	3.265430	0.9500092	0.9505286	0.9994536	0.0000000	
Parameter values							
PROB = 0.85	RHO = 0	BIGH = 1.4184			SMHL = 1.04		
Pair No.	SMH	SMK	PROB. COMP.	Risk level %	Risk% error		
1	1.42	1.416704	0.8500023	14.99977	2.264977E-04		
	XH	XK	PROB-JNT.	PROB-XH	PROB-XK	DIFF-PROBS	
	1.42	1.416704	0.8500023	0.9221961	0.9217153	0	
2	1.41	1.42685	0.8500074	14.99926	7.390976E-04		
	XH	XK	PROB-JNT.	PROB-XH	PROB-XK	DIFF-PROBS	
	1.41	1.42685	0.8500074	0.9207301	0.9231884	0	
3	1.40	1.437237	0.8500011	14.99989	1.072884E-04		
	XH	XK	PROB-JNT.	PROB-XH	PROB-XK	DIFF-PROBS	
	1.40	1.437237	0.8500011	0.9192433	0.9246748	0	
4	1.39	1.447966	0.8499946	15.00055	−5.483628E-04		
	XH	XK	PROB-JNT.	PROB-XH	PROB-XK	DIFF-PROBS	
	1.39	1.447966	0.8499946	0.9177356	0.9261868	0	

(continued)

Table 9.2 (continued)

Parameter values							
PROB = 0.85	RHO = 0	BIGH = 1.4184				SMHL = 1.04	
Pair No.	SMH	SMK	PROB. COMP.	Risk level %	Risk% error		
5	1.38	1.459138	0.8499983	15.00017	−1.728535E-04		
	XH	XK	PROB-JNT.	PROB-XH	PROB-XK	DIFF-PROBS	
	1.38	1.459138	0.8499984	0.9162067	0.9277364	−5.960465E-08	
6	1.37	1.470659	0.8499979	15.00021	−2.145767E-04		
	XH	XK	PROB-JNT.	PROB-XH	PROB-XK	DIFF-PROBS	
	1.37	1.470659	0.8499979	0.9146565	0.9293083	0	
7	1.36	1.482628	0.850003	14.9997	2.980232E-04		
	XH	XK	PROB-JNT.	PROB-XH	PROB-XK	DIFF-PROBS	
	1.36	1.482628	0.850003	0.9130851	0.9309133	0	
8	1.35	1.494953	0.8499997	15.00003	−3.576279E-05		
	XH	XK	PROB-JNT.	PROB-XH	PROB-XK	DIFF-PROBS	
	1.35	1.494953	0.8499997	0.911492	0.9325366	0	
9	1.34	1.507832	0.8500085	14.99915	8.46386E-04		
	XH	XK	PROB-JNT.	PROB-XH	PROB-XK	DIFF-PROBS	
	1.34	1.507832	0.8500085	0.9098773	0.9342012	0	
10	1.33	1.521074	0.8500038	14.99962	3.814697E-04		
	XH	XK	PROB-JNT.	PROB-XH	PROB-XK	DIFF-PROBS	
	1.33	1.521074	0.8500038	0.9082409	0.9358794	0	
11	1.32	1.534878	0.8500054	14.99946	5.364418E-04		
	XH	XK	PROB-JNT.	PROB-XH	PROB-XK	DIFF-PROBS	
	1.32	1.534878	0.8500054	0.9065825	0.937593	0	
12	1.31	1.549247	0.8500096	14.99905	9.536743E-04		
	XH	XK	PROB-JNT.	PROB-XH	PROB-XK	DIFF-PROBS	
	1.31	1.549247	0.8500096	0.9049021	0.9393387	0	
13	1.30	1.56399	0.8499921	15.0008	−7.987023E-04		
	XH	XK	PROB-JNT.	PROB-XH	PROB-XK	DIFF-PROBS	
	1.30	1.56399	0.8499921	0.9031996	0.94109	0	
14	1.29	1.579502	0.8499911	15.00089	−8.940697E-04		
	XH	XK	PROB-JNT.	PROB-XH	PROB-XK	DIFF-PROBS	
	1.29	1.579502	0.8499911	0.9014747	0.9428896	0	
15	1.28	1.595788	0.8500012	14.99988	−1.192093E-04		
	XH	XK	PROB-JNT.	PROB-XH	PROB-XK	DIFF-PROBS	
	1.28	1.595788	0.8500012	0.8997274	0.9447318	5.960465E-08	
No. of pairs		Mean PROB. DIFF.		STD. ERR. Mean PROB. DIFF.			
15		3.72529E-09		3.847463E-09			
Z-RAT. for mean		PROB. DIFF.	=0.968246	N.S.:- Not significant			
16	1.27	1.612656	0.8499981	15.00019	−1.907349E-04		
	XH	XK	PROB-JNT.	PROB-XH	PROB-XK	DIFF-PROBS	
	1.27	1.612656	0.8499981	0.8979577	0.9465904	0	

(continued)

9.2 Tables for Testing BIVNOR COMP. PROB. with JNT. PROB ...

Table 9.2 (continued)

Parameter values						
PROB = 0.85	RHO = 0	BIGH = 1.4184			SMHL = 1.04	
Pair No.	SMH	SMK	PROB. COMP.	Risk level %	Risk% error	
17	1.26	1.630307	0.8499963	15.00037	−3.695488E−04	
	XH	XK	PROB-JNT.	PROB-XH	PROB-XK	DIFF-PROBS
	1.26	1.630307	0.8499964	0.8961654	0.9484816	−5.960465E−08
18	1.25	1.64894	0.8500083	14.99917	8.285046E−04	
	XH	XK	PROB-JNT.	PROB-XH	PROB-XK	DIFF-PROBS
	1.25	1.64894	0.8500083	0.8943502	0.95042	0
19	1.24	1.66817	0.8499921	15.00079	−7.927418E−04	
	XH	XK	PROB-JNT.	PROB-XH	PROB-XK	DIFF-PROBS
	1.24	1.66817	0.8499921	0.8925123	0.9523589	0
20	1.23	1.688587	0.8499938	15.00062	−6.198883E−04	
	XH	XK	PROB-JNT.	PROB-XH	PROB-XK	DIFF-PROBS
	1.23	1.688587	0.8499938	0.8906514	0.9543507	0
21	1.22	1.710197	0.8500043	14.99957	4.291535E−04	
	XH	XK	PROB-JNT.	PROB-XH	PROB-XK	DIFF-PROBS
	1.22	1.710197	0.8500043	0.8887676	0.9563854	0
22	1.21	1.73281	0.8499985	15.00015	−1.549721E−04	
	XH	XK	PROB-JNT.	PROB-XH	PROB-XK	DIFF-PROBS
	1.21	1.73281	0.8499985	0.8868606	0.9584353	0
23	1.20	1.756822	0.8499986	15.00014	−1.430512E−04	
	XH	XK	PROB-JNT.	PROB-XH	PROB-XK	DIFF-PROBS
	1.20	1.756822	0.8499986	0.8849304	0.9605259	0
24	1.19	1.782434	0.8500072	14.99928	7.152558E−04	
	XH	XK	PROB-JNT.	PROB-XH	PROB-XK	DIFF-PROBS
	1.19	1.782434	0.8500072	0.8829768	0.9626608	0
25	1.18	1.809457	0.8499975	15.00025	−2.503395E−04	
	XH	XK	PROB-JNT.	PROB-XH	PROB-XK	DIFF-PROBS
	1.18	1.809457	0.8499975	0.8809999	0.96481	0
26	1.17	1.838678	0.850009	14.9991	9.000301E−04	
	XH	XK	PROB-JNT.	PROB-XH	PROB-XK	DIFF-PROBS
	1.17	1.838678	0.850009	0.8789996	0.9670188	0
27	1.16	1.869712	0.8499982	15.00018	−1.788139E−04	
	XH	XK	PROB-JNT.	PROB-XH	PROB-XK	DIFF-PROBS
	1.16	1.869712	0.8499982	0.8769756	0.9692381	0
28	1.15	1.903348	0.8499948	15.00052	−5.245209E−04	
	XH	XK	PROB-JNT.	PROB-XH	PROB-XK	DIFF-PROBS
	1.15	1.903348	0.8499948	0.8749281	0.9715024	0
29	1.14	1.940179	0.8500063	14.99937	6.318093E−04	
	XH	XK	PROB-JNT.	PROB-XH	PROB-XK	DIFF-PROBS
	1.14	1.940179	0.8500063	0.8728569	0.9738209	0

(continued)

Table 9.2 (continued)

Parameter values							
PROB = 0.85	RHO = 0		BIGH = 1.4184			SMHL = 1.04	
Pair No.	SMH	SMK	PROB. COMP.	Risk level %	Risk% error		
30	1.13	1.980015	0.8499935	15.00065	−6.496907E-04		
	XH	XK	PROB-JNT.	PROB-XH	PROB-XK	DIFF-PROBS	
	1.13	1.980015	0.8499935	0.870762	0.9761491	0	
31	1.12	2.024427	0.8499993	15.00007	−7.152558E-05		
	XH	XK	PROB-JNT.	PROB-XH	PROB-XK	DIFF-PROBS	
	1.12	2.024427	0.8499993	0.8686432	0.9785369	0	
32	1.11	2.073813	0.8499951	15.00049	−4.887581E-04		
	XH	XK	PROB-JNT.	PROB-XH	PROB-XK	DIFF-PROBS	
	1.11	2.073813	0.8499951	0.8665006	0.9809517	0	
33	1.10	2.130134	0.850003	14.9997	2.980232E-04		
	XH	XK	PROB-JNT.	PROB-XH	PROB-XK	DIFF-PROBS	
	1.10	2.130134	0.850003	0.864334	0.9834197	0	
34	1.09	2.19496	0.8500018	14.99983	1.728535E-04		
	XH	XK	PROB-JNT.	PROB-XH	PROB-XK	DIFF-PROBS	
	1.09	2.19496	0.8500018	0.8621435	0.9859168	0	
35	1.08	2.272206	0.8500079	14.99921	7.867813E-04		
	XH	XK	PROB-JNT.	PROB-XH	PROB-XK	DIFF-PROBS	
	1.08	2.272206	0.8500079	0.8599289	0.9884629	0	
36	1.07	2.36696	0.8499991	15.00009	−8.940697E-05		
	XH	XK	PROB-JNT.	PROB-XH	PROB-XK	DIFF-PROBS	
	1.07	2.36696	0.8499991	0.8576904	0.9910326	0	
37	1.06	2.491729	0.8499905	15.00095	−9.536743E-04		
	XH	XK	PROB-JNT.	PROB-XH	PROB-XK	DIFF-PROBS	
	1.06	2.491729	0.8499905	0.8554278	0.9936438	0	
38	1.05	2.680117	0.8500016	14.99984	1.609325E-04		
	XH	XK	PROB-JNT.	PROB-XH	PROB-XK	DIFF-PROBS	
	1.05	2.680117	0.8500016	0.8531411	0.9963201	0	
39	1.04	3.096978	0.8499984	15.00016	−1.66893E-04		
	XH	XK	PROB-JNT.	PROB-XH	PROB-XK	DIFF-PROBS	
	1.04	3.096978	0.8499984	0.8508301	0.9990225	0	

9.3 Tables Generated for Bivariate Normal Iso-Probable Quantile Pairs for TEST CASES

See Table 9.3.

9.3 Tables Generated for Bivariate Normal Iso-Probable ...

Table 9.3 Tables generated for bivariate normal iso-probable quantile pairs (test-cases)

Parameter values					
PROB = 0.91	RHO = 0.16		BIGH = 1.6742		SMHL = 1.35
Pair No.	SMH	SMK	PROB. level	Risk level%	Risk error%
1	1.68	1.668030	0.909994	9.000569	−0.000572
2	1.67	1.678020	0.909997	9.000319	−0.000322
3	1.66	1.688326	0.909993	9.000659	−0.000662
4	1.65	1.699146	0.910000	9.000021	−0.000024
5	1.64	1.710285	0.909997	9.000312	−0.000316
6	1.63	1.721942	0.909999	9.000069	−0.000072
7	1.62	1.734120	0.910005	8.999538	0.000459
8	1.61	1.746624	0.909995	9.000528	−0.000530
9	1.60	1.759849	0.909998	9.000176	−0.000179
10	1.59	1.773601	0.909997	9.000325	−0.000328
11	1.58	1.788079	0.910002	8.999819	0.000179
12	1.57	1.803284	0.910009	8.999073	0.000924
13	1.56	1.819026	0.910002	8.999788	0.000209
14	1.55	1.835696	0.910003	8.999676	0.000322
15	1.54	1.853298	0.910008	8.999234	0.000763
16	1.53	1.871834	0.910010	8.999032	0.000966
17	1.52	1.891309	0.910004	8.999592	0.000405
18	1.51	1.912115	0.910008	8.999229	0.000769
19	1.50	1.933865	0.909991	9.000885	−0.000888
20	1.49	1.957734	0.910010	8.999019	0.000978
21	1.48	1.982554	0.909992	9.000796	−0.000799
22	1.47	2.009890	0.910005	8.999454	0.000542
23	1.46	2.038966	0.909999	9.000147	−0.000149
24	1.45	2.070566	0.909992	9.000767	−0.000769
25	1.44	2.105475	0.910000	9.000039	−0.000042
26	1.43	2.143695	0.909997	9.000301	−0.000304
27	1.42	2.186405	0.909999	9.000116	−0.000119
28	1.41	2.234779	0.910006	8.999414	0.000584
29	1.40	2.289604	0.909996	9.000426	−0.000429
30	1.39	2.354788	0.910006	8.999432	0.000566
31	1.38	2.432682	0.909996	9.000379	−0.000381
32	1.37	2.531882	0.909991	9.000862	−0.000864
33	1.36	2.670364	0.910000	9.000004	−0.000006
34	1.35	2.901255	0.910001	8.999878	0.000119

(continued)

Table 9.3 (continued)

Parameter values					
PROB = 0.91	RHO = 0.16		BIGH = 1.6742		SMHL = 1.35
Pair No.	SMH	SMK	PROB. level	Risk level%	Risk error%
No. of pairs		Mean COMPUT. risk errors		STD. ERR.ERRORS.	
34		0.0000008515		0.0000919406	
Z-RATIO				Status of significance	
0.00926				N.S.—Not significant	
Parameter values					
PROB = 0.62	RHO = 0.68		BIGH = 0.6062		SMHL = 0.31
Pair No.	SMH	SMK	PROB. level	Risk level%	Risk error%
1	0.61	0.602424	0.620005	37.999540	0.000465
2	0.60	0.612464	0.620003	37.999750	0.000256
3	0.59	0.622942	0.620002	37.999780	0.000221
4	0.58	0.633876	0.620001	37.999880	0.000125
5	0.57	0.645285	0.619998	38.000250	−0.000250
6	0.56	0.657286	0.620006	37.999390	0.000608
7	0.55	0.669803	0.620006	37.999420	0.000584
8	0.54	0.682859	0.619994	38.000620	−0.000614
9	0.53	0.696676	0.620000	38.000010	−0.000012
10	0.52	0.711181	0.620002	37.999810	0.000185
11	0.51	0.726405	0.619995	38.000510	−0.000513
12	0.50	0.742574	0.620003	37.999680	0.000322
13	0.49	0.759624	0.620004	37.999580	0.000417
14	0.48	0.777689	0.620004	37.999620	0.000381
15	0.47	0.796908	0.620006	37.999410	0.000596
16	0.46	0.817323	0.619999	38.000130	−0.000131
17	0.45	0.839275	0.620006	37.999390	0.000608
18	0.44	0.862717	0.619999	38.000140	−0.000137
19	0.43	0.887999	0.619992	38.000830	−0.000829
20	0.42	0.915573	0.620003	37.999720	0.000286
21	0.41	0.945507	0.620001	37.999950	0.000048
22	0.40	0.978461	0.620005	37.999520	0.000477
23	0.39	1.014908	0.620003	37.999700	0.000304
24	0.38	1.055720	0.620000	38.000010	−0.000012
25	0.37	1.102169	0.620002	37.999780	0.000221
26	0.36	1.155929	0.620004	37.999560	0.000441
27	0.35	1.219469	0.619992	38.000780	−0.000775
28	0.34	1.298006	0.619991	38.000920	−0.000918
29	0.33	1.401070	0.619996	38.000370	−0.000370
30	0.32	1.594463	0.620002	37.999790	0.000215

(continued)

9.3 Tables Generated for Bivariate Normal Iso-Probable ... 633

Table 9.3 (continued)

Parameter values					
PROB = 0.62	RHO = 0.68		BIGH = 0.6062		SMHL = 0.31
Pair No.	SMH	SMK	PROB. level	Risk level%	Risk error%
31	0.31	1.893226	0.620001	37.999940	0.000060
No. of pairs		Mean COMPUT. risk errors		STD. ERR.ERRORS.	
31		0.0000705943		0.0000763388	
Z-RATIO				Status of significance	
0.92475				N.S.—Not Significant	
Parameter values					
PROB = 0.71	RHO = -0.12		BIGH = 1.0212		SMHL = 0.56
PAIR NO.	SMH	SMK	PROB. level	Risk level%	Risk error%
1	1.03	1.011889	0.709992	29.000800	−0.000793
2	1.02	1.021815	0.709997	29.000260	−0.000256
3	1.01	1.031938	0.709992	29.000750	−0.000751
4	1.00	1.042361	0.709996	29.000360	−0.000358
5	0.99	1.053090	0.710008	28.999170	0.000834
6	0.98	1.064034	0.710008	28.999230	0.000775
7	0.97	1.075200	0.709995	29.000550	−0.000548
8	0.96	1.086790	0.710005	28.999500	0.000507
9	0.95	1.098615	0.710001	28.999910	0.000095
10	0.94	1.110781	0.709999	29.000060	−0.000054
11	0.93	1.123296	0.709999	29.000070	−0.000066
12	0.92	1.136169	0.709999	29.000100	−0.000095
13	0.91	1.149406	0.709997	29.000290	−0.000286
14	0.90	1.163018	0.709992	29.000800	−0.000793
15	0.89	1.177112	0.709999	29.000150	−0.000143
16	0.88	1.191599	0.709998	29.000200	−0.000197
17	0.87	1.206587	0.710004	28.999560	0.000441
18	0.86	1.221990	0.709999	29.000090	−0.000089
19	0.85	1.237917	0.709996	29.000440	−0.000435
20	0.84	1.254378	0.709991	29.000900	−0.000900
21	0.83	1.271484	0.709996	29.000370	−0.000370
22	0.82	1.289150	0.709993	29.000670	−0.000662
23	0.81	1.307585	0.710006	28.999380	0.000626
24	0.80	1.326608	0.710002	28.999760	0.000244
25	0.79	1.346429	0.710004	28.999560	0.000447
26	0.78	1.367064	0.710006	28.999390	0.000614
27	0.77	1.388529	0.710001	28.999850	0.000149
28	0.76	1.411037	0.710008	28.999220	0.000781

(continued)

Table 9.3 (continued)

Parameter values						
PROB = 0.71	RHO = -0.12		BIGH = 1.0212		SMHL = 0.56	
PAIR NO.	SMH	SMK	PROB. level	Risk level%	Risk error%	
29	0.75	1.434411	0.709993	29.000680	-0.000674	
30	0.74	1.459060	0.709995	29.000470	-0.000471	
31	0.73	1.485006	0.710002	28.999780	0.000221	
32	0.72	1.512268	0.710002	28.999830	0.000173	
33	0.71	1.541067	0.710001	28.999900	0.000107	
34	0.70	1.571620	0.710003	28.999670	0.000334	
35	0.69	1.603953	0.709991	29.000930	-0.000930	
36	0.68	1.638680	0.709995	29.000530	-0.000530	
37	0.67	1.676022	0.710004	28.999640	0.000364	
38	0.66	1.716207	0.710002	28.999780	0.000221	
39	0.65	1.759852	0.709998	29.000210	-0.000209	
40	0.64	1.807772	0.709998	29.000180	-0.000173	
41	0.63	1.860784	0.709994	29.000560	-0.000560	
42	0.62	1.920295	0.709991	29.000910	-0.000906	
43	0.61	1.988494	0.709998	29.000190	-0.000185	
44	0.60	2.068159	0.710000	28.999960	0.000042	
45	0.59	2.164415	0.709997	29.000330	-0.000328	
46	0.58	2.287858	0.710003	28.999660	0.000346	
47	0.57	2.461590	0.709999	29.000110	-0.000107	
48	0.56	2.784104	0.710007	28.999320	0.000679	
No. of pairs		Mean COMPUT. risk errors		STD. ERR. ERRORS.		
48		-0.0000789457		0.0000693560		
Z-RATIO				Status of significance		
1.13827				N.S.—Not significant		
Parameter values						
PROB = 0.92	RHO = -0.69		BIGH = 1.7507		SMHL = 1.41	
Pair No.	SMH	SMK	PROB. level	Risk level%	Risk error%	
1	1.76	1.741449	0.919994	8.000601	-0.000602	
2	1.75	1.751400	0.920003	7.999742	0.000256	
3	1.74	1.761466	0.919991	8.000910	-0.000912	
4	1.73	1.772038	0.919991	8.000898	-0.000900	
5	1.72	1.783120	0.920001	7.999933	0.000066	

(continued)

9.3 Tables Generated for Bivariate Normal Iso-Probable ... 635

Table 9.3 (continued)

Parameter values					
PROB = 0.92	RHO = −0.69		BIGH = 1.7507		SMHL = 1.41
Pair No.	SMH	SMK	PROB. level	Risk level%	Risk error%
6	1.71	1.794517	0.920002	7.999808	0.000191
7	1.70	1.806428	0.920009	7.999110	0.000888
8	1.69	1.818658	0.920005	7.999540	0.000459
9	1.68	1.831406	0.920002	7.999826	0.000173
10	1.67	1.844675	0.919998	8.000178	−0.000179
11	1.66	1.858465	0.919992	8.000840	−0.000840
12	1.65	1.873171	0.920006	7.999378	0.000620
13	1.64	1.888208	0.919999	8.000136	−0.000137
14	1.63	1.904166	0.920005	7.999498	0.000501
15	1.62	1.920851	0.920009	7.999110	0.000888
16	1.61	1.938071	0.919994	8.000595	−0.000596
17	1.60	1.956609	0.920004	7.999563	0.000435
18	1.59	1.976079	0.920010	7.999021	0.000978
19	1.58	1.996483	0.920006	7.999408	0.000590
20	1.57	2.018213	0.920008	7.999194	0.000805
21	1.56	2.041274	0.920008	7.999170	0.000829
22	1.55	2.065668	0.919999	8.000058	−0.000060
23	1.54	2.091790	0.919991	8.000875	−0.000876
24	1.53	2.120423	0.920007	7.999343	0.000656
25	1.52	2.150789	0.919998	8.000183	−0.000185
26	1.51	2.184065	0.920000	8.000035	−0.000036
27	1.50	2.220644	0.920005	7.999486	0.000513
28	1.49	2.260920	0.920006	7.999444	0.000554
29	1.48	2.306068	0.920010	7.999003	0.000995
30	1.47	2.356875	0.920004	7.999563	0.000435
31	1.46	2.415687	0.920002	7.999778	0.000221
32	1.45	2.485633	0.920005	7.999528	0.000471
33	1.44	2.570624	0.919990	8.000958	−0.000960
34	1.43	2.683946	0.920004	7.999635	0.000364
35	1.42	2.849040	0.920004	7.999647	0.000352
36	1.41	3.186222	0.920009	7.999063	0.000936
No. of pairs		Mean COMPUT. risk errors		STD. ERR. ERRORS.	
36		0.0001862242		0.0000969860	
Z-RATIO				Status of significance	
1.92011				N.S.—Not significant	

9.4 Barycentric Coordinate Reading System: For Analysing Mixed Activities

In real-world scenario, any point located on the Barycentric coordinate system may lie anywhere on the equilateral triangle, depending on the per cent mix it represents of the three components that each vertex is required to represent, Upton and Cook (2005). Therefore, it would be relevant to develop a coordinate reading system for representation in terms of percentage of three components of the mixture assigned to three vertices of the triangle. The content of component is 100 % at the vertex assigned to it, but 0 % on the entire line falling opposite to it.

In the Barycentric coordinate system, developed on an equilateral triangle, each side is required to represent two scales, say A–B, i.e. the baseline with vertex A on the left side and vertex B on the right side. Then two lines A_1–B_1 and A_2–B_2, parallel to line A–B, are required to be drawn below line A–B. The percentage calibration on A_1–B_1 starts from A_1 to B_1 in decreasing sequence, i.e. to say at vertex A, A_1 is required to denote 100 % and as the calibration proceed from A_1 towards B_1, the next calibration point is to read 90 %. Then next in the sequel is to read 80 %, proceeding similarly as the calibration reaches B_1, corresponding to vertex B, the scale should read 0 %, representing decreasing concentration of the component of the mixture represented by vertex A. Thus, the scale line is required to be equally divided into 10 equal parts to represent the percentage of concentration of the components at a vertex. The other line for scale A_2–B_2 is required to represent calibration just in opposite direction, because it is required to represent the other component of the mixture, which has 100 % concentration at the vertex B and 0 % at the vertex A.

Similarly, two parallel lines of scale, i.e. B_1–C_1 and B_2–C_2, and percentage calibration are required to be indicated on them to denote concentration of the components represented by vertex B and C, respectively. Repeating the sequence, two more parallel lines C_1–A_1 and C_2–A_2 are required to be drawn to represent the scales for the concentration of the components represented by vertices C and A, respectively.

In fact, these percentages denote the dominance of components represented by the corresponding vertex. Nearer to the vertex, the product-mix, investment-mix, cropping-mix or soil-mix, greater would be the percentage of the component which the vertex is required to represent.

Thus, a mixture having coordinates at the centroid of the triangle is naturally composed of 33.33 % of each of the three components that vertices represent. Thus, it is evident that concentration of the component represented by the vertex A is zero on the line opposite to it, which is B–C. Similarly, concentration of the components represented by vertices B and C would, respectively, be zero on lines A–C and A–B but 100 % at the corresponding opposite vertices.

A Barycentric coordinate for any arbitrary point say X is read as below: drop perpendiculars from any arbitrary point X as indicated in Fig. 9.1 on each of the three lines of the triangle. Read the percentage points on each line from the

9.4 Barycentric Coordinate Reading System: For Analysing Mixed Activities

Fig. 9.1 Barycentric coordinate systems

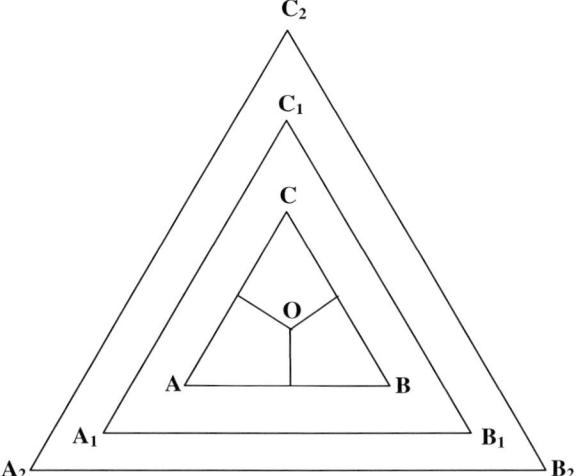

corresponding lines of the scales. For example, the perpendicular line from X to A–B reads 70 %, 0 % on line B–C and 65 % on line A–C. The coordinate point of X works out to (70 + 0 + 65)/3 = 45 % of the component A. Similarly, percentage of the component B in product mix represented by X is (30 + 38 + 0)/3 = 22.67 % and of the component C in the said mixture is (62 + 35 + 0)/3 = 32.33 %. The sum total of percentages for each component is therefore = 45.00 + 22.67 + 32.33 = 100 %. Similar analysis can be extended for four components through tetrahedral structure.

Reference

Upton, G., Cook, I.: Dictionary of Statistics. Oxford University Press, U.K (2005)

Chapter 10
Conclusions

This chapter, being the last chapter, contains five sections, starting with introduction in Sect. 10.1, conclusions in Sect. 10.2, caveat and caution in Sect. 10.3 and the ultimate questions with their answers in Sect. 10.4. The concluding Sect. 10.5 highlights Feller's dictum and Winston's aspiration with the remarks on achievement through this work.

10.1 Introduction

The primary aim of this book is to break away from unwanted assumptions of variable independence. In order to meet such requirements, tables of equi-quantiles BIGH values and 400 tables of biquantile pairs are placed in Chap. 8 as Table 8.2 of Chap. 8, for in Table 8.1 the combinations of probability levels 0.99 through 0.5 and correlation values between +0.95 and −0.95 through 0.[1] It is, therefore, mandatory to highlight their importance as well as cautions for their use. These are being presented in the following sections.

10.2 Conclusions

1. The crux of this book is to crack the age-old enigma of independence assumption in inductive inference of statistical method as primary element of uncertain decision-making processes, that is to induct at least one-stage dependence, also to play its role in such decision-making processes, as and when necessary.

---

[1]And their wide range of advantageous applications are presented in Chap. 7.

Further, as far as the dynamism is concerned, the same single-stage dependence of a variable is changed to advance or retard by a step in time dimension by replacing correlation with auto-correlation or with cross-auto-correlation. By such efforts, it could be possible to reduce the size of triangular base of the tetrahedral decision complexities.[2] The "instrument" to achieve that end is the development of tables of biquantile pairs and of equi-quantile values. The "assumption" behind such a process of decision-making is bivariate normal distribution which, in effect, is only the replacement of the univariate normal distribution. The "scope", therefore, naturally extends over those of the univariate normal distribution to those of bivariate normal distribution. This is for the reason that former is only a particular case of the latter obtainable on correlation being zero.

2. In the beginning, it traces back the history of decision enigma on account of its three components of complexities and of evolutionary process, which stand identified as uncertainty, dependence and dynamism, right from their origin during Vedic and post-Vedic era, to present the scenario, to the extent possible and contextual.
3. It provides the prospects for a paradigm shift in decision scenario to circumvent complexities on account of components of uncertainty and dependence, being faced by users and entrepreneurs, for two variable cases and also for pairwise correlated multiple sets of variables.
4. It provides methods and results of computing such biquantile pairs for different values of confidence probabilities, ranging from 0.99 to 0.5, and for correlation values from +0.95 to −0.95 through 0.0.[3]
5. Its programs have been tested for producing reliable results by formulating suitable test criteria and showing adherence to the same.
6. Computational errors have undergone rigorous test: the Z test of normality for each such generated tables of biquantile pairs and have been found to be far below the significant level. Every table presents such results at its bottom row. Such tests have not been performed for any table generated or computed so far. However, time has been the best tested for them. Such test establishes Hagen's hypotheses (theory of errors) for the tables generated for the text, making use of Herschel's hypothesis (hitting the bull's eye). Here, the bull's eye is the target probability, whereas the hypothesis is the same as that of Maxwell's, Rao (1973, 2006). Corresponding to Maxwell's variables u, v and w, variables here are the values of SMH, PROB and CORR or RHO.
7. Decision processes that existed during Vedic and post-Vedic period have been sketched. An outline of modern decision processes along with problems, being faced by a decision-maker, has been stated. Prospects for solution of some such problems by using biquantile pairs or equi-quantile values have been indicated.

---

[2]As presumed in Sect. 2.1 in Chap. 2.
[3]Through its tables in Chap. 8.

10.2 Conclusions

8. The application paradigms present multifaceted applications of biquantile pairs in undernoted fields of studies and of empirical research:

    (a) Comparison between simultaneous (joint) confidence intervals as are in use;
    (b) Discrimination and pattern recognition as applicable in diagnostics;
    (c) Optimization problem of joint chance-constrained programming;
    (d) Bivariate meteorological studies;
    (e) Bio-statistical data sets;
    (f) VaR (value at risk) for two correlated investment portfolio;
    (g) Copula model: the default correlation and biquantile pairs;
    (h) Bivariate quality control chart for correlated characteristics and for Six Sigma limits and a step towards;
    (i) Bivariate prediction with its joint confidence bound: a step in time dimension;
    (j) Problem of valid joint confidence interval for which the simulation has been waiting since long has been solved with the use of biquantile pairs/equi-quantile values;
    (k) Naïve attempt for the potential of reaching Higgs boson (so-called God Particle), by using equi-quantile value, has been indicated by providing condensed confidence interval on the basis of a hypothetical value of Bose–Einstein correlation (BEC), to its recent estimate of 125.3 ± 0.6 GeV/c^2 based on report published by Wikipedia® (op. cit.).
    (l) Rizopoulos (2009) paradox: problems put forward like those by Rizopoulos on their tangible modifications, solutions and their scope of generalization, for any value of correlation and extreme value of probability, by using biquantile pairs, which were considered not solvable otherwise.
    (m) Additional features of surrogating the level of one variable, for the other within realistic limits, where and whenever such exchanges are meaningful, but not available hitherto.

9. A realistic example of obtaining information gain at a farmer's fair, in quantitative terms, as a step towards multi-step pathway of personality development for decision orientation, towards any specific action.
10. While quoting citations from scriptures, it was repeatedly asserted that intention was neither to narrow down the deep and broad ethical thoughts and semantics underlying them to mere mundane interpretations nor to undermine the phenomenal achievements of generations' cumulative efforts made by mankind in making such advancements. But, only to assert that for such a class of cognition, mankind has been striving since Vedic and Vedantic epochs of which profuse evidences are available, usually not being referred and quoted hitherto.

Lastly but not the least important, it would be apt to quote *shloka-khanda* from mundakopanishad (celebrated preachings) Mundak 3':

"यं पश्यन्ति यतय: क्षीण दोषा: ||5||

meaning that as many times one keenly observes, so reductive are both the probability of committing error and its magnitude, one may encounter. This clearly reflects the existence of knowledge of iterative solution, or that of replications in an experiment in Vedant (the post-Vedic) period. And also that

"सत्यमेव जयति नानृतं सत्येन पन्था वित्ततो देवयान:|
येनाक्रमन्त्यृषयो ह्याप्तकामा यत्र त् सत्यस्य परमं निधानम् ||6||

Meaning: It is only the truth which prevails, not the falsehood or the error. The only way of emancipation is, therefore, to discover the truth and to eliminate the error, or as commonly conceived, rather confused with untruth.

10.3 Caveats and Cautions

A remarkable feature that could be observed was the fact of getting lower than trend values of both BIGH and SMHL, obtained for the grid point combinations of probability level 0.99 and correlation in between −0.89 and −0.94.[4] Such are the cases of extremely lower value of a risk level that is 0.01. That also for very high negative correlation around −0.9, which being rare has the least practical significance. Yet a caution needs to be taken in their use.

Therefore, the user of the tables is advised to avoid operating at such risk level of 0.01, i.e. of probability 1 %, when the correlation between −0.89 and −0.94 is observed. Instead, they should prefer to operate down to the risk level 0.02, when the data exhibit such a high negative correlation.

10.4 The Ultimate Question

What are the great gains of generating equi-quantiles or biquantile pairs? Its answers, the gains, are given as follows:

1. control over risk: as it offers the scope for choice of risk level;
2. formation of shortest possible joint confidence interval;
3. the expansion of decision alternatives, offering many choices for decision-making, like augmented risk management technique (ARMT) and INVEST-MIX[5], with features of:

---

[4]This may be seen from the Table 8.1 of Chap. 8.
[5]Reference may be made to Sect. 6.10 of Chap. 6 and Sect. 7.7 of Chap. 7.

10.4 The Ultimate Question

(i) no change in prefixed level of risk;
(ii) at no virtual cost, but with additional information about correlation which usually is available during bivariate data collection and processing;

4. freedom from the yoke of independence hypothesis at least for any numbers of pairs of variables;
5. and also for leap of one lag in time dimension: a step towards bivariate prediction and its joint confidence interval, and of all these;
6. power of generating any number of biquantile pairs has put the world to cross the first step of complexity boundaries of uncertainty, dependence and dynamism simultaneously;
7. For the greatest of among all the above, for some, it can be no. (3): greatness lies in its offering of multiple choices or alternatives to decision-makers for their chosen level of risk.
8. Yet, for others it could be no. (4), who may prefer freedom from yoke of independence hypothesis of higher human value than the former.

The question of preference of one over the other must not arise, because humanity is getting all of them at a time. They do not stand as alternative gains.

Such phenomenon opens a new vista for application and research in the field of multi-variable studies to serve the cause of all concerned, hitherto not available, Thus, opening scope for advancement and advantageous applications for not only bivariate but also multi-variate studies. Indeed, a recognizable addition to decision processes has the potential of being used even for precise prediction of Higgs boson, so-called God Particle, if the value of Bose–Einstein correlation (BEC) is made available in quantified terms, as reported by Wikipedia (2012) earlier.

10.5 Feller's Dictum and Winston's Aspiration

It is worth recalling Feller's (1972, 2009) dictum and his example made in connection with branching processes once again, which are given as follows:

1. **Feller's dictum**: "… stochastic independence is impossible". "… common inheritance and common environment are bound to produce similarities among brothers, which is contrary to our assumption of stochastic independence".
2. **Winston's aspiration**: Yet another author is Winston (2004) who has realized complication in determination of confidence interval, in situation of dependence in simulation data on account of correlation or auto-correlation between neighbouring, adjacent or time contiguous observations as reality. He there on expressed, "under such situation data are rarely if ever independent". Thereafter, he aspired "we must modify statistical methods to make proper inference from such simulation data".

Having generated and making use of biquantile pairs, it has been tried in the book to achieve a step of advancement towards solving problems raised through such a dictum and aspiration. But, of all who realized the handicap of "independent assumption" and are seeking, for "statistical methods" to circumvent such problems. Besides, efforts were made to predate some concepts of probability, statistics, psychological aspects of personality development and of decision processes, referring and quoting Indian scriptures and giving at places his own interpretations. All said and done what has been achieved is a contribution to decision processes facing complexities that arise due to uncertainty, dependence and dynamism. Whatever could be done, it appears to be the commandments of the Supreme Power, hence being submitted to Him.

त्वदीयं तुभ्यं समर्पयामि

References

Feller, W.: An Introduction to Probability Theory and its Applications, vol. 1. Wiley India, New Delhi (1972, 2009)
Mundkopanishad: Ishaadi Nov Upanishad. Gita Press, Gorakhpur
Rao, C.R.: Linear Statistical Inference and its Applications, 2nd edn. Wiley, New Jersey (1973, 2006)
Rizopoulos, D.: Quantiles for Bivariate Normal Distribution. http://www.R-project.org/posting-guide.html, http://www.r-project.org/posting-guide.html (2009)
Winston, W.L.: Introduction to Probability Models: Operations Research, vol. 2, 4th edn. Thompson Learning, Singapore (2004)
Wikipedia. Bose–Einstein Statistics: Higgs–Boson and Bose–Einstein Correlation (2012, 22nd July)

Index

A
Aasha, 107
Aavarna, 107
Ab initio recollection, 4
Action calculus, 34
Adaptive control processes, 37
Aggregate information, 188
Agni, 25, 26
Algorithmic approach, 37
Anrita, 10
Anti-symmetry, 93
Apex vertex of dynamism, 14
Application paradigm, 15, 53, 641
Artificial intelligence, 34
Association and dependence, 14
Associative memory, 4
Atigrahas, 27
Attitude, 6, 7, 9, 25, 104, 105, 107, 124
Augmented risk management technique, 129, 642

B
Barycentric coordinate, 16
Bayesian network, 34
Bhagavad Geeta, 104
Bhavitabhya, 22
Bhuma, 108
Bias, 23, 95, 100, 107
BIGH, 61, 75, 77, 98, 100, 140
Bio-statistical studies, 13, 16, 161
Biquantile pairs, 16
Biquantiles in optimization, 153
Bivariate meteorological prediction, 157
Bivariate normal distribution, 157
Bivariate normal integral, 157
Bivariate quality control charts, 13
Bivariate stochastic processes, 16, 173, 184
BIVTEST3, 95, 100
Bonferroni's interval, 83, 142, 144, 146, 160, 173, 180
Bose-Einstein correlation (BEC), 13, 16, 137, 177, 178, 641
Bose-Einstein statistics, 177
Boson, 13, 177, 178, 180, 643
Brahmasutra, 6, 7, 9, 22, 28
Branching processes, 39, 643
Brihadaranyako Upanishad, 10, 27
Brownian motion and Weiner process, 38

C
Causal algebra, 14, 30
Causal calculus, 30
Causal graph, 33
Causal structure, 32
Caveats and caution, 642
Central limit theorem, 47, 56, 155
Certainty equivalents, 33, 73, 105, 126, 130, 152
Chance-constrained version, 118
Chaotic state, 39
Characterization, 14, 48, 49, 61–63, 130
Charaiveti charaiveti, 35
Choice, 8, 9, 28, 69, 75, 79, 104, 113, 114, 116, 119, 123, 124, 127, 152, 153, 156, 164, 643
Cohesive bondage, 177
Components of rationality, 2, 8
Condensed confidence cover, 184, 185
Consensus algorithm, 131
Controlled Markov process, 36
Convex crystals or hyper-crystals, 156
Convexity problem, 154, 156

Convex structures, 156
Correlated Gaussian wavelet, 185
Correlation coefficient, 24, 30, 34, 63, 77, 85, 98, 137, 146, 147, 158, 168, 171, 177, 178
Crossover point, 184
Cumulative conscience, 104

D

Decision alternatives, 131, 137, 642
Decision complexities, 8, 13, 19, 20, 640
Decision function, 15, 114
Decision-oriented result, 7
Decision scenario, 15, 110, 131, 640
Default correlation
  copula, 14, 35, 165
Directed graph, 31, 186
Discriminant function, 14, 15, 116, 148
Discrimination and classification, 115
Discrimination and pattern recognition, 641
Dynamic programming, 36, 117
Dynamism, 1, 9, 12, 19, 20, 35, 40, 111, 640, 644

E

Eccentricity, 67, 158
Ellipticity, 68, 158
Empirical research, 113, 115, 641
Epsilon neighbourhood, 74
Equi-probability curves, 65, 74
Equi-quantile values, 16, 49, 64, 77, 79, 86, 103, 137, 152, 173, 640, 641
Exchangeability, 74, 168
Extreme paradigms, 184

F

Factor analysis, 14, 126
Feller's dictum, 17, 639, 643
Fiducial inference, 50
Filtering problems, 34
Fixed-lead prediction, 184
Fractile, 51
Fuzzy regression, 34, 132

G

Gaandiva, 105
Gaussian copula, 13, 16, 165, 166
GOD particle, 16, 177, 641, 643
Grahas, 27, 28, 115

H

Heuristic, 74, 147, 155
Heuristic algorithm, 13, 14, 74, 97, 147
Higgs boson, 177, 641, 643

High risk prone group, 15
Human mind, 7, 9, 111, 186
Human performance potential, 6

I

Inductive behaviour, 114
Info-sensors, 2
Integral table, 49, 56, 162, 165
Inverse Markov processes, 25
Iso-probable biquantile pairs, 14, 116, 126, 167
Iso-risk technology, 64, 119, 128
Iso probable ellipsoid, 72

J

Joint chance constraint, 12, 15
Joint confidence interval, 13, 15, 61, 64, 81, 139, 140, 143, 144, 146, 148, 159, 161, 170, 172, 174, 179, 642
Joint probability distribution function, 14, 114, 120, 165

K

Kaivalyam, 103, 107
Kalman filtrates, 37, 184
Kriti, 108

L

Large deviation, 39
Large hydrogen collider (LHC), 177
Law of error, 24, 49
Learning model, 2

M

Mal, 107
Markov chain and Bayesian inference, 36
Markovian decision processes, 36
Mati, 107
Memory ab initio, 4
MINMAD regression, 34
Moran's expression, 71
Multi-variate regression, 34

N

Nashadiya–sucta, 21
Nimitta karan, 7
Nistha, 108
Niyati, 22

O

Online stochastic combinatorial optimization (OSCO), 15, 130
Opulence, 125
Owen's computational scheme, 69
Owen's t-function, 79, 93, 100

Index

P
Path coefficients, 14, 31
Pattern recognition, 12, 33, 115, 117
Percentile, 50, 162
Pixel region, 159
Plexus, 4, 7, 186
Post-Vedic, 1, 2, 9, 14, 19, 22, 103, 640, 642
Prana, 28, 107
Predicate calculus, 5
Predicate logic, 5
Principal components, 32
Probabilistic casual algebra, 14
Prospect theory, 118, 119, 125
Psycho-kinetics, 1, 4, 19

Q
Qmvnorm, 86, 155, 180
Quantifiers, 5
Quantile, 12–14, 38, 39, 50–53, 55, 61, 63, 72, 73, 75, 78, 81, 85, 86, 91, 98, 101, 106, 119, 121, 126, 132, 141, 144, 147, 151, 155, 160, 163, 164, 171, 177, 178, 180, 184

R
Rama-charit manas, 22
Ramsalaka prashnavali, 22
Regression designs, 34
Regret algorithm, 131
Rig veda, 21, 26, 27, 110
Risk and uncertainty, 15, 123, 130
Risk aversions, 109, 124–126, 155, 156
Risk averting group, 15
Risk level, 56, 75, 76, 79, 122, 144, 155, 171, 642
Rizopoulos paradox, 13, 16, 87, 180, 641

S
Sambhavana, 22
Satya, 10, 107
Shifting condensed cover, 185
Shradha, 108
Shrinkage of joint confidence interval, 15, 140, 141, 143

Shrunken confidence cover, 184
Simultaneous joint confidence interval, 14, 16
Six sigma, 13, 16, 167, 641
Sliding window, 13, 184
Smarna, 107
SMH, 77, 78, 93, 95, 99, 151, 164, 183
Software BIVNOR, 14, 16, 75, 97
Software reliability, 15, 91, 97
Software testing criteria, 91
Stall effectiveness, 189
Standard software, 91
Stimulus, 4–6, 8, 186, 188, 189
Stochastic chain rule, 7
Stochastic differential equation, 14, 163
Stochastic kinetics, 4, 5
Stochastic modus ponens, 7
Sukha, 108
Synergic impact, 4, 8
Synergic pool, 104

T
Tri-quantile trios, 168

U
Upadaan kaaran, 6

V
Value at risk (VaR), 13, 16, 162, 641
Vedic, 1, 2, 13, 19, 22, 103, 640, 641
Vedic era, 20, 26, 640
Vertex of dependence, 14
Vertex of uncertainties, 19
Vic-chhepa, 107
Vigyana, 107

W
Well formed formulae, 7
Winston's aspiration, 17, 639, 643

Y
Yogavashistha, 22

Printed by Printforce, the Netherlands